Lecture Notes in Computer Science 12192

More information about this series at http://www.springer.com/series/7409

Pei-Luen Patrick Rau (Ed.)

Cross-Cultural Design

User Experience of Products, Services, and Intelligent Environments

12th International Conference, CCD 2020
Held as Part of the 22nd HCI International Conference, HCII 2020
Copenhagen, Denmark, July 19–24, 2020
Proceedings, Part I

 Springer

Editor
Pei-Luen Patrick Rau
Tsinghua University
Beijing, China

ISSN 0302-9743   ISSN 1611-3349 (electronic)
Lecture Notes in Computer Science
ISBN 978-3-030-49787-3   ISBN 978-3-030-49788-0 (eBook)
https://doi.org/10.1007/978-3-030-49788-0

LNCS Sublibrary: SL3 – Information Systems and Applications, incl. Internet/Web, and HCI

This Springer imprint is published by the registered company Springer Nature Switzerland AG
The registered company address is: Gewerbestrasse 11, 6330 Cham, Switzerland

Foreword

The 22nd International Conference on Human-Computer Interaction, HCI International 2020 (HCII 2020), was planned to be held at the AC Bella Sky Hotel and Bella Center, Copenhagen, Denmark, during July 19–24, 2020. Due to the COVID-19 coronavirus pandemic and the resolution of the Danish government not to allow events larger than 500 people to be hosted until September 1, 2020, HCII 2020 had to be held virtually. It incorporated the 21 thematic areas and affiliated conferences listed on the following page.

A total of 6,326 individuals from academia, research institutes, industry, and governmental agencies from 97 countries submitted contributions, and 1,439 papers and 238 posters were included in the conference proceedings. These contributions address the latest research and development efforts and highlight the human aspects of design and use of computing systems. The contributions thoroughly cover the entire field of human-computer interaction, addressing major advances in knowledge and effective use of computers in a variety of application areas. The volumes constituting the full set of the conference proceedings are listed in the following pages.

The HCI International (HCII) conference also offers the option of "late-breaking work" which applies both for papers and posters and the corresponding volume(s) of the proceedings will be published just after the conference. Full papers will be included in the "HCII 2020 - Late Breaking Papers" volume of the proceedings to be published in the Springer LNCS series, while poster extended abstracts will be included as short papers in the "HCII 2020 - Late Breaking Posters" volume to be published in the Springer CCIS series.

I would like to thank the program board chairs and the members of the program boards of all thematic areas and affiliated conferences for their contribution to the highest scientific quality and the overall success of the HCI International 2020 conference.

This conference would not have been possible without the continuous and unwavering support and advice of the founder, Conference General Chair Emeritus and Conference Scientific Advisor Prof. Gavriel Salvendy. For his outstanding efforts, I would like to express my appreciation to the communications chair and editor of HCI International News, Dr. Abbas Moallem.

July 2020   Constantine Stephanidis

HCI International 2020 Thematic Areas and Affiliated Conferences

Thematic areas:

- HCI 2020: Human-Computer Interaction
- HIMI 2020: Human Interface and the Management of Information

Affiliated conferences:

- EPCE: 17th International Conference on Engineering Psychology and Cognitive Ergonomics
- UAHCI: 14th International Conference on Universal Access in Human-Computer Interaction
- VAMR: 12th International Conference on Virtual, Augmented and Mixed Reality
- CCD: 12th International Conference on Cross-Cultural Design
- SCSM: 12th International Conference on Social Computing and Social Media
- AC: 14th International Conference on Augmented Cognition
- DHM: 11th International Conference on Digital Human Modeling and Applications in Health, Safety, Ergonomics and Risk Management
- DUXU: 9th International Conference on Design, User Experience and Usability
- DAPI: 8th International Conference on Distributed, Ambient and Pervasive Interactions
- HCIBGO: 7th International Conference on HCI in Business, Government and Organizations
- LCT: 7th International Conference on Learning and Collaboration Technologies
- ITAP: 6th International Conference on Human Aspects of IT for the Aged Population
- HCI-CPT: Second International Conference on HCI for Cybersecurity, Privacy and Trust
- HCI-Games: Second International Conference on HCI in Games
- MobiTAS: Second International Conference on HCI in Mobility, Transport and Automotive Systems
- AIS: Second International Conference on Adaptive Instructional Systems
- C&C: 8th International Conference on Culture and Computing
- MOBILE: First International Conference on Design, Operation and Evaluation of Mobile Communications
- AI-HCI: First International Conference on Artificial Intelligence in HCI

Conference Proceedings Volumes Full List

18. LNCS 12198, Digital Human Modeling and Applications in Health, Safety, Ergonomics and Risk Management: Posture, Motion and Health (Part I), edited by Vincent G. Duffy
19. LNCS 12199, Digital Human Modeling and Applications in Health, Safety, Ergonomics and Risk Management: Human Communication, Organization and Work (Part II), edited by Vincent G. Duffy
20. LNCS 12200, Design, User Experience, and Usability: Interaction Design (Part I), edited by Aaron Marcus and Elizabeth Rosenzweig
21. LNCS 12201, Design, User Experience, and Usability: Design for Contemporary Interactive Environments (Part II), edited by Aaron Marcus and Elizabeth Rosenzweig
22. LNCS 12202, Design, User Experience, and Usability: Case Studies in Public and Personal Interactive Systems (Part III), edited by Aaron Marcus and Elizabeth Rosenzweig
23. LNCS 12203, Distributed, Ambient and Pervasive Interactions, edited by Norbert Streitz and Shin'ichi Konomi
24. LNCS 12204, HCI in Business, Government and Organizations, edited by Fiona Fui-Hoon Nah and Keng Siau
25. LNCS 12205, Learning and Collaboration Technologies: Designing, Developing and Deploying Learning Experiences (Part I), edited by Panayiotis Zaphiris and Andri Ioannou
26. LNCS 12206, Learning and Collaboration Technologies: Human and Technology Ecosystems (Part II), edited by Panayiotis Zaphiris and Andri Ioannou
27. LNCS 12207, Human Aspects of IT for the Aged Population: Technologies, Design and User Experience (Part I), edited by Qin Gao and Jia Zhou
28. LNCS 12208, Human Aspects of IT for the Aged Population: Healthy and Active Aging (Part II), edited by Qin Gao and Jia Zhou
29. LNCS 12209, Human Aspects of IT for the Aged Population: Technology and Society (Part III), edited by Qin Gao and Jia Zhou
30. LNCS 12210, HCI for Cybersecurity, Privacy and Trust, edited by Abbas Moallem
31. LNCS 12211, HCI in Games, edited by Xiaowen Fang
32. LNCS 12212, HCI in Mobility, Transport and Automotive Systems: Automated Driving and In-Vehicle Experience Design (Part I), edited by Heidi Krömker
33. LNCS 12213, HCI in Mobility, Transport and Automotive Systems: Driving Behavior, Urban and Smart Mobility (Part II), edited by Heidi Krömker
34. LNCS 12214, Adaptive Instructional Systems, edited by Robert A. Sottilare and Jessica Schwarz
35. LNCS 12215, Culture and Computing, edited by Matthias Rauterberg
36. LNCS 12216, Design, Operation and Evaluation of Mobile Communications, edited by Gavriel Salvendy and June Wei
37. LNCS 12217, Artificial Intelligence in HCI, edited by Helmut Degen and Lauren Reinerman-Jones

38. CCIS 1224, HCI International 2020 Posters - Part I, edited by Constantine Stephanidis and Margherita Antona
39. CCIS 1225, HCI International 2020 Posters - Part II, edited by Constantine Stephanidis and Margherita Antona
40. CCIS 1226, HCI International 2020 Posters - Part III, edited by Constantine Stephanidis and Margherita Antona

http://2020.hci.international/proceedings

12th International Conference on Cross-Cultural Design (CCD 2020)

Program Board Chair: Pei-Luen Patrick Rau, Tsinghua University, China

- Kuohsiang Chen, Taiwan
- Zhe Chen, China
- Na Chen, China
- Yu-Liang Chi, Taiwan
- Wen-Ko Chiou, Taiwan
- Zhiyong Fu, China
- Sheau-Farn Max Liang, Taiwan
- Pin-Chao Liao, China
- Dyi-Yih Michael Lin, Taiwan
- Rungtai Lin, Taiwan
- Wei Lin, Taiwan
- Yongqi Lou, China
- Ta-Ping (Robert) Lu, China
- Liang Ma, China
- Alexander Mädche, Germany
- Chun-Yi (Danny) Shen, Taiwan
- Huatong Sun, USA
- Hao Tan, China
- Pei-Lee Teh, Malaysia
- Yuan-Chi Tseng, Taiwan
- Lin Wang, South Korea
- Hsiu-Ping Yueh, Taiwan
- Run-Ting Zhong, China

The full list with the Program Board Chairs and the members of the Program Boards of all thematic areas and affiliated conferences is available online at:

http://www.hci.international/board-members-2020.php

HCI International 2021

The 23rd International Conference on Human-Computer Interaction, HCI International 2021 (HCII 2021), will be held jointly with the affiliated conferences in Washington DC, USA, at the Washington Hilton Hotel, July 24–29, 2021. It will cover a broad spectrum of themes related to Human-Computer Interaction (HCI), including theoretical issues, methods, tools, processes, and case studies in HCI design, as well as novel interaction techniques, interfaces, and applications. The proceedings will be published by Springer. More information will be available on the conference website: http://2021.hci.international/.

General Chair
Prof. Constantine Stephanidis
University of Crete and ICS-FORTH
Heraklion, Crete, Greece
Email: general_chair@hcii2021.org

http://2021.hci.international/

Contents – Part I

Culture-Based Design

Cross-Cultural Behavior and Attitude

Cultural Facets of Interactions with Autonomous Agents and Intelligent Environments

Contents – Part II

Culture, Learning and Communication

Culture and Creativity

Cross-Cultural User Experience Design

A Service Design Framework for Brand Experience in the Creative Life Industry – A Case Study of the Millennium Gaea Resort Hualien in Taiwan

Shu-Hua Chang[1]([⊠]) and Rungtai Lin[2]

[1] Department of Arts and Creative Industries, National Dong Hwa University, Hualien, Taiwan
iamcsh0222@gms.ndhu.edu.tw
[2] Graduate School of Creative Industry Design, College of Design, National Taiwan University of Arts, New Taipei City, Taiwan
rtlin@mail.ntua.edu.tw

Abstract. This study combines the experience economy, customer-based brand equity and service design to explore the essential elements of the creative life industry. Data was collected from a single case study in Taiwan. The findings reveal that both the emotional and rational route should be designed into the realms of experience. First, the realms of experience reflected the main experiences offered by creative life enterprises. The results show that the respondents' perceptions of the various realms of experience through touchpoints could influence each other. Creative life enterprises should reinforce ways of triggering other experience realms if they want to provide tourists with comprehensive and high-quality experiences. Second, the tourists' perceptions of the realms of experience and of brand blocks mostly derived from specific touchpoints, such as the herb garden, the herbal remedy pools, the cuisine, the relaxation rooms, the staff, etc. The conclusions that can be drawn are that a one-source multiple-use strategy should be incorporated into the design of touchpoints, and peak experiences should be identified for build-up for tourists.

Keywords: Experience economy · Customer-based brand equity · Creative life industry · Service design

1 Introduction

In recent years, Taiwan has moved from being an original-equipment manufacturer and original-design manufacturer to being an original-brand manufacturer. "Brand Taiwan" has become a critical factor in stimulating the next wave of economic development in Taiwan. A development plan for the creative life industry was introduced and promoted by the Taiwanese government in 2002. The creative life industry is defined as an industry that uses creativity to integrate core knowledge of the life industry and provide in-depth experiences and high-quality aesthetics [1]. The Taiwanese government has proposed

© Springer Nature Switzerland AG 2020
P.-L. P. Rau (Ed.): HCII 2020, LNCS 12192, pp. 3–15, 2020.
https://doi.org/10.1007/978-3-030-49788-0_1

that the creative life industry orchestrates memorable events for their customers through specific products and services.

The "experience economy" refers to an emerging paradigm that encompasses a variety of industries, including tourism and hospitality [2, 3]. Experience industries can use services as a stage and products as props to spur individual consumers' participation [4]. Experience designers provide not only products or services but also experiences and memories elicited by the product or service. The various realms of experience of the experience economy were applied for the promotion of the creative life industry in Taiwan. The creative life industry is seen as a producer of experiences, wherein values can be created through various elements within related services, products, and activities.

A brand is a cognitive image in the mind, commitment, and experience of consumers. The goal of service design is to provide users with holistic services by integrating various design techniques, management styles, and engineering processes. This study examined how creative life enterprises can incorporate service design and realms of experience into the brand experience. This paper describes the attributes that define the framework of brand experiential service design using the Millennium Gaea Resort Hualien as a case study. Although brand equity has been considered a multidimensional construct by a number of authors [5–8], there are relatively few studies that examine brand experiences within the realms of experience developed by Pine and Gilmore [4] from the perspective of service design [9, 10] in the creative life industry.

Furthermore, few studies have examined these realms of experience from the perspective of a customer-based brand equity (CBBE) framework of designing customer experiences in the creative life industry. On the basis of the aforementioned studies, tools for designing tourist experiences and methods used in different disciplines should be integrated in order to provide an appropriate framework for experiential service design in the creative life industry. The following section provides a brief review of the literature on brand equity, the experience economy, service design, and the creative life industry. The first objective is to explore the elements of experience realms and the corresponding brand experience touchpoints. The second objective is to examine the crucial elements of brand experience through the various stages of service design. Understanding the components of an overall brand experience supports the efficacy of the service design of the brand experiential model for the creative life industry.

2 Theoretical Background

2.1 Customer Experience

Pine and Gilmore [4] developed the experience economy theory based on four realms of experience—education, escapism, aesthetics, and entertainment—and argued that including more realms would increase the likelihood of producing an experience "sweet spot". The following five steps were proposed to enable experience design—defining the theme, combining experiences with positive cues, eliminating negative cues, associating the experience with memorability, and integrating sensory stimulation [4]. These steps indicate that even after the realms of experience have been determined, experiential cues and sensory stimulation should be employed to enhance customer experiences.

Several studies have explored the experience economy and have applied the theory in different tourism contexts [1, 11–14]. The experience framework put forward by Pine and Gilmore is generally considered to be valid for measuring customer experiences in tourism industries. Quadri-Felitti and Fiore [15] pointed to the nature of destination-specific tourist experiences within the realms of experience in order to advance the experience economy theory. However, studies have shown that the best practices for the creation of tourism experiences remain a subject of debate in creative tourism. There is a need for a more comprehensive view of the experiential nature of creative tourism that draws on the tourist's perspective, especially with regard to what contributes to the creative tourism experience.

2.2 The CBBE Framework

The concept of customer-based brand equity (CBBE) was proposed by researchers in the field of marketing [5–7]. Brand equity was defined by Keller [6] as "the differential effect of brand knowledge on consumer response to the marketing of the brand". Keller considered the CBBE model misleading in terms of what customers learned, saw, felt, and heard about the brand as a result of their experiences. Marketers should strive to provide the right type of experiences through products and services and the accompanying marketing programs in order to build up a strong brand. Keller also suggested establishing six "brand building blocks" with customers that can be assembled in a pyramid through two routes, namely the rational route and the emotional route [7]. The CBBE model includes six blocks—salience, performance, imagery, judgements, feelings, and resonance. Brand performance and brand judgements represent the rational route, while brand imagery and brand feelings represent the emotional route.

According to Aaker [5], brand equity is "a set of brand assets and liabilities linked to a brand, its name and symbol that adds to or subtracts from the value provided by a product or service to a firm and/or to that firm's customers". The components of brand equity include brand loyalty, perceived quality, brand associations, brand awareness, and other proprietary brand assets (e.g., patents, trademarks, and channel relationships). To date, the CBBE framework has been applied in the tourism and hotel context [16–18].

Petromilli and Michalczyk [19] proposed a brand experience model in which the stages of brand experience were divided into pre-experience, during-experience, and post-experience. Brand experience represents not just customers' recognition of the brand's message, but also the process of experiencing product consumption to generate experience values of the brand. The CBBE framework provides a more comprehensive assessment of the creative life industry in terms of tourist perceptions than other frameworks, and will therefore be applied to the experience stages in the current study.

2.3 Service Design

Service design is the planning and designing of systems and processes through the integration of tangible and intangible media to provide customers with comprehensive and meticulous service experiences [10]. Service design must be developed from the customer's perspective to ensure that a service interface features differentiating characteristics that are feasible, useful, and effective, and which conform to customer expectations

[20]. Stickdorn and Schneider [10] suggested that service design should emphasize a user-centered, cocreative, sequencing, evidencing, and holistic principle.

Service design methods have been proposed that use various instruments, such as stakeholder mapping, customer journey mapping, cultural probing, user diaries, empathy mapping, and service blueprints [10, 21, 22]. Cook et al. [23] argued that to establish a competitive edge, service providers should emphasize interactive experiences at service touchpoints and focus on customer emotions, behaviors, and expectations through appropriate design and management. Previous studies have examined the generation and management of tourism experiences through the service design approach [9, 10, 24–26], and various service design methods, such as user diaries [11], personas, and service scenarios [24], have been applied in tourism experiences. This study used customer journey mapping to illustrate the findings.

3   Methodology

This case study applied the service design methods of observation, interviews, and customer journey mapping at the Millennium Gaea Resort Hualien in Taiwan. The subject of the case study is a creative life business certified by the Taiwanese government, and is located in the foothills of the central mountains. It comprises a collection of suites and villas in leafy surroundings, and offers spectacular views of the pristine mountains. The Millennium Gaea Resort Hualien boasts excellent dining, a tea house, special venue spots, a herbal boutique shop, and an organic agricultural shop, and emphasizes the cultivation of organic herbs and eco-friendly farming, for which it has made a name for itself over the past 30 years.

A case study is a suitable method for the present research because the issues examined and analyzed in this study lack relevant research and have exploratory value [27]. Case studies of a single target are suitable for the early stages of theory construction or the later stages of theory testing. Through participant observation, we depicted the relevant touchpoints and activities in a customer journey map, which served as the basis for focus group interviews. A focus group interview is a method that brings together 4–12 people for the purpose of providing thoughtful responses regarding a product, service, or concept. Focus group interviews are suitable for studies that involve different cultures, backgrounds and lifestyles, or for research into the behavior of complex social and psychological issues, motivation, attitudes, etc. [28].

This study explores the service design elements of brand experience. Since the respondents were expected to be different in terms of culture, background, lifestyle, behavior, motivation, attitudes, etc., a focus group interview was deemed to be suitable for this study. A focus group interview format was used to ascertain the participants' feelings and perceptions about the design elements of the brand experience service in the customer journey at the Millennium Gaea Resort Hualien. The research participants were purposeful tourists visiting the Millennium Gaea Resort Hualien. Purposeful tourists were chosen because they were more likely to have goals and expectations from their visits. Seven visitors were asked about—and confirmed—their willingness to participate in this study. After we compiled and organized the collected data, the data served as the basis for reviewing and modifying the observed customer journey map.

This case study applied the service design methods of observation and customer journey mapping through the format of focus group interviews conducted at the Millennium Gaea Resort Hualien in Taiwan. Based on previous studies, a research framework combining experience realms [2, 3] and brand building blocks [7] in the customer journey map [10] was put forward to explore the design elements of brand experience service in the creative life industry (Fig. 1).

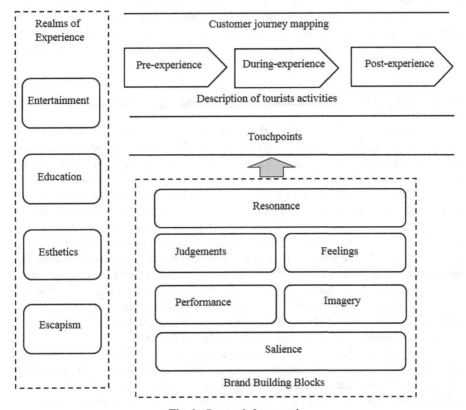

Fig. 1. Research framework

First, observations and documentary analysis were conducted to determine the customer journey map offered by the Millennium Gaea Resort Hualien. The documents used in the study included news reports, book interviews, and formal reports. These can be regarded as "mute evidence", and they provided useful information about event details [27].

Based on the depicted touchpoints in the customer journey map, the participants were asked to describe what they perceived during the various experience stages at the Millennium Gaea Resort Hualien, and how they assessed those perceptions. To safeguard the reliability of the data analysis and interpretation, we interpreted the exemplars individually prior to including them in the research, and explained the operational definitions of the study. For the data analysis, we selected excerpts from the participants' responses through a careful, thematic reading of their transcripts so that they would reflect perceptions and opinions that were relevant to the study.

4 Results

In the current study, we identified the behaviors, activities, and touchpoints of the respondents as they experienced their destination visit. For ease of interpretation, the results from the aforementioned research questions are summarized in Fig. 2. The customer journeys incorporated tourist experiences through the realms of experience and brand building blocks of all the touchpoints that the respondents had contact with during their stay at the Millennium Gaea Resort Hualien.

4.1 Elements of Experience Realms Relevant to the Customer Journeys

As shown in Fig. 2, the results reveal the realms of experience that were perceived by the respondents, and the corresponding touchpoints. Most of the respondents perceived escapism and aesthetic experiences, followed by education and entertainment experiences. Escapism is related to the privacy that is offered by architecture, a sense of home and being far away from the hustle and bustle, feeling undisturbed, and relaxing, which could be derived from the appearance of the resort, its herb gardens and herbal remedy pools, and its herb-related room facilities, etc. For example, a respondent stated that "it didn't look like a normal hotel, it looked more secluded and far away from the hustle and bustle outside". The aesthetic experiences derived from the resort's minimalism, its elegant environment and artistic atmosphere, and its well-designed facilities, including the herb garden landscape and the herbal remedy pools, as well as from the display of the rooms, the private outdoor soaking tubs, and the specially designed products, etc. Some of the respondents particularly appreciated the resort's herb garden, garden landscape, and artistic atmosphere in the rooms, which are elegant and minimalist.

The element of education derives from gaining knowledge about herbs from the guided tour of the herb garden, and from the essential oil extraction equipment and the kitchenettes. A respondent stated that "there were many kinds of herbs in the resort, and we learned about herbs through the guided tour". The entertainment experience is less obvious. Only one respondent stated that "I made tea and chatted with my friends in the tea house and enjoyed the beautiful garden landscape, which were good ways of relaxation".

4.2 Brand Building Blocks

Brand Salience. Brand salience is used to measure how often and how easily a brand is recalled or recognized by customers. Various types of cues that are necessary to customers are offered by enterprises. As shown in Fig. 2, in the pre-experience stage, word of mouth was the element of brand salience that was most often perceived by the respondents, accompanied by news, social media, and signboards. Business philosophy and product/service features in particular are important in enabling customers to recognize brand salience.

	Searching/ Booking	Arrival/ Check-in	Guided tour	Dining	Sleep	Shopping	Check Out
Activities	Searching/ Booking	Arrival/ Check-in	Guided tour	Dining	Sleep	Shopping	Check Out
Touchpoints	Official website Facebook	Parking lot Entrance Lobby Receptionist	Herb gardens Swimming pools Herbal remedy pools Essential oil extraction equipment Yoga Space Tea house Kitchenettes	Restaurant Herb and seasonal local & international cuisine	Herbal remedy facilities Private outdoor soaking tubs Herbal toiletries Spa	Boutique shop Organic agricultural products shop	Customer feedback questionnaire Parking lot
Realms of Experience Escapist		Privacy Home-style Far away from the hustle and bustle	Relaxation		Relaxation Not being disturbed		
Esthetics			Minimalism Elegant environment		Artistic atmosphere Minimalism	Well-designed	
Education			Professional guide Learning about herbs Picking of herbs Cooking				
Entertainment			Making tea and chatting				
Brand building blocks Salience	Word of mouth Social media TV Newspaper Signboard	Warm receptions Comfortable and pleasant space Far away from the hustle and bustle	Interior design	Gourmet food Warm service	Details of room facilities	Smallholder farmers' produce	
Performance	Word of mouth Impressions of the herb garden		Authenticity through the herbal remedy pools, essential oil extraction equipment, kitchenettes, Well-guided tour	Relaxed and healthy lifestyle Gourmet food Organic ingredients Staffs	Reliability, differentiation in hotel facilities, herbal remedy facilities, herbal toiletries	Reliability, differentiation in smallholder farmers, local specialties	
Imagery		Lobby atmosphere Beautiful scenery	Picking herb Learning about herbs Landscape of Yoga Space	Warm service	Herbal toiletries Relaxing space	Organic smallholder farmers Free food samples	
Judgments	Booking Shuttle service	Herbal plants	Well-guided tour Professional services Touch of herb Herbal remedy pools Quick response code	Description of the meal Healthy and delicious food High quality service	Herbal remedy facilities	Organic and healthy food	Inquire of guests about their stay
Feelings		Far away from the city Privacy of architecture		Warmth and security by service, interior design, and meals Gourmet food Receptionist Valuable service	Tranquility	Fun with smallholder farmers	
Resonance			Herbal remedy program Day tour	New dishes Special offers			

Fig. 2. Elements of realms of experience and brand blocks relevant to customer journey maps

In the experiential stage, a warm reception, a comfortable and pleasant environment and interior design, room facility design, organic meals, and smallholder farmers' products all contributed to the brand elements coming to mind. For example, one of the respondents shared the following experience: "When he arrived at the hotel, a dedicated receptionist warmly greeted him and introduced the room facilities in detail. If you want to get away from the hustle and bustle, it's a great place, and the overall design of the hotel allows you to relax completely. In particular, the hotel integrates organic ingredients from nearby small farmers to provide meals". Another respondent stated that she "loved the minimalist style and herb-themed devices beyond her imagination". The results reveal that dedicated service, the style of the hotel, meals that incorporated locally grown ingredients, etc., all contributed to the depth of brand awareness during the experiential stage.

Brand Performance. How did the customers view brand performance, such as reliability, durability, serviceability, style, and prices? In the pre-experience stage, positive word-of-mouth feedback and long-standing brand impressions were the most important elements of brand performance. A respondent mentioned that "this farm had a long history of running the herb garden, and when it was transformed into a resort, I thought its brand is reliable".

During the experiential stage, the respondents often made performance-related associations with the guided tour, and with the dining, accommodation, and shopping. As shown in Fig. 2, the results reveal three frequently mentioned factors within brand performance: authenticity, reliability, and differentiation. When the customers visited the herb garden and herb remedy pools, and engaged in essential oil extraction in the kitchenettes with help from the guides, they acknowledged the authenticity and reliability of the herb-themed resort. Most of the respondents stated that while experiencing the steam baths and enjoying the landscape, herbal meals, and kitchenettes, they perceived the relaxed and healthy life-style offered by the resort.

Brand Imagery. Brand imagery refers to the abstract way in which customers think about a brand, rather than what they think the brand actually does [7]. The respondents perceived that the brand imagery of the Millennium Gaea Resort Hualien included elements of quiet, intellectual pursuit, emotional tranquility, healing and health, and upper class. For example, one respondent stated that "the brand focuses on health care, full of the five senses of hearing, sight, smell, taste, touch, etc.".

Situations for which the respondents thought it was appropriate to use the brand included relaxing, having experiences with friends, traveling after retirement, entertaining guests, spending idle time luxuriously, and taking advantage of discount programs, etc. The pleasant memories which the respondents shared included well-served dining experiences, the spacious and elegant atmosphere, the herb-themed housing facilities, the herb garden and the essential oil extraction facilities for learning about herbs, the friendly staff, and the practical tasting experience at the organic agricultural products shop. A respondent mentioned that "from the entrance, the service staff gave a very enthusiastic reception, followed by a detailed description of the room facilities. I thought these services were very fine and be fitting of a herbal leisure-themed healthcare hotel".

Brand Judgements. As regards the customers' personal opinions about the brand, four types of judgements were made: quality, credibility, consideration, and superiority. Firstly, brand quality mostly reflected certain associations that mattered to the customers, such as the dining service, guided tours, herb-related remedy facilities, staff service quality, design, style and appearance, etc. One of the respondents stated that "the herb steam bath in the room and the outdoor vanilla spa pool were very special and met my expectations. The meal was delicious and professional, but it was a little less than is characteristic, but the quality of the service was good". Some of the respondents spoke about how "the guide's tone was relaxed, well-mannered, and informal, and tourists were encouraged to touch the herbs".

Secondly, regarding the level of credibility behind the brand, the respondents appreciated the fact that herbs were used widely in the resort, such as in some of the products and services, and that the QR code offered by the resort enhanced the customers' understanding of herbs. One of the respondents stated: "as regards the ecological environment of herbs and the use of ingredients produced by small holders in meals, the staff fully communicated the concept of healthy-living aesthetics to customers". Thirdly, to what extent did the customers view the brand as unique and superior to other brands? The organic herbs, the organic healthy food, the herbal recreational facilities, the relaxation spaces, and the geographical location all gave the resort an advantage over alternative destinations. For example, a respondent mentioned the "organic herbs designed into the holistic touchpoints, which was very special".

Brand Feelings. As regards the emotional responses and reactions to the Millennium Gaea Resort Hualien brand, the respondents perceived warmth and security, relaxation, fun, self-respect, tranquility, and privacy, all of which contributed to their feelings about the Millennium Gaea Resort Hualien brand.

Some respondents observed that the design and appearance of the building, the quality of service, and the meals gave them a feeling of warmth, security, and relaxation. One respondent stated: "the resort produced a feeling of self-respect through its dedicated service". The element of fun consisted of the feeling of a connection with the surrounding environment and with small-holder farmers, and the feeling that these co-creative values touch tourists' hearts.

Brand Resonance. As regards the dimensions of brand resonance—behavioral loyalty, attitudinal attachment, a sense of community, and active engagement—all the respondents expressed their willingness to revisit the brand, and expected new programs from the brand, such as one-day tours, herb planting, horticultural therapies, etc. Additionally, the respondents expected new service information about the cuisine, special offers, wedding programs, etc.

5   Discussion

This study explored realms of experience and brand blocks through service design in the creative life industry from the perspective of tourists. Through cross-analyses of perceptions during customer journeys, we proposed a design framework for brand experiential service in the service blueprint, as shown in Fig. 3. This study combined service blueprints with brand experiences from customer perspective to described the frontstage and backstage operations that support customer actions in the service interfaces.

First, the findings showed that the realms of experience perceived by the respondents were in relation to brand performance, imagery, judgement, and feelings. This finding

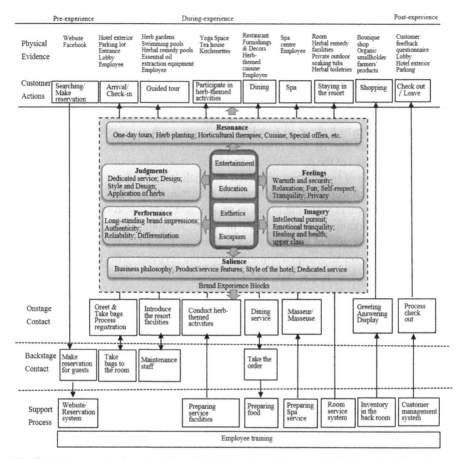

Fig. 3. A framework of service blueprint for brand experiences to the Millennium Gaea Resort

affirms that realms of experience can be regarded as a brand experience proposition and can be transformed into Keller's [7] "branding ladder". The elements of escapism corresponded to the touchpoints of brand performance (e.g., the herbal remedy pools, relaxation rooms, herbal remedy facilities, and herbal toiletries in the rooms), brand imagery (e.g., the herbal toiletries, and relaxation rooms), and brand feelings (e.g., being away from the city, and privacy of architecture). The elements of aesthetics (e.g., minimalism, elegant environment, and artistic atmosphere in the rooms) corresponded to the touchpoints of brand performance (e.g., the herb garden, landscape, and herbal remedy pools), brand imagery (e.g., the room furnishings), and brand feelings (e.g., the interior design). The elements of education (e.g., professional guide services, learning about herb gardens, and cooking) corresponded to the touchpoints of brand performance (e.g., the herbal remedy pools, essential oil extraction equipment, kitchenettes, and well-guided tours), brand imagery (e.g., picking herbs, and learning about herbs), and brand judgements (e.g., QR code). The elements of entertainment (e.g., chatting with friends) corresponded to the touchpoints of brand performance (e.g., the tea house). The findings of this study show that both the emotional and rational routes should be designed into the realms of experience.

Second, the current study extends the spectrum of escapism in the creative life industry [12, 14, 15] to privacy, a relaxing natural environment, and tranquility/not being disturbed. A lifestyle resort such as the Gaea offers a relaxing atmosphere with minimalist design and unique features, where people can unwind, not just in the rooms but throughout the resort. As regards the various stages of brand building, escapist experiences can be enhanced through theme-related facilities (e.g., herbal remedy pools, herbal remedy facilities in the rooms), guided tours, and products and services that improve the brand's performance, all of which represent a rational route. Brand feelings are also important in terms of shaping customers' escapist experiences, and this represents an emotional route. The finding shows that respondents perceived the resort was the distance from the hustle and bustle of life, as well as the relaxation of being in a place away from the cities.

The elements of esthetics were related to the beauty of nature (e.g., elegant and quiet surroundings), the beauty of the art (e.g., minimalist design, artistic atmosphere). Aesthetic experiences can be formed by meeting customers' functional needs in relation to brand performance (e.g., authenticity of the herb garden, landscape, herbal remedy pools) and in terms of emotional routes through imagery associations (e.g., room furnishings) and brand feelings (e.g., interior design). This study extends aesthetics from emotional to rational views.

The elements of education were in relation to the professional guides, learning about herb gardens, and cooking, which is consistent with the literature, since they focus on "learning experiences" [12, 14, 15]. Education experiences can and should be formed through the rational and emotional routes of brand performance (e.g., the herb garden, essential oil extraction equipment, kitchenettes) and brand imagery (e.g., picking up herbs).

Third, as regards brand salience, this study showed that most of the respondents did have a perception of brand awareness, in that they were willing to visit the branded resort. On the other hand, the depth of their brand awareness was insufficient; for example, the tourists not only expected to visit kitchenettes but also expected to experience cooking at kitchenettes. According to this case study, the best approach for improving the quality of visits in the creative life industry is to not only improve tourists' attitudes toward the brand but also to increase the depth of brand awareness, whereby more tourists would consider experiencing the brand.

Fourth, brand resonance focused on the ultimate relationship and level of identification that the tourists had with the brand [7]. For example, most of the respondents perceived escapism in terms of relaxation, privacy, etc., and stated that they would talk about the brand to others and revisit the place with family or friends. The findings of this study reveal that brand resonance resulted in greater realms of experience.

6   Conclusion and Implications

This study contributes to the literature on brand experience in the creative life industry, and develops a design framework for brand experiential service from the perspective of tourists. This study has both theoretical and managerial implications.

First, the realms of experience reflected the main experiences offered by creative life enterprises. Escapism appeared to be a relatively salient dimension of experiential values. In this study, aesthetic experience was associated with escapism, derived from the herb garden's recreational facilities. Educational experience was associated with aesthetics, in terms of enjoying and learning about herbs. The results revealed that the respondents' perceptions of the realms of experience through touchpoints could influence each other. Creative life enterprises should reinforce ways of triggering other experience realms in order to provide tourists with comprehensive and high-quality experiences.

Second, the tourists' perceptions of the realms of experience and brand blocks mostly derived from specific touchpoints, such as the herb garden, the herbal remedy pools, the cuisine, the relaxation rooms, and the staff, etc. A one-source multiple-use strategy should be incorporated into the design of touchpoints, and peak experiences should be identified for build-up for tourists.

This study focused on one small country—Taiwan. It therefore has limitations. The study used qualitative research methods for a specific creative life enterprise. Conclusions that are generalizable across the creative life industry and across countries cannot be easily drawn. The creative life industry has a wide range of variations. Future studies should replicate the current study in other creative life industry contexts in order to explore further realms of experience and CBBE and construct a universal model.

Acknowledgments. The authors gratefully acknowledge the support for this research provided by the Ministry of Science and Technology, Taiwan under Grants MOST 108-2221-E-259-004. The authors also wish to thank those who contributed to the research.

References

1. Chang, S.-H., Lin, R.: Building a total customer experience model: applications for the travel experiences in Taiwan's creative life industry. J. Travel Tourism Mark. **32**(4), 438–453 (2015). https://doi.org/10.1080/10548408.2014.908158
2. Gilmore, H.J., Pine II, B.J.: The Experience Is the Marketing. Brown Herron Publishing, Amazon.com eDoc (2002)
3. Gilmore, H.J., Pine II, B.J.: Differentiating hospitality operations via experiences: why selling services is not enough. Cornell Hotel Restaur. Adm. Q. **43**(3), 87–96 (2002)
4. Pine II, B.J., Gilmore, J.H.: The Experience Economy: Work is Theatre & Every Business a Stage. Harvard Business School Press, Boston (1999)
5. Aaker, D.A.: Managing Brand Equity. The Free Press, New York (1991)
6. Keller, K.L.: Conceptualizing, measuring, and managing customer-based brand equity. J. Mark. **57**, 1–22 (1993)
7. Keller, K.L.: Strategic Brand Management: Building, Measuring, and Managing Brand Equity, 2nd edn. Prentice Hall, Upper Saddle River (2003)
8. Yoo, B., Donthu, N.: Developing and validating a multidimensional consumer-based brand equity scale. J. Bus. Res. **52**(1), 1–14 (2001)
9. Stickdorn, M., Frischhut, B.: Service Design and Tourism: Case Studies of Applied Research Projects on Mobile Ethnography for Tourism Destinations. Books on Demand GmbH, Norderstedt (2012)
10. Stickdorn, M., Schneider, J.: This is Service Design Thinking: Basics, Tools, Cases. Wiley, New York (2012)

11. Lee, G., Tussyadiah, I.P., Zach, F.: A visitor-focused assessment of new product launch: The case of Quilt Gardens Tour^{SM} in Northern Indiana's Amish Country. J. Travel Tourism Mark. **27**(7), 723–735 (2010)
12. Hosany, S., Witham, M.: Dimensions of cruisers' experiences, satisfaction, and intention to recommend. J. Travel Res. **49**(3), 351–364 (2010)
13. Kastenholz, E., Carneiro, M.J., Marques, C.P., Loureiro, S.M.C.: The dimensions of rural tourism experience: impacts on arousal, memory, and satisfaction. J. Travel Tourism Mark. **35**(2), 189–201 (2018)
14. Oh, H., Fiore, A.M., Jeoung, M.: Measuring experience economy concepts: Tourism applications. J. Travel Res. **46**(2), 119–132 (2007)
15. Quadri-Felitti, D., Fiore, A.M.: Experience economy constructs as a framework for understanding wine tourism. J. Vacat. Mark. **18**(1), 3–15 (2012)
16. Boo, S., Busser, J., Baloglu, S.: A model of customer-based brand equity and its application to multiple destinations. Tour. Manag. **30**(2), 219–231 (2009)
17. Kladou, S., Giannopoulos, A.A., Mavragani, E.: Destination brand equity research from 2001 to 2012. Tourism Anal. **20**(2), 189–200 (2015)
18. Konecnik, M., Gartner, W.C.: Customer-based brand equity for destination. Ann. Tourism Res. **34**(2), 400–421 (2007)
19. Petromilli, M., Michalczyk, D.: Your most valuable asset. Mark. Health Serv. **19**, 4–9 (1999)
20. Mager, B.: Service design as an emerging field. In: Miettinen, S., Koivisto, M. (eds.) Designing Services with Innovative Methods, pp. 28–42. Taik Publications, Helsinki (2009)
21. Osterwalder, A., Pigneur, Y., Bernarda, G., Smith, A., Papadakos, T.: Value Proposition Design: How to Create Products and Services Customers Want. Wiley, Hoboken (2014)
22. IDEO: Human-Centered Design Toolkit. IDEO, San Francisco (2011)
23. Cook, L.S., Bowen, D.E., Chase, R.B., Dasu, S., Stewart, D.M., Tansik, D.A.: Human issues in service design. J. Oper. Manag. **20**(2), 159–174 (2002)
24. Trischler, J., Zehrer, A.: Service design: Suggesting a qualitative multi-step approach for analyzing and examining theme park experiences. J. Vacat. Mark. **18**(1), 57–71 (2012)
25. Tussyadiah, I.P.: Toward a theoretical foundation for experience design in tourism. J. Travel Res. **53**(5), 543–564 (2014)
26. Zehrer, A.: Service experience and service design: concepts and application in tourism SMEs'. Manag. Serv. Qual. **19**(3), 332–349 (2009)
27. Yin, R.K.: Case Study Research: Design and Methods, 3rd edn. Sage, London (2003)
28. Morgan, D.L., Krueger, R.A.: When to use focus groups and why. In: Morgan, D.L. (ed.) Successful Focus Group, pp. 3–19. SAGE, Newbury Park (1993)

Research on Sticker Cognition for Elderly People Using Instant Messaging

Cheih-Ying Chen[✉]

Department of Product Innovation and Entrepreneurship,
National Taipei University of Business, Taipei City, Taiwan, R.O.C.
c.y.chen@ntub.edu.com

Abstract. Elderly people neglected by relatives or friends have a poor social network and are thus lonely. Such loneliness can be alleviated through diverse Internet-based communication methods. Specifically, elderly people can effectively connect with their friends, relatives, children, and grandchildren if they can use instant messaging programs on smartphones. However, elderly people face health-related physical difficulties and loss of memory. Therefore, determining elderly people's needs and their ability to effectively understand instant messages is imperative. Accordingly, this study explored elderly people's perceptions and cognition of instant messaging software stickers. According to the results of a survey conducted in this study, stickers were designed to meet elderly people's communication requirements. Testing was conducted to verify whether the elderly could use the stickers appropriately.

First, this study explored the problem of social connection for elderly people from the perspective of human care and applied contextual inquiry to under-stand problems encountered by elderly people when using smartphones. Subsequently, the study designed instant messaging stickers. Sticker card sorting was used to study how elderly users use the designed instant messaging stickers. A semantic design method was used to design the stickers. This method entails assigning meanings to images and increases the ability of expressing emotions by using cartoon characters. The study results revealed that the semantic sticker design method could enable elderly users to effectively understand emotions conveyed by stickers without requiring textual descriptions.

Keywords: Cognitive ergonomics · Instant messaging · Semantic design · Sticker

1 Introduction

1.1 Research Background and Motivation

In addition to health concerns, lack of social connection is a problem encountered by elderly people. The number of divorced or widowed elderly people passing away alone is estimated to increase. Although elderly people are usually married having children, their children are often unable or unwilling to take care of them. Wildevuur, Dijk, Hammer-Jakobsen, Bjerre, Äyväri, and Lund stated that a person undergoes three major transitions

© Springer Nature Switzerland AG 2020
P.-L. P. Rau (Ed.): HCII 2020, LNCS 12192, pp. 16–27, 2020.
https://doi.org/10.1007/978-3-030-49788-0_2

in their later years: retirement, disability, and loss of a beloved [1]. These major transitions considerably affect elderly people's lifestyles. The transitions also break up their social connections, such as connections with colleagues, family members, and society on the whole, thus engendering negative emotions, such as sadness and loneliness. These conditions increase the risk of "social death"—which means losing connection with any form of social activity—among elderly people [2]. In addition to losing interest in all activities and being deprived of intimacy, care, interaction, identification capabilities, compassion, and dignity, elderly people are marginalized from society. They face crises in managing interpersonal relationships. Approximately half of the total elderly population is in contact with less than three family members or less than five friends. Disconnection with family and friends limits interpersonal relationships; elderly people are often forgotten by their relatives or by the people in their surroundings. As the gap in the connection between human beings increases, members of society eventually become lonely. Elderly people who stop engaging in social activities or in any occasion of interest to them experience reduced positive emotions. The aforementioned phenomena thus raise the following questions: How can elderly people live alone? How to make elders' mental prepared? The terms "alone" and "lonely" are not synonymous. Elderly people's children may not visit them probably because they do not have time to come home; moreover, elderly people may not meet their friends possibly because the friends do not have time to socialize. This thus raises the following question: How can the network of the elderly population be developed under this condition? Mausbach, Roepke, Depp, Patterson, and Grant suggested that the Internet or other communication technologies could solve the problem of loneliness faced by elderly people [3]. Communication technologies such as telephones, cellphones, and the Internet afford multiple means of communication. These technologies can enable individuals to share personal emotions, thus affording communication without time or space limitations. Elderly people can interact with others through various methods, such as chatting, telephone conversations, and texting. Furthermore, computer communication networks provide other options for interaction, such as calling through mobile phones, sending e-mails, or sending instant messages. Elderly people who can operate smartphones can connect with their relatives, friends, children, and grandchildren.

1.2 Research Purposes

Helping elderly people through the major transitions in their lives is crucial. Therefore, developing strategies, products, or services for satisfying their needs or for their habits is imperative. Maintaining elderly people's engagement in social activities, independence, and perceptions of connection with society through friends can help ensure that they live comfortable lives.

The advent of telephones, cellphones, and the Internet has enhanced people's connection, especially through communication networks. Several means of connection or communication have been established. Technologies such as e-mails, Facebook, and Line have increased the frequency and convenience of interaction among distant relatives. Technology-based interpersonal relationships have become more diverse and convenient. Accordingly, such technologies can enable elderly people to establish connections with relatives who are very far from them and thus have difficulty visiting to take care of them.

People aged older than 65 years who own a smartphone and know how to use the Internet can learn about one another's recent situation and share the joy of life through communication. The 1950s marked the beginning of the digital revolution; people born before 1950 are deemed too old to learn about computers, whereas those born after 1950 have learned the basics about computers before reaching their midlife years. Nevertheless, elderly people older than 65 years directly learn how to use a smartphone without learning how to use a computer. The traditional telephones and 2G cellphones have gradually exited Taiwan's electronic market, and most young individuals are netizens; therefore, elderly people are compelled or volunteer to replace their telephones with smartphones. People using instant messaging software can communicate through several media types such as text, images, sound, photos, and videos. Moreover, message stickers can be used to express emotions and pleasure in communication. Stickers represent anthropomorphic expressions and movements of cartoon characters, and some stickers include text representing emotions. However, elderly people may misinterpret the expressions and movements represented by stickers.

This study explored elderly people's perceptions and cognition of instant messaging software stickers to determine their preferences. The purpose of this study was to enable elderly people to use instant messaging stickers comfortably and to autonomously operate instant messaging systems. Elderly people's eyesight, learning ability, and cognition decrease gradually with age, a phenomenon that affects elderly their image recognition ability and eventually the type of stickers they prefer. Accordingly, to ensure comfortable and independent operation of communication systems for elderly people, the software interface and sticker designs of instant messaging software must be simplified.

2 Literature Review

2.1 Cognitive Ergonomics

The information processing model, also known as information processing psychology, constitutes the main theoretical framework of cognitive psychology. This model is used to explore the process of human perception, mental processing, and memory. Cognitive ergonomics is a new branch of human factors engineering that mainly entails highlighting cognitive processes required by operators in a modern working environment and subsequently researching, evaluating, and designing tasks for the operators. Human–computer interaction design, which focuses on cognitive information system design, is related to psychological processes such as sensation, perception, memory, attention, and motor response; this is because humans interact with other elements of a system during cognitive information processing [4]. Computer technology has been applied in many fields and has posed new challenges to human factors engineering. The primary challenge pertains to the interaction between humans and smartphones. Because of the diversity of information, multiple cognitive and emotional processes have been generated. Thus, human–computer interactions have become relatively complex. A user-centered design should be realized for a human–computer system in order to enhance users' experience.

The emotions and cognition of humans constitute a set of internal rules that can be applied daily to make decisions [5].

Stickers constitute a message mode in instant messaging. Sending or receiving stickers requires cognitive processes that are involved in processing and handling information, namely encoding, rehearsing, recalling, and transforming information in the human brain. The process of sending stickers in instant messaging can be categorized into two parts: encoding and decoding. Encoding involves sending stickers, and decoding involves receiving and interpreting stickers. Although the meanings of most stickers can be understood easily, the meanings of some stickers can be misunderstood. If a particular sticker is applicable in different contexts, at different times, or with different, users may have different interpretations of the meaning of the sticker. This problem can be attributed to erroneous decoding or encoding processes or to poor sticker design; accordingly, the sticker becomes meaningless. An effective sticker can directly and precisely express its meaning; by contrast, an invalid sticker can be completely misinterpreted by users. If a sticker from an instant messaging program can adequately express the sentiment of the sender and enable the receiver to understand the sentiment correctly, then the user's attitude toward and affinity with the sticker can improve. However, if the meaning of the sticker is unclear, the communication between the sender and receiver may involve cognitive errors. Accordingly, the meaning of stickers should be clear, and senders and receivers should have the same comprehension of a sticker; otherwise, stickers may cause confusion between users.

According to Table 1, stickers have a direct and obvious effect on the process of emotion expression. However, if the emotions represented by stickers are excessively exaggerated, elderly users may misunderstand the meaning of the stickers [6]. Initially, most stickers were composed of only images. With time, increasing numbers of stickers have included simple text to prevent cognitive errors. The addition of text to stickers may fill the generation gap between users of different ages. Accordingly, this study addressed problems pertaining to the design of stickers for elderly users and cognitive errors in communication among such users.

Table 1. Interpretation of stickers

Sticker				
The participants' interpretations	1.My apologies 2.Sorry 3.Can't recall 4.Acting cute	1.I am cool 2.It's you 3.You get me 4.I am confident on myself	1.I love you 2.I like it 3.Blowing a kiss 4.Over the moon;	1.Cool, isn't it? 2.I'm awesome! 3.I'm smart 4.Things are going well 5.Let me think

2.2 User Experience

This study explored the experience of elderly users while using instant messaging software in order to understand and address the difficulties encountered by such users. Accordingly, the following factors and approaches were considered on the basis of a literature review.

Contextual Inquiry. By studying users' product experience, product designers or engineers can under-stand customers' difficulties and requirements for using the product. Contextual inquiry is a qualitative research method. In this method, researchers can understand a question of interest through interviews, identify the problem, and devise strategies to solve the problem. In the context of product design, designers can conduct one-to-one interviews to understand users' experience of the product. During such interviews, designers can ask questions and observe every aspect of the problem encountered by users. Thus, designers and users can communicate in person, and the designers can comprehensively understand user difficulties and usage conditions [7]. In contextual inquiry, first-hand information can be collected, and the most direct reaction of the user can be noted. Data can be collected and organized to ensure their quality, thus enabling designers to modify product designs and improve product functionality.

Focus Group. The focus group method is a qualitative research tool that enables researchers to effectively, rapidly, and accurately understand a topic and collect data that can truly relate to the problem under investigation. Implementing this method still involves some challenges; nevertheless, the method is superior to and more effective than other existing research approaches. According to Mousa, participants, moderators, questionnaires, and other settings considered in a focus group should be appropriately configured according to a theme. The establishment of a theme determines the features of the focus group. The focus group method does not have specific standards, which increases the flexibility of the focus group during implementation and the value of the data obtained from the interviews [8].

Card Sorting. Card sorting is another experimental method that entails the use of cards as objects to enable respondents to perform classification actions. Qualitative and quantitative data can be obtained through card sorting, and researchers can select the required data according to the features of the research topic. Previously, card sorting themes mostly involved the classification of website architecture catalogs or topics related to information [9].

Different card sorting methods can be used for different testing purposes. The methods are roughly categorized into two types—closed and open. Closed card sorting can be used to determine how users classify content items that are set currently. Researchers can state the hypothesis of the current architecture and conduct sorting tests. Open card sorting entails establishing a neutral structure first. Subsequently, the tester must classify and confirm whether the content items are appropriate. Testers can focus on discussing the problems and attempt to meet the requirements of the interviewees to solve the problems.

2.3 Semantic Design

Cartoon characters, patterns, colors, and other visual elements are combined to design and develop stickers, engendering effects such as identification, description, and communication in the visual design field. The use of stickers is a trend currently. Stickers represent a "picture language" that can be easily remembered by any native language speaker. Appropriate stickers can replace language and can transcend language barriers. In sticker design, a set of interesting, visual, and lively designs are generated for conveying emotions during communication. The use of stickers is a crucial mode of instant communication (through a graphical language interface) between people. Stickers can be designed to convey specific feelings to users through cognitively understandable and interesting graphical language interfaces. This can thus enable users to break through language and cultural barriers, effectively convey their concepts or messages, and generate specific emotional attachments and actions to achieve fast and effective communication.

Advancements in cognitive linguistics and structural theory have largely promoted theoretical considerations of metaphors and metonymies [10]. A metaphor is a conceptual view in which one object is interpreted as another. In a metaphor, some qualities are transferred from a source domain to a target domain. For example, for two objects, namely A and B, a metaphor interprets A as B. A metonymy involves evoking the name of an object for another object that is associated with the first object; a common metonymic concept is "part for the whole". In a metonymy, one concept is substituted for another related concept. For example, A is substituted for B, where B has a salient association with A. An example of a metaphor is as follows: It is raining cats and dogs. Moreover, an example of a metonymy is as follows: It is an umbrella weather. In a metaphor, two elements are linked based on some specific analogy. However, in a metonymy, two elements are linked based on some understood association or contiguity. The major difference between a metaphor and a metonymy is that in a metaphor, two ideas are linked, whereas in a metonymy, a part is replaced for a whole [11].

3 Research Methods and Steps

On the basis of user experience, this study investigated the various cognitive communication problems faced by elderly people when using instant messaging application stickers through their smartphones. The steps of this study are described as follows. First, an elderly community center was visited to observe and interact with different elderly users to conduct user experience research. Second, observation results and contextual inquiries were collected, open-ended cognitive responses and preference were collected through questionnaires, and problems were summarized. Third, stickers were designed on the basis of the semantic design and survey conducted on elderly users. Finally, the instant messaging stickers were analyzed qualitatively and quantitatively to objectively discuss the problems experienced by the elderly users when using the stickers for social communication and determine the effects of the use of such stickers on communication.

3.1 Research Steps

Step 1. Observation

- Contextual inquiry: Research was conducted on the cognition and preferences of elderly users of instant messaging stickers.
- Focus Groups: Focus group interviews were conducted with elderly user of instant messaging stickers in Taiwan, and their experience of using stickers was acquired. After the interviews, specific research hypotheses were proposed.

Step 2. Texture design

- Semantic design: Stickers were designed on the basis of expressions, gestures, and actions. The designed stickers were supplemented with metaphorical and metonymical representations to effectively convey messages.
- Non-semantic design: Stickers were designed on the basis of expressions, gestures, and actions. However, the designed stickers were not supplemented with metaphorical or metonymical representations.

Step 3. Experimental design

- Card sorting: Card sorting was conducted using the survey data obtained from the elderly users to discuss the topic of sticker classification.
- Result analysis: The card sorting results and the effectiveness of the semantic design in enhancing communication were analyzed.

3.2 Focus Group Interviews

Focus group interviews were conducted with elderly users to collect qualitative information pertaining to the sticker design. The different opinions of the users could constitute a multifaceted basis for new design ideas. Six senior users participated in the interviews. The host provided the focus group statement and allowed every user to share their experiences or stories of using the communication software, discuss open-ended problems, present their opinions or requirements, discuss the commonly preferred functions and requirements, and elaborate on the type of stickers that can be understood and used in communication. Finally, the host collected answers and materials related to the topic (Fig. 1).

Fig. 1. Focus group meeting

3.3 Semantic Design of Stickers

Metaphors and metonymies are similar in various aspects. The major difference between these figures of speech is that in a metaphor, a concept is substituted for another for abstract analogy, and in a metonymy, ideas or concepts are substituted for related ideas or concepts. This study designed stickers on the basis of this difference. In the design of stickers representing metaphors, an element was substituted for another design element similar to the original element. In the design of stickers representing metonymies, an element is substituted for a related design element for contiguity. Specifically, stickers representing metaphors were designed by linking two ideas, whereas those representing metonymies were designed by replacing a part for a whole (Tables 2, 3 and 4).

Table 2. Metaphor design

Metaphor sticker	Meaning	Description
	Call back	The lightning symbol with the gesture of placing the phone on the ear to talk is used as a metaphorical representation of calling back.
	Good morning	The sunlight symbol with stretching exercises is used as a metaphorical representation of "good morning."
	Happiness	The heart symbol in the background of the sticker with the gesture of love and joy is used as a metaphorical representation of happiness.

Table 3. Metonymy design

Metonymy sticker	Meaning	Description
	Call back	A hand holding a mobile phone with the action of talking on the phone is a metonymical representation of calling back.
	Good morning	A hand holding a coffee cup and a nightcap with a greeting of "good morning" is a metonymical representation.
	Happiness	The firework symbol as the background with the gesture of sprinkles and joy is a metonymical representation of happiness.

Table 4. Non-semantic design

Non-semantic sticker
These stickers were designed to reflect greetings, happiness, and various other gestures of raising the hand to the ear, hinting expressions and actions such as "good morning," happy, and callback.

3.4 Card Sorting Meetings

This study designed cartoon-style stickers. Subsequently, 30 elderly users were invited and instructed to classify each of the stickers to one the following three categories: metaphor, metonymy, or Non-semantic. The classification was based on the subjective opinions of the users. The purpose of the classification task was to determine the ability of the elderly users to appropriately distinguish the stickers with different semantic designs; the classification results are described subsequently herein.

Elderly people may not understand the emotional expressions represented by stickers because of the generation gap. Thus, misunderstandings in meanings may occur. If such users can effectively understand stickers, they can easily and rapidly communicate and share their feelings with their friends.

In the card sorting process in this study, the compatibility between stickers' expected functionality and the users' experience was first established. Accordingly, the participants were asked to classify and confirm whether each content item was appropriate and to state whether their expectations were met by the stickers. This could thus reduce the possible influence of the researcher on the study findings.

(1) Questionnaire survey: The survey was essentially executed through a one-to-two method. Two enrolled users and the tester performed the test in the same space to reduce the influence of external factors. Information was collected systematically from the completed questionnaires.
(2) Explanation of card sorting to elderly users: The functions of the stickers were explained to the elderly users in a timely manner. The users were instructed to sort sticker cards and respond to raised questions through the questionnaires.
(3) Test result and feedback collection: After the test results were collected, the users were asked to provide feedback. Subsequently, a systematic analysis was conducted, and a complete solution was proposed.
(4) Classification of questionnaire data: The questionnaire data obtained were classified and analyzed to understand the users' perception of the readability, coherence, and attractiveness of the stickers (Fig. 2).

Fig. 2. Card sorting meeting

4 Results and Discussion

4.1 Sorting Results

Table 5 presents the results obtained after 30 elderly users sorted the sticker cards.

Regarding the "callback" sticker design, 15 users selected a metaphorical lighting sticker and 13 users selected a metonymic cell phone sticker. Moreover, for the remaining two stickers, each was selected by one user.

For the "good morning" sticker design, 12 users selected a metaphorical sunlight sticker and 7 users selected metonymic stickers displaying a coffee cup and a night cap image. The remaining 11 users selected among four other stickers.

Finally, regarding the "happiness" sticker design, six users selected a metaphorical heart sticker, eight users selected a metonymic sticker of fireworks and sprinkles, and six users selected a sticker depicting a laughing couple. The remaining seven users

Table 5. Results of the sticker card sorting

	Metaphor	Metonymy	Non-semantic				
Call back							
Frequency	15	13	1	1			
Good morning							
Frequency	12	7	4	3	2	2	
Happiness							
Frequency	5	5	5	2	1	1	1

selected among the remaining three stickers. In addition, three other users selected the metaphorical sunlight stickers.

4.2 Discussion

The semantic design method can be used to effectively convey emotions through stickers. The use of common elements makes it easier for users to identify the meaning of the relevant stickers and increases the readability, comprehensibility, and attractiveness of the stickers.

Among the aforementioned stickers, the lightning symbol is a metaphor for callback, the sunlight symbol is a metaphor for the sun rising in the morning, and the heart symbol is a metaphor for pleasure and joy. Metaphorical designs can indeed help elderly people to recognize the expression conveyed by stickers.

Moreover, the cellphone image in the stickers represents a metonymy for callback. The coffee and nightcap images represent a metonymy for the sun rising in the morning. The fireworks and sprinkles represent a metonymy for pleasure and joy. Metonymical designs can also help elderly users to recognize the expression conveyed by stickers.

Conveying messages through simple emotions, gestures, or actions of characters and shapes can easily cause misunderstanding. Semantic visual elements should be used effectively to break the barriers pertaining to sticker communication.

In design, semantic association enhances creativity. Moreover, metaphorical and metonymical design approaches are crucial cognitive methods. Stickers designed using metaphorical and metonymical design approaches can be better understood by elderly users.

In conclusion, semantic sticker design methods enable elderly people to easily understand and remember stickers without requiring any text description.

Acknowledgments. The author thanks the Yun-Cheng Lee at the Department of Urban Planning and Spatial Information, Feng Chia University, and Yun-Shu Lee at the Department of Information Management, National Sun Yat-sen University, for helping in the data collection and experimental procedures.

References

1. Wildevuur, S., van Dijk, D., Hammer-Jakobsen, T., Bjerre, M., Äyväri, A., Lund, J.: Connect: Design for an Empathic Society. BIS Publishers, Amsterdam (2014)
2. Mulkay, M., Ernst, J.: The changing profile of social death. Eur. J. Sociol. **32**(1), 172–196 (1991)
3. Mausbach, B.T., Roepke, S.K., Depp, C.A., Patterson, T.L., Grant, I.: Specificity of cognitive and Behavioral variables to positive and negative affect. Behav. Res. Ther. **47**(7), 608–615 (2009)
4. Falzon, P.: Cognitive Ergonomics: Understanding, Learning, and Designing Human-Computer Interaction. Academic Press, Cambridge (1990)
5. Rosenzweig, E.: Successful User Experience: Strategies and Roadmaps. Morgan Kaufmann, Burlington (2015)
6. Chen, C.-Y.: Using an eye tracker to investigate the effect of sticker on LINE app for older adults. In: Kurosu, M. (ed.) HCII 2019. LNCS, vol. 11567, pp. 225–234. Springer, Cham (2019). https://doi.org/10.1007/978-3-030-22643-5_18
7. Beyer, H.R., Holtzblatt, K.: Apprenticing with the customer. Commun. ACM **38**(5), 45–52 (1995)
8. Masadeh, M.A.: Focus group: reviews and practices. Int. J. Appl. Sci. Technol. **2**(10), 63–68 (2012)
9. Paul, C.L.: Analyzing card-sorting data using graph visualization. J. Usability Stud. **15**(1), 87–104 (2014)
10. Fass, D.: met*: a method for discriminating metonymy and metaphor by computer. Comput. Linguist. **17**(1), 49–90 (1991)
11. Chandler, D.: Semiotics: The Basics. Routledge, Abingdon (2007)

What Would Be the Next Design Evolution Under the Auspices of Industry 4.0?

Jyh-Rong Chou[✉]

Department of Creative Product Design, I-Shou University, Kaohsiung City 84001, Taiwan, ROC
jrchou@isu.edu.tw

Abstract. Industry 4.0 has recently been used as a synonym for the planned fourth Industrial Revolution, comparable to the radical transformations from the first Industrial Revolution toward "Mechanization" (Industry 1.0) and its subsequent second one toward "Electrification" (Industry 2.0) to the third Industrial Revolution toward "Digitization" (Industry 3.0). Product design accompanied by the industrial and technological developments has also evolved from "Design for mechanical production" (Design 1.0) and "Design for mass production" (Design 2.0) to a systemic paradigm shift toward "human considerations" and "environmental concerns" (Design 3.0). The concept of Industry 4.0 describes the trend toward digitalization, networklization, and intellectualization of the production environment, enabling industries to plan and project a product's entire value chain over its lifecycle. As such, a critical issue to both academia and industry is to ponder what will be the next design evolution (i.e., Design 4.0) in response to the initiative of Industry 4.0. Through a comprehensive review of current technological development, this study proposes a new perspective on total product-service design (TPSD), the purpose of which is to develop a practical methodology to facilitate the design and development of innovative products.

Keywords: Industry 4.0 · Design evolution · Total Product-Service design · Design methodology

1 Introduction

Industrialization plays a vital role in sustaining the advancement of human civilization, transforming our socioeconomic structure from an agrarian paradigm into an industrial one based on the manufacturing of goods. Historically, the first transformation (Industry 1.0) is known as the Industrial Revolution that began in Great Britain and quickly spread throughout Europe and North America in the mid-18th century with the shift toward mechanization. Individual manual labor was replaced by mechanical production through the employment of the steam engine and different types of machinery. The second period of radical transformation (Industry 2.0) started with the introduction of assembly lines and mass production with the application of electrical energy at the end of the 19th century. The division of labor is a key transition in production evolution for an assembly line that allows the separation of a work process into a number of tasks with

© Springer Nature Switzerland AG 2020
P.-L. P. Rau (Ed.): HCII 2020, LNCS 12192, pp. 28–45, 2020.
https://doi.org/10.1007/978-3-030-49788-0_3

each task performed by a separate worker or group. The advent of the second transformation opened the door to an age of affordable consumer products for mass consumption. In the mid-20th century, the use of digital electronics and information technology in industrial processes ushered in a new era of optimized and automated production. The computer-supported collaboration had emerged as a new way of industrial production in the third transformation (Industry 3.0), enabling the development of a digital and automated manufacturing environment as well as the digitization of the value chain. On the other hand, the consequences of mass production and mass consumption marked a major turning point in the Earth's ecology and humans' relationship with their environment. People had rethought profoundly the negative impact of industrialization on the environment. The concept of cleaner production had emerged as a preventive, industry-specific environmental protection initiative in the 1990s, focusing on the continuous application of integrated environmental strategies to products, processes, and services to increase efficiency and reduce risks to humans and the environment [1].

Since the first Industrial Revolution, subsequent transformations have resulted in radical changes in manufacturing and production, and industrial processes have become increasingly complicated and more difficult to co-ordinate. Moreover, current trends toward globalization, glocalization, demographic transition, and sustainable mobility associated with short product life cycles and diverse customer needs have challenged industries to achieve high flexibility, efficiency, and reliability for rapid product development and launch. Many countries attempt to rebuild their industrial structures and have proposed their own strategic initiatives in order to face the new challenges, such as the Advanced Manufacturing Partnership (AMP) by the U.S. Government in 2011, the German Industry 4.0 in 2012, the Japanese Industry Revival Plan in 2013, the Manufacturing Industry Innovation 3.0 Strategy by the South Korean Government in 2014, the Made in China 2025 Plan, and the Taiwan Productivity 4.0 Initiative in 2015. As an industrial giant worldwide, German Industry 4.0 is one of the most popular topics among industry and academia in the world and has been considered as the fourth Industrial Revolution with extreme impact on manufacturing in future [2–4]. It has been introduced as a paradigm shift in industry to describe the trend toward digitalization, networklization, and intellectualization of the production environment, leading to an intelligent, connected, and decentralized production and enabling industries to plan and project a product's entire value chain over its lifecycle [5, 6]. A core aspect of the Industry 4.0 is a continuous communication between humans, machines, and products during the production process enabled by cyber-physical systems (CPS) that integrate production facilities, warehousing systems, logistics, and even social requirements to establish the global value creation networks in which intelligent objects communicate and interact with each other [7, 8]. The overall aim is to increase cost- and time-efficiency and improve product quality, associated with the enabling technologies, methods, and tools [9].

Design is the conscious and intuitive effort to impose meaningful order by taking something from its existing state and moving it to a preferred state, while industrial design is a process of design applied to products that are to be manufactured through industrial production techniques. Accompanied by the industrial and technological developments, there are four design evolution phases corresponding to the four times of radical transformation in industry. At the early stage of the Industrial Revolution, people were not used to the machine-made products and some arguments had been proposed in response to the negative social and artistic consequences of the industrial production. For example, the Arts and Crafts Movement emerged in the mid-19th century, advocated production by traditional craftsmanship and attempted to revive earlier standards of workmanship and design for the handcrafted production. "Design for mechanical production" is the first evolution (Design 1.0) which sought to reform design processes to fit the industrial transformation in machinery and factory production. "Design for mass production" was a critical issue for the second industrial transformation (Design 2.0) in which normalized, standardized, and modularized design advanced the production of large amounts of consumer products via the processes of division of labor in an assembly line. The Bauhaus, founded in 1919, was the major influence on the second design evolution phase whose efforts contribute to the establishment of modern design fundamentals and the fulfillment of the idea that mass production could live in harmony with the artistic spirit of individuality. With the incredibly fast-paced advances being made in computer and information technologies, automation and digitization had emerged into the mainstream of the third industrial transformation (Design 3.0). There had been significant progress in industrial design and various design theories and approaches had been proposed in response to the wide application of such new technologies in products and production processes. In addition to computer-aided tools and techniques (e.g., CAD/CAM/CAE, concurrent and collaborative engineering etc.), human considerations had been involved in various aspects of product or system design. A large number of sophisticated human-focused theories and methodologies had been developed and employed, such as human factors and ergonomics (HF&E), human-computer interaction (HCI), usability engineering (UE), Kansei engineering (KE), user-centered design (UCD), interaction design (IxD), universal design (UD), and user experience (UX) design. On the other hand, environmental concerns have become strategically important to product design, development, and production since the 1990s. Various approaches had been developed and employed to address the issue of sustainable consumption and production, which responds to basic needs and brings a better quality of life while minimizing the use of natural resources and toxic materials as well as the emissions of waste and pollutants over the life cycle of the use of services or products so as not to jeopardize the needs of future generations [10]. These approaches include green design, eco-design, design for environment (DfE), design for sustainability (D4S), and various types of life cycle techniques (e.g., life cycle assessment/analysis (LCA), life cycle cost analysis (LCCA), and life cycle engineering (LCE)).

Design evolution is sequential and progressive associated with industrial and technological developments as well as human lifestyle needs. From the beginning of the 21st century onward, the world has changed radically in a few major steps. New industrial technologies and business models are emerging increasingly and rapidly. Robotics and artificial intelligence (AI), 3D printing, virtual/augmented/mixed reality (VR/AR/MR), social media, mobile networks, cloud platforms, and electronic commerce are changing our lifestyles and affecting the social, economic, and environmental aspects as well. The concept of Industry 4.0 is expected to radically transform industrial production and product value chains as well as business models, which implies that there tends to be a much closer relationship between manufacturing and service industries. Product design and manufacturing can no longer be the only source of competitive advantage and differentiation, whereas servitization of manufacturing in product firms has become one of the most active domains [11–13]. This domain is concerned with product firms shifting from designing, developing, and manufacturing products to innovating, selling, and delivering services [14–16]. Although current product-service systems (PSS) are regarded as a market proposition that extends the traditional functionality of a product by incorporating additional services to co-create value-in-use for customers [17–19], most existing PSS design methods support the conceptual design phase at the higher levels of abstraction and the technical design phase strongly lacks for methodical support. This can be explained by the existing challenge of coupling products and services in design processes since these two aspects focus on different perspectives and use specific models which appear as difficult to integrate [20]. In this context, this study proposes a new perspective on total product-service design (TPSD) beyond the existing product-service systems (PSS). The purpose of this study is to develop a practical methodology to facilitate the design and development of innovative products as well as to achieve the total performance of products.

2 Materials and Methods

2.1 Cyber-Physical Systems (CPS) and Internet of Things (IoT)

Industry 4.0 comprises a variety of state-of-the-art technologies (e.g., cyber-physical systems, RFID, wireless sensor networks, big data, cloud computing, Internet of things, etc.) to enable the development of a digital and automated manufacturing environment as well as the digitization of the value chain [21–23]. This results in improvements in product quality and a decrease of time-to-market as well as improvements in business performance [24]. Of these technologies, cyber-physical systems (CPS) and Internet of Things (IoT) are the two major ones, where information and communication technology-based (ICT-based) machines, systems and networks are capable of independently exchanging and responding to information to manage industrial production processes. A CPS can be regarded as a network of interacting cyber and physical elements in which natural and human-made elements (physical space) are tightly integrated with computation, communication and control elements (cyber space) to form the basis for further product and process optimization as well as complementary services [25, 26]. It consists of two main functional components: (1) the advanced connectivity that ensures real-time data acquisition from the physical world and information feedback from the cyber space; and (2) intelligent data management, analytics and computational capability that construct the cyber space [27]. CPS can produce intelligent computation such as autonomous predictive management, self-diagnosis/maintenance mechanism, and collaborative production planning for industry, business, and service performance. It can also be used as a data transfer platform to provide knowledge feedback for improving product design processes [28].

Internet of things (IoT), considered as an extension of the existing Internet, has been becoming a driving paradigm for the ICT evolution in recent years [29]. It envisions a future in which digital and physical entities can be linked by means of appropriate ICTs to enable a whole new class of applications and services that can bring tangible benefits to the society, economy, environment, and even individuals [30, 31]. The IoT is the internetworking of physical objects embedded with electronics, software, sensors, actuators, and network connectivity that enable these objects to share information across platforms as well as to communicate with each other and/or with humans in order to offer a given application or service [32]. This concept has captured the attention of both industry and research communities with a vision of extending Internet connectivity to a large number of "things" in the physical world [33, 34], where "things" can refer to a wide variety of devices (smart objects) which consist of sensors, actuators and embedded communication hardware, on demand storage and computing tools for data analytics, and visualization and interpretation tools which can be widely accessed on different platforms and can be designed for different applications. IoT is an essential driver for customer-facing innovation. As it is expected for smart objects to become ubiquitous on the market and omnipresent in our living environment, the need for creating new IoT-based products and services to improve people's everyday lifestyle will be inevitable [35, 36]. Although current IoT technology is still in its infant stage, the application of service-oriented IoT has driven the creation of new delivery channels for services and entirely new business models.

2.2 Business Models and Product-Service Systems (PSS)

The term "business models" has been the focus of substantial attention from both researchers and practitioners [37, 38]. A business model describes the rationale of how a company creates, delivers, and captures value through providing products and services [39, 40]. According to Mason and Spring [41], business models can be conceptualized as a three-dimensional construct, namely: (1) technologies used in structuring the product/service offerings and delivery management; (2) market offerings for structuring the producer-user interactions that generate the company's offering; and, (3) network architectures for structuring business activities of all buyers and sellers needed to make the market offering possible. Business models are often necessitated by technological innovation which creates both the need to bring discoveries to market and the opportunity to satisfy unrequited customer needs [42]. Nowadays the rapid *evolution of ICT-enabled commerce* is reducing entry barriers and opening new revenue streams to a range of individuals and companies. A number of new business models have emerged due to the widespread application of digital electronics and ICT-based technologies, such as electronic commerce (B2B, B2C, C2C, O2O, etc.), app store, metered service/pay-per-use, access economy, on-demand economy, circular economy, and sharing economy. A representative example is Apple's iPod that combines the iTunes software with the portable media player for offering audio/video media download service, which makes a great profit from both the product and service selling. The *integration of digital and physical experiences* is creating a new way for businesses to interact with customers, by using digital information to augment individual experiences with products and services. Customer demand is rising for *products and services that are free, intuitive, and radically user-oriented*. These will challenge companies to reinvent their business models from traditional firm-centric models to network-embedded ones to influence and co-create new value processes and business exchange patterns [38]. Moreover, recent research has stimulated discussion of business models for sustainability [43, 44]. It is also found that developing business models for sustainability serves as a solid basis for further research in the field [45].

The term product-service systems (PSS) is rooted in cleaner production as a new concept for businesses to improve their sustainability performance [46]. The key idea behind PSS is that customers do not have specific demand for a product per se, but rather are seeking the utility of the provided product and service. As such, PSS can be defined as the result of an innovation strategy, shifting the business focus from designing and selling physical products only, to selling a system of products and services which are jointly capable of fulfilling the same customers' demands with less environmental impact [47]. Various PSS models have been elaborated of which a frequently used categorization of PSS types includes product-orientated, use-orientated, and result-orientated PSS [48, 49]. Otherwise, Mont [50] classified PSS into five types, namely support, sale, use-based, maintenance, and end-of-life (EoL) services. A refined typology of PSS based on functional hierarchy modeling was also proposed by Van Ostaeyen et al. [51], who categorized PSS models according to two distinguishing features: the performance orientation of the dominant revenue mechanism and the degree of integration between product and service elements. Qu et al. [52] conducted a systematic literature review to analyze the state-of-the-art in the field of PSS design, evaluation, and operation methodologies. They concluded that design methodologies in PSS fell into six perspectives, namely customer perspective, modeling techniques, visualization methods, modularity, TRIZ, and system dynamics, of which the first two were the most commonly-used ones. Annarelli et al. [53] also conducted a comprehensive literature review on PSS to understand the origins, the current state-of-the-art, and the possible future research directions. They found that PSS design is one of the most attractive areas but there has been relatively little research on the exclusive design of PSS in literature. The design process of PSS involves not only the design of a service or a product, but also a whole system including a network of actors and support infrastructure [50]. Costa et al. [54] argued that the service design and PSS approaches require better coupling in order to co-create integrated solutions in manufacturing contexts. However, existing PSS methods and tools are mainly concerned with general PSS and service design, which cannot readily assist manufacturers of consumer products to implement and realize PSS solutions [55]. Moreover, most PSS design methods do not sufficiently specify engineering specifications to facilitate the design of physical products [56]. Developing novel product-service design methodology, therefore, has the potential to enable coupling products and services in design processes and to enhance the competitiveness of enterprises in the era of Industry 4.0.

3 Total Product-Service Design

The concepts of CPS and IoT pave the way for design trends toward cyber-physical products (CPP), also referred to as smart products or intelligent products. The notion of smart/intelligent product is still not established firmly, and there are numerous definitions available in the literature [57, 58]. The difficulties over the exact definition of smart/intelligent product are what we consider to be a product; why the product needs to be smart; how smart the product can perform; and, what are the elements of intelligence that may be associated to the product? In this study, the term cyber-physical products (CPP) is described as a physical product that combines ICT-based services with the product to perform required functionality and communicate information on its lifecycle for acquiring complementary service support. CPPs are a new type of consumer products embedded with smart components and connectivity to amplify the value and capabilities of the product, like monitoring the product's internal/external environment, optimizing its operations and usage, and even enabling some functions of the product to exist outside the physical device. For example, a novel refrigerator is able to detect what kinds of food ingredients are being stored inside the refrigerator and keep a track of the stock through barcode or RFID scanning whereby it collects the batch and manufacture detail directly from the Web.

Spring and Araujo [59] indicated that smart connected products capable of self-configuring can help achieve both business and sustainability objectives. Such products can also realize PSS implementation and business model innovations. As shown in Fig. 1, a CPP consists of three main components: (1) physical object made up of the product's mechanical and electrical parts to provide the product functionality, (2) cyber instrument used as a service-oriented IoT that allows the peer-to-peer owner-centric network connection for acquiring the complementary services anytime and anywhere, and (3) interface which is a shared boundary across the physical object and cyber instrument for exchanging information, by which the exchange can be between software, hardware, peripheral devices, humans, and combinations of these. A CPP can also include sensor and actuator of which the former is used to detect and respond to some type of input from the physical environment and the latter is to provide the corresponding output.

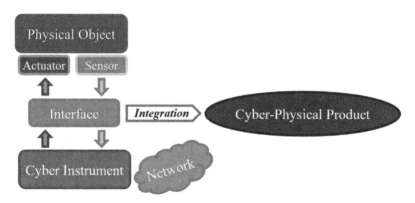

Fig. 1. Components of a cyber-physical product

In response to current design trends toward cyber-physical products, this study presents a new perspective on total product-service design (TPSD) to facilitate the design of such products. TPSD focuses on the design and development of a physical product associated with its possible service supports to fulfill customers' demands with less environmental impact. It is based on the construct that every aspect of product and service quality should be considered as a whole in the early design stage and from the entire product lifecycle perspectives [60]. As shown in Fig. 2, the fundamentals of TPSD are based

Fig. 2. Fundamentals of the total product-service design

on the 4 dimensions, namely functionality, usability, serviceability, and sustainability as described below:

Functionality: The particular use or set of uses for which a product is designed to fulfill customers' needs and expectations. Functionality is the essence of the product utility which comprises both physical functions and functional appearance, aiming at achieving the rationality of form-function relationship and increasing the attractiveness of a product to its potential consumers.

Usability: The extent to which a product can be used by specified users to achieve specified goals with effectiveness, efficiency, flexibility, and satisfaction in a specified context of use. Usability is often associated with the functionality of a product to assess how well the intended users can interact with the product to carry out the assigned activity.

Serviceability: The measure of and the set of the features that support the ease and speed of which the service can be conducted on a product. Serviceability focuses on the maintenance, extension, and enhancement of the product functionality with given resources and within a specified timeframe, identifying business opportunities to create value for both customers and enterprises.

Sustainability: The quality of a product-service development without significant deterioration of the environment and depletion of natural resources on which human well-being depends. Sustainability implies that a product associated with its complementary services should be economically viable, socially acceptable, and environmentally sensitive throughout the product lifecycle.

The implementation of TPSD requires a complex problem-solving process to provide products with descriptions of physical structures that enable essential form-function and specified service performances with the considerations of economic, social, and environmental sustainability. It combines physical product design, cyber service design, user interface design, and life cycle design into a holistic methodology to achieve the total performance of functionality, usability, serviceability, and sustainability (see Fig. 3). Many researchers argued that not only the product functions but also the product appearance can create a competitive advantage in the market [61, 62] and can influence its environmental sustainability [63, 64]. Physical product design aims to develop concepts and specifications of a tangible product that provides desired functions and rational appearance to customers. Cyber service design focuses on creating optimal service experiences to establish auxiliary applications supported by a built-in or an external ICT-enabled device for maintaining, extending, and enhancing the product's functional domains according to both the needs of customers and the competencies/capabilities of service providers. It could also create an innovative business model through the network-embedded service offering. The goal of user interface design is to maximize usability and the user experience, making the user's interaction as simple and efficient as possible in terms of accomplishing user goals. Life cycle design incorporates sustainability considerations into each phase of product and service design so the ultimate impacts of the product and service are minimized and optimized with economic, social, and environmental performance.

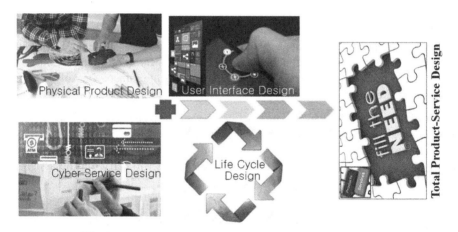

Fig. 3. Methodical support for the total product-service design

4   Proposed Methodology

Industry 4.0 highlights the need for digitalization, networklization, intellectualization, and servitization of manufacturing, while Design 4.0 focuses on the functionality, usability, sustainability, and serviceability of products. Traditional design methodologies do not integrate products and services into a holistic design process. Although various frameworks and methodologies for supporting PSS design processes have been proposed in the literature [65–67], most existing methodologies focus on a product's service system design rather than a methodical support for implementing entire product-service design and development. Based on the theoretical fundamentals given above, the total product-service design (TPSD) methodology is constructed under a horizontal timeline of product lifecycle (Design and development, Production, Use phase, and End of life) and vertical flow paths of the design processes as shown in Fig. 4. The implementation steps of the design processes are elaborated as follows:

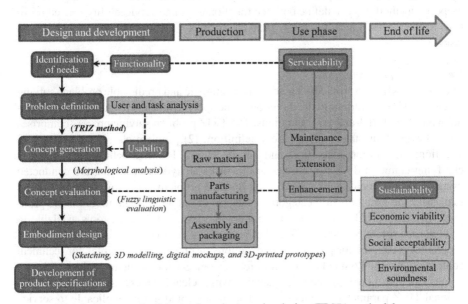

Fig. 4. Framework of the total product-service design (TPSD) methodology

Step 1. Identification of Needs

Total product-service design involves both the tangible product and intangible service designs in the entire lifecycle. The needs for the product's functionality and serviceability must be identified in order to search for a series of problems that should be resolved to fulfill customers' demands. The functionality aspect mainly focuses on the generation and improvement of the product utility including desired functions and appearance, while the serviceability aspect is centered on the maintenance, extension, and enhancement of the product functionality provided via cyber networks.

Step 2. Problem Definition

Product and service design is a goal-directed problem-solving activity aimed at providing practical solutions to satisfy customers' needs. Contradiction problems often occur in a product design process. Song and Sakao [68] also indicated that improvement or enhancement of one service attribute may cause the deterioration of another in the design process of product-service offerings. TRIZ has been recognized as a powerful method for dealing with contradiction problems with application of innovative solutions. The fundamental principle behind the TRIZ method is to find contradictions in a system and then eliminate them by using TRIZ tools. TRIZ has been widely used in the engineering and technical field for R&D and in the designing for tangible products [69]. The scope of TRIZ has also been extended to non-technical areas, such as marketing, sales, and advertising [70], service design [71, 72], maintenance support [73], eco-design [74], interaction design [75], cleaner production [76], and sustainable innovation [77]. As TRIZ is capable of providing methodical design support for the total performance of functionality, usability, serviceability, and sustainability, it is recommended as a technique for the proposed methodology to define both the functionality- and serviceability-related problems as well as the interdisciplinary problems between the product and service design tasks.

Step 3. Concept Generation

The aim of TRIZ is to provide designers with a strategic and predictable problem-solving process to advance further in the direction of the specific inventive solution with a minimum number of trial-and-error iterations. The TRIZ problem solving process comprises the following four stages: (1) problem definition; (2) problem classification and tool selection; (3) solution generation; and, (4) evaluation. Problem definition is aimed at clarifying common design questions and involves analytical tasks that are conducted using TRIZ analytical tools. After having finished problem definitions, designers have to classify the problems into categories and then select appropriate tools for the problems. For each problem category, there are some knowledge-based tools available to resolve the problems, such as 39 design parameters and the contradiction matrix, 11 separation principles for physical contradictions, 40 inventive principles for technical contradictions, and 76 standard solutions for substance-field (Su-field) analysis [78, 79].

Services must be designed in conjunction with a clear business case and model. Traditional TRIZ parameters, principles, and solutions could not be applicable to service design fields since TRIZ tools are originally developed for engineering applications to support technological innovation. To improve the strength of TRIZ application, there are numerous modified TRIZ tools available to resolve the service problems, such as Chang and Lu's service parameters [80], Zhang et al.'s modified inventive principles [81], Jiang et al.'s contradiction parameter options [82], and Gazem and Rahman's interpretation of the 40 inventive principles for performing services [72]. Moreover, Filippi and Barattin's interaction design guidelines (IDGL) [75] and Russo et al.'s iTree/ECO-TRIZ guidelines [74] can be used to link up the solutions with usability and sustainability aspects, respectively. Products and services require conscious adoption and interaction by users. TRIZ is capable of solving usability problems involving compromises and trade-offs. In this step, user and task analysis is also conducted to identify usability problems for the user interface design. Designers make choices about which solutions are heavily related to sustainability. By using the corresponding TRIZ tools, many possible solutions can be generated and then preliminarily evaluated to determine the best solutions to the problems of product, service, and both. The morphological analysis [83] is further used to systematically derive a set of possible combinations contained in the solutions of a multi-dimensional problem complex, which is used as the alternative concepts for the following concept evaluation step.

Step 4. Concept Evaluation

Concept evaluation is a critical step to identify promising concepts in the total product-service design process. It is a decision-making problem that consists of finding the most desirable concepts which can provide optimal functionality, usability, and serviceability with economic, social, and environmental sustainability. The evaluation is based on a hierarchical structure of the 3 sustainability criteria (Economic viability, Social acceptability, and Environmental soundness), each of which comprises 3 sets of sub-criteria, as shown in Table 1.

Table 1. List of sustainability criteria for the concept evaluation

Criterion	Sub-criterion	Description
C_1: economic viability	Manufacturing costs (c_{11})	The design is capable of economizing on manufacturing costs
	Assembly/packaging costs (c_{12})	The design is capable of driving assembly and packaging costs down
	Maintenance costs (c_{13})	The design is capable of reducing maintenance costs
C_2: social acceptability	Quality of life (c_{21})	The design is capable of improving the quality of life
	Health and well-being (c_{22})	The design is capable of providing health and well-being to humans
	Safety and security (c_{23})	The design is capable of ensuring safety and security throughout the product lifecycle
C_3: environmental soundness	Materials used (c_{31})	The design is capable of conserving the amount of materials used
	Energy saving (c_{32})	The design is capable of saving energy during production and use phase
	Waste reduction (c_{33})	The design is capable of expanding product lifespan and reducing waste with the objective of harmonizing end-of-life (EoL) treatments

To make the concept evaluation valid and effective, this study uses the fuzzy linguistic evaluation approach proposed by Chou [84, 85] to assess the concepts derived from the morphological analysis results. Given a set of linguistic ratings variables $\tilde{R} = [[\tilde{r}_{ij}]]$, and a set of linguistic variables of weights $\tilde{W} = [[\tilde{w}_{ij}]]$, the aggregation operator is formularized as

$$D_k\left(\tilde{R} \cdot \tilde{W}\right) = \frac{\sum_{i=1}^{3}[[\tilde{r}_{ij}]] \cdot [[\tilde{w}_{ij}]]}{\sum_{i=1}^{3}[[\tilde{w}_{ij}]]} = \frac{\int_{\alpha=0}^{1}(\tilde{r}_{i1} \times \tilde{w}_{i1})_\alpha + (\tilde{r}_{i2} \times \tilde{w}_{i2})_\alpha + (\tilde{r}_{i3} \times \tilde{w}_{i3})_\alpha}{\int_{\alpha=0}^{1}(\tilde{w}_{i1})_\alpha + (\tilde{w}_{i2})_\alpha + (\tilde{w}_{i3})_\alpha}$$
(1)

where

$D_k\left(\tilde{R}, \tilde{W}\right)$ represents the overall desirability of alternative concept k;

$\tilde{r}_{ij}$ represents the rating of the jth sub-criterion of criterion i; and,

$\tilde{w}_{ij}$ represents the importance (weight) of the jth sub-criterion of criterion i.

Linguistic variables $[[\tilde{r}_{ij}]]$ and $[[\tilde{w}_{ij}]]$ are fuzzy numbers. In this study, assessment is based on 7-point rating scales (ranging from VL to VH) described by the piecewise continuous triangular membership functions, $\mu A(x)$, shown in Fig. 5. The aggregation with continuous α-cuts ($0 \leq \alpha \leq 1$) is a combination of extended algebraic operations (addition, subtraction, multiplication, and division) based on interval arithmetic operations and requires that every fuzzy number is represented by a continuous membership function and can be completely defined by its family of α-cuts.

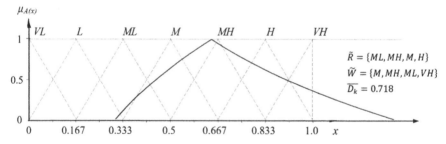

Fig. 5. Seven-point rating scales characterized by continuous triangular membership functions

The result of the arithmetic operations is a crisp set (interval) that represents the α-cut of the fuzzy set obtained by operating on fuzzy numbers $\tilde{r}_{ij}$ and $\tilde{w}_{ij}$. Through the aggregation operations, the family of α-cuts defined as the resultant membership function of the evaluated alternative can be presented as a convex and normalized fuzzy set, $D_k(x)$, which is also classified as a fuzzy number. Taking advantage of the center-of-gravity (COG) defuzzification method shown in Eq. (2), the quantitative value of the evaluated alternative, $\overline{D_k}$, can be derived. The higher the $\overline{D_k}$, value, the better is the evaluated alternative concept. An example of the aggregation and defuzzification result is also shown in Fig. 5.

$$\overline{D_k} = \frac{\int_s^t D_k(x) \cdot dx}{\int_s^t D_k(x)dx} \qquad (2)$$

By substituting the linguistic variables into Eq. (1) to perform the linguistic aggregation operation and further defuzzifying the resultant membership functions using Eq. (2), the rank of the alternative concepts can be obtained and thus the promising concepts can be identified.

Step 5. Embodiment Design
A product has a physical structure which is assembled from components, sub-assemblies, and parts into a workable whole. The embodiment design is concerned about the realization of the product concept associated with its service scheme. It is the process of uniting the imaginary separation between abstract concepts and concrete elements to verify the feasibility of the product and its complementary services. The physical embodiment is also used to visually communicate how an intentional function works via the physical structure. In this step, transformation from concepts to elements is accomplished by sketching, 3D modelling, digital mockups as well as 3D-printed prototypes that work for both visual and functional testing.

Step 6. Development of Product Specifications
According to the final product prototype, the product specifications for production engineering are developed to detail the product's bill of materials, packaging and other physical components and their assembly or relationship together. Moreover, the features of user interfaces and service offering are determined to specify how the product associated with its complementary services is to be used.

5 Concluding Remarks

The Industrial Revolution is a concept and a development that has fundamentally changed our society and economy. Industry 4.0 describes the planned fourth Industrial Revolution which leads to an intelligent, connected, and decentralized production paradigm and stands for a new level of organization and regulation of a product's entire value chain over its lifecycle. This study attempts to address the issue of Design 4.0 in response to the initiative of Industry 4.0 from the viewpoint of industrial and technological developments. It systematically analyzes the evolution of product design and categorizes the developed theories and methodologies in the existing literature corresponding to each of the four times of radical transformation in industry.

This study also proposes a new perspective on "total product-service design (TPSD)" beyond the existing "product-service systems (PSS)", which focuses on the implementation of entire product-service design and development instead of only supporting a product's service system design. The TPSD approach combines physical product, cyber service, user interface, and life cycle design into a holistic methodology to achieve the total performance of products. This methodology is expected to contribute to a new product design paradigm, facilitating the design and development of innovative cyber-physical products. Further research should conduct case studies to verify the applicability and effectiveness of the proposed methodology.

Acknowledgement. The author would like to thank the Ministry of Science and Technology, Taiwan, R.O.C., for financially supporting this research under grant number MOST 108-2221-E-214-003.

References

1. UNEP (United Nations Environment Programme): Cleaner Production: a training resource package. United Nations Environment Programme, Industry and the Environment, Paris (1996)
2. Li, L.: China's manufacturing locus in 2025: with a comparison of "Made-in-China 2025" and "Industry 4.0". Technol. Forecast. Soc. Change **135**, 66–74 (2018)
3. Sung, T.K.: Industry 4.0: a Korea perspective. Technol. Forecast. Soc. Change **132**, 40–45 (2018)
4. Muhuri, P.K., Shukla, A.K., Abraham, A.: Industry 4.0: a bibliometric analysis and detailed overview. Eng. Appl. Artif. Intell. **78**, 218–235 (2019)
5. Lasi, H., Fettke, P., Kemper, H.-G., Feld, T., Hoffmann, M.: Industrie 4.0. Bus. Inf. Syst. Eng. **6**(4), 239–242 (2014)
6. Qin, J., Liu, Y., Grosvenor, R.: A Categorical framework of manufacturing for industry 4.0 and beyond. Procedia CIRP **52**, 173–178 (2016)
7. Frazzon, E.M., Hartmann, J., Makuschewitz, T., Scholz-Reiter, B.: Towards socio-cyber-physical systems in production networks. Procedia CIRP **7**, 49–54 (2013)
8. Riedl, M., Zipper, H., Meier, M., Diedrich, C.: Cyber-physical systems alter automation architectures. Annu. Rev. Control **38**(1), 123–133 (2014)
9. Lu, Y.: Industry 4.0: a survey on technologies, applications and open research issues. J. Ind. Inf. Integr. **6**, 1–10 (2017)
10. UNCSD (United Nations Commission on Sustainable Development): Summary Report: The Symposium on Sustainable Consumption, Oslo, 19–20 Jan 1994. In: General Discussion on Progress in the Implementation of Agenda. New York: UNCSD, 2–5 (1994)
11. Gebauer, H., Gustafsson, A., Witell, L.: Competitive advantage through service differentiation by manufacturing companies. J. Bus. Res. **64**(12), 1270–1280 (2011)
12. Ostrom, A.L., Parasuraman, A., Bowen, D.E., Patrício, L., Voss, C.A., Lemon, K.: Service research priorities in a rapidly changing context. J. Serv. Res. **18**(2), 127–159 (2015)
13. Baines, T., Ziaee Bigdeli, A., Bustinza, O.F., Shi, V.G., Baldwin, J., Ridgway, K.: Servitization: revisiting the state-of-the-art and research priorities. Int. J. Oper. Prod. Manage. **37**(2), 256–278 (2017)
14. Gebauer, H., Edvardsson, B., Gustafsson, A., Witell, L.: Match or mismatch: strategy-structure configurations in the service business of manufacturing companies. J. Serv. Res. **13**(2), 198–215 (2010)
15. Ulaga, W., Reinartz, W.: Hybrid offerings: how manufacturing firms combine goods and services successfully. J. Mark. **75**, 5–23 (2011)
16. Kowalkowski, C., Gebauer, H., Oliva, R.: Service growth in product firms: past, present, and future. Ind. Mark. Manage. **60**, 82–88 (2017)
17. Baines, T., Lightfoot, H.W., Benedettini, O., Kay, J.M.: The servitization of manufacturing: a review of literature and reflection on future challenges. J. Manuf. Technol. Manage. **20**(5), 547–567 (2009)

18. Beuren, F.H., Gomes Ferreira, M.G., Cauchick Miguel, P.A.: Product-service systems: a literature review on integrated products and services. J. Clean. Prod. **47**, 222–231 (2013)
19. Kuijken, B., Gemser, G., Wijnberg, N.M.: Effective product-service systems: a value-based framework. Ind. Mark. Manage. **60**, 33–41 (2017)
20. Trevisan, L., Lelah, A., Brissaud, D.: New PSS design method of a pneumatic energy system. Procedia CIRP **30**, 48–53 (2015)
21. Chiarello, F., Trivelli, L., Bonaccorsi, A., Fantoni, G.: Extracting and mapping industry 4.0 technologies using Wikipedia. Comput. Ind. **100**, 244–257 (2018)
22. Alcácer, V., Cruz-Machado, V.: Scanning the industry 4.0: a literature review on technologies for manufacturing systems. Eng. Sci. Technol. Int. J. **22**, 899–919 (2019)
23. Kerin, M., Pham, D.T.: A review of emerging industry 4.0 technologies in remanufacturing. J. Clean. Prod. **237**, 117805 (2019)
24. Brettel, M., Friederichsen, N., Keller, M., Rosenberg, M.: How virtualization, decentralization and network building change the manufacturing landscape: an industry 4.0 perspective. Int. J. Mech. Aerosp. Ind. Mech. Manufac. Eng. **8**(1), 37–44 (2014)
25. Bagheri, B., Yang, S., Kao, H.-A., Lee, J.: Cyber-physical systems architecture for self-aware machines in industry environment. IFAC-PapersOnLine **48–3**, 1622–1627 (2015)
26. Leitaõ, P., Colombo, A.W., Karnouskos, S.: Industrial automation based on cyber-physical systems technologies: Prototype implementations and challenges. Comput. Ind. **81**, 11–25 (2016)
27. Lee, J., Bagheri, B., Kao, H.-A.: A Cyber-physical systems architecture for industry 4.0-based manufacturing systems. Manuf. Lett. **3**, 18–23 (2015)
28. Brandmeier, M., Bogner, E., Brossog, M., Franke, J.: Product design improvement through knowledge feedback of cyber-physical systems. Procedia CIRP **50**, 186–191 (2016)
29. Atzori, L., Iera, A., Morabito, G.: The internet of things: a survey. Comput. Netw. **54**(15), 2787–2805 (2010)
30. Miorandi, D., Sicari, S., De Pellegrini, F., Chlamtac, I.: Internet of things: vision, applications and research challenges. Ad Hoc Netw. **10**, 1497–1516 (2012)
31. Borgia, E.: The internet of things vision: key features, applications and open issues. Comput. Commun. **54**, 1–31 (2014)
32. Gubbi, J., Buyya, R., Marusic, S., Palaniswami, M.: Internet of things (IoT): a vision, architectural elements, and future directions. Future Gener. Comput. Syst. **29**(7), 1645–1660 (2013)
33. Want, R., Schilit, B.N., Jenson, S.: Enabling the internet of things. Computer **48**(1), 28–35 (2015)
34. Trappey, A.J.C., Trappey, C.V., Govindarajan, U.H., Chuang, A.C., Sun, J.J.: A review of essential standards and patent landscapes for the Internet of Things: a key enabler for Industry 4.0. Adv. Eng. Inf. **33**, 208–222 (2017)
35. Ouaddah, A., Mousannif, H., Elkalam, A.A., Ouahman, A.A.: Access control in the Internet of Things: big challenges and new opportunities. Comput. Netw. **112**, 237–262 (2017)
36. Stojkoska, B.R., Trivodaliev, K.: A review of Internet of Things for smart home: challenges and solutions. J. Clean. Prod. **140**(3), 1454–1464 (2017)
37. Zott, C., Amit, R., Massa, L.: The business model: recent developments and future research. J. Manag. **37**(4), 1019–1042 (2011)
38. Bankvall, L., Dubois, A., Lind, F.: Conceptualizing business models in industrial networks. Ind. Mark. Manage. **60**, 196–203 (2017)
39. Richardson, J.: The business model: an integrative framework for strategy execution. Strateg. Change **17**(5–6), 133–144 (2008)
40. Osterwalder, A., Pigneur, Y.: Aligning profit and purpose through business model innovation. In: Palazzo, G., Wentland, M. (eds.) Responsible Management Practices for the 21st Century, pp. 61–76. Pearson International (2011)

41. Mason, K., Spring, M.: The sites and practices of business models. Ind. Mark. Manage. **40**(6), 1032–1041 (2011)
42. Teece, D.J.: Business models, business strategy and innovation. Long Range Plan. **43**, 172–194 (2010)
43. Boons, F., Lüdeke-Freund, F.: Business models for sustainable innovation: state-of-the-art and steps towards a research agenda. J. Clean. Prod. **45**, 9–19 (2013)
44. Bocken, N.M.P., Short, S.W., Rana, P., Evans, S.: A literature and practice review to develop sustainable business model archetypes. J. Clean. Prod. **65**(15), 42–56 (2014)
45. Rauter, R., Jonker, J., Baumgartner, R.J.: Going one's own way: drivers in developing business models for sustainability. J. Clean. Prod. **140**, 144–154 (2017)
46. Boehm, M., Thomas, O.: Looking beyond the rim of one's teacup: a multidisciplinary literature review of product-service systems in information systems, business management, and engineering & design. J. Clean. Prod. **51**, 245–260 (2013)
47. Manzini, E., Vezzoli, C.: Product-Service Systems and Sustainability–Opportunities for Sustainable Solutions. UNEP (United Nations Environment Programme), Paris, France (2002)
48. Cook, M.B., Bhamra, T.A., Lemon, M.: The transfer and application of Product Service Systems: from academic to UK manufacturing firms. J. Clean. Prod. **14**(17), 1455–1465 (2006)
49. Reim, W., Parida, V., Örtqvist, D.: Product-Service Systems (PSS) business models and tactics-a systematic literature review. J. Clean. Prod. **97**, 61–75 (2015)
50. Mont, O.K.: Clarifying the concept of product-service system. J. Clean. Prod. **10**(3), 237–245 (2002)
51. Van Ostaeyen, J., Van Horenbeek, A., Pintelon, L., Duflou, J.R.: A refined typology of product-service systems based on functional hierarchy modeling. J. Clean. Prod. **51**, 261–276 (2013)
52. Qu, M., Yu, S., Chen, D., Chu, J., Tian, B.: State-of-the-art of design, evaluation, and operation methodologies in product service systems. Comput. Ind. **77**, 1–14 (2016)
53. Annarelli, A., Battistella, C., Nonino, F.: Product service system: a conceptual framework from a systematic review. J. Clean. Prod. **139**, 1011–1032 (2016)
54. Costa, N., Patrício, L., Morelli, N., Magee, C.L.: Bringing service design to manufacturing companies: integrating pss and service design approaches. Des. Stud. **55**, 112–145 (2018)
55. Yang, X., Moore, P., Pu, J.S., Wong, C.B.: A practical methodology for realizing product service systems for consumer products. Comput. Ind. Eng. **56**(1), 224–235 (2009)
56. Maussang, N., Zwolinski, P., Brissaud, D.: Product-service system design methodology: from the PSS architecture design to the products specifications. J. Eng. Des. **20**(4), 349–366 (2009)
57. Mühlhäuser, M.: Smart products: an introduction. In: Mühlhäuser, M., Ferscha, A., Aitenbichler, E. (eds.) AmI 2007. CCIS, vol. 11, pp. 158–164. Springer, Heidelberg (2008). https://doi.org/10.1007/978-3-540-85379-4_20
58. Meyer, G.G., Främling, K., Holmström, J.: Intelligent products: a survey. Comput. Ind. **60**, 137–148 (2009)
59. Spring, M., Araujo, L.: Product biographies in servitization and the circular economy. Ind. Mark. Manage. **60**, 126–137 (2017)
60. Liu, H., Chu, X., Xue, D.: An optimal concurrent product design and service planning approach through simulation-based evaluation considering the whole product life-cycle span. Comput. Ind. **111**, 187–197 (2019)
61. Chang, W.C., Wu, T.Y.: Exploring types and characteristics of product forms. Int. J. Des. **1**(1), 3–14 (2007)
62. Blijlevens, J., Creusen, M.E.H., Schoormans, J.P.L.: How consumers perceive product appearance: the identification of three product appearance attributes. Int. J. Des. **3**(3), 27–35 (2009)

63. Luchs, M.G., Brower, J., Chitturi, R.: Product choice and the importance of aesthetic design given the emotion-laden trade-off between sustainability and functional performance. J. Prod. Innov. Manag. **29**, 903–916 (2012)
64. Diego-Mas, J.-A., Poveda-Bautista, R., Alcaide-Marzal, J.: Designing the appearance of environmentally sustainable products. J. Clean. Prod. **135**, 784–793 (2016)
65. Cavalieri, S., Pezzotta, G.: Product-service systems engineering: state of the art and research challenges. Comput. Ind. **63**(4), 278–288 (2012)
66. Vasantha, G., Roy, R., Lelah, A., Brissaud, D.: A review of product-service systems design methodologies. J. Eng. Des. **23**(9), 635–659 (2012)
67. Song, W., Sakao, T.: A customization-oriented framework for design of sustainable product/service system. J. Clean. Prod. **140**, 1672–1685 (2017)
68. Song, W., Sakao, T.: Service conflict identification and resolution for design of product-service offerings. Comput. Ind. Eng. **98**, 91–101 (2016)
69. Chechurin, L., Borgianni, Y.: Understanding TRIZ through the review of top cited publications. Comput. Ind. **82**, 119–134 (2016)
70. Retseptor, G.: 40 inventive principles in marketing, sales and advertising. TRIZ J. April Issues (2005)
71. Chai, K., Zhang, J., Tan, K.: A TRIZ-based method for new service design. J. Serv. Res. **8**(1), 48–66 (2005)
72. Gazem, N., Rahman, A.A.: Interpretation of TRIZ principles in a service related context. Asian Soc. Sci. **10**(13), 108–130 (2014)
73. Vaneker, T., van Diepen, T.: Design support for maintenance tasks using TRIZ. Procedia CIRP **39**, 67–72 (2016)
74. Russo, D., Rizzi, C., Montelisciani, G.: Inventive guidelines for a TRIZ-based eco-design matrix. J. Clean. Prod. **76**, 95–105 (2014)
75. Filippi, S., Barattin, D.: Exploiting TRIZ tools in interaction design. Procedia Eng. **131**, 71–85 (2015)
76. Fresner, J., Jantschgi, J., Birkel, S., Bärnthaler, J., Krenn, C.: The theory of inventive problem solving (TRIZ) as option generation tool within cleaner production projects. J. Clean. Prod. **18**, 128–136 (2010)
77. D'Anna, W., Cascini, G.: Supporting sustainable innovation through TRIZ system thinking. Procedia Eng. **9**, 145–156 (2011)
78. Ideation International Inc.: An introduction to TRIZ: The Russian Theory of Inventive Problem Solving, Ideation International, Southfield, MI (1996)
79. Ilevbare, I.M., Probert, D., Phaal, R.: A review of TRIZ, and its benefits and challenges in practice. Technovation **33**, 30–37 (2013)
80. Chang, H.H., Lu, P.W.: Using a TRIZ-based method to design innovative service quality-a case study on insurance industry. J. Qual. **16**(3), 179–193 (2009)
81. Zhang, J., Chai, K.H., Tan, K.C.: 40 inventive principles with applications in service operations management. TRIZ J. December Issues (2003)
82. Jiang, J.C., Sun, P., Cheng, C.Y.: TRIZ contradiction parameter options for service system design (SSD). J. Stat. Manage. Syst. **13**(6), 1343–1361 (2013)
83. Zwicky, F.: Discovery, Invention, Research–Through the Morphological Approach. The Macmillan Company, Toronto (1969)
84. Chou, J.-R.: A linguistic evaluation approach for universal design. Inf. Sci. **190**, 76–94 (2012)
85. Chou, J.-R.: A Kansei evaluation approach based on the technique of computing with words. Adv. Eng. Inform. **30**, 1–15 (2016)

Envisioning the Future Scenario Through Design Fiction Generating Toolkits

Zhiyong Fu[(✉)] and Lin Zhu

Tsinghua University, Beijing, China
fuzhiyong@tsinghua.edu.cn, ifearlesslin@gmail.com

Abstract. Design with value orientation and World view are increasingly the basis of envisioning the future scenarios. Exploring the value of design from the perspective of society and humanity, further guiding the direction of science and technology have gradually become one of the future development directions of human-computer interaction field. Design concepts such as design fiction and speculative design have evolved into useful tools for thinking about future directions and reflecting on the problems posed by technology. However, the current researches mainly focus on provoking thinking and developing from the perspective of criticizing reality. And has not yet formed a mature practical method which can guide the design practice. Therefore, this essay mainly discusses how to integrate the concept of Design fiction with interaction Design and service Design. Make it more conducive to promoting the innovation of existing smart products or services, and help designers generate more innovative ideas. This essay develops the content of Design fiction and forms a structured Design toolkits: Future vision generating toolkits, to guide students in teaching and research to conduct future-oriented research in the early stages of Design. After the application, the feedback is relatively successful, which also has a good guiding and promoting effect on the further design practice.

Keywords: Design fiction · Future vision · Design toolkits · Smart city

1 Introduction

1.1 The Trend of the Smart City

Modern network communication technology, mobile Internet, social media, perception technology and so on. Making our life scene and environment become perceptive and interactive. With the growing trend of smart cities, the application of big data becoming increasingly extensive, citizens will be integrated to form a larger life system [1]. With the interaction of new ICT technologies, the urban environment has become more sensitive, interactive and flexible. The integration of the physical space of the city and the invisible network can support human activities at all levels and form a complex ecosystem.

"IT designers, design for/toward a future not far off, one that is plausible and even likely, but the company from our present - day this reality: Ubiquitous computing, smart cities, the Internet of Things, and wearable computing are all examples" [2]. Jeffrey

© Springer Nature Switzerland AG 2020
P.-L. P. Rau (Ed.): HCII 2020, LNCS 12192, pp. 46–59, 2020.
https://doi.org/10.1007/978-3-030-49788-0_4

Bardzell and Shaowen Bardzell proposed strategies for applying humanistic tendencies to design in the future. The construction of imagined scenarios - using any combination of design and fiction - serves as a cue for reflection and dialogue. Using a fully realized fictional character as a perspective to observe technological social variation.

The process of smart city is full of unknown and uncertainty, which requires to be promoted from multiple aspects and angles. Matthieu Cherubini has created three different algorithms for self-driving cars, each of which follows specific ethical principles/behavior Settings - and is embedded in a self-driving virtual car running in a simulated environment. Explore how self-driving cars will face various ethical dilemmas in these environments [3]. For integrating more social and cultural factors into our vision of the future and exploring how to overcome the constraints of technology in the field of combining Ai with human. So as to find a Preferable future more in line with people's expectations for the future. Therefore, we try to start from the perspective of Design innovation, explore the value of Design from the perspective of human-centered. And further think about society, humanity and technology. From such a perspective, we can see that Design innovation plays an important role in Design future. Further established Future vision generating toolkits: an integrated mindset that integrates cross-cutting factors to generate a vision of the Future from a broader perspective.

1.2   Regard Smart City as the Target of Design

Microsoft announced the "CityNext" smart city project at the WPC 2013 conference [4]. The CityNext smart city initiative will support innovation and encourage city leaders to leverage Microsoft partner networks and corporate technology solutions such as Azure, big data solutions, devices, and services to create sustainable cities. Through cloud computing, big data and Internet of things technologies, up to 30 solutions have been provided for the corresponding problems, and more than 100 global smart city cases have been published. Sidewalk Labs, a smart city subsidiary of Google's parent company Alphabet, published the Master Innovation and Development Plan (MIDP) in 2019 [5]. Transforming the Quayside area and Villers island on the Ontario shores east of Toronto, Canada, into a high-tech utopia. Sidewalk installed sensors in two smart neighborhoods and installed public wi-fi to collect "city data" as an important basis for improving housing quality and formulating smart transportation decisions. The goal is to use technological solutions to improve urban infrastructure and address issues such as the cost of living, efficient transportation and energy use to create a "smart city" of the future.

It is not hard to see that the technology companies focus on infrastructure construction and technological innovation in the research of Smart City. Just as the second principle proposed by Junichi ITO in "Whiplash: How to Survive Our Faster Future" [6]: pull is better than push. "Pull" refers to application, while "push" refers to science and technology. It points out that the development of science, technology and the declining cost of innovation can transfer power from the core to the periphery. And creating opportunities for innovators to inspire their own passion and potential. At its best, it helps people discover not only what they need, but what they don't know they need.

The traditional field of design is expanding in depth and breadth. In terms of "depth", the cross-boundary integration of social humanities and cutting-edge technologies has produced new ideas and new methods. The discipline of design will provide a new paradigm for a deeper understanding of human sensibility and social ethics. In terms of "breadth", the field and influence of design discipline are expanding, from user-centered interaction to more complex urban and social systems. Design discipline is becoming one of the most important means of community and urban transformation.

The development of Smart City requires not only short-term planning, but also long-term planning to achieve a more promising future. We propose "Future vision generating toolkits" as an experimental form of design and design education that combines design thinking with Future thinking, to align near-term design behavior with vision goals. While overcoming uncertainty and accelerating innovation toward the desired Future. As a practical basis for the long-term planning of Smart City based on the integration of design and technology, collaborative design is conducted for 2050 and beyond.

2 From Predicting Future to Creating Future

2.1 Related Theoretical Research

Almost everyone is talking about the future. In said, "It is an important space, a place where the future can be debated and discussed before It happens so that, at further in and found, the most desirable futures can be aimed for and the least desirable solutions." The past is history and can be used as a reference. What about the future where human imagination is limited? It's a nice debate which worthing contest for centuries [7].

Here are some design and futurology scholars' ideas and expectations about the future and design. Is the future necessarily going to be better and better? They come up with their own predictions and insights, showing us multiple models of the future. For the distant future, although we cannot predict the specific details, but through various means to outline the general outline.

"There are six basic concepts of futures thinking: the used future; The disowned the future; The alternative futures. Alignment; Models of social change; and uses of the future." In daily thinking, we do not include the future as part of the solution. When the future comes, we are often overwhelmed by lack of preparation, which is isolating ourselves [8]. The philosopher Nick Poststrom has described four possible models for the future of mankind. Peter Thiel concludes his book "Zero to One" [9] with a fourth scenario: the future accelerates. "no matter how many trends can be tracked, the future doesn't happen on its own. The future described by the singularity is no more important than the choice we face today: it is up to us to choose between the two most likely scenarios, to do nothing or to do our best. We cannot take for granted that the future will be better, but we must strive to create it today. Whether we achieve "singularities" on a cosmic scale may not matter as much as whether we seize unique opportunities to innovate in our daily lives.

In 2009, futurist Stuart Candy used a fascinating chart to illustrate various potential futures. [10] the core thinking mode is the future can be divided into four classes (4 p): a viable future Possible, technically Possible future Plausible, big probability (if there is no significant change) will implement the Probable future, idealized future Preferable. The RCA proposed a dream design and future oriented to do the design artifacts one design method [11]. Claims are bound by the existing thinking with its, be inferior to abandon the framework of rational thinking, Capturing an idealistic without scruple. Or the future is of the extreme (reality), and in this world with the product design to represent the performance of the alternative reality's thinking. Borrowing this to do an artifact of the reflection of today's society, to break the existing rigid thinking, promote the modern trend of progress.

2.2 Creating the Future Starting from Creating a New World Outlook

In imagining a true utopia, Eric Olin Wright described the social sciences of liberation as "the theory of the journey from the present to the possible future: the diagnosis and critique of society tells us why we should leave the world we live in; Choice theory tells us where to go; Transformation theory tells us how to get from here to there, how to make viable alternatives possible" [12]. For us, the realization of this journey to the future is highly unlikely if we plan the future as a blueprint. Instead, we believe that change must be achieved by opening people's imaginations and applying them to all areas of their lives. By generating alternatives, design can help people build a compass that guides them rather than a map that guides them straight to their destination.

William R. Huss argued that it was more important to develop the situation than to be visionary [13]. To find inspiration in design, we need to go beyond design to explore the methodologies of film, literature, science, ethics, politics and art; Explore, mix, borrow, and embrace many tools that can be used not only to make things, but also to conceive fictional worlds, cautionary tales, hypothetical scenarios, thought experiments, counter-factuals, reductive and absurd experiments, presuppositions of the future, and so on.

Instead of trying to predict the future, use design to open up possibilities that can be discussed, debated, and used together to define a better future for a particular group of people: from companies to cities to society. Designers can work with experts (including ethicists, political scientists, economists, etc.) Framing the future World view and discuss the type of future people's requirement, and gradually create a preferable future [14].

We believe that by speculating on more dimensions of future urban development and exploring other scenarios, the path from reality to a more hopeful future will become distinct. Although the future cannot be predicted, we can help to identify the factors of today that will increase the likelihood of a more ideal future. Similarly, factors that may contribute to an undesirable future can be identified early and addressed.

3 Future Vision Generating Toolkits

3.1 Originated from the Design Fiction

Although foresight is rooted in the social sciences, future-oriented design can combine science with fantasy to study a wide range of possibilities and speculate on radical scenarios. Future-oriented designers can conduct trend research and develop scenarios, and future-oriented designs quickly approach science fiction when the time scale tends to the distant future. In addition to the declining validity of the analytical data, the further out the "facts" of the future, the greater the change in value. Simon describes this phenomenon as "discounting the future" [15].

In 2009, Julian Bleecker's paper Design Fiction: A Short Essay on Design, Science, Fact and Fiction played A crucial role in the establishment of Design Fiction. "If design can be a way of telling a story by creating materials, what kind of story can it tell? What will be the style of the story?" [16]. Several important features of Design Fiction are mentioned here, namely Object and narrative, which can be extended to diegetic prototype and narratology respectively. This is what Design Fiction is all about: design-based storytelling. And Julian Bleecker emphasizes that fiction is based on fact. That is to say, such imagination and speculation is not "imagination", but imagination based on a certain real context.

When starting with creating our own Design Fiction, using this as a context to explore the shape of Design at a certain point in the future and embark on a unique journey of thinking. "Fiction constructs an imaginary world using text, whereas design fiction is the narrative of a story through design." This design prototype is an imagination based on reality, while this narrative or story is the reflection and speculation of the designer. Design fiction enables the conception, evaluation, and reflection of future product concepts and scenarios to be transmitted [17]. Within different users, physical environments and technologies. In addition, from the comparison with other qualitative research methods, the design novel has also attracted the attention of social issues for future research. In other words, it represents a special focus on the social aspects of the human and contextual feasibility of emerging technologies. With "situational" as the primary design outcome, as opposed to "technology-driven" and "enterprise-oriented" concepts for other design approaches to products. Accordingly, stakeholders can immerse themselves in the narrative of the design novel, instead of focusing on the non-textual conceptual products and services. Therefore, the social aspects of design proposals can be examined from a first-person subjective perspective [18].

3.2   The Framework of Future Vision Generating Toolkits

Future vision generating toolkits is an experimental form of representing design and design education that combines design thinking with Future thinking to align near-term design behavior with vision goals. While overcoming uncertainty and accelerating innovation toward the desired Future. The use of Future vision generating toolkits is divided into four stages: Generate Future Chronicles; Generate future vision from Chronicles from the perspective of an event node; Vision shaping creates Ai City Vision; Co-create. The first three stages are the designer-led vision generation process, and the fourth stage is to open up the future vision to the public and develop deeper design contents with stakeholders and other groups through co-creation (Fig. 1).

Vision generation process is mainly carried out in accordance with the consists of seven elements of template, respectively is: 1. The world outlook, 2. The theoretical foundation, 3. AI dielectric into the new world, 4. Roles, 5. Scene, 6. Props and metaphor, 7. The system diagram. The Ai city vision generating system is being applied to the teaching content of the interaction design course in 2020. The following is a more specific description of each basic module, combined with some students' assignments.

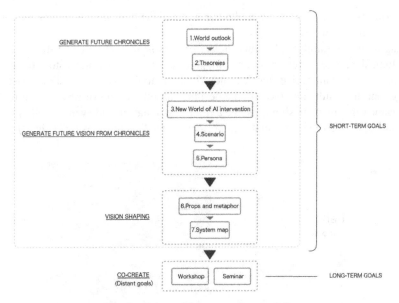

Fig. 1. Future vision generating toolkits

Generate Future Chronicles

Part 1. World view, from answering the question "what do you envision the future world view to be?" The problem begins.

Starting with building a framework for future development, and once you have a logical foundation, you can develop a schedule to improve every detail. Trying to start from an area that is related to or understood by the individual, divide the scope into multiple elements and integrate the elements to better stimulate the conception of the world view based on the existing cognition (Fig. 2).

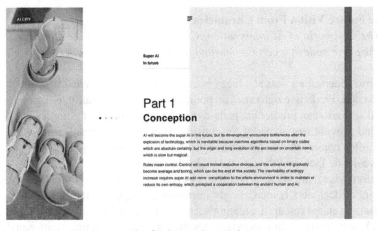

Fig. 2. Part 1. Conception

Part 2. Theory: search for theoretical basis and make a brief description based on the theory and my own world view;

Imagine the possibility of technological explosion in each era and the impact of technological change on future society. Imagine, speculate, construct a story by design. It is important to emphasize that it is not "imagination" and "speculation" based on nothing, but the results obtained by the designer based on a lot of research on reality. At the same time, using design to create a full of imagination and even alienation of the situation and narrative is particularly important for the "design illusion" (Fig. 3).

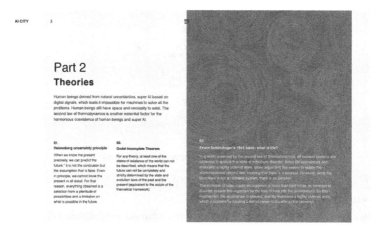

Fig. 3. Part 2. Theories

At this point, an appropriate future development framework can be formulated according to the project situation, and multiple development directions, such as health, education and transportation, can be set vertically with the future timeline as the abscissa. The perfect future world system is deduced step by step through the chronological way.

Generate Future Vision From Chronicles

Part 3. The new world of AI intervention. Creating the future of the chronicle for rod, making the plane node in a certain time and further describes the operating mechanism of the new world;

Scenario generation exists at a fuzzy boundary between creative design practice and strategic vision. Predictive outcomes do not represent one, but multiple futures. A well-developed scenario can predict the path, dynamics and sources of complex phenomena change, and provide new material for consideration [19, 20]. Evans argues that future scenarios often convey stories rather than facts [21]. The further we look into the future, the less we define facts and the less we understand them.

Therefore, visual information is still an essential element of communication here, and the strength of designers is increasingly important in the presentation of future vision. In most future-related research, creating scenarios requires human creativity and imagination. The data used to build the vision for the future is vague, uncertain, and difficult to process using scientific methods. In design research, by contrast, creativity abounds.

However, the usefulness of most design methods only applies to the immediate point of view: as designers move into the distant future, effectiveness declines dramatically (Fig. 4).

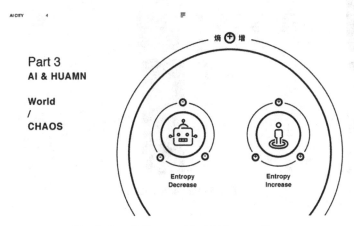

Fig. 4. Part 3. New world of Ai intervention

Part 4. Scenario. You can set up many different scenes, name each scene and expand to describe the story happened in the scene;

It's not just a scene, it's a story. A scenario is a reasonable description of the future based on a coherent set of assumptions. In fact, scenarios are more about extrapolation than invention, depending on the human understanding of a particular type of person. Usage scenarios typically contain user behavior, goals, and a set of activities and events, as well as relevant contextual information. The situational approach is a flexible design approach that can vary. Create at least one scenario for each character and explore the basic situation and status of the character. By describing people's experience of using products or enjoying services in a specific scenario, we can predict the future use of products from the perspective of users and finally obtain specific design ideas. So as to make the design project closer to the real life of users and create a design scheme suitable for target groups. In order to design more situational stories, we need to fully consider the problem from the perspective of the user and design a more satisfied state of people in the future (Fig. 5).

Part 5. Persona, set the main character for Ai in the new world, giving them names and identity information;

Personas based on the information in the previous sections can be used to render different user types associated with the project. A Persona is not a description of a real person, but a representative portrait of a user built from the information gathered. The number of typical user portraits should be 3 to 5. Helping the designer to consider the specific needs and expectations of different categories of target users and sort out the

Fig. 5. Part 4. Scenario

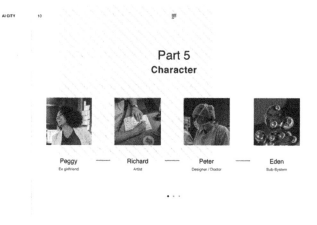

Fig. 6. Part 5. Persona

complicated design characters at hand layer by layer. It helps designers to step outside the scope of self-cognition and view the new World of Ai intervention from the perspective of users (Fig. 6).

Vision Shaping
Part 6. Tools and Metaphor, based on the created vision framework, further improve the props needed to realize the vision.

The biggest difference between fiction and movies is that "Diegetic Prototype of Design fiction" is supposed to be "immersive" and a first-person protagonist's point of view. Listing the future props needed to realize the preset vision, and further explain their functions and usage (Fig. 7).

Fig. 7. Part 6. Tools and metaphor

Part 7. System map. Sorting out the operation mechanism of the new world with Ai intervention in the current scene, and describing the operation mode of the new world by drawing system diagram (Fig. 8);

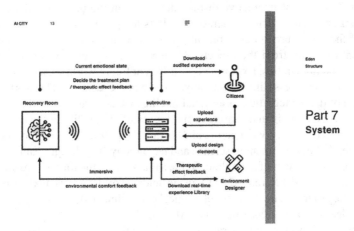

Fig. 8. Part 7. System map

Co-create

The variable value of the main factors creates a problem for future-oriented design (FOD): system design queries to be pushed over the medium to long term. In designing their current products, researchers aim to gather as much data as possible to illuminate existing systems and support short-term forecasting and decision-making. But the reality of the future is not singular. The researchers' subjective predictions about how external factors affect the system will follow an infinite number of trajectories. The core of the analysis used by the researchers required to be tested repeatedly [22].

The fourth stage is to open up the future vision to the public and develop deeper design contents with stakeholders and other groups through co-creation. This section will be created at the initial stage of Future vision by the designer inviting outsiders to co-create. Future vision generating toolkits the ideal operating process for Future vision generating toolkits is to follow the principle of reciprocating design. That is, to make repeated corrections in the four stages of the process in order to achieve a design that maximizes the positive value and minimizes the negative impact of direct and indirect stakeholders. At present, the first three stages have been well practiced in the teaching process and have received positive evaluation and feedback. The next step will be to showcase Future vision's generating toolkits production to more stakeholders through workshops and seminar. Further promote the deepening and implementation of the program.

3.3 Some Achievements of Future Vision Generating Toolkits

Future vision generating toolkits has been continuously introduced into interactive design curriculum as a teaching tool to stimulate students' creativity for Future design. And in the teaching activities of continuous accumulation and improvement. Some of these projects have been developed into landing design projects and are open to the public. The following "Ai City" project is one of the more mature works.

"AI City" is a set of interactive display devices with the theme of future AI City vision. Using the Future vision generating toolkits template framework. Based on the research of the future urban development trend, the author forecasts the future urban development direction from the aspects of science, technology, education, economy, environment and humanities. Exploring four different future world views, and critically thinks about urban access, life, community, mobile and education. The work explores the direction of deep integration of artificial intelligence and carries out the conception of AI fully participating in and sharing social scenes of human life in the next hundred years from 2022 to 2122. By collecting and sorting the political, economic, social, and the future trend of development in science and technology. The design team has written for the future development of a timeline. At the same time, in the form of crowdsourcing invited members added different AI CITY in the field of world view and text, respectively. Telling the story of the AI medical, psychological counseling and psychological comfort, AI scenario learning and transportation of AI scenarios.

The interactive system and operation manual composed of mobile phone scanning and multi-modal media interaction help the audience to build a complete exhibition experience. The access module is to enter the brain of the city and make the audience rethink "where will ai take the city?". Residential module is a kind of independent distributed virtual interconnection space, showing the future pattern of urban residents. The mobility module connects the city's past, future, derivation and convergence trends. The co-creation module provides a collaborative and exhibition platform for participating in the mapping of future cities (Fig. 9).

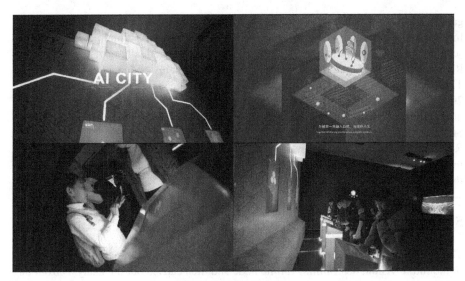

Fig. 9. "Ai City" project

4   Discussions and Future Works

We explored Toolkits that combine vision for the future with design to guide students through a process to produce a unique vision for the future. By setting the future timeline and deducing according to the development of time, we can better carry out the corresponding design. At the same time, we will also use future signs, future triangle, CLA and other tools for future design to reflect on the future trends and reflect on the creativity and development of design at the present stage. From a more unique perspective, I share with my classmates some visions of science fiction and the future. Guide students to integrate the experience of historical development and break through the shackles of the status quo. And explore some metaphorical stories to build a vision for the future.

We also introduce Future vision generating toolkits as teaching tools into four interactive design related classrooms in 2019–2020. It has greatly stimulated students' enthusiasm for thinking about the future. By sorting out students' feedback, we have also gained inspiration for improving the use and Settings of this tool. Some students said: "This is an interesting start, by creating a vision of the future rather than passively accepting it. I can choose to unfold my imagination from many angles". Some students showed their haze: "It seemed to me that there was a strong connection between the content and the sci-fi movies I had already seen. It was hard to jump out of the material and do something different." Some students pointed out that "Part 6. Tools and Metaphor. It's not hard to make up a story about the future, but it's hard to know where to start when it comes down to it."

We conclude that Future vision generating toolkits as teaching tools requires to be further improved as follows. 1. Some students mentioned that props and metaphors are difficult to understand. Indeed, creating a vision is a process of abstractly integrating existing knowledge. It depends on each person's past experiences and perceptions. For this, we are planning to build a Vision Library. Leverage the power of crowd creativity to gather vision knowledge across multiple domains to create the source story library. 2. As for how to find a unique perspective, we realize the importance of timeline. In fact, many scholars have used the method of creating a timeline to gradually envision the future. Richard Watson is a futuristic speaker and scenario thinker in London, published "MEGA TRENDS AND TECHNOLOGIES 2017-2050" [23]. The WBCSD's cornerstone Vision 2050 report calls for a new agenda for business laying out a pathway to a world in which nine billion people can live well, and within the planet's resources, by mid-century. Some charts of Pathway Toward a Sustainable development 2050 were drawn [24]. 3. Regarding the lack of possibility for ideas to be implemented, we envision introducing more tool support in future teaching. Put the idea into man-machine collaboration and make it more practical.

5   Conclusion

The essence of design is to subtly express an idea to the target. By making the chronicle of the future vision, determining a time node and then carried out the analysis from the political, economic, social, scientific and technological aspects. Using design thinking tools to describe the future city vision related to time nodes. In the future, the feedback will be communicated to more stakeholders in the form of workshops and seminars. And will be used to deepen the vision of improving the future city. So that participants can accept and have a sense of ownership, and at the same time enhance the thinking and innovation ability of designers, opening the door for the next round of deeper creative transformation.

Incorporating design fiction's approach to introducing future-oriented design, further shaping Future vision generating toolkits. To Establish a method to find a more comprehensive (social, political, emotional) design of the future direction of life and technology (method of envisioning new futures and technologies); A tool for communicating innovations to other researchers and to the general public; And a set of tools that provide inspiration and motivation for design to designers, engineers, inventors, etc. Achieving a better Future by using Future vision to generate toolkits for humanistic design projects. It emphasizes reflection on the design output to help people discover and determine the future they want. At the same time, hoping this toolkits can provide feedback to the present on the basis of speculation, encouraging designers to make revolutions from now on to influence the future.

Acknowledgements. The images used in this essay are from the works of students majoring in information design of Tsinghua University, and the feedback on the use of toolkits is from the students who participated in the interaction design course in 2020. We would like particularly to acknowledge all the members of the Interaction Design Course, along with He Xue, Zhao Zhexi, Lv Jiayi, Wang Jing, Wang Ke, Li Xiangyang, Yang Bolin, Zhou Wenxin, He Jiayi, Peng Chengyang, Zheng Zhi, Niu Yixuan, Xiang Chenzhuo and Li Hanxuan. Expressing our sincere thanks to their works.

References

1. Townsend, A.M.: Smart Cities: Big Data, Civic Hackers, and the Quest for a New Utopia. W. W. Norton & Company, New York (2013)
2. Bardzell, J., Bardzell, S.: Humanistic HCI. HCI, p. 25, March–April 2016
3. Cherubini, M.: Ethical autonomous vehicles 2013–2017 [EB/OL] (2017). http://mchrbn.net/ethical-autonomous-vehicles/
4. Microsoft. Smart cities: enhance citizen experiences, increase sustainability and resilience, and promote innovation for your city services [EB/OL], 11 July 2013. https://www.microsoft.com/en-us/industry/government/smart-cities
5. Sidewalk labs. Sidewalk labs is reimagining cities to improve quality of life [EB/OL], 25 June 2019. https://sidewalklabs.com
6. Joi, I., Howe, J.: Whiplash: How to Survive Our Faster Future, p. 51. MIT Media Lab: via Boston Herald, 3 April 2012
7. Dunne, A., Raby, F.: Speculative Everything: Design, Fiction, and Social Dreaming, p. 9. The MIT Press, Cambridge (2013)
8. Inayatullah, S.: Six Pillars: Futures Thinking for Transforming. Foresight **10**, 4–21 (2007)
9. Thiel, P.: Zero To One: Notes On Startups, Or How To Build The Future, p. 117. Crown Business, New York (2014)
10. For more on this, see Joseph Voros, "A Primer on Futures Studies, Foresight and the Use of Scenarios," Prospect, the Foresight Bulletin, no. 6, December 2001. http://thinkingfutures.net/wp-content/uploads/2010/10/A_Primer_on_Futures_Studies1.pdf. Accessed 21 Dec 2012
11. Dunne, A., Raby, F.: Speculative Everything: Design, Fiction, and Social Dreaming, p. 12. The MIT Press, Cambridge (2013)
12. Wright, E.O.: Envisioning Real Utopias, pp. 25–26 (2010)
13. Huss, W.R.: A move toward scenario analysis. Int. J. Forecast. **4**(3), 377–388 (1988). https://doi.org/10.1016/0169-2070(88)90105-7
14. Dunne, A., Raby, F.: Speculative Everything: Design, Fiction, and Social Dreaming, p. 15. The MIT Press, Cambridge (2013)
15. Simon, H.A.: The Sciences of the Artificial, 3rd edn, p. 157. MIT Press, Cambridge (1996)
16. Bleecker, J.: Design Fiction: A Short Essay on Design, Science, Fact and Fiction. Near Future Laboratory, p. 49 (2009)
17. Tamenbaum, J.: Design fictional interactions: why HCI should care about stories. HCI, p. 22, September–October 2014
18. Zhong, X.: Internet-of-Things Crisis Management. Delft University of Technology Faculty of Industrial Design, CE Delft (2015)
19. Schoemaker, P.J.H.: Multiple scenario development: its conceptual and behavioral foundation. Strateg. Manage. J. **14**(3), 193–213 (1993). https://doi.org/10.1002/smj.4250140304
20. Gausemeier, J., Fink, A., Schlake, O.: Szenario-Management: Planen und Führen Mit Szenarien. Hanser Fachbuch, Munich (1996)
21. Evans, M.: A design approach to trends and forecasting. In: Durling, D., De Bono, A., Redmond, J. (ed.) Proceedings of Future Ground, Design Research Society International Conference, p. 7. Monash University Press, Melbourne (2004)
22. Ritchey, T.: Wicked Problems—Social Messes: Decision Support Modelling with Morphological Analysis. Springer, Berlin (2011). https://doi.org/10.1007/978-3-642-19653-9
23. Watson, R.: Mega trends and technologies 2017–2050 [EB/OL] (2017). https://www.nowandnext.com/PDF/Mega%20Trends%20and%20Technologies%202017-2050%20(Web).png
24. World business council for sustainable development. Pathway toward a sustainable 2050 [EB/OL], 16 December 2010. https://www.wbcsd.org/Overview/About-us/Vision2050/Resources/WBCSD-s-cornerstone-Vision-2050-report

PIKAR: A Pixel-Level Image Kansei Analysis and Recognition System Based on Deep Learning for User-Centered Product Design

Yun Gong, Bingcheng Wang, and Pei-Luen Patrick Rau$^{(\boxtimes)}$

Tsinghua University, Beijing, China
rpl@tsinghua.edu.cn

Abstract. Can machines learn to perceive products like humans? Kansei Engineering has been developed to connect product design and human perception. While conventional Kansei Engineering Systems exhibit a high dependency on manual extraction of design elements and thereby are restricted to validity issues, we present a Pixel-level Image Kansei Analysis and Recognition (PIKAR) system that applies deep learning to extract and analyze the formation of human perception towards product designs automatically. Method validation is performed based on the evaluation of cosmetic packaging's kawaii. Two neural nets trained on 1,414 images, labeled by eight participants based on their perception of kawaii (1–5 Likert Scale), have achieved a better prediction than test persons. The implemented neuron analysis methodology for Kansei analysis points towards consistency with previous experimental kawaii studies and gives insight to individual differences. This work addresses the possibility of applying deep learning to support product design and user experience researches.

Keywords: Deep learning · Consumer product design · User experience · Kawaii

1 Introduction

The word "Kansei" originates in Japan and refers to the customers' feelings and perception towards a product. Kansei engineering has been developed as the translation technology between product design parameters and human perception and has maintained a significant role in the consumer-oriented product development paradigm [1]. For decades, the construction of Kansei Engineering System (KES) has relied on the manual definition and extraction of significant product attributes, i.e. the design elements [2–7]. Two typical types of element-level KES exist in this domain: a backward KES focuses on predicting Kansei from products and requires manual extraction of design elements prior to mathematical modelling; and a forward KES performs systematic analysis of design elements and identifies factors affecting Kansei. The classic KES can be summarized as an inductive system based on small samples and manual extraction, but it has the following problems to be solved [8]. Firstly, a typical KES usually only uses no more than 100 images of the product as a sample. Insufficient sample size and

© Springer Nature Switzerland AG 2020
P.-L. P. Rau (Ed.): HCII 2020, LNCS 12192, pp. 60–71, 2020.
https://doi.org/10.1007/978-3-030-49788-0_5

sample selection bias will lead to doubts about the validity of the conclusion. Secondly, the process of selecting key design elements often relies on expert advice, but human cognition has inevitable limitations. In addition, the selection of key features is often vague, so it is difficult to guarantee the comprehensiveness of the conclusions. Thirdly, the current system can only process manually processed images, not the original images. This means that the classic KES is difficult to apply on a large scale in the commercial field.

To circumvent the aforementioned constraints, we propose PIKAR system, a Pixel-level Image Kansei Analysis and Recognition system based on deep learning. Contrary to element-level KES, our method is an end-to-end, pixel-level approach that recognizes product Kansei from raw pixel distribution without prior feature handcrafting or extraction. Besides, we propose a deep learning based Kansei analysis method: the neuron analysis methodology incorporating net visualization techniques to identify features affecting Kansei for future design improvement. The method validation is performed based on cosmetic packaging's kawaii evaluation. In addition, autonomous extraction of product Kansei is crucial to customer-oriented product design and precision marketing. First, 1414 images of cosmetics labeled by 8 test persons according to their perception of kawaii (1–5 Likert scale) are used for training the deep neural nets KawaiiNet and KawaiiScoreNet. Results show that KawaiiNet achieves an accuracy of 84% in identifying kawaii design of cosmetic packaging; and KawaiiScoreNet predicts the kawaii values with an accuracy of 90.6% within a 1-point tolerance and 69.3% within a 0.5-point tolerance. The latter demonstrates a better prediction accuracy than the single test person (90.6% vs 56.1%, 69.3% vs. 28.3%). Second, we explore the deep representatives learned from images to identify design features affecting the perception of kawaii. Our results correspond with previous kawaii studies [9, 10] depicting the principal five factors causing kawaii feelings and give insight to subconsciously perceived product features and factor interaction.

There are mainly three contributions of this work. Firstly, for the first time, we propose a new method to extract subjective attribute "Kansei" from pixel-level images automatically. The system no longer uses design elements as the smallest input unit of the inductive engineering system, but directly uses image input to facilitate the direct use of big data. Secondly, we establish a neural net for predicting kawaii value of cosmetic packaging and a neuron analysis method for detecting features affecting Kansei within the deep learning framework, which is more insightful than the traditional method, and can also be applied to the study of individual differences. Thirdly, this work sheds light on a data-driven research paradigm in the field of product design and human factors, providing new perspectives to the current methodologies and theories.

2   Related Work

Extensive studies on product Kansei prediction are based on basic design elements such as the color combination [4], the texture pattern [2] or shape of a certain part [5]. Kansei Engineering System has been proved useful in industrial design [6, 7], packaging design [3, 4], and even interior design [11]. In comparison, less work has been focused on modelling product Kansei directly from raw images. Gong and Rau have proposed a system structure in this domain, but the system lacks further validation [8]. Another relevant study is Deepsentibank that propose a deep learning based, large-scale visual sentiment ontology using Adjective-Noun Pairs [12]. This model uses one million images to train convolutional neural nets and learns to recognize objects with their perceptual attributes. But their focus is rather on general items (e.g. flowers, faces) than on a specific product. Other studies perceive abstract attributes of images, such as photographic aesthetics [13], popularity of images [14], memorability [15] and artistic styles [16] and have provided strong evidence that deep learning is the most adaptive and suitable approach for recognizing vague image attributes.

In terms of Kansei analysis, the purpose is to discover the visual cues of a product that cause the formation of certain type of Kansei. Several research in the field of net visualization have revealed the connection between neuron activations and the appearance of certain visual cues. Girshick et al. point out that each neuron in a deep neural net captures one or two certain visual patterns [17]. Neurons on the lower-level layers (e.g. the first one or two hidden layers) represent fundamental visual properties such as color and edges, while high-level neurons fire to a specific shape, pattern or a category of object. Yosinski et al. demonstrate the "expectation" of each neuron by creating the images that maximize the activation of the certain neuron [18]. This inspires us that neuron activation patterns may be suitable for observing the visual characteristics of a certain image set. The contribution of design features to the formation of Kansei can theoretically be quantified, analyzed, and visualized within the framework of deep learning by interpreting the neuron activations. This system concept is briefly described in [8] and this study performs further implementation and validation of the concept.

3 Recognizing Kansei

This section introduces the construction of a pixel-level Kansei prediction system. In the following experiments, we choose the evaluation of cosmetic packaging's kawaii as the research object. The considerations for this choice are listed as follows: (1) Prone to compare and validate: Kawaii is a well-researched type of Kansei in comparison to other alternatives, especially among Japanese researchers (2) High practical value: Kawaii packaging of cosmetics has a significant effect on the purchase decision. Brands like Benefit, PAUL&JOE, Loretta, ETUDE are known for such design strategies. (3) Less noise of data: Cosmetics' images have favorable characteristics for image analysis, such as standard white background, single object of interest per image and adequate image resolution.

3.1 Method

For all experiments, we use a modified AlexNet [19] as the base model for transfer learning: we start with weights pre-trained on a larger dataset to increase model accuracy. Since we assume the dataset to be small in real practice (~1000 images in this study) and different from the original dataset, namely ImageNet, the following modifications are made to limit model overfitting:

- The classifier is trained from activations in the third pooling layer (pool3) since lower layers contain less dataset-specific features.
- Three fully connected layers with 40 nodes each are inserted before the classifier to obtain the feature vectors.

All images are resized to 256×256 pixels while retaining the original aspect ratio: in non-squared images and a white background is added. We perform random cropping to an image size of 227×227 pixels and random mirroring to augment dataset size. We use SoftmaxLoss as the loss function for kawaii classification (kawaii vs. not kawaii) and EuclideanLoss for regression (value of kawaii). The models are trained with Caffe [20].

3.2 Datasets

All cosmetics images are downloaded from the official websites of 25 international cosmetic companies. A total of 1,414 identical cosmetic images are collected, including products for skin care, body care, makeup for eyes, brows, face and lips, perfume and nail polish. All images have light-colored simple background with one product placed in the center of the image. The minimum resolution we accept is 230×230 pixels which is larger than the minimum input requirement of the network.

Next, 8 young females (Mean age: 24, SD: 1.51) representing the consumer demographics are recruited for the kawaii labelling task. All participants claim to have experiences in purchasing cosmetics from both online and shops. The number of the participants is decided based on a pilot experiment. During the pilot experiment, 15 participants are asked to label five cosmetics images' kawaii value using Likert-5 scale. Results show that the average kawaii values for all 5 test images change drastically when total number of testers is small but later become stable with relatively small differences from the total average scores (<0.5), starting from the 8th testers.

The 8 participants are given 2 h for labeling kawaii values for 1,414 images, allowed with brief breaks after sets of ~200 images to avoid fatigue. The labeling test is performed on a MatLab interface with 6 images per page, 5 radio buttons below each image and a next button. The images are displayed with an identical image size. Figure 1 shows the distribution of the final averaged kawaii values and the corresponding standard deviation of the dataset. The most and least kawaii samples are displayed on the right side.

Finally, the total dataset for the classification experiment comprise 500 most kawaii images and 500 least kawaii images while the regression experiment uses all 1,414 images. The dataset is divided into sub-divisions, namely the training set, validation set and test set, with a ratio of 7:1.5:1.5. The division is performed randomly.

3.3 Result of Experiments

Table 1 shows that KawaiiNet appears effective at predicting kawaii cosmetic designs (Overall accuracy: 84% vs. random: 50%). The columns report results on different subsets, showing a higher classification accuracy on images with obvious Kansei than the ambiguous ones (20 smallest SD: 85% vs. 20 largest SD.: 65%).

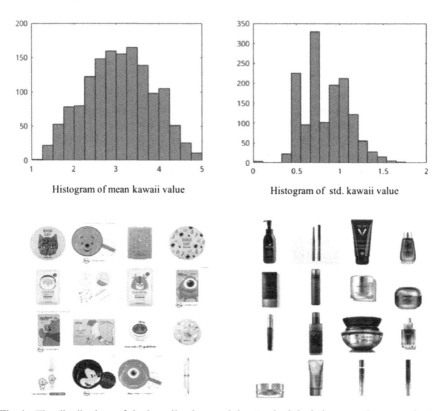

Histogram of mean kawaii value　　　　Histogram of std. kawaii value

Fig. 1. The distributions of the kawaii values and the standard deviations are shown on the top. The dataset is balances with kawaii and non-kawaii instances. The final kawaii value for each image is an average score of all participants. The mean kawaii value of all images is 2.9981 (SD = 0.79). Top kawaii samples and non-kawaii samples are shown on the bottom.

Table 1. Accuracy of *KawaiiNet*

	Overall	Top 20	Bottom 20	Large SD 20	Small SD 20	Random
Accuracy	84%	95%	90%	65%	85%	50%

Table 2 shows complete test results of KawaiiScoreNet. The predictions are compared with random guess and one human tester on four tolerance scales. Random guess is generated from a normal distribution (mean: 2.9981, SD: 0.79) close to the origin kawaii value distribution of the dataset. The human predictions are obtained from the test person who gives the closest scores to the average kawaii scores of eight participants. Results show that KawaiiScoreNet outperforms the single human predictor with a large margin in all scales, achieving a 90.6% confidence in providing prediction within 1-point tolerance and 69.3% within 0.5-point tolerance. Even after integer conversion of the predictions (since the regression results are not in Likert-5 scale in comparison to human), KawaiiScoreNet still gives kawaii scores that better represents the averaged Kansei than a single test person does.

Table 2. Accuracy of *KawaiiScoreNet*

Tolerance	Random guess	Best human predictor	Kawaii ScoreNet	Kawaii ScoreNet (Integer)
<0.1	6.2%	7.5%	15.1%	9.0%
<0.3	18.8%	23.1%	48.6%	20.6%
<0.5	30.8%	28.3%	69.3%	55.7%
<1	57.8%	56.1%	90.6%	85.3%

Figure 2 compares top 20% most and least kawaii images selected by PIKAR and all human testers, respectively. A preliminary Turing test conducted with a small group of people shows that most people we asked cannot distinguish the "machine-selected kawaii" from the "human-selected kawaii". Though it is only an early observation, we can see that PIKAR's ability in predicting kawaii and non-kawaii cosmetics is hardly distinguishable from that of a human.

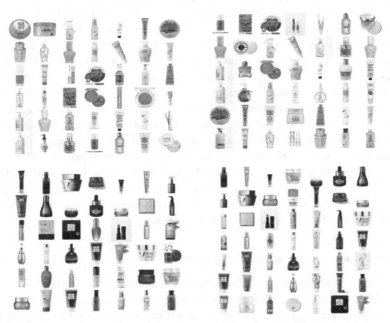

Fig. 2. The top 20% and bottom 20% images selected by human and by machine from the perspective of kawaii within the test image set (n = 212). The left ones are selected by human and the right ones are predicted by machine. Machine's ability in predicting kawaii cosmetics is hardly distinguishable from that of humans.

4   Analyzing Kansei

This section introduces our second application of deep learning in Kansei engineering. We present a neuron analysis approach to identify features that cause the formation of Kansei and support future product design.

4.1   Method

As recorded in previous studies, one neuron fires on to one or two certain visual modes [17, 18]. This means the total count of activations for a certain neuron indicates the appearance of certain visual cues in a collection of images. If these neurons are identified, which are more activated in image set A and, in comparison, far less activated in set B, they allow to describe the critical visual characteristics that distinguishes set A from set B. This briefly describes the core concept of our methodology. This concept is implemented and validated by the following two experiments. The first aims to detect kawaii factors and the second to understand individual differences in kawaii perception.

As our interest lies in medium-sized regional visual patterns (i.e. neither too general nor too detailed), we focus on neurons from the pool3 layer, the max-pooled output of the third convolutional layer with a receptive field of 99 * 99 pixels. A total of 13824 neurons are included in the selected layer. For the first experiment, 150 kawaii cosmetics images and 150 non-kawaii images are input to the KawaiiScoreNet separately for neuron activation computing and the total number of each neuron's activations is counted. Neuron activation count differences exceeding 50 between the datasets (top 0.7% large disparities) are selected as the crucial neurons for kawaii perception and further clustering is performed to reduce redundancy. Clustering operation is based on the distance of each neuron's activation vector. The activation vector is a 150-dimension vector composed of one certain neuron's activation values on the kawaii dataset. Second experiment, on the other hand, picks two testers who demonstrated different rating behaviors in the previous labelling task. The opposite image sets consist of 150 kawaii images each, but selected by different testers. We perform the same computation with the datasets as the first experiment except for the crucial neurons selection step. Since the disparities of individuals on perceiving the same type of Kansei is smaller than that of two opposite Kansei types, we adopt a less strict selection policy that captures neurons which are among the top 30% most activated neurons in image set A but fall below average in image set B. This policy results in a total number of 42 and 18 neurons representing the critical visual characters of each tester.

4.2   What Makes Cosmetic Packaging Kawaii?

Figure 3 shows the kawaii traits discovered by neuron analysis method. The graph exhibits the 100 crucial neurons identified to strongly affect the kawaii feelings and the corresponding clustering results. These crucial neurons are divided into 3 categories and 13 sub-categories. Typical images representing each sub-category are displayed below the graph. Each row contains 10 top-scoring images of one random neuron in the sub-category and demonstrates the dominant visual traits for which neuron is activated. Some rows share fewer clear characteristics than the others due to the fact that each neuron fires to more than one visual cues. This means factor interactions are considered in this method. Naming of the categories are performed by the authors. The resulting three principle factors affecting kawaii are pattern, color and shape. And the 13 sub-categories are named as follows:

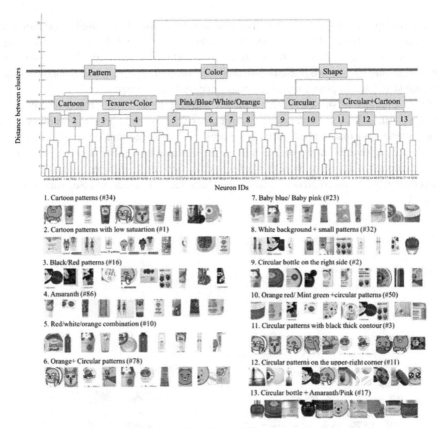

Fig. 3. Top 100 neurons affecting kawaii feelings are detected with neuron analysis method and are clustered into 3 categories and 13 sub-categories. Typical top-scoring regions corresponding to each cluster are displayed. The names of the categories are given according to the dominant traits.

- Cartoon patterns
- Cartoon patterns with low saturation
- Black/red patterns
- Amaranth
- Red/white/orange combination
- Orange + Circular patterns
- Baby blue/baby pink
- White background + Small patterns

- Circular bottle on the right side
- Orange red/mint green + circular patterns
- Circular patterns with black thick contour
- Circular patterns on the upper-right corner
- Circular bottle + Amaranth/pink

Comparing the results to the five principle factors identified to cause kawaii feeling, our method has covered three of them, namely, pattern, color and shape. The remaining two factors that are omitted in this study are size and tactile sensation. The reason for that is all testing images are presented virtually on a computer screen and displayed with the same size. This implies further labelling process improvements.

Other findings in sub-categories concord with the kawaii rules listed in previous studies [9, 10, 21], such as employing warm color scheme, high brightness colors, circular patterns and mammalian cartoon patterns. It is possible that our method cannot provide a comprehensive list of factors, but the dominant ones are detected. Besides, the deep representatives are more powerful in detecting subtle differences and even the subconsciously perceived product features and their factor interactions. See subcategory 8 and 11.

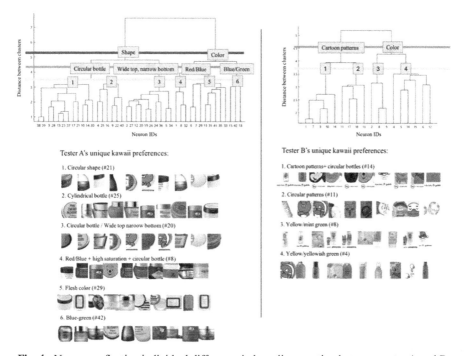

Fig. 4. Neurons reflecting individual difference in kawaii perception between tester A and B.

4.3 Individual Differences on the Perception of Kawaii

Figure 4 shows the most distinguishable visual cues detected with neuron analysis method to represent two selected test persons' kawaii preferences. 42 and 18 neurons are detected respectively, and clustering results show that tester A prioritizes shape while tester B focuses on patterns. More specifically, cosmetics with cylindrical bottles, and ones with wide top and narrow bottom can trigger the kawaii feelings in tester A. Tester B, in comparison, gives higher kawaii scores to the packaging printed with cartoon patterns. The disparities can be clearly observed from the top-scoring images. Additionally, two testers are found to have different reflection models between color and kawaii feelings. Flesh color, red/blue with high saturation and blue-green color are among the kawaii colors for tester A, while bright yellow, mint green and yellowish green are perceived as kawaii by tester B. The same behavior is observed in the labelling experiment. Though a more careful selection of parameters and policies can be used during the neuron selection process, the current results have shown the possibility of applying deep learning in researches such as individual differences in visual perception.

5 Conclusion

Can machines perceive products like humans? Our study is among the first attempts to answer this question using the deep learning technology. In this paper, we explore how low-level pixel distribution of product images can be interpreted to high-level human perception through deep learning. A case of evaluating the kawaii feelings of cosmetic packaging has shown the capability of machine in predicting abstract human perception. We further present how the neural nets can be analyzed to support design and provide insights into individual differences. The results correspond to previous experimental studies and exhibit interesting kawaii traits in the cosmetic packaging design. The proposed technology has possible applications in customized product recommendation systems, design-supporting systems and even in social robotics which mimic human perception.

Yet the answer to the aforementioned question is still open. Several directions remain for future researches. An important question is whether this method is applicable to variations in products and types of Kansei. To what extent can we reuse the neural nets trained with cosmetic packaging? And in what circumstances can we keep the same network structure? We have conducted several preliminary tests with small data size in this direction. The results suggest that a weight-level modification (i.e. keep the model structure and retrain the weights) may be required for cases such as "Luxury perception of cosmetic packaging" or "Kawaii perception of beverage bottles"; but a model-level modification (i.e. change the model structure) tents to be necessary for other applications such as "kawaii value of a girl's face". Neuron analysis for product design improvement is another interesting direction of exploration. Deep learning is well acknowledged as the powerful prediction tool, but this study has explored its ability in extracting visual information for design analysis.

Finally, this study has addressed the potential of applying deep learning in user-centered product design. Our method represents a new approach to gain customer insights with less restriction to the sample size and validity issues and is completed driven by data itself.

Acknowledgement. This work was funded by Tsinghua University Initiative Scientific Research Program 20193080010.

References

1. Nagamachi, M.: Kansei engineering: a new ergonomic consumer-oriented technology for product development. Int. J. Ind. Ergon. **15**, 3–11 (1995)
2. Matsubara, Y., Nagamachi, M.: Hybrid Kansei engineering system and design support. Int. J. Ind. Ergon. **19**, 81–92 (1997)
3. Tsuchiya, T., Ishihara, S., Matsubara, Y., Nishino, T., Nagamachi, M.: A method for learning decision tree using genetic algorithm and its application to Kansei engineering system. In: IEEE SMC 1999 Conference Proceedings. 1999 IEEE International Conference on Systems, Man, and Cybernetics (Cat. No. 99CH37028), pp. 279–283. IEEE (1999)
4. Nagamachi, M., Tachikawa, M., Imanishi, N., Ishizawa, T., Yano, S.: A successful statistical procedure on kansei engineering products. In: 11th QMOD Conference. Quality Management and Organizational Development Attaining Sustainability from Organizational Excellence to SustainAble Excellence, Helsingborg, Sweden, 20–22 August 2008, pp. 987–995. Linköping University Electronic Press (2008)
5. Chen, J.-S., Wang, K.-C., Liang, J.-C.: A hybrid kansei design expert system using artificial intelligence. In: Ho, T.-B., Zhou, Z.-H. (eds.) PRICAI 2008. LNCS (LNAI), vol. 5351, pp. 971–976. Springer, Heidelberg (2008). https://doi.org/10.1007/978-3-540-89197-0_93
6. Pritchard, S.C., Zopf, R., Polito, V., Kaplan, D.M., Williams, M.A.: Non-hierarchical influence of visual form, touch, and position cues on embodiment, agency, and presence in virtual reality. Front Psychol. **7**, 1649 (2016). https://doi.org/10.3389/fpsyg.2016.01649
7. Wang, K.-C.: A hybrid Kansei engineering design expert system based on grey system theory and support vector regression. Expert Syst. Appl. **38**, 8738–8750 (2011)
8. Gong, Y., Rau, P.-L.P.: DL-KES: a deep learning based kansei engineering system (特集 人工知能と感性). 感性工学 = J. Jpn Soc. Kansei Eng. **15**, 29–31 (2017)
9. Ohkura, M., Konuma, A., Murai, S., Aoto, T.: Systematic study for "kawaii" products (the second report)-commpmrison of "kawaii" colors and shapes. In: 2008 SICE Annual Conference, pp. 481–484. IEEE (2008)
10. Ohkura, M., Komatsu, T., Aoto, T.: Kawaii rules: increasing affective value of industrial products. In: Watada, J., Shiizuka, H., Lee, K.-P., Otani, T., Lim, C.-P. (eds.) Industrial Applications of Affective Engineering, pp. 97–110. Springer, Cham (2014). https://doi.org/10.1007/978-3-319-04798-0_8
11. Nomura, J., Imamura, K., Enomoto, N., Nagamachi, M.: Virtual space decision support system using Kansei engineering. In: Kunii, T.L., Luciani, A. (eds.) Cyberworlds, pp. 273–288. Springer, Tokyo (1998). https://doi.org/10.1007/978-4-431-67941-7_18
12. Chen, T., Borth, D., Darrell, T., Chang, S.-F.: Deepsentibank: visual sentiment concept classification with deep convolutional neural networks. arXiv preprint arXiv:1410.8586 (2014)
13. Lu, X., Lin, Z., Jin, H., Yang, J., Wang, J.Z.: Rapid: rating pictorial aesthetics using deep learning. In: Proceedings of the 22nd ACM International Conference on Multimedia, pp. 457–466 (2014)

14. Khosla, A., Das Sarma, A., Hamid, R.: What makes an image popular? In: Proceedings of the 23rd International Conference on World Wide Web, pp. 867–876 (2014)
15. Isola, P., Xiao, J., Torralba, A., Oliva, A.: What makes an image memorable? In: CVPR 2011, pp. 145–152. IEEE (2011)
16. Karayev, S., et al.: Recognizing image style. arXiv preprint arXiv:1311.3715 (2013)
17. Girshick, R., Donahue, J., Darrell, T., Malik, J.: Rich feature hierarchies for accurate object detection and semantic segmentation. In: Proceedings of the IEEE Conference on Computer Vision and Pattern Recognition, pp. 580–587 (2014)
18. Yosinski, J., Clune, J., Nguyen, A., Fuchs, T., Lipson, H.: Understanding neural networks through deep visualization. arXiv preprint arXiv:1506.06579 (2015)
19. Krizhevsky, A., Sutskever, I., Hinton, G.E.: Imagenet classification with deep convolutional neural networks. In: Advances in Neural Information Processing Systems, pp. 1097–1105 (2012)
20. Jia, Y., et al.: Caffe: convolutional architecture for fast feature embedding. In: Proceedings of the 22nd ACM International Conference on Multimedia, pp. 675–678 (2014)
21. Cheok, A.D.: Kawaii/cute interactive media. In: Cheok, A.D. (ed.) Art and Technology of Entertainment Computing and Communication, pp. 223–254. Springer, London (2010). https://doi.org/10.1007/978-1-84996-137-0_9

Technology Intelligence Practice for NTBFs in Developing Countries

Thi Ha Htun[✉], Weiguo Fang, and Yun Zheng

School of Economics and Management, Beihang University, Beijing 100083, China
{thihahtun,wgfang,yunzh}@buaa.edu.cn

Abstract. Technology intelligence (TI) is a process that allows technology-based companies to identify technological opportunities and threats that can affect the future growth and survival of companies. Therefore, the primary operational function of TI is the collection, analysis and dissemination of information for the development of knowledge about threats and technological opportunities. New technology-based firms (NTBFs) should be aware of the latest technological innovations if they want to take advantage of new opportunities and be mindful of potential threats. Since the needs of NTBFs may diverse, there may be different ways to design and implement TI activities depending on the business environment, the level of uncertainty, the strategy used and the resources they can get. The challenge of this study is what NTBF's capabilities should have to develop and how to implement TI processes for the firm. Therefore, this study considers a better understanding of how to develop and implement TI and its processes in practice for NTBFs in developing countries (Myanmar, China and India). Semi-structured interviews were conducted in this study with 19 technology-based organizations in the fields of Defense Avionics industry, ICT, pharmaceutical and electrical equipment located in Myanmar, China and India. 6 core questionnaires (60 detailed questions) were constructed for the interviews. Analysis and some managerial insights of the study are discussed. Finally, the general TI process cycle for NTBFs in developing countries is conceptualized with seven steps as a cycle for the continuation and integration of TI, and also comprehensively explained how it works.

Keywords: Technology intelligence · NTBFs · Implementation · Opportunities · Threats

1 Introduction

Technology intelligence (TI) is a process which allows technology-based companies to identify technological opportunities and threats that may affect the future growth and survival of companies. TI seeks data to collect, analyze and disseminate necessary technological information for better strategic planning and decision making associated with companies. The main operational function of TI is, therefore, to collect, evaluate and provide information in order to develop knowledge about threats and technological capabilities. For this reason, TI is characterized by a systematic model for

© Springer Nature Switzerland AG 2020
P.-L. P. Rau (Ed.): HCII 2020, LNCS 12192, pp. 72–90, 2020.
https://doi.org/10.1007/978-3-030-49788-0_6

compiling, analyzing and distributing information from the technological environment in order to find new opportunities. Kerr described TI as the collection and transfer of technological information as part of the process by which an organization develops an awareness of technological threats and opportunities [1]. TI responds to a wide range of decision-making needs (from strategic to operational), as it helps a company to recognize important advances in technology [1].

The challenge for TI is to take advantage of potential opportunities and defend against potential threats by quickly delivering relevant information on technology trends in the environment of the technology-based firms. Thus, the activity in the field of TI can support decision-making on technological and general management issues by timely preparation of relevant information about technological facts and trends through the collection, analysis and dissemination [2]. In other words, TI is the potential used by technology companies to support the decision-making process through the collection and provision of information about new technologies. Technology-based firms employed TI to identify new technologies, exploit potential opportunities, and also discover problems related to the current threat in the technology environment of the firms.

In order to identify technological opportunities and threats in the company's environment, many companies have implemented TI processes, which aim at the timely learning of technological trends [3]. However, studies have shown that many approaches to TI have failed in the past [4]. Nevertheless, several technology-based companies systematized their search processes by creating additional TI activities to acquire technologies due to the growing acquisition of external technologies [5].

Flexibility and responsiveness are crucial of new technology-based firms (NTBFs), and knowledge base has to be properly structured and carefully cultivated to support sustainable growth [6]. Therefore, it is of vital importance to understand and classify the knowledge management activities of NTBFs concerning the development of appropriate processes, methods, and tools. So, information on market/product/technology changes must be observed:

- from industrial and commercial fairs
- from specialized or business publications
- through links with (lead) suppliers or others business partners, governmental organizations and research centers, and consultants
- directly from market observation through personal contacts, and recent and prospective customers

For this purpose, numerous companies have established a so-called intelligence system (technology, business or competitive intelligence) to manage these monitoring activities, to analyze the information received, to improve the quality of decision making, and to communicate insight [6]. Since NTBFs' requirements may be diverse, there may be many ways depending on the commercial environment for the development and implementation of TI activities. One of the important factors that affect TI requirements for NTBF depends on the stages of the life cycle of the firms. In the initial stages (initiation, survival and growth phase), NTBFs faced particularly with technology management activities that change rapidly and, therefore, needed to change TI accordingly [6].

The effectiveness of technology management is fundamentally influenced by observations made on current and future technological trends [7]. Consequently, TI (observation and evaluation of technological trends) is one of the core activities of technology management.

NTBFs should be aware of the latest technological innovations if they want to take advantage of new business opportunities and be aware of potential threats. Many worldwide established firms have already implemented and developed TI systems, designed to capture information on emerging technologies and trends, as well as implemented to provide information obtained in a usable form for decision-makers. However, as an NTBF, creating this type of system can difficult and requires the ability to analyze and evaluate large amounts of data, identify what is relevant and transfer it to those who need it.

Since the needs of NTBFs may vary, there may be different ways to design and implement TI activities depending on the business environment, the level of uncertainty, the strategy used and the resources they can get. Therefore, the objective of this study is to provide a better understanding of the development and implementation of TI and its processes for NTBFs in developing countries.

2 Literature Review

Having effective TI capabilities in hand is becoming increasingly important as business become global, and technology life cycles are shrinking. While the literature is being widened, one word can be found as various terms to characterize the activity of analysis and evaluating TI, such as Technology intelligence [1, 8], Technology forecasting [9–11], Technology monitoring [12, 13], Technology scouting [14] and so on. It seems that the term TI is widely diffused and commonly accepted in modern literature. Savioz (2002) expressed that technology monitoring, technology forecasting, technology scouting and competitive intelligence are summarized under TI [15]. Byungun (2008) also agreed that TI is one of the important and critical processes of technology management that includes technology forecasting, monitoring and assessment concepts [16]. The new era is characterized by an economy based on knowledge and rapid technological change. Therefore, TI and forecasting future-technologies must be active in the face of rapid technological change and generating new innovations is important in this complex environment. That is why TI is a vital process to sustain the firms' competitiveness.

TI includes activities that support technological and general management using the timely preparation of relevant information about technological facts and trends (opportunities and threats) in the organization's environment through the collection, analysis and dissemination [15]. In other words, TI covers activities related to the collection, evaluation and transfer of relevant information on technological trends to support technological and more general decisions of firms. According to this definition, TI includes observation and analysis of individual competitors, as well as universities and start-up firms. In addition, TI performs additional functions, such as organizational learning [17, 18].

In the face of increased competition in a globalized environment, companies receive a huge burden to improve their TI capacity. For this reason, in recent years, there has been a growing interest in a systematic TI approach in many technology-intensive companies

[4], and its application in different technological fields is widely accepted [19, 20]. Initial work in TI mainly suggested establishing a centralized intelligence unit, but recent work has shown that TI processes require more complex solutions [17, 21, 22]. In order to arrive at a comprehensive view of TI management, three complementary organizational mechanisms can coexist within any given firm. First, the firm can rely on structural organization, that is, particular organizational structures to coordinate intelligence [23]. Second, a firm can use a project-based organization, which refers to temporary projects to coordinate intelligence activities [24]. Third, different intelligence tasks can be carried out informally.

In recent days, to increase market share and profitability, and to survive in the competitive environment, firms essentially necessitate to develop, devise and produce high-technology products with high added value, and in relation to this fact, developing countries oriented to establish new technology-based firms (NTBFs) due to their flexibility and speed of response. They are at the forefront of technological change due to their ability to innovate, and they have resurfaced as an important part of modern economies [9]. Therefore, this paper explores the question of which capabilities NTBFs should have for the development and implementation of TI process in developing countries.

3 Methodology

To date, literature has not presented many practical examples of TI practice in full for NTBFs concerning the developing countries. For this study, therefore, 19 technology-based organizations in the field of Defense Avionic Industry, ICT, Pharmaceutical and Electrical Equipment located in China, India and especially in Myanmar have been investigated to gain a broad perspective on TI. Table 1 gives some examples of high-technology industries, and these industries are considered in this study.

Table 1. High-technology industries considered in this study

High-technology industries
Aerospace
Automotive
Biotechnology
Chemicals
Defense
Electrical equipment
ICT
New materials technology
Medical technology
Pharmaceutical

This study aims to contribute to the purpose of:

- Understanding how TI systems are currently operationalized in NTBFs located in developing countries
- Acquiring knowledge of TI implementation issues

The analysis is based on the intelligence framework of Kerr et al. [1]. The framework describes two approaches to connect sources of information to decision-makers: Top-down (the information consumers ask the intelligence system to gather information); Bottom-up (the information collecting system knows who would be interested in receiving particular information acquired). Besides that, the framework consists of:

- Decision-makers
- Actions (reasons for performing intelligence)
- Intelligence Streams (MI-Market Intelligence, CI-Competitive Intelligence, TI)
- Types of information acquired and information sources

In this study, semi-structured interviews were held with 19 technology-based organizations to analyze six main questionnaires concerned understanding how TI systems are currently operationalized:

1. For what reasons the firms get to establish TI?
2. How about the organizational capabilities for organizing and integrating TI (budget, staffing, infrastructure, etc.)?
3. How about the TI activities of the firms according to the intelligence framework [1]?
4. What sources (published or non-published information sources) the firms use to collect data in the TI process? (Collecting Data)
5. How do the firms evaluate the information collected from the TI search process? (Analyzing Information)
6. How about the application of intelligence knowledge derived from the TI evaluation process? (Disseminating Information)

The first question consists of 6 items concerning starting phase questions that address the motivation elements for firms to do TI (making TI for realizing what goals). The second question consists of 5 items which analyze structural elements that the firms created to organize and integrate TI processes (budgeting, staffing, creating infrastructure). The third question involves 14 activities according to the intelligence framework [1]. The fourth question generates six published sources and 11 non-published sources for collecting data to evaluate as information for decision-makers in a usable form. The fifth question sets up eight evaluation methods at which the firms analyze the information through TI collecting process. The last question consists of 10 items concerning the application of the results derived from the TI evaluation process. In order to construct the questionnaires in accordance with the purpose of the study, nine managers working in the technology-based companies located in Myanmar advised research methods on the viability and sufficiency about data collection. In the beginning, it was planned to send questionnaires to 27 NTBFs located in Myanmar, China and India, but only 19

NTBFs accepted to take part in this study. General description of case studies in this study is expressed in Table 2.

Table 2. General description of case studies

Case	Fields/Industries	Total employees	Interviewee's role	Countries
1	Defense (Avionics)	140	Head of Department	Myanmar
2	ICT	20	Technical Manager	Myanmar
3	ICT	500	CEO	Myanmar
4	ICT	25	Technical Manager	Myanmar
5	ICT	142	Chief Operation Officer	Myanmar
6	ICT	21	Chief Technical Officer	Myanmar
7	ICT	80	Project Manager	Myanmar
8	ICT	10	Head of the Centre	Myanmar
9	ICT	100	Project Leader	Myanmar
10	ICT	60	Project Manager	Myanmar
11	ICT	24	Senior Project Manager	Myanmar
12	CT	140	Director	Myanmar
13	ICT	15	Project Manager	Myanmar
14	ICT	51	Project Manager	Myanmar
15	Pharmaceutical	200	Project Manager	Myanmar
16	Pharmaceutical	40	Project Manager	Myanmar
17	Electrical Equipment	35	General Manager	Myanmar
18	ICT	400	Game Content Manager	China
19	ICT	500	Operations Manager	India

The various types of firms located in Myanmar, China and India were selected to study in order to cover a wide range of sectors and firm types with the condition of the strong interest in technological developments. The interviews were conducted at the firm's premises. The questionnaires are sent to interviewee one day ahead before the interview for better understanding the questionnaires. The organizations' name and interviewees' anonymity were protected. The interviewees' position in the organization were displayed to ensure that the survey instrument was completed by the appropriate and complete informant. After the analysis, the interviewees were asked to check the case report for any inconsistencies.

4 Analysis and Results

For the 1st question of the study, the six reasons [25] why the firms establish TI are expressed in Table 3, and the analysis results are given in Fig. 1.

Table 3. Analysis of the reasons to establish TI

No.	Reasons
Q1-1	To satisfy the customer's new requirements and expectations
Q1-2	To increase the quality and performance of the existing products
Q1-3	To find out new ways in order to decrease the design and production costs
Q1-4	To be aware of new commercial and technological opportunities
Q1-5	To take advantage of the capabilities and resources of other institution/companies and to establish partnerships (such as R&D partnership)
Q1-6	To watch potential and existing rivals/products

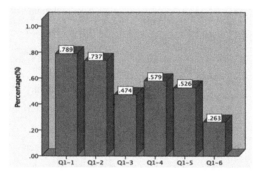

Fig. 1. Analysis result of the reasons to establish TI

According to the result, as shown in Fig. 1, most of the NTBFs in this study have a motivation to perform the TI process except watching potential and existing rivals/products (Q1-6). The firms mostly focused on satisfying the customer's new requirement and expectation (Q1-1), and to increase the quality and performance of the existing products (Q1-2). Only a few firms (Case 3, 9, 15, 16, 19) aim to watch potential and existing rivals/products. This is because, to date, there is still a lesser number of NTBFs which have been established in Myanmar compared with other developing countries, so competition between NTBFs is also not too high, and still no need to worry about competitive advantages. Nevertheless, potential and existing rivals/products should be watched for long term competitive advantage.

For the 2nd question of the study, five items [25] concerning organizational capabilities to integrate TI process are selected in Table 4, and the analyzed results are given in Fig. 2.

According to the result, as shown in Fig. 2, among five items concerning organizational capabilities to organize and integrate TI process, two items (Q2-2 and Q2-4) are more operationalized in most of the firms than other three items (Q2-1, Q2-3 and Q2-5). Several researches argue that TI cannot be delegated fully to dedicated units. Nevertheless, according to recent empirical studies, the most technology-intensive companies in the world have internal TI steering groups or personnel.

Table 4. Analysis of organizational capabilities to integrate TI

No.	Items
Q2-1	Technology intelligence process is defined in work procedures
Q2-2	Responsible employees are in charge of technology intelligence
Q2-3	Specialized internal technology intelligence unit/department exists
Q2-4	Information technology and networks for technology intelligence are provided
Q2-5	Dedicated budget for technology intelligence is allocated

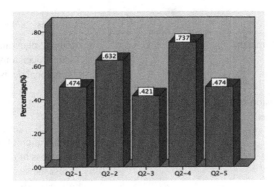

Fig. 2. Analysis result of organizational capabilities to integrate TI

For the 3rd question of the study concerning intelligence framework [1], Table 5 and Fig. 3 reports the results whether the firms work Top-down (following the TI user's requests) or Bottom-up (feeding information to users without a specific request).

Table 5. Analysis of top-down and bottom-up approach

No.	Approach
Q3-1	Top-down
Q3-2	Bottom-up

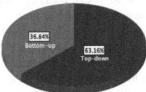

Fig. 3. Analysis results of two approaches operation in organizations

According to the analysis result of two approaches to connect sources of information to decision-makers, most of the NTBFs in this study operate their firms frequently with a Top-down approach. This result could be related to the limited sources and size of NTBFs, and owner-manager factor of the firms. Top management prefers to make inferences based upon their experience. On the other hand, the strongest form of Top-down approach was found for target activities, where intelligence followed a list of pre-identified technologies. The Top-down approach dominated in small groups and where decision-makers knew to whom to address the information requests.

For the 3rd question of the study, Table 6 is drawn to survey the TI activities according to the intelligence framework [1], and the results are shown in Fig. 4.

Table 6. Analysis of the TI activities according to the intelligence framework [1]

No.	Factors	Activities
Q3-3	Decision makers	Strategic planning
Q3-4		Tactical decisions
Q3-5		Operational decisions
Q3-6	Are decision-makers part of the technology intelligence?	Yes
Q3-7		No
Q3-8	Actions	Identification of opportunities
Q3-9		Awareness of threats
Q3-10		Assess the state of the art
Q3-11		Profile trends
Q3-12	Intelligence streams	Market intelligence
Q3-13		Competitive intelligence
Q3-14		Technology intelligence

The result of the analysis of activities performing in organizations indicate that most of the decision-makers in Myanmar emphasize strategic planning, but just a few decision-makers from Myanmar and two firms from China and India emphasize all of the strategic planning, tactical and operational decision.

Nearly all of the decision-makers of the firms participated in this study are part of the TI process, as most of them operate TI in their firms with Top-down approach.

The firms in this study mostly observe to identify opportunities, to be aware of threats and profile trend, but lack to assess the state of the art in the firms' environment.

The analysis shows that TI is the first priority compared with MI and CI for NTBFs in Myanmar, but the two firms from China and India emphasize all of three intelligence streams. Although the firms are based on technology, market and competitive intelligence should be acted for long term goals. As the development approaches commercialization, market and competitive intelligence can provide manufacturing costs for effective pricing, performance analysis of competitive products, and provide as early indicators of technology abandonment. Each firm must examine the collected information about technologies, products and markets using TI, CI and MI. TI can introduce

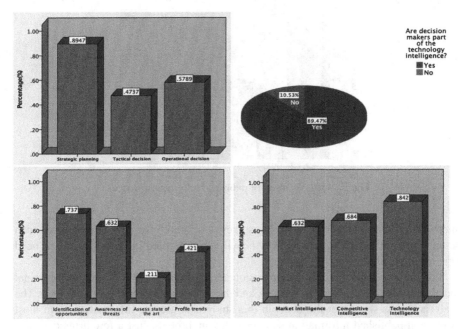

Fig. 4. Analysis result of activities performing in organizations

innovation processes. Every company should have an MI activity, typically in marketing departments. CI can be considered as part of MI or TI.

For the 4th question of the study, there is a distinction between *published information* and *non-published information*.

For the *published information*, the acquisition of data is analyzed in Table 7, and the results of the analysis are given in Fig. 5.

Table 7. Analysis of published information sources

No.	Published information sources
Q4-P-1	Internet/websites
Q4-P-2	Patents
Q4-P-3	Field publications
Q4-P-4	Non field publications
Q4-P-5	External intelligence reports
Q4-P-6	Governmental foresight studies

The most popular sources of published information applying NTBFs in this study were internet websites (P-1), and the 2nd popular sources were field publications (P-3). Internet search engines are used to identify new sources of information related to current

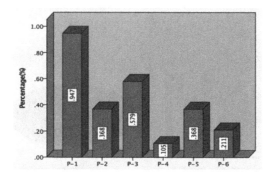

Fig. 5. Analysis result of published information sources

interests. Internet searches need to choose keywords carefully, and online sources can be unreliable. Although TI should rely heavily on published information, it will need time for publication.

For the *non-published information*, collecting information sources from external social networks are analyzed in Table 8, and the analysis results are given in Fig. 6. The analysis results show that the firms in this study also acquire information from most of the non-published information sources listed in Table 8. Just a few firms can seek information out from acquaintances, friends and relatives (N-5), government agencies (N-7), scientific conferences (N-10) and start-up fairs (N-11).

Table 8. Analysis of non-published information sources

No.	Non published information sources
Q4-N-1	University contacts/projects
Q4-N-2	Product fairs
Q4-N-3	Commercial conferences
Q4-N-4	Suppliers
Q4-N-5	Acquaintances, friends and relatives
Q4-N-6	Co-operations
Q4-N-7	Government agencies
Q4-N-8	Customers
Q4-N-9	Consultants
Q4-N-10	Scientific conferences
Q4-N-11	Start-up fairs

The 5th question expresses evaluation methods [25] that the firms examine the information collected by the TI search process to achieve the objectives of the firms. The findings of the analysis for the 5th question are given in Table 9 and Fig. 7.

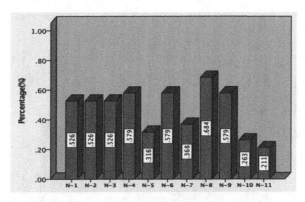

Fig. 6. Analysis result of non-published information sources

Table 9. Analysis of evaluation methods for information

No.	Evaluation methods
Q5-E-1	Patent analysis
Q5-E-2	Bibliometric analysis
Q5-E-3	Technology trend analysis
Q5-E-4	Roadmap analyses
Q5-E-5	Benchmarking processes
Q5-E-6	Technology development analysis, S-curves
Q5-E-7	Quality Function Deployment (QFD) analysis
Q5-E-8	Some heuristics and creative problem-solving methods are utilized depending on employee experience

The instruments and tools that are used to analyze and evaluate the technological information in NTBFs are mostly based on Technology trend analysis (E-3). And also, roadmap analysis (E-4), technology development analysis and S-curves (E-6), quality function deployment analysis (E-7) and, some heuristics and creative problem-solving methods (E-8) are normally used as evaluation methods in the firms. Patent analysis (E-1), Bibliometric analysis (E-2) and benchmarking process (E-5) are the least preferred tools. It has appeared that only two firms located in China and India in this study use bibliometric analysis (E-2) to evaluate information.

For the 6th question, the application of intelligence knowledge derived from the TI evaluation process is surveyed in Table 10 [25] and the results of the application of intelligence knowledge are expressed in Fig. 8.

The results show that the firms mostly apply the intelligence knowledge which came out from TI evaluation process by regularly reporting to the top management (A-1) (because decision-makers are part of TI process) and by inputting for strategic goals

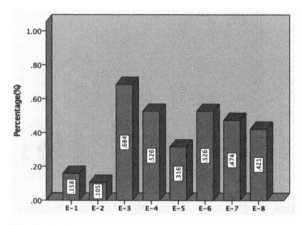

Fig. 7. Analysis result of evaluation methods for information

Table 10. Analysis of the application of intelligence knowledge

No.	Items
Q6-A-1	Technology intelligence results are regularly reported to top management
Q6-A-2	Technology intelligence results are kept in the database that is accessible for whole organization
Q6-A-3	Technology intelligence results are discussed in daily meetings
Q6-A-4	Technology intelligence results are shared within whole organization via e-mail, minutes of meeting, etc.
Q6-A-5	Technology intelligence results are input for selecting new business partners
Q6-A-6	Technology intelligence results are shared with the project or business partners
Q6-A-7	Technology intelligence results are input for technology transfer decisions
Q6-A-8	Technology intelligence results are integrated to R&D and new product development process
Q6-A-9	Technology intelligence results are the input for project selection
Q6-A-10	Technology intelligence results are input for strategic goals development process

development process (A-10). It is because the firms participated in this study are operating with Top-down approach, top management usually makes inferences based upon their knowledge and experience, and they emphasize strategic goals. Only four firms kept the TI results in the database that is accessible for the whole organization (A-2). According to the analysis results, it can be seen that NTBFs in this study exploit the intelligence knowledge in most of the ways listed in Table 10.

In questionnaires of this study, 60 detailed items are constructed concerning the understanding of TI process. Among the 19 cases/firms participated in analyzing TI

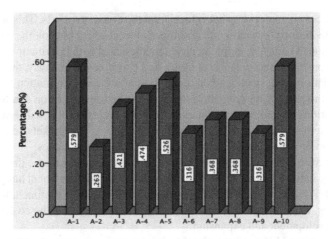

Fig. 8. Analysis result of application of intelligence knowledge

activities, case 15 (Pharmaceutical industry) is mostly working out (54 items). According to the interview and also the real situation, this firm has a well-structured type of organization and possesses successful competitive advantages in its industry environment in Myanmar. Among the cases, it is clearly seen that larger organizations (Case 1, 15, 18, 19) can manage budget and employees to integrate TI process inside the organization and exploit TI activities effectively than other smaller ones. According to the interview, most of the firms from Myanmar participated in this study are operating TI process, but there is still no comprehensive and systematized TI operation within organizations, although specialized internal TI units/department exists.

5 Discussion

All the factors listed in the above tables were considered while analyzing the TI systems of 19 technology-based companies located in Myanmar, China and India. Although being aware of the limited number of case studies and limited types of industries in this study, the participated companies performed TI process in line with the listed aspects of 6 main questionnaires, although specific approaches varied and not all aspects were implemented by all companies. These listed aspects represent approaches to how TI works and can help companies to review their current intelligence capabilities. The more aspects the firms can act, the more comprehensive TI systems they can get because these aspects are the essential foundation for the development and implementation of the TI system.

The NTBFs should align market, technology and competitive intelligence, and communicate well between them. The firms should manage the interface between decision-makers and intelligence to enable the efficient dissemination of information requirements and intelligence. The firms should formalize the intelligence activities without reducing their flexibility and capability. The firms should dedicate progressively more resources to TI development by understanding that these are important assets for the firm.

The TI process is to decide what information the company needs. This will depend not only on the industry context and current product estate but also on the firm's future plans. In addition, TI must be supplemented and integrated with information about other aspects of the business, such as market trends and competitors. In a way, the TI process is the accumulation, assessment and dissemination of relevant information for decision-makers about the firm's environment in terms of markets, competition and technology. Often information about technology, market and competitors is collected separately. It is important to encourage connection between these three areas, as market and competitive intelligence can provide valuable information about new technologies.

The successful TI process adds alternative and new ways to achieve the same goals that are already being pursued domestically. As technology development has proceeded, more focused TI efforts will help avoid the potential surprise of launching competitive technology. Effective intelligence can reduce uncertainty over the costly period of technology development and expansion.

For an effective TI process, it should closely work with users to define specific intelligence collection goals, so it can increase the likelihood that results will be valued and used (for example, what decisions will affect according to the information gathered, what is now known, what should be studied or confirmed, when information is needed and to what degree of detail). There are often a large number of technologies with potential interest. The one thing to do is to integrate technical, market and competitive information. This will help determine what technologies currently have the greatest impact on products or processes, and also provides clues to what new technologies are likely to be important in the future. No single collection method is suitable for all situations, and each of them has its own advantages and disadvantages. Firms must consider a variety of different methods, be sensitive to the purpose of information gathering, as well as time and resources.

Based on literature reviews and this case study, the general TI process cycle for NTBFs in developing countries is conceptualized with seven steps as a cycle for the continuation and integration of TI as follow:

- *Organization and Direction*: Firstly, planning is carried out for TI objectives, for organizing TI, and for directing actions concerning TI (for example, what are the objectives? Who will conduct the TI team and what are their roles? What other elements should be involved? And what direction should be acted and how?). The objectives of TI should be: to provide early warning of external technical developments or the movement of rivals; to evaluate information concerning new products, processes, or collaboration prospects created by external technical activities; and to anticipate and understand the changes or trends associated with science and technology in the competitive environment for organizational planning.
- *Formulating the needs (technology/competitive/market-intelligence)*: Successful efforts in the field of TI are clear purposes, focus on high impact technologies and the use of several methods. In order to formulate the needs, it is necessary to consider what information the firm needs, what others do, and how it can affect. It should be considered to collect data including TI, CI and MI. As mentioned earlier, technological information, market information and competitive information are often collected separately. Therefore, it is important to encourage communication between the three

areas, since both market and competitive intelligence can provide valuable information on new technologies.

- **Searching (formal/informal) and Data Collection**: For knowledge management issue and approaches, it is applied to **Mine** and **Trawl** for storing and retrieving intelligence information. Firm trawl internal knowledge that is not explicitly stored and inaccessible in the **Mine** mode. **Target** is organized around the main technology groups, where the distribution of tasks follows a hierarchy that reflects the organizational structure. **Target** activities are coordinated by people with proven technical experience and knowledge of corporate technology policy. They interface TI with MI and CI, as well as with decision-makers who identify new technologies for investigation.

The model (Fig. 9) demonstrates the string of search functions through common top-down and bottom-up processes that connect decision-makers with information sources. Although each of the functions is necessary, the balance between them varies depending on the specific needs of the firm. The **Scan** is mainly operated due to rapid changes in technology and environment of NTBFs in developing countries. There are three steps – focus, compile and test - to approach to collect timely information effectively.

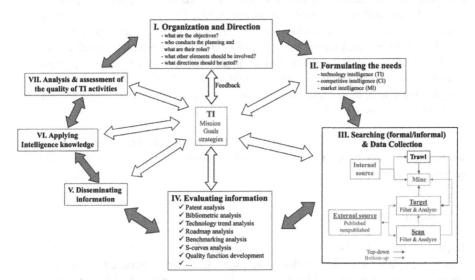

Fig. 9. General TI process cycle for NTBFs in developing countries

As shown in Fig. 9, all the information from **Target**, **Trawl**, and **Scan** are fed into the **Mine** function. The first step is to find information in the organization. When it is not clear where it is, the **Trawl** function is applied to the repositories and finally redirects the **Target** function. In the case of the Bottom-up approach, no input is fed for searching. Using the **Target** and **Scan** functions, information is collected from published and nonpublished sources (as shown in Tables 7 and 8), and then filtered and analyzed to reduce data overload, and converted data to information/knowledge with a strategic value (For example, according to **Scan**, the question "Is this technology interesting?" is *filter,* and "Does this technology meet the needs of the firm? " is ***analysis***. According to ***Target***, the question "How is this technology different from what is known?" is the *filter,* and "What is the value of this technology?" is ***analysis***).

- ***Evaluating information***: At this stage, the firm examines the information collected in the TI search process to achieve the goals of the firm and to continue disseminating information (as shown in Table 9). Evaluation of information should be added in TI reports being part of the decision-making process. The key factors are awareness, risk reduction, the necessary development, innovation and cooperation, as well as in line with the organizational goals.
- ***Disseminating information***: The information examined in the evaluation stage is disseminated to those who need it, or to decision-makers as a usable form in order to adapt the knowledge development about technological threats and opportunities to the decision-making process.
- ***Applying intelligence knowledge***: The intelligence knowledge gained as a result of the TI process is applied to the missions, goals and strategies of the firm (as shown in Table 10).
- ***Analysis and assessment of the quality of TI activities***: The quality of TI activities is needed to analyze (quantitatively and/or qualitatively) and assess (activity-based measure and/or outcome-based measure [26]). TI requires resources, competences and capabilities which may possess by not all companies, especially small and newly established firms. Consequently, TI can be prepared to reorganize, to re-plan and to upgrade in order to match the present situation of the firm's environment. Accuracy, depth, relevance, responsiveness and timeliness also have to be considered for the whole TI process of the firm.

The one important thing that needs to be understood is that there is always a feedback action at each stage to align performance with mission, goals and strategies of the TI system of the firm.

6   Conclusion

To date, no model has been developed to introduce TI for NTBFs in developing countries in accordance with their various conditions, lack of resources, and nature of their organization still being in the formation and being in a rapidly changing environment. TI supports decision making at many levels. TI outputs can be used for operational, tactical and strategic decisions. TI process also reduces development time and provides innovation processes. TI can be used to identify technology marketing opportunities. TI processes are geared toward creating opportunities rather than solving problems.

New ideas in science and technology are sought for further development that fits the company's business. All technologies are pre-commercial, although there are still no products or processes. When NTBFs are located in developing countries, in order to invest in technology, it must be watched locally and globally what others are doing and understood how it could affect them. If technical progress is missed, the opportunity may be lost forever, or the position may be lost in the marketplace. If a firm invests in the wrong technology or at the wrong time, the investment cannot be recovered. For this, TI can provide valuable information about new technologies to invest.

This TI study provides real case studies of TI process, general TI system model and challenges for understanding which capabilities NTBFs should have for development and implementation of TI process. Future research will investigate how best to deploy the TI system of NTBFs in developing countries.

Acknowledgements. The authors would like to thank all the companies and individuals who contributed to this research and special thanks should go to ACE Data System in Myanmar and two ICT companies in China and India. We are very grateful to those who helped us throughout the duration of the course.

References

1. Kerr, C.I.V., Mortara, L., Phaal, R., Probert, D.R.: A conceptual model for technology intelligence. Int. J. Technol. Intell. Plann. **2**(1), 73–93 (2006)
2. Savioz, P., Tschirky, H.: Technology intelligence systems: benefits and roles of top management. In: Bringing Technology and Innovation into the Boardroom (2004)
3. Alencar, M., Porter, A., Antunes, A.: Nanopatenting patterns in relation to product life cycle. Technol. Forecast. Soc. Chang. **74**, 1661–1680 (2007)
4. Lichtenthaler, E.: Third generation management of technology intelligence processes. R&D Manag. **33**(4), 361–375 (2003)
5. Huston, L., Sakkab, N.: Connect and develop: inside Procter & Gamble's new model for innovation. Harvard Bus. Rev. **84**, 58–66 (2006)
6. Savioz, P., Luggen, M., Tschirky, M.: Technology Intelligence: structuring it into the new technology-based firm (NTBF). Tech Monitor, July–August 2003. Special Feature: Technology Road-Mapping, pp. 41–46 (2003)
7. Iansiti, M.: How the incumbent can win: managing technological transitions in the semiconductor industry. Manag. Sci. **46**(2), 169–185 (2000)
8. Porter, A.L.: QTIP: quick technology intelligence processes. Technol. Forecast. Soc. Chang. **72**(9), 1070–1081 (2005)
9. Savioz, P., Blum, M.: Strategic forecast tool for SMEs: How the opportunity landscape interacts with business strategy to anticipate technological trends. Technovation **22**(2), 91–100 (2002)
10. Daim, T.U., Rueda, G., Martin, H., Gerdsri, P.: Forecasting emerging technologies: use of bibliometrics and patent analysis. Technol. Forecast. Soc. Chang. **73**(8), 981–1012 (2006)
11. Robinson, D.K.R., Huang, L., Guo, Y., Porter, A.L.: Forecasting Innovation Pathways (FIP) for new and emerging science and technologies. Technol. Forecast. Soc. Chang. **80**(2), 267–285 (2013)
12. Nosella, A., Petroni, G., Salandra, R.: Technological change and technology monitoring process: evidence from four Italian case studies. J. Eng. Technol. Manag. JET-M **25**(4), 321–337 (2008)
13. Lee, C., Jeon, J., Park, Y.: Monitoring trends of technological changes based on the dynamic patent lattice: a modified formal concept analysis approach. Technol. Forecast. Soc. Chang. **78**(4), 690–702 (2011)
14. Rohrbeck, R.: Harnessing a network of experts for competitive advantage: technology scouting in the ICT industry. R&D Manag. **40**(2), 169–180 (2010)

15. Savioz, P.: Technology Intelligence in Technology-based SMEs: Conceptual Design and Implementation. ETH Zürich, B-327422, Verlag Industrial Organization (2002)
16. Byungun, Y.: On the development of a technology intelligence tool for identifying technology opportunity. Expert Syst. Appl. **35**, 124–135 (2008)
17. Lichtenthaler, E.: Managing technology intelligence processes in situations of radical technological change. Technol. Forecast. Soc. Chang. **74**, 1109–1136 (2007)
18. Karaoz, M., Albeni, M.: Dynamic technological learning trends in Turkish manufacturing industries. Technol. Forecast. Soc. Chang. **72**, 866–885 (2005)
19. Behkami, N., Daim, T.: Research forecasting for health information technology (HIT), using technology intelligence. Technol. Forecast. Soc. Chang. **79**(3), 498–508 (2012)
20. Dereli, T., Altun, K.: A novel approach for assessment of candidate technologies with respect to their innovation potentials: quick innovation intelligence process. Expert Syst. Appl. **40**(3), 881–891 (2013)
21. Majidfar, F., Salami, R.: Impacts of technology intelligence cycles on university-industry networking. In: Proceedings of the Fourth National Conference on Management of Technology, Tehran, Iran (2010)
22. Laursen, K., Salter, A.: Open for innovation: the role of openness in explaining innovation performance among U.K. manufacturing firms. Strateg. Manag. J. **27**, 131–150 (2006)
23. Makadok, R., Barney, J.: Strategic factor market intelligence: an application of information economics to strategy. Manag. Sci. **47**, 1621–1638 (2001)
24. Slater, S., Narver, J.: Intelligence generation and superior customer value. J. Acad. Mark. Sci. **28**, 120–127 (2000)
25. Kilic, A., Cakmak, T., Eren, H., Sakarya, F.: Technology intelligence process in technopark firms; an empirical research in Turkey. Int. J. Econ. Financ. Stud. **8**(1), 74–100 (2016)
26. Loh, Y.W., Mortara, L.: How to measure technology intelligence? Int. J. Technol. Intell. Plann. (2017)

Application of Design Thinking to Optimize Change Management Procedures with a Case Study on Reference Book Stores

Yu-Lun Huang[⊠] and Ding-Hau Huang

Institute of Creative Design and Management, National Taipei University of Business, Taipei,
Taiwan, ROC
{1076D002,huahuang}@ntub.edu.tw

Abstract. To deal with the constant changes in the social environment today, companies must adopt appropriate changes to face different challenges. Some change model was developed, however merely following the steps and procedures of these change models is not enough. Being able to change the thinking of the team is also a crucial factor. In recent years, many companies have used design thinking methods to engage in successful innovation and transition. Design thinking is a human-centered approach that focuses on interdisciplinary communication and the establishment of innovative thinking and implementation. The purpose of this study was to use design thinking to optimize the unfreeze step of change, develop a model, and determine whether this model can effectively help teams reach a consensus and set suitable visions and goals so that they can successfully initiate change. We used change in small physical bookstores (reference book stores) as an example in our case study. The results indicated that by integrating design thinking and change management, physical bookstore owners can efficiently define the visions of their companies when they are making transitions so that they can make preliminary plans for the change. Future studies can use the proposed model to examine similar small retail companies that are in need of change and innovative transitions as the times change.

Keywords: Design thinking · Change management · Bookstore

1 Introduction

Today, many companies are facing difficult challenges due to globalization, rapid technological progress, and changes in social development and the economic environment. If they do not change the way they operate, they will be unable to survive in this intensely competitive environment. As the external environment changes swiftly, they must understand what state the external environment is in and devise a set of strategies based on their internal conditions so as to achieve sustainability [1, 2].

With the growing importance of change to companies, a number of researchers are investigating issues associated with change management and proposing relevant change models. One of the most widely known change models is that presented by

© Springer Nature Switzerland AG 2020
P.-L. P. Rau (Ed.): HCII 2020, LNCS 12192, pp. 91–103, 2020.
https://doi.org/10.1007/978-3-030-49788-0_7

the founder of social psychology, Kurt Lewin, which involves three steps: unfreezing, changing, and refreezing. Another is the eight-step change model for business success proposed by change management guru John Kotter, which includes creating a sense of urgency, building a guiding coalition, forming a strategic vision and initiatives, enlisting a volunteer army, enabling action by removing barriers, generating short-term wins, sustaining acceleration, and instituting change. John Kotter has significant successful experience in business. The procedures he proposed are much clearer and easier to understand than other change models, and as a result, this model received more favor than other change models later on [3]. However, the content of Kotter's change model is an ideal framework created based on personal practical experience; it lacks external citations and evidence, and not every step has concrete implementation methods. This change model can only provide reference when change procedures are beginning to be executed, and there is no guarantee that implementing this model will lead to success [4].

Many past studies pointed out that over half of organizations implementing change do not succeed [5–8]. During the process of change, members within the organizations may resist the changes and prevent the changes from being successful due to self-interests, lack of trust in the changes, uncertainty, different goals or assessments of the changes, and their personal needs [9]. Researchers have also pointed out that true change is not achieved via a single approach but via an organizational way of thinking. Simply following the steps of a procedure is not enough to stimulate organizational innovation, and changing team thinking is essential.

With regard to changing team thinking, an innovative thinking model has become widely used in various fields in recent years: design thinking. Design thinking is a type of iterative exploration. It uses human-centered spirit and methods to explore the potential needs and behaviors of people while taking technological and commercial possibilities into account [10]. In addition to being human-centered, the features of design thinking include assisting in the creation of interdisciplinary teams, enhancing interdisciplinary communication, promoting strategy formulation, and aiding people in exploring more ideas and goals. IDEO founder Tim Brown said in his book, Change by Design, "It takes a systematic approach to achieve organization-wide change." He also advocated that all employees, regardless of their level or department, should take part in design thinking [10]. However, few existing studies have explored whether the features of design thinking can be used to solve the difficulties that arise during organizational change due to lack of proper communication or consensus in teams.

In the initial stages of effecting change, a team to lead the change must be established, as well as goals and a vision. However, it can be difficult to find a suitable vision for the future or the right strategies and plans without systematic guidance [11].

2   Change Management

Change is a crucial factor that promotes innovation and sustainability in companies. To avoid being eliminated by the ever-changing times, companies must continuously innovate and search for new goals and directions and periodically ascertain whether the organization's visions, competitiveness, management structure, and staffing lead towards the new goals. During this stage, the leaders play a pivotal role and guide the company in the new direction. They must constantly make plans and have discussions with organization members. Change involves the behaviors, culture, structure, and strategies of organization members [12]. With current technological advances and fierce global competition, we know that change must be ongoing [13].

Change is triggered by changes in the internal and external environment or situation of the company, or by the discovery of a crisis. Companies with strong organizational culture often have difficulty breaking conventions when implementing change or have problems pinpointing the real problem and defining suitable innovative methods when making plans for change. Not knowing what it will be like after the change or distrusting and rejecting the change can make organization members uneasy and agitated. If they cannot accept the change, it can ultimately prevent the change from succeeding. Kotter [14] indicated that the key to success of change lies in the methods of the leadership. This shows how important leaders are in change. This chapter examines the two change management models that are often used by companies implementing change and the issues that they encounter.

2.1 Eight-Step Change Model

This model was proposed by Kotter in the Harvard Business Review. Investing resources in change does not always work, and Kotter listed common reasons why transformation efforts fail: allowing too much complacency, not creating a powerful enough guiding coalition, underestimating the power of vision, under-communicating the vision by a factor of ten, not removing obstacles to the new vision, not systematically planning for and creating short-term wins, declaring victory too soon, and not anchoring changes in the corporation's culture [15]. To help companies solve these problems, Kotter proposed an eight-step model for success: create a sense of urgency, build a guiding coalition, form a strategic vision, enlist a volunteer army, enable action by removing barriers, generate short-term wins, sustain acceleration, and institute change [14] (See Table 1).

The change management model proposed by Kotter is the most popular change management framework and the most widely used [16]. It assists in making fundamental changes to business operations to deal with new, more challenging market environments [15]. The superiority of Kotter's model over other change models lies in it being clearer and easier to understand and based on Kotter's personal successful experience in business [3].

2.2 Three-Step Change Model

Initially, social psychologist Kurt Lewin proposed a three-step change model based on his Force Field Analysis, the three steps being unfreezing, changing, and refreezing. It applies the driving and restraining forces within organizations to reduce resistance to change and ease internal pressure as much as possible [17]. Lewin's three-step change model is the first model proposed for change. The steps and contents of the three-step change model are as shown in Table 2.

This study mainly investigates the early unfreezing step in the change process. It is the initial preparation stage of change, and if any problems appear in this step, the

Table 1. Eight-step change model

Step	Content
Create a sense of urgency	Understand market trends and ensure that organization members understand the urgency and necessity of change for the organization. Without creating a sense of urgency, there is no chance for change to take place
Build a guiding coalition	Build a coalition with members of adequate status and credibility so that they can communicate with, coordinate with, and influence organization members during the process of change
Form a strategic vision & Initiatives	Form plans and goals for the change, which are of upmost importance to explaining the methods and direction of change
Enlist a volunteer army	This step focuses on communication to ensure that all members can quickly understand the content of the change in detail and understand or communicate the definition of the vision and goals
Enable action by removing barriers	Expand the coalition implementing the change plans to the entire organization and give members responsibilities and tasks to obtain more acceptance and involvement
Generate short-term wins	To prompt more members to get involved in the change, announce that these changes will be implemented once a certain amount of change has been completed. Provide the entire organization with a change model and demonstrate the effectiveness of the changes
Sustain acceleration	Once the changes that have been implemented are successful, further promotion is needed to increase implementation and acceptance
Institute change	Integrate all of the changes so that they are inseparable, and anchor them into the organization so that members will understand that the new goals have become part of organizational operations

difficulty of the subsequent steps will only increase. During this step, the change needs, change objectives of the organization, the possible sources of resistance to change, and suitable change instruments and techniques are defined to begin the first step of change. However, lack of trust in upper management, lack of a sense of involvement in the change, and previous experience are common issues that cause organization members to become uneasy and agitated [18], to the point that the change fails. In the next section, we compile the solutions that have been suggested by existing studies.

Table 2. Three-step change model

Step	Content
Unfreeze	Identify current issues in the organization, arouse determination for change, and take apart existing organizational structures or management models
Change	Develop new behaviors, values, and attitudes and change the current conditions of the company to achieve new standards
Refreezing	Incorporate the new model into the organization

During the process of change management, team collaboration and communication are crucial. In 2013, Calber presented six methods for the unfreeze step: communication, training, employee involvement, stress management, negotiation, and coercion [17] (See Table 3). After examining 200 cases of organizational change, Jones et al. [19] named three major factors that can influence change management in organizations and promote successful change: effective communication, effective guidance, and the importance of teamwork/employee involvement. We can thus infer that team collaboration and communication are vital factors in the unfreeze step of change.

Table 3. Methods to promote unfreeze stage

Method	Content
Communication	Give employees a better understanding of the change situation
Training	Provide the knowledge and techniques needed to achieve the goal
Employee involvement	Allow employees to participate in change decisions
Stress management	Give employees the opportunity to discuss
Negotiation	Offer employees incentives that will persuade them to support the change
Coercion	Give employees an ultimatum to change or give up

3 Design Thinking

Design thinking is a concept that combines both theory and practice. David Kelley, founder of design consultancy IDEO, proposed the design thinking process, and in recent years, many companies have made successful innovations and transitions using design thinking methods and procedures. Design thinking is a set of methods to implement innovative thinking. Emphasized as being human-centered, design thinking converts observations into insight and finally produces a set of new solutions. Design thinking can also be used to stimulate creative thinking and implement concrete actions, all the while learning, discovering, defining, solving problems, and promoting interdisciplinary teamwork. Design thinking can be used in any subject [10]. It can not only aid in product innovation but also be used for organizational innovation [20].

Design thinking is an iterative design process including inspiration, ideation, and implementation. Compared to linear ways of innovation, design thinking is not a standardized process but a framework integrating creativity, reasoning, and analysis [21]. In commercial applications, design thinking is a way of solving the uncertainty problem within organizations. Tim Brown pointed out that if today's corporate leaders want innovation to bring differentiation and competitive advantages, they must incorporate design thinking in all stages of innovative processes for better effectiveness. So far, however, few existing studies have focused on the use of design thinking in companies and organizations [10].

Design thinking does not have any specific steps. Adjustments can be made flexibly depending on the needs. Design thinking is an iterative process of divergence, convergence, discovery, definition, and development. One widely used design thinking process is the five-step process proposed by Stanford d. school [22] below:

Empathy. Convert sympathy into empathy. Design with the users as the focus. Understand and get to know the users via various methods. Assist the design thinkers. From the users' perspective, find the real needs and problems of the users.

Define. Defining needs means using information collected in the empathy step to remake deeper definitions of the problems. Like exploring an iceberg under the surface of the water, delve deeper to find the read needs of the users and give them a brief definition.

Ideate. The objective of the ideate step is to develop multiple solutions to solve the problems identified in the define step. By following the rules of brainstorming, ideating can produce countless ideas. A truly suitable solution can then be found using different voting standards. The rules of brainstorming are defer judgment, one conversation at a time, stay on topic, build on the ideas of others, be visual, encourage wild ideas, go for quantity, and headline your ideas.

Prototype. Creating a simple prototype during the design process can make the abstract concrete. A concrete presentation can serve as tool for communication within the team or with users. The prototyping process can also clarify thinking with hands-on thinking.

Test. The actual test uses the prototype created in the previous step to communicate with users. With scenario simulations, the suitability of the prototype for users is tested. How the users use and respond to the prototype is also observed. Based on user responses, the needs are redefined or the solution is modified. A more in-depth understanding of the users is also obtained.

Carlgren et al. [23] conducted semi-structured interviews with employees on various levels in different departments and presented five core themes in design thinking: diversity, problem framing, empathy, visualization, and experimentation. They indicated that these core themes are closely linked to one another and can be used in different orders during the process of design thinking. In the next section, we look further into problem framing, which is associated with the discovery and definition of problems and team cohesion, and diversity, which involves interdisciplinary cooperation.

Problem Framing. The problem framing stage is crucial. Problem framing for a large scope can create more room for solutions. This stage requires various tools and methods to reconstruct the most fundamental problems, build the scope of ideation, and develop room for problems and strategies. The problem definition stage is repeatedly questioned and reconstructed until a problem with an appropriate scope is defined.

Diversity. This stage involves innovative interdisciplinary cooperation. What is unique about these teams is that they comprise members of different levels, ages, and genders, or different companies and departments. They may also include both internal and external stakeholders [24]. Establishing a diverse interdisciplinary team provides different perspectives for joint decision-making.

During change, the means by which leaders alter the preconceived perceptions and behaviors of organization members is a factor that is critical to the success of change. Finding ways to establish communication between leaders and organization members and achieve a favorable consensus are important to the implementation of change.

The characteristics of design thinking include interdisciplinary teamwork and discovering and solving problems. Design thinking can also help companies incorporate their strategies and vision into their strategic goals [11]. This study used a design thinking model to design a change management and design thinking workshop to help implement change in small physical bookstores and investigated whether the characteristics of interdisciplinary teamwork and discovering and solving problems in design thinking can facilitate team communication and vision and goal establishment in change management and lower the possibility of change fails.

4 Methodology of Research

Rapid technological progress and changes in the social environment have caused many industries to shrink in recent years. One example is the retail industry. The appearance of new commercial models and technology have radically changed the retail industry over the past decade, and part of the retail industry includes bookstores. In the past, bookstores mainly operated out of physical stores, but they have been continuously dwindled due to the appearance of new commercial models and technology and changes in consumer book-buying habits. Thus, if physical bookstores want to survive, it is inevitable that they make corresponding innovations.

Among small physical bookstores, bookstores dedicated to selling reference books are unique and are in urgent need of change due to the shrinking of the industry. We therefore conducted a case study using reference book stores. Using critical case sampling, we chose reference book stores located near schools and a train station as the targets of this study. It have been in business for over 20 years and have high local brand awareness. We speculated that the results obtained from this case can be logically applied to other similar cases. Through in-depth interviews, we gained an understanding of the current status of the industry and the willingness and attitude of the interviewees with regard to change. We interviewed the owners of the reference book stores, both of whom have been in the business for more than 30 years and have had their own brands for more than 20 years. Locally, their brands are highly recognized. The interviewees have many years of experience in the industry as well as a thorough understanding of the changes in the industry. Through literature review, we broadened our understanding of the characteristics and objectives of each step in change management and used design thinking tools and methods to optimize the unfreezing step of change management. We then compiled the content of the in-depth interviews and constructed an unfreezing model. Figure 1 displays the process of this study.

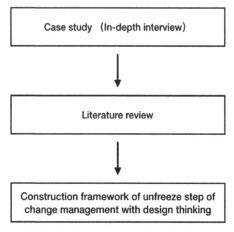

Fig. 1. Research process

5   Result

5.1   Results of In-depth Interviews

The final results of this study were subjected to selective coding based on the two primary focuses: understanding of the current status of the reference book store industry and willingness and attitude with regard to change (See Table 4).

With regard to the current status of the industry, the reference book industry is intertwined with education in Taiwan. The education system in Taiwan was conventionally focused on further education and obtaining educational degrees. However, it is now transitioning to support diversified learning and development among students, and that nothing is more important than studying is no longer believed. The times have changed, and parents are beginning to support their children's learning in various aspects and no longer believe that only getting good grades in academic subjects is important. The books that reference book stores specialize in are focused on academic subjects; they do not sell books that help stimulate student interest or build their skills in other aspects. This may also be a reason why the reference book store industry is dwindling.

With regard to willingness to change, our interviews revealed that the interviewees were positive about changing and willing to try it. They felt that any methods that were innovative and worth trying could be observed and attempted. The interviewees both indicated that although the reference book store industry is already a sunset industry,

Table 4. Selective coding

Core category	Code
Understanding of current status of industry	Business model
	Impact on industry
	Major differences between high and low seasons
	Service = Expertise
Willingness and attitude with regard to change	Willingness to change
	Brand extension
	Industrial advantages

there is still the possibility that it will not disappear. However, they had seen failed transitions before and thus had no specific ideas of what a successful transition would be like.

The in-depth interviews with the reference book store owners revealed that they have already become aware of the changes in the industry. The changes in the education in Taiwan from traditional education to supporting diverse learning have exerted an impact on reference book stores, which purely sell books for academic purposes. If the owners want their businesses to survive, they will have to make changes. They presented positive attitudes toward change, but how and in what direction these changes should be made are what troubles them.

5.2 Construction Framework

Our literature review revealed that during the process of change, the lack of effective communication in teams and insecurities about the changes brought by the change often prevent change from going smoothly. Thus, with Kotter's eight-step change model and Lewin's three-step change model as the foundations of change implementation, we optimized the change process of the unfreeze step using design thinking (See Fig. 2). Below, we use the first four steps of Kotter's eight-step change model for in-depth application of design thinking.

Create a Sense of Urgency. This is the first step of implementing change. To create a sense of urgency in change management, one must understand the current market trends and ensure that organization members understand the urgency and necessity of change for the organization. Thus, in this step we employ three procedural tools in design thinking. Through internal discussion and four major macro-environmental factors, we search for preliminary opportunity gaps. Next, we use a stakeholder map to provide an understanding on the internal and external stakeholders of the industry and establish the foundation of the workshop in the next step.

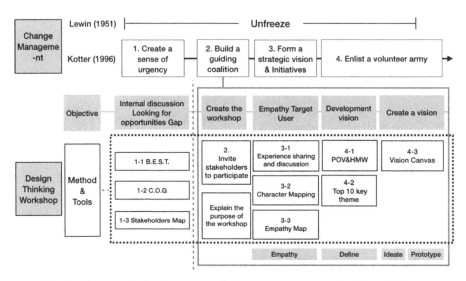

Fig. 2. Framework of unfreeze step of change management with design thinking.

Step 1-1 B.E.S.T. To gain a better understanding of the case, we modify the PEST strategy analysis model by changing "political" to "business" for macro-environmental analysis. To find a suitable direction for change, macro-environmental factors in the business, economic, social, and technological aspects can help companies understand and analyze the main external environmental factors that impact them in various aspects. Participants can brainstorm in each aspect, write down the keywords or issues they think of on post-its, and then categorize them under business, economic, social, or technological aspects.

Step 1-2 C.O.G (Change Opportunity Gaps). Based on the macro aspect trends developed in the previous step, specific opportunity gaps can be identified. For this step, we use the POG (Product Opportunity Gaps) concept mentioned in the book, *Creating Breakthrough Products*. To bring it closer to the issue of change, we change "product" to "change". Combining the analysis results of the business, economic, social, and technological aspects in the previous step with the keywords or issues in different aspects, we then deduce the possible Change Opportunity Gaps. This step can be conducted via discussion among the participants.

Step 1-3 Stakeholder Maps. Using the Change Opportunity Gaps established in Steps 1-1 and 1-2, aspects that the change may be aimed at can be discussed from the macro perspective. Using a stakeholder map, companies can clarify the relationships among internal employees and managers, external consumers and competitors, and other stakeholders. The connections between different roles can clarify their interactions for analysis and give a detailed understanding of all the people related to the company from the inside out. The Change Opportunity Gaps identified in Step 1-2 can serve as the starting point for Step 2, which relevant personnel can be invited to take part in.

Build a Guiding Coalition. Building a guiding coalition in change management means forming a team to execute the change. The members comprising the team must have adequate status and credibility so that they can communicate with, coordinate with, and influence organization members during the process of change. The stakeholder map created in the previous step can give an understanding of all the people related to the company from the inside out. Choose a representative for each type of stakeholder, and invite related people to participate in the next step.

Form a Strategic Vision. Forming a strategic vision means formulating plans and objectives for the change. Setting appropriate visions and objectives is important in change, but this can be a particularly difficult step for companies. The design thinking tools experience sharing and discussion, character mapping, and empathy map can help companies understand how to define objective groups and understand the needs of target groups in discovering problems under empathy.

Step 3-1 Experience Sharing and Discussion. In this step, the goal of the workshop and the current status of the industry are described. The internal and external stakeholders participating in the workshop are then asked to briefly explain their own opinions of the current conditions and issues. During this process, have all of the participants write down the keywords and issues that they hear on post-its and then discuss them together.

Step 3-2 Character Mapping. Using the character mapping tool, set the character images of multiple target groups. Create several characters for each group, name the characters, and then define their characteristics (e.g., age, gender, occupation, educational background, income, etc.) based on the theme. The contents should be linearly written in a form based on level and size. Select a representative character for each group by putting it to a vote, and then proceed to the next step.

Step 3-3 Empathy Map. Empathize with the representative characters established in the previous character mapping step using an empathy map for different roles. This tool is divided into two phases: (1) divergent phase – brainstorm and categorize in a form what the target groups think and feel, hear, see, and say and do; (2) convergent phase – categorize all of the elements in the divergent phase as pains or gains.

Enlist a Volunteer Army. This step requires a significant amount of communication to ensure that all members have a detailed understanding of the contents of the change. Focusing and brainstorming in this step is conducted using the convergence-divergence process.

Step 4-1 POV (Point of View) & HMW (How Might We). Creativity comes from constant observation and discovery. After empathizing with the empathy map, systematic, well-founded, and creative brainstorming is performed. Based on the content established in the previous empathy map step, the needs of the target groups are discovered, and different keyword combinations on the empathy map can be used to develop point of view (POV) insight to provide direction for brainstorming. Based on the POV insight content, the executive aspect, How Might We (HMW)…, is established to convert ideas into actions and add ideas with regard to execution.

Step 4-2 Top 10 Key Themes. Using the POV and HMW analysis results derived in the previous step, participants can share and discuss five to ten possible visions for the future and list them in a form.

Step 4-3 Vision Canvas. In this step, participants vote for the issue developed from the key issues in the previous step with the most consensus and then use the Vision Canvas tool mentioned in the book *Design a Better Business* to brainstorm about vision [25]. Using the Vision Canvas tool, it is predicted that if the company is to achieve the change objectives, what concrete plans will be needed, what challenges will be encountered, and what solutions will be used to overcome them. The Vision Canvas tool can help companies make preliminary plans for their visions for the future.

6 Discussion

We conducted a case study to understand the willingness and attitude of reference book store owners with regard to change. In-depth interviews revealed that they held positive attitudes toward change but had difficulty setting appropriate objections and visions for the change.

Our literature review showed us that change is a crucial factor that promotes innovation and sustainability in companies. With the growing importance of change to companies, researchers have proposed various change models. However, as the change is initiated, lack of proper communication with employees, lack of consensus, and lack of a sense of involvement in the change among employees often result in less than half of changes succeeding.

This study developed an unfreeze procedure for change and employed methods and tools developed based on the interdisciplinary teamwork and discovering and solving problems characteristics of design thinking to facilitate team communication and vision and goal establishment in change management and lower the possibility of change fails. The purpose of this study was to use the case study to gain a preliminary understanding of the current status of the industry and the willingness and attitude of the interviewees with regard to change. In addition, we used a literature review to understand the difficulties that companies and organizations encounter during the unfreeze step of change and employed design thinking tools and methods to optimize the unfreeze step. However, the validity has yet to be verified, which may be done in future studies.

References

1. Ramanujam, V., Varadarajan, P.: Research on corporate diversification: a synthesis. Strateg. Manag. J. **10**, 523–551 (1989)
2. Schaltegger, S., Wagner, M.: Sustainable entrepreneurship and sustainability innovation: categories and interactions. Bus. Strategy Environ. **20**(4), 222–237 (2011)
3. Wentworth, D.K., Behson, S.J., Kelley, C.L.: Implementing a new student evaluation of teaching system using the Kotter change model. Stud. High. Educ. **45**, 1–13 (2018)
4. Appelbaum, S.H., Habashy, S., Malo, J.L., Shafiq, H.: Back to the future: revisiting Kotter's 1996 change model. J. Manag. Dev. **31**(8), 764–782 (2012)

5. Beer, M., Nohria, N.: Breaking the Code of Change. Harvard Business School Press, Boston (2000)
6. Burke, W.W.: A perspective on the field of organization development and change: the Zeigarnik effect. J. Appl. Behav. Sci. **47**(2), 143–167 (2011)
7. Jacobs, G., Zeyse, J.C.: A theoretical framework of organizational change. J. Organ. Change Manag. **26**(5), 772–792 (2013)
8. Michel, A., Burnes, B.: The limitations of dispositional resistance in relation to organizational change. Manag. Decis. **51**(4), 761–780 (2013)
9. Daft, R.L., Steers, R.M.: Organizations: A Micro/Macro Approach. Scott Foreman & Company, Glenview (1986)
10. Brown, T.: Design thinking. Harvard Bus. Rev. **86**(6), 84–92 (2008)
11. Kumar, V., Ward, A., Runcie, E., Morris, L.: Embedding innovation: design thinking for small enterprises. J. Bus. Strategy **30**, 78–84 (2009)
12. Carnall, C.A.: Managing Change in Organizations. Pearson Education, London (2016)
13. Armenakis, A.A., Harris, S.G.: Reflections: our journey in organizational change research and practice. J. Change Manag. **9**(2), 127–142 (2009)
14. Kotter, J.P.: Leading Change. Harvard Business School Press, Boston (1996)
15. Kotter, J.P.: Leading change: why transformation efforts fail. Harvard Bus. Rev. (1995). https://hbr.org/1995/05/leading-change-why-transformation-efforts-fail-2
16. Todnem By, R., Hughes, M., Ford, J.: Change leadership: oxymoron and myths. J. Change Manag. **16**(1), 8–17 (2016)
17. Calder, A.M.: Organizational change: models for successfully implementing change. Underduate Hoonors Capstone Project, Utah State University (2013)
18. Brickman, J.: How to get health care employees onboard with change. Harvard Bus. Rev. (2016). https://hbr.org/2016/11/how-to-get-health-care-employees-onboard-with-change
19. Jones, J., Firth, J., Hannibal, C., Ogunseyin, M.: Factors contributing to organizational change success or failure: a qualitative meta-analysis of 200 reflective case studies. In: Evidence-Based Initiatives for Organizational Change and Development, pp. 155–178. IGI Global, Pennsylvania (2019)
20. Brown, T., Katz, B.: Change by design. J. Prod. Innov. Manag. **28**(3), 381–383 (2011)
21. Liedtka, J.: Perspective: linking design thinking with innovation outcomes through cognitive bias reduction. J. Prod. Innov. Manag. **32**(6), 925–938 (2015)
22. Stanford d school (2010). https://dschool.stanford.edu
23. Carlgren, L., Rauth, I., Elmquist, M.: Framing design thinking: the concept in idea and enactment. Creativity Innov. Manag. **25**(1), 38–57 (2016)
24. Shapira, H., Ketchie, A., Nehe, M.: The integration of design thinking and strategic sustainable development. J. Clean. Prod. **140**, 277–287 (2017)
25. Piji, P., Lokitz, P., Solomon, J., Kay, L.: Design a Better Business: New Tools, Skills, and Mindset for Strategy and Innovation. Wiley, New Jersey (2016)

Infiltration of Sustainable Environmental Space Service Design in a Cross-Cultural Context

Xing Ji[1,2(✉)], LiuYing Huang[1,2(✉)], ChengYao Cai[1,2(✉)], and Jie Tang[1,2(✉)]

[1] Beijing Technology Institute, Zhuhai, People's Republic of China
54619184@qq.com, 477067869@qq.com, 93713953@qq.com,
4448286@qq.com
[2] Studying School for Doctor's Degree, Bangkok Thonburi University, Bangkok, Thailand

Abstract. In our modern society, the global information has been integrated. The destiny of a country's economy and society depends on the social and cultural resources, as well as the cultural innovation and creativity. As the inevitable theme of this era, sustainable development integrates the innovative creative ability into the sustainable architectural design, urban planning and design, public environment design, and private space design, etc. This can solve the environmental problems, social problems, and development problems, existing in some cities, or even in some countries. What's more, the sustainable design of the environment space takes the environmental factors, socioeconomic factors, and cultural factors into account, which has a positive influence to the ecological sustainable development, and promote the country's economic and social development. Thereby, it can help us to achieve our goal of sustainable design.

There are many ways to achieve sustainable design of environmental space. For example, the use of flexible decorative materials and variable spatial structures in the design can make the space have growth properties, which has no need for the repeated demolition to meet new demands, but can achieve the goal of low carbon environmental protection and recycling development in the constant changing of space. From the perspective of ecological environment, the sustainable design of environmental space can improve the ratio of space utilization, thereby improving the utilization of urban space and the sustainable utilization of land, which has a positive effect on the rational use and integration of land resources. It can satisfy people's current needs without threatening the future development of mankind.

Service design is an effective way to achieve sustainable development in environmental space. Under the cross-cultural background, we have to take the differences in thinking pattern, values system, customs and religious beliefs as prerequisites first, then let's apply the concepts and methods of sustainable design, which comes from life and has the goal to serve the society, so we can meet the needs of environmental space in different regions.

Keywords: Sustainable design · Cross-cultural · Environmental space · Design service

© Springer Nature Switzerland AG 2020
P.-L. P. Rau (Ed.): HCII 2020, LNCS 12192, pp. 104–114, 2020.
https://doi.org/10.1007/978-3-030-49788-0_8

1 Introduction

In recent years, the development of China's urban construction and housing industry has made a historical leap. However, while enjoying the comfortable life that brought by the advancement of human science & technology and urban development, we are also facing a series of risks, such as the serious damage to the earth environment, resource depletion, etc. The deterioration of the ecological environment not only hinders the development of human society, but also directly threatens the human survival [2]. Therefore, we must seek a new way of survival and development, which could guide our current construction of the space environment. This is the common goal that chosen and pursued by all our human beings—sustainable development.

The World Commission on Environment and Development (WCED) states: "Sustainable development is the development that not only meets the needs of contemporary people, but does not pose a danger to future generations' ability to meet their needs." The concept of sustainable development shows that after experiencing various environmental problems, we human beings have begun to consider the path of coordinated development between man and nature, and between man and the environment. This is a feedback that is conducive to human survival, and also a way in which the relationship between environment and nature is restored to equilibrium [11].

Under different historical backgrounds, the different regions, will not only have differences in political systems, economic systems, development stages, development levels, values, etc., but also have diversities in culture, resources of leisure landscape, and urban service design functions [3]. A modern city should have an international standard of quality environmental life. International standards of living standards include four aspects: interactive living standards, GEN ecological standards, cross-cultural standards and healthy living standards. The sustainable development of the design for the environmental service in the public space should not only meets the needs of international urban development, but also considers the different cultural backgrounds, the ways of thinking, the values, and also the customs and religious beliefs in the cross-cultural field (see Fig. 1). Under the goal of serving the society with the design which inspired by living life, we should consider all-round ways of coordinated development between people and people [12], between people and nature, and between people and the environment. Only in this way, we could achieve the sustainable development of a cross-cultural environment space. It is a multi-dimensional, all-round integrated design that integrates the aspects of ecological environment system, social environment system and economic environment system.

2 Research Value

2.1 Academic Value

Service design is also called humanized design. The humanized design of public space environment is not only based on the functional needs and comfort of humanities, but also embodies the protection of ecological balance and the natural environment [12]. "Green design" and "sustainable development" are important topics of humanized design at this stage.

Fig. 1. Schematic diagram of International Green Paradigm New City Standard 3.0

The most primitive "humanized" pursuit of our ancestors for the public space environment was sheltering the wind and rain, keeping them away from the beasts (see Figs. 2, 3 and 4). They could not imagine the splendid civilization and the city in ten thousand years after. However, once they met the basic needs for survival, they began to pursue and create ideal living conditions. Human beings today have come out from darkness and cold; have changed their surroundings with smart minds and wisdom; have established a human society which is independent from nature, and have created an increasingly comfortable artificial environment [2, 6]. However, it is this increasing artificial environment that makes humans more and more comfortable and do what they want. They neglected the desertification of land, the drying rivers, the resource depletion, and the violent climate (see Figs. 5, 6 and 7). The beauty and gentle of the natural environment is swallowing by the industrial civilization which is created by our own.

In the 21st century, the environment no longer exists only as the human beings' boat. It is tolerant, giving, but also sensitive and fragile. It shares the same fate with us. It awakens humanity's awareness of protecting the nature. "Humanized design", "Cross-cultural design" and "sustainable design" have become important topics in the global design field.

Fig. 2. Living space when humans appeared

Fig. 3. Ancient house

Fig. 4. Lived where humans lived in Korea about 6000 years ago

Fig. 5. Land desertification

Fig. 6. River flow

Fig. 7. Resource depletion

2.2 Application Value

By our side, the so-called ecological, green, and sustainable buildings & public spaces are not rare, but we should also see that now many housing communities are promoting their selling by publicizing the small bridges, rocky vegetation, etc. (see Figs. 8 and 9). But due to the design defects or inadequate input costs, the ponds eventually become stinky gutters. It's not just a community that has such cases, but also a city. Due to the early urban planning was not forward-looking enough, the early buildings could

Fig. 8. Landscape stinking gutter

Fig. 9. No water in the landscape

Fig. 10. Due to unreasonable planning, the new community was demolished

Fig. 11. Due to unreasonable planning, the new community was demolished

not meet the needs of follow-up living and travelling. Some buildings were blasted and demolished in just a few years, causing serious waste of resources (see Figs. 10 and 11).

"Sustainable design" is more than just a slogan. It should not just be a gimmick for business propaganda. It should be a practical combination of the application of science and technology and the increasing awareness of human beings in the process of sustainability. When the government, the society and the public have a deeper understanding of the concept of "sustainable design", the goal of "sustainable design" may be thoroughly implemented and realized [7]. There are many ways to realize the sustainable design of public space environment. For example, the use of flexible decorative materials and variable space structure in the design can make the space have growth properties without the need for repeated disassembly to meet new requirements. From that, we can meet the requirements of low-carbon environmental protection and cyclic development in the continuous change of space. From the perspective of ecological environment, the sustainable design of public space environment can improve the utilization of space, and then improve the utilization of urban space and sustainable utilization of land, which has a positive effect on the rational use and integration of land resources. It can meet people's current needs without threatening human development in the future. On the basis of sustainable design, humanized service design can promote public space to sustainability [8].

3 Research Status

At present, the design of sustainable services for public space environment has become a consensus to all countries in the world and a strong voice of the times. It marks a major turning point in the history of human civilization, and is also an important milestone in transforming traditional design models and developing modern innovative designs. Sustainable development has become a common goal of all countries in the world. In addition, we must also see that the concept of sustainable development is based on the integration of environment and development. It has insight into the causes of environmental and ecological problems hidden in the development philosophy and lifestyle of industrial civilization [1]. Therefore, it not only requires the prevention and interception of environmental problems from the development mechanism, but it also advocates the renewal and change of human civilization.

3.1 Research Status in China

The environmental design of public space in China is not only the environmental design in the field of architecture. It also has a certain difference from the environmental design in foreign countries [2]. China's environmental design is not only an integrated design of art such as urban planning, gardens, interior design, etc. It also incorporates rich multicultural connotations, which contains the meaning of life. Therefore, environmental design is a discipline system with rich connotations and a wide range. Although it has been studied and developed in China for more than thirty years, and has undergone many practices, until now its discipline theory construction and sustainable design concepts have not been perfect and need to be further improved.

As the world's largest developing country, China actively responds to the international community's call for sustainable design of environmental space service. In 1994, China took the lead in promulgating the China's Agenda 21. In accordance with the new situation and the new requirements for sustainable development, the Chinese government promulgated the China's Action for Sustainable Development in the Early 21st Century in 2003. In addition, for the first time, the Third Plenary Session of the 16th CPC Central Committee explicitly put forward a scientific development concept, which is people-oriented, comprehensive, coordinated and sustainable; the Fourth Plenary Session of the 16th CPC Central Committee proposed a socialist harmonious society where people and nature live in harmony. The Fifth Plenary Session of the 16th CPC Central Committee proposed to accelerate the construction of a resource-saving and environment-friendly society, conforming to the current trend of world development, and reflecting the latest development concept of the contemporary world. These have created a good domestic environment for the research and practice of China's sustainable development and innovative design [11].

3.2 Research Status in Foreign Country

Compared with it in China, the foreign research on the development awareness of sustainable design for public space environmental service is earlier and the content involves more fields. The reason is that in the process of economic development, foreign countries, especially developed countries, frantically pursue economic growth because of their indifference to environmental pollution and ecological damage. They are the first to experience the revenge from the nature. Such as the world's eight major environmental pollution incidents. Among them, the United States occupies two seats and Japan occupies four seats. The constant emergence of ecological and environmental disasters has alerted the West, especially the developed countries or society, to start rethinking the relationship between man and nature, and explore new paths suitable for the development of the country and the world [4]. Judging from the generation history of sustainable development environmental design, it emerged in the process that western scholars continue to study and explore how human society chooses the development path or direction when facing environmental and ecological crisis: Optimists are represented by future research in the United States. They believe that human beings have survived millions of years without talking about environmental protection and will continue to survive in the future.

Judging from the most direct embodiment of the sustainable design for the public space environmental service—humanistic sustainable space design, the humanistic design consciousness in the West, especially in developed countries, is generally higher. There are many reasons for this. The role of environmental protection books that have important influences around the world cannot be ignored, such as "Silent Spring" (translated by Rachel Carson, translated by Lu Ruilan, Li Changsheng, Jilin People's Publishing House, I997), "The Limits of Growth" ((US) Dennis · Midus et al., Translated by Li Baoheng, Jilin People's Publishing House, 1997), "Our Homeland—Earth" (by Schlidas Ravl, Translated by Xia Fangbao, etc., China Environmental Science Press. 1993) "The Green Sabbath" (by Thor Herald, translated by Zhao Huiqun, Chongqing Press, 2005), "The Green Mountains" (by Ishizaka Yojiro, translated by Yu Lei, Foreign Literature Press, 1983), etc. The establishment of environmental awareness and the shaping of sustainable development awareness are of great inspiration or educational significance.

Foreign researches on sustainable environmental space service design mostly focus on the life cycle research, including product life cycle analysis, assessment tools, and life cycle management. In addition, there are studies related on materials and service systems [4]. China's sustainable environmental space service design also focuses on product life cycle assessment, and there are relatively many researches on sustainable design theory. Research on sustainable environmental design is still focused on academic research, but has not been widely used in practice, especially in China. Although some environmentally friendly materials, low-carbon products and publicity slogans have appeared, they have not improved the waste of resources to meet humanity. Therefore, while conducting theoretical research on the design of sustainable environmental space services, it should also be in line with reality, and the sustainable thinking should be widely used in social life to benefit future generations.

4 Improvement Measures

4.1 Human Oriented; Embody the Humanization of the Humanistic Spirit

The first is to strengthen the experience of public space design. Take the various urban residents as the basis for research. In the specific service design of public spaces, user experience should be the main focus, and art, interaction design, and digital media design should be involved. In this way, the humanistic spirit of environmental design will be more prominent (see Fig. 12).

Fig. 12. Strengthen the experience of public space design

The second is to strengthen the social experience. In the design of public space, it is necessary to pay attention to the role of space in social aspects, so that the environmental design of public space will have more practical significance (see Fig. 13).

The third is event experience. Regarding the social experience as a normal state of urban public space, then the experience activities in the space will become events that can be regularly activated in the space. By combining the event experience and the space experience, the space and vitality, as well as the charm will be maintained for a long time [9].

The fourth is the customized experience, which is the concreteness based on the human-oriented. The designers and planners of space take the special needs of people's physiology and psychology into consideration, so as to provide a design plan for the public space environment with individual needs. By this way, it can meet the individual needs of more people (see Fig. 14).

Therefore, in the service design of public space environment, we should insist the human-oriented, highlighting the experience, and embodying the humanistic spirit into the environment design. In this way, the space environment design will be more humane

Fig. 13. Strengthen the social experience

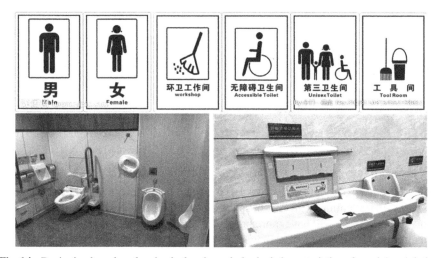

Fig. 14. Designing based on the physical and psychological characteristics of special populations

4.2 Refer to the "Regional Culture"; Embody the "Cross-Cultural" of Public Space

The modern cities are facing the unprecedented development of economic internationalization, and its resident population also has the characteristics of internationalization. While designing the public space environment, it must meet the groups of different races, different countries, different beliefs and different ages. It should cover the service design into special groups such as the elderly, the young children, and the disabled [12]. By understanding the levels and characteristics of service design requirements from the foreign groups and the special groups, we should build a service innovation system in order to integrate the dynamics of service design requirements for such groups, and then show the characteristics of security, interaction, flexibility and humanization of service design. By this way, we can realize the innovative research in the field of cross-cultural design, and promote the development of urban internationalization (see Figs. 15, 16 and 17).

Fig. 15. Intercultural design should take into account different countries, different races and different cultures. The world's three major religions are Buddhism, Christianity, and Islam.

Fig. 16. Intercultural design should take into account different countries, different races and different cultures. The world's three major religions are Buddhism, Christianity, and Islam.

Fig. 17. Intercultural design should take into account different countries, different races and different cultures. The world's three major religions are Buddhism, Christianity, and Islam.

4.3 The Ecological Awareness Based on the Principle of Resource Protection Reflects "Sustainability"

First, the principle of sustainable development effects the survival of human beings and nature as a whole. This is not about personal interests or the existence of individuals, but about maintaining the existence of human beings and all living things. The key point to the sustainable design is to coordinate the harmonious development of man and nature. The relationship between man and nature has always reflected the interaction between human civilization and natural evolution. How to learn from existing foreign experience to develop a circular economy has become a strategic issue in China [3, 5].

At present, the humanized design under the principle of sustainable development has many international examples that we can learn from and refer to. Having a certain understanding of foreign design is a necessary accumulation for designers. But we cannot copy it 100% as what it is exactly. It is necessary to combine the specific characteristics of local geography and humanities, then take our own path of sustainable development. This is also a deeper level of humanized design.

5 Conclusion

The environmental design of public space must be in line with the service design, cross-cultural concepts and sustainable concepts. With the assistance of sustainable design, we can improve the quality of local residents' life. By collecting relevant survey data at home and abroad; proposing suggestions objectively; studying and analyzing in-depth the needs for public space service design of different people in modern cities; mastering the feasibility of service design for public space environment in different cultural backgrounds; analyzing its advantages and insufficient; and exploring in-depth in which way that the sustainable design can be applied to the service design of public space environment, we can not only make an active service plan for the geographical location and human attributes of modern cities, and improve the quality of public space environmental design, but also promote the humanistic care of the discipline in the real society and create well-being for its urban residents.

The sustainable development of public space environmental design is an inevitable trend, and it is also a key point and breakthrough of the environmental design industry in the new era. In the future, as the environmental design majors continue to mature on the study of innovative projects in service design and sustainable design, the development of the environmental design industry will rapidly advance in the direction of humanization and internationalization, improving the well-being of human settlements in worldwide.

References

1. Gao, C.L.: Sustainable space design under the concept of green ecology—review of introduction to ecological environment protection. J. Saf. Environ. 2242 (2019)
2. Li, X.Y.: Urban development studies, pp. 27–31. China Urban Science Research Association, Beijing, July 2019
3. Li, T.: Creation and design of urban green public living space. Beauty Times 80–81 (2019)
4. Zhang, Y.L., Zhang, D.Y.: Analysis of the urban public space design practice of Chicago in 1909. Archit. Cult. 144–145 (2019)
5. Zhang, L.H., Yang, G.M.: Sustainable space design under the concept of green ecology—review of introduction to ecological environment protection. Creation Des. 66–70 (2015)
6. Tang, J.: Research on flexible space design of human living environment based on sustainable development. Ind. Des. 99–100 (2019)
7. Zhao, R.Z., Chen, Q.: Urban spatial planning for a sustainable high-density environment. World Archit. 52–57+116 (2017)
8. Yang, F.R.: Analysis on sustainable development of urban space environment design. Fujian Archit. Constr. 4–5+3 (2007)
9. Zhang, L., Lv, W.L.: Exploration of public cultural space design in interactive urban communities. Beauty Times 62–63 (2019)
10. Lu, J.J.: Research and application of modern cultural space design. Art Sci. Technol. 138–139 (2014)
11. Bergman, D.: Sustainable Design: A Critical Guide. Jiangsu Phoenix Science and Technology Press, Nanjing (2019)
12. Polaine, A., Reason, B.: Service Design from Insight to Implementation. Tsinghua University Press, Beijing (2015)
13. Peng, F.: Cross-Cultural Design. Electronic Industry Press, Beijing (2016)

A Preliminary Study on the Game Design of Pokémon GO and Its Effect on Parent-Child Interaction

Hsuan Lin[1]([✉]), Kuo-Liang Huang[2], and Wei Lin[3]

[1] Department of Product Design, Tainan University of Technology, Tainan, Taiwan
te0038@mail.tut.edu.tw
[2] Department of Industrial Design, Sichuan Fine Arts Institute, Chongqing, China
shashi@scfai.edu.cn
[3] School of Architecture, Feng Chia University, Taichung, Taiwan
wlin@fcu.edu.tw

Abstract. In today's society characterized by information explosion, the smart phone has become an indispensable convenience in modern people's lives, facilitating information inquiry, shopping, basic needs, as well as entertainment; meanwhile, it has changed people's living habits. Over the past ten or more years, there have been many game-related studies which mainly focused on online games for personal computers (PC). However, with the popularity of smart phones, mobile games have entered the new mainstream of the market. Pokémon GO is an online game that combines a mobile device with augmented reality (AR). Through their smartphone screens, players can search for cartoon and animation characters in the real world. Ranging from small children, teenagers, and middle-aged people to old people over 70 years of age, all players try to capture virtual creatures, called Pokémon. It is difficult for ordinary mobile online games to attract players of different age groups, i.e., young, middle-aged, and old players at the same time, while Pokémon GO enables a wide variety of players to participate in the game simultaneously. Therefore, this study first analyzed the game design of Pokémon GO and then explored the interactive factors involving the context where parents and children play together. The semi-structured interview was employed, with three participants interviewed. Through the interview, it was found that the game design of Pokémon GO includes the following features: 1) a user-friendly interface, 2) going outdoors, 3) association with places of interest, 4) combining games with exercise, 5) cooperative gaming with the community, and 6) highlighting festivals to attract players. As for its impact on parent-child interaction, Pokémon Go boasts the characteristics below: a) a common topic for the family, b) family members doing their cooperative gaming, c) training children to classify things and use logic, and d) fostering children's independence. Due to the foregoing factors, parent-child interaction can improve a lot. The results of this study can be referred to by game designers.

Keywords: Appeal · Child · Senior citizen · Game design · Cooperative gaming

© Springer Nature Switzerland AG 2020
P.-L. P. Rau (Ed.): HCII 2020, LNCS 12192, pp. 115–127, 2020.
https://doi.org/10.1007/978-3-030-49788-0_9

1 Introduction

1.1 Background

As an online game, Pokémon GO blends a mobile device with augmented reality (AR). Players can search for cartoon and animation characters in the real world through their smartphone screens. Since its launch in July 2016, this game has been downloaded more than 500 million times worldwide [1, 2]. Players depend on the Global Positioning System (GPS) to move around in real life, explore maps, capture Pokémon, and fight with other factions to battle for a Gym. Since being released in Taiwan, Pokémon GO has been highly popular. The majority of its players are young people born after 1980, who had watched the Pokémon cartoon when they were young children. But this online game has kept players of all ages attached to it. Players, whether children, middle-aged people, or even old people over 70, join the treasure hunt.

1.2 Research Purposes

Mobile games, or virtual games, have long been considered to be one of the exclusive pastimes for modern-day youngsters. With the popularity of smart phones, mobile games have become one of the pastimes for users of different ages. Although the impact of online games on parent-child relationships is growing constantly, only a few studies have evaluated issues about parent-child cooperative gaming (co-op). If the appeal of family cooperative games as well as the effect on parent-child interaction can be identified, the developers of such games will be greatly helped. Consequently, besides interviewing parents who were involved in the parent-child co-op of Pokémon GO, this study explored the impact of this online game on parent-child interaction. The findings will be helpful to future research into the influence of online games on parent-child interaction.

In the past, online games could only be run on a personal computer. With the rapid development of smartphones and the Internet, smartphones have become the main carrier of online games. Mobile games have been a battleground on the game market, so many young game designers and software engineers have engaged in game development. It is difficult for ordinary mobile online games to attract children and players of the other age groups at the same time. Unlike its counterparts, Pokémon GO leads people of all ages to participate in it. Therefore, this study first analyzed the game design of Pokémon GO to understand the factors that attract its players. Next, it was explored how the online game combined with augmented reality affects parent-child interaction concerning family cooperative gaming. The findings will be beneficial to the developers of online games. The purposes of this study are as follows: 1) to summarize the features of Pokémon GO's game design, 2) to understand the relationship between Pokémon GO and parent-child cooperative gaming, and 3) to probe the impact of Pokémon GO on parent-child interaction.

2 Literature Review

2.1 Online Game

Online games have developed into one of the primary leisure activities done by modern people. There are many different motives for players' being attracted to the games

[3, 4]. The most important component of online games lies not in the games themselves but in players; specifically, online games and players are inseparable. After researching the motivation of online gamers, Rouse pointed out that in the process of game-playing, players usually regard it as the most interesting part to manifest their creativity and that this dynamic process is named interaction [5]. In other words, whenever players exert their imagination to the full and try new techniques or strategies, they will feel particularly fulfilled and moved. This is what makes video games unique and appealing. Yee classified the motivations of video gamers into three types: achievement, sociability, and immersion [6]. Achievement-type players put emphasis on completing game tasks, upgrading quickly, and accumulating game resources, all of which are in line with the avatar's desire to demonstrate his/her power. Sociability-type players care about maintaining online relationships with others, for the avatar's name can present many aspects of a player, such as his/her qualities and personality. Concerning immersion-type players, they become immersed in the story through the designated avatar. Therefore, the avatar's design as well as the background or style of the avatar's story will affect the player's immersion. Games like Pokémon Go have huge potential, for virtual reality (VR) games exert a positive impact on players' health. Players have an excuse for going out, not only increasing physical activity but also improving psychological quality and social skills [1].

2.2 Virtual Community

A virtual community refers to a group of people who communicate with and contact each other through electronic media, such as telephones, faxes, radios and computers [7]. That is, the virtual community means a kind of social gathering on the Internet. When there are enough users to perform long-term and full participation, communication and management, a body of social aggregation maintaining a social network is formed in the cyberspace [8]. In the past, people had to convey messages through face-to-face conversations, telephone calls, faxes, and letters. Today, interpersonal relationship is free from geographical obstacles, so people in different regions, cities, and countries can have a dialogue without being restricted by space. What's more, people can use different virtual identities to perform various activities at the same time, transcending the limits of time and space.

The rise of virtual communities can be attributed to the booming development of the Internet. Due to the popularity of the Internet and its users' habits, it has created the dynamics of a new social domain along with a space reflecting the nature of society [9]. With the swift progress of online communication, Internet users are increasing continuously who have changed the group behavior in society because of frequent interactions. People gather and communicate on this online platform, gradually forming a new social phenomenon, which has brought virtual communities into existence.

The popularity of smartphones, improvement in Internet speed, and continuous invention or innovation of applications (APP) have greatly influenced and changed the way people interact with each other. In the past, the telephone was invented to help people communicate with each other without the hindrance of distance. Later, the emergence of mobile phones made communication more immediate. Nowadays, such everyday activities as communication, shopping, consumption, entertainment, social activities, and documenting daily life depend on smartphones, which appeal to users of all ages. As the elderly are beginning to keep up with this trend, their lives, social circles, and ways of interaction have changed. Through online communities like Line and Facebook, old people are using smartphones. Mobile games have also evolved with this trend in diverse directions, combining community and interaction. The games enable players, their friends and relatives to experience the fun of teamwork simultaneously by completing game tasks.

2.3 About Pokémon GO

The release of Pokémon GO in 2016 started a growing wave of sensation all over the world. Pokémon GO is an online game that combines a mobile device with augmented reality (AR) [2]. While walking in the real-life world with the help of the Global Positioning System (GPS), players follow the virtual map, constantly exploring and capturing Pokémon. Also, in order to earn Pokécoins, players can battle with their counterparts to seize or defend a gym. A new way of gameplay in Pokémon GO was added in June 2017, under which players can participate in Raid Battles with others, attacking Legendary Pokémon. Only after defeating the Legendary Pokémon are players given the chance to catch rare Pokémon. Moreover, because of the above feature, many players who abandoned this game have resumed it. In March 2018, two new ways of gameplay were introduced, i.e., Special Research and Field Research.

Taking up Special Research, players must complete missions at different stages to get rewards. The later tasks are much more difficult than the earlier ones, challenging players' stamina, skill, and luck. As long as players complete the tasks, they are qualified to capture Legendary Pokémon, such as Mew and Celebi, with their success ensured. Field Research means that players need to collect topics, or tasks, from a Poke Stop, and there will be a reward for the completed tasks. In July 2019, Team GO Rocket was launched to capture all Poke Stops in various places. Players are likely to catch Shadow Pokémon after overwhelming the Rockets' Shadow Pokémon. After being purified, the captive Shadow Pokémon will be changed back to normal Pokémon. Pokémon GO has been one of the very successful interactive online games in recent years. The rapid growth of online APP games has resulted in the development of new physical activities, which produces a certain effect on schools, families and the entire society [10, 11].

3 Methodology

By means of interviews, this study aimed to understand the characteristics of Pokémon GO's game design and the effect of cooperative gaming (co-op) on parent-child interaction. The interviewees were three parents/guardians involved in co-op of Pokémon GO together with their children. Before the interview, the purposes and procedures of this experiment were explained to the interviewees. Then, the interviewees' information was confirmed; after that, the formal interview was started. Lastly, the collected interview records were organized for further analysis and discussion.

3.1 Interviewees

Targeting at parents/guardians who did much cooperative gaming of Pokémon GO with their children, this study interviewed three participants. The participants' average age is 53 (from 41 to 65), and all of them reached the game level 36 or above, as shown in Table 1.

Table 1. The interviewees' information

Code name	Parenthood/ guardianship	Age	Gender	Number of children	Child's age
M1	Father	41	Male	One	7
M2	Father	53	Male	Two	9 and 13
F1	Grandma	65	Female	One	5

3.2 Selecting a Form of Interview

There are various forms of interview, including open-ended, structured, semi-structured, and focus group interviews [12]. The difference of the first three forms lies in the interviewer's control and understanding of the questions before the interview. As for focus groups, the interviewer guides a group of participants to discuss specific topics. This study adopted the semi-structured interview, which combines the characteristics of structured and open-ended interviews, with open and closed questions included. In this way, interviewees will answer questions from easy to difficult ones, gradually entering the core of the topics.

3.3 Interview Procedures

This study first developed an interview outline which was followed to guide different interviewees to answer the same topics. As the interview was performed, the interviewer asked questions in accordance with the outline at first. Later, the interviewee was asked deeper questions, having enough time to think and talk, so that he/she could elaborate his/her answers and relevant ideas. The interviewee didn't move on to the next topic until he/she could no longer reveal new information [13].

The interviewees were requested to answer related questions about the game design of Pokémon GO's and its effect on parent-child interaction connected with cooperative gaming. The interview procedures are set forth below. First, the research purposes were explained to the interviewees. Next, the basic information of the interviewees was confirmed. After that, the formal interview started, with its details shown in Table 2.

Table 2. Interview procedures

Procedure	Topic	Question
1) Explanation		The interviewer explained the research purposes
2) Confirming interviewees' information	Background information	Game level, child's age, and gaming time per week
3) Formal interview	Game design of Pokémon GO The effect of co-op on parent-child interaction	1) When did you start to play Pokémon GO? 2) What led you to start playing Pokémon GO? 3) What features of Pokémon GO appealed to you? 4) What motivated you to continue playing Pokémon GO? 5) In what way did Pokémon GO affect your interaction with your child/children? 6) Do you think whether Pokémon GO can improve parent-child interaction? If so, give some examples

4   Results and Discussion

This study was directed at two domains, namely, the game design of Pokémon GO and its effect on parent-child interaction connected with cooperative gaming. The interview results found that the game design of Pokémon GO incorporates the following features which lead its players to continue playing: 1) the user-friendly interface, 2) going outdoors, 3) association with places of interest, 4) combining games with exercise, 5) cooperative gaming with the community, and 6) highlighting festivals to attract players. As for its impact on parent-child interaction, Pokémon Go shows the features below: a) becoming a common topic for the family, b) the family members doing their cooperative gaming, c) training children to classify things and use logic, and d) fostering children's independence.

4.1   Features Concerning Game Design of Pokémon GO

1) User-friendly interface

Immediately after being launched, Pokémon GO created a wave of craze, so a large number of players were often seen to capture Pokémon characters in many public places or at roadsides. Players in this game range from youngsters and middle-aged people to senior citizens, namely, peoples of all ages. Its simple gameplay and user-friendly interface are two factors that appeal to a wide variety of players. For example, when a player clicks on a Pokémon to be caught and slides his/her finger from bottom to top, which means throwing a virtual Pokémon Ball, he/she has the chance to capture Pokémon characters. In addition, Pokémon is a cartoon character that accompanied young people born around 1980 in their process of growing up, which makes them willing to download and try playing the newly-released game. The interviewee M2 stated that a news report led him to play Pokémon GO in the beginning. After downloading the game, he was captivated by augmented reality (AR) technology [14]. The interviewee M1 said, *"Seniors and children join the game mainly because its interface is easy to operate. The former do not need to spend time memorizing the game functions, and the text has fewer explanatory words."* Therefore, Pokémon GO keeps older people obsessed with it. By contrast, because other mobile games are more complicated, players have encountered difficulties right from the start; in the meantime, players need a lot of learning and textual descriptions to understand the rules of those games. Faced with the unfriendly kinds of gameplay, many inexperienced players feel frustrated and give up playing due to confusion, let alone lead children and seniors to participate in the games.

2) Going outdoors

The main feature of Pokémon GO is that players who previously played video games at a fixed place or on a computer must go outside. It changes players' past behavior, that is, to play online games at home. The gameplay design of Pokémon GO demands that players should move constantly in the real world so that they can move synchronously in the virtual world. The global positioning system (GPS) on mobile phones can accurately locate the locations in the real and virtual worlds. This mechanism forces players to go outdoors one after another, which is one of the factors that make many old players attached to this game.

3) Association with places of interest

In addition to going outdoors, Pokémon Go has another feature and important component, Poke Stop. Players must visit a nearby Poke Stop to collect items so that they may obtain Pokémon Balls to catch Pokémon. The landmark pictures of Poke Stops are numerous, including statues, parks, artistic works, shops, and paintings. What's more, special items or patterns at roadsides can be chosen as Poke Stops. Players can randomly get Pokémon Balls or other items by turning a Poke Stop. Furthermore, the Poke Stop is designed to be updated every 5 min, so players will move on to the next Poke Stop instead of staying in the same place and waiting for an update. Since the game presents real-life scenes, players just click on the Poke Stop. As shown in Fig. 1, the picture in the circle is the actual presentation of the landmark. Before resupply, the Poke Stop is blue, and it turns purple after resupply. The interviewee M1 commented, *"I have lost sight of many valuable things at roadsides. Only when I saw the pictures on the Poke Stops did I discover the things that I had long neglected."*

Additionally, Pokémon GO has still another feature, Safari Zone. From October 3 to 6, 2019, Taiwan hosted a Safari Zone event at the Metropolitan Park in New Taipei City, staring from 10:00 till 17:00. The interviewee M1 said, *"The Metropolitan Park of New Taipei City, in the eyes of residents in other counties or cities, was practically unknown. Amazingly, it became a well-known scenic spot overnight after the mass media reported about it. According to my brother-in-law, the players in their hometown rented a tour*

(a)Before resupply   (b) After resupply

Fig. 1. Poke Stop, taken from a screenshot of the game

bus and traveled northward to the event site, or the Metropolitan Park, in order to catch rare Pokémon". As shown in Fig. 2, there were many Pokémon-related decorations and characters for tourists to take pictures of.

Pokémon GO cooperates with places of interest all over the world through the Safari Zone event and at the same time promotes tourism, making rare Pokémon found easily or exclusively in a designated site. Many players scramble to reach the site so that they may catch those rare Pokémon. After the disastrous earthquake in March, 2011, many villages in Northeastern Honshu, Japan were severely destroyed by tsunamis. In the past few years, in order to attract tourists as well as accelerate reconstruction and revival in the quake-stricken regions, the local governments have employed Pokémon GO to organize various activities. For instance, Lapras appeared in large numbers at Ishinomaki in Miyagi Prefecture; as a result, many Pokémon fans swarmed into Ishinomaki to capture Lapras.

4) Combining game with exercise

Unlike other online games, Pokémon Go boasts one more feature, namely, helping to increase players' amount of exercise. Through augmented reality (AR), this game gets players out of their homes, requiring those who usually stay at home or walk less to go outdoors. In consequence, people who have been accustomed to interacting on the Internet return to the real world, engaging in actual exercise. The interviewee F1 said, *"Playing Pokémon can increase the amount of physical activity and help me to stay in shape."* For the game to go smoothly, the player is compelled to move constantly; meanwhile, there is a speed limit for egg-hatching, which is used to calculate mileage.

Fig. 2. A Safari Zone event at Metropolitan Park in New Taipei City, Taiwan

Therefore, players who want to upgrade and capture Pokémon creatures must walk, run, or ride a bicycle in real life. However, if the mileage of travel exceeds 30 km, the figure will not be accepted by the program. The craze for Pokémon GO has led many people to go outdoors to do their treasure-hunting and egg-hatching. According to Howe's findings, in the first week after downloading Pokémon Go, its players moved 955 more steps per day on average than those who did not play the game [1]. In the first six weeks, players' amount of exercise indeed increased greatly; nevertheless, after six weeks, other more creative ways had to be found to help the players continue their habit of exercise.

5) Community co-op
Pokémon GO is a game that can be shared for pleasure with family members and friends [2]. It is mainly for the above reason that a lot of players continue playing it [14]. Since Pokémon creatures will appear anywhere randomly and disappear after a certain period of time, it is generally difficult to control the situation far away. Many players didn't know each other before but often met each other when playing Pokémon GO nearby, so they ended up becoming friends and creating a community. When a special Pokémon appeared or a group battle (Raid) occurred, the message was immediately sent to the community; soon afterwards, lots of players came to the scene to capture Pokémon or fight a Raid. In the eyes of an unsuspecting person, a group of people suddenly appeared and gathered somewhere, and then, after a short time, they dispersed equally suddenly, as if they were flash mob.

6) Highlighting festivals
Downloading mobile games through the online APP has expanded from the regional market to the global market. It is much more convenient than before for players to choose and acquire the games. How to keep players fascinated and attached has been a

major challenge for game developers. Online games are often combined with festivals and relevant celebrations to attract as many players as possible. In addition to the special festive atmosphere in the real-life environment, players can create game scenes in the virtual world with the same festive layout and decoration, and even dress the virtual characters in festive costumes. Pokémon GO is no exception, for there are special festive celebrations in addition to the regular community day every month. From October 18, 2019 to November 1, 2019, the Pokémon GO Halloween 2019 event was launched. The brand-new Shadow Pokémon, costumes (avatar items) and Shiny Pokémon appeared in the game. More Ghost- and Dark-type Pokémon appeared in the wild, Eggs, and Raids. To capture special Pokémon, which are only available during special festivals, many players showed up on such occasions.

4.2   Effect of Parent-Child Co-op on Interaction

1) Pokémon as a common topic

Pokémon has become a topic among the family members. Its game design mainly enables players to move around, find Pokémon characters randomly, and use the Pokémon Ball to conquer them for the purpose of collecting Pokémon as well as opening the book. The interviewee F1 said, *"After collecting Pokémon, my granddaughter felt very happy and fulfilled. Also, she shared and discussed her experiences with her peers."* The interviewee M1 remarked, *"After my child came back home, he directed all his talks at Pokémon. Right after returning home, he came up to me, asking what special character I had captured, where I caught it, what its combat power (CP) was, and whether its individual values (IV) were high."* Many seniors cannot remember Pokémon's name, but children can. The interviewee F1 added, *"My granddaughter explained to me the characteristics of a particular Pokémon, which is so powerful as to spray water. Then, I asked her whether it was the one I suspected. If not, I would request her to talk more about it. We talked and talked in vain for 30 min without getting bored."* Pokémon GO has become not only a topic among friends but also a bridge connecting parents with children.

2) Cooperation in Pokémon-hunting

In a modern society characterized by hustle and bustle, parents are usually busy with work while children need to digest heavy schoolwork. As a result, parents and children spend a meager amount of time getting together, with the opportunities for parent-child interaction decreasing. The interviewee M2 commented, *"I started playing Pokémon GO just to keep up with the trend, thinking the news report seemed to be amusing. Immediately after the game was released, my children and I started to play it, using the same account. We did our cooperative gaming most of the time, discussed the progress of the game, and studied our strategies. In this way, we would walk together in the park near our home, catching Pokémon at the same time."* Some parents had not played online games themselves, but after starting to play Pokémon GO, they were willing to stroll outdoors, chat with their children, and improve the parent-child relationship. Some of their relatives and friends participated in the game together, and they took their children to a specific site to join the Community Day. Thus, opportunities for parent-child interaction increased.

3) Training children to classify and use logic

How to upgrade and rear highly powerful Pokémon in the shortest time is also one of Pokémon GO's features. On Pokémon Event Day, the chance for special Pokémon to appear is very good; besides, different events are accompanied by different doubled rewards. For example, the candy, experience points, and Star Dust awarded to players are doubled to increase their willingness to capture Pokémon creatures. Pokémon GO Halloween in 2019 doubled the candy reward, and many players turned to Pinap Berry to have a better chance of getting candy. The interviewee M1 said, *"When my child and I first started playing Pokémon GO together, he would use Berry for whatever he saw; that is to say, his timing was incorrect. After taking my advice, he knew when to save Berry and when to use it. On some occasions, as the item bag was full, he needed to consider which items to throw away and which ones to keep. Those matters needed to be explained to him clearly."*

In addition to capturing Pokémon outdoors, it is necessary to frequently tidy the limited Pokémon zone and item zone. The former zone is no longer capable of capturing Pokémon when it is full; likewise, the latter is no longer capable of holding added items when it is full. For example, as each Pikachu (Pokémon's name) is randomly captured, its individual values (IV) and combat power (CP) are also randomly assigned. With such a big difference, it is necessary to spend some time considering which Pikachu should be kept or traded for candy. The interviewee M1 added, "I explained to my seven-year-old child how to identify those Pikachus' IV and CP and then to rank them in order of the two values. Five Pikachus were kept at one time, with extra Pikachus traded for candy. In this way, the Pikachus with high IV and CP were reserved, so a lot of Pokémon space was freed."

The interviewee M1 went on adding, *"When a combat is fought in Pokémon GO, the case of conflicting types frequently occurs. My child used to fight inefficiently when playing against his opponent. After he had suffered a number of defeats in the one-to-one combat, I explained to him the importance of types. Now, my child can remember the type of each Pokémon and equip it with an appropriate attribute during the combat. I often ask my child which Pokémon can be dispatched to defeat its enemy."*

4) Fostering children's autonomy

The main reason why many parents object to children playing online games is that children are easily addicted to the games. The interviewee F1 remarked, *"The main reason for me to play Pokémon GO with my granddaughter is to cultivate her autonomy. If a child is stopped from playing online games at an early age, he/she will easily become addicted after owning his/her mobile phone. It is better to accompany the child and guide him/her to play online games in his/her childhood than to prevent him/her from playing."* The interviewee M1 said, *"The main benefit of Pokémon GO is that its players must walk around outdoors to play the game. Such being the case, children no longer demand to play online games at home. Moreover, children's merit points can serve as a reward. If children perform well, take them out for fun on a holiday or an outing."* Modern people spend more and more time playing games online. It is very difficult for adults to show self-control, not to mention children. It is suggested that children's autonomy should be cultivated by means of online games.

5 Conclusions

This study explores the game design of Pokémon GO and its impact on parent-child interaction, discovering several features of its game design. Its game interface is easy to operate, appealing to a wide range of players, from 5 to 65 years old. Unlike other online games, it compels players to go outdoors. Players must walk in order to play this game, which increases their amount of exercise and wins widespread praise. To give players an unceasing sense of novelty, the game company host different activities every week, combining places of interest and festivals to keep players attached.

Pokémon GO has become a common topic in family life. All the family members participate in Pokémon-hunting together under the influence of a common interest. The time for parent-child interaction has increased, and so parent-child relationship has also improved. Through cooperative gaming, parents can try to understand their children's interests, cultivate common interests, train children's classification and logical abilities, and foster children's autonomy. When parents put too much blame on their children for being addicted to online games and neglecting studies, they push their children farther away. Parents can respect their children's ideas by understanding and even participating in the latter's game together. When children notice their parents caring about what they care about, the conversations in a family are no longer just parents' order-like instructions or worries about schoolwork, but discussions about the game are frequently heard. This game has changed parent-child relationships. Specifically, children teach parents how to play the game, reversing their roles in the family. Parents do not play the role of educators at this time. When facing new things, parents can consult with or learn from their children. Children achieve a sense of achievement in the teaching process. In the end, more interactions between parents and children lead to a closer relationship.

Players are reminded to pay attention to the following when playing online games. They should spend a moderate amount of time playing, manage their money wisely, be always alert to safety, never play on the road, and never indulge in games.

Acknowledgments. The author is grateful to the Ministry of Science and Technology of the Republic of China for supporting the research under grant No. MOST 108- 2221-E-165-001.

References

1. Howe, K.B., Suharlim, C., Ueda, P., Howe, D., Kawachi, I., Rimm, E.B.: Gotta catch'em all! Pokémon GO and physical activity among young adults: difference in differences study. BMJ **355**, i6270 (2016)
2. Paavilainen, J., Korhonen, H., Alha, K., Stenros, J., Koskinen, E., Mayra, F.: The Pokémon GO experience: a location-based augmented reality mobile game goes mainstream. In: Proceedings of the 2017 CHI Conference on Human Factors in Computing Systems, pp. 2493–2498. ACM (2017)

3. Jansz, J., Tanis, M.: Appeal of playing online first person shooter games. CyberPsychol. Behav. **10**, 133–136 (2007)
4. Lucas, K., Sherry, J.L.: Sex differences in video game play: a communication-based explanation. Commun. Res. **31**, 499–523 (2004)
5. Rouse III, R.: Game Design: Theory and Practice. Jones & Bartlett Learning, Burlington (2010)
6. Yee, N.: Motivations for play in online games. CyberPsychol. Behav. **9**, 772–775 (2006)
7. Cerulo, K.A., Ruane, J.M., Chayko, M.: Technological ties that bind: media-generated primary groups. Commun. Res. **19**, 109–129 (1992)
8. Rheingold, H.: The Virtual Community: Finding Commection in a Computerized World. Addison-Wesley Longman Publishing Co., Inc., Boston (1993)
9. Mitchell, W.J.: City of Bits: Space, Place, and the Infobahn. MIT Press, Cambridge (1996)
10. Gao, Z., Chen, S., Pasco, D., Pope, Z.: A meta-analysis of active video games on health outcomes among children and adolescents. Obes. Rev. **16**, 783–794 (2015)
11. Gao, Z., Chen, S.: Are field-based exergames useful in preventing childhood obesity? A systematic review. Obes. Rev. **15**, 676–691 (2014)
12. Fontana, A., Frey, J.: The Art of Science. In: The Handbook of Qualitative Research, pp. 361–376 (1994)
13. Rogers, Y., Sharp, H., Preece, J.: Interaction Design: Beyond Human-Computer Interaction. Wiley, Hoboken (2002)
14. Rasche, P., Schlomann, A., Mertens, A.: Who is still playing Pokemon Go? A web-based survey. JMIR Serious Games **5**, e7 (2017)

Teaching to Find Design Opportunities for Behavior Change Through Causal Layered Analysis

Peter Scupelli[(✉)]

School of Design, Carnegie Mellon University, Pittsburgh, PA 15213, USA
scupelli@cmu.com

Abstract. Change is exponential. Products and services are developed faster, hold a shorter shelf-life disrupted by new offerings, and exist within global challenges such as climate change and sustainability. We live in the Anthropocene Era with human-induced Climate Change and the sixth mass extinction. Design challenges for product/service creation are shifting from a paradigm of customer-centric focus linear system to a new paradigm that includes a broader planetary context with a circular economy. The *Human-Centered Design Thinking* process popularized by IDEO/Stanford d.school identifies human-centered design opportunities for products and services through five steps: empathize, define, ideate, prototype, and test. Unfortunately, a mere customer-centric view of product or service is insufficient for planetary-scale challenges requiring a broader focus that includes all life forms and the planet's health. Thus, moving forward, design for the 21st century requires broader perspectives; design educators are challenged to overcome the limitations of *Human-Centered Design Thinking*. The point of this paper is that *Human-Centered Design Thinking* needs to be used in conjunction with new design methods that take into account the larger planetary context over time. In this paper, I discuss how *futures thinking* methods can augment *design thinking* methods to overcome limiting human-centered worldviews, epistemologies, and ontologies. In particular, how Causal Layered Analysis (CLA) can augment designers' creative responses to behavior change challenges such as rapid-decarbonization through four layers (e.g., litany, systems, worldview, myths/metaphors). We provide a case study on CLA from design courses taught to both undergraduate and graduate design students.

Keywords: Dexign Futures · Design Thinking · Futures thinking · Causal Layered Analysis · Case study · Design pedagogy

1 Introduction

It seems that so much of design education in recent decades has been focused on training designers to use *Human-Centered Design Thinking* to design products, services, or environments for a company or organization. Yet the challenges to human long term survival on planet earth are in peril due to human-induced climate change. Such a shift

© Springer Nature Switzerland AG 2020
P.-L. P. Rau (Ed.): HCII 2020, LNCS 12192, pp. 128–144, 2020.
https://doi.org/10.1007/978-3-030-49788-0_10

in the scale of design problems from artifacts/services to larger systems and societal level problems means that design students wanting to engage with such larger-scale societal challenges need to learn new: tools, design methods, systems perspectives, worldviews, and myths and metaphors to operate at the planetary scale that such problems entail. For example, Tony Fry describes the shift from *defuturing* (Fry 1999) to *futuring* (Fry 2008). He describes *defuturing* as engaging in actions in the present that take away our collective future and *futuring* as actions in the present that give us a collective future.

An example of *defuturing* is lack of action towards rapid decarbonization. Climate change and sustainability are global challenges that must be addressed quickly in the next decade. In 2018, the Intergovernmental Panel on Climate Change (IPCC) issued the 1.5 °C report that clearly states that rapid decarbonization is needed by the year 2030 to avoid climate change catastrophe (UN-IPCC 2018). This means that action needs to be taken quickly on a global scale. Since most things within the consumerist world play a large role in the carbon emissions causing Climate Change (UN-IPCC 2018), we posit that design education needs to change quickly around the world.

Two design moves that Fry identifies as necessary for *design futuring*: to *design in time* and a *values-based* design perspective (Fry 2008). To *design in time* with attention to the *defuturing* impact of design has ethical and moral value based implications. To operate in the present in a way that gives us a future, designers must acknowledge that they need to align their design practice with a much larger context as defined by systems, worldviews, perspectives, and epistemologies. Next, we briefly describe shifts in historical understandings of the scale of design, and shifts in values.

1.1 Design Challenges Are Shifting in Scale and Size

Historically, design disciplines have engaged with broad societal challenges, examples of which include architecture for humanity (2006), industrial design (e.g., Papanek 1972), communication design for good (e.g., Berman 2009), organizational change (e.g., Brown and Kātz 2009), and instructions for continued life on planet earth (Fuller 1969), among others.

The Scale of Design is Changing. On one hand, some of the design problems are getting smaller and more specialized, while on the other hand, the problems on which designers work are getting bigger (e.g., Brown and Kātz 2009), connected to global supply-chains, with global social, economic, and environmental impact. Designers are engaged with systems-wide implications of designed products and services (e.g., Mau 2005). Such changes influence professional practice and require design schools to adapt their courses and curricula to better prepare for such changing environments. Teaching students to engage with larger-scale problems requires an emphasis on collaboration between multiple disciplines and methods of design research that work at small and larger scales. In particular, models and frameworks that allow designers to describe and articulate the relationships between small scale and larger scale problems are particularly helpful for such collaborations.

1.2 Design Evolves to Engage with Complex Large-Scale Problems

There is a long tradition within design research and design methods communities to problematize narrow conceptions of design when addressing larger societal problems. Many thought-leaders have articulated shifts in the field of design over time. Next, we mention a few:

Design Beyond Craft. John Chris Jones described four levels of design – components, products, systems, and community – to advocate for new design methods that go beyond craft to cover new challenges such as traffic congestion and air quality (Jones 1992).

Levels of Complexity for Product Design. Jay Doblin described three levels of complexity: (a) *products* – the simplest form of design; (b) *unisystems* – coordinated products and the people that operate them; and (c) *multisystems* – the sets of competing *unisystems* (Doblin 1987).

Levels of Complexity for Communications Design. Meredith Davis explains that complexity expands within the field of communication design according to breadth of system and resulting human experience (Davis 2008). As complexity and human experience increase, communication design goes from simple logo design to corporate identity, to branding, to service design.

Orders of Design. Richard Buchanan introduced four orders of design to contrast the traditional understandings of the disciplines of communication design (symbol), industrial design (product), interaction design (action), and systems design (thought) with new understandings of design that blur the distinctions between types of design (Buchanan 1992).

Different Kinds of Design Thinking. Elizabeth Pastor (2013) co-founder of Humantific with G.K. VanPatter, articulated the differences between four types of design that shift as levels of complexity increase: Design 1.0 traditional design thinking, Design 2.0 Product/Service Design thinking, Design 3.0 as Organizational Transformation Design thinking, and Design 4.0 as Social Transformation Design Thinking.

DesignX was a nomenclature created to get beyond the number of designs (e.g., Norman 2014) to a broader version of design that shifts from a focus on products and services to a broader range of complex societal issues.

Exponential Design. Arnold Wasserman (2011) describes four versions of design to include design 1.0 as artifact-centric (e.g., making and selling); design 2.0 as human-centric (e.g., strategic field building and embedding; design 3.0 as Socio-centric (e.g., changing the world); and Design 4.0 as the post-anthropocene (e.g., sustainable prosperity @ one planet). The term sustainable prosperity @ 1 Planet was coined by Chris Luebkman in 2009 in the ICSID 2009 conference (International Council Societies Industrial Design).

Wasserman points out that from a set theory perspective the different kinds of design are nested with the lower order of design being contained by the higher orders. Therefore, the higher-order design does not replace the lower-levels of design. To design at the higher levels of design, all the lower levels of design are necessary. More importantly, alignment between lower and higher order of design is needed to accomplish the larger-scale ambitions such as Sustainable Prosperity @ One Planet, UN Sustainable Development Goals.

An obvious corollary to Wasserman's previous point is that many of the problematic aspects of the Anthropocene Era are a direct result of lack of acknowledgment on the larger *defuturing* aspects of design at the lower levels as described by Tony Fry (1999). For example, focusing on the human-centered needs at the detriment of the other life forms on the planet. Hence, a shift in paradigms from *defuturing* to *futuring* is needed to accomplish goals such as Sustainable Prosperity @ One Planet.

We note that a decade has passed since the term "Sustainable Prosperity @ 1 Planet" was coined, and popular discourse in 2019 shifted to "Extinction Rebellion" with a much more urgent plea (Extinction Rebellion 2019). The ongoing effects of Climate Change experienced currently such as record: heatwaves, hurricanes, tornadoes, and floods (e.g., Shaftel 2019) has shifted the discourse from sustainability to climate adaptation and mitigation. As the 2018 IPPC 1.5 °C report makes abundantly clear, decisive action is increasingly more urgent to avoid the catastrophic effects of Climate Change (i.e., 50% decarbonization by year 2030, and 100% by year 2050).

In short, design shifts as the world shifts; design thinking and design methodologies evolve to address the complexity of ongoing and emerging societal challenges. The current challenges involve designing products and services within the context of a swift transition to a decarbonized world.

1.3   Design Evolves to Engage with Complex Large-Scale Problems

Challenges such as societal-level sustainability require new thought, temporalities, and action. In the past decades, design temporalities have shortened dramatically. There is a tension between ever shortening design cycles and long-term thinking. Examples in the design industry include fast-fashion (e.g., Claudio 2007), continuous beta (e.g., O'Reilly 2005), and lean start-up (Ries 2011).

Design on one time horizon is insufficient. For example, in large technology companies, designed products might be conceived for three timescales: three months, three years, and seven years. The three-time horizons model referenced was developed by Baghai, Coley, and White, from McKinsey Consulting (2000).

Much of traditional craft-based design education combined with *Design Thinking* focuses on teaching students to craft products or services that can be made in a short time horizon. Design thinking often focuses on finding problems and opportunities through empathy with customers about their current needs. In other words, problems that can be solved, manufacture, and sold quickly in the first time horizon.

Instead, plans for societal-level sustainability instead are grounded in long term horizons, such as 2050 or even 2100 (e.g., WBCSD 2010; IFTF 2020). Design for change requires aligning near-term design actions to long-term visions. Our current efforts are focused on developing pedagogies to teach these new skills.

Next, we describe how specific methods from the field of futures studies can help designers bridge that gap in thinking about long time horizons and aligning the present with possible futures.

Scupelli, Wasserman, Brooks, et al. have published extensively about the Dexign Futures courses taught with the flipped classroom pedagogy at the School of Design School at Carnegie Mellon University. In this paper, I describe one specific method in detail and describe how it was used in three design courses offered to both undergraduates and graduate students (Dexign Futures). First, we describe the Dexign Futures case study, followed by a description of the futures studies method Causal Layer Analysis.

2 Case Study: Dexign Futures

A Dexign Futures class was introduced in the design studies spine as part of a new undergraduate curriculum launched in 2014. The futures studies related addition was spurred by a recognition that emerging designers will, in their lifetimes, confront pervasive challenges such as sustainability and climate change.

The futures class at the School of Design at Carnegie Mellon University, required for all third-year design students, was introduced to help prepare them for these grand challenges. Design educators addressing such larger-scale concerns confront an inherent tension between covering traditional artifact-centered approaches and the systems perspectives addressing societal level concerns (Irwin et al. 2015; Kossoff 2011; Scupelli et al. 2016a, b, 2017, 2018a, b). Case studies describe ways to bridge such tensions (Scupelli 2019).

Sustainability is often framed in terms of long-range challenges unfolding over periods of a generation or more, for example looking to a specified multi-decadal time horizon like the year 2050 (WBCSD 2009). Dexign Futures explicitly focused on aligning near-term design action with sustainable futures. The "X" in Dexign was originated by Wasserman (2013) to signify an experimental form of design and design education combining design thinking with futures thinking to align near term design action with long-range vision goals – while navigating uncertainty and accelerating innovation toward desired futures (Wasserman et al. 2015; Scupelli et al. 2016a, b).

Design studies courses at the School of Design at Carnegie Mellon University focus on design research methods, explorations into design culture, and new topics (e.g., systems, placing, cultures, futures, persuasion). The design studies courses are required of all undergraduate design students in one class (approximately 40–50 students) in 80-min classes twice a week. Design studies classes are taught as lecture classes with hands-on activities to apply key concepts. Instead, required studio courses are taught in three different tracks (i.e., Products, Communications, Environments) with approximately 16–20 students with 3-h sessions twice a week.

The Dexign Futures course was a required 15-week course for all third-year design students in 2016–2017. The course covered different approaches to constructing and critiquing futures. There were four modules: Futures Narratives and People, Critiquing Alternative Futures Scenarios, Critiquing Normative Futures Scenarios, and Making Experiential Futures. In 2018, a different version of Futures was taught (Scupelli et al. 2019a, b). In 2019, for design students, the Dexign Futures course was redesigned to be a 7-week course called Futures 1. It had three modules: Futures Narratives, Alternative Futures Over Time, and Linking Pasts, Presents, and Futures. Futures 1 was followed by a 7-week course called Futures 2 focused on experiential futures. An elective course version Dexign Futures was taught to both undergraduate and graduate students as a 7-week course.

In 2016, 2017, and 2019 the Dexign Futures courses were taught as flipped classroom and had two main parts: (a) online pre-exposure to concepts that help students prepare for (b) in-class hands-on application activities. The class meets twice a week for 80-min sessions. The flipped classroom pedagogy as an alternative to the traditional lecture approach (Scupelli and Brooks 2018).

"Flipped" courses shift new-content exposure to pre-class work and use class time for hands-on application activities (Bergmann and Sams 2012). Pre-class work included online readings, videos, and interactive questionnaires providing immediate feedback; as well as a mechanism for students to submit questions to the instructor ahead of each session. Weekly reflections asked students to explain how they might integrate futures methods into design practice.

The online platform, Open Learning Initiative (OLI), included an information dashboard highlighting the top five questions that students had answered incorrectly in the pre-work, so that the instructor could address student misconceptions. The discussion then paved the way to active engagement with hands-on individual and group activities, during which the instructor provided just-in-time guidance.

McCarthy (2016) lists six potential benefits and limitations to the flipped classroom model: (a) students can learn at their own pace, assuming access to online resources at home; (b) they are introduced to productive self-directed and independent learning, as well as the benefits of collaborative, group-oriented learning; although some may come unprepared to class, ignore class discussion, or rely on better-prepared team members; (c) teachers may have more insight into student performance and learning challenges, but interactive in-class materials take significantly more time to prepare; (d) students have 24/7 access to rich, customized, interactive course materials, assuming that instructors can muster the time, skill, and effort to create them; (e) class time can be used more effectively with peer interaction, provided that students are on task during class, and not doing other coursework or browsing the internet; (f) students enjoy the new technologies used in the flipped classroom and find it motivating, except for those who prefer familiar teaching methods (McCarthy 2016).

Scupelli and Brooks (2018) suggest three further potential benefits and limitations: (g) the flipped class format positions students to actively engage the course material, but active learning is more effortful than passive or distracted learning (e.g., "listening" to a lecture while surfing the web); (h) online homework can be automatically and easily graded (e.g., done/not done), but some students may click through exercises without watching the videos, engaging with the questions, or reasoning about feedback received; (i) the many scaffolded in-class activities provide opportunities to practice and apply new knowledge encountered during the pre-work, but the course instructor needs to provide timely feedback on in-class assignments to foster learning opportunities.

The Dexign Futures courses were based on the premise that students need a broad introduction to futures, the goal being to help them bring these methods of longer-term, pluralistic thinking into applied contexts. In the next section, I describe what Causal Layered Analysis is and how it was taught in the Dexign Futures courses.

2.1 What Is Causal Layered Analysis?

The saying "We don't see things as they are; we see them as we are" by novelist Anaïs Lin captures a challenge that many designers have when required to look critically beyond their own assumptions, lifestyles, worldviews, and myths. For example, many design students struggle to understand why their own carbon-intensive lifestyles seem impossible to change. It is difficult to see that which challenges our own lifestyles, assumptions, worldviews, and stories.

Design research methods such as *Human-Centered Design Thinking* are often unable to meaningfully address sustainability-related behavior change questions due to several limitations such as lacking broader multi-disciplinary perspectives (e.g., DiSalvo et al. 2010). To help students to begin to think about such complex questions in a structured and meaningful way students are introduced to Causal Layered Analysis (CLA).

Sohail Inayatullah defines CLA as a "theory of knowledge and a methodology for creating more-effective policies and strategies" composed of four levels: litany, social systems, worldview, and myth/metaphor (1998, 2004, 2015).

- **Litany.** The first level, is the most obvious and visible aspect of an issue the quantitative data and trends, problems, often exaggerated by the news media.
- **Social Causes.** The second level involves a systems perspective. Quantitative data is interpreted by experts as: social causes, including economic, cultural, political and historical factors.
- **Discourse/Worldview.** The third deeper level of analysis regards the structure and the supporting discourse/worldview that provides cultural legitimacy. Different lenses frame how people understand an issue (e.g., economic, religious, cultural).
- **Myth/metaphor.** The fourth layer of analysis involves metaphor or myth. It provides an emotional level understanding to the worldview under inquiry. Often expressed as "deep stories, collective archetypes, or unconscious dimensions of the problem or the paradox" (Inayatullah 1998).

CLA draws from different epistemologies drawing from poststructuralism, macrohistory, and postcolonial multicultural theory (Inayatullah 2004). The power of CLA lies in the ability to make explicit such different perspectives in a coherent manner. The method is most frequently used to analyze current situations and to explore possible futures scenarios. It can be used to generate and explore alternative scenarios as well.

2.2 Challenges Teaching Causal Layered Analysis to Design Students

Human-Centered Design Thinking is implicated in our current unsustainability (e.g., Fry 1999). Among a new wave of design educators, there is a shift from the product/service focused *human-centered design thinking* in the present time frame to a new paradigm based on two ideas: (a) designing in time (b) for specific values (Fry 2008). Designing in time involves at a minimum, aligning short-term design action with long-term sustainable development goals. Designing for specific values such as sustainability (or sustainment) entails not taking away our collective future (*defuturing*). For example, as research centered in sustainability makes clear, it is a long term problem that requires a values-based design approach. Designers committed to a sustainability values-centered design approach take *futuring* actions that give us a collective future (e.g., Fry 2008).

In general, design students typically struggle to learn to design in time from *futures thinking* methods for three reasons: (a) shifting paradigms and worldviews from short-term design action to include long-term design thinking and action, (b) learning a new epistemology about time, and (c) learning to transfer futures thinking to core design studio courses and design practice (Scupelli et al. 2019a, b).

As mentioned previously, most of *human-centered design thinking* operates at the litany level (i.e., empathize, define, ideate, prototype, and test). Empathizing often takes place with designers observing and interacting with customers to notice needs and challenges. Design students become proficient in collecting data about such everyday difficulties at the expense of the planet. Likewise, defining, ideating, prototyping, and testing product/services opportunities often focus on immediate needs time-frames.

Students struggled to apply to design some of the ideas from design studies courses added to the curriculum (e.g., Systems, Cultures, Futures, Persuasion) described previously. Next, I mention some student challenges with transferring core ideas from such courses to the futures course applied to design practice. Through Systems, students learn to map systemic aspects around design but often struggle to use such systems knowledge to drive design action. In the Cultures course, students learn key ideas related to social causes/worldview but struggle to apply worldviews such as feminism and gender theory to specific design of products and service opportunities. In the Persuasion course, students learn of the powers of persuasion by design but lack the agility to think more broadly about how to shape designed products and services through the stories, myths, and metaphors that people, organizations, communities, and countries may live by.

As described above, the design studies courses provide students with key theoretical and practical building blocks to engage with the four levels of CLA. Yet, CLA. It was difficult for students to link concrete situations to the abstract ideas on the social causes/systems level, worldview level, and myth/metaphors level.

Furthermore, CLA requires students to think about different points in time. Unfortunately, design history is not taught to students in the new curriculum. Instructors are expected to embed some design history into the design studio classes they teach. Unfortunately, the study of history is more complex than mentioning relevant form exemplars from the past (e.g., what). Historical studies also involve explaining critical frameworks as to why certain ideas, products, and services emerged in different points in time (e.g., why, how, where), and so forth. Consequently, students lacking in design history, and how to think through the conditions that shaped design history, lack the historical perspectives necessary to explore the very social, technological, economic, environmental, and political forces (STEEP) that shaped the past, present, and futures of designed products/services. Lacking a deep understanding of how the past was shaped by STEEP forces, makes it difficult to notice STEEP forces shaping the present times. Being unable to notice how times changed due to STEEP forces makes it more difficult to notice how the zeitgeist of the times changed, and how worldviews and core narratives changed.

Many of the examples of CLA in the literature and videos describe government level or organizational level challenges with regards to strategic decisions. Students struggled to connect the CLA method to design type problems. The goal of applying CLA type thinking to a design problem was to provide three new views that designers often ignore. At times students seemed perplexed by what to make of the worldview and myth/metaphor levels.

3 Two Causal Layered Analysis Design Exercises

Next, I describe two assignments linked to CLA in the Dexign Futures classes: personal futures, and weekly reflection. First, I describe the exercises followed by student examples.

3.1 Personal Futures

To lessen the learning curve in the Dexign Futures class students explored possible alternative futures regarding their career choices. Design students know their own career choice considerations, compared to a public policy issue. Through a guided step-by-step exercise, students created a simple 2×2 matrix with critical career choices they faced when considering career choices[1].

For example, if a student was considering careers in medicine or design those might be two ends of one axis. The other axis might include professional practice or research. The four quadrants define four possible career paths: clinical physician, medical researcher, design researcher designer, and professional designer[2]. Students then pick one alternative career that they want to explore more in-depth through CLA. In the next section, we describe some examples from student work in the Dexign Futures courses.

Weekly Reflection on CLA. Each week students were asked to reflect on a specific question related to the futures thinking methods explored. The week on personal futures and CLA they responded to the following prompt:

> *You've explored alternative futures based on two critical uncertainties this week focused on career choice and other related uncertainty (e.g., research/practice, corporate/freelance, government/NGO). For one such future scenario, you conducted at Causal Layered Analysis to describe the four layers (i.e., litany, social system & structure, worldview, and myths & metaphors). CLA is a structured analysis tool. What did you learn from conducting such an exercise? On what kind of projects might you use CLA type analysis as a designer?*
>
> *Please post one comment (150-word max) by midnight Sunday.*

In the next section, I provide examples of student work to illustrate the two exercises described previously: personal futures, and student reflections on CLA.

---

[1] All course materials are available online at https://dexignfutures.org.

[2] Another example might be with the first axis as: art - design and the second axis as individual practice commercial practice. The resulting four career choices might be: artist, small studio design owner, corporate designer, corporate art director.

3.2 Personal Futures Assignment

The students were given the opportunity to explore a personal future related to their career choices. The assignment asked students to explore in a structured way career decisions that they were considering. In the personal futures assignment, students explored different carrer options they were considering and how each possible career might differ in substantial ways. Students began by imagining four alternative careers according to two dimensions.

Students explored on one axis different carrers such as design vs. other career (Fig. 1). On the second axis they explored different types of jobs, such as, corporate vs. freelance or professional practice vs. research.

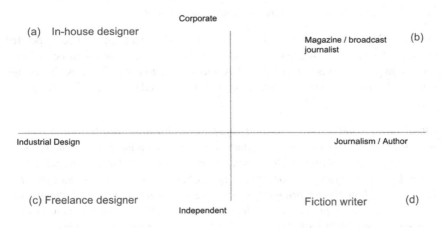

Fig. 1. One undergraduate student's mapping of four alternative career scenarios differing based on two dimensions: type of employment (e.g., corporate or independent) and type of job.

Once the students identified four carreer scenarios, they picked one they wanted to explore in more depth through the lens of CLA. They were asked to describe the four layers of litany, social systems, worldview, and myths and metaphors. For example one student picked five images to illustrate of the litany level work for an Interaction Designer living in San Francisco: one image was of the Golden Gate bridge, two images showed interaction designers working in a project room with post-it notes and whiteboard sketches, two more images showed wireframe sketches on a whiteboard and on a computer screen. For the other three levels of CLA analysis students mostly wrote text based descriptions.

3.3 Student Weekly Reflections on CLA

At the end of the week students wrote 100–150 word reflections on the CLA analysis of their personal futures exercise. Below I provide examples from themes that emerge from three undergraduate and graduate students followed by a brief discussion by theme.

Opening up the Design Space. One graduate student described how CLA could inform the design of user experiences from the digital touchpoints to include the other layers of social systems, worldviews, and myths/metaphors.

CLA may be a good method to use when creating a new product. As a UX designer, I tend to focus more on the details of the interactions and often forget to view the impact of my designs from a higher level. By using CLA, I can take a step back and learn how the design of a new product impacts society at different levels (litany, social system & structure, world view and myths & metaphors). This is especially relevant in today's world, because technology is more intertwined in our daily lives, a new product can have a major impact on users. I believe it is our responsibility, as designers, to create products that positively impact users. Furthermore, we should be designing products for a future we want to be a part of.

Below, a design undergraduate student reflects on how her design process operated at the litany level, and as she learned about other layers how those might be included in her design process. It is evident that CLA allowed her to critically assess her design process and lay out an area for change. In short, CLA expanded opportunities for design thinking.

I think I tend to get bogged down and only focus on the lower levels (litany) when it comes to my design process. As a student it is easy to do this because the scope of projects we do and their scale are often much smaller than the 'real world' and it can be easier to just focus on what is most obvious. Since my freshman year I have noticed how much more my eyes have been opened to considering the other layers. In my future projects I can definitely see myself using this framework to help ensure considerations of all the layers. This is especially important in projects that have a really large audience and the people it will reach are outside of my own worldview and direct social systems. If I can take the time to really consider all four layers then the outcome will hopefully be more successful.

The first graduate student and the undergraduate describe how CLA allowed them to focus on the big picture in addition to the details of a design or interface. It is interesting how both students comment on how CLA allowed them to notice how the details of a design have larger implications. Those larger views are typically missing from *Human-Centered Design Thinking*.

Noticing Relationships in the Whole Picture. A second graduate student described what he perceived to be the relationships between the layers of CLA as a series of nested "Why" questions where each layer goes deeper. An open question remains about how to shift into a different set of nested logics.

CLA, while feeling a little too obtuse to use as an explanation to others, seems to be at least an interesting way to structure more general critical thinking. Each layer essentially acts as a "why" to the previous: litany is sort of daily life and what things you do, the next layer is why you do those things–what social systems and structures influenced your decisions and methods? The third layer asks why those systems and structures exist, what is your worldview that causes you to interpret them that way? Finally, myths & metaphors asks why you have the worldview that you do? What cultural influences and 'legends' do you use to justify that view?

Similar to the comment above, a design undergraduate student speculated on how CLA opened up a series of opportunities to look at the deeper aspects of a perceived design problem by moving past the litany and exploring deeper causes. She connects the broader context of CLA to a wider range of possible solutions.

Learning about CLA taught me about how I could approach design problem from the inside out. I think that a lot of the time we will be presented with a problem and look for solutions to that problem without really looking at the deeper causes. For example, let's say the problem is there's a lot of trash on the street. A "solution" might be get more trash cans or punish people who litter; however, that is just looking at the litany. What is the deeper cause? What causes this excess of trash? Why do we create trash at all? Looking at these other layers helps you approach a problem from a different touch point creating better, or at least different, solutions. This would be very helpful when looking at wicked problems.

Both of these examples illustrate how the CLA exercise allowed students to notice the logic that connects each layer of CLA, and how innovative design solutions can emerge by looking at the broader context in a structured manner.

Shaping and Questioning Reality. A third graduate student describes how CLA can inform human-centered design processes by using the higher layers to reinforce behavior in the litany. She questioned if the litany layer can challenge the myths and metaphors?

From a practical standpoint, CLA is useful for the human-centered design process because it triangulates human motivations and how we've constructed our litany. If the designer wishes to increase product-market fit, social structure, worldview, and myth are all useful factors to consider. However, I am curious how the designer might create a product or experience which successfully challenges one of these layers. For instance, what are the conditions that led Uber/Lyft to succeed? Was it that users were so unwilling to wait for a taxi (myth: time is money, 21st century services should be on demand) that they were willing to trust strangers? When does challenging these layers reveal opportunity vs. go against deeply-held beliefs?

Personal Insights from CLA. An undergraduate student gained insights from her personal reflection through CLA about how the science and technology based worldview embedded in her tier one research university helped to shape the carreer choices of students after graduation.

Causal Layered Analysis helped me to reflect on myself, and look at the path of my own future through the lens of another perspective. Most people here at CMU (Carnegie Mellon University) plan go into the real world after graduation doing jobs related to their current studies, mostly somehow tech-related. This is what people think of immediately at a litany level. When we think of it from a cause and worldview level (from a deeper, bigger picture), we can understand that it is because of the influence of the information age that we are living under that's shaping this trend, especially when CMU is a very STEM focused school. CLA also reminded me of how as a designer, it is important for us to not only think in one direction for the most obvious solution, but also consider views on other macro and micro levels for the best result for better understanding of the problem and to build empathy through this understanding.

4 Discussion

The previously mentioned student projects and reflections can be interpreted to illustrate that students are using CLA to notice relevant aspects of their design process, school environment, and personal lives that previously went unnoticed or spent little time thinking about.

Some students are framing the problems in new ways and noticing new design opportunities. The four layers of analysis likely helped students to move past some of the obvious aspects of the litany to ask deeper questions.

One student described the process as following a series of nested "why" questions linking across layers from behavior to a social/systems perspective, a worldview, and myths/metaphor. It is quite uncommon for design students to be able to articulate such a holistic explanations unprompted. This is a testament to the power of the scaffolding of CLA as a structuring device for designerly thinking.

4.1 Personal Futures

The goal of the personal futures brief exercise was to allow students to practice mapping the four levels of CLA to a personally meaningful future related to a possible career choice. The exercise presented challenges for the students. For example, one graduate student's exploration of the worldview layer described what she imagined to be a tension between the perceived worldview of the interaction designers job in industry (technology as commodity) and the student's personal values (desire to make a positive impact on society). The tension in worldview associated with the workplace and personal desires created an interesting dilemma for the student.

Another student decided to explore her career options in Denmark. She realized that she did not know what a Danish design firm worldview and myths and metaphors might be. The CLA exercise became a way for her to realize what her known-unknowns were about a specific career path. Articulating her knowledge gaps was valuable because it made it possible for her to overcome such lacunae.

Some students' explorations of the social systems, worldviews, and myths/metaphors levels read as hurried, superficial explorations done to turn in an assignment. One might too quickly negatively judge such rushed work. However, if one considers that the goal of this assignment was to allow a first attempt to apply an abstract framework close to student's own experiences, the exercise succeeded.

Obviously, to develop a deep understanding and profound insights more practice is needed. That is why students were given repeated opportunities to engage with CLA in the

Dexign Futures course through the flipped classroom pedagogy (e.g., online exercises, in class discussion, hands on exercise started in class and finished for homework, followed by an end of week reflection, and in class discussion). The student reflections discussed in the next section, demonstrate that even repeated feeble attempts at CLA yielded deep insights cumulatively.

4.2 CLA Reflections

Next, I describe themes that emerge from students' written reflections: opening up the design space, noticing relationships holistically, shaping and questioning reality, and personal insights.

Opening up the Design Space. Students reflected on how CLA opened up the design space past the superficial litany to include broader views and product opportunities by the social systems, worldview, and myths/metaphors levels.

Noticing Relationships Holistically. Students describe the relationships between the four levels of CLA and how the framing of a design opportunity is linked to the context within which a problem is considered. These comments illustrate how CLA allows designers to move past problems located in the litany to consider a broader context described in other layers of CLA.

Shaping and Questioning Reality. A graduate student explained how CLA informed human-centered design thinking by using the other layers to reinforce the litany layer. She noticed how customer and company fit can be improved by manipulating the levels of CLA appropriately. She wondered how products and services might question and challenge current prevailing worldviews, myths/metaphors? Sohail Inayatullah, the inventor of CLA, suggests that the behavior in the litany is unlikely to shift social systems, worldview, and myths/metaphors within an organization due to the misalignment between values/rewards and observed behavior. Within an organization, it is clear why the established cultural values drive rewards and bon uses liked to behavior and not vice versa.

The example of Lyft/Uber that the student raises can be explained as the Uber/Lyft product/service offering operates within a specific behavior, social system, worldview, and myth/metaphor. People can opt into the service offering of on-demand rides but to do so, they need to also opt into the specific social system (e.g., create an Uber/Lyft account, accept the sharing economy, on demand technological, GPS location tracking, surveillance capitalism worldview, and the associated myths/metaphors). Opting into the Uber/Lyft worldview might be at odds with other worldviews and myths/metaphors based on company taxis, government licenses, and so forth.

The challenge for the designers is to create a product/service/interface that allows people to opt into such a multi-layered reality; when the traditional transportation and taxi cabs does not require app accounts, worldview shifts, and myth/metaphor changes. In short, the design problem for companies like Uber/Lyft is to shift people from one layered version of reality to a new one. The same is true for companies like AirBnB, and so forth. The salient point being made is that: *Human-Centered Design Thinking* that focuses merely on the litany level, lacks the theoretical basis that CLA offers to account for the existance and importance of shifting people's behavior and underlying layered reality.

One might argue that while our parents told us "not to accept rides from strangers" the affordances designed into the Uber/Lyft interface and underlying systems allow riders to trust strangers (and drivers to trust unknown riders). Riders likely believe they can trust drivers because drivers are vetted by the Uber/Lyft companies. The drivers and riders are held accountable for their behavior through online rating, review, and reputation system. Furthermore, the Uber/Lyft companies have mechanisms to resolve driver and customer complaints. These design details help people to trust strangers. But to trust strangers on the Uber/Lyft platform, people need to tell themselves different stories at the myth/metaphor level, adopt a worldview, and opt into a social system. In other words, the different levels of CLA can inform in a strategic way to design the problem at hand.

Personal Insights. CLA allowed an undergraduate student to better understand how the technology focus of Carnegie Mellon University shaped students' ideas about careers worth pursuing through a STEM related worldview. The student also noticed that she should consider both the micro and macro levels.

5 Summary

The world is changing at an accelerating rate. Design educators are struggling to keep up with such change. Tools and methodologies like *Human Centered Design Thinking* are showing their limitations when used for problems from the Anthropocene Era like human induced Climate Change and Sustainable Development. In this paper, I've suggested that *Human-Centered Design Thinking* alone is insufficient to address problems like Climate Change and Sustainable Development. Other methods that allow to design in time and designing for specific values are needed. The field of Futures Studies and Futures Thinking methods can help designers overcome such challenges.

I described work done in a class called Dexign Futures that teaches students to combine Design Thinking with Futures Thinking to create Dexign Futures (or Dexign Thinking). I provide examples of how designers can benefit from using a Futures Studies method called Causal Layered Analysis (CLA) developed by Futurist Sohail Inayatullah. The CLA method claims that reality has four layers (i.e., litany, social system, worldview, myth/metaphor). CLA allows designers to move past the superficial level of *Human Centered Design Thinking* that operates at the litany level to include a broader context (e.g., systems, worldview, and myths/metaphors). I provided two design exercises that both undegraduate and graduate students did to illustrate how CLA expanded on their design abilities. Future work is needed to further illustrate how futures thinking can expand on design thinking.

References

Architecture for Humanity: Design Like You Give a Damn: Architectural Responses to Humanitarian Crises. Metropolis Books, New York (2006)

Baghai, M., Coley, S., White, D.: The Alchemy of Growth: Practical Insights for Building the Enduring Enterprise. Perseus Publishing, Cambridge (2000)

Berman, D.B.: Do Good: How Designers Can Change the World. AIGA, Berkeley (2009)

Bergmann, J., Sams, A.: Flip Your Classroom: Reach Every Student in Every Class Every Day. International Society for Technology in Education, Eugene (2012)

Brown, T., Kātz, B.: Change by Design: How Design Thinking Transforms Organizations and Inspires Innovation. Harper Business, New York (2009)

Buchanan, R.: Wicked problems in design thinking. Des. Issues **8**(2), 5–21 (1992)

Claudio, L.: Waste couture: environmental impact of the clothing industry (2007)

Davis, M.: Why do we need doctoral study in design? Int. J. Des. **2**(3) (2008)

DiSalvo, C., Sengers, P., Brynjarsdóttir, H.: Mapping the landscape of sustainable HCI. In: Proceedings of the SIGCHI Conference on Human Factors in Computing Systems, pp. 1975–1984. ACM, April 2010

Doblin, J.: A short, grandiose theory of design. STA Des. J. Anal. Intuit. **6**, 16 (1987)

Extinction Rebellion: This Is Not a Drill: The Extinction Rebellion Handbook. Penguin Books (2019)

Fry, T.: Design Futuring: Sustainability, Ethics and New Practice. Berg, Oxford (2008)

Fry, T.: A New Design Philosophy: an Introduction to Defuturing. UNSW Press, Sydney (1999)

Fuller, R.B.: Operating Manual for Spaceship Earth. Southern Illinois University Press, Carbondale (1969)

Inayatullah, S.: Causal layered analysis: poststructuralism as method. Futures **30**(8), 815–829 (1998)

Inayatullah, S.: The Causal Layered Analysis (CLA) Reader: Theory and Case Studies of an Integrative and Transformative Methodology. Tamkang University Press, Tamsui, Taipei (2004)

Inayatullah, S.: Causal layered analysis: an integrative and transformative theory and method. Futures Research Methodology, Version, 3 (2009)

Inayatullah, S., Milojevic, I.: CLA 2.0: transformative research in theory and practice (2015)

Institute for the Future. http://www.iftf.org/our-work/global-landscape/sustainability/. Accessed 20 Jan 2020

Irwin, T., Kossoff, G., Tonkinwise, C., Scupelli, P.: Transition Design. Carnegie Mellon School of Design, Pittsburgh (2015)

Jones, J.C.: Design Methods. Van Nostrand Reinhold, New York (1992)

Kossoff, G.: Holism and the reconstitution of everyday life: a framework for transition to a sustainable society. Doctoral dissertation, University of Dundee, Centre for the Study of Natural Design, Dundee, Scotland (2011)

Luebkman, C.: ICSID World Design Council Singapore 2009: Design2050 Studios. Commissioned by DesignSingapore Council a department within the Ministry of Information, Communications and the Arts, Singapore (2009)

Mau, B.: Massive Change: Bruce Mau and the Institute Without Boundaries. Phaidon, London (2005)

McCarthy, J.: Reflections on a flipped classroom in first year higher education. Issues Educ. Res. **26**(2), 332–350 (2016). http://www.iier.org.au/iier26/mccarthy-j.html

Norman, D.: DesignX: A Future Path for Design, 4 December 2014. https://www.linkedin.com/pulse/20141204175515-12181762-designx-a-future-path-for-design. Accessed 25 Jan 2020

O'Reilly, T.: What Is Web 2.0, 30 Sept 2005. https://www.oreilly.com/pub/a/web2/archive/what-is-web-20.html?page=4. Accessed 31 Jan 2020

Papanek, V.: Design for the Real World: Human Ecology and Social Change. Thames and Hudson, London (1972)

Pastor, E.: The OTHER Design Thinking, 7 November 2013. https://issuu.com/humantific/docs/theotherdesignthinking/1. Accessed 25 Jan 2020

Ries, E.: The Lean Startup: How Constant Innovation Creates Radically Successful Businesses. Portfolio Penguin, London (2011)

Scupelli, P.: Teaching to transition design: a case study on design agility, design ethos, and dexign futures. Cuadernos del Centro de Estudios de Diseño y Comunicación, no. 73, pp. 111–132 (2019)

Scupelli, P.: Designed transitions and what kind of design is transition design? Des. Philos. Pap. **13**(1), 75–84 (2016)

Scupelli, P., Fu, Z., Zheng, Y., Brooks, J.: Teaching to dexign futures in China: a vision for a blended learning pedagogy to be deployed at scale. In: 9th International Conference the Future of Education, Florence, Italy (2019a)

Scupelli, P., Candy, S., Brooks, J.: Teaching futures: trade-offs between flipped classroom and design studio course pedagogies. In: IASDR 2019: Design Revolutions, 2–5 September 2019, Manchester, UK (2019b)

Scupelli, P., Brooks, J.: What features of a flipped course improve design student learning experiences? In: 21st DMI: Academic Design Management Conference, Next Wave, 1–2 August 2018, London, UK (2018)

Scupelli, P., Wasserman, A., Wells-Papanek, D., Brooks, J.: The futures of design pedagogy, learning, and education. In: 21st DMI: Academic Design Management Conference, Next Wave, 1–2 August, 2018, London, UK (2018)

Scupelli, P., Wells-Papanek, D., Wasserman, A., Brooks, J.: Opening a design education pipeline from university to K-12 and back. In: IASDR 2017 Re: Research, 31 October–3 November 2017, Cincinnati (2017)

Scupelli, P., Brooks, J., Wasserman, A.: Making dexign futures learning happen: a case study for a flipped, open-learning initiative course. In: Design Educators IDSA International Conference 2016: Making Things Happen, 17–20 August, Detroit, MI, USA (2016a)

Scupelli, P., Wasserman, A., Brooks, J.: Dexign futures: a pedagogy for long-horizon design scenarios. In: Proceedings of DRS 2016, Design Research Society 50th Anniversary Conference, 27–30 June 2016, Brighton, UK (2016b)

Shaftel, H. (ed.): Climate change: how do we know? NASA Global Climate Change. Earth Science Communications Team at NASA's Jet Propulsion Laboratory, 26 December 2019

Wasserman, A.: Thinking about 50 years of design thinking (2011). https://vimeo.com/60342260. Accessed 20 Jan 2020

Wasserman, A., Scupelli, P.: Dexign the Future: Human Centered Innovation for Exponential Times (2013). https://dexignthefuture.com/. Accessed 30 Jan 2020

Wasserman, A., Scupelli, P., Brooks, J.: Learn! 2050 and design futures: lessons learned teaching design futures. In: Design Educators IDSA International Conference 2015: Future of the Future, 19–22 August, Seattle, WA (2015)

United Nations, IPCC: Global Warming of 1.5 °C (2018). https://www.ipcc.ch/sr15/

WBCSD - World Business Council for Sustainable Development (2009). https://www.wbcsd.org/Overview/About-us/Vision2050/Resources/Vision-2050-The-new-agenda-for-business. Accessed 18 Jan 2020

Webster, K.: The Circular Economy—A Wealth of Flows. Ellen MacArthur Foundation (2015)

Impacts of Employee Turnover on the Integrated Inventory Model When Overtime Occurs

Xuefang Sun and Renqian Zhang[(✉)]

School of Economics and Management, Beihang University, Beijing 100191, China
zhangrenqian@buaa.edu.cn

Abstract. In this paper, we first consider a single-vendor single-buyer integrated inventory model. The vendor's daily production rate is assumed to be less than the daily demand. To meet the demand, the vendor adopts overtime strategy to increase the production capacity. However, when overtime occurs, the number of departing employees for the vendor is not constant but varies with overtime. We obtain the optimal ordering quantity, overtime working hours, the number of shipments and overtime days by minimizing the daily total cost of the model. Then, we extend the integrated model by considering that a vendor supplies multiple buyers. In this model, we assume that the vendor's production cycle equals to the sum of production cycles for all buyers. Based on mathematical analysis of the integrated system, a lipschitz iterative method is proposed to obtain the optimal ordering and overtime strategy. Numerical example and sensitivity analysis about this case have been drawn to present some insights into parameter effect on the optimal ordering and overtime strategy.

Keywords: Integrated inventory model · Employee turnover · Overtime · Transportation cost

1 Introduction

The integrated inventory management is an important way to improve the system performance and to gain economic advantages in the competitive global markets. In recent decades, most integrated inventory systems have focus on the integration between vendor and buyer. The basic integrated inventory model was proposed by Goyal [1]. Subsequently, the extension of the integrated model are studied to show how the integrated inventory management evolved in different directions.

One research extension of the integrated inventory problem is considering insufficient production capacity. For the production capacity problem, readers are referred to Hariga [2], Sinha and Sarmah [3] and Banerjee [4]. Among the extensive literature in this area, overtime, as an effective measure, has become one of enterprises' preferred options to temporarily expand production capacity. Bahll and Ritzman [5] took overtime strategy into lot-sizing model when the manufacturer has production capacity constraints. Hariga [2] proposed an production-ordering model in which a combination of subcontracting and overtime is considered when the production capacity is insufficient to satisfy the

© Springer Nature Switzerland AG 2020
P.-L. P. Rau (Ed.): HCII 2020, LNCS 12192, pp. 145–159, 2020.
https://doi.org/10.1007/978-3-030-49788-0_11

annual demand. In the model, it proves that the mixed policy of Hariga [2] outperforms the policy of Goyal and Gopalakrishnan [5]. Recently, Zhang and Sun [6] introduced overtime and maintenance time into the integrated model when the vendor's production rate is less than demand rate. In their model, maintenance time is assumed to be a polynomial function of the vendor's production cycle. The optimal production-delivery policy is presented by minimizing the joint cost of their model using proposed solution procedure. Then, Sun and Zhang [7] extended the model of Zhang and Sun [6] by relaxing the assumption of deterministic demand and considered that the lead time demand is stochastic and follows a normal distribution.

In most literature about overtime problem, enterprises' employee turnover is ignored or it assumes that there is no employee turnover is very common phenomenon. Meanwhile, modern enterprise management theory suggests that employee turnover makes technology exchange and cultural exchange between enterprises become possible. With the rapid growing of turnover phenomenon, employee turnover problem is increasingly concerned by managers and researchers. For this problem, researchers analyzed reasons and influence factors of employee turnover in all aspects.

Porter and Steer [8] proposed met expectation theory and analyzed the relationship between turnover and various factors in work situation. The met expectation theory in their model asserts that job satisfaction depends on how closely a job fulfills employee's initial job expectation. Dalton et al. [9] challenged the assumption about dysfunctional turnover and introduced functional turnover into the analyses about the data on employees of Western bank branches. Krackhardt and Porter [10] investigated the relationship between turnover and communication networks and the separated analyses in their paper pointed out that turnover does not occur randomly throughout a work group but is concentrated in patterns delineated by role similarities in a communication network. While, Kacmar et al. [11], Call et al. [12] and Hausknecht et al. [13] explored mechanism between turnover rates and organizational performance and moderators of these effects. Recently, Chen et al. [14] and Liu et al. [15] estimated how job satisfaction trajectories predict turnover intention, respectively. Ramesh and Gelfand [16] found that person-job fit was more important to predict turnover in the United States than in India, and Hom and Xiao [17] examined the influence of guanxi networks on employee retention in China. Based on Penrose's growth theory, Pernille and Christina [18] studied the relationship between early employment growth and long-run survival and pointed out that employee turnover has an effect on joint experience in the firm, the development of the productive opportunity set and long-term utilization of early employment expansion. Other studies concerned about turnover can be referred to Shapiro et al. [19], Rothausen et al. [20], An [21].

In this paper, we attempt to introduce the combination of overtime and employee turnover into the integrated inventory model. The vendor adopts overtime strategy to increase daily output when limited production capacity existing. Meanwhile, as overtime increases, the number of turnover employees for the vendor is not constant but varies with overtime. We assume in this paper that the number of departing employees is a linear function of the overtime working hours. First, we introduce overtime and employee turnover into an integrated model with a vendor and a buyer. The objective of this model is to determine ordering quantity, overtime working hours, the number of shipments and overtime days, which minimizes the daily total cost of the integrated system. Then, we extend the integrated model by considering that the vendor supplies multiple buyers. In this extended model, the vendor's production cycle equals to the sum of production cycles for all buyers. To obtain the optimal ordering and overtime strategies, we analyze the mathematical properties of the daily cost in the integrated model and a lipschitz iterative method is proposed. Finally, numerical example is conducted to verify the effectiveness of the proposed solution procedure and sensitivity analysis is developed to present how the main parameters affect the optimal ordering and overtime policy.

The organization of this paper is as follows. Section 2 briefly describes the running process of the single-vendor single-buyer integrated model by considering overtime and employee turnover when the vendor's daily production rate is less than the daily demand. The assumptions are provided throughout this paper. Section 3 formulates the related constrained optimization model and Sect. 4 analyzes the propositions of the daily total cost function and proposes a solution procedure to solve the optimal strategy. The extended model with a vendor supplying multiple buyers is considered in Sect. 5. To obtain the optimal ordering and overtime strategy, we probe the mathematical properties of the objective function and a lipschitz iteration method is provided in Sect. 5.2. Numerical example and sensitivity analysis are developed to illustrate the effectiveness of the proposed iterative method in Sect. 5.3. Finally, we give conclusions and future research of the paper in Sect. 6.

2 Model Description

In this section, we consider an integrated inventory model with a vendor and a buyer. The buyer orders nQ units product from the vendor at each setup and ships Q units product to the buyer over n times. Therefore, the cycles of the buyer and the vendor respectively are Q/D and nQ/D, where D is the daily demand. To avoid excessive holding cost, the vendor's production quantity should be equal to the buyer's order quantity during each production cycle. In this integrated model, we assume that the daily production rate of the vendor is smaller than the daily demand D. In order to meet the large demand, the vendor increases the production capacity by overtime.

For the vendor, there are M work positions and each position only need one employee. The normal and overtime working hours of employees are t_{norm} and t_{over}, respectively. However, when overtime occurs, the number of departing employees Num increases as t_{over} grows. We assume that the number of departing employees has a linear relationship with the overtime working hours, i.e., $Num(t_{over}) = kt_{over} + C_{con}$, where k is a random number and C_{con} is a constant. What's more, the unit producing cost of the product increases from c to c_o when overtime occurs. Since each work position need one employee, a count of new employees will be recruited to fill the vacant positions. Therefore, the number of departing employees equals the number of new recruited employees. For convenience, we also denote Num as the number of new recruited employees.

For the vendor, the fixed cost is S when recruiting new employees. The new employees are trained after recruited and the training cost per employee is c_{tr}. Hence, the total cost of recruiting new employees is $S + c_{tr}Num$. For skilled employees, each one manufactures P_S units product per hour. While, each new employee produces P_N units product per hour. Obviously, $P_S > P_N$. Therefore, the daily total output of product with overtime, $\overline{P}$, is

$$\overline{P}(t_{over}) = P_S(t_{norm} + t_{over})M - (P_S - P_N)(t_{norm} + t_{over})(kt_{over} + C_{con}), \quad (1)$$

and the daily total output of product without overtime, $\underline{P}$, can be expressed as

$$\underline{P} = P_S t_{norm}M - (P_S - P_N)t_{norm}C_{con}. \quad (2)$$

Following the assumption of Zhang and Sun [6], the vendor reserves maintenance time Δt to maintain and repair equipment when finishing each lot. To ensure the validity of maintenance time for the vendor in integrated model, we assume that $\Delta t/nT \geq \alpha$, where α is a constant. The specific operating mode of this model is shown in Fig. 1. To develop the proposed integrated model, we have the following assumptions.

1. There is no initial inventory and shortages are not allowed.
2. The system operates for infinite time horizon.
3. The vendor's daily production rate under normal work is less than the daily demand, i.e., $\underline{P} < D$. In order to make overtime meaningful for the vendor, the daily production rate under overtime work should be not less than the daily demand, i.e., $\overline{P} \geq D$.

3 Mathematical Formulation

Based on description in Sect. 2, the buyer orders the product according to the economic order quantity (EOQ) model. The total cost for the buyer consists of setup cost, holding cost and transportation cost. For the buyer, the ordering cost per order is A_b and the buyer's daily ordering cost is written as $A_b D/Q$. The buyer's daily holding cost per unit is h_b and the buyer's daily holding cost can be expressed as $\frac{h_b Q}{2}$. The transportation cost from the vendor to the buyer is charged by the buyer. The buyer's transportation cost per

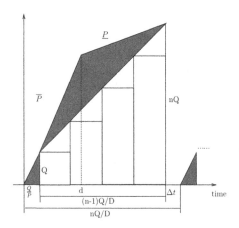

Fig. 1. The vendor's inventory level against time for $n = 5$.

cycle is $c_t Q$, where c_t is transportation cost per unit. Therefore, the buyer's daily total cost $TC_b(Q)$ is written as

$$TC_b(Q) = \frac{A_b D}{Q} + \frac{h_b Q}{2} + c_t D. \tag{3}$$

For the vendor, the setup cost per setup is A_v and the vendor's daily setup cost is $A_v D / nQ$. As mentioned above, the vendor adopts an overtime strategy to increase daily output when the daily production rate is less than the daily demand. The unit producing costs of the product under normal work day and overtime day are c and c_o, respectively. The vendor overtimes for d consecutive days during each production cycle. Hence, the vendor's total producing cost during each production cycle can be calculated as $c_o \overline{P} d + c\underline{P}[\frac{Q}{\underline{P}} + (n-1)\frac{Q}{D} - d]$. Accordingly, the vendor's daily producing cost is $\frac{(c_o \overline{P} - c\underline{P})dD}{nQ} + \frac{c\underline{P}[D/\overline{P} - (n-1)]}{n}$.

Based on the operating model shown in Fig. 1, the vendor's accumulated inventory quantity during each production cycle is calculated as $\frac{dQ(n-1)}{2}(\frac{\overline{P}}{D} - 1) + \frac{nQ^2}{2\overline{P}} - \frac{(n-1)Q^2}{2D}$ and the vendor's total holding cost in a lot is $h_v[\frac{dQ(n-1)}{2}(\frac{\overline{P}}{D} - 1) + \frac{nQ^2}{2\overline{P}} - \frac{(n-1)Q^2}{2D}]$, where h_v is the vendor's holding cost per unit product. Therefore, the vendor's daily holding cost is $h_v[\frac{(d\overline{P} - dD - Q)(n-1)}{2n} + \frac{QD}{2\overline{P}}]$.

Besides, according to the description in Sect. 2, the total cost of recruiting new employees during each production cycle is $S + c_{tr} Num$ and the daily cost of recruiting new employees is $\frac{(S + c_{tr} Num)D}{nQ}$. The vendor's total cost consists of setup cost, producing cost, holding cost and recruited cost. Consequently, the vendor's daily total cost $TC_v(Q, n, d, t_{over})$ is

$$TC_v(Q, n, d, t_{over}) = \frac{A_v D}{nQ} + \frac{(c_o \overline{P} - c\underline{P})dD}{nQ} + \frac{c\underline{P}[D/\overline{P} - (n-1)]}{n} + h_v[\frac{(d\overline{P} - dD - Q)(n-1)}{2n} + \frac{QD}{2\overline{P}}] + \frac{(S + c_{tr} Num)D}{nQ}, \tag{4}$$

and the daily total cost of the integrated system $TC(Q, n, d, t_{over})$ can be expressed as

$$TC(Q, n, d, t_{over}) = \frac{A_b D}{Q} + \frac{h_b Q}{2} + c_t D + \frac{A_v D}{nQ} + \frac{(c_o \overline{P} - c\underline{P})dD}{nQ} + \frac{c\underline{P}[D/\overline{P} - (n-1)]}{n}$$
$$+ h_v[\frac{(d\overline{P} - dD - Q)(n-1)}{2n} + \frac{QD}{2\overline{P}}] + \frac{(S + c_{tr} Num)D}{nQ}. \tag{5}$$

In order to make overtime strategy feasible in this model, we previously assumed that $\overline{P} \geq D$, i.e., $P_S(t_{norm} + t_{over})M - (P_S - P_N)(t_{norm} + t_{over})(kt_{over} + C_{con}) \geq D$. After some algebra, it yields

$$-k(P_S - P_N)t_{over}^2 + [P_S M - (P_S - P_N)(kt_{norm} + C_{con})]t_{over} - (D - \underline{P}) \geq 0. \tag{6}$$

Besides, working long hours everyday not only effects employees' health, but also reduces the quality of their work. In order to ensure stability of production line for the vendor, the number of departing employees Num should be not larger than the vendor's workshop capacity. Therefore, the length of overtime should be too long. We assume that $t_{over} \leq \tilde{t}_{over}$, where $\tilde{t}_{over}$ is a constant.

Based on the above argument, we can write down the mathematical formulation of this integrated model as follows.

$$\text{Minimize} \quad TC(Q, n, d, t_{over}) \tag{7}$$

$$\text{Subject to:} \; -k(P_S - P_N)t_{over}^2 + [P_S M - (P_S - P_N)(kt_{norm} + C_{con})]t_{over} - (D - \underline{P}) \geq 0, \tag{8}$$

$$t_{over} \leq \tilde{t}_{over}. \tag{9}$$

4 Theoretical Solution

In order to avoid excessive holding cost, the vendor's production quantity should equal to the buyer's order quantity during each production cycle, i.e., $\overline{P}d + \underline{P}[\frac{Q}{\overline{P}} + (n-1)\frac{Q}{D} - d] = nQ$. And the buyer's order quantity can be expressed as

$$Q = (\overline{P} - \underline{P})d / [n - \frac{P}{\overline{P}} - \frac{(n-1)\underline{P}}{D}]. \tag{10}$$

As shown in Eq. (10), the vendor's order quantity Q is decided by overtime t_{over}, overtime days d and the number of shipments n. Meanwhile, since the order quantity Q is a non-negative number, it deduces that $n - \frac{P}{\overline{P}} - \frac{(n-1)\underline{P}}{D} > 0$, i.e., $\overline{P} > \underline{P}/[n - \frac{(n-1)\underline{P}}{D}]$. Clearly, this inequality holds when following the assumption $\overline{P} \geq D$. The daily total cost of the model be can rewritten as

$$TC(n, d, t_{over}) = (A_b + \frac{A_v + S + c_{tr}Num}{n}) \frac{D}{(\overline{P} - \underline{P})d}[n - \frac{P}{\overline{P}} - \frac{(n-1)\underline{P}}{D}] + c_t D + \frac{c\underline{P}[D/\overline{P} - (n-1)]}{n}$$
$$+ \frac{(c_o\overline{P} - c\underline{P})D}{n(\overline{P} - \underline{P})}[n - \frac{P}{\overline{P}} - \frac{(n-1)\underline{P}}{D}] + \frac{(h_b + \frac{D}{\overline{P}} - \frac{n-1}{n})(\overline{P} - \underline{P})d}{2[n - \frac{P}{\overline{P}} - \frac{(n-1)\underline{P}}{D}]} + h_v \frac{(\overline{P} - D)(n-1)d}{2n}. \tag{11}$$

As described in Zhang and Sun [15], the vendor reserves maintenance time Δt to maintain and repair equipment when finishing each lot, and $\Delta t / nT \geq \alpha$. It can be deduced that $n \leq \frac{1}{\alpha}(1 - \frac{D}{\overline{P}})$. Combined Eq. (1) with Constraints (8) and (9), the lower and upper bound of overtime are obtained, and they, for convenience, are denoted $\underline{t_{over}}$ and $\overline{t_{over}}$ respectively. Hence, $n \leq \frac{1}{\alpha}(1 - \frac{D}{\max\{\overline{P}(t_{over})\}})$. Since the number of shipments during each production cycle n is a positive integer, we derive that the lower and upper bounds of n are 1 and $n \leq \lfloor \frac{1}{\alpha}(1 - \frac{D}{\max\{\overline{P}(t_{over})\}}) \rfloor$, respectively.

For given n and t_{over}, we obtain the first and second derivatives of $TC(n, d, t_{over})$ with respect to d as follow.

$$\frac{\partial TC(n, d, t_{over})}{\partial d} = -(A_b + \frac{A_v + S + c_{tr}Num}{n}) \frac{D[n - \frac{P}{\overline{P}} - \frac{(n-1)\underline{P}}{D}]}{(\overline{P} - \underline{P})d^2} + (\frac{h_b}{2} + \frac{D}{2\overline{P}} - \frac{n-1}{2n}) \frac{(\overline{P} - \underline{P})}{[n - \frac{P}{\overline{P}} - \frac{(n-1)\underline{P}}{D}]} + h_v \frac{(\overline{P} - D)(n-1)}{2n}. \tag{12}$$

$$\frac{\partial^2 TC(n, d, t_{over})}{\partial d^2} = 2(A_b + \frac{A_v + S + c_{tr}Num}{n}) \frac{D[n - \frac{P}{\overline{P}} - \frac{(n-1)\underline{P}}{D}]}{(\overline{P} - \underline{P})d^3}. \tag{13}$$

Since $\frac{\partial^2 TC(n,d,t_{over})}{\partial d^2} > 0$ for given n and t_{over}, $TC(n,d,t_{over})$ is a convex function in d. Setting $\frac{\partial TC(n,d,t_{over})}{\partial d} = 0$, we obtain

$$d = \sqrt{\frac{(A_b + \frac{A_v + S + c_{tr}Num}{n})D[n - \frac{P}{\overline{P}} - \frac{(n-1)P}{D}]}{(\frac{h_b}{2} + \frac{D}{2\overline{P}} - \frac{n-1}{2n})\frac{(\overline{P}-P)^2}{[n - \frac{P}{\overline{P}} - \frac{(n-1)P}{D}]} + h_v\frac{(\overline{P}-D)(n-1)(\overline{P}-P)}{2n}}}. \tag{14}$$

Substituting Eq. (14) into Eq. (11), the mathematical model can be rewritten as

$$\text{Minimize} \quad TC(n,d,t_{over}) \tag{15}$$

$$\text{Subject to:} \quad \underline{t_{over}} \le t_{over} \le \overline{t_{over}} \tag{16}$$

For given n, $TC(n,t_{over})$ is a function with respect to t_{over}. It is impractical to solve the model by analyzing the first and second derivatives of $TC(n,d,t_{over})$ with respect to t_{over}. Therefore, we attempt to find out a method for solving the optimal solution and the following proposition of the function $TC(n,t_{over})$ is provided.

Proposition 1. For given n, when $t_{over} \in [\underline{t_{over}}, \overline{t_{over}}]$, $\underline{TC} < TC(n,t_{over}) < \overline{TC}$, where

$$\underline{TC} = \sqrt{2D(A_b + \frac{A_v + S + c_{tr}C_{con}}{n})(h_b + \frac{D}{P(\overline{t_{over}})} - \frac{n-1}{n})} + c_tD + c\underline{P}\frac{D}{n\overline{P}(\overline{t_{over}})} - \frac{c\underline{P}(n-1)}{n},$$

and

$$\overline{TC} = \sqrt{2D(A_b + \frac{A_v + S + c_{tr}M}{n})[h_b + \frac{1}{n} + \frac{[n - \frac{(n-1)P}{D}]h_v(n-1)}{n}]} + \frac{(c_o + c)D}{2n}[n - \frac{(n-1)P}{D}] + c_tD + \frac{cD}{n} - \frac{c\underline{P}(n-1)}{n}.$$

Proof. For given n, $TC(n,t_{over}) > \sqrt{2D(A_b + \frac{A_v + S + c_{tr}C_{con}}{n})(h_b + \frac{D}{\overline{P}} - \frac{n-1}{n})} + c_tD + c\underline{P}\frac{D}{n\overline{P}} - \frac{c\underline{P}(n-1)}{n}$.

Let $F(n,t_{over}) = \sqrt{2D(A_b + \frac{A_v + S + c_{tr}C_{con}}{n})(h_b + \frac{D}{\overline{P}} - \frac{n-1}{n})} + c_tD + c\underline{P}\frac{D}{n\overline{P}} - \frac{c\underline{P}(n-1)}{n}$. Based on the above derivation, it implies that $TC(n,t_{over}) > F(n,t_{over})$ for given n. Obviously, $F(n,t_{over})$ is a decreasing function in t_{over}. Then, for any $t_{over} \in [\underline{t_{over}}, \overline{t_{over}}]$, $TC(n,t_{over}) > F(n,\overline{t_{over}})$ for given n, i.e., $TC(n,t_{over}) > \sqrt{2D(A_b + \frac{A_v + S + c_{tr}C_{con}}{n})(h_b + \frac{D}{P(\overline{t_{over}})} - \frac{n-1}{n})} + c_tD + c\underline{P}\frac{D}{n\overline{P}(\overline{t_{over}})} - \frac{c\underline{P}(n-1)}{n}$. For simplicity, let $\underline{TC} = \sqrt{2D(A_b + \frac{A_v + S + c_{tr}C_{con}}{n})(h_b + \frac{D}{P(\overline{t_{over}})} - \frac{n-1}{n})} + c_tD + c\underline{P}\frac{D}{n\overline{P}(\overline{t_{over}})} - \frac{c\underline{P}(n-1)}{n}$. Hence, $TC(n,t_{over}) > \underline{TC}$ for given n.

Likewise, $TC(n,t_{over}) < \overline{TC}$ for any $t_{over} \in [\underline{t_{over}}, \overline{t_{over}}]$ under given n, where $\overline{TC} = \sqrt{2D(A_b + \frac{A_v + S + c_{tr}C_{con}}{n})[h_b + 1 - \frac{n-1}{n} + h_v\frac{(n-1)[n - \frac{(n-1)P}{D}]}{n}]} + \frac{(c_o + c)D[n - \frac{(n-1)P}{D}]}{2n} + c_tD + \frac{cD}{n} - \frac{c\underline{P}(n-1)}{n}$. $\square$

For given n, the intervals of overtime and the daily total cost are evenly divided into Z sub-intervals respectively as shown in Fig. 2. Accordingly, this generates Z^2 grid points

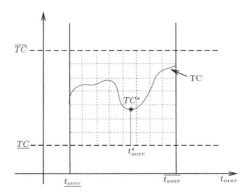

Fig. 2. The range of overtime and the daily total cost.

and Z^2 grids in Fig. 2. The optimal daily total cost TC^* is in one of these grids and the optimal solution t^*_{over} is in corresponding sub-interval. Given Proposition 1 and Fig. 2, the following proposition of $TC(n, t_{over})$ can be obtained.

Proposition 2. For given n, $TC(n, t_{over})$ is a lipschitz function in t_{over} on the interval $[\underline{t_{over}}, \overline{t_{over}}]$.

Proof. Under given n, for any $t_{over,1}, t_{over,2} \in [\underline{t_{over}}, \overline{t_{over}}]$, when $|t_{over,1} - t_{over,2}| \leq \varepsilon$, it can conclude that $t_{over,1}$, $t_{over,2}$ and their corresponding costs $TC(n, t_{over,1})$ and $TC(n, t_{over,2})$ are in the same grid when ε is sufficiently small. Therefore, $|t_{over,1}-t_{over,2}| \leq \frac{\overline{t_{over}}-\underline{t_{over}}}{Z}$ and $|TC(n, t_{over,1})-TC(n, t_{over,1})| \leq \frac{\overline{TC}-\underline{TC}}{Z}$. Let $\varepsilon = \max\{\frac{\overline{t_{over}}-\underline{t_{over}}}{Z}, \frac{\overline{TC}-\underline{TC}}{Z}\}$. We deduce that $|TC(n, t_{over,1})-TC(n, t_{over,1})| \leq \varepsilon$ when $|t_{over,1}-t_{over,2}| \leq \varepsilon$. Therefore, for given n, $TC(n, t_{over})$ is a lipschitz function with respect to t_{over} on the interval $[\underline{t_{over}}, \overline{t_{over}}]$. $\square$

Based on Proposition 1 and 2 and the analysis above, we propose the following solution procedure to find the optimal policy.

Solution Procedure

Step 1: Given a sufficient integer Z and set $n := 1$. Let $TC_{min} := +\infty$.
Step 2: If $n \leq \left\lfloor \frac{1}{\alpha}(1 - \frac{D}{\max\{\overline{P}(t_{over})\}}) \right\rfloor$, go to step 3; otherwise, go to step 8.
Step 3: Set $i := 0$ and go to step 4.
Step 4: If $i \leq Z$, go to step 5; Otherwise, go to step 7.
Step 5: Let $t_{over} = \underline{t_{over}} + \frac{i}{Z}(\overline{t_{over}}-\underline{t_{over}})$. Substitute n and t_{over} into Eq. (15), obtain the corresponding cost TC. If $TC < TC_{min}$, let $TC_{min} := TC$. Then, go to step 6.
Step 6: Let $i := i + 1$ and go to step 4.
Step 7: Let $n := n + 1$ and go to step 2.
Step 8: Regard TC_{min} as the optimal daily cost and denote it as TC^*, and the corresponding solutions n and t_{over} as n^* and t^*_{over}. Substitute n^* and t^*_{over} into Eqs. (14) and (10), we obtain the optimal overtime days d^* and the optimal ordering quantity Q^*.

5   The Extended Model with Multiple Buyers

5.1   The Extended Model

In this section, we extend the integrated model by considering the vendor supplying N buyers. For the convenience of vendor management, all buyers order the product from the vendor at the same time during each production cycle. In this extended model, we assume that buyer i has a deterministic daily demand D_i and orders Q_i units product from the vendor independently. To make overtime strategy feasible, we assume that $\underline{P} < \sum_{i=1}^{N} D_i$ and $\overline{P} \geq \sum_{i=1}^{N} D_i$. When buyer i places an order, it incurs an ordering cost A_i per order and incurs an inventory holding cost h_i per unit per hour. Meanwhile, when the vendor delivers to buyer i, the transportation cost c_i per unit is incurred and is paid by buyer i. The daily total cost BC_i for buyer i can be expressed as

$$BC_i(Q_i) = A_i D_i / Q_i + h_i Q_i / 2 + c_i D_i. \tag{17}$$

Since the vendor delivers the product to N buyers, it is not easy to find out the production cycle. In this extended model, the vendor finishes a lot at a period T_v, and we assume that the vendor's production cycle T_v is $T_v = \sum_{i=1}^{N} \frac{Q_i}{D_i}$. Based on this assumption, the vendor's producing cost in a lot is $c_o \overline{P} d + c\underline{P}(\sum_{i=1}^{N} \frac{Q_i}{D_i} - d)$ and the daily producing cost can be written as $(c_o \overline{P} - c\underline{P}) d / \sum_{i=1}^{N} \frac{Q_i}{D_i} + c\underline{P}$. As the ordering quantity equaling the total production quantity in a lot, the daily producing cost is rewritten as $\frac{(c_o \overline{P} - c\underline{P})}{(\overline{P} - \underline{P})}(\sum_{i=1}^{N} D_i - \underline{P}) + c\underline{P}$. In a lot, the vendor temporarily stores the produced product with average accumulated inventory quantity $\frac{\overline{P} d^2}{2} + \frac{\underline{P}}{2}(\sum_{i=1}^{N} \frac{Q_i}{D_i} - d)^2 + \overline{P} d(\sum_{i=1}^{N} \frac{Q_i}{D_i} - d)$, and the daily holding cost is $\frac{h_v}{2(\overline{P} - \underline{P})} \sum_{i=1}^{N} \frac{Q_i}{D_i}[\overline{P}(\sum_{i=1}^{N} 2D_i - \underline{P}) - (\sum_{i=1}^{N} D_i)^2]$, where h_v is the daily holding cost per unit. Because the vendor incurs setup cost during each production cycle and recruiting cost when recruitment happening, the hourly costs of setup and recruiting new employees are $A_v / \sum_{i=1}^{N} \frac{Q_i}{D_i}$ and $(S + c_{tr} Num) / \sum_{i=1}^{N} \frac{Q_i}{D_i}$, respectively. Since the vendor's ordering quantity is equal to the total production quantity during each production cycle, we have $d = (\sum_{i=1}^{N} \frac{Q_i}{D_i})(\sum_{i=1}^{N} D_i - \underline{P}) / (\overline{P} - \underline{P})$. The daily total cost VC for the vendor is given

$$VC(tover, Q_1, \cdots, Q_N) = \frac{A_v + S + c_{tr} Num}{\sum_{i=1}^{N} \frac{Q_i}{D_i}} + \frac{(c_o \overline{P} - c\underline{P})}{\overline{P} - \underline{P}}(\sum_{i=1}^{N} D_i - \underline{P}) + c\underline{P}$$

$$+ \frac{h_v}{2(\overline{P} - \underline{P})}(\sum_{i=1}^{N} \frac{Q_i}{D_i})[\overline{P}(2 \sum_{i=1}^{N} D_i - \underline{P}) - (\sum_{i=1}^{N} D_i)^2], \tag{18}$$

and the daily total cost $JC(t_{over}, Q_1, \cdots, Q_N)$ of the single-vendor multi-buyer integrated model is expressed as

$$JC(t_{over}, Q_1, \cdots, Q_N) = \sum_{i=1}^{N} (\frac{A_i D_i}{Q_i} + \frac{h_i Q_i}{2} + c_i D_i) + \frac{A_v + S + c_{tr}(kt_{over} + C_{con})}{\sum\limits_{i=1}^{N} \frac{Q_i}{D_i}}$$

$$+ \frac{(c_o \overline{P} - c\underline{P})}{\overline{P} - \underline{P}} (\sum_{i=1}^{N} D_i - \underline{P}) + c\underline{P} + \frac{h_v}{2(\overline{P} - \underline{P})} (\sum_{i=1}^{N} \frac{Q_i}{D_i}) [\overline{P}(2 \sum_{i=1}^{N} D_i - \underline{P}) - (\sum_{i=1}^{N} D_i)^2].$$

$$(19)$$

Following the assumptions about production capacity and overtime in last section, the constraints of the model are given: $-k(P_S - P_N)t_{over}^2 + [P_S M - (P_S - P_N)(kt_{norm} + C_{con})]t_{over} - (\sum_{i=1}^{N} D_i - \underline{P}) \geq 0$ and $t_{over} \leq \tilde{t}_{over}$. And the lower and upper bounds of overtime in this model, t'_{over} and t''_{over}, are solved, i.e., $t'_{over} \leq t_{over} \leq t''_{over}$. Hence, mathematical formulation of the single-vendor multi-buyer integrated model is as follows.

$$\text{Minimize } JC(t_{over}, Q_1, \cdots, Q_N)$$

$$\text{Subject to } t'_{over} \leq t_{over} \leq t''_{over}$$

5.2 The Solution Procedure for Determining the Optimal Policy

The problem for the integrated model is to find the optimal overtime t_{over} and all buyers' ordering quantities $Q_1, \cdots, Q_N$ that minimizes the system's daily total cost $JC(t_{over}, Q_1, \cdots, Q_N)$.

There is no other constraints for each ordering quantity Q_i except $Q_i > 0$, $i = 1, \cdots, N$. And it can easily proved that $JC(t_{over}, Q_1, \cdots, Q_N)$ is a convex function in $(Q_1, \cdots, Q_N)$, because hessian matrix of $JC(t_{over}, Q_1, \cdots, Q_N)$ in $(Q_1, \cdots, Q_N)$ is positive definite matrix.

$$\nabla = \begin{pmatrix} -\frac{A_1 D_1}{Q_1^2} + \frac{h_1}{2} - \frac{A_v + S + c_{tr}(kt_{over} + C_{con})}{(T_v)^2 D_1} + \frac{h_v}{2(\overline{P}-\underline{P})D_1}[\overline{P}(2\sum\limits_{i=1}^{N} D_i - \underline{P}) - (\sum\limits_{i=1}^{N} D_i)^2] \\ -\frac{A_2 D_2}{Q_2^2} + \frac{h_2}{2} - \frac{A_v + S + c_{tr}(kt_{over} + C_{con})}{(T_v)^2 D_2} + \frac{h_v}{2(\overline{P}-\underline{P})D_2}[\overline{P}(2\sum\limits_{i=1}^{N} D_i - \underline{P}) - (\sum\limits_{i=1}^{N} D_i)^2] \\ \vdots \\ -\frac{A_N D_N}{Q_N^2} + \frac{h_N}{2} - \frac{A_v + S + c_{tr}(kt_{over} + C_{con})}{(T_v)^2 D_N} + \frac{h_v}{2(\overline{P}-\underline{P})D_N}[\overline{P}(2\sum\limits_{i=1}^{N} D_i - \underline{P}) - (\sum\limits_{i=1}^{N} D_i)^2] \end{pmatrix}$$

$$\text{Hessian} = \begin{pmatrix} \frac{2A_1 D_1}{Q_1^3} + 2\frac{A_v + S + c_{tr}(kt_{over} + C_{con})}{(T_v)^3 D_1^2} & 2\frac{A_v + S + c_{tr}(kt_{over} + C_{con})}{(T_v)^3 D_1 D_2} & \cdots & 2\frac{A_v + S + c_{tr}(kt_{over} + C_{con})}{(T_v)^3 D_1 D_N} \\ 2\frac{A_v + S + c_{tr}(kt_{over} + C_{con})}{(T_v)^3 D_1 D_2} & \frac{2A_2 D_2}{Q_2^3} + 2\frac{A_v + S + c_{tr}(kt_{over} + C_{con})}{(T_v)^3 D_2^2} & \cdots & 2\frac{A_v + S + c_{tr}(kt_{over} + C_{con})}{(T_v)^3 D_2 D_N} \\ \vdots & \vdots & \ddots & \vdots \\ \frac{A_v + S + c_{tr}(kt_{over} + C_{con})}{(T_v)^3 D_1 D_N} & 2\frac{A_v + S + c_{tr}(kt_{over} + C_{con})}{(T_v)^3 D_2 D_N} & \cdots & \frac{2A_N D_N}{Q_N^3} + 2\frac{A_v + S + c_{tr}(kt_{over} + C_{con})}{(T_v)^3 D_N^2} \end{pmatrix}$$

Therefore, the optimal ordering policy is acquired when $\nabla = \mathbf{0}$ and the following equations are derived.

$$\frac{A_1 D_1}{Q_1^2} + \frac{A_v + S + c_{tr}(kt_{over} + C_{con})}{(T_v)^2 D_1} = \frac{h_1}{2} + \frac{h_v}{2(\overline{P} - \underline{P})D_1}[\overline{P}(2\sum_{i=1}^{N} D_i - \underline{P}) - (\sum_{i=1}^{N} D_i)^2] \tag{20}$$

$$\vdots$$

$$\frac{A_N D_N}{Q_N^2} + \frac{A_v + S + c_{tr}(kt_{over} + C_{con})}{(T_v)^2 D_N} = \frac{h_N}{2} + \frac{h_v}{2(\overline{P} - \underline{P})D_N}[\overline{P}(2\sum_{i=1}^{N} D_i - \underline{P}) - (\sum_{i=1}^{N} D_i)^2] \tag{21}$$

After some algebra, the ordering quantity for buyer i can be expressed

$$Q_i = \frac{D_i \sqrt{A_i} Q_1}{\sqrt{(D_1 h_1 - D_i h_i)Q_1^2/2 + A_1 D_1^2}}, i = 1, \cdots, N. \tag{22}$$

Combined Eq. (22) with Eq. (20), the optimal ordering policy is obtained and it easily be proved that the ordering strategy is related to overtime strategy. Therefore, the ordering quantity for buyer i, Q_i, can be denoted as $Q_i(t_{over})$. For convenience, $Q_i(t_{over})$ and Q_i will be used interchangeably.

Based on the analysis about the ordering policy, the daily total cost for the integrated model can be rewritten as a function with respect to t_{over}. Proceeding the same fashion about overtime, we derive the following proposition.

Proposition 3. $JC(t_{over}, Q_1, \cdots, Q_N)$ is a lipschitz function with respect to t_{over} on the interval $[t'_{over}, t''_{over}]$.

Proof. Following the proof of Proposition 2. □

Proposition 3 depicts the lipschitz condition of $JC(t_{over}, Q_1, \cdots, Q_N)$ and the optimal overtime can be derived using the lipschitz method when the optimal ordering policy is given. Give the following points: $t'_{over}, t'_{over} + \frac{t''_{over} - t'_{over}}{L}, t'_{over} + \frac{2(t''_{over} - t'_{over})}{L}, \cdots, t''_{over}$, and we obtain the corresponding daily costs. The minimum of these costs is regarded as the optimal cost and its relevant overtime is the optimal overtime value.

5.3 Numerical Example

Example 1. We consider an integrated system with three buyers. The vendor's workshop can accommodate 30 employees and each employee normally works 8 h per day. On average of 2 employees leave when no overtime happen. While, when overtime occurs, the number of departing employees increases as the daily overtime working hours surge and the number of separated employees Num is $Num(t_{over}) = kt_{over} + 2$, where k is a random number between 2 and 5. Besides, we set $\tilde{t}_{over} = \min\{(M - C_{con})/k, 8\}$. Other related parameters in this case are denoted as follows. $A_v = 200$ \$/setup, $S = 100$ \$/setup, $c = 2$ \$/unit, $c_o = 3$ \$/unit, $c_{tr} = 50$ \$/employee, $P_S = 60$ units/hour, $P_N = 50$ units/hour, $h_v = 0.002$ \$/unit. For each buyer, we assume the parameters used in this model as follow: $D_i = 5000$ units/hour, $A_i = 100$ \$/order, $h_i = 0.002$ \$/unit, $c_i = 0.003$ \$/unit.

The correlation coefficient k in the relationship between turnover employees and overtime working hours is a random number. To demonstrate the effects of key parameters on the optimal overtime and delivery policy, we temporarily assume that k is a constant. The values of other parameters remain the same as in the above case, while k varies. Sensitivity analysis with respect to k is shown in Table 1.

Table 1. The computational results under different overtime constraints

k	$\bar{t}_{over} = \min\{(M - C_{con})/k, 8\}$		
	Q_i^*	t_{over}^*	JC^*
2	1.4734e + 05	8.0000	3.3586e + 04
3	1.4734e + 05	8.0000	3.3704e + 04
4	1.4830e + 05	7.0000	3.3987e + 04
4.5	1.4950e + 05	6.2222	3.4216e + 04
5	1.5089e + 05	5.6000	3.4463e + 04

As depicted in Table 1, the daily total cost of the integrated system grows under different values of k. When the parameter k is small, employees are not sensitive to overtime and the increase of *Num* is small as overtime increases. In this situation, the vendor requires employees to overtime as long as possible. Besides, from Table 1, it concludes that the ordering quantity for each buyer is proportional to the parameter k and is inversely proportional to overtime t_{over} when k rises. Conversely, employees are more sensitive to overtime when k grows. Hence, the vendor slashes overtime to ensure avoiding excessive cost and too many departing employees as k increases. Despite that, each buyer still orders higher ordering quantity, because the total training cost for new recruited employees is less than the additional cost incurred by overtime.

Figure 3 provides an numerical analysis ($k = 2$) to illustrate the relationship between training cost c_{tr} and the daily total cost JC. It shows that the daily cost of the integrated model is a monotonous increasing function of the training cost. While, as shown in above case, the training cost is relatively tiny compared with setup cost and producing cost. So, fluctuation in training cost of new recruited employees has no effect on the ordering strategy and the ordering quantity for each buyer remains unchanged under different c_{tr}. It implies that enterprises should plan training program in advance to prevent unnecessary expenses.

In addition, we provide the effect of the unit producing cost c_o on the daily total cost of the integrated model in Fig. 4. Obviously, JC is a linear function with respect to the unit producing cost c_o. The vendor's producing cost adds as c_o increases. The same trend as applies to the daily cost of the model JC. This is in accordance with the actual business activities. Hence, managers adopt standardized products, improved technology, large-scale production and other effective measures to reduce producing cost. While, employees for the vendor is relatively insensitive to overtime in this case ($k = 2$) and overtime policy stays the same as c_o varies.

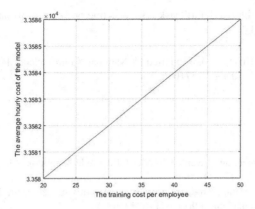

Fig. 3. The daily total cost on varying training cost c_{tr}.

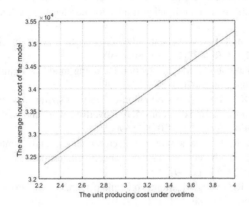

Fig. 4. The daily total cost with different values of c_o.

6 Conclusion

In this paper, we first consider an integrated inventory model, which consists of a vendor and a buyer. For the vendor, there are several work positions and each position only need one employee. The vendor's daily production rate is assumed to be less than the daily demand. To meet the demand, the vendor adopts overtime strategy to increase the production capacity. However, when overtime occurs, the number of departing employees for the vendor is not constant but varies with overtime. Then, we extend the integrated model by considering that a vendor supplies to multiple buyers. In this extended model, we assume that the vendor's production cycle equals to the sum of production cycles for all buyers. By analyzing the mathematical property, a lipschitz iterative method is proposed to obtain the optimal ordering and overtime strategy. Numerical example and sensitivity analysis have been drawn to present some insights into the parameter effect on the optimal ordering and overtime strategy.

In future research on this problem, it would be interesting to extend this integrated model by introducing stochastic demand, multiple buyers and shortage, etc. Besides, considering trade credit and default risk into overtime problem also can be the extension of the integrated model.

Acknowledgements. This work is supported by National Nature Science Foundation of China under Grants No. 71971010 and 71571006.

References

1. Goyal, S.K.: An integrated inventory model for a single supplier-single customer problem. Int. J. Prod. Res. **15**, 107–111 (1976)
2. Hariga, M.A.: Economic production-ordering quantity models with limited production capacity. Prod. Plann. Control **9**(7), 671–674 (1988)
3. Sinha, S., Sarmah, S.P.: Supply-chain coordination model with insufficient production capacity and option for outsourcing. Math. Comput. Model. **46**, 1442–1452 (2007)
4. Banerjee, A.: A joint economic-lot-size model for purchaser and vendor. Decis. Sci. **17**, 292–311 (1986)
5. Bahl, H.C., Rizman, L.P.: A cyclical scheduling heuristic for lot sizing with capacity constraints. Int. J. Prod. Res. **22**(5), 791–800 (1984)
6. Goyal, S.K., Gopalakrishnan, M.: Production lot sizing model with insufficient production capacity. Prod. Plann. Control **7**(2), 222–224 (1996)
7. Zhang, R.Q., Sun, X.F.: Integrated production-delivery lot sizing model with limited production capacity and transportation cost considering overtime work and maintenance time. Math. Probl. Eng. https://doi.org/10.1155/2018/1569029
8. Sun, X.F., Zhang, R.Q.: The single-manufacturer single-retailer integrated production-delivery lot sizing model with production capacity under stochastic lead time demand. Procedia CIRP **83**, 528–533 (2019)
9. Porter, L.W., Steer, R.M.: Organization, work, and personal factors in employee turnover and absenteeism. Psychol. Bull. **80**, 151–176 (1973)
10. Datlon, D.R., Krackhardt, D.M., Porter, L.W.: Functional turnover: an empirical assessment. J. Appl. Psychol. **66**, 716–721 (1981)
11. Krackhardt, D., Porter, L.: The snowball effect: turnover embedded in communication networks. J. Appl. Psychol. **71**, 50–55 (1986)
12. Kacmar, K.M., Andrews, M.C., Van Rooy, D.L., Chris Steilberg, R., Cerrone, S.: Sure everyone can be replaced… but at what cost? Turnover as a predictor of unit-level performance. Acad. Manag. J. **49**(1), 133–144 (2006)
13. Call, M.L., Nyberg, A.J., Ployhart, R.E., Weekley, J.: The dynamic nature of collective turnover and unit performance: the impact of time, quality, and replacements. Acad. Manag. J. **58**(4), 1208–1232 (2015)
14. Hausknecht, J.P., Trevor, C.O., Howard, M.J.: Unit-level voluntary turnover rates and customer service quality: implications of group cohesiveness, newcomer concentration, and size. J. Appl. Psychol. **94**, 1068–1075 (2009)
15. Chen, G., Ployhart, R., Thomas, H., Anderson, N., Bliese, P.: The power of momentum: a new model of dynamic relationships between job satisfaction and turnover intentions. Acad. Manag. J. **54**, 159–181 (2011)
16. Liu, D., Mitchell, T.R., Lee, T., Holtom, B., Hinkin, T.: When employees are out of step with coworkers: How job satisfaction trajectory and dispersion influence individual-and unit-level voluntary turnover. Acad. Manag. J. **55**, 1360–1380 (2012)

17. Ramesh, A., Gelfand, M.J.: Will they stay or will they go? The role of job embeddedness in predicting turnover in individualistic and collectivistic cultures. J. Appl. Psychol. **95**, 807–823 (2010)
18. Hom, P.W., Xiao, Z.: Embedding social networks: how Guanxi ties reinforce Chinese employees' retention. Organ. Behav. Hum. Decis. Process. **116**, 188–202 (2011)
19. Gjerløv-Juel, P., Guenther, C.: Early employment expansion and long-run survival: examining employee turnover as a context factor. J. Bus. Ventur. **34**(1), 80–102 (2019). S0883902616303925
20. Shaprio, D., Hom, P., Shen, W., Agarwal, R.: How do leader-departures affect subordinates' organizational attachment? A 360-degree rational perspective. Acad. Manag. Rev. **41**, 479–502 (2016)
21. Rothausen, T.J., Henderson, K.E., Arnold, J.K., Malshe, A.: Should I stay or should I go? Identity and well-being in sensemaking about retention and turnover. Journal of Management **43**(7), 2357–2385 (2017)
22. An, S.H.: Employee voluntary and involuntary turnover and organizational performance: revisiting the hypothesis from classical public administration. Int. Public Manag. J. **1**, 1–26 (2019)

New Digital Media Technologies Put Forth Cross-domain Design

Jie Tang[1,2]([✉]), ChengYao Cai[1,2], LiuYing Huang[1,2], and Xing Ji[1,2]

[1] Beijing Technology Institute, Zhuhai, People's Republic of China
4448286@qq.com, 93713953@qq.com, 477067869@qq.com,
54619184@qq.com
[2] Studying School for Doctor's Degree, Bangkok Thonburi University, Bangkok, Thailand

Abstract. The phenomenon "global village" is exhibited in the field of design as not only cultural fusion, but also the increasingly important and prominent inter-disciplinary cooperation. This is particularly true of the emergence and rapid development of new digital media technologies, which will become a new medium that enables new integration and unification of all design disciplines and interdisciplinary knowledge. Design is no longer a marginal discipline. It will play an active role in various fields to show its importance, and will lead the development and progress of society and human civilization. New interdisciplinary integration and unification of design has not only epoch-making significance for the development of modern design, but also a huge impact on the current mode of design thinking. The fundamental purpose of all designs is to meet human needs rather than research and manufacture the products. Such purpose is being fulfilled in a slow manner due to technical limitations, despite some remarkable achievements. Today, the idea can be better developed and brought into innovation by means of new digital media technologies, making interdisciplinary design more reasonable and feasible. Therefore, this article aims to establish a framework for contemporary designers for interdisciplinary design in the future, to explore the continuity and originality of modern design by combining public needs with the development of new digital media technologies.

Keywords: Modernist design · New digital media technologies · Integrated interdisciplinary design · Interdisciplinary learning of design

1 Introduction

Various disciplines have gradually emerged as people acquire more and more skills and knowledge and their living skills become more and more detailed. In the modern society where the division of labor is relatively sound, most people believe that, as long as they learn a discipline well, they can find a foothold in society. Therefore, universities have further divided disciplines for some majors, in the hope that their students will be able to survive in society or give full play to the skills they have learned in the workplace after they learn a discipline well.

© Springer Nature Switzerland AG 2020
P.-L. P. Rau (Ed.): HCII 2020, LNCS 12192, pp. 160–168, 2020.
https://doi.org/10.1007/978-3-030-49788-0_12

However, owing to the further development of electronic information technology, especially the emergence of the Internet which covers all fields, people around the world are able to communicate with each other without obstacles. This paves the way for a "global village" that dissolves cultural differences and distance. It is precisely because of these Internet information technologies that social development has become very efficient. In many industries, there are no absolute differences between disciplines, with the boundaries of various knowledge increasingly blurred. In the new society, there is a demand for interdisciplinary talents instead of specialized talents with knowledge of a single discipline [1, 5]. Excellent talents should be equipped with specialized knowledge in depth and breadth, as well as the ability to transform and update knowledge within the specialized field. Design as a discipline has seen the debate over art and design, as well as form and function since the beginning of modern design theory established by Bauhaus of Germany a century ago. This long-term debate may be better balanced and developed as new digital media technologies advances.

2 Discussion on the Debate Over Art and Design

It is generally believed that design as a discipline originated in the field of art, while modernist design came from Bauhaus in Germany. When it comes to art and design, there are always people trying to determine which is better. At the same time, many people think that what Bauhaus represents is a functionalist style, yet with disagreement over whether it is aimed at pursuing function or form. In fact, both art and design are ultimately oriented towards human needs. What we need to understand is the fundamental significance of the difference between art and design in terms of human needs.

2.1 Art is Based on Sensibility

Art as a discipline has been regarded by the public as "mysterious", because it pursues a perceptual way of expression. Emotion dominates the entire artistic process. It can reflect reality with representation or abstraction, but it features more typical social ideology than reality. Also an important supplement to language, it triggers people's thinking or emotional activity. In many cases, artistic expressions are created with personal sensibility as the starting point, which allows other people to resonate in ideology and directly provides perceptual cognition to the senses [1]. Therefore, art is also considered as a form of expression for one's pursuit of ego. From the perspective of the viewers, it is very difficult for many art works to resonate with them, because only the artists themselves can grasp the origin of their art. Art usually exists in a small group of people [3], and art works are not cheap. Throughout the history of art around the world, outstanding works of art serve the royal family, the ruling class, the nobility and a small number of people. It is difficult for the general public to have the need for or enjoy the pleasure brought by art works. According to Maslow's theory on the hierarchy of needs, most people mainly fulfill their physiological and safety needs in the course of history. Therefore, it is generally assumed that art is people's ability to pursue ego after the physiological and safety needs are met. In essence, it is expensive and luxurious.

2.2 The Core Spirit of the Bauhaus Style

Traditionally, people think that design is a purposeful creative act based on art. However, the core spirit of the modernist design that affects society today is to design not only to serve all the people, but also more importantly to meet the needs of people, with affordable prices [1], practical functions and beautiful styles. These ideas can be summarized when we analyze the establishment of the Bauhaus Design School in Germany in 1919 and its contributions over the past century [5].

2.3 Design is Based on Rationality

The Bauhaus Manifesto, drafted by Walter Gropius, is an important document of modern design. It expresses the efforts of artists to integrate design with art and technology. For affordability and practicality, they put their thoughts into new materials and new processes and try to solve problems on the basis of mass industrial production. However, traditional aesthetics was built on artistic techniques of expression that limited the exploration into affordability and practicality. This was one of the main reasons why modernist design was called "utopianism" at the time.

Many well-known artists at the time taught at the Bauhaus Design School. Among them, Kandinsky, Itten, and Dubsburg tried to capture artistic sensuality with rational words through continuous thinking of traditional art works and exploration of new styles, like abstractionism and Dutch style as well as how artistic emotions could be expressed in lines, colors, and geometric forms. In the end, these artists summarized how emotions were described in rational words, developed a theory and put it into practice, thereby laying the theoretical foundation for education on modern design [1, 5]. It is beautiful and practical to use lines, colors, and geometric forms to show traditional artistic aesthetics. Besides, the geometric forms can further reduce the cost of design for affordability. Although the theory was based on architectural design, in the later exploration, all theoretical knowledge served and paved the way for design as the main discipline, with no sub-discipline in the strict sense.

2.4 Root of Debate Over Art and Design

Because the modernist design style has made art very rational, it has been controversial since its inception, lasting for a century. Artists believe that excessively rational words deprive people of their reverie and soul, and human emotions should not be restrained by rules and regulations but be allowed greater freedom and pursuit of ego. So in order to oppose the indifference and simplification brought about by modernist design, post-modernist design was born in the 1980s. However, just like the French Art Deco movement [3] that emerged at the same time as the Bauhaus, it decayed within less than thirty years, failing to replace the dominance of modernist design represented by the Bauhaus in society today [2, 5].

The design represented by the Bauhaus style is aimed to serve the public, which is set to be violated by the sensibility brought by art. The postmodernist design brings perceptual colors and restores design to its essence—art, which is luxurious and decorative and serves a small number of people (see Fig. 1). It was difficult to bring sensibility and rationality together under design with technologies in the past, while accommodating affordability, practicality and aesthetics. Yet today the rapid development [3, 6, 7] of new digital media technologies has shown artists and designers the possibility of integration.

Fig. 1. Changsha meixi lake international culture and art center designed by Zaha Hadid

3   History and Development of New Digital Media Technologies

New digital media technologies consist of many technologies in many fields, but what is mainly referred to in this paper is Augmented Reality (AR) and Virtual Reality (VR), which can improve design and design services [4]. In recent years, such technologies have become more and more sophisticated. People are offered a thorough experience of the functional practicability of products, and obtain emotional enjoyment and sensual stimulation.

AR is the image technology superimposed by computer on real-life images and inserts the virtual world into the real one to interact with it for the purpose of "augmenting" reality. VR is to generate a three-dimensional space on a computer to provide users with a virtual experience [4, 6, 7] about sight, hearing, feeling and other senses, making users feel as if they are personally on the scene. New digital media technologies date back to the first Augmented Reality (AR) and Virtual Reality (VR) devices in the 1950s and 1960s. After more than half a century of development, AR and VR are embracing their golden age. Now they are prevalent in our lives, most widely seen in the field of games (see Fig. 2) and children's picture books (see Fig. 3).

Fig. 2. Pokemon go (Internet Game)

Fig. 3. Children's picture book

4 Interdisciplinary Design

Generally, a discipline often has its own methodological traditions. Researchers who have been working in the discipline for a long time, because of path dependency, often find it difficult to bring design into innovation. As the boundaries between the specializations of design become increasingly blurred [4, 8], knowledge continues to merge. Since there are similar needs in multiple dimensions, such as science and technology, design, art, and humanities, students are required to dig into the knowledge and conduct interdisciplinary and cross-domain cooperation.

4.1 Mercedes-Benz F015 Luxury in Motion

At CES in 2015, Mercedes-Benz introduced a concept car F015 Luxury in Motion which was inspired by the "cross-cultural discussions" among Mercedes designers, experts in auto-driving technology, car manufacturers, and interior futurists (see Fig. 4).

Fig. 4. F 015 Luxury in Motion

Mercedes-Benz is committed to creating the experience where passengers can immerse themselves in their own world. When they are in the car, they can interact with the car through gestures, eye tracking or touch control. The drivers only need to wave their hands or roll their eyes to control the car. The car integrates six display screens, which keeps passengers informed of the current information and allows them to interact with the vehicle (see Fig. 5). For those who are keen on driving or traveling, the front windshield of the car can be used to change the scene outside with AR [8], such as the F1 track or the beautiful scenery. But for most people who are not passionate about driving, it is more valuable to save the time of driving for work and reading, etc. Mercedes-Benz believes that the road ahead will be shared by all pedestrians and vehicles. Cars will be able to give precedence to the pedestrians, and even project the pedestrian crossings in front of pedestrians.

Fig. 5. Interior (F 015 Luxury in Motion)

4.2 SPONtech

The German company SPONtech has developed a design for spinal examination and surgery with the help of commonly used electronic equipment, a design built upon new digital media technologies (see Fig. 6, 7). The design simulates human behaviors on electronic devices based on patients' spinal data from the hospital's big data. It enables diagnosis of the patients' spinal problems through simulation and comparison. Then the doctors will simulate the surgical process to obtain the best surgical solution [6, 7]. Interviews with members of the design team reveal that it was inspired by the interdisciplinary conflicts between designers and the medical field. Through discussions and communications with the doctors, the design team applied new digital media technologies to finally achieve interdisciplinary integrated and innovative design.

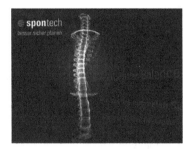   

Fig. 6. SPONtech   **Fig. 7.** SPONtech (Simulation operation)

It can be seen from the above cases that the design not only concerns visual communication design, product design, interaction design and space design, but also comes with interdisciplinary integration through new digital media technologies. In this sense, design is no longer about single-disciplinary study, but needs more assistance from in-depth research and interdisciplinary knowledge of many disciplines.

5   How to Make Design Interdisciplinary?

To make design interdisciplinary with new digital media technologies, firstly, a personal system of design knowledge should be set up and research orientations should be specified. When a new field is involved, a problem will emerge as to how to deal with the old definitions of the knowledge system. Knowledge of a new field tends to be closely related to the knowledge of some specific disciplines. Detailed division of the knowledge system helps to free designers from the constraints of the knowledge of a single discipline for more creative thinking.

5.1   The Core of Interdisciplinary Learning

The core of using new digital media technologies for interdisciplinary learning is interdisciplinary thinking. With flights of thinking, learning and exposure to new knowledge over a long time, it is easier to resonate and rethink to some extent in the face of new knowledge and fields. The learning process should not be bound by the traditional methods of any discipline [1]. We should be bold in borrowing and transplanting methods of other disciplines. Of course, such methods are not limited to specializations related to design. We should make good use of knowledge of different disciplines to explain the problem we are studying and connect explanations from different fields to solve the problem.

Interdisciplinary thinking also encourages interdisciplinary discussions and exchanges. It takes time for design innovation to be recognized. During this period, the knowledge, the premise of design, and the methods leading to the final design project will be questioned. Open and rational academic discussions will not only help to improve the knowledge, but also promote more reliable theoretical prerequisites and clarify the design and research methods adopted, so that knowledge can be transferred from the creator to the academic community. The German Association of Craftsmen was clearly aware of this a hundred years earlier, as it was made up of people from various fields for discussions and exchanges, laying the foundation for the Bauhaus School.

5.2 Different Countries to Interdisciplinary Learning

German design colleges have always adhered to interdisciplinary cooperative learning, which can be confirmed by the Bauhaus textbooks and curricula. The Royal College of Art in the U.K. is recognized as the best art institution. It has always emphasized the philosophy of interdisciplinary learning. In the United States, Carnegie Mellon University is a mecca for many design students, because its projects are all interdisciplinary and it has also highlighted interdisciplinary learning on its official website [6, 9]. Many art colleges around the world aim at comprehensive training of students, integrating art, design and technology in their teaching. Many of the best design colleges in China have also begun to carry out interdisciplinary learning and research. Among them, the School of Art and Design, Guangdong University of Technology, has introduced digital media technologies to the undergraduate graduation design projects, which will definitely turn design from a single sub-discipline to a complete discipline [1]. The Bauhaus already proposed that design was the new integration and unification of art and technology a hundred years ago.

6 Conclusion

It is precisely because of the difficulty in expressing complex emotions and thoughts in words that, a hundred years ago, it was impossible to meet more emotional needs through rational design. But now we can integrate new digital media technologies into a fixed thing that serves as the carrier [1, 7], so that people in the virtual environment and space can get a sensual experience. In the future, in addition to vision, technology will also satisfy other senses like smell and taste.

The fundamental purpose of design is to serve the needs of people rather than products. In the future, with the advent of the faster 5G network and the assistance of new digital media technologies, design that combines sensibility and rationality will play an active role in various fields due to human needs, leading the development and progress of society and human civilization [9].

References

1. Wang, S.: History of Modern Design in the World. China Youth Press, Beijing (2002)
2. Cui, B., Ren, Y.: The design of Zaha Hadid was abolished in Tokyo. Arch. Des. Manag. **09**, 94–96 (2015)
3. Guo, Q.: Decorative art movement and its style. Ornament **09**, 96–97 (2004)
4. Yang, M., Guo, X., Kong, C.: Augmented reality and its application and development in various fields. Electron. World **20**, 49–50 (2017)
5. Zong, M., Wang, R.: Fragments of history—glimpse of the formation of german industrial design thoughts. Design **01**, 50–59 (2015)
6. Ma, Z.: User experience under augmented reality. Art Technol. **07**, 14–19 (2015)
7. Hu, B., Yang, Y.: The impact of VR interactive design based on internet on art communication. Electron. Technol. Softw. Eng. **24**, 04–05 (2019)
8. Zeng, L.: Design of human-computer interaction interface based on context awareness. Intern. Combust. Engine Parts **02**, 167–168 (2019)
9. Wei, Y.: Development of virtual reality and augmented reality in the 5G era. Guangxi Commun. Technol. **01**, 04 (2019)

Design of Vibrotactile
Direction Feedbacks on Wrist
for Three-Dimensional Spatial Guidance

Jo-Hsi Tang[1]($\boxtimes$), Giuseppe Raffa[2]($\boxtimes$), and Liwei Chan[3]($\boxtimes$)

[1] National Taiwan University, Taipei, Taiwan
tanya76567@gmail.com
[2] Intel, Santa Clara, USA
giuseppe.raffa@intel.com
[3] National Chiao Tung University, Hsinchu, Taiwan
liweichan@cs.nctu.edu.tw

Abstract. A wrist-worn vibrotactile interface was previously studied but was aimed at low-resolution navigation tasks such as driving. The previous design had achieved up to six directions for three-dimensional navigation. We argue that the expressivity of vibrotactile navigation on the wrist has not been fully explored, and we address how three-dimensional direction cues can be packed into a wrist-form tactile interface. We present an 8-tactor cuboid worn in wrist form to generate high-density three-dimensional direction feedback around the wrist. This sparse arrangement of 8 vibrotactors allows up to 26 directions to be presented, when benefitting from phantom illusion. We conducted a study with 36 participants to inform the effective design of the interface regarding two factors: the cuboid shape (e.g., the length along the wrist), by comparing 4-cm, 6-cm, and 8-cm configurations, and the direction feedback, which includes point stimuli and motion stimuli. The results show that 6 cm strikes a balance between form and recognition rate. The direction feedbacks made with motion stimuli (80.2%) are generally more discernible than those made with point stimuli (69.6%).

Keywords: Haptic guidance · Spatial guidance · Vibrotactile feedback

1 Introduction

Vibrotactile stimulation provides an unobtrusive way to access information while keeping it confidential; moreover, the tactile channel is generally less overloaded in complex scenarios. These benefits have been studied and applied to HCI systems such as warning systems, navigation systems, and tactile displays and support applications including learning motor skills [13,15,16] and rehabilitation [5,7,8]. Among other uses, a wrist-worn interface embedding tactile information is of particular interest to due to the sufficient skin region suggested by a highly acceptable form (i.e., smartwatch). Although wrist vibrotactile interfaces have been extensively studied for navigation systems [3], previous research has been limited to a small set of directions (4 to 6) [2,5,9,15,16,18] and thus to

© Springer Nature Switzerland AG 2020
P.-L. P. Rau (Ed.): HCII 2020, LNCS 12192, pp. 169–182, 2020.
https://doi.org/10.1007/978-3-030-49788-0_13

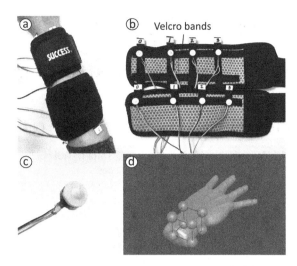

Fig. 1. The wristbands and tactors for our experiments. (a) The wristbands worn by a participant. (b) The arrangement of tactors in the wristbands to from a cuboid shape around wrist. (c) The tactor with plastic caps to enhance tactile sensation. (d) The cuboid shape around wrist, where each green sphere indicates a tactor.

low-resolution navigation. While this might be sufficient in simple tasks such as driving navigation, which primarily involves turning left or right, it would remove the opportunity to navigate more complicated environments.

This work explores how 3-dimensional direction cues can be packed into a wrist-form vibrotactile interface. Inspired by EdgeVib [12], which exploited a square design formed by 2×2 tactors on a wrist back to deliver alphanumeric patterns (i.e., $10 + 26$ recognition items) with spatiotemporal vibrotactile cues, we sought to conduct an explorative study, extending EdgeVib to a cubic design formed by $2 \times 2 \times 2$ tactors around the wrist to generate 3D direction cues.

In this study, we exploit two vibrotactile wristbands (Fig. 1), each containing 4 tactors, which together form an adjustable cubic tactor around the wrist, allowing up to 26 directions for 3D navigation. To determine the effective distance between the 2 wristbands and the appropriate vibration feedback, 36 participants were involved in the controlled lab experiments with 3 distance configurations (4 cm, 6 cm, and 8 cm) and two forms of vibration feedback (point stimuli and motion stimuli). The result reported that 6 cm separation strikes a balance between form and recognition rate. Direction cues made with motion stimuli is generally more discernible than with point stimuli.

2   Related Works

This work presented a study on wrist-based vibrotactile spatial guidance. For brevity, we will address the previous haptic research on hand-based and wrist-based interfaces.

Chen et al. [2] exploited tactor localization using 3×3 grids of vibration motors on both the dorsal and volar sides of the wrist, and they concluded that only 2 motors could be reliably distinguished on either the dorsal or the volar side. Lee et al. [9] revealed that different parameters, such as tactor type, sensory saltation, and locus of stimuli, will affect the performance when transmitting tactile information, which was explored with a 3×3 grid of tactors on the wrist. Other applications, such as delivering alphanumeric characters [12], enabling eyes-free interaction for a wristwatch [14], producing alerts to on-the-go users [10], and motion guidance [5], are all enhancing the interactivity and expressiveness of wrist-worn vibrotactile devices. In 2011, as a follow-up to Schätzl's work, Weber et al. [18] compared wristband configurations with 4 and 6 tactors and evaluated the users' ability to perceive a vibration signal in one of the predefined directions. Similarly, Hong et al. [4] evaluated 32 vibrotactile directions using 4 and 8 tactors in 1 wristband for 2D guidance. In this work, the absolute movement error in interpreting and executing the direction haptic signal is about 25°. This error indicates low resolution of direction cues on the wrist. In addition, this study was confined to 2D guidance.

Most of these works used a 4-tactors wristband with a small set of directions (4 to 6) for only 2-dimensional spatial guidance. Three-dimensional direction patterns are needed to enhance the performance of spatial guidance. For our work, we use an 8-tactor cuboid layout with 2 wristbands to achieve 26 directions for 3-dimensional spatial guidance.

3   Apparatus and Method

We implement an experimental system with 2 vibrotactile wristbands connected to an Arduino Mega microcontroller and a Unity program as an interface for presenting the experimental tasks and communicating with the Arduino.

The two wristbands, as shown in Fig. 1, use a total of 8 tactors, with 4 tactors in each band, around the wrist. The tactors used to generate tactile stimuli were a 10-mm Precision Microdrive 310-117[1], one of the circular eccentric rotation mass (ERM) type, with rated operating voltage of 3 V, which can easily be integrated into wristbands. Following a previous study [9], we attached a hemispherical plastic cap with 4 mm diameter to each vibration tactor to enhance stimulus with a smaller contact area.

For each band, we placed 2 tactors on the dorsal and 2 tactors on the volar side. As a result, the layout of the vibrotactors is in a cuboid (Fig. 1d), allowing it to intuitively map the spatial direction with the tactors around the wrist. The 2-band tactors were attached to adjustable Velcro wrist guards, to accommodate varying wrist sizes of the users. Also, to cover the 2 bands separated by 3 distances (4 cm, 6 cm, and 8 cm) for the study, we used 1 wrist guard to cover all 8 tactors in the 4-cm configuration (Fig. 2a) and 2 wrist guards for the 6-cm and 8-cm configurations (Fig. 2b,c).

---

[1] https://www.precisionmicrodrives.com/product/310-117-10mm-vibration-motor-3mm-type.

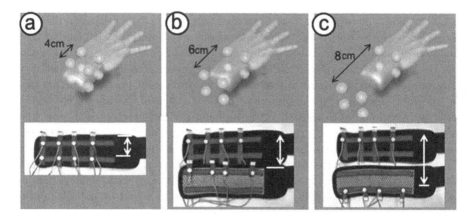

Fig. 2. Three distances, 4 cm, 6 cm, and 8 cm, separating the two wristbands are compared in our study.

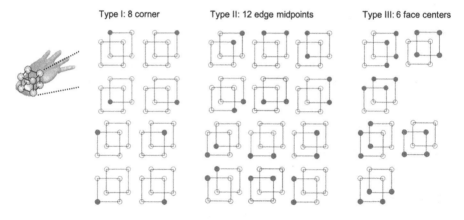

Fig. 3. The 26 nodes consist of three types of tactile groups. Type I: 8 nodes at corners (with one tactor); Type II: 12 (virtual) nodes at edge midpoints (formed by two tactors); Type III: six (virtual) nodes at face centers (formed by four tactors).

3.1 Design of the Direction Feedback

Our goal was to generate dense directions using sparse motors (here, 8 tactors) around the wrist. To address the explorative nature of this research, we set the goal to have 26 overwhelming directions. Using 1-1 mapping, however, the 8 tactors around the wrist can only indicate up to 8 directions. To increase the number of actuating nodes around the wrist, we introduce *virtual points* using a phenomenon called *phantom illusion*. A phantom illusion simulates a sensation of single vibration between multiple vibrotactors, when tactors in close proximity actuate simultaneously.

We simulated 18 virtual nodes using the 8 tactors. The 12 virtual nodes were at the midpoints of the cuboid's 12 edges when actuating the 2 tactors at the edge's 2 ends. The 6 virtual nodes are at the midpoints of the cuboid's 6 faces with the four tactors of the face's corners, as illustrated in Fig. 3.

Point Stimuli. With the 26 nodes (8 physical tactors plus 18 virtual tactors), which point to 26 directions from the central wrist, we defined the point stimuli pattern using 1-1 mapping, as shown in Fig. 4. However, in a pilot test, we realized that a point stimulus as a single-end effector may lead to user confusion in decoding the direction, due to the lack of a start effector (e.g., the central wrist in the case of point stimuli). As such, a point stimulus, as shown in Fig. 5, may indicate a set of nearby directions. We sought to resolve this issue with motion stimuli, as discussed below.

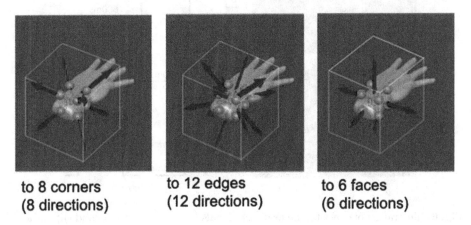

to 8 corners
(8 directions)

to 12 edges
(12 directions)

to 6 faces
(6 directions)

Fig. 4. The 26 directions are formed by pointing to 26 nodes around the wrist from the wrist central.

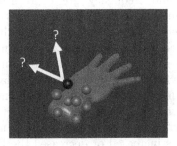

Fig. 5. Directional ambiguity in point stimuli.

Motion Stimuli. For the motion stimuli, we added a start effector as the reference point for users to make up the direction with the end effector used for point stimuli. The start effect is in the opposite location of the end effector along the intended direction. Therefore, each direction cue in motion stimuli consists of two sets of tactors acting sequentially, with each set acting as the start and end effectors, respectively, causing a motion stroke sensation, known as apparent motion [1]. We argued that motion stimuli would have higher recognition rates than point stimuli. However, motion stimuli would suffer from a comparatively lower refresh rate.

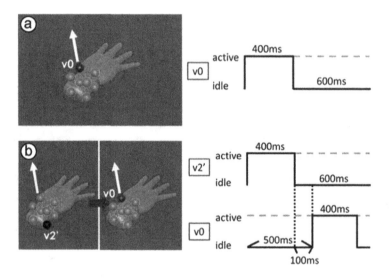

Fig. 6. Illustration of exact tactile feedback signals for (a) point stimuli and (b) motion stimuli.

3.2 Haptic Feedback

The tactors' amplitude and frequency are controlled by the pulse-width-modulation (PWM) through Arduino. The range of the vibration amplitude in the experiment was between 1.8 g and 2.5 g, and the corresponding vibration frequency was between 220 Hz and 250 Hz. To ensure the same perceptual intensity of vibrotactile feedback in different locations on the skin, we conducted a simple sensitivity test with 8 participants from our lab. In each form of distance (e.g., 4 cm, 6 cm, and 8 cm), we used the right motor on the volar side of the wrist in the first wristband, with a fixed 4.0 V reference motor and the 7 remaining motors of the wristbands being the adjusting motors. For each adjusting motor, the participants increased or decreased the PWM values until their intensity was rated the same as that of the reference motor, following one-up one-down adaptive procedures [11]. Their differences were compensated for in the official study.

The active duration of each single vibration was 400 ms. The delimiter between consecutive direction cues was 600 ms. Therefore, for point stimuli, a single direction cue took 400 ms, and its refresh rate was once per second, as shown in Fig. 6a. As for motion stimuli (Fig. 6b), the 2 sets of vibrations for the start and end effectors took 900 ms, with a 100-ms divider of effectors. The same 600-ms delimiter was applied to consecutive motion stimuli.

4   User Study

The experiment was designed to find the *effective distance* for two wristbands and *direction feedbacks*.

4.1   Experiment Design

There were 2 variables: distance and feedback. Feedback refers to 2 various cue designs: motion stimulus and point stimulus. Distance investigates the separation between 2 wristbands that are 4 cm, 6 cm, or 8 cm. The experiments followed a mixed design, in which distance was a between-subject and feedback was a within-subject variable, in order to reduce the term of study for each participant. The conditions in the respective variables were presented in counterbalanced order.

4.2   Procedure

For each participant, we arranged the tactors on the 2 wristbands to fit the study condition and the participant's wrist size. We then helped the participant to put on the wristbands with proper tightness. Before the study started, all of the vibrotactile directional patterns were fully explained. Each tactor was independently actuated to ensure that the feedback was correctly perceived. The participants were instructed to maintain their posture without rotating the wrist

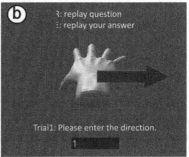

Fig. 7. (a) Environmental setup. (b) The interface where the red arrow represents the input answer form the participant.

and without having the wrist contact the table during the study, as shown in Fig. 7a. To input a direction, the participants selected from a sheet of schematic vibrotactile directions placed on the table in front of them. The participants wore a headset playing pink noise to block the noise caused by the vibrators.

There were 26 directional patterns. For each trial, the program randomly played 1 of the 26 directions. The participant could replay the vibration as many times as he or she wanted. In the training phase, the correct answer was displayed after the participant's input, while in the testing phase, this information was not provided. The interface of the experiment is shown in Fig. 7b.

In the training phase, each participant completed 52 training trials–each direction in the pre-defined 26 vibrotactile patterns was selected twice–and the order of the directions was randomized. The recognition rate in the training phase needed to be above 50% before the participant could proceed to the testing phase. This threshold was used to ensure that the participant fully understood the vibration patterns. The testing phase contained 78 trials: all 26 directions appeared 3 times, and the directions were randomized. During the testing trials, the participants were asked to replay each trial as little as possible and complete each trial as quickly and accurately as possible. There was a 5-min break between each 26 trials. The time of each break could be extended according to the participant's condition. The study took approximately 120 min per participant.

4.3   Participants

We recruited 36 participants (18 males) from our university aged 24.4 on average, ranging from 20 to 30 years old. Only right-handed participants were recruited.

Table 1. Average recognition rate and result of pairwise t-tests in distance configurations.

Distance	Motion stimuli (SD)	Point stimuli (SD)	T-test
4 cm	80.24% (7.74%)	70.95% (10.51%)	0.0124 (*)
6 cm	80.19% (14.13%)	69.58% (9.87%)	0.0048 (**)
8 cm	88.47% (6.28%)	82.95% (4.84%)	0.0522

4.4   Results

We considered 1 participant's data in the 6-cm motion condition who reported sensitivity fatigue after the study. Two participants in the 8-cm point stimulus condition were also considered as outliers when using Tukey's fences [17]. Their data were removed from the analysis.

Recognition Rate. A two-way mixed ANOVA was applied. The average recognition rates for each condition in the different distances are shown in Table 1. The standard deviation and p-value of the t-test are also indicated in the table. The results revealed a significant main effect of recognition rate on distance, with $F(2, 33) = 5.452$, $p = 0.009 < 0.05$ and $\eta_p^2 = 0.248$, and on Feedback, with $F(1, 33) = 12.862$, $p < 0.05$, and $\eta_p^2 = 0.28$ (strong effect). Motion stimuli (83.0%) are on average more discernible than point stimuli (74.5%). Pairwise Tukey HSD tests further showed significant differences between the 8-cm and 6-cm configurations (85.7% vs. 74.9%) and the 8-cm and 4-cm (85.7% vs. 75.6%) configurations (both $p < 0.05$), but there was no difference between the 6-cm and 4-cm configurations.

According to Fig. 8, the motion stimuli were more discernible than the point stimuli across all of the distances. The lower recognition rates in the point-stimuli conditions may have been caused by the direction ambiguity from the point stimuli (see Fig. 5). Also, the higher recognition rates for the motion stimuli may have been due to double information, with the start point and end point providing participants more information with which to assist their the location judgment.

To gain more insight about the interaction between feedback and distance, we ran paired t-tests comparing the 2 feedback conditions for each distance. The results showed significant differences in the 4-cm ($p < 0.05$) and 6-cm ($p < 0.05$) conditions but no difference in the 8 cm conditions (see Table 1). This implies that the enhancement of recognition rates for motion stimuli over point stimuli decreases when the distance between 2 wristbands increases. In addition, the t-test results for 8 cm were on the borderline of statistical significance, which indicates that for a distance above 8 cm between 2 vibrotactile wristbands, the difference in recognition rates between 2 feedback conditions might decrease, and

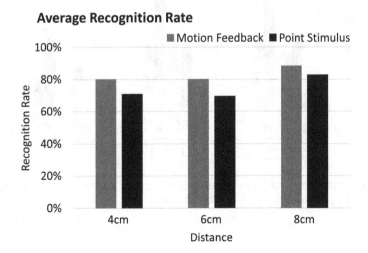

Fig. 8. The average recognition rate in motion stimuli and point stimuli with respect to different wristband distances.

the direction ambiguity caused by point stimuli might be compensated for by a greater distance between the wristbands.

Cross-Type Error. According to the number of tactors around the wrist, the 26 vibration patterns can be divided into 3 types, as shown in Figs. 3 and 4. We further characterized the errors by classifying them into *cross-type error* and *within-type error*. Cross-type errors occur in a different type of direction than the type of the correct direction, and within-type errors are those mistaken within the same type as the correct direction. As Fig. 9 shows, for point stimuli, a higher error rate is caused by cross-type error as compared to for motion stimuli. This difference may be caused by the direction ambiguity when using point stimuli.

Angular Error. Figure 10 shows the angular errors across all trials. The angular error is defined as the angle between a direction recognized by the user and the corresponding correct direction. Most of the angular error was found below 90°. This indicates that users might not be precise in telling the exact direction below 90°, although they still have a good idea about the general direction. Moreover, in general, the motion stimuli caused error below 45° less frequently than point stimuli did, revealing that motion stimuli attribute a stronger cutaneous direction cue around the wrist than point stimuli do.

Fig. 9. Cross-type error rate and within-type error rate in motion stimuli and point stimuli with respect to different wristband distances.

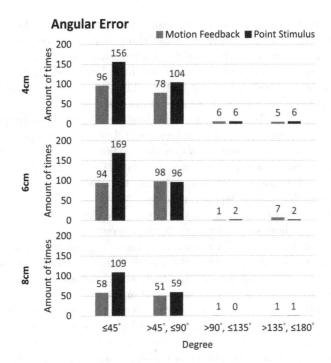

Fig. 10. Angular errors with respect to difference wristband distances and feedback types.

5   Discussion and Limitations

With the different distances between the 2 vibrotactile wristbands, the average recognition rate for our 26 directional vibration patterns was higher with motion stimuli than with point stimuli in the 4-cm and 6-cm conditions. This suggests different wearable forms and haptic direction feedback designs.

For instance, when a form of sports wristband interface for 3D guidance is preferable (e.g., 4 cm or 6 cm long), motion stimuli should be used to ensure a high recognition rate. However, the displaying of motion stimuli has a discrete nature, requiring consecutive motion stimuli to be divided with a delimiter, leading to a lower refresh rate than for point stimuli. On the other hand, when a wearable form of sleeve cover is an option, using point stimuli can be good way to keep up the recognition with a higher refresh rate than that of motion stimuli.

We analyzed the cross-type error rate and the angular error for each feedback and distance condition. The results reveal that a higher cross-type error rate and a larger amount of angular errors under 90° occurred in the point stimulus than in the motion feedback conditions. This result echoed our argument that point stimuli are easily attributed to direction ambiguity when dense direction cues are desirable. This ambiguity is greatly reduced in motion stimuli, due to the clear direction cue being contributed by the phantom effect from two tactors acting sequentially.

In the current study, ERM vibrotactors were adopted for the prototype, which are known to have a lower response time than linear-resonant-actuator (LRA) vibrotactors. Some of the participants reported a residual vibration during the experiments. This residual vibration may have been due to the long response time (about 100 ms) of our tactors. This effect can be mitigated with an LRA vibrotactor, whose short response time allows for sharper vibrations.

In addition, ERM vibrotactors whose vibration intensity and frequency are bound with the voltage do not allow us to separately evaluate the potential influences of the 2 parameters. However, we set the voltage to generate vibration from 210 Hz to 250 Hz, which was previously proved effective for cutaneous perception [6]. Further investigation into their interaction requires actuators such as LRA vibrotactors, which afford independent control.

Due to the explorative nature of this study, the participants were in a stationary situation; they were seated with their wrist not contacting anything. It is unclear how users being in mobile situations (e.g., standing and walking) would have mattered for the results. We observed that when users rotated their wrist, the tactors would be relocated, which would distort the intended cubic geometry, due to uneven skin being stretched around the wrist. This effect was well-controlled in our experiment with a fixed wrist posture. This relocation can be corrected by selectively using a high density of tactors to keep the cubic geometry in accordance with the wrist orientation, or potentially by compensating for the relocation with phantom sensations that fine-tune the perceived actuating locations. Further studies are required to determine the potential side effects. In addition, our study was conducted with a mixed design to reduce the overhead per participant. A complete within-subject design is invited to confirm the results.

6   Conclusion

This work presented a comparative study to inform the effective design of a wrist-worn haptic interface for 3D directional cues. While previous research has revealed various designs for sparse direction feedbacks (e.g., 4 to 6 directions), we were interested in how much we could boost the direction resolution by using sparse tactors around wrist. Our 36-participant study with an 8-tactor cuboid layout on the wrist suggested that motion stimuli are generally more discernible than point stimuli across all distances; however, the former's benefits are suppressed with an increased space between the wristbands. In future work, we would like to extend the study to mobile situations such as standing and walking. In addition, we learned some distinct strengths of motion and point stimuli. A mixed design using both stimuli interactively can strike a balance among wearable form, recognition rate, and feedback types.

Acknowledgment. This research was supported in part by the Ministry of Science and Technology of Taiwan (MOST 108-2633-E-002-001), National Taiwan University, Intel Corporation, and Delta Electronics.

References

1. Anstis, S.M., Mackay, D.M.: The perception of apparent movement [and discussion]. Philos. Trans. R. Soc. Lond. B Biol. Sci. **290**(1038), 153–168 (1980)
2. Chen, H.-Y., Santos, J., Graves, M., Kim, K., Tan, H.Z.: Tactor localization at the wrist. In: Ferre, M. (ed.) EuroHaptics 2008. LNCS, vol. 5024, pp. 209–218. Springer, Heidelberg (2008). https://doi.org/10.1007/978-3-540-69057-3_25
3. Ho, C., Tan, H.Z., Spence, C.: Using spatial vibrotactile cues to direct visual attention in driving scenes. Transp. Res. Part F: Traffic Psychol. Behav. **8**(6), 397–412 (2005)
4. Hong, J., Stearns, L., Froehlich, J., Ross, D., Findlater, L.: Evaluating angular accuracy of wrist-based haptic directional guidance for hand movement. In: Proceedings of the 42nd Graphics Interface Conference, GI 2016, pp. 195–200. School of Computer Science, University of Waterloo, Waterloo, Ontario, Canada. Canadian Human-Computer Communications Society (2016)
5. Jin, Y.S., Chun, H.Y., Kim, E.T., Kang, S.: VT-ware: a wearable tactile device for upper extremity motion guidance. In: The 23rd IEEE International Symposium on Robot and Human Interactive Communication, pp. 335–340, August 2014
6. Johnson, K.O.: The roles and functions of cutaneous mechanoreceptors. Curr. Opin. Neurobiol. **11**(4), 455–461 (2001)
7. Kapur, P., Jensen, M., Buxbaum, L.J., Jax, S.A., Kuchenbecker, K.J.: Spatially distributed tactile feedback for kinesthetic motion guidance. In: 2010 IEEE Haptics Symposium, pp. 519–526, March 2010
8. Karime, A., Al-Osman, H., Gueaieb, W., El Saddik, A.: E-glove: an electronic glove with vibro-tactile feedback for wrist rehabilitation of post-stroke patients. In: 2011 IEEE International Conference on Multimedia and Expo, pp. 1–6, July 2011
9. Lee, J., Han, J., Lee, G.: Investigating the information transfer efficiency of a 3x3 watch-back tactile display. In: Proceedings of the 33rd Annual ACM Conference on Human Factors in Computing Systems, CHI 2015, pp. 1229–1232. ACM, New York (2015)
10. Lee, S.C., Starner, T.: Buzzwear: alert perception in wearable tactile displays on the wrist. In: Proceedings of the SIGCHI Conference on Human Factors in Computing Systems, CHI 2010, pp. 433–442. ACM, New York (2010)
11. Leek, M.R.: Adaptive procedures in psychophysical research. Percept. Psychophys. **63**(8), 1279–1292 (2001)
12. Liao, Y.C., Chen, Y.L., Lo, J.Y., Liang, R.H., Chan, L., Chen, B.Y.: Edgevib: effective alphanumeric character output using a wrist-worn tactile display. In: Proceedings of the 29th Annual Symposium on User Interface Software and Technology, UIST 2016, pp. 595–601. ACM, New York (2016)
13. Lieberman, J., Breazeal, C.: Tikl: development of a wearable vibrotactile feedback suit for improved human motor learning. IEEE Trans. Rob. **23**(5), 919–926 (2007)
14. Pasquero, J., Stobbe, S.J., Stonehouse, N.: A haptic wristwatch for eyes-free interactions. In: Proceedings of the SIGCHI Conference on Human Factors in Computing Systems, CHI 2011, pp. 3257–3266. ACM, New York (2011)
15. Schönauer, C., Fukushi, K., Olwal, A., Kaufmann, H., Raskar, R.: Multimodal motion guidance: techniques for adaptive and dynamic feedback. In: Proceedings of the 14th ACM International Conference on Multimodal Interaction, ICMI 2012, pp. 133–140. ACM, New York (2012)

16. Sergi, F., Accoto, D., Campolo, D., Guglielmelli, E.: Forearm orientation guidance with a vibrotactile feedback bracelet: on the directionality of tactile motor communication. In: 2008 2nd IEEE RAS EMBS International Conference on Biomedical Robotics and Biomechatronics, pp. 433–438, October 2008
17. Tukey, J.W.: Exploratory Data Analysis. Addison-Wesley, Boston (1977)
18. Weber, B., Schätzle, S., Hulin, T., Preusche, C., Deml, B.: Evaluation of a vibrotactile feedback device for spatial guidance. In: 2011 IEEE World Haptics Conference, pp. 349–354, June 2011

How Design with Intent Cards Facilitate Behavioral Design Ideation for Humanities, Design, and Engineering Students

Yuan-Chi Tseng[1,2(✉)]

[1] Institute of Service Science, National Tsing Hua University, Hsinchu, Taiwan
Yuanchi.Tseng@gmail.com
[2] Industrial Engineering and Engineering Management, National Tsing Hua University, Hsinchu, Taiwan

Abstract. The behavioral design not only requires people with a background in the humanities and behavioral sciences to understand behavioral strategies, but also requires designers to translate these behavioral strategies into functional, system, or product details. It also requires engineering technologists to develop design concepts into real products. This study investigates the ideation ability of humanities, design, and engineering students, as well as the perspectives of their ideas. This study explores whether the Design Intent (DwI) cards, a card-based design toolkit designed to promote social and environmental behavior, can help students from different disciplines (humanities, design, and engineering) to generate better ideas to help target audiences change behavior. The empirical results show that students from design and humanities disciplines have better initial ideas. In addition, with the use of DwI cards, the number of ideas from students from all three disciplinary backgrounds, the rate of ideation from humanities and design students, and the quality of ideas from engineering students has been improved. However, quantitative data show that the DwI cards cannot alleviate differences in ideation among students from the three disciplines. On the other hand, the qualitative results of the post-interview show that after using DwI cards, the perspectives of ideas have become broader and more homogeneous among students from the three disciplines. Finally, this study discusses insights into future research and design practices for multidisciplinary collaborative teams, involving how to use the card-based toolkit DWI to allow team members to have a common communication language and idea base to facilitate the ideation stage of the design process.

Keywords: Design ideation · Design education · Engineering education · Disciplinary backgrounds · Design with Intent · Behavioral change · Communication

1 Introduction

Ideation is one of the stages of the design process, which usually follows the identification and definition of problems, opportunities, or needs of target users. This is a creative

© Springer Nature Switzerland AG 2020
P.-L. P. Rau (Ed.): HCII 2020, LNCS 12192, pp. 183–199, 2020.
https://doi.org/10.1007/978-3-030-49788-0_14

process that can generate new ideas individually or collaboratively to solve problems that have been discovered (Basadur et al. 1982; Daly et al. 2016; Faste et al. 2013). A good idea is key to many successful design projects, including behavioral design, which helps individuals make behavioral changes. However, generating innovative ideas is not simple, especially when it comes to solving complex problems, such as those related to social or environmental issues. Designing persuasive technology systems to promote behavior change often requires multidisciplinary knowledge, including humanities, design, and engineering (Agapie et al. 2016; Chou and Tseng 2018; Consolvo et al. 2009; de Vries et al. 2016; Fogg 2002; He et al. 2010; Hekler et al. 2013; Klasnja 2011; Kuo et al. 2019; Li and Tseng 2018; Tseng et al. 2018; Tseng and Chiu 2017) to ensure the optimal balance is achieved between desirability, feasibility, and viability (Brown 2009; Kandachar et al. 2010; Smith 2007) when developing products, services, and technologies that promote people's behavior change. An HCI professional, usually from the humanities, design, and engineering disciplines, is well-suited for collaboration on behavioral design (behavioral change design) projects. However, perspectives, terminologies, and value propositions are very different among disciplines. While the diversity of disciplinary knowledge can make the team come up with more and better ideas (Chan et al. 2012), the disciplinary differences due to different disciplinary training may have negative effects on the ideation performance of the multidisciplinary team (Diehl and Stroebe 1987; Jehn et al. 1999; Pearsall et al. 2008; Pelled et al. 1999; Price et al. 2006; Rafols and Meyer 2006).

Many card-based design tools have been developed to help designers generate innovative ideas to solve problems in their design projects (Roy and Warren 2019; Wölfel and Merritt 2013). Recently, Design with Intent method toolkit, a card deck developed by Lockton and colleagues (Lockton et al. 2009; Lockton et al. 2008; Dan Lockton et al. 2010a, b), is particularly designed to stimulate ideas to promote environmentally and socially beneficial behavioral changes (see Fig. 1). These cards provide real-word examples of behavioral design, drawn from knowledge in many different disciplines related to behavior change. DwI cards may also be effective in helping people generate more and better behavior change ideas. More specifically, it may also be useful in mitigating disciplinary differences (e.g., different terminologies, different idea perspectives, different value propositions, etc.) in the ideation process.

Therefore, the purpose of this study is to investigate the ideation disciplinary differences of humanities, design, and engineering students, and to investigate whether DwI cards can alleviate such differences, thereby allowing students from different disciplines to produce highly homogeneous and equal-quality ideas. This study conducted a two-stage ideation experiment to measure quantitative ideation performance, including the number of ideas, the rate of ideation, and the quality of idea. This study also conducted post-interviews to investigate whether participants' ideas were derived from their original disciplines or inspired by DwI cards. More importantly, this study also evaluated the characteristics of their ideas to examine whether their ideas are homogenous. The empirical findings show that the initial ideation performance varies among humanities, design and engineering students. It seems that our design students performed better than engineering students. DwI cards had different effects on the ideation performance of students in the three disciplines. It boosted the number of ideas in all three disciplines.

It helped our student participants in the humanities and design majors to generate ideas faster and helped student participants in engineering to generate better quality ideas. The different effects of DwI cards on the humanities, design and engineering students' idea generation suggest that in the future, card-based design tools should be tailored to the different needs of different disciplines to eliminate barriers to interdisciplinary collaboration.

Fig. 1. An example card of Design with Intent (Lockton et al. 2010b). Each card of DwI has a question phrase to stimulate users' ideas about behavior change. A sample image is shown next to the question to let users quickly grasp the concept.

The results of the evaluation of idea characteristics and analysis of post-experiment interviews also show that DwI cards had an impact on expanding the ideation perspectives of humanities, design and engineering students. DwI cards particularly helped humanities students use the product and technology-oriented solutions, helped design students use social and behavioral strategies and technology in their ideas, and helped engineering students extend their ideas from using their domain knowledge, such as ICT techniques, to use behavioral intervention strategies learned from the DwI cards. The DwI allowed students in each discipline to extend their ideation perspectives beyond their original knowledge, and as a consequence, the ideation perspectives of the three disciplines became more identical. The results show that using DwI cards is an effective and cheap way to alleviate the influence of disciplinary differences in generating behavioral change ideas.

2 Related Works

2.1 Disciplinary Difference in Ideation

Due to the different emphasis given to core training in different disciplines, the performance of ideation, problem-solving and the ideation perceptive varies among humanities, design, and engineering students (Jablokow et al. 2015; Lawson 1979). For example, Lawson (1979) found distinct problem-solving strategies when experimentally comparing the behavior of senior undergraduate architecture and science students. Lehman et al. (1988) also found that different postgraduate training results in different performance on different problem-solving tasks. Ideation behavior has also been found to be related to personality traits (Batey et al. 2010) and cognitive styles (Aggarwal and Woolley 2013). For example, Ansburg and Hill (2003) found that cognitive traits are related to the performance in creative problem-solving.

The diversity of individual members is often seen as a double-edged sword in team collaboration (Gilson et al. 2013; Milliken and Martins 1996). Individual and disciplinary differences may harm ideation performance in interdisciplinary collaboration (Rafols and Meyer 2006). For example, Rafols and Meyer (2006) proposed that there is a cost in the integration of disciplines in team collaboration. A team needs to put additional efforts into coordinating and communicating different knowledge of various disciplines. Disciplinary differences also result in culture barriers, which can hinder the effectiveness and efficiency of groups in the short term (Jehn et al. 1999). To ensure the success of teamwork, finding a way to alleviate the impact of the individual and disciplinary differences on ideations has become one of the most important issues in interdisciplinary collaborations.

2.2 Design with Intent Toolkit for Behavior Design Ideation

Many card-based design tools has been developed to stimulate idea generation, structure design or help users and designers communication during the design process (Beck et al. 2008; Deng et al. 2014; IDEO, 2003; Lafrenière et al. 1999; Lockton et al., 2009; Lockton et al. 2008; Lockton et al. 2010a, b; Lucero and Arrasvuori 2010; Roy and Warren 2019; Wölfel and Merritt 2013). Among these, Design with Intent (DwI) toolkit (Lockton et al. 2008; Lockton et al. 2010a, 2b) was developed to help designers design behavioral interventions to promote behavior change. Lockton et al. (2008) found that it is difficult for designers to directly apply persuasive technologies and theories to design projects that change user behavior. Therefore, Lockton and colleagues proposed the Design with Intent method and develop a card-based toolkit to help designers generate ideas. Designers who are unfamiliar with the related theories of behavior change and potential persuasive technologies can learn from the examples shown on the DwI cards to generate broader behavior design ideas. Design with Intent has eight lenses, including, Architectural, Errorproofing, Interaction, Ludic, Perceptual, Cognitive, Machiavellian, and Security lenses. These lenses are derived from many knowledge related to behavior change, such as environmental and ecological psychology, affordance, persuasive technology, gamification, product semantics, semiotics, gestalt psychology, behavioral economics, decision science, and marketing, etc. As shown in Fig. 1, each DwI card has an illustrative example and a sentence or a question related to behavior change to stimulate behavior design ideation. For example, the personality card asks a question, "Can you give your system a personality or character that engages users becoming a 'social actor", and shows a real example that responds to the question (see Fig. 1). Studies have shown that design-by-analogy, accessing and transferring previously obtained examples to support current design problems, is the center of innovative design and product development (Ball and Christensen 2009; Ball et al. 2004; Casakin 2004; Casakin and Goldschmidt 1999; Christensen and Schunn 2007; Moreno et al. 2014; Visser 1995). From this perspective, DwI cards that use real examples of behavior change designs have the potential to broaden ideation perspective and, as a result, improve ideation.

3 Experiment

To study students' ideation performance in the humanities, design and engineering disciplines, whether the DwI cards can help students generate more and better ideas, and whether it can alleviate the disciplinary differences, this study conducted an experiment with two ideation stages (pre-card: initial stage and post-card: card stimulation stage). A post-interview was conducted after each stage to understand how our student participants came up with their ideas and where these ideas came from.

3.1 Method

Experiment Design

This study used a mixed-model design, comprising discipline (humanities, design, and engineering) and card (DwI card, placebo card) as a between-subjects factor, and ideation stage (pre-card, post-card) as within-subjects factors. Every participant participated in both ideation stages. The ideation experiment was conducted in the Department of Industrial Design at the National Cheng Kung University, and ethical approval was granted by the Human Research Ethics Committee. Informed consent was obtained from all participants before the experiment.

Participants

Twelve humanities, twelve design and Twelve engineering students from the University of National Cheng Kung University aged 20–25 years, volunteered to participate in the experiment. Six participants from each disciplinary background were randomly assigned to the DwI card group, while the other half were assigned to the placebo card group (the Ripple card was used as a placebo in this experiment). They received 150 TWD ($4.73) as compensation for their time.

Materials

A question sheet listed three ideation questions was provided to each participant. Unlimited answer sheets were provided. Each answer sheet had two ideation answer boxes (see Fig. 2). There were three ideation questions: (1) how to help people save water at home; (2) how to prevent people from staying up late and help them go to bed regularly; and (3) how to help people do more exercise during leisure time. All of these design issues are related to behavior change design. Participants had to generate ideas about products, services or technology systems to help people build up pro-environmental behaviors, healthy behaviors, and lifestyles. Student participants were asked to write their ideas on the answer sheets. They were allowed to draw to help them explain their ideas. They were informed that any content in an answer box would be counted as an idea.

Fig. 2. An example of questions for ideation. Participants were allowed to write or draw their ideas on the answer sheet.

There were two kinds of cards used in this experiment. (1) Design with Intent cards (Lockton et al. 2010b), which is developed to help designers to generate ideas for behavior change, was used to examine whether participants could generate more and better ideas. Two graduate research assistants translated DwI cards into the Chinese version. This study conducted a pilot study with 5 student participants to examine the comprehension of our Chinese version DwI cards (see Fig. 3a). As long as one participant claimed that the card was not easy to understand, we modified the card until all the cards were easy to understand. (2) Ripple card (Chu and Tan 2003), which is developed to help people to think positively, were used as a placebo in the experiment (see Fig. 3b). We expected that it would not have an effect on the idea generation in the experiment. It is written in Chinese so we did not need to translate it. Both cards used in this experiment were the Chinese version.

Procedure
The experiment had two ideation stages. The first ideation stage was the pre-card stage, during which participants generated ideas without any external help. Participants were given three questions and unlimited answer sheets. All participants were told they had

(a)

(b)

Fig. 3. a. The Chinese version of DwI cards. They were translated from the original Design with Intent toolkit (Lockton et al. 2010b); b. An example of Ripple Card. Each Ripple card has a motivational message to encourage people to think positively. It is not designed for promoting social or environmental behavior change.

unlimited time to come up with their ideas. They were told to answer one question at a time. After they thought they could no longer generate any more answers for a question, they could ask to give up answering the question and started to answer the next question. After answering all three questions, a semi-structured post-interview was conducted to ask participants how each idea was generated and where it came from.

The second ideation stage was a post-card stage (i.e. card intervention stage). The experimental procedure was the same as the first stage. A post-interview was conducted after the ideation. The only difference was that in the second stage, participants could use the assigned cards (either DwI cards or Ripple cards) to generate ideas. They were free to adopt their ideation strategy and were not forced to use the assigned cards. They were also allowed to ignore the assigned cards if they thought the cards were useless. The three questions used in the second stage were the same as the three questions used in the first stage. The reason this study used the same questions in both stages was that with this design, this study could examine whether DwI cards could help people come up with more solutions to the question they thought they couldn't generate more. This study assumes that the DwI cards can expand people's ideation perspectives, and therefore it should be able to help stimulate more ideas to solve the questions that have been abandoned in the first stage.

Data Analysis

The number of ideas was calculated as follows. All writing and drawing in an answer box were converged to an idea. Ideas to all three questions were summed together to get the number of ideas. The rate of ideation was calculated by dividing the number of ideas by time (hour) that participants spent on generating ideas.

Following Dean et al. (2006), the quality of each idea was evaluated based on three dimensions, including novelty, workability and relevance. Therefore, each idea would receive a score for these three dimensions on the 5-point Likert scale. Two graduate research assistants were recruited and trained to independently evaluate all ideas. First, they evaluated 93 ideas on these three idea quality dimensions. The inter-evaluator reliability was 77.4%, 87%, and 87% for novelty, workability, and relevance, respectively. The overall inter-evaluator reliability was 83.9%. In addition, the two evaluators met together to discuss the scoring criteria and principles they used. They reviewed these criteria and principles together to obtain more general scoring criteria and principles in each quality evaluation dimension. Then, they evaluated the total of 249 ideas with new criteria and principles. Their inter-evaluator reliability became 90.4%, 90.4%, and 88% for novelty, workability, and relevance, respectively. The overall inter-evaluator reliability was 89.2%. When the idea of scoring a difference of more than 2 points on any of the dimensions, the study recruited a third graduate student for judgment. This third evaluator also discussed with the first two evaluators to learn the scoring criteria. Two of the three scores which were closer were retained. Finally, this study averaged the three dimension scores for each idea. The individual's quality of ideas was the average quality of each idea generated by the individual participant.

To understand the experimental participants' ideation perspective, two post-interviews were conducted following each of the ideation section. The post-interview transcripts were codified, analyzed and categorized by thematic analysis (Braun and Clarke 2006). The members of the code team were three graduate research assistants. There were two steps in this process: (1) The code team reviewed the post-interview data and marked the representation of their ideas, and gave these marks descriptive names. (2) the code team grouped concepts with similar descriptive name into a category (i.e., perspective). This study then investigated the difference in ideation perspectives in the pre- and post-ideation stages to investigate how DwI cards can change the ideation perspectives of humanities, design and engineering students.

4 Results and Discussion

Table 1 shows the results of the number of ideas, the mean rate of ideation, and the mean quality of ideas of each participant in the pre-card and post-card ideation stages. There were six groups: Humanities with DwI, Humanities with Ripple, Design with DwI, Design with Ripple, Engineering with DwI and Engineering with Ripple.

4.1 The Initial Ideation Performance

Because one of the purposes of this study is to examine whether students' initial ideation performance was different in the three disciplines, one-way ANOVA was used to compare the number of ideas, the rate of ideation and the quality of ideas in the different discipline in the pre-card ideation stage. Post-hoc comparisons were conducted using Fisher least significant difference (LSD) corrections. This study used data from DwI and Ripple group (N = 12) in the first ideation stage because the two card groups were in the same condition during the pre-card ideation stage. During the pre-ideation stage, neither group was intervened by the card.

Table 1. The results of the number of ideas, the mean rate of ideation, and the mean quality of ideas of each participant in the pre-card and post-card ideation stages. The ideation card (DwI & Ripple), disciplinary background (Humanities, Design, Engineering) were between-subjects variables. There were six groups: Humanities with DwI, Humanities with Ripple, Design with DwI, Design with Ripple, Engineering with DwI and Engineering with Ripple). Each group had 6 participants.

	Humanities				Design				Engineering			
	DWI		Ripple		DWI		Ripple		DWI		Ripple	
	Pre	Post	Pre	Post	Pre	Post	Pre	Post	Pre	Post	Pre	Post
Number of idea	3	21	14	6	24	24	34	6	16	19	16	3
	32	44	9	5	22	25	12	3	13	17	3	3
	15	15	20	8	12	18	24	18	9	11	10	10
	21	18	24	13	22	24	22	6	14	16	20	4
	9	12	21	10	20	60	24	18	14	17	11	3
	28	22	28	11	8	6	30	10	17	14	10	3
Mean	18	22	19.33	8.83	18	26.17	24.33	10.17	13.83	15.67	11.67	4.33
SD	11.14	11.4	6.86	3.06	6.45	18.05	7.53	6.46	2.79	2.8	5.82	2.8
Rate of ideation	4.77	16.97	22.75	12.78	21.35	26.58	29.26	11.31	26.62	15.81	24.63	20.73
	23.94	30.4	18.42	7.79	23.94	30.34	13.02	2.67	21.63	14.14	4.35	3.49
	23.77	19.79	40.34	13.27	12.12	13.51	21.18	11.77	6.37	6.17	18.91	11.73
	22.35	37.92	25.12	16.6	19.79	16.04	20.6	7.54	13.44	16.86	28.45	10.26
	6.8	7.77	24.8	11.41	16.44	18.3	18.17	12.64	15.42	23.51	23.19	6.42
	19.3	22.75	22.11	25.71	10.31	6.76	34.07	18.87	20.81	16.1	14.41	3.96
Mean	16.82	22.6	25.59	14.59	17.33	18.59	22.72	10.8	17.38	15.43	18.99	9.43
SD	8.73	10.53	7.61	6.15	5.35	8.65	7.66	5.41	7.15	5.57	8.65	6.45
Quality of idea	3.06	3.15	3.42	3.33	3.17	3.18	3.13	3.29	2.98	3	2.73	2.39
	3.25	3.2	2.94	2.61	2.9	3.27	3.11	3.12	2.97	2.95	3.33	2.89
	3.04	3.29	3.1	2.68	3.07	3.26	3.33	3.36	3.04	3.04	3.27	3.19
	3.21	3.21	3.2	3.07	3.24	3.31	3.24	3.42	2.95	3.26	2.76	2.44
	2.98	3.23	2.97	3.18	3.11	3.32	3.11	3.38	2.95	3.12	2.89	2.78
	2.81	2.76	2.82	2.78	3.34	3.26	3.02	3.19	3.06	3.25	2.94	2
Mean	3.06	3.14	3.07	2.94	3.14	3.27	3.16	3.29	2.99	3.11	2.99	2.62
SD	0.16	0.19	0.21	0.29	0.15	0.05	0.11	0.12	0.05	0.13	0.26	0.42

The Number of Initial Ideas

Discipline had a significant main effect on the number of Initial ideas, $F(2, 33) = 4.366$, $p = 0.021$. In post hoc comparisons, the mean number of ideas of students in the design discipline ($M = 21.167$, $SD = 7.46$) was significantly greater than that of the engineering students ($M = 12.75$, $SD = 4.49$), $p = 0.007$, while humanities students ($M = 18.67$,

SD = 8.85) were not significantly different in the number of ideas from design and engineering students.

The Rate of Initial Ideation

Regarding the initial conception rate, there was no significant main effect of discipline, $F(2, 33) = 0.445$, $p = 0.645$. Post hoc comparisons revealed no significant differences between the three disciplines. This shows that without the help of tools, the speed of ideation was about the same between the three disciplines.

The Quality of Initial Ideas

Discipline did not have a significant main effect on the initial performance of the quality of ideas, $F(2, 33) = 2.863$, $p = 0.071$. Post hoc comparisons showed that the mean quality of ideas of design students (M = 3.15, SD = 0.13) was significantly higher than that of engineering students (M = 2.99, SD = 0.18), $p = 0.023$, while humanities students (M = 3.07, SD = 0.18) was not significantly different from students in design and engineering.

These three results show that the initial ideation of behavioral change design was not the same among students majoring in humanities, design, and engineering. In particular, design students could generate more ideas and better ideas than engineering students. However, students in all three disciplines had the same rate of ideation.

4.2   The Effect of DwI Cards on Ideation

Figure 4 illustrates the mean number of ideas, the mean rate of ideation and the mean quality of ideas for humanities, design, and engineering students. The mean number of ideas, the rate of ideation and the quality of ideas were analyzed within a $3 \times 2 \times 2$ mixed model analysis of variance (ANOVA) with discipline (humanities, design, and engineering) and card (DwI and Ripple) as between-subjects factors, and ideation stage (pre-card stage and post-card stage) as a within-subjects factor to examine whether the DwI cards have an effect on improving the ideation in all three disciplines.

The Number of Ideas. The main effect of discipline ($F2,30 = 4.229$, $p = 0.024$), card ($F1,30 = 5.982$, $p = 0.021$), and ideation stage ($F1,30 = 4.205$, $p = 0.049$) were significant. The significant main effect on card factor was as the expectation that the DwI cards, which is designed for inspiring behavior change design, had an effect on generating more ideas, and the Ripple card was only used for placebo.

More importantly, the interaction between card and ideation stage was significant ($F1,30 = 24.8$, $p < 0.001$). These results indicate that the DwI cards had an effect on generating more ideas than the placebo card after used in the post-card ideation stage, where participants answered the same questions that they had given up in the pre-card ideation stage. This study then further examined the interaction effect of each discipline between card and ideation stage. The results show that the interaction between card and ideation stage was significant in the humanities ($F1,10 = 12.141$, $p = 0.006$), design ($F1,10 = 8.186$, $p = 0.017$) and engineering ($F1,10 = 10.240$, $p = 0.009$). These results suggest that DwI cards had an effect on generating more ideas (on average, 4 more

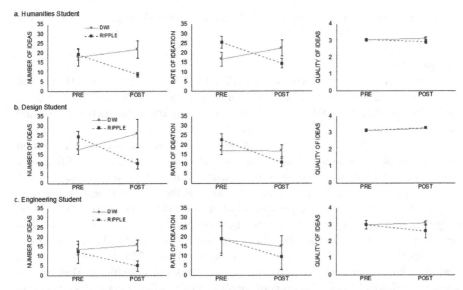

Fig. 4. The mean number of ideas, rate of ideation and quality of ideas in the pre-card and post-card ideation stage, with the DwI and Ripple cards, for (a) humanities, (b) design, and (c) engineering students. Error bars represent the sem.

ideas for humanity students, 7.5 more ideas for design students, and 2.92 more ideas for engineering students) for all three disciplines, after it was used in the post-card ideation stage (see subfigure: number of ideas in Fig. 4a, b and c).

The Rate of Ideation
The main effect of discipline and card were not significant. The main effect of the ideation stage on the rate of ideation was significant ($F1,30 = 16.868$, $p < 0.001$). This result indicates that the rate of ideation became lower in the post-card ideation stage (M = 15.1183, SD = 8.03938) than the pre-card ideation stage (M = 19.8036, SD = 7.77984). This result is consistent with our intuition that those questions that have been answered and given up required more time to generate novel ideas. In addition, using the cards in the second stage required more time to comprehend the contents of the cards.

The interaction between card and ideation stage was also found significant, $F(1,30) = 28.931$, $p < 0.001$. This result shows that participants could generate ideas faster through external stimuli from the DwI cards than when using the placebo card in the post-card ideation stage. Figure 4 shows that the placebo group using Ripple card had a lower rate of ideation than the pre-card ideation stage, while the DwI group using DwI cards had a higher rate of ideation than the pre-card ideation stage. This shows the positive effect of DwI cards on the rate of ideation. This study then further examined the interaction effect between card and ideation stage in each discipline. The results show that the interaction between card and ideation stage was significant in humanities ($F1,10 = 11.320$, $p = 0.007$) and design ($F1,10 = 27.305$, $p < 0.001$), but not in engineering ($F1,10 = 3.542$, $p = 0.089$). These results suggest that while the DwI cards can help students in humanities and design discipline to generate ideas faster (see subfigure: rate of ideation in Fig. 4a and b), it could not help engineering students generate ideas faster (see subfigure: rate of ideation in Fig. 4c).

The Quality of Ideas

The main effect of discipline factor was significant ($F_{2,30} = 7.609$, $p = 0.002$), but that of card factor and ideation stage factor were not significant ($F_{1,30} = 3.008$, $p = 0.093$, and $F_{1,30} = 0.061$, $p = 0.807$, respectively). The 2-way interaction between card and ideation stage ($F_{1,30} = 12.765$ $p = 0.001$) and the 2-way interaction between discipline and ideation stage ($F_{2,30} = 5.617$ $p = .008$) were significant. The three-way interaction between discipline, card, and ideation stage was also significant ($F_{2,30} = 5.062$, $p = 0.013$). The significant interaction between card and ideation stage shows that using DwI cards could help participants generated better ideas in the post-card ideation stage (M = 3.17, SD = 0.15) than in the pre-card ideation stage (M = 3.06, SD = 0.14). This study then further examined the interaction effect between card and ideation stage in each discipline. The results show that the interaction between card and ideation stage in humanities ($F_{1,10} = 3.839$, $p = 0.079$) and design ($F_{1,10} = 0.020$, $p = 0.890$) were not significant, but in engineering ($F_{1,10} = 12.276$, $p = 0.006$) was significant. These results suggest that although the DwI cards could not help students from humanities and design to generate better ideas, it could help engineering students not generate lower-quality ideas after using the DwI cards.

4.3 Sources of Idea

The results of post-interviews show that there were two main sources of ideation. (1) ideas were drawn from personal life experience. Since all three ideation questions were closer to everyday experiences, participants often came up with ideas from their previous life experiences. For example, many participants said that they transferred the way they used to reserve water in the past to support current ideation. (2) ideas came from similar products or services they are familiar with. Participants often used concepts read from newspapers, magazines, social media or technical reports to generate new ideas. Many participants said that their design ideas were directly derived from similar products or concepts that exist in real worlds. Or, they turned part of existing concepts of current products or services into their new design ideas that can solve the design questions of this study. These results are consistent with previous findings, suggesting that the ideation is drawn from past design work, other design projects or products (Ball et al. 2001; Cross 2006)

4.4 The Effect of DwI Cards on Ideation Perspective

The results show that the DwI cards, having many examples of behavior change design from multiple perspectives, had various effects on expanding the design concept space and changing the ideation perspective of students in three disciplines. Next, we describe the results one discipline after another.

Humanities Student

The results of the pre-card ideation post-interview listed in Table 2 show that humanities participants were good at using their previously acquired social and behavioral science knowledge, such as policy promotion and behavior strategies, to addressed the behavior design issues. In the post-card ideation post-interview, all humanities participants agreed that the examples showed on the DwI cards had become a very important knowledge base for their ideation. This finding is consistent with previous studies, showing that people's ideation is usually driven by previous examples (Ball and Christensen 2009; Ball et al. 2004; Casakin 2004; Casakin and Goldschmidt 1999; Christensen and Schunn 2007; Moreno et al. 2014; Visser 1995). They said that after using DwI cards, they not only used the behavior strategies such as reward and punishment or policy promotion but also used products to promote behavior change. DwI cards allowed them to broaden their ideation perspectives. Pictures of real products and technology systems showed on DwI cards motivate students in the humanities to use similar ways to solve behavior change problems.

Table 2. Ideation perspectives for students in the humanities, design and technology discipline in the pre- and post-ideation stage. Bold typeface indicates new perspectives in the post-card ideation stage.

Discipline	Pre-card ideation stage	Post-card ideation stage
Humanities	Policy promotion Reward & punishment	Policy promotion Reward & punishment **Products** **Behavior change strategies**
Design	Products App	Products App **Behavior change strategies** **Reward & punishment**
Engineering	Sensing technology ICT App	Sensing technology ICT App **Behavior change strategies**

Design Student

Design students in the pre-card ideation stage of this experiment preferred to generate ideas using products and mobile applications to solve problems. Sometimes they would also use reward and punishment mechanisms. After using the DwI cards, students in design majors started using reward and punishment or more extensive behavior change strategies to promote behavior change, although the major ideas in this stage were still related to products or mobile applications (see Table 2).

Engineering Student

This study found that the ideation perspectives from engineering students were more limited to technology. They preferred to monitor the user's physical condition and provide feedback to the user and solve the user's problems through sensing and mobile technology. Some participants had experience in mobile application programming, so their ideas were majorly based on mobile applications. After using DwI cards, participants in the engineering background started using behavior change strategies to solve the problem. Their ideas had more perspectives (see Table 2). They also showed that they would be happy to apply the real examples showed on the DwI cards to their new design ideas.

All students from three disciplines combined the ideas they generated in the pre-card ideation stage with the strategies shown on the DwI cards to generate new ideas in the post-card ideation stage and, as a consequence, supplement the initial ideation perspectives. As students in each discipline had more and broad perspectives, the ideas of students from different disciplines became more intersecting and homogeneous.

5 Conclusion

Ideation is a creative process that generates ideas that solve problems. Individual differences arising from different training across academic disciplines may harm team collaboration in ideation. To understand whether the design with intent method card, a card-based design tool developed to promote socially and environmentally beneficial behavior, can help students in the humanities, design and engineering disciplines generate more and better ideas and whether it can increase more and broaden ideation perspectives and alleviate disciplinary differences, an experiment was conducted. The results show that the initial ideation performances were different in quantity and quality in three disciplines (humanities, design, and engineering). This study found that DwI cards can help students in all three disciplines generate more ideas. However, it can only help students in the humanities and design majors to generate ideas faster, and it can only help engineering students generate higher-quality ideas. It seems that the DwI cards cannot quantitatively alleviate the disciplinary differences in ideation. The different impacts of DwI cards on humanities, design and engineering students imply that DwI itself or how DwI cards are used should be tailored to different disciplines to alleviate the impact of disciplinary differences and, as a consequence, improve the ideation in multidisciplinary teamwork. Nevertheless, this study shows that DwI cards can help students increase and broaden their perspectives and generate ideas that exceed the limits of their knowledge. Moreover, it helps participants from different disciplines to generate intersecting and homogeneous ideas, thereby improving communication in multidisciplinary teamwork.

Acknowledgments. Yuan-Chi Tseng gratefully acknowledges two grants from the Ministry of Science and Technology (Grant No. 107-2221-E-007-114 and 107-2410-H-007-093-MY3). I also gratefully acknowledge the time spent at National Cheng Kung University.

References

Agapie, E., Avrahami, D., Marlow, J.: Staying the course: system-driven lapse management for supporting behavior change. Paper presented at the Proceedings of the 2016 CHI Conference on Human Factors in Computing Systems (2016)

Aggarwal, I., Woolley, A.W.: Do you see what I see? The effect of members' cognitive styles on team processes and errors in task execution. Organ. Behav. Hum. Decis. Process. **122**(1), 92–99 (2013)

Ansburg, P.I., Hill, K.: Creative and analytic thinkers differ in their use of attentional resources. Pers. Individ. Differ. **34**(7), 1141–1152 (2003)

Ball, L.J., Christensen, B.T.: Analogical reasoning and mental simulation in design: two strategies linked to uncertainty resolution. Des. Stud. **30**(2), 169–186 (2009)

Ball, L.J., Lambell, N.J., Reed, S.E., Reid, F.J.: The exploration of solution options in design: a 'naturalistic decision making' perspective, pp. 79–93. Designing in Context, Delft University Press, Delft, The Netherlands (2001)

Ball, L.J., Ormerod, T.C., Morley, N.J.: Spontaneous analogising in engineering design: a comparative analysis of experts and novices. Des. Stud. **25**(5), 495–508 (2004)

Basadur, M., Graen, G.B., Green, S.G.: Training in creative problem solving: Effects on ideation and problem finding and solving in an industrial research organization. Organ. Behav. Hum. Perform. **30**(1), 41–70 (1982)

Batey, M., Chamorro-Premuzic, T., Furnham, A.: Individual differences in ideational behavior: can the big five and psychometric intelligence predict creativity scores? Creat. Res. J. **22**(1), 90–97 (2010)

Beck, E., Obrist, M., Bernhaupt, R., Tscheligi, M.: Instant card technique: how and why to apply in user-centered design. Paper presented at the Proceedings of the Tenth Anniversary Conference on Participatory Design 2008 (2008)

Braun, V., Clarke, V.: Using thematic analysis in psychology. Qual. Res. Psychol. **3**(2), 77–101 (2006)

Brown, T.: Change by design (2009)

Casakin, H.: Visual analogy as a cognitive strategy in the design process: expert versus novice performance. J. Des. Res. **4**(2), 124 (2004)

Casakin, H., Goldschmidt, G.: Expertise and the use of visual analogy: Implications for design education. Des. Stud. **20**(2), 153–175 (1999)

Chan, J., Paletz, S.B., Schunn, C.D.: Analogy as a strategy for supporting complex problem solving under uncertainty. Mem. Cogn. **40**(8), 1352–1365 (2012)

Chou, P.-Y., Tseng, Y.-C.: GreenAction: a service design for promoting pro-environmental behaviour. Paper presented at the Persuasive technology 2018, Waterloo, Canada (2018)

Christensen, B.T., Schunn, C.D.: The relationship of analogical distance to analogical function and preinventive structure: the case of engineering design. Mem. Cogn. **35**(1), 29–38 (2007)

Chu, P., Tan, Y.-M.: Ripple Cards. Senchi Tarot Publishing House, Taiwan (2003)

Consolvo, S., McDonald, D.W., Landay, J.A.: Theory-driven design strategies for technologies that support behavior change in everyday life. Paper presented at the Proceedings of the SIGCHI Conference on Human Factors in Computing Systems (2009)

Cross, N.: Designerly ways of knowing. Springer, London (2006). https://doi.org/10.1007/1-84628-301-9

Daly, S.R., Seifert, C.M., Yilmaz, S., Gonzalez, R.: Comparing ideation techniques for beginning designers. J. Mech. Des. **138**(10), 101108 (2016)

de Vries, R.A., Truong, K.P., Kwint, S., Drossaert, C.H., Evers, V.: Crowd-designed motivation: motivational messages for exercise adherence based on behavior change theory. Paper presented at the Proceedings of the 2016 CHI Conference on Human Factors in Computing Systems (2016)

Dean, D.L., Hender, J.M., Rodgers, T.L., Santanen, E.: Identifying good ideas: constructs and scales for idea evaluation. J. Assoc. Inf. Syst. **7**(10), 646–699 (2006)

Deng, Y., Antle, A.N., Neustaedter, C.: Tango cards: a card-based design tool for informing the design of tangible learning games. Paper presented at the Proceedings of the 2014 Conference on Designing Interactive Systems (2014)

Diehl, M., Stroebe, W.: Productivity loss in brainstorming groups: Toward the solution of a riddle. J. Pers. Soc. Psychol. **53**(3), 497 (1987)

Faste, H., Rachmel, N., Essary, R., Sheehan, E.: Brainstorm, Chainstorm, Cheatstorm, Tweetstorm: new ideation strategies for distributed HCI design. Paper presented at the Proceedings of the SIGCHI Conference on Human Factors in Computing Systems (2013)

Fogg, B.: Persuasive technology: using computers to change what we think and do. Ubiquity 2002, 5 (2002)

Gilson, L.L., Lim, H.S., Luciano, M.M., Choi, J.N.: Unpacking the cross-level effects of tenure diversity, explicit knowledge, and knowledge sharing on individual creativity. J. Occup. Organ. Psychol. **86**(2), 203–222 (2013)

He, H.A., Greenberg, S., Huang, E.M.: One size does not fit all: applying the transtheoretical model to energy feedback technology design. Paper presented at the Proceedings of the SIGCHI Conference on Human Factors in Computing Systems (2010)

Hekler, E.B., Klasnja, P., Froehlich, J.E., Buman, M.P.: Mind the theoretical gap: interpreting, using, and developing behavioral theory in HCI research. Paper presented at the Proceedings of the SIGCHI Conference on Human Factors in Computing Systems (2013)

IDEO, I.: Method cards: 51 ways to inspire design, Palo Alto (2003)

Jablokow, K., Teerlink, W., Yilmaz, S., Daly, S., Silk, E., Wehr, C.: Ideation variety in mechanical design: examining the effects of cognitive style and design heuristics. Paper presented at the ASME 2015 International Design Engineering Technical Conferences and Computers and Information in Engineering Conference (2015)

Jehn, K.A., Northcraft, G.B., Neale, M.A.: Why differences make a difference: a field study of diversity, conflict and performance in workgroups. Adm. Sci. Q. **44**(4), 741–763 (1999)

Kandachar, P.: Designing for global sustainable solutions. In: Ceschin, F., Vezzoli, C., Zhang, J. (eds.) Sustainability in Design: Now! Challenges and Opportunities for Design Research, Education and Practice in the XXI Century, pp. 60–75. Greenleaf Publishing Limited, London (2010)

Klasnja, P., Consolvo, S., Pratt, W.: How to evaluate technologies for health behavior change in HCI research. Paper presented at the Proceedings of the SIGCHI Conference on Human Factors in Computing Systems (2011)

Kuo, H.-C., Tseng, Y.-C., Yang, Y.-T.C.: Promoting college student's learning motivation and creativity through a STEM interdisciplinary PBL human-computer interaction system design and development course. Think. Ski. Creat. **31**, 1–10 (2019)

Lafrenière, D., Dayton, T., Muller, M.: Variations of a theme: card-based techniques for participatory analysis and design. Paper presented at the CHI'99 Extended Abstracts on Human Factors in Computing Systems (1999)

Lawson, B.R.: Cognitive strategies in architectural design. Ergonomics **22**(1), 59–68 (1979)

Lehman, D.R., Lempert, R.O., Nisbett, R.E.: The effects of graduate training on reasoning: Formal discipline and thinking about everyday-life events. Am. Psychol. **43**(6), 431 (1988)

Li, J.-T., Tseng, Y.-C.: The effect of doing and messaging pro-environmental behavior on fostering the behavior. Paper presented at the 2018 CHI Conference Extended Abstracts on Human Factors in Computing Systems, Montreal, Canada, USA (2018)

Lockton, D., Harrison, D., Holley, T., Stanton, N.A.: Influencing interaction: development of the design with intent method. Paper presented at the proceedings of the 4th international conference on persuasive technology (2009)

Lockton, D., Harrison, D., Stanton, N.: Design with intent: persuasive technology in a wider context. In: Oinas-Kukkonen, H., Hasle, P., Harjumaa, M., Segerståhl, K., Øhrstrøm, P. (eds.) PERSUASIVE 2008. LNCS, vol. 5033, pp. 274–278. Springer, Heidelberg (2008). https://doi. org/10.1007/978-3-540-68504-3_30

Lockton, D., Harrison, D., Stanton, N.A.: The design with intent method: a design tool for influencing user behaviour. Appl. Ergon. **41**(3), 382–392 (2010a)

Lockton, D., Harrison, D., Stanton, N.A.: Design with intent: 101 patterns for influencing behaviour through design: Equifine (2010b)

Lucero, A., Arrasvuori, J.: PLEX cards: a source of inspiration when designing for playfulness. Paper presented at the Proceedings of the 3rd International Conference on Fun and Games (2010)

Milliken, F.J., Martins, L.L.: Searching for common threads: Understanding the multiple effects of diversity in organizational groups. Acad. Manag. Rev. **21**(2), 402–433 (1996)

Moreno, D.P., et al.: Fundamental studies in design-by-analogy: a focus on domain-knowledge experts and applications to transactional design problems. Des. Stud. **35**(3), 232–272 (2014)

Pearsall, M.J., Ellis, A.P., Evans, J.M.: Unlocking the effects of gender faultlines on team creativity: Is activation the key? J. Appl. Psychol. **93**(1), 225 (2008)

Pelled, L.H., Eisenhardt, K.M., Xin, K.R.: Exploring the black box: an analysis of work group diversity, conflict and performance. Adm. Sci. Q. **44**(1), 1–28 (1999)

Price, K.H., Harrison, D.A., Gavin, J.H.: Withholding inputs in team contexts: member composition, interaction processes, evaluation structure, and social loafing. J. Appl. Psychol. **91**(6), 1375 (2006)

Rafols, I., Meyer, M.: Knowledge-sourcing strategies for cross-disciplinarity in bionanotechnology. Paper presented at the 2006 Annual Conference, Technology Transfer Society: Next Generation Innovation: New Approaches and Policy Designs (2006)

Roy, R., Warren, J.P.: Card-based design tools: a review and analysis of 155 card decks for designers and designing. Des. Stud. **63**, 125–154 (2019)

Smith, C.E.: Design for the Other 90%. Cooper-Hewitt, National Design Museum, Smithsonian Organization (2007)

Tseng, Y.-C., Chang, H.-Y., Yen, S.-W.: The different effects of motivational messages and monetary incentives on fostering walking behavior. Paper presented at the 2018 CHI Conference Extended Abstracts on Human Factors in Computing Systems, Montreal, Canada, USA (2018)

Tseng, Y.-C., Chiu, Y.-C.: The influence of fictitious community trend on individual physical activity. Paper presented at the Proceedings of the 2017 CHI Conference Extended Abstracts on Human Factors in Computing Systems (2017)

Visser, W.: Use of episodic knowledge and information in design problem solving. Des. Stud. **16**(2), 171–187 (1995)

Wölfel, C., Merritt, T.: Method card design dimensions: a survey of card-based design tools. Paper presented at the IFIP Conference on Human-Computer Interaction (2013)

Study on Innovative Gestures Applicable to the Elderly

Ming-Hong Wang[✉] and Shuo-Fang Liu

Department of Industrial Design, National Cheng Kung University, Tainan, Taiwan (R.O.C.)
wming0403@gmail.com

Abstract. This study aims to provide the elderly with a more convenient gesture operation mode of smartphones. It first conducted field investigation and interviews to identify the gestures that are deemed difficult by the elderly. Through literature review and patent research on operation gestures, this study found that the action of smartphone use can be completed with one-finger operation. This study compared the existing gestures with innovative operation gesture modes, and found that the gesture requiring the operation of two fingers is difficult for the elderly. On the other hand, innovative operation gestures, compared with traditional operation gestures, have more significance in performance. Therefore, this study suggested that the innovative gesture with single finger operation could replace current operation gestures, and provide solutions to the inconvenience of the elderly or individuals with upper extremity disabilities in operating smart mobile devices.

Keywords: Elderly · Smart mobile device · Innovative gesture operation · Touch gesture

1 Introduction

Under the development and prevalence of high-tech products, people use computers, mobile phones, televisions, home appliances, and vehicles in their daily life. As the technology products are designed to bring convenience and social progress, among them, the mobile products feature the fastest development and high elimination and replacement rate. The continuous development technology can make the mobile products more compact, yet be equipped with many functions, such as audio control, online searching, listening to radio or watching TV shows, tabulation, calendar for daily planning, photographing, sending e-mail, etc. The design of interface develops from physical keys to touch screen. The development trend reveals that the previous designs were aimed to satisfy the usage habit of the masses in pursuit of convenience. As long as the device can be portable, other audio and video devices and equipment would no longer be in demand. In the design of these products, the designers added the interface suitable for the usage of the elderly, but met the problem of "over-simplicity" design, which actually presented a kind of stereotype. The designers often unselfconsciously eliminate "the elderly" from the users of high-tech devices, and propose to lower the complexity of technology devices

© Springer Nature Switzerland AG 2020
P.-L. P. Rau (Ed.): HCII 2020, LNCS 12192, pp. 200–211, 2020.
https://doi.org/10.1007/978-3-030-49788-0_15

to increase the acceptability of the group. With the advent of the aging society, relevant topics have emerged, and received attention from countries across the world. The aging tendency of society is a topic deserving attention in the course of development. The aging society is characterized by the sharp increase in the population the elderly every year, thus making the satisfaction of elderly users a top priority. Currently, when people walk on the street or wait at the bus stops, they are often using smart mobile devices, some even carry over two models at a time, which is much different from the stereotype that the elderly either take walks in parks or live alone. The elderly can use the smart mobile devices as proficiently as the young adults. Therefore, despite the increase in senior populations across the world, they have begun to use these high-tech products on a daily basis, and started intimate interaction in the usage process, thereby developing a sense of safety and reliability. In this way, they have gained autonomous right that allows them to use the personal smart mobile device independently and conveniently without being dependent on the guidance of a younger family member. They are becoming the potential users of the new technology.

This study focuses on gesture operation, proposes operation modes more suitable for the elderly, and attempts to develop an innovative gesture design method for smartphone operation gesture. This study discusses the usability of mobile phone, and reviews the existing interface design from the perspectives of the users. The usability principle is taken as the evaluation standard. Through the experiment method, the interface usage and usability in interactive operation are discussed, taken as reference for designing single-point touch interface software. At present, most smartphones adopt multiple-point touch operation. Different from past physical touch provided by the hardware interface, the operation is carried out by the gesture made by fingers directly on the screen. Without the need for press action, actions including gently knocking, long press, dragging can be performed to complete interface operation. All users must adapt to the new form of operation. Gestures are a natural and powerful tool for communication among people. Piper mentioned in the "Exploring the accessibility and appeal of surface computing for older adult health care support" that comparing the multiple-point touch device and traditional devices, the former is closer to the usage habit of users [1]. Researches on gesture operation in Taiwan and abroad mentioned that relevant gestures can be interpreted as direct and indirect touch. For the elderly, direct operation is more efficient than indirect operation. If intuitive gestures can be applied, it can shorten the time for elderly users to learn new gestures, and increase the interaction and sense of security between products and the elderly users.

2   Literature Review

Touch devices serve as standard operation interface of 3C products. They are widely applied in public facilities and transportation system (e.g., ATM, iBon multifunction machine, high-speed rail ticket machine) [2], medical health system machines in hospitals [3], and social network platform [4]. The touch operation is a design product under the concept of direct manipulation [5]. Its advantages include fast learning, better hand-eye coordination, less occupation of hardware space, and high availability in public places. However, without adjusting the hand gesture and angle of view, users may easily feel fatigue [6]. In terms of gesture control, there are a variety of technologies, which can be divided into two-dimensional gesture, which refers to the operation by using fingers or hands on touch screen, and three-dimensional gesture, which is defined as operation by using movement space in more free forms. For two-dimensional operation gesture, this study aims to develop better smartphone operation to assist the elderly in operation freely. Among relevant literature on the degradation of exercise performance of the elderly, Chaparro pointed out the doubt in the adaptability of finger operation for the elderly in the research of "Range of motion of the wrist: implications for designing computer input devices for the elderly" [7]. If the wrist joint operation is included in gesture operation, the flexibility of wrist of the elderly should be included into the evaluation, so as to determine whether a gesture suits the elderly. For various kinds of smartphones in the market, this study still doubted whether the interface gestures are suitable for the usage of the elderly. To date, relevant topics on smartphone gesture design among the middle-aged and the elderly are still rare. This study team attempts to develop innovative ways for smartphone interface gesture operation.

2.1 Cognitive Psychology of Aging

When people age, they face physical degeneration, which greatly lowers their mobility, memory, and learning ability. The changes in visual function are the most evident. Due to the diminution of vision, the elderly find it hard to see things clearly and the adaptation to brightness and acceptance of sensitivity and color are affected. The life of the elderly is therefore affected. The physical degeneration further leads to the degeneration of thinking ability and judgment. From another point of view, after the elderly retreat from the workplaces, they have less exposures to social contacts and self-improvement, which inevitably results in a relative distance from the society, as well as a gap between their real life and cognition. The influence on mentality is also on the rise. In terms of recognition, understanding and learning, the elderly are subject to degeneration of message processing as they age, such as decreased attention, narrowed recognition on time and space, lowered distinguishing ability of instant message, lowered complex information processing ability, slowed response to stimulation, which impacts the record and memory of event details. The changes in recognition of the elderly are mainly reflected in handling inductive inference, fluency in semantics and language, and figure and space concept. Those changes affect their communication effect of information reading and emotion reaction of advertisement [8]. From the point of view of social psychology, the elderly connect with the society through media, and rely on news, entertainment and message to look for familiar roles to compensate their diminishing social networks. This kind of media contact features the function of "quasi-social role" [9]. The elderly use media for the purpose of looking for happiness, killing time or pursuing leisure. They bridge the gap in social experience via media or learn new roles for self-affirmation.

2.2 Research on the Usage of Man-Machine Interaction

The intuition-based operation allows users to interact with the devices with direct fingers operation. Although it attracts elder computer users, there are still problems to be explored from the perspective of theory or practice [10]. For example, with thumb as fixing point, the index finger is used to make a loop. One typical case is that the gesture for viewing photo is intuition-based, but complex editing operations are needed, such as selection, rotation, or copy-paste become too complicated for figure gestures. McLaughlin and Charness proposed that the effect of interactive device is based on a set of precise data calculation [11]. For the elderly, choosing interactive devices that are compatible with operation mode is the key factor to ensure performance. Barker studied the use of smartphones by the elderly, and proposed eight strategies: 1) priority is given to click and slide gestures; 2) demonstration gesture; 3) enlarge touch target; 4) place the target in eye-catching places; 5) reserve enough space between targets; 6) interaction with target is to provide feedback; 7) limit text input if possible; 8) use clear font with proper size [12]. Caprani et al. found that the elderly perform better when using touch screen than other input devices [13]. Compared with other input devices, the touch screen is preferred by the elderly.

2.3 Welfare Design

With the progress in medical and technology, the elderly population is able to enjoy active social participation and independent living ability. From the perspective of technology, how to apply various technology aids to help the elderly who suffer from gradual physical degeneration to enjoy a more active, healthy, comfortable, and safe life is an important subject. In 1992, the First International Congress on Gerontechnology was held in Europe, which established research structure for enhancing technology and the aged well-being. Graafmans and Bouma from Eindhoven University of Technology of the Netherlands defined the welfare technology for the elderly as follows: based on the knowledge on aging progress, the welfare technology for the elderly is engaged in the research and development of technology and related products in the hope of providing the elderly with better life and work environment, as well as supporting medical care. In September 1997, the International Society of Gerontechnology was established in Europe with the purpose to design technology and environment that could help the elderly to live independently, healthily, comfortably and safely, and to able to participate in social activities [17, 18].

3   Methods

The research objectives are divided into three parts, and the research methods include literature review and interviews. First, this study measured the key points of operation design from four aspects of ergonomics (physical and mental status of the elderly, HMI and operation gesture), and reviewed literature on the ergonomic design principle. It then developed the ergonomic engineering principle evaluation form of touch gesture. In the second part, the ergonomic engineering principle evaluation form of usability was developed to evaluate gesture operation status. Based on the scores, individual improvement suggestions were provided. In the third part, the field evaluation was conducted. of the problems identified in the assistant measure design of smartphone for the usage of the elderly were identified, and overall improvement suggestions were proposed. This study also proposed the guiding principle for gesture operations suitable for the elderly. Nielsen and Tullis and Albert pointed out that the usability of evaluation system or user interface is not one-dimensional, but consists of the following five criteria: (1) efficiency, (2) errors, (3) memorability; (4) learnability; (5) satisfaction [14–16].

3.1 Collection of Current Products and Sample Selection

This study took the best-selling mobile phone brand products in Taiwan as samples. The smartphones selected by this study were under five brands, include HTC, Samsung, Sony, Apple and ASUS (as shown in Fig. 1). During experiment process, these brand names and enterprise logos were not shown to eliminate the existing impression of participants on some brands. According to literature review, this study summarized the common intuitive gesture designs, including zoom with two fingers, one finger as the fixing and the other finger for rotation, screenshot viewed images by pressing two function keys. The common gesture operation at present is indeed challenging to the elderly.

Fig. 1. Popular mobile phone brands

3.2 Group Positioning and User Observation

This study observed the elderly who use smartphones more often in daily life, and found that the gesture operation is difficult for elderly users. According to previous researches, when using the touch interface of smartphones, the elderly may have problems due to physical degeneration, such as blurred vision, amnesia, insensitivity of touch due to increase of wrinkles. The elderly may have difficulty in recognizing the function pattern on the interface, and find it hard to target the object when operating due to lowered mobility. Moreover, when the elderly were browsing images, they would continuously use two fingers to zoom in/out to adjust the screen due to failure to focus resulted from blurring of vision. They would also touch the screen repeatedly due to insensibility of finger touch, which easily led to error in pressing and mistakenly switching the current viewing pages, thus affecting smoothness of browsing. The operation gesture affected the comfort of smartphone usage. Therefore, the gesture designed for the elderly population should focus on easy comprehension, and simple operation as priority.

3.3 Product Item Analysis (Item Category)

This study took five popular smartphone brands for gesture analysis. Actual measurement, observation and operation were carried out on length, width, thickness, weight, main body shape, color, materials, Logo, operation methods according to appearance (as shown in Table 1).

Table 1. Brand model and operation methods

Model / Operation	Moving gesture	Zooming gesture	Rotating gesture
HTC ONE M8			Click on editing rotating function key separately for usage
Samsung NOTE4			Click on editing rotating function key separately for usage
Sony Xperia			Gently touch screen to display toolbar and gently touch function selection bar and select rotate. The photo would be stored in new direction
iPhone 6s			
Zenfone			Start editing function under album or photo mode

3.4 Field Interview

This study recruited 30 elderly people aged over 65 with experience in using smart phones. For those who suffer from nearsightedness, farsightedness and presbyopia, corrective therapy or the wearing of an assistant device such as glass should be carried out to ensure the ability in identification and reading. Before the interview, the researcher first introduced the procedures, and explained the current operation gesture of smartphones. The participants were led to experience the zoom-in, zoom-out, rotation, and other operation gestures to view images. During the operation process, the researcher recorded the feedbacks of the participants on their operation gesture experiences. The interview lasted for about 30 min. Each participant was given enough time to think and organize their thoughts. The interviews were expected to identify the operation gestures that satisfy the needs of the elderly users. This study focused on identifying the reason for the usage difficulty of the elderly, and developing the operation method that is most convenient for the elderly (Fig. 2).

Fig. 2. Interactions of hospital volunteers/participants

3.5 Existing Operation Gesture Design Mode in the Market

Based on literature and interview results, this study found that the operation gesture that is difficult for the elderly participants is rotation and zoom-in & zoom out (as shown in Fig. 3, Fig. 4). The reason mainly lies in the insensibility of skin touch of two-finger operation and easy action repetition caused by shaking during operation process. Therefore, it's difficult for the elderly to use a gesture with over one finger.

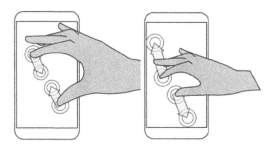

Fig. 3. Zoom in and zoom out with two fingers

Fig. 4. The rotation effect with thumb as rotating fixed point, index finger as radius (Traditional Rotation, TR)

3.6 Results of Developing New Design Gesture Mode

This research invited product designer and HMI developers to discuss the innovative design of operation gestures and observed the experiments and studied gesture operation suitable for the elderly. In addition, some past studies showed that due to physical degeneration and natural degeneration, the elderly tend to be slow and inflexible when performing two-finger action. According to actual observation and review of past studies, this research initially developed the prototype of gesture operations suitable for the elderly (as shown in Fig. 5 and Fig. 6).

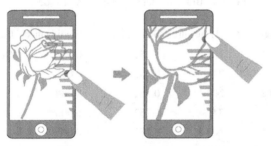

Fig. 5. Innovative design to achieve zoom-in and zoom-out with one-finger operation (New Design, ND1)

Fig. 6. Innovative design to achieve rotation with one-finger operation (New Design, ND2)

4   Methods

This study evaluated the two operation gestures under innovative design and availability of existing rotating operation gesture (TR). The relationships of various usability evaluation items are as follows: (1) Learning ability: the gesture operation of ND1 is superior to ND2 and TR; to evaluate learning ability, when learning and becoming familiar with the operation, the participants indicated that is it easier to understand of icon meaning of ND1 operation gesture; (2) Efficiency: to evaluate efficiency, this study invited the participants to open one image in the folder, and move the images by innovative zoom-in and zoom-out gesture; as seen from the time recorded, the usage of ND1 operation gesture is more efficient than the usage of ND2 and TR; (3) Memorability: in general, the ease of learning is related to memory to some extent; most of the participants can understand the meaning of ND1 gestures in the interface; during the test, after the participants used smartphones and operation gesture of different brands, they still knew how to use the innovative operation gestures proposed by this study and their meaning; (4) Errors: the error rate is related to efficiency to some extent; this study set one image in the folder, and measured the time spent by the participants to open it, in order to evaluate the efficiency; it found that the longer time taken led to a higher error rate; (5) Overall satisfaction: compared with existing operation gestures, the innovative operation gestures designed by this study received positive feedback (as shown in Table 2)

Table 2. Brand model and operation methods

Variable	(I) gesture	(J) gesture	Standard error	Significance	Comparison result
Learnability	ND 1	ND 2	0.152	0.000	ND1 > TR > ND2
		TR	0.152	0.001	
	ND 2	ND 1	0.152	0.000	
		TR	0.152	0.006	
	TR	ND 1	0.152	0.001	
		ND 2	0.152	0.006	
Efficiency	ND 1	ND 2	0.155	0.000	ND1 > ND2 > TR
		TR	0.155	0.000	
	ND 2	ND 1	0.155	0.000	
		TR	0.155	0.911	
	TR	ND 1	0.155	0.000	
		ND 2	0.155	0.911	

(continued)

Table 2. (*continued*)

Variable	(I) gesture	(J) gesture	Standard error	Significance	Comparison result
Memorability	ND 1	ND 2	0.166	0.487	ND1 > ND2 > TR
		TR	0.166	0.002	
	ND 2	ND 1	0.166	0.487	
		TR	0.166	0.060	
	TR	ND 1	0.166	0.002	
		ND 2	0.166	0.060	
Errors	ND 1	ND 2	0.136	0.000	ND1 > TR > ND2
		TR	0.136	0.000	
	ND 2	ND 1	0.136	0.000	
		TR	0.136	0.474	
	TR	ND 1	0.136	0.000	
		ND2	0.136	0.474	
Satisfaction	ND 1	ND 2	0.172	0.000	ND1 > ND2 > TR
		TR	0.172	0.000	
	ND 2	ND 1	0.172	0.000	
		TR	0.172	0.001	
	TR	ND 1	0.172	0.000	
		ND 2	0.172	0.001	

5 Discussions and Conclusions

In the past studies, there was a preliminary finding that the demand and recognition of the operation of smart devices would change to some extent as the users age. On the contrary, the younger generation had better control over technology products either in familiarity in usage or response to product operation, as compared with the elderly users. Based on the discussion of the human factors and the cognitive, previous studies obtained the effective response time and feedback time data of the elderly users during operation. The results were used to promote the production of smart products with perception, memory, decision making, attention, response execution, and feedback mechanism, in order to meet the needs of lifelong mobility of the elderly users and help them to operate smart devices more conveniently. According to previous literature, the main obstacles faced by the elderly include: deterioration of vision, memory and cognition. The cognition part is the mental factor that caused the elderly users to reject the use of computers. In their opinion, they are too old to use high-tech products or are afraid of breaking the computer during the learning process (Zajicek 2001). Under the social structure tending to aging in the future, for the elderly and those with disability in hands, the intuitive operation and work efficiency should be more valued in product operation. The findings of this study are expected to contribute to the elderly group or those in needs. The topic deserves further attention. In relevant reports about the elderly and technology, the high-tech mobile devices such as mobile phone and tablets are mentioned. The continuously advancing cloud services in recent years has contributed to the fast development in smart mobile devices, and the service providers continue to launch new services to satisfy the usage habit of users. The service providers could reduce learning cost and efficiency for time saved. The entry of more service providers into the market has brought more vigorous development of mobile devices. With understanding of the various factors that influence the touch gestures of the elderly, the service providers could design operation gestures that meet the needs of the elderly users, thus benefiting the elderly population. This will also point a direction for the future development of various touch mobile devices. It is expected that the touch gestures for elderly users can retain the accuracy of the original gestures, while reducing the difficulty in re-learning and memorizing due to human factors of the exiting touch control gestures. The elderly could access the information provided by mobile devices in a more efficient and pleasant way.

References

1. Albert, W., Tullis, T.: Measuring the user experience: collecting, analyzing, and presenting usability metrics. Newnes (2013)
2. Bhalla, M.R., Bhalla, A.V.: Comparative study of various touchscreen technologies. Int. J. Comput. Appl. 6(8), 12–18 (2010)
3. Caprani, N., O'Connor, N.E., Gurrin, C.: Touch screens for the older user. In: Assistive technologies: IntechOpen (2012)
4. Carstensen, L.L.: The influence of a sense of time on human development. Science 312(5782), 1913–1915 (2006)
5. Chaparro, A., Rogers, M., Fernandez, J., Bohan, M., Sang Dae, C., Stumpfhauser, L.: Range of motion of the wrist: implications for designing computer input devices for the elderly. Disabil. Rehabil. 22(13–14), 633–637 (2000)
6. Charness, N., Boot, W.R.: Aging and information technology use: potential and barriers. Curr. Dir. Psychol. Sci. 18(5), 253–258 (2009)
7. Hertzum, M., Hornbæk, K.: How age affects pointing with mouse and touchpad: a comparison of young, adult, and elderly users. Int. J. Hum. Comput. Interact. 26(7), 703–734 (2010)
8. Hollinworth, N., Hwang, F.: Investigating familiar interactions to help older adults learn computer applications more easily. Paper presented at the Proceedings of the 25th BCS Conference on Human-Computer Interaction (2011)
9. McLaughlin, A.C., Rogers, W.A., Fisk, A.D.: Using direct and indirect input devices: attention demands and age-related differences. ACM Trans. Comput. Hum. Interact. (TOCHI) 16(1), 2 (2009)
10. Nielsen, J.: Usability Engineering. Elsevier, Amsterdam (1994)
11. Piper, A.M., Campbell, R., Hollan, J.D.: Exploring the accessibility and appeal of surface computing for older adult health care support. Paper presented at the Proceedings of the SIGCHI Conference on Human Factors in Computing Systems (2010)
12. Roy, A., Harwood, J.: Underrepresented, positively portrayed: Older Adults in Television Commercials (1997)
13. Sutter, C., Müsseler, J.: User specific design of interfaces and interaction techniques: what do older computer users need? Paper presented at the International Conference on Universal Access in Human-Computer Interaction (2007)

14. Tsai, T.-H., Chang, H.-T., Wong, A. M.-K., Wu, T.-F.: Connecting communities: designing a social media platform for older adults living in a senior village. Paper presented at the International Conference on Universal Access in Human-Computer Interaction (2011)
15. Wang, M.-H., Chang, Y.-C., Liu, S.-F., Lai, H.-H.: Developing new gesture design mode in smartphone use for elders. Paper presented at the International Conference on Human Aspects of IT for the Aged Population (2015)
16. Wilson, M.L., Kules, B., Shneiderman, B.: From keyword search to exploration: designing future search interfaces for the web. Found. Trends® Web Sci. 2(1), 1–97 (2010)
17. Wood, E., Willoughby, T., Rushing, A., Bechtel, L., Gilbert, J.: Use of computer input devices by older adults. J. Appl. Gerontol. 24(5), 419–438 (2005)
18. Wright, P., et al.: Using audio to support animated route information in a hospital touch-screen kiosk. Comput. Hum. Behav. 26(4), 753–759 (2010)

Cultural Discourse in User Interface Design: Investigating Characteristics of Communicators in Microsoft Word

Chunyan Wang[1](✉) and Xiaojun Yuan[2]

[1] Xiamen University of Technology, Xiamen 361024, Fujian, People's Republic of China
ahwangchunyan@126.com
[2] University at Albany, State University of New York, Albany, NY 12222, USA
xyuan@albany.edu

Abstract. This paper addressed cultural factors in user interface design. Specifically, it took the words and symbols in the user interface as a discourse, and analyzed the characteristics of the communicator who created such discourse under the frame of cultural discourse studies created by Shi-xu (2015). The characteristics of communicators were revealed diachronically in a series of Microsoft Word versions through the content, forms of communication, and historical relations of the user interface discourse. The influence of Microsoft Word user interface design on other word processors such as Jinshan WPS was also investigated. Results showed that the typical words and/or symbols in the Microsoft user interface were the most frequently elements of English documents and were listed according to western thinking method. It seems that the word processors in other cultures tended to copy the form and content of the design of Microsoft Word without taking their cultural differences into consideration. How to use cultural discourse studies to design a user-centered interface was discussed.

Keywords: Cultural discourse · Discourse study · Discourse analysis · Cultural discourse studies · User interface · Communicator · Microsoft Word · Word processor

1 Introduction

Cultural HCI studies have been studied for several decades since 1990s (Heimgärtner 2017). Hofstede's cultural dimensions theory (Hofstede 1980) addressed cultural differences that influenced on the internationalization and localization of user interface design. However, to our knowledge, few cultural studies have focused on discourse perspective.

In cultural discourse studies, "discourse" was defined as a situated communicative event through which people can achieve social (inter)action using linguistic and other symbolic ways (Shi-xu 2015). Shi-xu (2015) emphasized that the nature of a discourse is culture. To reveal the cultural nature of a discourse, he proposed four key categories – communicators, forms, mediums of communication, purposes – and two relationships including historical and (inter)cultural elements, to describe, analyze and evaluate who

© Springer Nature Switzerland AG 2020
P.-L. P. Rau (Ed.): HCII 2020, LNCS 12192, pp. 212–227, 2020.
https://doi.org/10.1007/978-3-030-49788-0_16

communicated what information by which medium and the reasons behind the phenomenon. Based on this theory, a user interface (UI) that applies words or symbols to communicate between users and software applications (e.g. Microsoft Word) can be considered as an instance of a discourse. A communicator means a specific person or social organization form (such as media, units and associations) that uses language symbols to generate meaning in a specific context. A communicator or subject of discourse is the most creative factor in the process of discourse, the driving force to participate in society and change the social practice of the world, and the core component of society and culture (Shi-xu 2010).

Since the communicators in a user interface are usually a group of people under the name of a company, it is challenging to point out who used the language symbols to express their ideas. However, we can collect their characteristics from what information they provided (Shi-xu 2010), how they arranged it, and how they think about it when they switched from the old version to the newer version of Word (Shi-xu 2010). In that way, whether a UI follows the internationalization or localization rules or not can be revealed.

In this paper, we aim at employing different Microsoft Word versions (including version 1.x for DOS, 2.0 for Windows, 2003, 2007, 365) as a case study, and carrying out a cultural UI discourse study to investigate the characteristics of communicators, and then to evaluate the cultural nature of these communicators.

2 Research Methodology

We adopted qualitative method to discuss the nature of the discourse of Microsoft Word and other widely used word processors.

In China, Microsoft's Microsoft Office products dominate the PC market with its pre-installed advantage of Windows operating system, with a penetration rate of 97.31%, according to CCW research, an ICT research institute in China as of 2017 (CCW 2017). That is the major reason why we chose Microsoft Word as the case study here. In addition, Microsoft Office has been integrated into other widely used software, such as SDL Trados Studio, a famous computer-aided technology application. Translators were required to install Microsoft Office first before they could use pre-SDL Trados 2007 (SDL 2020; Baidu 2020). Since SDL Trados Studio 2009, Microsoft Office has not been a necessity (Baidu 2020), but still an important part if certain functions work correctly. It is not surprising that for students and researchers in China, Word has become a basic and necessary tool to learn.

To reveal the characteristics of communicators in Microsoft Word, both diachronical and synchronical descriptions were employed. A diachronical investigation is the study of a discourse through the course of its history (Hu 2011). Specific Word versions (Microsoft Word 1.x for DOS, Microsoft Word 2.0 for Windows, Word 2003, Word 2007) and their related documents were selected for the following considerations: Word 1.x for DOS introduced Word's basic functions. Word 2.0 for Windows, with its nine menus, the Standard toolbar and the Formatting toolbar, set the basic UI structure from Word 2.0 to Word 2003 (Harris 2005a). Word 2007 changed its appearance based on collected user data, such as the frequency of commands and the keyboard shortcuts (Harris 2005b), then introduced ribbon, contextualization and galleries to reconstruct the

UI (Harris 2005c). We will analyze whether Word UI design took internationalization and localization into consideration or not in the Sect. 3.

Synchronical description used 2019–2020 as the point of observation (Hu 2011). The influence of Microsoft Word UI design on other word processors such as Jinshan WPS, and 126.com (a popular Chinese e-mail tool) is also discussed.

3 Characteristics of Communicators in Microsoft Word

According to localization practice, language resources in software usually contain strings, menus and dialogues (Wang et al. 2005). Menus used here as discourse analysis object not only refer to list of options from which a user can choose (Microsoft 2019) but also menu buttons and certain keys with graphic or signals of text, or letters. Strings and dialogues are not the focus of this study.

Six study materials are included in this research: (1) versions of Microsoft Words from 1.x to Word 365; (2) help documents of Word 2007; (3) more than 150 blogs in Microsoft website written by Word 2007 research executive Jensen Harris (Harris 2006a); (4) comments on Microsoft Word older versions (Manes 1984); (5) the authors' teaching experiences and observances; and (6) 109 Microsoft Style Guides, including Simplified/Traditional Chinese.

In the following, three steps were taken to outline the characteristics of communicators of MS Word, that is, the discourse content, forms of communication, and historical relations between 2003 and 2007 versions, and then followed by a comparison between Microsoft Word and other word processors.

3.1 Discourse Content

The content of discourse in UI include the typical buttons/menus chosen and the reasons why they, not others, were chosen. To compete with other word processors, such as WordStar, SuperWriter, PFS: Write, Multimate, Perfect Writer, and WordPerfect, MS Word showed its strength of "What You See Is What You Get" (WYSIWYG) (which was thought to be originated by WordStar) (Bergin 2006). MS Word provided two methods for users to apply the formatting, that is, the formatting commands and the formatting keys. Nine-character formats and 11 paragraph formats were assigned keys preceded by Alt (Microsoft 1983). The character keys are listed in Table 1. Two functions—boldface, and italics—were emphasized (Microsoft 1983) (WordStar advertised on *Personal computing* in January 1981 to focus on underlining and boldfacing functions) (Manes 1984). From MS Word 1.x for DOS, bold and Italic have often been used in Microsoft Word for DOS, Windows and Apples (Fig. 1, Fig. 2).

Word 2.0 for Windows outlined the basic UI structure for the pre-2003 versions. One typical characteristic was its nine menus, the same as in Word 2003 (Fig. 2, Fig. 3).

The other characteristics were its two basic toolbars: the Standard toolbar and the Formatting toolbar, which became much longer and contained more commands in 2003 (Fig. 3).

To track the user's habits, the Microsoft Office Customer Experience Improvement Program was used in Office 2003. Anonymous data was collected if the user chose

Table 1. Nine-character formatting keys

No.	Format	Key	No.	Format	Key
1	Normal character	Alt + Space	6	Small capitals	Alt + K
2	Boldface	Alt + B	7	Strikethrough	Alt + S
3	Italic	Alt + I	8	Superscript	Alt ++
4	Underline	Alt + U	9	Subscript	Alt +−
5	Double underline	Alt + D			

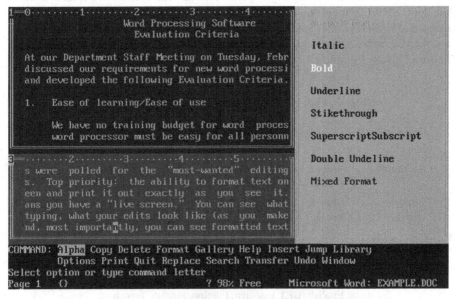

Fig. 1. Microsoft Word 1.15 (DOS) (Winworldpc 2019)

Fig. 2. Word 2.0 for Windows (English version) (Harris 2005a)

Fig. 3. Microsoft Word 2003 (Simplified Chinese version, Basic Edition), accessed by 2020/1/4

"Help Make Office Better". During 2005, about 1.3 billion sessions and over 352 million command bar clicks in Word were collected from January to April (Harris 2005b). In Word 2003, Paste, Save, Copy, Undo, and Bold were the top 5 frequently used orders by all users (Harris 2005d). Italic was not listed here, but it was among the first 8 (Harris 2005c). Was that really the data of all users or only English-speaking users? Bold and italics were commonly used in English documents, but were they also frequently used in other languages? Though Harris did not mention it, the Microsoft Style Guides in other languages told a different story.

There were 108 valid versions of Localization Style Guides in Microsoft Language Portal (Microsoft 2019a) which had been updated from December 2008 to February 2019 besides Microsoft Writing Style Guide, and was last revised in September 2019 (Microsoft 2019b). Four documents – Basque, Bosnian (Cyrillic), Malay (Brunei Darussalam), and PASHTO, provided no information about Italic or Bold. Four out of 108 have listed specific different usages between English Italic and their own languages. They are Chinese (Simplified), Chinese (Traditional), Georgian and Korean (Table 2).

Table 2. Bold or Italic usages in Localization Style Guides

No.	Usage of Bold or Italic	Language	Number
1	No information	Basque, Bosnian (Cyrillic), Malay (Brunei Darussalam), PASHTO	4
2	Different usages of Italic	Chinese (Simplified), Chinese (Traditional), Georgian and Korean	4
3	Same usages of Bold or Italic	Afrikaans, Albanian, Amharic, Arabic, Armenian, Assamese, Azerbaijani, Bangla (Bangladesh), Bangla (India), Belarusian, Bosnian (Latin), Bulgarian, Catalan, Central Kurdish, Croatian, Czech, Danish, Dari, Dutch, English (UK), Estonian, Filipino, Finnish, French (Canadian), French (France), Galician, German, Greek, Gujarati, Hausa, Hebrew, Hindi, Hungarian, Icelandic, Igbo, Indonesian, Inuktitut, Irish, IsiXhosa, IsiZulu, Italian, Japanese, Kannada, Kazakh, Khmer, K'iche', Kiswahili, Konkani, Kyrgyz, Lao, Latvian, Lithuanian, Luxembourgish, Macedonian, Malay (Malaysia), Malayalam, Maltese, Maori, Marathi, Mongolian (Cyrillic), Nepali, Norwegian, Norwegian Nynorsk, Odia, Persian, Polish, Portuguese (Brazilian), Portuguese (Portugal), Punjabi, Quechua, Romanian, Russian, Scottish Gaelic, Serbian (Cyrillic), Serbian (Latin), Sesotho sa Leboa, Setswana, Sindhi (Pakistan), Sinhala, Slovak, Slovenian, Spanish (Mexico), Spanish (Spain), Swedish, Tajik, Tamil, Tatar, Telugu, Tigrigna, Thai, Turkish, Turkmen, Ukrainian, Uyghur, Urdu, Uzbek (Latin), Vietnamese, Welsh, Wolof, Yorùbá	100
Total			108

Chinese culture prizes upright posture. Microsoft takes this into consideration in Font localization. For example, In Chapter 3 of the Chinese (simplified) Style Guide, the special 3.1.7 Font localization requires users to follow the general rules (Microsoft 2019c): "

a. Please do NOT use *italic* or **bold** font style for words or sentences that are italic or bold in the source files....
b. When localizing UI terms within software messages/descriptions, please enclose UI terms with double byte double quotation marks ("").
c. As for UI terms in documentation and online help, move the UI out of the <bold>, <italic> or <ui> tags, and enclose it with double byte double quotation marks ("")."

It is similar in 3.1.7 Font localization of traditional Chinese version. It says, "**Bold** and *Italic* style should be avoided for Traditional Chinese characters" (Microsoft 2019d). It also lists a few exceptions to keep the bold style, like for headings, emphasis, user input, replacement of Italic style. It is obvious that bold and Italic styles are not used often in traditional Chinese. But in the localization of the UI, B and I were not only put in the outstanding place in the normal Chinese user interface but also listed as the first two buttons in the right-click menu, labeled as the Mini Toolbar to afford super-efficiency for mouse access (Harris 2006) (Fig. 4).

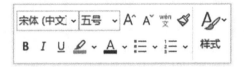

Fig. 4. Upper part of Right-click menu of Word (Office 365 accessed by 2019/12/5 Chinese version, personal edition.)

Georgian Style Guide points out in Quotation marks of 3.1.13 Punctuation that "As opposed to US English usage of italic formatting for titles or foreign words, Georgian tends to use quotation marks in this context as well" (Microsoft 2019e). Korean Style Guide suggests in 3.1.5 Font localization that "bold font can be applied same as English but italic font is recommended to remove for better readability" (Microsoft 2019f).

It is not surprising that buttons in Word were first chosen typically from the English documents and suited the habits of users with English as the first language. Though Microsoft knew the different usages of other languages, it laid down rules to avoid cultural conflicts in the documents, except its UI design. The proportion 3.7% (4 out of 108) is less than 4%, which is too small to deserve special attention. However, the total number (19.87% of all the world, by 2019 mid-year estimates) and the number of Internet users (20.13% of all the world, by June 30, 2019) (Miniwatts 2020) of Chinese-speaking, Korean-speaking, and Georgian-speaking people are large enough for being taken into consideration.

3.2 Discourse Form

Form refers to the graphic or text shape here. In MS Word, not all the buttons are graphic, and many of them are letters (Fig. 5). As can be seen in English version, most letters were originally short for English letters, such as B for bold, I for italics, U for underline, and the first letter A in English alphabet is an example to increase or decrease font size, change case, and clear all formatting, also as a phonetic guide, character border, text effects and typography, font color, and character shading. The only exception among the English letters under the Home ribbon is that a Chinese character 字 is used for function Enclose Characters to emphasize a character by placing a circle or box around it. If the English interface is localized or only translated into other languages, the localization rules require that target languages should be used as many as possible. Ideally, B, I, U are to be translated into the target language respectively. In this paper, we downloaded all the available language packs of newer Office versions in the website of Office and install them in Office 365, in order to research what they really show in Word interface.

Fig. 5. Microsoft Word 365 (English version, personal edition), accessed by 2019/12/23

Our approach is first to open a Word document, click File, choose Options at the end of the list, and then a Word Options dialogue pops up. The sixth of the narrower column is Language. Click Language, we can Set the Office Language Preferences, which has two parts, "Office display language" and "Office authoring languages and proofing." The former says "Buttons, menus, and other controls will show in the first available language on this list", and a box with languages in it for selection. Below there is a link, "Install additional display languages from Office.com", which links another dialogue box labeled as "Install a display language" (Fig. 6).

As soon as people select the language they want to install, they will be directed to Language Accessory Pack for Office on the support.office.com webpage where they can install the specific language. Sometimes the link is opened to the single chosen language installment page, sometimes it is a general language pack page, where we should choose the language again. No matter which page was directed, the website of the support page is automatically shown, in our case, in Chinese. As we analyze the languages in English, to keep the language name consistent, we use a general language pack installment page shown in English instead (Microsoft 2020).

On the page, 2 steps were listed. Step 1: Install the language accessory pack, choose Newer versions. After downloading the language, run the exe file and then follow the directions. If the pack includes UI language, the preferred language will appear in the

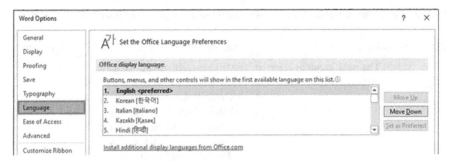

Fig. 6. Word Options of Microsoft Word 365, last accessed 2020/1/28

box under "Office display language". Then follow the directions in step 2 to configure Office language, or just set the chosen language as Preferred, close and reopen Word, then Word is displayed in the assigned language.

However, as can be seen in the 102 respective languages, the directions were difficult to follow. To solve the problem, we managed to install English version first and record step by step the positions of the clicked buttons according to their relatively fixed places (Fig. 6). For example, it was helpful to remember that Language is often the sixth from top to bottom in the left or narrower column. As to the Arabic, its related languages, or other languages which are read from right to left, the Language should be in the right narrower column.

Microsoft Office 365 provides 102 language packs. Among them, 91 packs include selected display language. As the rest 11 packs have a spell checker only, they will not be taken into consideration (Table 3).

These languages in Table 3 not only refer to the languages of nations, but also refer to those used in a specific area. Taking the Portuguese as an example. There are Portuguese (Brail) and Portuguese (Portugal). However, for English, there is no difference between American English, British English, Australian English and Canadian English.

We took screenshots of the first two parts (Clipboard and Font, as shown in Fig. 5) of the 91 localized user interfaces under ribbon Home. Then we listed all the corresponding translations of English letters B, I and U. Table 4 displays the form of languages in UI. As can be seen, 70 languages keep all the three letters of B, I, U in the UI as English does. While 3 languages—French, Portuguese (Brazil), Portuguese (Portugal)—single out I to stand for Italic, 4 languages—Danish, German, Norwegian Bokmal, Swedish keep U for underline, only one language Malay (Latin) keep both, 13 packs can change all these three letters into totally different letters. Those languages, like Italian, Spanish, practiced in European countries, and some part of Asia, especially those employed near European, such as Kazakh, tend to display a different interface.

There are two ways to localize B, I, U symbols. One common strategy is to change B, I, U respectively into their own languages, which is adopted by Estonian, Hungarian, Italian (Fig. 7), Kazakh, Latvian, Russian, Serbian (Latin, Serbia), Slovenian, Spanish, Turkish, and Ukrainian (Table 5). Most of them are the first letter of related words, except Russian Ж, Ч, Spanish K for Cursiva and Turkish T for İtalik. There are two Ķs in Kazakh, two Ps in Serbian (Latin, Serbia) which have different meanings.

Table 3. Localization extent of the language packs used in newer version of Microsoft Office

L10n extent	Description	Languages	Numbers
1	The pack includes selected display language	Afrikaans, Albanian, Amharic, Arabic, Armenian, Assamese, Azerbaijani, Bangla (Bangladesh), Bangla (Bengali India), Basque (Basque), Belarusian, Bosnian (Latin), Bulgarian, Catalan, Chinese (Simplified), Chinese (Traditional), Croatian, Czech, Danish, Dari, Dutch, English, Estonian, Filipino, Finnish, French, Galician, Georgian, German, Greek, Gujarati, Hebrew, Hindi, Hungarian, Icelandic, Indonesian, Irish, Italian, Japanese, Kannada, Kazakh, Khmer, KiSwahili, Konkani, Korean, Kyrgyz, Latvian, Lithuanian, Luxembourgish, Macedonian (North Macedonia), Malay (Latin), Malayalam, Maltese, Maori, Marathi, Mongolian (Cyrillic), Nepali, Norwegian Bokmal, Norwegian Nynorsk, Odia, Persian (Farsi), Polish, Portuguese (Brazil), Portuguese (Portugal), Punjabi (Gurmukhi), Quechua, Romanian, Russian, Scottish Gaelic, Serbian (Cyrillic, Serbia), Serbian (Cyrillic, Bosnia & Herzegovina), Serbian (Latin, Serbia), Sindhi (Arabic), Sinhala, Slovak, Slovenian, Spanish, Swedish, Tamil, Tatar (Cyrillic), Telugu, Thai, Turkish, Turkmen (Latin), Ukrainian, Urdu, Uyghur, Uzbek (Latin), Valencian, Vietnamese, Welsh	91
2	The pack includes a spell checker	Hausa, Igbo, IsiXhosa, IsiZulu, Kinyarwanda, Pashto, Romansh, Sesotho sa Leboa, Setswana, Wolof, Yorùbá	11

Table 4. Form of languages in UI

Forms	Languages	Numbers
B, I, U kept	Afrikaans, Albanian, Amharic, Arabic, Armenian, Assamese, Azerbaijani, Basque (Basque), Bangla (Bangladesh), Bangla (Bengali India), Belarusian, Bosnian (Latin), Bulgarian, Catalan, Chinese (Simplified), Chinese (Traditional), Croatian, Czech, Dari, Dutch, English, Filipino, Finnish, Galician, Georgian, Greek, Gujarati, Hebrew, Hindi, Icelandic, Indonesian, Irish, Japanese, Kannada, Khmer, KiSwahili, Konkani, Kyrgyz, Luxembourgish, Macedonian (North Macedonia), Malayalam, Maltese, Maori, Marathi, Mongolian (Cyrillic), Nepali, Norwegian Nynorsk, Odia, Persian (Farsi), Polish, Punjabi (Gurmukhi), Quechua, Romanian, Scottish Gaelic, Serbian (Cyrillic, Serbia), Serbian (Cyrillic, Bosnia & Herzegovina), Sindhi (Arabic), Sinhala, Slovak, Tamil, Tatar (Cyrillic), Telugu, Thai, Turkmen (Latin), Uyghur, Urdu, Uzbek (Latin), Valencian, Vietnamese, Welsh	70
B replaced	Danish, Estonian, French, German, Hungarian, Italian, Kazakh, Korean, Latvian, Lithuanian, Malay (Latin), Norwegian Bokmal, Portuguese (Brazil), Portuguese (Portugal), Russian, Serbian (Latin, Serbia), Slovenian, Spanish, Swedish, Turkish, Ukrainian	21
I replaced	Danish, Estonian, German, Hungarian, Italian, Kazakh, Korean, Latvian, Lithuanian, Norwegian Bokmal, Russian, Serbian (Latin, Serbia), Slovenian, Spanish, Swedish, Turkish, Ukrainian	17
U replaced	Estonian, French, Hungarian, Italian, Kazakh, Korean, Latvian, Lithuanian, Portuguese (Brazil), Portuguese (Portugal), Russian, Serbian (Latin, Serbia), Slovenian, Spanish, Turkish, Ukrainian	16
Fully replaced	Estonian, Hungarian, Italian, Kazakh, Korean, Latvian, Lithuanian, Russian, Serbian (Latin, Serbia), Slovenian, Spanish, Turkish, Ukrainian	13

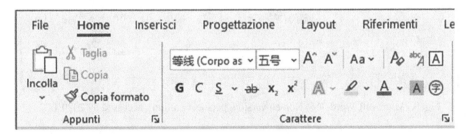

Fig. 7. Microsoft Word 365 (Italian version, personal edition), accessed by 2020/1/20

The other strategy is to choose one character to replace these three letters, as Korean and Lithuanian did (Table 6). While the Lithuanian language shares the same first letter in three different expressions, Korean picks the first character of their Alphabet chart to represent these three meanings. This way will not cause much confusion, as different styles help users to distinguish the languages from each other (Fig. 8).

Table 5. Fully localization of B, I, U in different letters

No.	Language	B	Bold	I	Italic	U	Underlined
	Language	B	Bold	I	Italic	U	Underlined
1	Estonian	P	Paks	K	Kursiiv	A	Allakriipsutus
2	Hungarian	F	Félkövér	D	Dőlt	A	Aláhúzott
3	Italian	G	Grassetto	C	Corsivo	S	Sottolineato
4	Kazakh	Қ	Қалың	Қ	Қиғаш	A	Астын сызу
5	Latvian	T	Treknraksts	S	Slīpraksts	P	Pasvītrots
6	Russian	Ж	Полужирный	K	Курсив	Ч	Подчеркивание
7	Serbian (Latin, Serbia)	P	Podebljano	K	Kurziv	P	Podvučeno
8	Slovenian	K	Krepko	L	Ležeče	P	Podčrtano
9	Spanish	N	Negrita	K	Cursiva	S	Subrayado
10	Turkish	K	Kalın	T	İtalik	A	Altı çizili
11	Ukrainian	Ж	Жирний	K	Курсив	П	Підкреслення

Table 6. Fully localization of B, I, U in same letters

No.	Language	B	Bold	I	Italic	U	Underlined
1	Korean	가	굵게	가	기울임꼴	가	밑줄
2	Lithuanian	P	Paryškintasis	P	Pasvirasis	P	Pabrauktasis

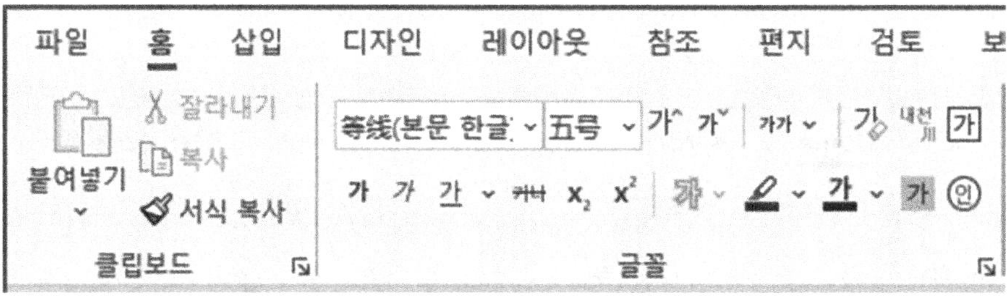

Fig. 8. Microsoft Word 365 (Korean version, personal edition), accessed by 2020/1/21

Specifically, B is replaced by 9 letters, half of them are the first three highly used F, N and P, while G, K, T, Ж, 가, and Қ share the other half. As Italic is based on the early Italian cursives (Etymonline 2020; Microsoft 2017), or spelled as Kursive, K and C are major alternatives for I, the others, D, L, P, S, T, 가 and Қ appears once. The most important substitutions of U are S, A and P, the other three 가, Ч, and П are rarely used (Table 7).

Table 7. Total substitutes of B, I, U in UI

	1	2	3	4	5	6	7	8	9	Total
B replaced by:	F	N	P	G	K	T	Ж	가	K	
Numbers	5	3	3	2	2	2	2	1	1	21
I replaced by:	K	C	D	L	P	S	T	가	K	
Numbers	9	1	1	1	1	1	1	1	1	17
U replaced by:	S	A	P	가	Ч	П				
Numbers	5	4	4	1	1	1				16

Translating these letters is not easy, because they are not only the first alphabet of those commands, but also part of the shortcut key. For example, Italic command can be realized by shortcut Ctrl + I, Bold by Ctrl + B. Ideally, the first letter of the substitutes should also do so, but it is not always the case. The letter I is translated into C, short for Corsivo as an Italian command, but is still used in shortcut key: Ctrl + I. The first ones in Grassetto (Bold) and Sottolineato (Underlined) are part of the shortcuts Ctrl + G and Ctrl + S (Table 8). G is employed in so many shortcut keys as Ctrl + G, because it is exchangeable with Ctrl + B.

Table 8. Shortcut keys of localized UI

	Ctrl + B	Total	Ctrl + I	Total	Ctrl + U	Total
1	Ctrl + G	49	Ctrl + K	10	Ctrl + S	7
2	Ctrl + N	7	Ctrl + T	1	Ctrl + A	1
3	Ctrl + F	4	STRG + UMSCHALT + K	1	STRG + UMSCHALT + U	1
4	Ctrl + I	1				
5	Ctrl + K	1				
6	STRG + UMSCHALT + F	1				
		63		12		9

As can be seen, Microsoft Word was originally designed to use English words and their short forms to imitate the printed format of English documents. Consequently, shortcut keys, operated by keyboard, are often referred to the English letters. However, when the Microsoft Word was developed into an international application, it has integrated the international factors, such as letters in French, Italian, Spanish, and Chinese characters, into the design of its user interface. In the localization process, it has only selected a few to localize, and kept the English short form as international symbols.

3.3   Historical Relations

Word 2007 (Fig. 9) has adopted groups to manage the buttons, which was different from the 2003 (Fig. 10) and previous versions. As is shown from Fig. 9, three main groups in the middle part under Home tab from left to right are Font, Paragraph, and Styles, which emphasize the individual elements the first, holistic modules the second. Again, this is another example that cultural factors were not considered in the design of UI since the thinking habit of western focus on the individual object first, while the eastern is famous for their holistic view (Nisbett 2003).

Fig. 9. Microsoft Word 2007 (Simplified Chinese version, Home and Student Edition), accessed by 2020/1/4

Fig. 10. Microsoft Word 2003 (Simplified Chinese version, Basic Edition), accessed by 2020/1/4

3.4   Intercultural Relations

Besides Microsoft Word, Jinshan WPS (Word Processing System) is a popular and free word processor in China. It seconds to Microsoft Office in Chinese PC market with a penetration rate of 61.74% as of 2017 (CCW 2017). It has been internationalized and installed over 1.2 billion times worldwide (Jinshan 2019) (Fig. 11, 12).

The Microsoft Office clearly had an impact on Jinshan WPS. In the simplified Chinese version of WPS 2019 PC personal edition, the English letters B, I and U were almost in the same position as they were in the Microsoft Word 2007 to 365, with no absence in the upper part of right-click button. These B, I and U English symbols were supposed to be available to help people who seldom use them.

The text edit in email, such as NetEase's 126.com, was also modeled by following the distinct Microsoft style (Fig. 13).

Generally speaking, the basic features of the word processors in China were not designed based on Chinese documents. Moreover, Chinese users' habits were not reflected in the design of the UI.

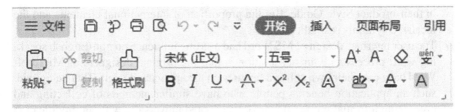

Fig. 11. Jinshan WPS 2019 (PC Simplified Chinese version, personal edition, free of charge), accessed by 2019/12/23.

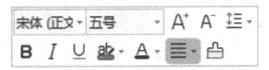

Fig. 12. Upper part of Right-click menu of WPS 2019 (PC English version, personal edition, free of charge), accessed by 2020/1/7.

Fig. 13. Messenger writing styles of www.126.com.

4 Discussions and Conclusion

In this paper, we took the words and symbols in the user interface as a discourse, and analyzed the characteristics of the communicator. By investigating the discourse content, discourse form, and historical relations of different versions of Microsoft Word, we found that the characteristics of the user interface of MS Word are:

(1) MS Word was originally designed mainly for English-speaking people, in order to help them edit English documents on computer. Therefore, it extracted typical elements, such as bold, italic, from English documents, and expressed them respectively in English short forms.

(2) As the number of function keys increased and the traditional display methods were not effective to help users find the needed functions quickly, MS Word divided their functions in groups by following western culture.

(3) When MS internationalized their MS Office software, they integrated other languages' elements, for example, Chinese characters, to enrich their user interfaces.

(4) In the localization process, MS Word changed a small part of their English symbols into target language short forms, and listed different usages of target languages

in their product Style Guide. But the proportion of international elements and the localized symbols is small.

(5) The user interface design of MS Word had a strong influence on non-English speaking word processors. Our observation is that the original designers of Microsoft Word have cast their own cultural features on the application itself. That being said, such an application benefits people who have similar methods of collecting and processing information, but it cannot be adopted universally without modification or with only slight modification.

It would be challenging if we expect a user interface to take every culture into fully consideration. To better adapt a UI design to existing applications, we should learn the basic word processor design strategies from the three perspectives, that is, what distinctive features should be collected, how their own traditional symbols can be employed, and what thinking methods and using habits are often practiced. We are further examining how eastern-thinking-method guides the users to finish their document style in our immediate next step.

References

Hofstede, G.: Culture's Consequences: International Differences in Work Related Values. Sage Publications, London (1980)

Heimgärtner, R.: Using converging strategies to reduce divergence in intercultural user interface design. J. Comput. Commun. 5, 84–115 (2017)

Shi-xu: Cultural discourse studies. Int. Encycl. Lang. Soc. Interact. 1(1), 1–9 (2015)

Shi-xu: Communication of Contemporary China: Studies in a Cultural Discourse Studies Perspective. Peking University Press, Beijing (2010)

CCW Research: Office software industry faces new change after 30 years (2017). http://www.ccw research.com.cn/view_point_detail.htm?id=753664. Accessed 2 Dec 2019

SDL Trados Studio (2020). https://www.sdltrados.cn/cn/resources/infographics/history-timeline. html. Accessed 13 Jan 2020

Baidu Baike: TRADOS (2020). https://baike.baidu.com/item/TRADOS/10916594?fr=aladdin. Accessed 13 Jan 2020

Hu, Z.: Linguistics: A Course Book, 4th edn. Peking University, Beijing (2011)

Harris, J.: Ye Olde Museum of Office Past (Why the UI, Part 2) (2005a). https://blogs.msdn.micros oft.com/jensenh/2005/10/03/ye-olde-museum-of-office-past-why-the-ui-part-2-2/. Accessed 6 Jan 2020

Harris, J.: Inside Deep Thought (Why the UI, Part 6) (2005b). https://blogs.msdn.microsoft.com/ jensenh/2005/10/31/inside-deep-thought-why-the-ui-part-6-2/. Accessed 28 Dec 2019

Harris, J.: Enter the Ribbon (2005c). https://blogs.msdn.microsoft.com/jensenh/2005/09/15/enter-the-ribbon/. Accessed 28 Dec 2019

Wang, H., Cui, Q., et al.: Software Localization. Publishing House of Electronics Industry, Beijing (2005)

Microsoft Language Portal (2019). https://www.microsoft.com/zhcn/language/Terminology. Accessed 23 Oct 2019

Harris, J.: An Office User Interface Blog (2006a). https://blogs.msdn.microsoft.com/jensenh/2008/ 03/13/table-of-contents/. Accessed 23 Oct 2019

Manes, S.: The unfinished word. PC Mag. 3(3), 192–195 (1984)

Bergin, T.J.: The origins of word processing software for personal computers: 1976–1985. IEEE Ann. Hist. Comput. **28**(4), 32–47 (2006)

Microsoft Word: Word Processing Program for IBM Personal Computer (1983). http://toastytech.com/manuals/MS%20Word%201.00%20for%20DOS%20Manual.pdf. Accessed 23 Dec 2019

Microsoft Word 1.x (DOS). https://winworldpc.com/product/microsoft-word/1x-dos. Accessed 28 Dec 2019

Harris, J.: No Distaste for Paste (Why the UI, Part 7) (2005d). https://blogs.msdn.microsoft.com/jensenh/2005/11/07/no-distaste-for-paste-why-the-ui-part-7-2/. Accessed 5 Dec 2019

Microsoft: Localization Style Guides (2019a). https://www.microsoft.com/en-us/language/StyleGuides. Accessed 23 Dec 2019

Microsoft: Microsoft Writing Style Guide (2019b). https://docs.microsoft.com/zh-cn/style-guide/welcome/. Accessed 13 Jan 2020

Microsoft: Microsoft Chinese (Simplified) Style Guide (2019c). http://download.microsoft.com/download/4/c/a/4ca95933-3496-4793-9d77-a89b60a8312c/zho-chn-StyleGuide.pdf. Accessed 23 Dec 2019

Microsoft: Microsoft Chinese (Traditional) Style Guide (2019d). http://download.microsoft.com/download/0/0/c/00c8bd4a-d56a-4a87-9737-154d44fc3712/zho-twn-StyleGuide.pdf. Accessed 23 Dec 2019

Harris, J.: Not So Set In Our Ways After All (2006). https://blogs.msdn.microsoft.com/jensenh/2006/02/27/not-so-set-in-our-ways-after-all/. Accessed 28 Dec 2019

Microsoft: Georgian Style Guide (2019e). http://download.microsoft.com/download/8/0/0/8004AA92-9A6B-4225-AF09-520781BD8011/kat-geo-StyleGuide.pdf. Accessed 13 Jan 2020

Microsoft: Korean Style Guide (2019f). http://download.microsoft.com/download/6/3/9/639fc431-8d7d-482d-b80e-ae4971dc5270/kor-kor-StyleGuide.pdf. Accessed 23 Dec 2019

Miniwatts Marketing Group: Internet Usage in Asia (2020). https://www.internetworldstats.com/stats3.htm. Accessed 14 Feb 2020

Microsoft: Language Accessory Pack for Office (2020). https://support.office.com/en-us/article/language-accessory-pack-for-office-82ee1236-0f9a-45ee-9c72-05b026ee809f?ui=en-US&rs=en-US&ad=US. Accessed 19 Jan 2020

Etymonline: italic (2020). https://www.etymonline.com/search?q=italic. Accessed 21 Jan 2020

Microsoft: Monotype Corsiva font family (2017). https://docs.microsoft.com/en-us/typography/font-list/monotype-corsiva. Accessed 19 Jan 2020

Nisbett, R.E.: The Geography of Thought: How Asians and Westerners Think Differently… and Why. Free Press, Michigan (2003)

Jinshan Corporation: About Kingsoft Office. https://www.wps.com/about-us. Accessed 7 Jan 2020

Jinshan Corporation: WPS Office 2019. https://pc.wps.cn/. Accessed 7 Jan 2020

Netease. www.126.com. Accessed 5 Dec 2019

The keys of products used in this paper are listed as follows:

Microsoft Office 2003: QC77B-823CD-DCW2X-RW4K3-DWMXQ
Microsoft Office 2007: P6TQW-X9DQ4-WFKB2-J8MF6-GFYW8
Microsoft Office 365: DRHR9-WVF46-M67VX-WKGGT-4XJVZ

Service Design for Proposing an Innovative Intergenerational Interaction Platform

Sheng-Ming Wang(✉), Sara Sanchez, Yu-Chen Wang, Wei-Jie Pan,
and Cheng-Yen Lin

Department of Interaction Design, National Taipei University of Technology, Taipei, Taiwan
ryan5885@mail.ntut.edu.tw

Abstract. This research uses the service design method to propose an innovative intergenerational interaction platform to provide solutions to the growing gaps and the interaction problems between generations. This research begins with an introduction on the prototype of an intergenerational communication, collaboration, and time management application called ShareCare. ShareCare proposes an interdisciplinary application that uses design thinking and service design to analyze the problems and provide some solutions. Secondly, we used Service Design methods to extend ShareCare and propose an Intergenerational Service Innovation Framework (ISIF) that integrates virtual and physical services with the application of the Internet of Things (IoT) technologies. Then, this research evaluated the ISIF using the Kano Model to analyze and classify the proposed functions and their related satisfaction factors. At the end of this research, we provided future development and implementation strategies of ISIF. As concluded in this research, technology and information development, smart infrastructures, and innovative services are a promising platform to support and develop practical solutions that increase the possibilities for interaction between generations. The ISIF proposed in this research provides a possible solution for solving the problems of the intergenerational gap and improve the interaction between ages. However, for future work, it's necessary to fully develop the system platform prototype, the implementation of the service framework, and the intergenerational user experience evaluation of the ISIF proposed in this research.

Keywords: Service design · Intergenerational interaction · Internet of Things · Data-driven storytelling · Kano Model

1 Introduction

The world is undergoing rapid population change. Countries are experiencing growth in the number and proportion of older people. The global trend shows that, while the number of surviving generations in a family may have increased, these generations are more likely to live separately today. This demographic and family change means there are fewer opportunities for generations to interact. Mutual benefits are clear, but today it's not easy to get people of different age groups to interact with each other. Our society is becoming segregated increasingly by age. Loneliness and isolation are not only

© Springer Nature Switzerland AG 2020
P.-L. P. Rau (Ed.): HCII 2020, LNCS 12192, pp. 228–244, 2020.
https://doi.org/10.1007/978-3-030-49788-0_17

characteristics of the aging society but also of younger generations. Research shows the apparent demand for the design of innovative goods and services and solutions to help generations connect.

This research proposes a framework for integrating virtual and physical service innovation as a possible solution to the identified problem of the intergenerational gap. We presented the objectives of the study below:

1. Introduce design thinking and service design to analyze the problem and create a service design mechanism to provide a solution for the intergenerational gap.
2. Based on the pilot study project "ShareCare," a communication and time-management platform that connects elderly and youngsters through task services and collaboration, the research is extended to develop a further service design mechanism to connect generations.
3. Propose an Intergenerational Service Innovation Framework (ISIF) that extends from the pilot study ShareCare, and that integrates virtual and physical services with the application of the Internet of Things (IoT) technologies.
4. Evaluate the Intergenerational Service Innovation Framework (ISIF) using the Kano Model to analyze and classify the proposed functions and their associated satisfaction factors.
5. Propose the future development and implementation strategies of ISIF.

This research proposed two limitations in future studies.

6. The first limitation concerns the geographic scope of participants. The group of people that participated in the focus groups was mainly Westerners.
7. The second limitation concerns the focus of the mechanism on services and the innovation proposition. It was beyond the scope of this research to outline the specific, technical details of a prototype and the system implementation details.

2   Literature Review

In recent years, researchers, social businesses, and non-profits have focused on investigating traditional and non-traditional ways to develop and implement programs that promote the interaction between generations. Much research has been done about intergenerational service-learning as well. Intergenerational service-learning aims at developing significant relationships between different age groups [1]. Related projects frequently include students engaging with the elderly to decrease the generation gap. Several non-profit organizations and volunteer organizations have also been trying to close this gap. In the years 2000 and 2003, a nursing home in cooperation with an organization that runs several kindergartens in Vienna started a unique project, a kindergarten integrated into a retirement home [2]. Another example is the development of the business model and service "myBestHelper," a digital tool that offers families the easiest way to book helpers for child care based on the parents' needs (myBestHelper 2018). Another service provided by a for-profit company is "Papa." Papa [3] – Grandkids on-demand – connects university students with older people who need support with senior services like transportation and household chores.

Internet of Things (IoT) and smart infrastructures are the next frontiers of technology that transform the way people live and work. IoT represents a future vision of a sophisticated ecosystem where all physical objects, virtual objects, and smart objects that interconnect to each other. Another use of IoT to close the intergenerational gap was a conceptual study done in Portugal in which IoT was used to develop a playful and immersive cultural heritage tourism project to promote the social, cultural, and economic development of Portuguese rural areas. LOCUS IoT system [4] supports intergenerational communication to avoid isolation and contribute to healthy aging. The integration of IoT and Artificial Intelligence (AI) is a hybrid technology named Ambient Intelligence (AmI). The emergence of ambient intelligence technologies portrayed to be a significant potential for in-home assistance and elderly support, providing an environment that senses and response to the presence of people. Zhou et al. did research to investigate a case-driven ambient intelligence system [5] with the purpose of sensing, predicting, reasoning, and acting as a response to older people's daily living activities in their homes. Results show that ambient intelligence is a promising technology for developing smart and cost-effective solutions for elderly-care applications. A study done in 2011 explores notions of accessibility to these technologies for older generations. Still, it extends the research ideas to broaden the way to consider older people in age-related research. This study focuses on the importance of the intergenerational context as an emphasis on the design of innovative technologies.

The so-called "Millennials" – those born between the early 1980s to the early 2000's – stand out for their high-rate of technology use. But elderly generations also embrace digital life when they perceive the technology as not complex and beneficial for them. A study done by Pew Research Center [6] shows how Millennials have developed rapid adoption habits and therefore created opportunities for product and service technology development. Surveys also found that seniors are moving towards more digitally connected lives. However, they still face unique barriers to adoption, ranging from physical challenges to a lack of comfort and familiarity with technology.

Design Thinking is a user-oriented, prototype-driven methodology for innovation. Many kinds of research prove the usability of the tool for service innovation. A study done in 2018 showed how these tools promote creative design within the organization and understand how to design for users in an efficient way [7]. Service Design's holistic approach, on the other hand, allows it to be applied not only to services but to products as well. An application of Service Design to products showed a "smart carrying interactive chair designed for improving emotional support and parent-child interactions to promote sustainable relationships between elderly and other family members" [8]. The Kano Model screens important service function attributes and studies the nature of customer needs by providing a way for a better classification of customer needs. The Quality Function Deployment (QFD), on the other hand, is defined as a "method for developing design quality aimed at satisfying the consumer and then translating the consumers' demand into design targets" [9]. A study done in Taiwan researched how to integrate and use the Kano Model analysis, and the Quality Function Deployment for quality improvement in healthy fast-food chain restaurants. The Kano Model was used to understand how customers perceive service characteristics. The QFD was employed to describe the relationship among the most crucial service characteristics and corresponding improvements as well as to recognize the priority of those improvements [10].

3 Research Methodology

In this section, we provide the plan and related methodologies used in this research, which extended from the ShareCare pilot studies. The chapter begins by showing the results of the pilot study, ShareCare. We'll detail and discuss the design thinking and service design tools and how to extend the ShareCare and propose the ISIF. Finally, we'll include the Kano Model for evaluating the ISIF. Based on an interdisciplinary service design process, this research presents a pilot study of a prototype for intergenerational and time-sharing management platform, ShareCare. We designed ShareCare to be a platform that connects older adults with youth for task management and collaboration.

We used the design thinking approach will to understand the problem and suggest an innovative solution. To gain empathic understanding with the prospective users, we observed and surveyed two groups with the following characteristics. The idea chosen for development was a platform that supports older adults that need help to complete daily life tasks – like grocery shopping or gardening – by connecting them with youth that applies on the platform to complete these tasks in exchange for some kind of reward. In everyday speech, older people get support, and youth receive reward points in exchange for helping older people. We used the tools persona analysis, customer journey map, service blueprint, among others, to understand the users who were the targets of our designs: elderly and young people.

During the prototype phase, we produced an inexpensive scaled-down version of the product or service. The prototype would be useful to share and test the tool to collect information from the users. Then, we built the prototype using a high-fidelity fast-prototyping tool called Flinto that allowed designers to create an app testable on mobile devices. For the test and validation phase, we validated the built prototype with the same two groups identified during the empathy process. Insights on how they used their apps were going to be collected. The design thinking process was extended later in this research to define and introduce the Intergenerational Service Innovation Framework (ISIF). The prototype was evaluated first by executing a business model. The figure below shows the evaluation (Fig. 1).

Using Peter Morville's user experience honeycomb required hands-on testing with the users to evaluate how they used the service. For this phase, we used the same group we interviewed for the empathy process. The value of the pilot study was to validate a service that solves the intergenerational gap by connecting young and elderly users.

ShareCare presents some advantages and disadvantages. One advantage focuses on the platform as a service mechanism with a proven model behind it. Another advantage is that this service mechanism can bring generations together to communicate with each other. A disadvantage of the ShareCare pilot study is that the solution is suitable for young people but not for elderly users. Specifically, younger users are more comfortable using mobile applications on their phones. Older people struggle with complex tasks on their phone, including filling out forms or registering a simple job. Therefore, the pilot

Fig. 1. Pilot study ShareCare business model canvas (Source: this research)

study suggests a need for a framework that integrates virtual and physical technologies to develop cutting-edge technologies for the elderly in a way that they can be supported and later connected with young people. Because of this need, this research proposed the development of an Intergenerational Service Innovation Framework (ISIF) that integrates virtual and physical services with the application of the Internet of Things technologies (Fig. 2).

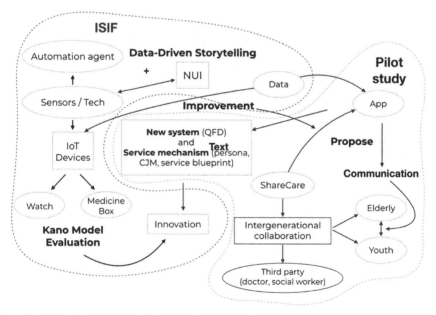

Fig. 2. Proposed intergenerational service innovation framework (ISIF) (Source: this research)

The previous research was developed to generate an intergenerational service innovation framework (ISIF), included below. The ISIF proposed architecture for creating innovative intergenerational interaction. Thus, provide solutions to solve the identified problem of the intergenerational gap. This research extends to the proposal of a framework that integrates virtual and physical services with the application of the Internet of Things (IoT) technologies.

Extending the research meant running more interviews and executing an ideation workshop, which allowed us to approach the study with different perspectives.

Focus group users were from the U.S. or Westerners currently living in Taiwan. We were also able to validate the third-party group as part of the extensive research. This group was composed of two insurance companies whom we contacted and offered us the opportunity to talk to three different doctors from the organizations. The questions during the interview were open-ended questions focused on collecting more data about habits, lifestyles, needs, and technology use for elderly and young people. During the meeting for the third-party (in this case, doctors in insurance companies) questions, we're also open-ended and focused on collecting information about the company's needs and perspective in services related or offered to older people.

After the interviews, a workshop focused on ideation and brainstorming would be executed. The purpose was to understand what users could do with data and the possibilities of using virtual and physical IoT technologies as part of the service. The proposed ISIF is evaluated later on with the use of the Kano model to analyze and classify the intended function and their related satisfaction factors.

4   The Proposition of Intergenerational Innovation Service

This research extends by continuing the Design Thinking methodology and the Design Thinking tools applied to the proposed Intergenerational Service Innovation Framework (ISIF). The ISIF integrates virtual and physical services with the application of the Internet of Things (IoT) technologies. In this section, we'll include the results obtained from the empathic phase and extensive qualitative research collected through the focus group interviews and provided details of the ideation workshop results. Finally, we gave the service design tools like the persona analysis supported by empathy maps, customer journey maps, and service blueprints.

This part of the research presents the insights obtained from the open questions that were part of the extensive research. The ideas collected from the focus group interviews allowed us to understand that a combination of virtual and physical solutions could serve both elderly and younger users. The elderly could use virtual technologies that were not intimidating or invasive, and that could understand their behavior for relevant use cases like health and medicine intake. Younger users could interact with a more complex set of virtual technologies that feel native to their typical mobile app experiences. Together, this set of virtual and physical technologies support the value proposition of the ISIF by connecting generations. The findings also suggest that third parties like insurance companies, doctors, and other health professionals could benefit from this virtual and physical integration since they would have more information to make better decisions for their clients. Data harvested from the set of services could become a competitive advantage by allowing these actors to optimize their interactions with their clients.

After running the focus group interviews and collecting insights that provided us with an understanding of users' behavior and needs, we ran a workshop. This workshop was developed with the purpose of ideating and developing ideas to understand the possibilities we could design depending on those identified needs.

The questions presented during the activity aimed to understand the opportunities we had to integrate storytelling with data. Since the ISIF wants to integrate virtual and physical technologies, including IoT, we knew that we would generate essential data. Findings and results of the ideation workshop allowed us to understand how users could interact with the data collected from sensors and IoT technologies.

After doing the interviews, the persona analysis was extended and updated with the information collected from the validation and extensive research. This part of the research allowed us to organize the information collected about our users and continue with the extensive research using design tools in the proposed ISIF in this research. The persona analysis would allow us to examine the process and workflow that the user would prefer to use in a proposed solution to achieve their objectives (Table 1).

Table 1. Persona analysis according to proposed ISIF (Source: this research)

Persona	Description
Mr. Johnson **Age:** 69 years old **Gender:** Male **Hobbies:** Read, watch TV, play solitaire	Mr. Johnson is in his 70's. He doesn't understand technology well, but he is interested in using it and uses his phone to stay connected with his children, who often send him pictures of his grandchildren. Mr. Johnson thinks technology can be useful for healthcare. He tries to stay active, but sometimes he lacks physical activity, or he forgets taking his medicine
Alex **Age:** 23 years old **Gender:** Male **Hobbies:** Listen to music, play videogames, chatting online, play with new technologies	Alex is in his early 20's. He is a student of the Industrial Design program. Alex spends most of his time at the university campus. He rarely sees his family because he lives in the city. Alex enjoys spending time with his friends. He is tech-savvy and uses his phone almost 40% of his time during the day
Dr. Parken **Age:** 45 years old **Gender:** Male **Hobbies:** read, research, exercise	Dr. Parken works for an insurance company. He is part of a team that works with data to understand patients (specifically elderly) behavior related to healthcare. He noticed the company has been trying to incorporate innovative strategies into the company to improve the value they offer to their clients

A Customer Journey Map (CJM) helps to systematically understand users' interactions and the touchpoints that lead them to particular experiences. A CJM was developed based on the ISIF and the information collected during the empathic process of extensive research (Fig. 3).

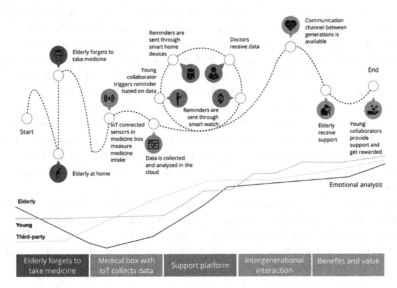

Fig. 3. Customer journey map of proposed ISIF (Source: this research)

In the following service blueprint, we show the relationships between the different service components and processes that are directly related to the touchpoints in the specific customer journey [11] based on the proposed ISIF. Stakeholders are highlighted in bold to identify their roles during the process (Fig. 4).

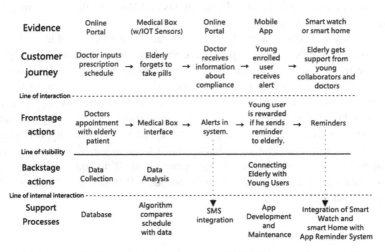

Fig. 4. Service blueprint based on proposed ISIF (Source: this research)

After the pilot study ShareCare, the research was extended based on the evaluation of the pilot study. The pilot study gave information about the challenges of the elderly while using technology. One crucial insight was that even though they feel that technology can improve their lives, it's tough for them to interact with sophisticated technologies. Therefore, as we extended the research, we developed the ISIF, and we extended the design thinking and service design tools to define the intergenerational service innovation framework better. The ISIF proposes using virtual and physical technologies that integrate IoT. The ISIF proposes a new system and service mechanism that connect across generations. IoT promises to support humans by providing passive monitoring through the collection of data with sensors. By identifying a problem faced by older people – forgetting to take their medicine – we propose a medical box that uses sensors connected to networks enabled by IoT to support medicine intake. The medical box would use that collected data to support the elderly user through third parties like insurance companies and younger generations that help them remember to take their medicine.

The benefits of this innovation do not restrict to the elderly. Young people would also find benefits by interacting with an older generation and getting rewarded for their knowledge of technology. Based on the information above, we propose a Macro and Microanalysis of the service mechanism that allows us to define the functions and user behavior operations needed for the ISIF.

5   Research Evaluation and Analysis

This chapter focuses on the evaluation of the ISIF proposed in this research. It starts by defining the macro and micro analysis to define the functions that the Kano model used for evaluation based on an identified scenario. Macro and Microanalysis allow us to have a holistic but detailed perspective about a problem we're trying to solve. During this research, we implement Macro and Microanalysis to understand the service mechanism functions and the user behavior operational needs.

We analyzed the scenario according to the identified situation in which older people forget to take their medicine. Young people are comfortable using technology, especially on their mobile phones. Older people struggle with these tasks. Young people can support older people's compliance with medicine intake. Third parties need information about older people's compliance with treatment to understand better how to support their elderly user's health. The ISIF proposes the integration of virtual and physical services with the application of Internet of Things technology as a solution to support the elderly with easy-to-use and non-invasive technologies that can be handled by young people and third parties. In this section, we'll enumerate the identified parts of each analysis. Below we enumerate the macro analysis of the proposed ISIF based on the service design results and the analyzed scenario.

Elderly:

- Older people have difficulties executing physical tasks and often forget to complete personal tasks like taking their medicine.
- Older people don't like sophisticated technologies.
- Older people are frequently isolated.
- Older people have a positive perception of technology and believe technology can usefully support their healthcare.

Younger people:

- Younger people are tech-savvy.
- During their studies, young people are not able to commit to a part-time or full-time job.
- Younger people want to earn money and acquire skills.
- With the development of new technologies, younger people are also isolated from communities and interactions with other age groups.

Insurance Companies:

- Insurance companies need to gain a competitive advantage through innovation.
- Insurance companies want to introduce more technologies into their products and operations.
- Insurance companies want to collect more data about their clients to make better decisions.

   Below we address the microanalysis of the proposed ISIF based on the service design results and the analyzed scenario.

Elderly:

- Older people prefer to interact with a non-screen interface, like voice-controlled assistants and other products that use a Natural User Interface.
- The elderly are more likely to share their data with a service that supports their health and shares data with specific, trusted third parties like doctors or family members.
- Measure medicine intake without having to calculate or write down anything.
- Elderly users want help remembering to comply with their treatment and don't mind receiving notifications like SMS, phone calls, or other prompts.

Young People:

- Young people see the internet as offering ways to earn money.
- Young people want a service that allows them to make money on a flexible schedule.
- A specific subset of young people enjoys the company of older people, especially young people who may not have access to an extended family nearby and want a medium to connect with them.

Insurance Companies:

- Insurance companies want to monitor factors like blood pressure, changes in heart rate, daily exercise, and emotional state.
- Be able to remind and prompt clients to comply with treatment, using SMS, email, or Natural User Interfaces.
- Visualize elderly medicine intake daily, without relying on self-reported data.
- Insurance companies want to offer information about the behavior, premium, payments, and other details of an elderly family members' insurance plan since family members may not live with the elderly family members but want to monitor their health.

Based on the information collected on the extended design thinking and service design analysis, we used the Quality Function Deployment (QFD) tool to define the functions and obtain the base for the development of the questionnaire. The QFD allowed us to score them and assign them an order of priority by designers' subjective analysis. Based on the quality function deployment analysis at the physical level, the priority order is 1) smart watch; 2) smart medical box, and 3) smart home device. At the practical level, the priority order is: 1) reminding information, 2) screen interfaces, and 3) Voice-controlled interfaces. In terms of functions, the priority order is 1) measure vital signs, 2) SMS reminders, and Phone call reminders. These last two score the same, therefore they share the same level of priority. At last, in terms of services, the priority order is 1) practical challenges (streak), 2) smart medicine stocking, and chat with users of other ages. These last two score the same, therefore they share the same level of priority. The importance of this evaluation relies on having the information relevant to the design of services that meet users' needs based on the priority score (Table 2).

Kano's model is used in innovative development to identify potential user needs and to evaluate service functions and users. The questionnaire in this research refers to the quality elements of the proposed functions and their related satisfaction factors based on the ISIF. We can measure customer satisfaction by using a customer satisfaction coefficient. The coefficient number increases when the design met the requirement and increases satisfaction and vise Versa. By calculating an average satisfaction across customers, we can derive a useful overview of our service design and isolate the effects of individual service requirements. The calculation of this coefficient is as follows (Table 3).

Equation 5.1 Kano model analysis satisfaction coefficient (Source: Zacarias 2015)

$$Satisfaction\ Coefficient = \frac{A + O}{A + O + M + I}$$

Equation 5.2 Kano model analysis dissatisfaction coefficient (Source: Zacarias 2015)

$$Dissatisfaction\ Coefficient = -\frac{O + M}{A + O + M + I}$$

Table 2. Quality function deployment analysis of ISIF functions (Source: this research)

<table>
<tr><td rowspan="3"></td><td rowspan="3"></td><td colspan="4">Physical</td><td colspan="3">Virtual</td><td rowspan="3">Score</td><td rowspan="3">Priority</td></tr>
<tr><td>Smart medical box</td><td>Smart home device</td><td>Smart watch</td><td>Medicine tracker</td><td>Reminding information</td><td>Screen interfaces</td><td>Voice-controlled interfaces</td></tr>
<tr></tr>
<tr><td rowspan="7">Functions</td><td>SMS reminders</td><td></td><td>○</td><td>◎</td><td></td><td>◎</td><td>△</td><td></td><td>27</td><td>2</td></tr>
<tr><td>Smart watch alerts</td><td></td><td></td><td>◎</td><td></td><td>◎</td><td></td><td>○</td><td>24</td><td></td></tr>
<tr><td>Email reminders</td><td></td><td></td><td>△</td><td></td><td>◎</td><td>◎</td><td></td><td>21</td><td></td></tr>
<tr><td>Phone call reminders</td><td></td><td>○</td><td>◎</td><td></td><td>◎</td><td></td><td>△</td><td>27</td><td>2</td></tr>
<tr><td>Voice reminders</td><td></td><td></td><td>△</td><td></td><td>◎</td><td></td><td>◎</td><td>21</td><td></td></tr>
<tr><td>Measure vital signs with medical box</td><td>◎</td><td></td><td></td><td></td><td>◎</td><td>○</td><td>○</td><td>30</td><td>1</td></tr>
<tr><td>App download</td><td></td><td></td><td>○</td><td></td><td>△</td><td>◎</td><td></td><td>18</td><td></td></tr>
<tr><td rowspan="7">Service</td><td>Online community</td><td>◎</td><td>◎</td><td></td><td></td><td>○</td><td></td><td>○</td><td>30</td><td>2</td></tr>
<tr><td>Doctor's online access to patient's compliance</td><td>◎</td><td></td><td>△</td><td></td><td>○</td><td>◎</td><td></td><td>27</td><td></td></tr>
<tr><td>Online visualization of healthcare data</td><td></td><td>△</td><td>○</td><td></td><td>○</td><td>◎</td><td></td><td>24</td><td></td></tr>
<tr><td>Virtual challenges (streak)</td><td>◎</td><td>○</td><td>◎</td><td></td><td>○</td><td>◎</td><td></td><td>39</td><td>1</td></tr>
<tr><td>Smart medicine stocking</td><td>◎</td><td>○</td><td></td><td></td><td></td><td></td><td>○</td><td>30</td><td>2</td></tr>
<tr><td>Chat with users of other ages</td><td></td><td></td><td>◎</td><td></td><td>○</td><td>○</td><td>◎</td><td>30</td><td>2</td></tr>
<tr><td>Healthcare online portal for third-parties</td><td>◎</td><td></td><td></td><td></td><td>◎</td><td></td><td></td><td>18</td><td></td></tr>
<tr><td colspan="2">Score</td><td>54</td><td>30</td><td>60</td><td>0</td><td>105</td><td>60</td><td>39</td><td></td><td></td></tr>
<tr><td colspan="2">Priority</td><td>2</td><td>3</td><td>1</td><td></td><td>1</td><td>2</td><td>3</td><td></td><td></td></tr>
</table>

"A positive customer satisfaction coefficient ranges in value from zero to one; the closer to one the value is, the greater the influence on customer satisfaction. The negative customer satisfaction operates in the same way. A value of zero signifies that this feature does not cause dissatisfaction if it is not met" (Qiting, Uno & Kubota 2013). We describe the customer satisfaction coefficients for the ISIF in the table below. This information allowed us to understand which functions and services if met or not during the design of the service, would bring satisfaction or dissatisfaction to the user.

Table 3. Kano model customer satisfaction coefficient results (Source: this research)

Function	Quality element	SC	DC
Smart medical box	A	0.7	− 0.25
SMS reminders	M	0.5	−0.8
Reminders through smart watch	O	0.65	− 0.6
Online community	A	0.7	−0.35
Voice-controlled interfaces	O	0.8	− 0.65
Smart home device	O	0.85	−0.6
App download	R	0.2	− 0.2
Doctor's online access to patient's compliance	O	0.85	− 0.65
Smart watch	O	0.75	− 0.6
Reminders through smart homes	O	0.75	−0.55
Online visualization of healthcare data	O	0.9	− 0.65
Virtual challenges (streak)	A	0.79	−0.32
Healthcare online portal for insurance/doctors	O	0.8	−0.55
Medicine tracker	O	0.75	−0.7
Reminders through email	1	0.85	−0.65
Measure vital signs with medical box	M	0.4	− 0.6
Chat with users of other ages	A	0.75	−0.4
Smart medicine stocking	A	0.75	−0.35
Screen interfaces	O	0.67	−0.78
Phone calls reminders	M	0.35	−0.6

There were 20 questionnaires done completed in this research. The studied group was the same one that completed the second round of interviews detailed before. By identifying a problem faced by older people – forgetting to take their medicine – we proposed a medical box that uses sensors connected to networks enabled by IoT to support medicine intake. The medical box would use that collected data to support the elderly user through third parties like insurance companies and younger generations that help them remember to take their medicine. We evaluated the functions to understand the related satisfaction factors. Functions that imply a direct action from the user are perceived as "less attractive" than the ones that use natural user interfaces to support user interactions with technology to solve a problem.

With the information collected before, we can understand how to design an innovative service for intergenerational connections between ages better. As technology keeps advancing fast, such as technology adoption in different generations. As it shows in Table 4, nowadays people take for "granted" functions like SMS and phone calls. And even though older people don't like to interact with sophisticated technologies, they're interested in how technologies with devices, sensors, and voice-controlled functionalities can support them in their daily life and health. Table 4 shows the final result based on the Kano model analysis and quality elements axis and the relationship with the virtual and physical functions and services of the ISIF.

Table 4. Kano model analysis results (Source: this research)

Attractive requirements (excitement needs)	One dimensional requirements (expected needs)	Must-be requirements (basic needs)
• Smart medical box • Online community • Virtual challenges (streak) • Chat with users of other ages • Smart medicine stocking	• Reminders through smart watch • Voice-controlled interfaces • Smart home device • Doctor's online access to patient's compliance • Smart watch • Reminders through smart homes • Online visualization of healthcare data • Medicine tracker • Healthcare online portal for insurance/doctors • Screen interfaces	• SMS reminders • Measure vital signs with medical box • Phone calls reminders

6 Conclusions and Future Works

This research reaches the following conclusions:

1. **Results of Pilot Study**: Results of the pilot study included at the beginning of this research provide us information about an initial service designed to connect generations based on task service and a time-sharing platform, ShareCare, for elderly and young people. Results from a Service Design and Design Thinking analyses helped us to understand if we solve the users' needs. The evaluation of the pilot study helped us to discover that young people have an easier time using technology, in particular, mobile apps. Older people found it challenging to interact with the mobile application interface. Based on this, we extended the research and proposed the Intergenerational Service Innovation Framework (ISIF), which proposes the integration of virtual and physical services that use natural user interfaces and IoT technologies to support services design for intergenerational interactions.
2. **Extended Research and ISIF Proposal:** The extension and comprehensive analysis of the ISIF with an ongoing application of Design Thinking and Service Design tools provided us with information to define the macro and micro perspectives of another,

more specific scenario. This scenario focuses on the problem of elderly medicine intake. Older people often forget to take their medicine. The integration of virtual and physical services with IoT technologies and the connection between generations can serve as a system to solve problems faced by both young and elderly.

3. **Macro and Micro Analysis:** The results of the service design analysis provided information to execute a micro and macro analysis. We were able to define the functions of the system that could add satisfaction to the user experience. We used the Quality Function Deployment to understand the relevance of the system functions according to users' needs of the ISIF.

4. **Quality Function Deployment Evaluation:** Based on the Macro and Microanalysis, we executed a Quality Function Deployment (QFD), which allowed us to categorize the insights collected into categories of services and function. Each of these two categories is then divided into a physical and virtual version of the service of function. The QFD then allowed us to establish the priority of each category. The decision criteria for the QFD was the designer's subjective analysis.

5. **Kano Model Evaluation:** We evaluate the proposed ISIF using the Kano model. Based on the functions identified, we run a questionnaire to understand how users value functions and how those functions satisfy specific needs. As seen in the data showed in the previous chapter, some functions that users considered attractive are: smart medical box, online community, practical challenges, the opportunity to chat with others, and smart medicine stocking. Attractive features are those that please the user when present, but that has no adverse effect when missing. Some of the functions considered one-dimensional are reminders through a smart watch, voice control interfaces, smart home devices, doctors' online portal for patient compliance, reminders through the smart home, online visualization of healthcare data, online healthcare portal for insurance and doctors, and screen interfaces. By one-dimensional, we mean those that cause satisfaction to the user when it does not meet their expectation but has a positive effect when it exceeds them. The functions identified as must-be were related to necessary technologies like SMS reminders, phone call reminders, and the measure of vital signs. Those attributes are unnoticeable by a customer when they present but have substantial adverse effects when missing. At last, we defined one function as indifferent, and one fell under reverse. The indifferent functions were reminders through email, which allowed us to understand that emails are not an essential medium for seniors. We categorized the function of mobile application downloads as the reverse. These results are from the complex challenges that the download process implies.

6. **Value of Research:** The evaluation of the ISIF becomes relevant for future research related to the design for services for intergenerational design. The results of this research suggest that future services aimed at bridging the international gap should incorporate both physical and virtual services to benefit stakeholders who work together.

Future works were considered during research and could explore the following:

7. By extending the pilot study and proposing the ISIF, we layout the conceptualization of a system that aims to connect generations through virtual and physical interfaces that use IoT technologies.
8. For practical validation, we recommended developing a prototype for implementation. It is also essential to develop a prototype of the proposal to understand the technical possibilities and feasibility of the proposed system. The analysis provides valuable insights that influence the end-user experience of the service design.
9. Another consideration of the research was concerning the participation of a third party in the system. In the case of this study, we focused on insurance companies, but this can extend to private medical practices, physicians, and even family members who often are also in charge of an older person's health.
10. Another consideration is that the product design fleshed out in more detail based on the ISIF. Product design would allow us to, for example, focus on the medical box in more detail. The tracking system included in the box, the friendliness and intuitiveness of the product could also be explored and understood better in the context of the ISIF.
11. Based on the results of the Quality Function Deployment, we obtained the category of services. Nonetheless, we did not explore in depth in this research. Future works could focus on researching in detail the service aspect to discuss what to do in terms of design and how services like this could add value and can create future business models for intergenerational services.

References

1. Kogan, L.R., Schoenfeld-Tacher, R.M.: Participation in an intergenerational service learning course and implicit biases. Educ. Gerontol. **44**(2–3), 90–98 (2018)
2. Amaro, A., Oliveira, L.: IoT for playful intergenerational learning about cultural heritage: the LOCUS approach. In: Present in 5th International Conference on Information and Communication Technologies for Ageing Well and e-Health, Heraklion, Crete, Greece, 2–4 May 2019 (2019)
3. Zhou, F., Jiao, J.R., Chen, S., Zhang, D.: A case-driven ambient intelligence system for elderly in-home assistance applications. IEEE Trans. Syst. Man Cybern Part C (Appl. Rev.) **41**(2), 179–189 (2011)
4. Anderson, M., Perrin, A.: Technology Use Among Seniors. Pew Research Center, Washington, DC (2017)
5. Czaja, S.J.: The role of technology in supporting social engagement among older adults. Public Policy Aging Rep. **27**(4), 145–148 (2017)
6. Yu, E., Sangiorgi, D.: Service design as an approach to implement the value cocreation perspective in new service development. J. Serv. Res. **21**(1), 40–58 (2018)
7. Tseng, W.S.-W., Hsu, C.-W.: A smart, caring, interactive chair designed for improving emotional support and parent-child interactions to promote sustainable relationships between elderly and other family members. Sustainability **11**(4), 961 (2019)
8. Kuo, H.-M., Chen, C.-W.: Int. J. Innov. Comput. Control **7**(1), 253–268 (2011)

9. Chen, K.J., Yeh, T.M., Pai, F.Y., Chen, D.F.: Integrating refined Kano model and QFD for service quality improvement in healthy fast-food chain restaurants. Int. J. Environ. Res. Public Health **15**, 7 (2018). https://doi.org/10.3390/ijerph15071310

10. Pinto, S., Cabral, J., Gomes, T.: We-care: an IoT-based health care system for elderly people. In: 2017 IEEE International Conference on Industrial Technology (ICIT), Toronto, ON, pp. 1378–1383 (2017)

Trends on Design Tools Under Futurology

Qing Xia and Zhiyong Fu[✉]

Tsinghua University, Beijing, China
fuzhiyong@tsinghua.edu.cn

Abstract. This paper proposes a structured research for design tools from the perspective of futurology, especially the tool research that can help future scenarios, and aims to comprehensively understand the current research situation. This paper displays a framework to help summarize the current research focus, using design, futurology and tool research as the three keywords for discussion, and taking the "worldview-scenario-product" of future design as the three levels of content, carrying design process with four dimensions of "look back-insight-construction-reflection". On this basis, this article discusses views of design future and its corresponding design methods and tools, bringing new perspectives and content to design tools.

Keywords: Futurology · Design future · Design method · Design tool

1 Introduction

Nowadays, with the prosper of human-computer interaction, in order to quickly seize the market, many products and services are developed and launched rapidly, and are iterated mainly relying on user feedbacks. This is actually a trend in which designers replace their speculation with users' feedback. However, when giving feedback, users pay more attention to their own interests, rather than the long-term and widespread impact of the products and services. On this occasion, we try to use the design tools to implant future sensitivity into the creating activities, so that designers can identify whether the future state presented is what we want when designing the current form of products or services. The juxtaposition of design and futurology has two aspects. One is to establish future thinking, and uses the future-oriented research methods to improve the existing design methods. One is to use the practice-oriented characteristics of design to promote the futurology from phenomenon research to practically application on shaping the future.

Through the search and analysis, this paper studies the development direction of futurology, the methods and contents of designing future, especially the existing methods and tools of futurology and design. Because of the uniqueness of futurology research, with the typical characteristics of forward-looking, it is very helpful to discover the current social focus and future view of social disciplines. This article analyzes the keywords of literatures to show the research focus of Futurology in the past 2 years to help locate possible trends for design research. At the same time, according to the framework of future study and the model of design thinking, the process of designing future is set, that

P.-L. P. Rau (Ed.): HCII 2020, LNCS 12192, pp. 245–254, 2020.
https://doi.org/10.1007/978-3-030-49788-0_18

is, "looking back-insight-construction-reflection", and 3-level content of design future is developed, that is "value-scene-product". Under this framework, the existing future study tools and design tools are cross-analyzed. The framework demonstrates the possibility of design tool development from the futuristic view, bringing new perspectives and contents to design method and design tool research.

2   Futurology and Design

The term futurology was first proposed and used by the scholar Ossip Flechtheim in 1943. Futurology (also known as futures studies and futurism) is a discipline that studies the possible, plausible, probable, and preferable futures, the worldview behind, and operating laws. Futurology attempts to understand what may continue and what may change. Therefore, part of the discipline is to seek a systematic, model-based understanding of the past and present, and to measure the likelihood of future events and trends.

Futurology developed after World War II, and in the early stages of development, it mainly studied around the sociology and politics. Later, researches in socio-economic, scientific and technological fields was involved. It has now expanded to interdisciplinary, including design, to jointly solve the Wicked Problem (Table 1).

Table 1.   The developmental stage of futurology.

Late 1940s–1950s	1960s	After 1970s	Now
Concept generation Theoretical and academic discussion Sociological category Strong political tendencies	Focus on social and economic development	Focus on science and technology Various schools of futures studies have emerged	Intersects with many fields, including design

This article analyzes the hot topics of the Journal of Futures Studies [1],, and summarizes the hotspots of futures studies in the past two years from three parts of the theory, the fields involved, and the specific practice (Table 2).

Because of the forward-looking characteristics of futures studies, we can see future research challenges from the hot issues that other social disciplines may face, including design disciplines. From the analysis, now the main areas of intersection between design and futures studies include practical service design, speculative design focusing on method reflection, and design fiction focusing on expression of ideas.

From the perspective of design discipline, design itself also includes exploration of the future. Design deconstructs current problems and creates future scenarios by creating products or services. Starting from the future concept video "Microsoft 2020", we see a series of new design concepts of the future development trend proposed based on technology, showing the future life scenarios and new models of human-computer interaction. The development of technology has also led to humanistic thinking, about the changes in

Table 2. Research hotspots of futurology in 2017–2019

Topic of study	Theme	Research direction
Theory		Criticism and epistemology, holisms and integrative theories, future-proofing, experiential futures
Fields	Politics	Policy making, governance, leadership, peaceful future, globalization, global governance, planetary mega-crisis, war
	Society and culture	Future consciousness, racial equality, racial spirit, gender equality, cross-culture/cross-civilization, demography, education, ethics, macro history and time change, community building, social governance, urbanization
	Humanities	Metaphor, narratology, philosophy
	Economy	Enterprise and vision management, experiential future, participatory networks
	Technology	Communications, innovation, media, modeling, simulation and games, nuclear
	Natural	Climate change, environmental protection, ecological environment, energy, evolution, health and sustainability
	Design	Service design, design fiction, speculative design
Practice	Specific areas	Cross-environmental applications, Africa, China, India, Iran, USA
	For special groups	Children, the elderly, women

human society's lifestyles and changes in values. We hope to provide some references for the future development of science and technology from a humanistic perspective.

As far as human-computer interaction disciplines are concerned, there is a tendency to rely too much on product feedback. When a product is designed, it will be released immediately to occupy the market and then it adjusted by user feedbacks. This is actually a technical thinking-oriented, data-driven product design development. But the design method is in fact giving up some designers' speculations to user feedbacks. This model brings new problems. Users are relatively short-sighted in feedback and pay more attention to the benefits available in the near future, rather than the long-range development of the product. And, there is a lack of thoughts about the meaning and the value of products and services. So, we have to re-emphasize the speculative level future thinking in design.

Design tools are approaches that inspire thinking. On the surface layer, they are simple tool papers or sheets (some tools only have rules and no corresponding material objects), but in fact tools are the materialization of a specific way of thinking. We study design tools under Futurology, aiming to present future thinking in the form of tools to help designers raise concerns about the future and train designers' future sensitivity. With these tools, we hope that designers can think more deeply about the future, lead the direction of future development by designing products or services, and reach a preferable future.

3   The Process and Content of Designing Future

Designing future takes the future as a design object, and design its concepts (worldview), scenarios, products and services. It is the practice of futurology in the design field. In order to sort out the process of designing future, the process of futurology and design is used as a reference here. On the process of futures studies, this paper adopts *The six pillars of futures studies* by Sohail Inayatullah [2], which is mapping - anticipation - timing - deepening - creating alternatives - transforming. One the process of design, the paper adopts *Design thinking* process proposed by IDEO for its wide range of influences, which is empathy - define - ideation - prototype - test. We conduct a comparative analysis of the two models to find the combination of the futurology creative process and the design process, helping to establish the basic process of designing future (Table 3).

Table 3. Process of futures studies and design

Futures studies	Design Thinking
Mapping	
	Empathy
Anticipation	Define
Timing	
Deepening	Ideation
Creating alternatives	
Transforming	Prototype
	Test

By simplifying and merging, we propose a designing future model of lookback - insight - construct - rethink. In this model, we retain mapping which is very futuristic, but instead of taking the full mapping connotation of the past, present, and future, we focus on a review of the past to help designers analyze products or services by the trajectory that have happened. Insight includes anticipation and timing in future research, as well as empathy and define in design, which means identifying and determining the issues to study. Construct includes deepening, creating alternatives, and transforming in future research, as well as ideation and prototype in design, which means creating and detailing solutions. Rethink is a very important part. Although it is not reflected in futures studies, the speculative design method commonly used in futurology emphasizes taking the future form of objects designed by present, to think over their current choices [3–5] (Fig. 1).

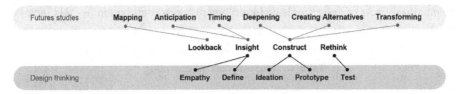

Fig. 1. Process of designing future

If we consider the process of designing future as action steps built on the timeline, the object that runs through the entire process is the content of designing future.

We divide the content of designing future into three levels, namely, the object, the environment, and the intention. The three levels represent as the product, the scenario, and the value in terms of design entities, and represent as the appearance, the system, and worldview behind at the metaphor level. Through layered research on content of designing future, it can help to explore the opportunities of design tools and check the comprehensiveness of tool design [6, 7] (Fig. 2).

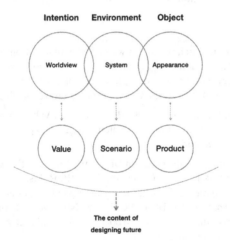

Fig. 2. The content of designing future

In future-oriented design, this research emphasizes the shaping of macro values, that is, the overall operating law of the future world, and then passes the values it contains down to the specific product or service design. With this top-down approach, the products or services are designed to be appropriate to the preferable world.

4 Design Tools and Tools of Futurology

We compile the commonly used futures study and design tools in order to compare and find the differences and connections between the two, to help analyze the possible development direction of design tools from the perspective of futurology, and explore the possible effects of design tools on futurology research contribution [8–22] (Tables 4 and 5).

Table 4. Common tools for futurology

Process	Tool
Mapping	Shared history, Futures triangle maps, Futures landscape
Anticipation timing	Emerging issues analysis, Futures wheel
Deepening	Causal layered analysis, 4-quadrant mapping
Creating alternatives	Nuts and bolts, Scenarios, Articulates scenario archetypes, Scenario writing, Four dimensions scenario
Transforming	Questioning, Concept video, Analytic scenario, Backcasting, Transcend method

Table 5. Common tools for design

Process	Tool
Empathy	Secondary researches, Graffiti walls, Stakeholders, Conversation starters, User observation, Shadowing, Fly-on-the-walls, Guided tours, Simulation Exercises, User interviews, Focus groups, Intercept interviews, Questionnaire, Personal Inventories, Diary studies, Culture probes, Crowdsourcing, Card Sorting
Define	Personas, Empathy maps, User Journey maps, Value Proposition Maps, HMW How might we? Point-of-view statements, Territory maps, Mood boards, Image Boards, Design briefs
Ideation	Brainstorming, Lotus Blossom Diagram, Mindmap, Design Mash-up, 6 Thinking Hat, Fast Idea Generator, SCAMPER, Fishbone, Provocation, Affinity diagram, 2 × 2 Matrix, Business Model Canvas, Lean Canvas, PEST, SWOT, Analysis, 5W1H, PAPCM, How Now Wow Matrix, KANO Model, Value Opportunity Analysis, $100 Test, Weighted Matrix, Dot Voting, Possession tool, Speed dating
Prototype	Tomorrow headline, Scenarios, Storyboard, Paper prototype, Flexible prototype, Technical Documentation, Video Visualization, Role-playing, Wizard of Oz, Experience Prototype, Blueprint, Business Origami
Test	Concept Testing, Usability Testing, Heuristic Evaluation, A/B Testing, 5-Second Usability Test, Eye Tracking, Web Analytics, Net Promoter Score, Semantic Differential Scale, Feedback Capture Grid, 5E Model, The Love Letter & The Breakup Letter, System Usability Scale, Testing Design Using the Competition, Snap Tests
Cross process	Design Sprint, Design Jam, Creative Workshop, World Café

In order to visually understand each tool in the stages, we put the futurology research tools and design tools listed above the designing future process model according to their functions. We separate the two types of tools into two ends of the progress line. The curve on background represents the divergence (high) and contraction (low) of the thoughts (Fig. 3).

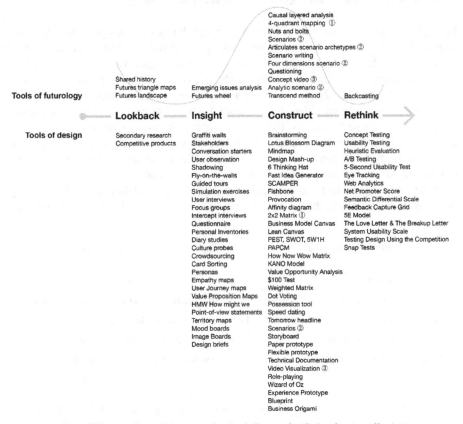

Fig. 3. Futurology research tools vs. design tools (Color figure online)

By comparing and analyzing futurology research tools and design tools in the process of designing future, we find that compared to design tools (about 120 commonly used, the figure shows only some of the tools that are highly relevant to futuristic research), the number of futurology research tools relatively small. There are cases of cross-use between the two or mutual variation (marked in red), such as the scenario method, the 2 × 2 quadrant method, and the use of video. It verifies that there is no huge gap between the two, and they can learn from each other or even use directly. At the same time, although some design tools have not been used for futures studies, their properties and functions are similar to futurology research tools and can be applied directly (blue part), such as the culture probes, a tool to probe group culture through diegetic objects which is very suitable for studying cultural phenomena in specific cultural environment; the fishbone, a tool for digging the root of a problem in a progressive way through inferencing; the tomorrow headlines, a tool that simulates news coverage to show future concept that is difficult to implement immediately, etc. It is worth trying to directly use these tools in futurology research and verify their effectiveness in practice.

5 Opportunities of Design Tool Under Futurology

Through a comparative analysis of futurology research tools and design tools, this study finds that future-oriented design tool research opportunities include: 1) There are few design tools in the existing "lookback" part, which have typical futuristic research features. The design tools can learn from the characteristics of this stage to increase the diversity of design tool; 2) The two have a very large mutual reference or even direct application. Relatively speaking, futures study tools are more focused on speculative, design tools are more focused on the smooth process, the two can be merged to evolve each other and generate new design tools; 3) Design tools which have the properties of futures study tools can be directly applied to future design, and researchers can summarize the data of using on futures through experimental results.

In addition, we conduct a cross analysis from the process and the content of designing future in order to examine the possible development direction of design tools (Table 6).

Table 6. The cross-analysis of designing future process and content

	Product	Scenario	Value
Lookback	Trajectory of products existed	Scenario in history	Existence value verification
Insight	Future needs	Perceptive scenario	Internal desires
Construct	Fictional products	Futuristic scenarios	Empathy creation
Rethink	Reflect on products	Reflect on scenarios	Reflect on values

In the cross analysis of process and content of designing future, the details presented by each cross grid may become the directions of design tool development (this form is only an example, designers can fill in more possible answers according to the prompts of the horizontal and vertical nouns). The cross-analysis table shows some of the possible trends on design tools under futurology. Here taking the cross space between product and lookback as an example, we can design a tool that helps designers review and resolve existing products and their development trajectories, to help similar products avoid risks and better control the future development of their own products.

6 Reflection

This research mainly focuses on the research of design tools, that is, how to help designers improve the sensitivity of the future by creating tools, so there is no specific product or service involved. And, the article is about trend research, showing the overall direction and potential of designing future tools, no specific design tool and usage as the example in detail.

In the subsequent research, the main research direction will be to create design tools under futurology to verify and improve the designing future process and content, and help designers implant future-considered genes into the design process. At the same time, through the practice of tool design and application research, the features and design methods of designing future tools will be explored to help tool designers.

7 Conclusion

Based on the literature research, this paper extracts the keywords of future research in the past 2 years to show the trends of futures studies, and puts forward the three levels of "value - scenario - product", with the corresponding metaphorical layer "worldview - system - representation" as design tools research content. In terms of the designing future process, this paper proposes "looking back - insight - construct - rethink" as the four dimensions of design advancement. On this basis, this paper summarizes the existing major futurology research tools and design tools, and further analyzes the two types of tools in the progress axis of the designing future process, demonstrates that the two can learn from each other, and shows possible further development of design tool development areas.

Relevant research of next step will be the use of the three levels of designing future content and the four dimensions of process as a basic framework to design a set of design tools to promote future design, and give out verification cases and associated design implications.

The paper mainly displays structured research for the design tools from the perspective of futurology, and taps the existing research opportunities to provide a basic framework and perspective for the research of tools for future scenarios. This research analyzes the existing literature of futurology, futurology research tools and design tools, summarizes key words, compares and analyzes existing futurology research methods and design tools, and proposes models of designing future process and content. On this basis, this article discusses possible future views of design and corresponding design methods and tools, bringing new perspectives and content to design tools as a response to the perspective of human culture in the vision of future scenarios. At present, the framework still needs to be further improved, which is will be used as a theoretical basis and model to provide more appropriate guidance for specific tool design and experiments.

References

1. Journal of Futures Studies homepage. https://jfsdigital.org. Accessed 1 Mar 2020
2. Inayatullah, S.: Six pillars: futures thinking for transforming. Foresight **10**(1), 4–21 (2008)
3. Boulding, E., Boulding, K.: The Future: Images and Processes. Sage, London (1995)
4. Chopra, D.: Synchrodestiny. Rider & Co., London (2005)
5. Dator, J.: The Futures of Cultures and Cultures of the Future. Academic Press, New York (1979)
6. Marsella, T., Ciborowski, T., Tharp, R. (eds.): Perspectives on Cross Cultural Psychology. Academic Press, New York (1979)

7. Inayatullah, S.: Editor of Special issue on Macrohistory: contributing authors include Anthony Judge, Philip Daffara, James Dator, Yongseok Seo, Jay Weinstein, Walter Truett Anderson, William Halal, Anodea Judith, and Jay Earley, J. Futures Stud. **9**(1) (2004)

8. Inayatullah, S.: The causal layered analysis (CLA) reader. Theory and Case Studies of an Integrative and Trans-formative Methodology, Tamkang University Press, Tamsui (2004)

9. Inayatullah, S.: Alternative futures of occupational therapy and therapists. J. Futures Stud. **11**(4), 41–58 (2007)

10. Karoly, L.A., Panis, C.: The 21st century at work: forces shaping the future workforce and workplace in the United States. Report prepared for the US Dept of Labor, Rand, Santa Monica, CA (2004)

11. Molitor, G.: The Power to Change the World: The Art of Forecasting. Public Policy Forecasting, Potomac (2003)

12. Stone, H., Stone, S.: Embracing Our Selves: The Voice Dialogue Manual. New World Library, Novato (1989)

13. Voros, J.: Nesting social-analytical perspectives: an approach to macro-social analysis. J. Futures Stud. **11**(1), 1–22 (2006)

14. Dunne, A., Raby, F.: Speculative Everything. Design, Fiction, and Social Dreaming. The MIT Press, Cambridge (2013)

15. Martin, B., Hanington, B.M.: Universal Methods of Design: 100 Ways to Research Complex Problems, Develop Innovative Ideas, and Design Effective Solutions. Rockport Publishers, Beverly (2012)

16. Damle, A., Miller, T.: Influence of design tools on conceptually driven processes. In: Proceedings of C&C 2011: The 8th ACM Conference on Creativity, pp. 327–328 (2011)

17. Spaulding, E., Faste, H.: Design-driven narrative: using stories to prototype and build immersive design worlds. In: Proceedings of CHI 2013: The SIGCHI Conference on Human Factors in Computing Systems, pp. 2843–2852 (2013)

18. Wang, H., Rosé, C.P., Cui, Y., Chang, C., Huang, C., Li, T.: Thinking hard together: the long and short of collaborative idea generation in scientific inquiry. In: Proceedings of CSCL 2007: The 8th International Conference on Computer Supported Collaborative Learning, pp. 754–763 (2007)

19. Zimmerman, J., Stolterman, E., Forlizzi, J.: An analysis and critique of research through design: towards a formalization of a research approach. In: Proceedings of DIS 2010: The 8th ACM Conference on Designing Interactive Systems, pp. 310–319 (2010)

20. Kantola, N., Jokela, T.: SVSb: simple and visual storyboards. Developing a visualisation method for depicting user scenarios. In: Proceedings of OZCHI 2007: The 19th Australasian conference on Computer-Human Interaction: Entertaining User Interfaces, pp. 49–56 (2007)

21. Blythe, M., Encinas, E., Kaye, J., Avery, ML., McCabe, R., Andersen, K.: Imaginary design workbooks: constructive criticism and practical provocations. In: Proceedings of the 2018 CHI Conference on Human Factors in Computing Systems (paper No. 233), Montréal, QC, Canada (2018)

22. Bella, F., Fletchera, G., Greenhillb, A., Griffithsa, M., McLeanc, R.: Science fiction prototypes: visionary technology narratives between futures. Futures **50**(6), 5–14 (2013)

A Literature Review of the Research on the Uncanny Valley

Jie Zhang[1,2], Shuo Li[3], Jing-Yu Zhang[1,2], Feng Du[1,2], Yue Qi[1,2(✉)], and Xun Liu[1,2]

[1] CAS Key Laboratory of Behavioral Science, Institute of Psychology,
Chinese Academy of Sciences, Beijing, China
[2] Department of Psychology, University of Chinese Academy of Sciences, Beijing, China
qiy@psych.ac.cn
[3] Department of Psychology, Nankai University, Tianjin, China

Abstract. Depend on the development of science and technology, the demands for robots are not only limited to the use of functions but also pay more attention to the emotional experience brought by the products. However, as the robot's appearance approach human-likeness, it makes people uncomfortable, which is called the Uncanny Valley (UV). In this paper, we systematically review the hypothesis and internal mechanisms of UV. Then we focus on the methodological limitations of previous studies, including terms, assessment, and materials. At last, we summarize the applications in interaction design to avoid the uncanny valley and propose future directions.

Keywords: Uncanny valley · Humanoid robots · Human-computer interaction · Affective design · Human-likeness

1 Introduction

With the boom of computer technology and the development of related hardware facilities, robots have been used more and more widely in human society and provided many conveniences to people's life [1]. In the past 20 years, social robots have developed fast and been used to interact with humans in many places, such as homes, hospitals, and shopping malls [1]. In order to improve human-robot interaction, engineers have designed robots that resemble humans highly [2]. There is a positive relationship between the human-likeness of robots and feelings of comfort with them. However, it has a steep dip in comfort and felt eeriness when robots looked almost but not entirely human, which called the "uncanny valley" [3].

The concept of "uncanny valley" was first proposed by Mori in 1970 [4]. In his paper, he envisioned people's reactions to robots that looked and acted almost like a human and took some examples to verify his thought. He proposed that the level of affinity for the robot increased up with its appearance becoming more humanlike until people perceived the faces as eerie suddenly. However, as the robot's human-likeness went on increasing, the eeriness reverted to likeability. This concept is useful to design a robot and works as a guide to improve human-robot interaction.

© Springer Nature Switzerland AG 2020
P.-L. P. Rau (Ed.): HCII 2020, LNCS 12192, pp. 255–268, 2020.
https://doi.org/10.1007/978-3-030-49788-0_19

This paper systematically combs the explanation and internal mechanisms of the uncanny valley, the problems and deficiencies in existing research, and its practical applications in interaction design. The paper has the following structure. In Sect. 2, we describe different explanations of the uncanny valley. In Sect. 3, we present the defects of existing research, including the terms, assessments, and materials. In Sect. 4, we summarize the application of the phenomenon in the design of robots to avoid the uncanny valley.

2 Explanations of the Uncanny Valley

Researchers have proposed a variety of explanations to account for the uncanny valley phenomenon [2]. These hypotheses can be mainly divided into two categories. One category explains the phenomenon from an evolutionary psychology perspective that the uncanny feeling comes from facial features themselves, including the Threat Avoidance hypothesis [2, 3, 5, 6] and the Evolutionary Aesthetics hypothesis [2, 5, 6]. The other category interprets the phenomenon based on cognitive conflicts, including the Mind Perception hypothesis [1, 7], the Violation of Expectation hypothesis [1–3, 5], and the Categorical Uncertainty hypothesis [1, 2, 8]. Most related empirical studies focus on the latter because the cognitive response is easy to quantify and manipulate. However, the hypothesis of evolutionary psychology has little empirical research.

2.1 Explanations Based on Evolutionary Psychology

Threat Avoidance Hypothesis. Mori [3] first pointed out that the UV phenomenon "may be important to our self-preservation". During the process of evolution, diseases and death are two main threats to human beings. Thus, there are two explanations for the uncanny valley stemming from the avoidance of threat. The first explanation is called pathogen avoidance, which indicates that when people perceive the imperfections of humanoid robots, they will associate the defects with diseases [2]. Moreover, because of the high human-likeness, people may consider that humanoid robots are genetically close to humans and are likely to transmit diseases to humans [2, 5, 6]. However, this hypothesis is just an inference based on Rozin's theory of disgust and has not been tested directly [2, 5]. Another explanation named mortality salience was proposed based on the terror management theory. Hanson [9] indicated that the flaws of humanoid robots combined with a humanlike appearance could remind us of mortality. From the aspect of this explanation, the uncanny feeling is the anxiety for mortality and the fear of death triggered by humanoid robots. People may be reminded of death and consider humanoid robots as dead individuals who come alive [2, 5]. However, there is only one study testing the hypothesis directly and found that the sensitivity to the vulnerability and impermanence of the physical body was significantly correlated with eerie ratings of android [10].

Evolutionary Aesthetics Hypothesis. The hypothesis pays attention to the attractiveness of physical features and regards the uncanny feeling as an aversion to unattractive individuals. By morphing the images of abstract robots and realistic robots or real humans, Hanson's research [9] found that the high-attractive images were consistently rated low in eeriness. Attractiveness is judged based on specific external characteristics that humans are sensitive to, such as bilateral symmetry, facial proportions, and skin quality [6]. These traits are associated with health, fertility, and other aspects that are close to the reproduction, and we inherit the preference for these traits from our ancestors who successfully reproduced under the selection pressure [2, 5, 6]. In a word, aesthetic properties are shaped by natural selection and determine the feeling of humanoid robots potentially.

These hypotheses explain the uncanny valley from the perspective of evolutionary psychology. Although they focus on various mechanisms to suggest the explanations, the essence is to achieve self-preservation and successful reproduction, which is the core of evolutionary psychology. However, the empirical studies supporting these hypotheses are still insufficient [2].

2.2 Explanations Based on Cognitive Conflicts

Mind Perception Hypothesis. Gray and Wegner [7] proposed that humanoid robots are uncanny because they are so realistic that people may ascribe to them the capacity to feel and sense. However, this capacity is considered as the unique characteristic of humans, which is not expected to emerge on the robots [2, 7]. People are happy to have robots do works as human, but not have feelings like humans.

Violation of Expectation Hypothesis. This hypothesis expands the mind perception hypothesis and believes that people will elicit specific expectations of the humanoid robots whose appearance resembles that of humans. For example, humanoid robots are expected to perform movements or speak as smoothly as humans. However, the robots often violate these expectations: the movements may perform mechanically, and the voice may be synthetic [2, 5]. The mismatch between expectations and reality results in negative emotional appraisal and avoidance behaviors, and leads to the feelings of eeriness and coldness [1, 11].

Categorical Uncertainty Hypothesis. The hypothesis emphasizes that the feeling of eeriness is caused by the ambiguous boundary of categories [2, 5, 6]. There are many empirical studies on this hypothesis, but the results are quite controversial. Some studies support the Mori's uncanny valley that the most humanlike robots are perceived as the robots. This perception blurs the category boundary between humans and machines to the greatest extent [12]. However, Ferrey, Burleigh, & Fenske's study [13] employed human-robot and human-animal morphing images, and found that the negative peak is not always close to the human end (Line 1 in Fig. 1). The perceptual ambiguity was maximum at the midpoint of each continuum (Boundary 1 in Fig. 1). Furthermore, recent research found that the location of the category boundary did not coincide with the classic uncanny valley either (Boundary 2 in Fig. 1), and the negative peak was near the machine end (Line 2 in Fig. 1) [14].

These hypotheses interpret the uncanny valley based on cognitive conflicts. The conflict may exist between deduction and stereotype, between expectation and reality, or between different categories. Although there are many related empirical studies because the cognitive response is easy to quantify and manipulate, the explanation of the uncanny valley is still controversial.

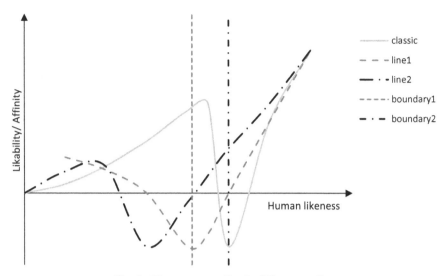

Fig. 1. The uncanny valley in different studies

3 Defects of Existing Research

At present, the related research of the uncanny valley involves computer science, psychology, material science, and other fields. Researchers studied the feelings of eeriness from various groups of users [15, 16], and explore the methods to improve the design of androids or computer-animated characters [14, 17, 18]. However, there are some problems in the existing studies, which may lead to inconsistent findings.

The classic uncanny valley is proposed by Mori. Line 1 is proposed by Ferrey, Burleigh, & Fenske (2015). Line 2 is proposed by Mathur, Reichling, & Lunardini, et al. (2020). Boundary 1 and 2 exhibit the category boundary in Ferrari et al. and Mathur et al.'s study, respectively.

3.1 Terms

Firstly, the absence of a clear definition of uncanny feelings may be a major cause of the controversial findings [19–21], especially the inconsistency of the translation [1]. Mori [4] used "shinwakan" or "bukimi" to represent the feelings when people faced different human replicas (e.g., androids or artifacts), when the feelings changed against human-likeness [22]. The original Japanese term "bukimi" was translated clearly into eeriness. However, the word "shinwakan" was first translated into familiarity, which was not equivalent and proved complex to define, partly because of its two meanings in English-a sense of closeness or lack of novelty [22–25]. Thus, it is no surprise that Mori's original items have been extended to various interpretations and used in numerous studies. Realizing that, Mori et al. [3] revised the translation of familiarity into affinity, which refers to novelty or strangeness. Unfortunately, according to the literature review recently, although affinity has been used in some research, it is still not accepted and used consistently (Table 1).

Moreover, the same term can be explained as different connotations in various studies. For instance, "likability" is interpreted as friendly and enjoyable [14, 26], or aesthetic or pleasant appearance of the character [21]. Distinct instructions result in complicated comprehension.

One more reason for the dilemma may be that a single concept could not cover the uncanny feeling. Ho et al. [27] verified that uncanny feeling includes several kinds of emotions, such as fear, disgust, nervousness, dislike, and shock. Future research is encouraged to adopt a universal definition of the original term "shinwakan", such as affinity [28], as well as confirm its boundaries and content compositions.

Table 1. List of items used in different studies

Original item	Item	Author & Year
Positive Shinwakan	Acceptability	Hanson, Olney, & Prilliman et al., 2005
	Affinity	Mori, MacDorman, & Kageki, 2012; Zibrek, Kokkinara, & McDonnell, 2018; Kätsyri, Gelder, & Takala, 2019, Study 2 & 3
	Appeal	Hanson, 2005
	Attractiveness	Ho & MacDorman, 2010; Burleigh, Schoenherr, & Lacroix, 2013, Study 1; Destephe, Zecca, & Hashimoto et al., 2014; Ho & MacDorman, 2017
	Familiarity	Hanson, 2005; MacDorman & Ishiguro, 2006; MacDorman, 2006; Bartneck, Kanda, & Ishiguro et al., 2009; Cheetham, Wu, & Pauli et al., 2015, Study 2; Chattopadhyay & MacDorman, 2016; MacDorman & Chattopadhyay, 2017; Schwind, Floerke, & Ju et al., 2018; Pütten, Krämer, & Maderwald et al., 2019
	Pleasantness/Pleasure	Seyama & Nagayama, 2007; Ho & MacDorman, 2010; Burleigh, Schoenherr, & Lacroix, 2013, study 2
	Likability	Bartneck, Kanda, & Ishiguro et al., 2007; Ferrey, Burleigh, & Fenske et al., 2015; Zlotowski, Sumioka, & Nishio et al., 2015; Mathur & Reichling, 2016; Kätsyri, Mäkäräinen, & Takala, 2017; Pütten, Krämer, & Maderwald et al., 2019; Mathur, Reichling, & Lunardini et al., 2020
	Valence and arousal	Cheetham, Suter, & Jäncke, 2011; Cheetham & Jancke, 2013; Cheetham, Wu, & Pauli et al., 2015, Study 1
	Warmth	Ho & MacDorman 2010; MacDorman & Entezari, 2015; Chattopadhyay & MacDorman, 2016; MacDorman & Chattopadhyay, 2016; Ho & MacDorman, 2017

(*continued*)

Table 1. (*continued*)

Original item	Item	Author & Year
Negative Bukimi	Eeriness	Hanson, 2005; MacDorman, 2006; MacDorman & Ishiguro, 2006; Bartneck, Kanda, & Ishiguro et al., 2009; Ho & MacDorman, 2010; Burleigh, Schoenherr, & Lacroix, 2013; Destephe, Zecca, & Hashimoto et al., 2014; Strait & Scheutz, 2014; Zlotowski, Sumioka, & Nishio et al., 2015; Chattopadhyay & MacDorman, 2016; Koschate, Potter, & Bremner et al., 2016; MacDorman & Chattopadhyay, 2016; Kätsyri, Mäkäräinen, & Takala, 2017; MacDorman & Chattopadhyay, 2017; Strait, Floerke, & Ju et al., 2017; Buckingham, Parr, & Wood et al., 2019; Kätsyri, Gelder, & Takala, 2019, study 1; Appel, Izydorczyk, & Weber et al., 2020

3.2 Assessments

Self-report questionnaires are widely used in previous studies. Gray and Wegner [7] used the Likert scale to collect the participants' feelings of uneasy, unnerved, and creepy. Meanwhile, different scales were employed, such as a visual analog scale [14, 26], single-target IAT [29], and semantic differential scale [30, 31]. However, there are several potential limitations. Firstly, the construct validity of these questionnaires and scales are still questioned. For example, some dimensions include only one item, and some dimensions are highly correlated [2, 26, 32, 33]. Secondly, there are few suitable external calibrations to test whether the items measure the putative inner constructs (emotions). The assessment of uncanny feelings is subjective and lacks objective indexes [2]. Thirdly, psychometric noise will also bring an impact on the effectiveness of subjective rating [2]. Subjects may also give socially desirable responses [33].

Recently, objective indicators with high sensitivity, such as reaction times, pupillary responses, EMG (facial electromyography), and brain activity (ERPs and fMRI), are gradually adopted in this area [34–39]. For example, an fMRI study found that VMPFC (the ventromedial prefrontal cortex) integrates likability and human-likeness to an explicit UV reaction [39]. The fMRI technology used to explore uncanny feelings could go back to 2011 [40], while eye-tracking data collected firstly to study monkeys' uncanny feelings in 2009 [41]. Thus, objective indexes and measurements are expected to determine the occurrence and operation mechanisms of uncanny feelings.

3.3 Materials

The selection criteria of experimental materials are not consistent [10, 26]. Similar to uncanny feelings, human-likeness is also a complex variable without a unified definition [24]. Therefore, various stimuli used in previous research induce irrelevant variables that may lead to confounding results. Table 2 shows the stimuli used in the experiments which aim to verify the UV effect in the past five years.

Table 2. Stimuli used in the past five years

Author & Year	Medium				Display				Serialization		Artificiality			Response	Amount of stimuli
	Video	Graph	Vignette	Others	Face	Head	Whole body	Others	Independent or scattered	Series	Nature	Both nature & artificial	Artificial		
Mathur, Reichling, & Lunardini, et al., 2020		•			•					•	•			On-line	80 face pictures
Appel, Izydorczyk, & Weber et al., 2020, Study 1			•					•	•				•	On-line	3 short descriptions of robot
Villacampa, Ingram, & Rosa, 2019		•			•				•		•			Laboratory	5 faces
Pütten, Krämer, & Maderwald et al., 2019		•					•		•			•		Laboratory	36 pictures of 6 stimulus categories
Kätsyri, Gelder, & Takala, 2019, Study 1		•			•					•		•		On-line	60 faces pictures of 6 actors
Reuten, Dam, & Naber, 2018		•			•					•			•	Laboratory	8 face pictures

(*continued*)

Table 2. (*continued*)

Author & Year	Medium				Display				Serialization		Artificiality			Response	Amount of stimuli
	Video	Graph	Vignette	Others	Face	Head	Whole body	Others	Independent or scattered	Series	Nature	Both nature & artificial	Artificial		
MacDorman & Chattopadhyay, 2017		•			•					•		•		On-line	7 face pictures
Strait, Floerke, & Ju et al., 2017		•						•	•		•			Laboratory	60 half-body pictures
Ho & MacDorman, 2017, Study 4	•							•	•		•			On-line	12 video clips of 12 characters
Kätsyri, Mäkäräinen, & Takala, 2017	•						•		•		•			Laboratory	60 video clips of 15 movies
Wang & Rocha, 2017, Study 1		•			•				•		•			Laboratory	89 face pictures
Mathur & Reichling, 2016, Study 1		•				•			•		•			On-line	80 face pictures

Note: "Medium" means the form of stimuli presentation. "Display" means which part of the stimuli could be observed. "Serialization" means whether the stimuli are series of many components, such as a series of morphing images. "Artificiality" means whether the stimuli were morphed, for example, a photo of robot from Google or filmed is natural.

Participants were asked to make evaluations based on different forms of stimuli, such as videos, pictures, descriptions, words, or even interactions [14, 17, 21, 24, 29, 42, 43]. However, few studies compared the uncanny feelings evoked by these various mediums directly. Moreover, it is also difficult to infer whether people had similar feelings when they only see a part of the robots (e.g., face, head, or body), even all of them are displayed as static graphs [14, 26, 39, 44]. Furthermore, a small number of discontinuous stimuli could not reflect the continuous axis of human-likeness correctly. Bartneck et al. [24] got a result against Mori's prediction, but the author pointed out that by using one human and his robotic copy as the stimuli was unable to confirm or disconfirm the Mori's hypothesis. If the stimuli are arbitrarily or subjectively selected, then researchers would not be possible to obtain reliable conclusions of the UV effect [25].

Additionally, morphing artifact becomes one of the common methods to manipulate stimuli [20]. Following the guidelines that endpoint images should be similar to each other to reduce morphing artifacts [45], using similar source images of humans and robots for morphing restrict the generated range of human-likeness [31]. Even if the morphing artifacts controlled perfectly, it is still questioned whether the objectively manipulated human-likeness percentages are equal to perceived human-likeness [31, 45].

4   Practical Applications

The relevant research results of the uncanny valley, which involve users' attitudes and concepts towards humanoid robots, play a significant role in the field of human-computer interaction, especially in interaction design. The development and innovation of humanoid robot design are trying to reduce the negative impact of the uncanny valley. From the perspective of a human, the question is whether the individual differences among the users can predict sensitivity to the uncanny valley and acceptance to the humanoid robots [10, 46]. From the view of the robot, the question is what kind of design is more acceptable to the majority of users [9, 46]. Therefore, in order to avoid the uncanny valley, there are two directions to improve the design of robots.

One way is to pursuit a nonhuman design deliberately so that the robots can lie at the first peak of affinity. Find a moderate degree of human likeness and a considerable sense of affinity, rather than taking the risk to increase the degree of human likeness continually [3]. There are two suggestions:

(1) *Keep the balance between humanness and machine-like.* The existence of the nose, eyelids, and mouth can increase the perception of humanness. Several design suggestions are proposed, for example, four or more features on the head, wide head with wide eyes, details in the eyes, or complex curves in the forehead [47].
(2) *Design the robots for target users.* For example, children rate human-machine like robots as the most positive, and they prefer cartoon-like and mechanical features, such as exaggerated facial features and wheels [48–50]. Elderly users have their preferences as well [46].

The other way is to reach the second peak and increase the level of human-likeness to step over the uncanny valley. The main idea of this way is to narrow the gap between robots and humans from various aspects:

(1) *Make robots alive.* Hanson [9] indicated that people feel unease because robots seem partly-dead. For example, robots shut down instead of going to sleep like humans. Thus, it is better to remove these flaws to make robots alive, friendly, and attractive.

(2) *Express emotions.* The addition of emotion display (e.g., emotional expressions, gait, voice, or gestures) can decrease the sense of uncanniness successfully [18, 51]. These emotion displays narrow the gap between the expectation the design raises about human nature and the perception of it, achieving a harmonious interaction.

5   Conclusions and Future Directions

Robots are becoming increasingly prevalent in everyday life. Humanoid robots are expected to be used more friendly and experienced more comfortably. Therefore, how to define and design the best appearances of humanoid robots is a critical question to be answered. In summary, decades of research develop two main explanations of the uncanny valley effect from the views of evolutionary psychology and cognitive conflict. The inconsistency of previous studies may be due to the absence of a unified definition, robust measure, and the representativeness of materials. Practically, pursuit a nonhuman design and increase the rate of human-likeness as high as possible are both helpful to avoid uncanny feelings. Future research is encouraged to reach a consensus on how to define the uncanny feelings, no matter it is a single item or complex emotions. Moreover, creating a sizeable and diverse database of images (or videos) covers a continuous series of human-likeness, as created by Mathur et al. [14], could avoid manipulation defects such as heterogeneous or discontinuous stimuli. Finally, considering most of the previous studies focus on young adults, future research is expected to test the uncanny valley in a more diverse user group.

Acknowledgments. Jie Zhang and Shuo Li made equal contributions to this manuscript. This research is supported by fund for building world-class universities (disciplines) of Renmin University of China. Project No. 2018, the Beijing Natural Science Foundation (5184035), and CAS Key Laboratory of Behavioral Science, Institute of Psychology.

References

1. Broadbent, E.: Interactions with robots: the truths we reveal about ourselves. Annu. Rev. Psychol. **68**, 627–652 (2017). https://doi.org/10.1146/annurev-psych-010416-043958
2. Wang, S., Lilienfeld, S.O., Rochat, P.: The uncanny valley: existence and explanations. Rev. Gen. Psychol. **19**(4), 393–407 (2015). https://doi.org/10.1037/gpr0000056
3. Mori, M., MacDorman, K., Kageki, N.: The uncanny valley [from the field]. IEEE Robot. Autom. Mag. **19**(2), 98–100 (2012). https://doi.org/10.1109/mra.2012.2192811
4. Mori, M.: The uncanny valley. Energy **7**(4), 33–35 (1970)
5. MacDorman, K.F., Ishiguro, H.: The uncanny advantage of using androids in cognitive and social science research. Interact. Stud. **7**(3), 297–337 (2006). https://doi.org/10.1075/is.7.3.03mac
6. MacDorman, K.F., Green, R.D., Ho, C.C., Koch, C.T.: Too real for comfort? Uncanny responses to computer generated faces. Comput. Hum. Behav. **25**(3), 695–710 (2009). https://doi.org/10.1016/j.chb.2008.12.026

7. Gray, K., Wegner, D.M.: Feeling robots and human zombies: mind perception and the uncanny valley. Cognition **125**(1), 125–130 (2012). https://doi.org/10.1016/j.cognition.2012.06.007

8. Yamada, Y., Kawabe, T., Ihaya, K.: Categorization difficulty is associated with negative evaluation in the "uncanny valley" phenomenon. Jpn. Psychol. Res. **55**(1), 20–32 (2013). https://doi.org/10.1111/j.1468-5884.2012.00538.x

9. Hanson, D.: Expanding the aesthetic possibilities for humanoid robots. In: IEEE-RAS International Conference on Humanoid Robots (2005)

10. MacDorman, K.F., Entezari, S.O.: Individual differences predict sensitivity to the uncanny valley. Interact. Stud. **16**(2), 141–172 (2015). https://doi.org/10.1075/is.16.2.01mac

11. MacDorman, K.F., Chattopadhyay, D.: Reducing consistency in human realism increases the uncanny valley effect; increasing category uncertainty does not. Cognition **146**, 190–205 (2016). https://doi.org/10.1016/j.cognition.2015.09.019

12. Ferrari, F., Paladino, M.P., Jetten, J.: Blurring human-machine distinctions: anthropomorphic appearance in social robots as a threat to human distinctiveness. Int. J. Soc. Robot. **8**(2), 287–302 (2016). https://doi.org/10.1007/s12369-016-0338-y

13. Ferrey, A.E., T.J. Burleigh, M.J. Fenske: Stimulus-category competition, inhibition, and affective devaluation: a novel account of the uncanny valley. Front. Psychol. **6**(249) (2015). https://doi.org/10.3389/fpsyg.2015.00249

14. Mathur, M.B., Reichling, D.B., Lunardini, F., et al.: Uncanny but not confusing: multisite study of perceptual category confusion in the uncanny valley. Comput. Hum. Behav. **103**, 21–30 (2020). https://doi.org/10.1016/j.chb.2019.08.029

15. Buckingham, G., Parr, J., Wood, G., et al.: Upper- and lower-limb amputees show reduced levels of eeriness for images of prosthetic hands. Psychon. Bull. Rev. **26**(4), 1295–1302 (2019). https://doi.org/10.3758/s13423-019-01612-x

16. Destephe, M., Zecca, M.., Hashimoto, K., Takanishi, A.: Uncanny valley, robot and autism: perception of the uncanniness in an emotional gait. In: Proceedings of the 2014 IEEE International Conference on Robotics and Biomimetics, pp. 1152–1157. IEEE, Bali (2014). https://doi.org/10.1109/robio.2014.7090488

17. Ho, C.-C., MacDorman, K.F.: Measuring the uncanny valley effect. Int. J. Soc. Robot. **9**(1), 129–139 (2017). https://doi.org/10.1007/s12369-016-0380-9

18. Koschate, M., Potter, R., Bremner, P., Levine, M.: Overcoming the uncanny valley: displays of emotions reduce the uncanniness of humanlike robots. In: 11th ACM/IEEE International Conference on Human-Robot Interaction (HRI), pp. 359–366. IEEE, Christchurch (2016). https://doi.org/10.1109/hri.2016.7451773

19. Olivera-La Rosa, A.: Wrong outside, wrong inside: a social functionalist approach to the uncanny feeling. New Ideas Psychol. **50**, 38–47 (2018). https://doi.org/10.1016/j.newideapsych.2018.03.004

20. Katsyri, J., Forger, K., Makarainen, M., Takala, T.: A review of empirical evidence on different uncanny valley hypotheses: support for perceptual mismatch as one road to the valley of eeriness. Front. Psychol. **6**(390) (2015). https://doi.org/10.3389/fpsyg.2015.00390

21. Katsyri, J., Makarainen, M., Takala, T.: Testing the 'uncanny valley' hypothesis in semirealistic computer-animated film characters: an empirical evaluation of natural film stimuli. Int. J. Hum Comput Stud. **97**, 149–161 (2017). https://doi.org/10.1016/j.ijhcs.2016.09.010

22. Ho, C.-C., MacDorman, K.F.: Revisiting the uncanny valley theory: developing and validating an alternative to the Godspeed indices. Comput. Hum. Behav. **26**(6), 1508–1518 (2010). https://doi.org/10.1016/j.chb.2010.05.015

23. Bartneck, C., Kanda, T., Ishiguro, H., Hagita, N..: Is the uncanny valley an uncanny cliff? In: Proceedings of the 16th IEEE International Conference on Robot & Human Interactive Communication, pp. 368–373. IEEE, Jeju (2007)

24. Bartneck, C., Kanda, T., Ishiguro, H., Hagita, N.: My Robotic Doppelgänger – a critical look at the uncanny valley. In: 18th IEEE International Symposium on Robot and Human Interactive Communication, pp. 269–276. IEEE, Toyama (2009)
25. Lay, S., Brace, N., Pike, G.: Circling around the uncanny valley: design principles for research into the relation between human likeness and eeriness. i-Perception 1–11 (2016). https://doi.org/10.1177/2041669516681309
26. Mathur, M.B., Reichling, D.B.: Navigating a social world with robot partners: a quantitative cartography of the Uncanny Valley. Cognition **146**, 22–32 (2016). https://doi.org/10.1016/j.cognition.2015.09.008
27. Ho, C.-C., MacDorman, K.F., Pramono, Z.A.D.: Human emotion and the uncanny valley: a GLM, MDS, and isomap analysis of robot video ratings. In: 3rd ACM/IEEE International Conference on Human-Robot Interaction (HRI), pp. 169–176. IEEE, Amsterdam (2008)
28. Wang, S., Rochat, P.: Human perception of animacy in light of the uncanny valley phenomenon. Perception **46**(12), 1386–1411 (2017). https://doi.org/10.1177/0301006617722742
29. Villacampa, J., Ingram, G.P.D., Corradi, G., Rosa, A.O.-L.: Applying an implicit approach to research on the uncanny feeling. J. Artic. Support Null Hypothesis **16**(1), 11–22 (2019)
30. MacDorman, K.F., Chattopadhyay, D.: Categorization-based stranger avoidance does not explain the uncanny valley effect. Cognition **161**, 132–135 (2017). https://doi.org/10.1016/j.cognition.2017.01.009
31. Katsyri, J., de Gelder, B., Takala, T.: Virtual faces evoke only a weak uncanny valley effect: an empirical investigation with controlled virtual face images. Perception **48**(10), 968–991 (2019). https://doi.org/10.1177/0301006619869134
32. MacDorman, K.F.: Subjective ratings of robot video clips for human likeness, familiarity, and eeriness: an exploration of the uncanny valley. In: Proceedings of the ICCS/CogSci 2006 Long Symposium 'Toward Social Mechanisms of Android Science', pp. 26–29 (2006)
33. Bartneck, C., Kulić, D., Croft, E., Zoghbi, S.: Measurement instruments for the anthropomorphism, animacy, likeability, perceived intelligence, and perceived safety of robots. Int. J. Soc. Robot. **1**, 71–81 (2009). https://doi.org/10.1007/s12369-008-0001-3
34. Saygin, A.P., Chaminade, T., Ishiguro, H., Driver, J., Frith, C.: The thing that should not be: predictive coding and the uncanny valley in perceiving human and humanoid robot actions. SCAN **7**, 413–422 (2012). https://doi.org/10.1093/scan/nsr025
35. Strait, M., Scheutz, M.: Measuring users' responses to humans, robots, and human-like robots with functional near infrared spectroscopy. In: 23rd IEEE International Symposium on Robot and Human Interactive Communication, pp. 1128–1133. IEEE, Edinburgh (2014)
36. Cheetham, M., Wu, L., Pauli, P., Jancke, L.: Arousal, valence, and the uncanny valley: psychophysiological and self-report findings. Front. Psychol. **6**(981) (2015). https://doi.org/10.3389/fpsyg.2015.00981
37. Strait, M., Vujovic, L., Floerke, V., Scheutz, M., Urry, H.: Too much humanness for human-robot interaction. In: Proceedings of the 33rd Annual ACM Conference on Human Factors in Computing Systems - CHI 2015, Seoul, pp. 3593–3602 (2015). https://doi.org/10.1145/2702123.2702415
38. Reuten, A., van Dam, M., Naber, M.: Pupillary responses to robotic and human emotions: the uncanny valley and media equation confirmed. Front. Psychol. **9**(774) (2018). https://doi.org/10.3389/fpsyg.2018.00774
39. Rosenthal-von der Puetten, A.M., Kraemer, N.C., Maderwald, S., Brand, M., Grabenhorst, F.: Neural mechanisms for accepting and rejecting artificial social partners in the uncanny valley. J. Neurosci. **39**(33), 6555–6570 (2019). https://doi.org/10.1523/jneurosci.2956-18.2019
40. Cheetham, M., Suter, P., Jäncke, L.: The human likeness dimension of the "uncanny valley hypothesis": behavioral and functional MRI findings. Front. Hum. Neurosci. **5**(126) (2011). https://doi.org/10.3389/fnhum.2011.00126

41. Steckenfinger, S.A., Ghazanfar, A.A.: Monkey visual behavior falls into the uncanny valley. Proc. Natl. Acad. Sci. U.S.A. **106**(43), 18362–18366 (2009). https://doi.org/10.1073/pnas. 0910063106

42. Ramey, C.H.: An inventory of reported characteristics for home computers, robots, and human beings: applications for android science and the uncanny valley. In: Proceedings of the ICCS/CogSci 2006 Long Symposium 'Toward Social Mechanisms of Android Science' (2006)

43. Appel, M., Izydorczyk, D., Weber, S., Mara, M., Lischetzke, T.: The uncanny of mind in a machine: humanoid robots as tools, agents, and experiencers. Comput. Hum. Behav. **102**, 274–286 (2020). https://doi.org/10.1016/j.chb.2019.07.031

44. Strait, M.K., Floerke, V.A., Ju, W., et al.: Understanding the uncanny: both atypical features and category ambiguity provoke aversion toward humanlike robots. Front. Psychol. **8**(1366) (2017). https://doi.org/10.3389/fpsyg.2017.01366

45. Cheetham, M., Jancke, L.: Perceptual and category processing of the uncanny valley hypothesis' dimension of human likeness: some methodological issues. Jove J. Visual. Exp. (76) (2013). https://doi.org/10.3791/4375

46. Prakash, A., Rogers, W.A.: Why some humanoid faces are perceived more positively than others: effects of human-likeness and task. Int. J. Soc. Robot. **7**(2), 309–331 (2015). https:// doi.org/10.1007/s12369-014-0269-4

47. DiSalvo, C.F., Gemperle, F., Forlizzi, J., Kiesler, S.: All robots are not created equal: the design and perception of humanoid robot heads. In: Proceedings of the 4th conference on Designing Interactive Systems: Processes, Practices, Methods, and Techniques, pp. 321–326. ACM Press, London (2002). https://doi.org/10.1145/778712.778756

48. Woods, S.: Exploring the design space of robots: children's perspectives. Interact. Comput. **18**(6), 1390–1418 (2006). https://doi.org/10.1016/j.intcom.2006.05.001

49. Woods, S., Dautenhahn, K., Schulz, J.: The design space of robots: investigating children's views. In: 13th IEEE International Workshop on Robot and Human Interactive Communication, pp. 47–52. IEEE, Kurashiki (2004). https://doi.org/10.1109/roman.2004.137 4728

50. Lin, W., Yueh, H.-P., Wu, H.-Y., Fu, L.-C.: Developing a service robot for a children's library: a design-based research approach. J. Assoc. Inf. Sci. Technol. **65**(2), 290–301 (2014). https:// doi.org/10.1002/asi.22975

51. Jizheng, Y., Zhiliang, W., Yan, Y.: Humanoid robot head design based on uncanny valley and FACS. J. Robot. (2014). https://doi.org/10.1155/2014/208924

52. Hanson, D., et al.: Upending the uncanny valley. In: The Twentieth National Conference on Artificial Intelligence and the Seventeenth Innovative Applications of Artificial Intelligence Conference, Pittsburgh, Pennsylvania, USA, 9–13 July 2005 (2005)

53. Zibrek, K., Kokkinara, E., McDonnell, R.: The effect of realistic appearance of virtual characters in immersive environments - does the character's personality play a role? IEEE Trans. Visual. Comput. Graphics **24**(4), 1681–1690 (2018). https://doi.org/10.1109/tvcg.2018.279 4638

54. Burleigh, T.J., Schoenherr, J.R., Lacroix, G.L.: Does the uncanny valley exist? An empirical test of the relationship between eeriness and the human likeness of digitally created faces. Comput. Hum. Behav. **29**(3), 759–771 (2013). https://doi.org/10.1016/j.chb.2012.11.021

55. Chattopadhyay, D., MacDorman, K.F.: Familiar faces rendered strange: Why inconsistent realism drives characters into the uncanny valley. J Vis. **16**(11), 7 (2016). https://doi.org/10. 1167/16.11.7

56. Schwind, V., Leicht, K., Jaeger, S., Wolf, K., Henze, N.: Is there an uncanny valley of virtual animals? A quantitative and qualitative investigation. Int. J. Hum Comput Stud. **111**, 49–61 (2018). https://doi.org/10.1016/j.ijhcs.2017.11.003

57. Seyama, J.I., Nagayama, R.S.: The uncanny valley: effect of realism on the impression of artificial human faces. Presence Teleop. Virt. Environ. **16**(4), 337–351 (2007). https://doi.org/10.1162/pres.16.4.337

58. Zlotowski, J., Sumioka, H., Nishio, S., Glas, D.F., Bartneck, C., Ishiguro, H.: Persistence of the uncanny valley: the influence of repeated interactions and a robot's attitude on its perception. Front. Psychol. (2015). https://doi.org/10.3389/fpsyg.2015.00883

Quantification of Natural Multimodal Interaction Capacity

Jian Zheng⑩, Pei-Luen Patrick Rau⁽✉⁾ ⑩, and Jingyu Zhao⑩

Department of Industrial Engineering, Tsinghua University, Beijing, China
`rpl@mail.tsinghua.edu.cn`

Abstract. In multimodal interaction, information is presented to users through multiple channels, e.g., sight, sound, touch, smell, and taste. Too much information delivered in a short time, however, may result in information overload that overflows people's information processing capacity. We summarized the methods of quantifying the capacity by categorizing them into the span of storage or the speed of processing. The span of storage mainly includes short-term memory and working memory capacity and multiple object tracking capacity. Working memory is required in many intellectual functions, and its capacity could be tested with change detection tasks, self-ordered tasks, and complex span tasks. Whether different modalities have separate capacities, whether objects or features are stored, and whether the capacity works as discrete slots or a continuous resource pool were discussed. The speed of processing could be calculated as the information transfer rate with the stimuli and responses matrix; Entropy is used for more complex stimuli such as languages. The relative capacity of multitasking, which is often incorporated in multimodal interaction, could be calculated with the capacity coefficient. The application of these methods to the non-traditional modalities in human-computer interaction, e.g., touch, smell, and taste, was discussed.

Keywords: Multimodal output · Visual attention · Working memory

1 Introduction

Multimodal output uses sight, sound, touch, and even smell and taste to transfer information [1]. Multimodal output is more accessible to people with disabilities, and is especially useful in virtual and augmented reality to enhance the feeling of immersion. Wearable devices [2] and mid-air haptics [3, 4] can present rich haptic information. Although digitizing the chemical senses is challenging [5], there are already initial works have used smell [6, 7] and taste [8] as outputs. Even distributed over multiple senses, the amount of information that people can process is still limited. If too much information is pushed to the user in a short time, the user may get information overload [9]. For example, students cannot absorb and digest everything if they skim several thick books on the night before an exam.

Capacity is the upper limit of human's processing capability, beyond which information cannot be accurately processed [10, 11]. In extreme conditions, the brain's capacity

© Springer Nature Switzerland AG 2020
P.-L. P. Rau (Ed.): HCII 2020, LNCS 12192, pp. 269–283, 2020.
https://doi.org/10.1007/978-3-030-49788-0_20

of understanding language is limited to just one word at one time [12]. The existence of capacity, or bottleneck, has been demonstrated for various stimuli, tasks, and modalities. However, despite the increasing popularity of multimodal interaction, most studies on capacity have focused on uni-modal information processing. Investigation of multimodal capacity will contribute to our knowledge of human's cognition capability and provide guidance for multimodal interface design.

Information processing functions in the brain are complex with multiple bottlenecks limiting the efficiency. Based on the theory of visual attention [13], we categorized relevant studies as addressing either the span of storage or the speed of processing. The span of storage, which includes concepts of visual attention span, short-term memory and working memory capacity, and multiple object tracking capacity, is usually denoted as the parameter K and is measured in discrete units such as items or chunks. The speed of processing is usually reflected in response time data and measured in items or bits per second (items/s and bits/s respectively).

In this paper, we aim to focus on methods of capacity quantification. Because the calculation is closely related to the experimental paradigms, we briefly introduce the three basic experimental paradigms to be used frequently in later sections, i.e., detection, identification, and discrimination tasks. In detection tasks, participants answered whether an object, a feature, or a change was present or absent. In identification tasks, participants reported the identity of the object. In discrimination tasks, participants classified objects into several pre-defined categories.

2   Span of Storage

2.1   The Parameter K

The maximum number of visual objects than can be perceived simultaneously is defined as the visual span [14]. It is often used interchangeably with visual short-term memory and working memory. All these concepts share the tradition of using K to denote capacity. The capacity of visual span, visual working memory, and multiple (visual) object tracking are all believed to be about four items [13, 15].

2.2   Working Memory

Generic working memory is defined as "the ensemble of components of the mind that hold a limited amount of information temporarily in a heightened state of availability for use in ongoing information processing" [16]. The relationship between short-term memory and working memory is historically complex. Some studies used the two terms interchangeably, whereas others defined short-term memory as a subset of working memory [16]. According to Daneman and Carpenter [17], working memory must be measured with tasks including both storage and processing of information, whereas short-term memory tasks involve only information storage.

Paradigms
Researchers have used several paradigms to measure working memory capacity. We

focus on three of them: change detection tasks, self-ordered tasks, and complex span tasks.

Change Detection Tasks

In each trial of a typical change detection tasks, participants saw a few items, usually four to eight, with different colors for a short time (e.g., 500 ms). Then the items disappeared, and after an interval (e.g., 1000 ms), one item re-appeared at the same location with the same or a different color. Participants were required to answer whether the color had changed [18]. The working memory capacity could be calculated with the probability of a hit and a correct reject [19]: $K = N(H + CR - 1)$, where N is the set size, H is the probability of a hit, and CR is the probability of a correct reject. If we allow some mistakes arising from participants not paying attention and some correct responses from lucky guessing, the calculation becomes more complex. The probability that the change is detected, $d = min\ (K/N, 1)$. H and CR could be denoted as:

$$H = a[d + g(1 - d)] + (1 - a)g \tag{1}$$

$$CR = a[d + (1 - g)(1 - d)] + (1 - a)(1 - g) \tag{2}$$

Where a is the attention parameter, and g is the guessing parameter. With maximum likelihood estimation, parameters K, a, and g could be estimated. For example, in [20], the K was estimated to be 4.39 items. For later development of the calculation, please see [21].

In change detection tasks, each object is defined by two features. In visual change detection tasks, for example, participants need to remember the location and color of each item. In auditory change detection tasks, participants need to remember the value and voice of spoken digits presented serially [22]. Change detection tasks could be generalized to other modalities as long as the objects in that modality could be defined by two features.

Self-ordered Tasks

In each trial of typical self-ordered tasks, eight three-dimension simple line drawings were presented in 3×3 grids, with one blank grid. Participants needed to select all the drawings each one time in a self-determined order, in which selecting any drawing twice was considered an error [20]. In the first step, participants could select any object, which was always correct. After the selection, the objects were re-arranged within the grids, and the grid of the first selection became blank. In the second step, the participants needed to select any object except the one selected in Step One. In the third step, they needed to avoid the two objects selected previously, and so on. For more details, please see [20, 23]. The number of items maintained in memory on step S is $m = min\ (S - 1, K)$. The probability of making an error in that step is

$$E = a\frac{S - 1 - m}{N - m} + (1 - a) \tag{3}$$

Where a is the attention parameter. Maximum likelihood estimation could be used to estimate the parameters a and K. For example, in [20], K was estimated to be 6.20

items. The self-ordered tasks involve complex spatial information, which is difficult for modalities other than vision to present.

Complex Span Tasks

In each trial of complex span tasks, participants needed to first decide whether a sentence made sense, and then remembered a letter. After three to seven sentence-letter pairs, participants needed to select the letters in the correct order from a pool of 12 letters [24]. The calculation is similar to that of the self-ordered tasks but more complex, please see [20]. K was estimated to be 5.36 items in [20]. The stimuli, i.e., sentences and letters, could be presented in visual or auditory channels.

Results of correlation analysis and exploratory factor analysis suggested a separate capacity for the complex span tasks from the change detection and self-ordered tasks [20]. The capacity complex span tasks, which rely on both storage and processing, is more closely linked with the fluid intelligence [25]. Recently, it has been proposed that the working memory capacity had been underestimated: The capacity would be larger if we consider the features besides and the ensemble information of objects and treat the capacity as a continuous resource pool rather than as slots.

Objects or Features

Working memory capacity is usually measured in units of items or chunks, supported by evidence that we can only remember four visual objects no matter how many features are contained in each object [18, 26]. Some researchers have argued that the capacity is also limited in features in addition to objects, and the information load of items should be considered [27, 28]. If so, the capacity should be measured in bits rather than items.

Slots or Resource

Another influential assumption about working memory is that it works as discrete slots. However, recent studies have proposed that working memory capacity should be conceptualized as a limited resource pool that could be distributed flexibly across much more than four or seven items [29]. According to the variable precision model of working memory, the amount of capacity resource devoted to a single item generally decreases as the set size increases but might vary across items and trial [30]. The precision of memory and the capacity resource devoted to an item follows a power law [31]. The slots or resource debate is ongoing [32].

Structure and Layout

Instead of randomly placed objects, the natural visual world often comprises structure and regularity [33]. Our visual perception could make use of statistics to summarize information across groups of objects. Objects not attended focally were not reported correctly in classic working memory tasks but still were included in the ensemble representation [34]. Observers have some knowledge of the entire visual field. The capacity of object layout was estimated as 13.4 items [35], and the capacity of objects with structure was estimated as much as 24 items [36, 37].

2.3 Multiple Object Tracking

People often need to copy with moving objects, e.g., cars, beside static ones. The number of objects they can track correctly is the multiple object tracking capacity. In multiple object tracking tasks, participants needed to monitor moving objects of varying set sizes. After a random interval, the objects stopped moving and were covered by grey discs. Participants were asked to identify the location of a random object, e.g., the red one, by clicking on the disc.

In the classic multiple object tracking tasks, participants had only one chance to click, and the probability of a correct click is $d = min (K/N, 1)$. K value calculated with this method is rather small, i.e., 1.4–2.6 items [38]. In the multiple object awareness tasks [39], however, participants had more than one chance of click. They kept clicking different discs until the target was found, although they were required to make as few clicks as possible. Even if participants did not remember the exact location of the target, they still have some knowledge as "it should be around here". The expected number of clicks could be calculated as

$$C = \frac{K}{N} + \frac{(N - K)(N - K + 1)}{2N}$$
(4)

With the observed data of clicks, the value of K could be calculated. The multiple object tracking capacity is estimated to be 7–10 items with the multiple object awareness paradigm [39].

3 Speed of Processing

3.1 Processing Speed of Visual Attention

The visual processing speed reflects the amount of visual information can be processed in one second [40]. Within the framework of the theory of visual attention, the typical experiment paradigm to test the speed is partial-report tasks. In Shibuya and Bundesen's study [41], for example, participants were presented an array of digits and letters simultaneously for a varying exposure time and then were required to report as many digits correctly as possible.

The number of correctly reported digits was predicted by a model with four parameters, including the processing speed and the span K, and with the number of digits, the number of letters, and exposure time as input. The processing speed was estimated to be 45 items/s [41]. The parameters can be estimated with a maximum-likelihood algorithm in MATLAB program packages such as LIBTVA [42, 43]. The processing speed calculated in this way could be used to quantify individual difference in attentional ability [44].

3.2 Information Transfer Rate

One behavioristic approach of quantifying the capacity of processing speed is to treat the brain as a black box and focus on stimuli and responses. The amount of mutual information between the input, i.e., the stimuli, and output, i.e., the responses, of sense channel(s) is defined as information transfer. Information transferred in one second is information transfer (IT) rate, and the maximum IT rate that can be achieved through certain channel(s) is the capacity of that sensory channel(s) [45, 46].

Stimulus-Response Confusion Matrices

Assume that a discrimination task contains L alternative stimuli and M alternative responses. The IT of one stimulus-response pair can be calculated with the matrix of all the possible stimuli and responses:

$$IT = \sum_{j=1}^{M} \sum_{i=1}^{L} \frac{n_{ij}}{n} log_2\left(\frac{n_{ij}n}{n_i n_j}\right) \tag{5}$$

Where n is the number of all the trials, n_{ij} is the number of trials presenting S_i and responded with R_j, n_i is the number of trials presenting S_i, and n_j is the number of trials responded with R_j. The maximum of IT is log_2 min(L, M). In identification tasks, L equals M (each stimulus is associated with one correct response), and the maximum of IT equals log_2L. Information transfer rate in bits/second could be calculated as the product of information transfer in bits/item and the presentation rate in items/second. The capacity of bulb light pattern calculated with Eq. 5 was 10.5 bits/s [47].

The presentation rate, i.e., how many items are presented in one second, is limited by reaction time: if participants take an average of 250 ms to respond, the presentation rate cannot surpass four items/second; otherwise the participants will miss some of the stimuli. This limitation could be overcome with the AXB paradigm, in which the participants only need to respond to the middle of three consecutive stimuli. The IT is calculated with information in stimuli (IS) and error rate (e): IT $=$ IS$(1 - 2e)$.

$$IS = -\sum_{i=1}^{L} P(S_i)log_2 P(S_i) \tag{6}$$

Where $P(S_i)$ is the probability of the stimuli taking the value of S_i. The capacity of tactile signal with masking is estimated to be 21.9 bits/s [48].

Entropy and Reading

The information contained in each stimulus is measured in entropy. It is the unpredictability of a piece of message. For stimuli with known distribution, the entropy could be calculated with Eq. 6. For stimuli with (almost) infinite possibilities, the calculation is complex and not covered in this paper. According to Shannon [49], the entropy of each letter in English is 1.5 bits. The capacity of processing visual and auditory semantic information is about 40–60 bits/s [45].

However, in natural interaction, stimuli flow such as language is not just items put together randomly: a stimulus usually follows, continues, and depends on its precedent. For example, if the first word of a sentence is "I", the second word probably is "am" and should not be "is". Depending on how many preceding words are known, the zero-, first-, second-order entropy of English words, for example, are estimated as 11.47, 6.06, and 2.01 bits respectively [50]. Other results reported a range between 0.6 and 1.3 bits per letter [49] or 5.7 bits per word [51]. The entropy of Chinese characters in meaningful text ranges from 4 to 5.31 bits per character [51–56]. Entropies of other languages can be found in [51].

The capacity of reading could be calculated with the language entropy and reading speed: $IT = S \times E \times R$, where S is reading speed, E is the language entropy, and R is the comprehension rate. Anne Jones, the six-time World Speed Reading champion, could read over 4,200 words per minute with a comprehension rate of 67%, corresponding to at least 140 bit/s; college students typically operate scanning processes at rates over 600 wpm, which is equal to 20 bit/s [57, 58]. There are also methods of calculating the entropy of images [59, 60] and sounds [61, 62], though they are not ready to be incorporated into human processing capacity.

3.3 Visual Search Slopes

People often need to find a target in a complex visual environment full of non-targets and distractors. The processing speed is also reflected in the efficiency of performing visual search tasks [63, 64], in which participants were asked to detect a target defined by one or several features among distractors of varying set size. Plotting the response time against the set size, we can calculate the slope and intercept of the visual search. The search slopes can reflect the search efficiency.

According to the feature integration theory [65], visual selective attention controls a pre-attentive, parallel stage and a serial stage. Visual search involves both serial and parallel processing: items enter the processing unit serially, but multiple items can be processed parallelly [63]. Some attributes, such as color, size, and motion, could be processed in parallel [66–68]. Visual search for targets defined according to those features is efficient, with search slopes near zero. Visual search for other features and feature conjunctions is inefficient, and the search slopes for target-present trials are about 20–30 ms/item, and the search slopes for target-absent trials are a bit more than 40–60 ms/item [69], which is equal to 17–25 items/s.

4 Multimodal Capacity

Multimodal capacity often takes a relative form as the ratio of the multi-task performance to the single-task performance. It has no units but is compared with certain boundaries, e.g., zero and one. Again, there are span of multimodal storage and speed of multimodal processing.

4.1 Multimodal Working Memory Capacity

There is still controversy about whether the working memory has a single component [70] or multiple components [71]. Multimodal working memory capacity could shed light on this debate. To measure the visual-auditory multimodal capacity, for example, participants needed to perform a visual change detection task and an auditory change detection task at the same time [72]. The dual-task performance would be compared with the single task performance. The capacity of dual-task is denoted as K_{DT}, and it is the sum of the capacity of each task in the dual-task condition. The capacity of single task is denoted as K_{ST}.

There are several ways to calculate the multimodal capacity. First, it could be calculated with maximum capacity method [22], $C = K_{DT}/\max(K_{ST1}, K_{ST2})$. Alternatively, if the capacity is assumed to be shared equally between two modalities, the capacity should be calculated as with the normalized measures, which is the average ST capacity method [72]: $C = K_{DT}/(K_{ST1} + K_{ST2})/2$. C values larger than one indicate intermodal saving and thus separate capacities for multiple modalities; C equal to one indicates one central capacity. If the capacity could be shared flexibly between two modalities, ΔK could be calculated [15]:

$$\Delta K = \left(\frac{K_{ST1} - K_{DT1}}{K_{ST1}} + \frac{K_{ST2} - K_{DT2}}{K_{ST2}} \right)/2 \times 100 \tag{7}$$

ΔK equal to 50 indicates one limited central capacity, ΔK equal to zero indicates completely separate capacities for different modalities, and ΔK of a value between 0 and 50 indicates a hybrid of both models. The value of ΔK in [72] ranged from 15 to 35, supporting that the working memory capacity is constrained by both the central and the modality-specific limits.

4.2 Multimodal Processing Speed

The typical paradigm of measuring multimodal processing speed is the redundant target detection tasks, in which multiple targets were presented in different locations or modalities [73]. In the OR tasks, participants respond positively when they detect any of the targets. In the AND tasks, participants respond positively only when they detect all the targets. The response times in conditions of two targets and one target would be compared.

Based on the systems factorial technology [74, 75], the capacity coefficient for OR tasks (C_{OR}) and the capacity coefficient for AND tasks (C_{AND}) could be calculated with the distribution of response time:

$$C_{OR}(t) = \frac{\ln(1 - F_{AB}(t))}{\ln((1 - F_A(t))(1 - F_B(t)))} \tag{8}$$

$$C_{AND}(t) = \frac{\ln(F_A(t)F_B(t))}{\ln(F_{AB}(t))} \tag{9}$$

Where F_A and F_B are the cumulative distribution function of response time when only one target is present. F_{AB} is the cumulative distribution function when both targets are present. The capacity coefficient could be estimated with the sft R package [74].

C values less than one indicate limited (or fixed) capacity, and the processing of multimodal information is slower than that of uni-modal information. C equal to one indicates unlimited capacity, and the processing of multimodal information is perfectly parallel and equals to the uni-modal processing speed. C larger than one indicates a super capacity, and the processing speed of multimodal information is faster than that of uni-modal information.

An alternative parametric approach based on the linear ballistic accumulator decision model [76] to measure the processing speed [77] has been proposed. The distribution of response time was used to estimate the drift rate v for dual task v_{DT} and single task v_{ST}. The capacity coefficient could be calculated simply as the ratio, $C = v_{DT}/v_{ST}$, and the interpretation of the value is the same as in the paragraph above.

5 Capacity of Touch, Smell, and Taste

5.1 Touch

The capacity of communication through touch calculated with the information transfer rate was estimated to be 10–18 bits/s [46]. Because vibrations of different strengths could be presented to different body parts, it is possible to design change detection tasks with haptic information. However, participant could not detect the change reliably [78]. Future studies may consider presenting the stimuli to the fingers, which are more sensitive [79]. Alternatively, haptic working memory tasks could be designed with real objects placed in different positions [80].

5.2 Smell

The smell capability, or olfactory sensitivity, could be measured with Sniffin Sticks [81]. The test suite includes detection threshold, discrimination, and identification. The detection of a certain odor is faster than the identification, and the identification is faster than judging the valence and edibility of the odor [82]. Processing of olfactory stimuli is generally slow compared with visual and auditory processing [83]. As a result, to make participants perceive visual and olfactory stimuli at the same time, the olfactory signal should be released (362 ms on average) before the visual signal [84]. It is also possible to present spatial information with smell: participants could tell apart stimuli presented to the left or the right nostril reliably [85, 86]. It is possible to design change detection tasks with two smell stimuli of different locations and odors.

5.3 Taste

The basic tastes are sour, sweet, bitter, salty, and umami. They have different receptors in the taste bud and thus different processing speeds [87]. The reaction times to the most condense solution are about 400 ms for salty and sour and about 700 ms for sweet and bitter [88]. The reaction time generally decreases as the concentration increases [88]. It is faster to detect salty and sour tastes than to discriminate them, whereas there is no such difference in sweet and bitter tastes [89]. Spatial information could be presented with tastes. Participants could tell apart stimuli presented to the left, right, and both sides of the tongue with an accuracy over 90% [90]. Taste change detection tasks could be designed with two stimuli of different locations and tastes.

The taste capability, or more specifically, the gustatory sensitivity, could be measured with Taste Strips [91]. The procedure of testing taste capability with taste strip consists of 32 trials, four taste (sweet, sour, bitter, and salty) with four concentration on both left and right sides of the tongue. The final score is simply the number of trials got correct responses. The average scores for age groups of 18–40, 41–60, and over 60 years were respectively 25.5, 21.4, and 19.4, respectively, showing a clear declining in taste sensibility with age.

6 Summary

Because there are reliable response-time experiments for all the modalities, it should be safe to say that all the modalities and their combination have a limited processing speed. The information transfer rate could be calculated for all the modalities. Nevertheless, the set size of stimuli in the identification tasks should be large enough, at least larger than four or five as used in current gustatory experiments. As for the span, there might be working memory modules for each modality, or the information of all the information could be all limited by a semantic capacity module: the information from touch, smell, and taste might be all stored as semantic information.

In earlier studies of capacity, the calculation of capacity is rather simple and could be done manually, sometimes it is even just counting numbers. In later studies, the models began to consider more parameters, such as the attention parameter and the guessing rate, and became more complicated. Software like MATLAB or R are employed to estimate the parameters including capacity. Uni-modal capacity usually has a unit, such as bits, bits/s, items, items/s, or chunks. Multimodal capacity, more often than not, comes in the form of a ratio or percentage and has some meaning boundaries like zero and one. The latter is a relative capacity.

Parallel or serial processing is an issue of great importance to capacity estimation. Many theories or models, e.g., the feature integration theory, the theory of visual attention, and the guided search model of visual search, all have underlying assumptions of parallel or serial processing. Systems factorial technology provides an approach to determine whether a process is parallel or serial with the survivor interaction contrast [70, 71]. Some simple visual information, e.g., objects of different colors, could be processed parallelly. Information of other modalities, however, tends to be easily mixed with or masked by other information within the same or from other modalities [92]. The parallel processing of information in modalities other than vision might be difficult.

Many of paradigms mentioned above involve spatial information. The location of objects needs to be processed in change detect tasks, self-ordered tasks, multiple object tracking tasks, and visual search tasks. Visual objects could be presented at different spatial locations. Other modalities have limited appropriateness of presenting spatial information. People can tell apart sounds coming from different directions [93] but not very reliably [94]. Similarly, people can also tell apart stimuli presented to different fingers, but the discrimination of stimuli to other body parts, especially the torso, is less efficient [79]. The possible locations for smell and taste seem to be just two: the left and right nostrils and left and right sides of the tongue. It will remain difficult to combine those paradigms involving spatial information with modalities other than vision until further evidence and more flexible devices are provided.

Acknowledgement. This study was funded by the National Key Research and Development Plan 2016YFB1001200 and Tsinghua University Initiative Scientific Research Program 20193080010.

References

1. Obrist, M., et al.: Touch, taste, & smell user interfaces: the future of multisensory HCI. In: Proceedings of the 2016 CHI Conference Extended Abstracts on Human Factors in Computing Systems, pp. 3285–3292. ACM (2016)
2. Pacchierotti, C., Sinclair, S., Solazzi, M., Frisoli, A., Hayward, V., Prattichizzo, D.: Wearable haptic systems for the fingertip and the hand: taxonomy, review, and perspectives. IEEE Trans. Haptics **10**, 580–600 (2017). https://doi.org/10.1109/TOH.2017.2689006
3. Carter, T., Seah, S.A., Long, B., Drinkwater, B., Subramanian, S.: Ultrahaptics: multi-point mid-air haptic feedback for touch surfaces. In: Proceedings of the 26th Annual ACM Symposium on User Interface Software and Technology, New York, NY, USA, pp. 505–514. ACM (2013). https://doi.org/10.1145/2501988.2502018
4. Obrist, M., Subramanian, S., Gatti, E., Long, B., Carter, T.: Emotions mediated through mid-air haptics. In: Proceedings of the 33rd Annual ACM Conference on Human Factors in Computing Systems, New York, NY, USA, pp. 2053–2062. ACM (2015). https://doi.org/10.1145/2702123.2702361
5. Spence, C., Obrist, M., Velasco, C., Ranasinghe, N.: Digitizing the chemical senses: possibilities & pitfalls. Int. J. Hum. Comput. Stud. **107**, 62–74 (2017). https://doi.org/10.1016/j.ijhcs.2017.06.003
6. Dmitrenko, D., Maggioni, E., Obrist, M.: I smell trouble: using multiple scents to convey driving-relevant information. In: Proceedings of the 2018 on International Conference on Multimodal Interaction, pp. 234–238. ACM (2018)
7. Maggioni, E., Cobden, R., Dmitrenko, D., Obrist, M.: Smell-O-Message: integration of olfactory notifications into a messaging application to improve users' performance. In: Proceedings of the 2018 on International Conference on Multimodal Interaction, pp. 45–54. ACM (2018)
8. Vi, C.T., Arthur, D., Obrist, M.: TasteBud: bring taste back into the game. In: Proceedings of the 3rd International Workshop on Multisensory Approaches to Human-Food Interaction, p. 1. ACM (2018)
9. Eppler, M.J., Mengis, J.: The concept of information overload: a review of literature from organization science, accounting, marketing, MIS, and related disciplines. Inf. Soc. **20**, 325–344 (2004). https://doi.org/10.1080/01972240490507974
10. Hsia, H.J.: The information processing capacity of modality and channel performance. AV Commun. Rev. **19**, 51–75 (1971)
11. Moray, N.: Where is capacity limited? A survey and a model. Acta Physiol. **27**, 84–92 (1967). https://doi.org/10.1016/0001-6918(67)90048-0
12. Strother, L.: A neural basis of the serial bottleneck in visual word recognition. PNAS **116**, 9699–9700 (2019). https://doi.org/10.1073/pnas.1905456116
13. Bundesen, C.: A theory of visual attention. Psychol. Rev. **97**, 523–547 (1990). https://doi.org/10.1037/0033-295x.97.4.523
14. Bosse, M.-L., Tainturier, M.J., Valdois, S.: Developmental dyslexia: the visual attention span deficit hypothesis. Cognition **104**, 198–230 (2007). https://doi.org/10.1016/j.cognition.2006.05.009
15. Fougnie, D., Marois, R.: Distinct capacity limits for attention and working memory: evidence from attentive tracking and visual working memory paradigms. Psychol. Sci. **17**, 526–534 (2006). https://doi.org/10.1111/j.1467-9280.2006.01739.x
16. Cowan, N.: The many faces of working memory and short-term storage. Psychon. Bull. Rev. **24**, 1158–1170 (2017). https://doi.org/10.3758/s13423-016-1191-6
17. Daneman, M., Carpenter, P.A.: Individual differences in working memory and reading. J. Verbal Learn. Verbal Behav. **19**, 450–466 (1980). https://doi.org/10.1016/S0022-5371(80)90312-6

18. Luck, S.J., Vogel, E.K.: The capacity of visual working memory for features and conjunctions. Nature **390**, 279–281 (1997). https://doi.org/10.1038/36846
19. Cowan, N.: The magical number 4 in short-term memory: a reconsideration of mental storage capacity. Behav. Brain Sci. **24**, 87–114 (2001). https://doi.org/10.1017/S0140525X01003922
20. Van Snellenberg, J.X., Conway, A.R.A., Spicer, J., Read, C., Smith, E.E.: Capacity estimates in working memory: reliability and interrelationships among tasks. Cogn. Affect Behav. Neurosci. **14**, 106–116 (2014). https://doi.org/10.3758/s13415-013-0235-x
21. Rouder, J.N., Morey, R.D., Cowan, N., Zwilling, C.E., Morey, C.C., Pratte, M.S.: An assessment of fixed-capacity models of visual working memory. PNAS **105**, 5975–5979 (2008). https://doi.org/10.1073/pnas.0711295105
22. Saults, J.S., Cowan, N.: A central capacity limit to the simultaneous storage of visual and auditory arrays in working memory. J. Exp. Psychol. Gen. **136**, 663–684 (2007). https://doi.org/10.1037/0096-3445.136.4.663
23. Curtis, C.E., Zald, D.H., Pardo, J.V.: Organization of working memory within the human prefrontal cortex: a PET study of self-ordered object working memory. Neuropsychologia **38**, 1503–1510 (2000). https://doi.org/10.1016/S0028-3932(00)00062-2
24. Unsworth, N., Heitz, R.P., Schrock, J.C., Engle, R.W.: An automated version of the operation span task. Behav. Res. Methods **37**, 498–505 (2005). https://doi.org/10.3758/BF03192720
25. Cowan, N., et al.: On the capacity of attention: its estimation and its role in working memory and cognitive aptitudes. Cogn Psychol. **51**, 42–100 (2005). https://doi.org/10.1016/j.cogpsych.2004.12.001
26. Vogel, E.K., Woodman, G.F., Luck, S.J.: Storage of features, conjunctions and objects in visual working memory. J. Exp. Psychol. Hum. Percept. Perform. **27**, 92–114 (2001)
27. Alvarez, G.A., Cavanagh, P.: The capacity of visual short-term memory is set both by visual information load and by number of objects. Psychol. Sci. **15**, 106–111 (2004). https://doi.org/10.1111/j.0963-7214.2004.01502006.x
28. Brady, T.F., Konkle, T., Alvarez, G.A.: A review of visual memory capacity: beyond individual items and toward structured representations. J. Vis. **11**, 4 (2011). https://doi.org/10.1167/11.5.4
29. Ma, W.J., Husain, M., Bays, P.M.: Changing concepts of working memory. Nat. Neurosci. **17**, 347–356 (2014). https://doi.org/10.1038/nn.3655
30. van den Berg, R., Shin, H., Chou, W.-C., George, R., Ma, W.J.: Variability in encoding precision accounts for visual short-term memory limitations. PNAS **109**, 8780–8785 (2012). https://doi.org/10.1073/pnas.1117465109
31. Bays, P.M., Husain, M.: Dynamic shifts of limited working memory resources in human vision. Science **321**, 851–854 (2008). https://doi.org/10.1126/science.1158023
32. Luck, S.J., Vogel, E.K.: Visual working memory capacity: from psychophysics and neurobiology to individual differences. Trends Cogn. Sci. **17**, 391–400 (2013). https://doi.org/10.1016/j.tics.2013.06.006
33. Geisler, W.S.: Visual perception and the statistical properties of natural scenes. Annu. Rev. Psychol. **59**, 167–192 (2007). https://doi.org/10.1146/annurev.psych.58.110405.085632
34. Alvarez, G.A.: Representing multiple objects as an ensemble enhances visual cognition. Trends Cogn. Sci. **15**, 122–131 (2011). https://doi.org/10.1016/j.tics.2011.01.003
35. Sanocki, T., Sellers, E., Mittelstadt, J., Sulman, N.: How high is visual short-term memory capacity for object layout? Atten. Percept. Psychophys. **72**, 1097–1109 (2010). https://doi.org/10.3758/APP.72.4.1097
36. Cohen, M.A., Dennett, D.C., Kanwisher, N.: What is the bandwidth of perceptual experience? Trends Cogn. Sci. **20**, 324–335 (2016). https://doi.org/10.1016/j.tics.2016.03.006
37. Cohen, M.A.: What is the true capacity of visual cognition? Trends Cogn. Sci. **23**, 83–86 (2019). https://doi.org/10.1016/j.tics.2018.12.002

38. Horowitz, T.S., Klieger, S.B., Fencsik, D.E., Yang, K.K., Alvarez, G.A., Wolfe, J.M.: Tracking unique objects. Percept. Psychophys. **69**, 172–184 (2007). https://doi.org/10.3758/BF0319 3740

39. Wu, C.-C., Wolfe, J.M.: Your hidden capacity revealed! the multiple object awareness (MOA) paradigm. J. Vis. **18**, 1019 (2018). https://doi.org/10.1167/18.10.1019

40. Lobier, M., Dubois, M., Valdois, S.: The role of visual processing speed in reading speed development. PLoS ONE **8**, e58097 (2013). https://doi.org/10.1371/journal.pone.0058097

41. Shibuya, H., Bundesen, C.: Visual selection from multielement displays: measuring and modeling effects of exposure duration. J. Exp. Psychol. Hum. Percept. Perform. **14**, 591–600 (1988). https://doi.org/10.1037//0096-1523.14.4.591

42. Kyllingsbæk, S.: Modeling visual attention. Behav. Res. Methods **38**, 123–133 (2006). https://doi.org/10.3758/BF03192757

43. Dyrholm, M.: A MATLAB/C/C++ library for modeling visual attention with Bundesen's Theory of Visual Attention (2012). http://www.machlea.com/mads/libtva.html

44. Duncan, J., Olson, A., Humphreys, G., Bundesen, C., Chavda, S., Shibuya, H.: Systematic analysis of deficits in visual attention. J. Exp. Psychol. Gen. **128**, 450–478 (1999). https://doi.org/10.1037/0096-3445.128.4.450

45. Reed, C.M., Durlach, N.I.: Note on information transfer rates in human communication. Presence Teleop. Virt. Environ. **7**, 509–518 (1998)

46. Tan, H.Z., Reed, C.M., Durlach, N.I.: Optimum information transfer rates for communication through haptic and other sensory modalities. IEEE Trans. Haptics **3**, 98–108 (2010)

47. Klemmer, E.T., Muller, P.F.: The rate of handling information: key pressing responses to light patterns. J. Mot. Behav. **1**, 135–147 (1969)

48. Tan, H.Z., Reed, C.M., Delhorne, L.A., Durlach, N.I., Wan, N.: Temporal masking of multidimensional tactual stimuli. J. Acoust. Soc. Am. **114**, 3295–3308 (2003)

49. Shannon, C.E.: Prediction and entropy of printed English. Bell Syst. Tech. J. **30**, 50–64 (1951)

50. Bell, T.C., Cleary, J.G., Witten, I.H.: Text Compression. Prentice-Hall Inc., Englewood Cliffs (1990)

51. Montemurro, M.A., Zanette, D.H.: Universal entropy of word ordering across linguistic families. PLoS ONE **6**, 1–9 (2011)

52. Feng, Z.: Ultimate entropy of Chinese characters. Chin. Inf. 53–56 (1996)

53. Huang, X., Wu, L., Guo, Y., Liu, B.: Computation of the entropy of modern Chinese and the probability estimation of sparse event in statistical language model. Acta Electronica Sinica **28**, 110–112 (2000)

54. Sun, F., Sun, M.: Statistical estimation of ultimate entropy of Chinese characters. Presented at the 25 Anniversary of the Chinese Information Society Conference Proceedings (2006)

55. Wu, J., Wang, Z.: The entropy of Chinese and the perplexity of the language models. Acta Electronica Sinica. 24 (1996)

56. Xu, B., Wu, L.: The LZW data compression algorithm for the Chinese language. J. South China Univ. Technol. (Nat. Sci.) **17**, 1–9 (1989)

57. Carver, R.P.: Rauding theory predictions of amount comprehended under different purposes and speed reading conditions. RRQ. 205–218 (1984)

58. Carver, R.P.: Reading rate: theory, research, and practical implications. J. Read. **36**, 84–95 (1992)

59. Sheikh, H.R., Bovik, A.C.: Image information and visual quality. IEEE Trans. Image Process. **15**, 430–444 (2006)

60. Sheikh, H.R., Bovik, A.C., De Veciana, G.: An information fidelity criterion for image quality assessment using natural scene statistics. IEEE Trans. Image Process. **14**, 2117–2128 (2005)

61. Cox, G.: On the relationship between entropy and meaning in music: an exploration with recurrent neural networks. Presented at the Proceedings of the Annual Meeting of the Cognitive Science Society (2010)

62. Le Bot, A.: Entropy in sound and vibration: towards a new paradigm. Proc. R. Soc. A Math. Phys. Eng. Sci. **473** (2017)
63. Wolfe, J.M.: Guided search 4.0: current progress with a model of visual search. In: Integrated Models of Cognitive Systems, pp. 99–119. Oxford University Press, New York (2007)
64. Wolfe, J.M.: Visual search revived: the slopes are not that slippery: a reply to kristjansson (2015). Iperception **7** (2016). https://doi.org/10.1177/2041669516643244
65. Treisman, A.M., Gelade, G.: A feature-integration theory of attention. Cogn. Psychol. **12**, 97–136 (1980). https://doi.org/10.1016/0010-0285(80)90005-5
66. Treisman, A.: Preattentive processing in vision. Comput. Vis. Graph. Image Process. **31**, 156–177 (1985). https://doi.org/10.1016/S0734-189X(85)80004-9
67. Treisman, A.: Features and objects in visual processing. Sci. Am. **255**, 114–125 (1986)
68. Treisman, A.M., Gormican, S.: Feature analysis in early vision: evidence from search asymmetries. Psychol. Rev. **95**, 15 (1988)
69. Wolfe, J.M.: What do 1,000,000 trials tell us about visual search. Psychol. Sci. **9**, 33–39 (1998)
70. Cowan, N.: Working Memory Capacity, Classic edn. Routledge, London (2016)
71. Baddeley, A.: Working Memory, Thought, and Action. OUP, Oxford (2007)
72. Fougnie, D., Marois, R.: What limits working memory capacity? evidence for modality-specific sources to the simultaneous storage of visual and auditory arrays. J. Exp. Psychol. Learn. Mem. Cogn. **37**, 1329–1341 (2011). https://doi.org/10.1037/a0024834
73. Townsend, J.T., Nozawa, G.: Spatio-temporal properties of elementary perception: an investigation of parallel, serial, and coactive theories. J. Math. Psychol. **39**, 321–359 (1995). https://doi.org/10.1006/jmps.1995.1033
74. Houpt, J.W., Blaha, L.M., McIntire, J.P., Havig, P.R., Townsend, J.T.: Systems factorial technology with R. Behav. Res. **46**, 307–330 (2014). https://doi.org/10.3758/s13428-013-0377-3
75. Houpt, J.W., Townsend, J.T.: Statistical measures for workload capacity analysis. J. Math. Psychol. **56**, 341–355 (2012). https://doi.org/10.1016/j.jmp.2012.05.004
76. Brown, S.D., Heathcote, A.: The simplest complete model of choice response time: linear ballistic accumulation. Cogn. Psychol. **57**, 153–178 (2008). https://doi.org/10.1016/j.cogpsych.2007.12.002
77. Eidels, A., Donkin, C., Brown, S.D., Heathcote, A.: Converging measures of workload capacity. Psychon. Bull. Rev. **17**, 763–771 (2010). https://doi.org/10.3758/PBR.17.6.763
78. Gallace, A., Tan, H.Z., Spence, C.: Tactile change detection. In: First Joint EuroHaptics Conference and Symposium on Haptic Interfaces for Virtual Environment and Teleoperator Systems. World Haptics Conference, pp. 12–16 (2005). https://doi.org/10.1109/WHC.2005.122
79. Dim, N.K., Ren, X.: Investigation of suitable body parts for wearable vibration feedback in walking navigation. Int. J. Hum. Comput. Stud. **97**, 34–44 (2017). https://doi.org/10.1016/j.ijhcs.2016.08.002
80. Paz, S., Mayas, J., Ballesteros, S.: Haptic and visual working memory in young adults, healthy older adults, and mild cognitive impairment adults. In: Second Joint EuroHaptics Conference and Symposium on Haptic Interfaces for Virtual Environment and Teleoperator Systems (WHC 2007), pp. 553–554 (2007). https://doi.org/10.1109/WHC.2007.64
81. Hummel, T., Sekinger, B., Wolf, S.R., Pauli, E., Kobal, G.: 'Sniffin'sticks": olfactory performance assessed by the combined testing of odor identification, odor discrimination and olfactory threshold. Chem. Senses **22**, 39–52 (1997)
82. Olofsson, J.K., Bowman, N.E., Gottfried, J.A.: High and low roads to odor valence? A choice response-time study. J. Exp. Psychol. Hum. Percept. Perform. **39**, 1205–1211 (2013). https://doi.org/10.1037/a0033682

83. White, T.L., Prescott, J.: Chemosensory cross-modal stroop effects: congruent odors facilitate taste identification. Chem. Senses **32**, 337–341 (2007)
84. Höchenberger, R., Busch, N.A., Ohla, K.: Nonlinear response speedup in bimodal visual-olfactory object identification. Front. Psychol. **6** (2015). https://doi.org/10.3389/fpsyg.2015.01477
85. Spence, C., Ketenmann, B., Kobal, G., Mcglone, F.P.: Selective attention to the chemosensory modality. Percept. Psychophys. **62**, 1265–1271 (2000). https://doi.org/10.3758/BF03212128
86. Spence, C., Kettenmann, B., Kobal, G., McGlone, F.P.: Shared attentional resources for processing visual and chemosensory information. Q. J. Exp. Psychol. Sect. A **54**, 775–783 (2001). https://doi.org/10.1080/713755985
87. Roper, S.D., Chaudhari, N.: Taste buds: cells, signals and synapses. Nat. Rev. Neurosci. **18**, 485–497 (2017). https://doi.org/10.1038/nrn.2017.68
88. Yamamoto, T., Kawamura, Y.: Gustatory reaction time in human adults. Physiol. Behav. **26**, 715–719 (1981). https://doi.org/10.1016/0031-9384(81)90149-9
89. Wallroth, R., Ohla, K.: As soon as you taste it: evidence for sequential and parallel processing of gustatory information. eNeuro **5** (2018). https://doi.org/10.1523/ENEURO.0269-18.2018
90. Andersen, C.A., Alfine, L., Ohla, K., Höchenberger, R.: A new gustometer: template for the construction of a portable and modular stimulator for taste and lingual touch. Behav. Res. **51**, 2733–2747 (2019). https://doi.org/10.3758/s13428-018-1145-1
91. Landis, B.N., Welge-Luessen, A., Brämerson, A., Bende, M., Mueller, C.A., Nordin, S., Hummel, T.: "Taste strips"–a rapid, lateralized, gustatory bedside identification test based on impregnated filter papers. J. Neurol. **256**, 242 (2009)
92. Bolton, M.L., Edworthy, J., Boyd, A.D.: A formal analysis of masking between reserved alarm sounds of the IEC 60601-1-8 international medical alarm standard. In: Proceedings of the Human Factors and Ergonomics Society Annual Meeting (2018). https://doi.org/10.1177/1541931218621119
93. Parise, C., Spence, C.: 'When birds of a feather flock together': synesthetic correspondences modulate audiovisual integration in non-synesthetes. PLoS ONE **4**, e5664 (2009). https://doi.org/10.1371/journal.pone.0005664
94. Van Erp, J.B.F., Kooi, F.L., Bronkhorst, A.W., van Leeuwen, D.L., van Esch, M.P., van Wijngaarden, S.J.: Multimodal interfaces: a framework based on modality appropriateness. In: Proceedings of the Human Factors and Ergonomics Society Annual Meeting, vol. 50, pp. 1542–1546 (2006). https://doi.org/10.1177/154193120605001606

Culture-Based Design

Towards the Ethnic Understanding of Taiwanese Indigenous Peoples: A Mashup Based on Semantic Web and Open Data

Yu-Liang Chi[1]([✉]), Han-Yu Sung[2], and Ying-Yuan Lien[1]

[1] Department of Information Management, Chung Yuan Christian University, No. 200,
Zhongbei Road, Zhongli District, Taoyuan City 32023, Taiwan (R.O.C.)
maxchi@cycu.edu.tw
[2] Department of Allied Health Education and Digital Learning,
National Taipei University of Nursing and Health Sciences, No. 365, Mingde Road,
Beitou District, Taipei City 11219, Taiwan (R.O.C.)

Abstract. The purpose of this study is utilizing semantic web technologies to mix diverse datasets of Taiwanese indigenous peoples to facilitate ethnic understanding. Because indigenous peoples have been in a lower social grade for centuries, indigenous culture and corpus are relatively rare in the modern society. Insufficient information not only enlarge the gap of misunderstanding but lead to the emotional alienation between ethnics. To integrate multiple data sources, a data mashup is a way to achieve the integration of data. Furthermore, the result of mashups may offer new insights to discover interesting knowledge. To implement data mashup on the Web, three components are identified: (1) collecting available open data related to indigenous people; (2) performing a data conversion process to convert non-RDF to RDF data; (3) developing SPARQL federated queries to mix data from diverse endpoints. Furthermore, a web-based application based on an interactive map is utilized to facilitate ordinary users to find information from the general information of tribes to cultural artworks. The experimental results show that data mashup design can help users quickly understand essential information of indigenous peoples.

Keywords: Data mashup · Indigenous peoples · RDF · SPARQL

1 Introduction

The Taiwanese indigenous peoples usually refer to their ancestors who settled the island before the 17th century. Thereafter, lots of Han Chinese emigrated from mainland China. From 17th century to 19th century, the Dutch, Spanish, and Japanese successively occupied Taiwan. During the past centuries, indigenous peoples suffered colonial rule, discrimination, and marginalization. For centuries, ruling government carried out the policy of cultural assimilation that resulted in disappear of indigenous languages and loss of original cultural identity. In short, destruction of cultural heritage of indigenous peoples

© Springer Nature Switzerland AG 2020
P.-L. P. Rau (Ed.): HCII 2020, LNCS 12192, pp. 287–297, 2020.
https://doi.org/10.1007/978-3-030-49788-0_21

is common without sense of guilty. Since the end of 20th century, most people recognize that cultural genocide is mistakes and leads to irreversible damages. Governmental authorities started to reform and remedy historical mistakes by making a series of legislation and cultural revival activities. To integrate government and social resources, Taiwan has established a central government-level department, the Council of indigenous peoples (CIP)[1], to manage revitalization programs and support system development since 1996. The reviving programs, such as cultural identity, language revitalization, and restoration of tribal lifestyle, are under progress. Today, approximately 16 known tribes and 26 languages of Taiwanese indigenous peoples are recognized. However, more than half of languages are now extinct or endangered. Therefore, we should accelerate revival activities and encourage public to participate the understandings of multicultural societies and indigenous cultures.

To facilitate the understandings of relevant affairs of indigenous peoples, tangible and intangible assets must be collected, rebuilt, and activated. Web technologies provide a new way related to digital asset management and preservation. It further gives us the opportunities to effectively activate various applications such as dissemination and data integration. In recent years, the open data movement is widely promoting to diverse industries. Open government data (OGD) is one of the most valuable and authoritative data sources because the data is collected by public sector [1, 8, 12]. On the other hand, the concept of data mashup is mixing data together to create something new and more useful knowledge. Furthermore, semantic web is an extension of traditional web technology to construct a framework for making Internet data machine-readable [7]. World Wide Web Consortium (W3C) is responsible for developing a series of standards to realize various types of applications. Fundamental standards of semantic web such as RDF (a metadata data model), RDF Schema (RDF vocabularies for structuring RDF resources), and SPARQL (an RDF query language) offer the solution of data editing and searching across the Internet. Simply put, open data and semantic web technologies have presented developers with new opportunities to create web-based mashups.

This study, therefore, focuses on a mashup of open datasets of indigenous people by using semantic web technologies. The mashup performs mixing on general information of tribes, population distribution, heritages, and cultural artworks. In addition, we developed a web-based application with interactive design to increase user interests of knowledge exploration. Consequently, the benefits of data mashup can facilitate user engagement of indigenous affairs and ethnic integration.

2   Literature Review

2.1   Open Data and Linked Open Data

The idea of open data is promoted again and is realized because the Internet is widely used. There is no doubt that governments are the important driver of the open data movement because they take the lead role in releasing data. Open government data (OGD) becomes one of the major sources that promote transparency and value creation by making government data available to everyone [11]. Todays, more than 50 countries

---

[1] Council of indigenous peoples (CIP), https://www.apc.gov.tw/portal/index.html.

have launched open data initiatives. Furthermore, lots of organizations and institutions increasingly join the open data community to release huge quantities of data. The DBpedia is an example. Because people have unbounded creativity, some innovative impactful solutions can be proposed by combining available open data. In order to provide the usability of applications, open data usually utilizes structured formats, such as CSV, JSON, XML, RDF, etc. In short, open data has become a worldwide movement and has facilitated the idea of data sharing and reuse.

In order to integrate datasets from dispersed sources, open data fuses linked data becoming powerful Linked Open Data (henceforth LOD). LOD has significant advantages including open sources, RDF graph, and cross-domain links for achieving data sharing and reuse. The LOD Cloud (https://lod-cloud.net) is a Knowledge Graph that demonstrates a Semantic Web of Linked Data. DBpedia is a typical example of LOD and the core in LOD Cloud [5, 9]. Developers can save big efforts in organizing multiple data sources [10]. From open data to LOD, there is a rating scheme of data quality, i.e., a 5-star scheme, proposed by Sir Tim Berners-Lee. The scheme can be briefly explained as follows: 1 star is data available on the web with an open license; 2 stars are machine-readable data; 3 stars are data with non-proprietary format; 4 stars are data with RDF standards; and finally, 5 stars are exactly LOD.

2.2 Semantic Web Technologies

Conventional World Wide Web (known as the web) coded in HTML is built for human browsing rather than machine understanding. The term Semantic Web, coined by Tim Berners-Lee, aims to provide a common framework that allows data reuse across web applications [3]. The World Wide Web Consortium (W3C) gives a series of interoperable specifications to unify terms, protocols, and architectures to create the standard environment for applications. From a theoretical perspective, the Semantic Web consists of a set of specifications such as syntax using XML, data interchange using Resource Description Framework (RDF), and ontology using Web Ontology Language (OWL). The famous diagram of Semantic Web Stack (e.g., https://en.wikipedia.org/wiki/Sem antic_Web_Stack) illustrated the hierarchy of technologies, where each layer exploits the capabilities of the layers below. Because the very bottom layer of the stack is built on the foundation of a Uniform Resource Identifier (URI), Unicode, and HTTP, the Semantic Web can be regarded as an extension rather than a replacement of the current web. Some technologies of the Semantic Web and important terms used in this study are simply introduced as follows.

RDF is a metadata model for expressing information about resources. It codifies a statement about resources through triples in the form (subject, predicate, and object). A triple is a set of three entities that is similar to entity-attribute-value model. Thus, the three entities can be seen as resource identifier (subject), property name (predicate), and property value (object). In a set of triples, both a subject and an object are represented by URIs that unambiguously identify a particular resource. The predicate is also a URI to denote relationships between the subject and the object. The collection of sets of triples is usually called an RDF dataset. There are several common serialization formats used to encode RDF datasets such as N3, RDF/XML, and Turtle. Because Turtle is a relatively concise and readable format, all RDF examples in this study will use it.

Furthermore, for better performance in data access, a database-like triplestore is used to store RDF datasets. A comparison of available triplestores can be found online at en. wikipedia.org/wiki/Comparison_of_triplestores. This study employs Apache Jena (an open source can be found at https://jena.apache.org/) to develop a triplestore. Finally, to prevent confusion of data meanings, the predicate is suggested using common or published vocabulary sets. We can take advantage of a namespace lookup site (e.g., http://prefix.cc) to find common vocabulary sets. In contrast, linked data is data based on RDF and contains links to other resources to discover more things [7]. Tim Berners-Lee proposed four principles of linked data; they are listed as follows [5]: (1) Use URIs to identify things. (2) Use HTTP URIs so that people can look up those names. (3) When someone looks up a URI, provide useful information, using the standards (RDF, RDFS, and SPARQL). (4) Include links to other URIs, so that they can discover more things. Therefore, any RDF dataset matching these principles can be regarded as linked data.

SPARQL is the W3C standard query language for RDF. Its syntax and functionality are similar to SQL used in relational databases. Through coding a SPARQL query, one can retrieve RDF data from a triplestore. The queries are performed by a specific SPARQL processor that can search endpoints and return results. An endpoint is a website that is capable of receiving and processing SPARQL protocol requests. The basic structure of a SPARQL query comprises prefix declarations (abbreviating URIs), result clause (using SELECT to identify what information to return from the query), and query pattern (using WHERE to specify what to query for in the underlying dataset). A simple example is as follow. In this example, the query return all the values of "*?name*". The query pattern of SPARQL looks like RDF Turtle.

```
PREFIX  foaf: <http://xmlns.com/foaf/0.1/>
   SELECT
   WHERE{
       ?Subject  foaf:name  ?name.
}
```

3   Research Design

To develop a mashup of available open data, three components are proposed as shown in Fig. 1. The first component is collecting machine readable data related to indigenous

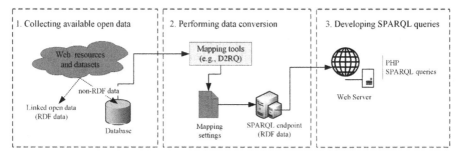

Fig. 1. Research design of data mashup

peoples. There are two types of the collected data including linked open data (i.e., RDF data) and non RDF data. The second component performs a data conversion process to convert non RDF data sources to RDF datasets. The process is mainly making a mapping definition between data formats. The last component develops SPARQL queries to implement mashups across the web.

3.1 Collecting Available Open Data Related to Indigenous Peoples

There is relatively little data regarding Taiwanese indigenous peoples, due to long-term destruction and lack of maintenance. Thus, each data is a precious asset for reviving indigenous history and cultures. Because a dataset is a structured collection of data, we are actually collecting available datasets. As shown in the most left block of Fig. 1, there are two types of collected data. The first type of data is linked open data (LOD), which is interlinked with other data, allowed public access, and formatted to RDF. Two LOD datasets are utilized in this study as follows:

The "TaiUC-anthropology-cultural heritages of Taiwanese indigenous people" is a set of LOD datasets to preserve the digitized items of cultural heritages and handicraft works. The set is upgraded from the digital contents of National Taiwan museum. The set has more than one hundred thousand triples and the major dataset of the set is presented by 42 attributes. The LOD is available of the website of Datahub at http://old.datahub.io.

DBpedia is a well-known linked open data and has been widely used in diverse disciplines [2, 5, 9]. The DBpedia corpus extracted from semi-structured Wikipedia allows users to search for context by a SPARQL request. In this study, some general information are pulled from DBpedia, such as the brief introduction of tribes and administrative divisions.

Table 1. Datasets of Taiwanese indigenous peoples

Dataset names	Fields descriptions
Townships of indigenous peoples	Using 4 attributes to present the townships of Taiwanese indigenous peoples
Population of indigenous peoples by townships	Using 24 attributes to present the demographics of each tribe group by administrative division
Museums of indigenous peoples	Using 8 attributes to present general information of indigenous museums
Statistics of reserve land ownership of indigenous peoples	Using 15 attributes to present the land area and amount categorized by the sectors of public, private, state, county, and township respectively
Ancient roads of indigenous peoples	Using 14 attributes to present cultural landscapes and special natural landscapes by indigenous tribe
Tourist routes of indigenous peoples	Using 8 attributes to present the travel itinerary of tribe reservation

On the other hand, the second type of data is non-RDF open data that is formatted by other machine readable formats, such as JSON, CSV, and excel. The collected non-RDF datasets are obtained from an official website at http://data.gov.tw. Most datasets are released by Taiwan CIP. Table 1 listed 6 datasets participated in the experiment of this study.

3.2 Performing Data Conversion

Because a database can import most machine readable data, it plays an important role in data conversion of non-RDF to RDF. As shown in the first block of Fig. 1, non-RDF datasets are first imported to a database. Data conversion is then the process of mapping database corpus to RDF data. To assist the mapping from a relational database (RDB) to RDF data, W3C has published two specifications, including:

- RDB to RDF Mapping Language[2], developers first edit the mappings of data structure and vocabulary based on the specification, and then convert database corpus to RDF data.
- A Direct Mapping of Relational Data to RDF[3], a database corpus is directly converted to initial RDF data, developers then edit the mapping settings to correct the conversion.

As shown in the second block of Fig. 1, a D2RQ tool (available on the website at http://d2rq.org) is utilized to data conversion. The database corpus is virtually presented as RDF data according to the settings in a mapping file. The tool consists of two main components: (1) D2RQ engine, it can generate an initial mapping file corresponding to a database corpus; (2) D2R Server, it provides an endpoint to allow access by using SPARQL requests.

The initial mapping file comprises a set of settings that defines how a table schema in a database converts to RDF data model (i.e., subject, predicate, and object). The settings do the data conversion right, however, conversion result may not achieve the final goal. For example, a predicate of an RDF triple borrows a column name of a database table as its temporary vocabulary. Developers usually need modify the definition of settings for getting meaningful data conversion. The modification is performed by using D2RQ mapping language. Three typical types of setting modifications are worth to note as follows.

- Modifying the settings of a subject of an RDF statement. In D2RQ mapping language, the vocabulary of *d2rq:uriPattern* is utilized to declare the subject of a set of RDF triples. The initial settings use the combination of a database table name and its primary key to naming a subject of a RDF statement. Developers should utilize a proper URI to naming the subject.
- Modifying the settings of a predicate of a RDF statement. The vocabulary of *d2rq:property* is utilized to naming a predicate of a RDF statement. The initial settings

---

[2] R2RML, https://www.w3.org/ns/r2rml.

[3] Direct Mapping, https://www.w3.org/TR/rdb-direct-mapping/.

simply uses column names of a table as initial predicates. According to W3C specifications, a predicate of a RDF statement are suggested to use a defined vocabulary that applications can understand its meanings. The defined vocabularies are usually wrapped together into a namespace. Online applications can use defined vocabularies by importing a namespace. Some formal sets of vocabularies (e.g., RDF, RDFS, OWL, and etc.) and common sets of vocabularies (e.g., FOAF, vcard, Dublin core, and etc.) are widely used to naming predicates.

- Modifying the settings of an object of a RDF statement. In initial mapping file, a data value of a column are mapping to an object of a RDF statement. Thus, the object is literals rather than an URI. To achieve the idea of linked data, developers can replace a literal to a proper URI. For example, using the URI of http://dbpedia.org/resource/Taipei instead of the literal of *Taipei*.

3.3   Developing SPARQL Queries to Implement Data Mashup

After all datasets are ready, we implement data mashup by editing SPARQL requests followed the specification of SPARQL federated query[4]. The specification describes how a query can retrieve data from multiple remote endpoints. The *SERVICE* keyword commands a federated query processor to invoke a portion of a SPARQL query against a remote SPARQL endpoint. A simple example shown as follow. The SERVICE keyword put a subquery inside of the query that shows how to pull data from a remote SPARQL endpoint and join the returned data with the data from the local RDF dataset. The description about the names of cities (variable *?city*) is retrieved from the endpoint of http://dbpedia.org/sparql.

```
PREFIX  dbo: <http://dbpedia.org/ontology/>
SELECT  ?abstract
WHERE {
        ?subject  rdf:type  ?city .
    SERVICE <http://dbpedia.org/sparql>
        {
        ?city dbo:abstract  ?abstract .
        }
}
```

In this study, we focus primarily on the federation over SPARQL endpoint infrastructure. As shown in Fig. 2, three major sources joined mashups including the datasets of open government data, a museum dataset, and DBpedia. SPARQL queries are capable to link and merge data together across endpoints. For example, related cultural artworks of each tribe can be took by entities having the same identity in the datasets of OGD and artworks respectively. Furthermore, all Taiwanese tribes in the dataset of OGD can be connected with tribes in DBpedia by the vocabulary of *owl:sameAs* to get more description. Also, administrative districts in the dataset of OGD can get further explanation in DBpedia by query two entities having the same identity.

---

[4] SPARQL 1.1 Federated Query, https://www.w3.org/TR/sparql11-federated-query/.

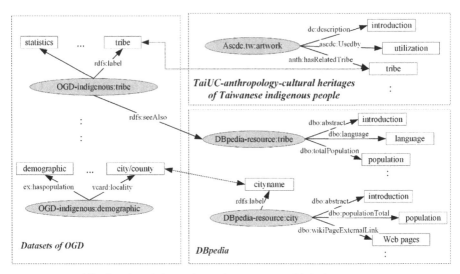

Fig. 2. The relations of a mashup among multiple data sources

4   System Implementation

To implement the proposed design of a data mashup, this study develops a web-based application built up by a web server (XAMPP) and a SPARQL server (Apache Jena Fuseki[5]). The application is programmed on the server and wraps SPARQL query inside a PHP function to carry out a search on dataset endpoints. As shown in the upper part of Fig. 3, it listed 16 officially recognized indigenous tribes. Users can simply click on the name of a tribe to browse the brief pulled from DBpedia endpoint. The lower part of Fig. 3 shown messages of historic trails pulled from government open datasets, such as hunting path and ancient roads. Each historic trail is a restored heritage that is described by using the original name, location, related ethnicities, history, and influence. This page provides general concepts of indigenous tribes for publics.

Due to insufficient information and cultural barriers, we have only a superficial knowledge of indigenous people. For example, we may aware of that someone is indigenous peoples but hard to tell the name of tribe to which someone belongs. Furthermore, detailed information, such as population distribution and cultural characteristics, are even difficult. Figure 4 presented an interactive design using geographical map of Taiwan to trigger a data mashup. The left-hand side of Fig. 4 is a map contained 22 administrative divisions. Users can simply click on an area to enable query. The right-hand side of the figure presents multiple information including the introduction of geographical area, a list of museums, population distribution, an introduction by indigenous tribe, and cultural artworks. In the example of Fig. 4, a user who clicked on the area of "*New Taipei city*" got five types of data. Each type of data represented by the marks of (1) to (5) inside the figure is described as follows.

---

[5] Apache Jena Fuseki, https://jena.apache.org/documentation/fuseki2/.

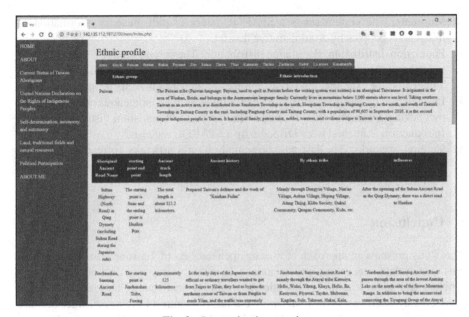

Fig. 3. System implementation

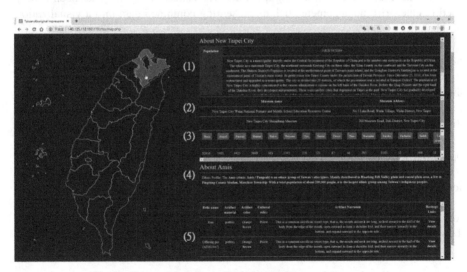

Fig. 4. Geographical map of Taiwan to trigger a data mashup

1. Geographical area: The system sends a SPARQL query to related endpoints to get populations of indigenous peoples and an administrative district respectively. The first row presented the populations by two numbers of "*55858/3972204*". Furthermore, the query pull description of an administrative district from DBpedia as shown in the second row.

2. Museums: The system sends a SPARQL query to pull museum information of indigenous peoples by administrative district.
3. Population distribution: Population distribution: The system sends a SPARQL query to fetch population data by indigenous tribe within an administrative district. In New Taipei city, for example, the largest number of all tribes is Amis people.
4. Introduction by indigenous tribe: The introduction of an indigenous tribe is invoked by clicking a name of tribe located in the previous "Population distribution". The Introduction is fetched from DBpedia by a SPARQL request.
5. Cultural artworks: The last part presented cultural artworks corresponding to an indigenous tribe which we clicked in previous step. The information is fetched from the linked data of "TaiUC-anthropology-cultural heritages".

5   Conclusion

This study presents an approach of mixing open datasets of Taiwanese indigenous peoples to stimulate inter-ethnic understanding. In recent years, open data pushed by government and some leading organizations to support data sharing and reuse. A single open data by itself can be difficult to give a larger picture of the knowledge, because the purpose of its creation is simple and limited. Mixing multiple data therefore can provide a different view or new insight of data. The concept of mashup has been widely used in enterprise management such as presenting combined data into a visualization dashboard.

In this study, we implement mashups of open datasets based on semantic web technologies. All collected datasets are linked data or open datasets that can be converted to RDF data. Through SPARQL queries, a mashup can be easy implemented across the Internet. For example, in Fig. 4, a user clicks on an area inside the map to enable federated query to multiple endpoints. The result of mashup provides users to gather diverse information, such as population distribution among tribes, a brief description of indigenous tribes, cultural artworks, and etc. In short, data mashup provides a great opportunity to get know better the indigenous peoples for publics. Future work will focus on extending the range of data sources and modifying an interactive application for better user experience.

Acknowledgement. This research was supported by the Ministry of Science and Technology, Taiwan, Grant Number: MOST 108-2410-H-033-036 and MOST 106-2511-S-227 -005 -MY2.

References

1. Attard, J., Orlandi, F., Scerri, S., Auer, S.: A systematic review of open government data initiatives. Gov. Inf. Q. **32**(4), 399–418 (2015). https://doi.org/10.1016/j.giq.2015.07.006
2. Auer, S., Bizer, C., Kobilarov, G., Lehmann, J., Cyganiak, R., Ives, Z.G.: DBpedia: a nucleus for a web of open data (2007)
3. Berners-Lee, T.: Linked Data - Design Issues (2006). http://www.w3.org/DesignIssues/LinkedData.html. Accessed 12 Dec 2019

4. Berners-Lee, T.: Putting Government Data online (2009). http://www.w3.org/DesignIssues/GovData. Accessed 12 Dec 2019
5. Bizer, C., Heath, T., Berners-Lee, T.: Linked data - the story so far. Int. J. Semant. Web Inf. Syst. **5**(3), 1–22 (2009). https://doi.org/10.4018/978-1-60960-593-3.ch008
6. Bizer, C., et al.: DBpedia - a crystallization point for the web of data. Web Semant. Sci. Serv. Agents World Wide Web **7**(3), 154–165 (2009). https://doi.org/10.1016/j.websem.2009.07.002
7. Heath, T., Bizer, C.: Linked data: evolving the web into a global data space. Synth. Lect. Semant. Web Theory Technol. **1**(1), 1–136 (2011). https://doi.org/10.2200/s00334ed1v01y201102wbe001
8. Heise, A., Naumann, F.: Integrating open government data with stratosphere for more transparency. Web Semant. Sci. Serv. Agents World Wide Web **14**, 45–56 (2012). https://doi.org/10.1016/j.websem.2012.02.002
9. Lehmann, J., et al.: Dbpedia–a large-scale, multilingual knowledge base extracted from wikipedia. Semant. Web **6**(2), 167–195 (2015). https://doi.org/10.3233/sw-140134
10. Musto, C., Basile, P., Lops, P., de Gemmis, M., Semeraro, G.: Introducing linked open data in graph-based recommender systems. Inf. Process. Manag. **53**(2), 405–435 (2017). https://doi.org/10.1016/j.ipm.2016.12.003
11. Shadbolt, N., et al.: Linked open government data: lessons from data.gov.uk. IEEE Intell. Syst. **27**(3), 16–24 (2012)
12. Veljković, N., Bogdanović-Dinić, S., Stoimenov, L.: Benchmarking open government: an open data perspective. Gov. Inf. Q. **31**(2), 278–290 (2014). https://doi.org/10.1016/j.giq.2013.10.011

The Craft Aesthetics of the Cultural Relics Collection in the Qing Dynasty

Mei-Jin Hsiao[✉] and Shu-Hua Hsueh

Fuzhou University of International Studies and Trade, No. 28, Yuhuan Road,
Shouzhan New District, Changle District 350202, Fujian, People's Republic of China
Xiaomeijin@fzfu.edu.cn, iec@fzfu.com

Abstract. Since China had been the emperor system, all folk treasures would be collected by the aristocracy of the palace. Emperor Qianlong, the emperor Gao Zong of the Qing Dynasty loved cultural relics by nature and took great care of his treasures. Some treasures are from the former dynasties, some are imported from foreign tribute, and some are from domestic officials or folks. In order to protect these ancient relics, many famous artists such as painters, sculptors, carpenters, and stone smiths were recruited, So the brilliant craft aesthetics was well developed. This study mainly aims to explore the collection of cultural relics from 1735 to 1795 during the reign of Emperor Gao Zong of the Qing Dynasty and introduce various delicate boxes created by the Qing artisans. In other words, through these ancient cultural relics, such as ancient jade, calligraphy, books, literature, and other exquisite cultural relics, in this paper the author presents the aesthetics of the Qianlong emperor's collection and the effort of Qing artisans who shaped the relics of the Royal Collection, shaping various combinations, storage and display, as well as exquisite boxes such as Doubger and Borg.

Keywords: Qianlong · Qing Dynasty · Cultural relics · Craft

1 Introduction

China had been an emperor system since the founding of the country, and Chinese emperors had a hobby for collecting cultural relics. Therefore, all the folk treasures would be collected by the nobles of the palace. In particular, this kind of collection prevailed in the Qing Dynasty, especially during the Qianlong period due to the vast territory, the state treasury, political stability, developed economy, and prosperous cities. For example, there were a variety of crafts produced in Guangdong and all kinds of crafts and paintings imported from Europe. They were all landed from Guangzhou, sold through commercial banks, and resold to Beijing, Yangzhou and other major cities.

At that time, the prosperity of commercial trade had opened the eyes of artisans to learn from each other from imitation to innovation, and had improved local traditional craftsmanship. The economic prosperity also allowed the Emperor Qianlong to accumulate a large amount of wealth for the treasury, which laid a solid and luxurious material foundation for the development of the "House of the Palace of Internal Affairs".

© Springer Nature Switzerland AG 2020
P.-L. P. Rau (Ed.): HCII 2020, LNCS 12192, pp. 298–312, 2020.
https://doi.org/10.1007/978-3-030-49788-0_22

According to historical records, the manufacturing office was in its heyday, with 42 professional workshops underneath, and the manufacturing office at that time was specialized in making daily necessities for the imperial court. In addition to being closely related to the royal residence, it included repairs, imperialism, and imperialism. There was everything from supplies, drawings, even leisure, antiquities and furnishings, and the "old relics and new gadgets" collected by Emperor Qianlong had reached their peak. Each workshop brought together the first-class elites from all over the country, and also created the superb craftsmanship aesthetics of the Qing Dynasty. Emperor Gaozong of the Qing Emperor Qianlong liked the relics by nature, and took great care of his treasures. Some treasures were from the former dynasty, some were imported from foreign tribute, and some were from domestic officials or folks. In order to protect these owned cultural relics, many well-known figures such as painters, sculptors, carpenters, and stone smiths had been recruited. So, the brilliant craft aesthetics was well developed. This study mainly aims to explore the collection of cultural relics from 1736 to 1795 during the reign of Emperor Gao Zong of the Qing Dynasty and introduce various delicate boxes created by the Qing artisans. In other words, through these ancient cultural relics, such as ancient jade, calligraphy, books, literature, and other exquisite cultural relics, in this paper the author presents the aesthetics of the Qianlong emperor's collection and the effort of Qing artisans who shaped the relics of the Royal Collection, shaping various combinations, storage and display, as well as exquisite boxes such as Doubger and Borg. For example, the Qianlong Emperor's collection of "Carved Red Sandalwood and Dragon Square Hundred Pieces" was carved in red sandalwood dragon square box with flexible space design, widely used in decorative techniques and materials, is the representative of the hundred pieces in the Qianlong period. And "Red Sandalwood and Bamboo Silk Turntable Lattice" (Fig. 1a).

(a)   (b)

Fig. 1. National Palace Museum

Figure 1b was recorded in Qianlong's Thirty-one Year (1766) "Live Plan File" (National Palace Museum 2017).

At that time, the Emperor Qianlong requested the combination of wooden seats and hand scroll albums. The appearance is decorated by the lotus flower pattern of bamboo silk collage. When combined, it shows a cylindrical shape. When opened, the column is divided into four equal parts. Craftsmen cleverly used a quarter of the cylindrical space, staggered into interesting layers of cells. All the jade articles in the small grid were matched with suitable wooden bases, arranged in space, one high and one low, intricate, and there were patterned carvings on the pillars, showing the aesthetic interest of the Qianlong period, the ingenuity of the craftsman, and superb craftsmanship. This box is the best portrayal of the display, storage and packaging of palace relics.

In addition, during the Qianlong period, they relied on their cultural relics. Therefore, the maintenance of cultural relics was specially tailored. Various types of crafts were designed and produced by the "making office", the shape of materials, the use of materials. Whether it is sitting, standing or lying, the cultural relics display the most beautiful posture, making the collected cultural relics a beautiful piece of art, reflecting the royal etiquette and aesthetic needs, the inheritance of these palace craftsmanship. It also represents the highest achievement of the craft aesthetics of the Qing Dynasty.

2   Overview of the Antiquities Collection

Collection is a natural instinct of human beings and an act that has existed since ancient times. All over the world, People's love for collection is all the same.

The so-called collection refers to the habit of collecting, storing, sorting and maintaining items. In general, collectors usually collect valuable antiques, but they can also be small things. Therefore, of course, the collection of items will vary according to the individual's emotions, interests, and tastes. But it is worthwhile to be sure that the items that the collector collects are things that he really likes. For example, the Medici family in Florence during the Renaissance loved collecting art and was also a patron of the artist. Personal collectors can trace back to Rudolph II. He specializes in collecting the treasures of Renaissance art. This tradition continues to today's art collectors.

In addition, many museums in the world, such as the U.S.-New York Metropolitan Museum and the Spain-Madrid Thyssen Museum, are donated by collectors and have become places for people to enjoy art in the future. And China is no exception. For generations, noble literati also loved ancient relics and art collections. For example, the famous collector in the history of China, Xiang Yuanzhang (1525–1590) of the Ming Dynasty, became a well-known collector in China because he was good at appreciating ancient and elegant art works, and good at managing his collections. He was the leader of private collectors of all ages in history.

Emperor Qianlong of the Qing Dynasty had always been very interested in art collection, and had established the largest collection, even affecting the study of Chinese art as a whole (Clunas 2012).

During the reign of Emperor Qianlong, he loved tasting cultural relics, so he collected ancient relics on a large scale, and studied the sources and allusions of these ancient relics, and even systematically stored, organized, and furnished them. The love of history had enabled him to properly manage, classify, store and maintain antiquities. Therefore, the Emperor Qianlong can be regarded as the best representative of collection and storage.

The interpretation of "relics" according to "Ci Hai" is: historical and cultural relics that remain in society or buried underground. "Relics" broadly refers to the national etiquette and music regulations. The cultural relics protected by the state in China must have the following conditions: (1) ancient cultural sites, ancient tombs, ancient buildings, cave temples, stone carvings, and murals with historical, artistic, and scientific value;

(2) with major historical events, revolutionary movements modern, important modern historical sites, and physical representative buildings that are related to famous people and have important commemorative, educational, or historical value; (3) precious works of art and arts and crafts in various eras in history; (4) historical important revolutionary materials of the era, manuscripts and books with historical, artistic, and scientific value; (5) representative real objects that reflect the social system, social production, and social life of all times, ethnic groups in history (Chi 2009). In addition, ancient vertebrate fossils and ancient human fossils of scientific value are protected by the state just like cultural relics. Therefore, the objects on the surface are antiques and old objects. In essence, cultural relics are relics and relics of historical value left by human beings in social activities. They are precious historical and cultural heritage of human beings and are miniatures of public sentiment under the witness of historical culture.

The collection of folk cultural relics can be described for a long time. According to historical records, folk cultural relics have been collected since the Sui and Tang Dynasties. Due to political stability, social harmony had been brought about, and the exchange of commercial trade had become alive. Therefore, the economy has also developed accordingly. According to Tang notes and historical records, the antique market at that time had a variety of antique markets in addition to the sale and purchase of artworks such as jewelry, paintings and calligraphy. In Tang Xuanzong, there were also merchants who specialized in collecting calligraphy and antiques for the royal family. In the tenth year of Kaiyuan, Wang Xizhi's "Lanting Collection Preface" extension edition alone cost tens of thousands of yuan, which shows that the folks were very popular for collecting antiques at the time (Chi 2009).

However, the real rise of cultural relics collection should start from the Song Dynasty. As people's material living standards improved, the ancient cultural relics market would also flourish and flourish. In the glorious collection of cultural relics, the Song Dynasty, it is recorded in the picture that there were only over 500 ancient artifacts of various sizes in collectors, and then the number reached to be more than 6,000 ancient artifacts during the Zhenghe period. Also in Wuxian County's Customs and Records: "The rich and wealthy house collects antiques, or paintings, or characters, or utensils such as a bottle and a bowl. The size is small but the price is pretty high." (Lai 2009) From the above, we can understand the enthusiasm and love of cultural relics from the royal to the folk in the Song Dynasty. Since then, the collection of cultural relics has gone through the Yuan Dynasty, the Ming Dynasty and the Qing Dynasty, and in particular, the collectors of the Qing Dynasty have become more universal.

Until today, the collection of antiquities has become a common practice. Due to social stability and the open era, the economy is booming, and people in the market have a better chance to enjoy the rare treasures in museums, art galleries, and antique markets. The enjoyment of culture and art has not reached Official patents and literati patents, therefore, the number of people who like to collect more and more in the society has also promoted the diversification of the collection of antiquities. At the same time, it has also enhanced people's tastes and appreciation of life, so that people get spiritual satisfaction from arts.

3   Arts and Crafts of the Qing Dynasty

The arts and crafts of the Qing Dynasty experienced four stages of development: the first stage was the growth period of the early Qing Dynasty; the second stage was the prosperity period of Kangxi and Qianlong; the third stage was the decline period of Jia, Tao and Xian; The fourth stage referred to the recovery period of Tongzhi, Guangxu, and Xuantong. This article will use the "Kangxi and Qianlong prosperity" era to explain the development of various crafts in the Qing Dynasty.

3.1   Ceramic Craft

The production site of Qing Dynasty ceramics mainly came from Jingdezhen, and the Qing royal family kiln factory was also located here. The heyday of Emperor Kangxi and Qianlong was the highest example of Qing Dynasty porcelain burning art and porcelain art. The Qing Dynasty vigorously copied the four famous kiln of the Guan Ware, Ding Ware, Jun Ware and Ge ware in the Song Dynasty; in addition, many new monochrome glazes were created. Emperor Yongzheng loved monochrome glazed porcelain not only for its pure style but also for its delicate shape. Yongzheng porclain changed from the thick and awkward style of the Kangxi period to the handsome and exquisite, the glaze was fine, light and exquisite, and the lines were elegant and soft.

The three generations of Kangxi, Yongzheng, and Qianlong reached their heyday in the firing of monochrome glazed porcelain. After firing a single-color glaze, it presents different single colors, such as cyan glaze, red glaze, green glaze, blue glaze, etc. Although the color of the single-color glazed porcelain is single, the color is light and elegant, but in a certain degree, it is beautiful and elegant (LNCS). The process of ceramic decoration is divided into "on-glaze" and "under-glaze". The glaze is made by painting the burned plain porcelain and then grilling it at low and medium temperatures. Because the decoration is attached to the glaze, it is called glaze. In the Kangxi period, the on-glaze color inherited the multicolored Ming dynasty, and also introduced the western enamel color, and later created the pastel color, which made the Qing Dynasty glaze colorful.

Twenty-three years after Emperor Kangxi relaxed the Sea Ban, western enamel wares were continuously imported into Guangzhou, where missionaries and local governments paid tribute to the emperor. In the 58th year of Kangxi (1719), the missionary Jean Baptise Gravereau entered the court to guide the painting of enamel, so the court had its own paintings of enamel (National Palace Museum 2011) In the late Kangxi period, porcelain enamels, also known as enamels, appeared on the porcelain tyres produced in Jingdezhen. They belong to new varieties of glazed colors, such as glazed red color, multicolored and pastel colors. Three stages of enamel color were development in the Qing Dynasty. The Kangxi period was rigorous and neat (Fig. 2a). The enamel colors of the Yongzheng period were more colorful and richer than the previous period (Fig. 2b). The Qianlong period enamel colors became more matured and the colors were more vivid (Fig. 2c).

Under-glaze is a main decoration method of ceramics. Various colored patterns are drawn on the formed and dried pins, and then covered with white transparent glaze or light-colored glaze, and fired at one time. The pattern after firing is covered by a

(a)   (b)   (c)

Fig. 2. National Palace Museum (Color figure online)

transparent glaze film. The surface is bright and soft, smooth and does not protrude, which makes the ceramic more transparent. Blue and white porcelain is the most representative of underglaze. The blue and white flowers of the Kangxi period were pure blue, with a bright color. The blue-and-white pattern was made lighter and darker by rendering (Fig. 3a). The blue and white flowers of Yongzheng and Qianlong emulated the blue and white flowers of Xuande, and also achieved good artistic achievements (Fig. 3b).

(a)   (b)

Fig. 3. National Palace Museum (Color figure online)

3.2 Jade Craft

Jade in the Qing Dynasty can be divided into antique jade and Shi Zuoyu. There are two types of antique jade, one is antique Yi and the other is Han Yu. The antique Yi is a jade made from the shapes and patterns of the Shang and Zhou bronzes. It inherits the antique Yi of the Ming Dynasty. It is more comprehensive in the shape of the device than in the Ming Dynasty, and is not limited to the imitation of jade. Imitation Han jade refers to the broad antiquity, not specifically to the Han Dynasty jade. During the period of Qianlong, imitation of Han jade was very popular. Qianlong highly appreciated the sapphire two-eared ear cups, and had the article to record its events.

Such as (Fig. 4) The production of antique utensils in the Qianlong period became common practice. This piece, regardless of the ornamentation and shape, has many antique styles, and pursues the antiques of ancient jade. Shizuo jade has various shapes and patterns, but through analysis, it is not difficult to find that Shizuo jade can be summarized into two categories: one is advocating fine carvings, focusing on the exquisite craftsmanship, the complexity of the patterns and the shape gorgeous, this is the characteristic of jade in the Qing Dynasty; the other is simple decoration, simple carving, fully showing the beauty of the texture of the jade.

Fig. 4. XiLingYinShe

3.3 Lacquer Craft

Lacquerware of the Qing Dynasty is a continuation of the Ming Dynasty. The development of lacquer ware is mainly reflected in the restoration of red lacquerware and the promotion of imitation lacquerware. The main production areas of lacquerware and lacquerware are in Suzhou. Suzhou in the Qing Dynasty was an important town for handicrafts and an important place for officials to tribute. In the third year of Qianlong (1738), the famous master of life-bamboo carving carved a prototype and sent it to Suzhou for mass production, so that the red picking device had been restored. In addition, during the Qianlong period, another red picker was produced in Yangzhou. Suzhou and Yangzhou's tick red paint industry was destroyed due to the Taiping Heavenly Kingdom battle.

As for the development of tick red in Beijing in the late Qing Dynasty and the beginning of the 20th century, it was mainly based on the Suzhou red tick like Qianlong (Fig. 5). Imitation lacquerware began in the Ming Dynasty. Suzhou imitation Japanese lacquer lacquerware used a large amount of gold powder, lacquered gold, and gold tracing in the Qing Dynasty. It can still achieve the brilliant artistic effect of Japanese lacquer lacquerware, so it was affected by the emperor Favor. The Forbidden City of Qing Dynasty has a large collection of Suzhou lacquered furniture and lacquer boxes (Fig. 6) you can see the glory of the lacquerware craft in the middle of Qing Dynasty.

Fig. 5. XiLingYinShe
(Color figure online)

Fig. 6. National Palace Museum

3.4 Metal Craft

In the Qing Dynasty, metal technology flourished and was divided into gold, silver, copper, iron, tin and so on. Most cities and towns in the Qing Dynasty opened gold and silver craft workshops. The transparent enamel on silver jewelry was generally popular among women in the lower middle class because of its low price. Gold crafts belong to precious metal craftsmanship and have a long tradition. The silver wares in the palace were also made by the Yangxindian Manufacturing Office. There were three types of bronzes in the Qing Dynasty: bronze, copper and brass.

For example, during the tenth year of Qianlong (1745) to sixty years (1795), a batch of furnaces, cylinders, bottles, and furnishings such as lions, turtles and cranes were cast in half a century. The gilt plating was undertaken by the casting furnace of the Yangxindian Manufacturing Office. Its craftsman division was very fine. The Qing Dynasty imitation of Xuande furnace, exquisite craftsmanship was very popular (Fig. 7).

Fig. 7. XiLingYinShe

The famous production areas of tin ware in the Qing Dynasty include Suzhou Wuzhong Mudu, Lingnan Guangzhou, and southern Yunnan. The tin ware belongs to metal craftsmanship, but it is not as splendid as gold and silver, and it is not as loud as bronze wares, so it is unknown in history books. It was formally favored by people around the beginning of the Ming Dynasty, and it was not until the Qing Dynasty that it shined in utensils. The tin pots in the tin ware were famous all over the world made by famous craftsmen. The shapes of the tin ware are simple and elegant.

3.5 Enamel Craft

Enamel in the Qing Dynasty was generally developed, with Kangxi and Qianlong being the highest peaks. In the 32nd year of the reign of Emperor Kangxi, the establishment of the Yangxin Hall was expanded and a workshop was set up. Enamels are generally divided into filigree enamels, hammer tire enamels, transparent enamels, and painted enamels.

Filament enamel should be the main product undertaken by the craftsmen of the early Kangxi enamel workshop. During the Qianlong period, the production scale of filigree enamel was huge, the output soared, the product was widely used, the style was unique, and it had a profound impact on future generations. In addition to the production center of Yangxindian, the silk enamel also made in Beijing, Yangzhou, Guangzhou and other cities. In the late Qing Dynasty, Beijing became the only place where silk enamel was produced in China, and its workmanship was more refined. Hammer tire enamel was found in the Qianlong period of the Qing Dynasty. It was formed by hammering out uneven patterns on gold, silver, copper and other metal tires, and then blue, baked and gold-plated in the pattern. The jade inlaid on it is mostly used for making Buddhist supplies. Transparent enamel was coated with transparent enamel on artistically processed metal tires, and the patterns on the tires were exposed after baking and firing. Generally, enamel materials were used in four colors: purple, blue, green, and yellow. Monochrome and multicolor can be used. There are two types of enamel: soft and hard. Soft enamel and burnt blue on jewelery work can be produced all over the country. Hard and transparent enamel was only produced in Guangzhou. Painting enamel was a device for painting on metal tires with enamel. Chinese painting enamel began in the Kangxi reign of the Qing Dynasty. In 55 years (1716), Guangzhou enamel maker Pan Chun and Yang Shizhang Jinjing brought the rose-red enamel recipe to the office enamel. Chen Zhongxin, a French enamel craftsman, entered the court in the 58th year of Kangxi (1719) to fire painted enamels; missionaries Ma Guoxian and Lang Shining also painted. On the basis of the techniques of painting enamel during the Kangxi and Yongzheng periods, painting enamel during the Qianlong period had greater achievements and development. The enamel painting style in the court is dignified and well-proportioned. Its pattern is rigorous and smooth as well as the painting is delicate and exquisite with the color extremely rich and bright (Fig. 8a,b).

3.6 Weaving and Embroidery Craftsmanship

In the Qing Dynasty, weaving and embroidering were developed on the basis of Ming dynasty skills. Not only were weaving and embroidery a wide variety, patterns and colors were also more abundant. Suzhou, Hangzhou, Nanjing and other places had formed weaving and embroidery production centers. The Qing royal family set up weaving officers in Suzhou, Hangzhou, and Nanjing to weave official satin. Three-woven weaving and embroidery products represent the highest level of weaving and embroidering

(a)   (b)

Fig. 8. National Palace Museum

technology in the Qing Dynasty. Suzhou imitation Song Jin, continued production from the Ming Dynasty to the early Qing Dynasty, to the Kangxi and Qianlong periods, its weaving technology and variety of colors had reached unprecedented heights. In Kangxi, Yongzheng, and Qianlong, Nanjing Yunjin had been further developed. The pattern is rich and large as well as the color is strong. The flat gold hem was used to make the color bright and bright, and it had a brilliant decoration effect. From the early Qing Dynasty to Qianlong, this period was the most brilliant era in the history of silk.

The royal family was the largest user of filigree, mainly produced in Suzhou, Wuxian and Beijing. Works of filigree included ornamental scrolls such as scrolls, albums, and hanging screens; and practical scrolls such as clothes, accessories, and fans. It can be said that the method of filigree is a collection of filigree silk of the past dynasties, comprehensive use of various filigree silk, with the assistance of painting, dots, dyeing, etc. This is one of the characteristics of the filigree craft in the Qianlong period of the Qing Dynasty. With the expansion of the scope of embroidery, embroidery crafts have become increasingly commercialized, so embroidery is blooming everywhere.

Jiangsu Su embroidery, Guangdong Yue embroidery, Hunan Xiang embroidery, Beijing Jing embroidery and other major schools had been formed successively. Su embroidery ranked first in the embroidery art world and had a profound influence on the national embroidery. Su embroidery was meticulous and elegant in the court, and folk embroidery was also very fine. Cantonese embroidery was rooted in the folk. Yue embroidery workers were male workers, mostly from Guangzhou and Chaozhou. Yue embroidery was most famous for gold embroidery with floating pads. Mascots such as Long Fei and Feng Wu were popular themes. Xiang embroidery was originated in Changsha.

After the middle of the Qing Dynasty, there was a great demand for embroidery, so it came into being and became an independent embroidery school. The characteristics of color matching in Hunan embroidery were mainly shades and black and white, so elegant as ink painting. On the basis of northern folk embroidery, Beijing Embroidery was formed under the influence of court embroidery and painting academies. Beijing embroidery patterns were realistic, mostly based on Gongbi painting, with subjects such as flowers and fruits, grass insects, and small gardens. The color of Beijing embroidery was fresh and beautiful, and its color was similar to the pastel and enamel in porcelain. Obviously, Beijing embroidery follows the court's artistic expression in many aspects, such as color and composition.

3.7   Bamboo and Wood Carving Crafts

Bamboo carving wood carving art is indispensable in the history of Chinese arts and crafts. The history of bamboo carving in China has a long history, and the Ming and Qing dynasties were the prosperous period of small sculptures. Literati dominated the status of bamboo and wood carving in the Qing Dynasty and became an indispensable companion for many literati lives. Literary writers and painters of the Qing Dynasty loved the carving of Yaxing. Therefore, the bamboo and wood carvings had the elegant atmosphere of scrolls. At the beginning of the Qing Dynasty, the four famous masters of Wu Zhizheng, Feng Xilu, Zhou Hao, and Pan Xifeng were all literati who were good at painting and calligraphy as well as bamboo and wood carving, which had a certain impact on the development of bamboo and wood carving crafts during the Qianlong period.

In addition, the combination of bamboo and wood carving with calligraphy and painting had also become a unique feature of the carving process during the Qianlong period, and at the same time pushed the bamboo and wood carving craft of the Qing Dynasty to its peak. The Ming and Qing Dynasties was a glorious period of woodcarving art. Woodcarving includes architectural woodcarving, furniture woodcarving, and craft woodcarving. It requires technical excellence, and the materials used are mostly rosewood, Huanghuali rosewood, ebony, agarwood, and Nanmu precious hardwood.

In Hanmo Piaoxiang's traditional Chinese culture, the bamboo-carved pen holder was an important tool in the study room. "I'm in the study room using my pen holder, it is my model for learning and I'm going to live with it together." This word described the humble vigor of bamboo, which is the spiritual realm that scholars and scholars aspire to (Fig. 9a,b).

(a)   (b)

Fig. 9. XiLingYinShe

4   Emperor Qianlong's Collection from Bogug

Duobaoge is also called "Baibaoge" or "Bogug". It is a display grid for displaying antique objects, and it can also be used for furnishing antique objects. Duobaoge is unique in that it will create vertical and horizontal spaces, uneven heights, and uneven spaces. Duobaoge is generally small, mostly placed on the table, and belongs to the category of crafts. Bogug racks are generally tall, placed on the ground and belong to the furniture category. Duobaoge, Baibaoge, Boguge, and Bogujia flourished in the Qing Dynasty and demonstrated the Qing style furniture such as existing in the entire palace in Beijing's Forbidden City (Fig. 10). It looks dazzling, just like the contemporary window design, it is really full of decoration.

Fig. 10. National Palace Museum

The materials of Duobaoge include sandalwood, Huanghuali, rosewood, elm, and alder. In terms of craftsmanship, there are gold paint inlays such as inlaying, painting, carving and filling. From the shape, there are rectangle, round, semi-circular, octagonal and other shapes. Some can be covered with a large multi-shelf on a wall, and some can be placed on a table or pocket-type multi-hob (National Palace Museum 2017).

According to the "Internal Affairs Office's Production Office's preparations for keeping records", "Forbidden City Item Check Report" and "The Qing Dynasty's Office's Production Office's Archives According to the General Assembly", the requirements of the Emperor Qianlong for distribution and assignment were recorded in detail. The collection, induction and management of cultural relics had certain regulations. The Emperor Qianlong handled the antiques by himself. Each item in the collection had detailed records of the dimensions, texture (materials), decoration and modeling characteristics. He was extremely particular about each link, and had his style and products for large and small collections. Therefore, during the reign of the Qianlong Emperor, from time to time, artisans were required to make various packaging boxes for cultural relics, and even to re-examine and adjust them. For example, some articles were placed in a single box, and some items were placed in a box. The big ones were such as "Qinglong Emperor Qianlong Carving a Red Sandalwood Nine Dragon Box Hundred Boxes and 100 Pieces" as the (attached 44 pieces of play) Qing dynasty, Qianlong reign (1736–1795) Square red sandalwood curio box with carved dragon decoration (contains 44 curio pieces) (Fig. 11a). The younger one is as the Qing dynasty (1644–1911) Magnificent Standards box with ten bronzes (Fig. 11b), not only must the antiques be exquisite, but even the matching seat and storage had strict requirements to achieve practical and aesthetic functions. The Jade, Bronze and Porcelain in <Bogug> all had their own seats. In particular, the shape of jade ran through ancient and modern times, with various shapes and different sizes. Therefore, in the Qianlong period, artisans coordinated the collection and placement of antiquities. Or stand, such as the Elephant-shaped stand of the Han (Fig. 12a), or a wooden pavilion, such as the Dazai Bell in the Spring and Autumn Period (with a rosewood wooden stand in the Qianlong period of the Qing Dynasty) Spring and Autumn period (770–476 BCE) Zhutaizai bell, with red sandalwood frame made in Qianlong reign, Qing dynasty (Fig. 12b), or lying, such as Gong drinking vessel of the Tang (Fig. 12c), or screen insert, such as Qijia Culture Jade (with a wooden insert screen stand in the Qianlong period of the Qing Dynasty) Qijia culture, c. Millennium BCE Bi disk, with wood frame stand made in Qianlong reign, Qing dynasty (Fig. 13a) Qing dynasty, 18th century Four-tiered lacquer box with encircled flower decoration (contains 32 jade pieces) (Fig. 13b), some boxes such as the Qianlong Dragon Fragrant Imperial Ink (with lacquer box and Lacquer box) Qing dynasty, Qianlong reign (1736–1795) Imperial stick, with lacquer case (Fig. 14), and other packaging shapes for furnishings, etc. All of the above crafts had the royal collection and the box containing the antiquities shine together and shape out Beautiful packaging art in history (National Palace Museum 2017).

(a)   (b)

Fig. 11. National Palace Museum

(a)   (b)   (c)

Fig. 12. National Palace Museum (Color figure online)

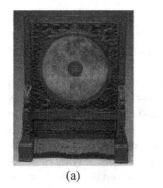
(a)

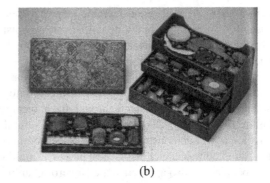
(b)

Fig. 13. National Palace Museum (Color figure online)

Fig. 14. National Palace Museum

5   Conclusion

In the 19th century, German philosopher Hegel regarded art as the spirit of mankind, and his ideas were reflected in the thoughts of Eastern China, which were completely consistent. The Emperor Qianlong had to deal with a lot of complicated government affairs every day, working day and night, and outside the government, but there was no lack of interest in life. Because of his natural love for cultural relics, he had a special room full of rare treasures, and used art to enjoy his leisure time in order to relieve heavy stress. He often appreciated, examined, organized, and stored all of his collections. From this, a variety of packaging designs such as "treasure chests", "best pieces", "multiple grids", "baibaoges", "Baibao Ge" and other packaging designs with different sizes, both for viewing and furnishing were produced.

Without extensive knowledge and considerable cultivation, people cannot engage in collections, especially cultural relics. In turn, through the process of collection, people will gradually accumulate rich knowledge and improve their cultivation. Emperor Qianlong always studied ancient books in his spare time, tasting antiquities (taste and identification of antiquities), viewing ancient calligraphy and paintings (appreciating ancient calligraphy and paintings), and learning about the historical context of cultural relics, the production of cultural relics and the use of materials. During the Qianlong period, there were also considerable achievements in establishing a reasonable work space for handicrafts, including craftsmanship outside the court and the establishment of workshops in the court, all of which were superb crafts produced under this kind of background of craftsmanship.

In addition, from the ancient relics in the treasure chest of the Emperor Qianlong, one can also understand the emperor's intention to reconstruct the historical trajectory and dialogue with the ancients through the ancient relics, and use history as a mirror to build a royal family full of cultural heritage. Cultural relics are not just static furnishings anymore. They demonstrate different aspects of thinking and aesthetic interest. Through the "hundred treasure chests", "best pieces", "multi treasures", sitting or lying or standing, people can also appreciate the emperor Qianlong's emphasis on the collection and combination of cultural relics, as well as the innovation and creativity of Qianlong's art and culture. All in all, we may say that he is the originator of modern window and packaging design.

References

Clunas, C.: Art in China. A Division of Shanghai Century Publishing, Shanghai (2012)

Kleutghen, K.: Imperial Illusions: Crossing Pictorial Boundaries in the Qing Palaces, pp. 271–274 (2015)

Kleutghen, K.: Chinese occidenterie: the diversity of "western" objects in eighteenth-century China. Eighteenth-Century Stud. 47(2), 117–135 (2014)

Ledderose, L.: Ten Thousand Things: Module and Mass Production in Chinese Art. SDX Joint, Beijing (2012)

Lai, H.: Japanese products and everyday life in Suzhou, 1736–1795, no. 63, pp. 1–48. Institute of Modern History, Academia Sinica, ©Academia Sinica Institute of Modern History (2009)

LNCS. https://www.sothebys.com/zh-hant/2020/1/25

National Palace Museum: Story of a Brand Name the Collection and Packaging Aesthetics of Emperor Qianlong in the Eighteenth Century, pp. 17–18, 230–231. National Palace Museum, Taipei (1997)

National Palace Museum: The Emperor Kangxi and the Sun King Louix XIV, pp. 17–18, 276. National Palace Museum, Taipei (2011)

Shi, C.: Perspectives of Qing court on the wooden nesting cups from the "western ocean". J. Art Stud. 21, 30–43 (2017)

Wu, R., Wang, D.: A preliminary study on the consumption and collection of local goods in Qianlong dynasty, no. 89, pp. 1–41. Institute of Modern History, Academia Sinica, ©Academia Sinica Institute of Modern History (2015)

Yu, P.: The appeal of connoisseurship: eighteenth century ceramic manuals and realated question. Natl. Palace Museum Res. 22, 133–166 (2004)

Zhou, J.: Aesthetic cultural features of Ming and Qing crafts from porcelain and furniture. J. Zibo Univ. 4, 84–88 (2000)

Transforming Chinese Cultural Features into Modern Product Design

Yukun Hu[1(✉)], Suihuai Yu[1], Yafang Ju[1], Dengkai Chen[1], Weiwei Wang[2], Yanpu Yang[3], and Chen Chen[1]

[1] Northwestern Polytechnical University, Xi'an, People's Republic of China
huyukun0315@163.com
[2] Shaanxi University of Science and Technology, Xi'an, People's Republic of China
wangjiarui1202@163.com
[3] Chang'an University, Xi'an, People's Republic of China

Abstract. With the advent of experience economy, people are more pursuing the cultural meaning of products and spiritual satisfaction brought to them. The combination of design and culture is a trend of economic development. In this context, products with cultural characteristics have started hitting the market. Chinese traditional culture, with its beautiful and primitive visual arts and crafts, provides great potential for improving design value and gaining recognition in the global market. There is evidence that Chinese traditional culture is likely to become an important cultural element in future design applications. The purpose of this paper is to explore the connotation of Chinese traditional culture and explore its cultural characteristics. This article attempts to describe how to transform Chinese cultural features into modern product design to provide a good user experience and meet the users' needs.

Keywords: Experience economy · Cultural experience · Cultural product design · Cultural application · Chinese traditional culture

1 Introduction

Experience economy is gradually developing into the fourth stage of human economic life [1, 2]. Under the this background, customers are no longer satisfied with the basic material enjoyment brought by products, but pay more attention to the cultural connotation and artistic value of products to experience the emotional satisfaction. The emotional change of users is the most direct manifestation of user experience [3]. In order to meet this new demand, it is more and more important to improve the cultural connotation of products. Through the development and redesign of cultural resources, the product can be endowed with cultural attributes and the connotation of design can be enriched to meet the requirements.

At the same time, designers recognize the importance of linking products with cultural characteristics to provide unique value, and incorporate local culture into the design in order to create product identification strategies in the global market [4, 5]. Industrial

P.-L. P. Rau (Ed.): HCII 2020, LNCS 12192, pp. 313–324, 2020.
https://doi.org/10.1007/978-3-030-49788-0_23

design plays an important role in this process, embedding cultural elements into products and enhancing their cultural value in a competitive global product market. Therefore, it can be said that designing products with local characteristics to emphasize their cultural value has become a key issue in the design process [6, 7].

Therefore, this paper proposes a method to integrate culture into products. Starting from user experience, this method conceives how to transform cultural elements into product design elements based on the three levels of user experience, and carries out cultural design of products according to the conception. We perform a user study to evaluate our method. The evaluation and results confirm the convenience of our method for cultural product design.

In the remainder of the paper, Sect. 2 reviews the way how the culture is integrated into the product. Section 3 explains the motivation of proposing the method of this paper. Section 4 explains the implementation process. Section 5 presents a case study to demonstrate its effectiveness. We conclude and describe the future work in Sect. 6.

A central idea of our work is trying to find a way to transform cultural elements into product design elements according to the three levels of user experience, and assist designers to design cultural products, so as to create a good cultural experience for users. We hope the methods and steps mentioned in this paper can help designers interpret and sort out culture elements, and put it into the design practice. The method can not only provide reference for the design of cultural products, but also enhance the cultural attribute of products. Moreover, products designed in this way would provide users with multi-dimensional cultural experience.

2 Related Works

Culture plays an important role in the field of design, and it will become a design trend in the global market to incorporate more cultural elements into products. We need to better understand our own culture, not only to participate in global markets, but also to develop local designs [8–14]. The importance of cultural research has been repeatedly demonstrated in various fields of technical design and researchers have also made many explorations. Wu et al. put forward a design process, which includes collecting related materials, analyzing materials, brainstorming, hand drawing, printing a 3-D model, confirming the final design, based on the proposed appearance–behavior–culture (ABC) theory. And the effectiveness is verified through the design case of glass teapot, proving that the process can be used to enhance cultural factors in product design [15]. Huang et al. emphasized the impact of "benevolence" and "propriety" on kids' home products design and proposed some suggestions, which explored how to apply the essence of traditional Chinese cultural values in user experience design of kids' home products [16]. According to the characteristics of the development of creative museum cultural products, Song's article discussed the whole development process of the "Ceramics and Wood" hanger series of Nanjing museum. The development process included product issues, concept visualization, and design commercialization as well as exploring the product lifecycle [17]. Designers [18] obtain important cultural elements from collecting and analyzing native cultural information and then design their products after brainstorming for the correlation between cultural elements and popular daily products.

Instead of directly applying cultural into design practice, some researchers [19] first translate cultural knowledge into useful design information and then apply it in the three steps of modern product design. They design the incense burner based on the image of Mazu.

In recent years, researchers have been gradually considering the experience aspect of the design of cultural creative products. Yang's article introduces the perspective of emotion reflection to design experience based Museum Cultural creative products. From the cultural emotion connotation of cultural creative products in museums, this paper discussed the value and application of emotional reflection in the design process [20]. Sun et al. introduced a design method to optimize products from the perspective of user experience. Based on the experience design theory, traditional culture is applied to modern product design, taking cultural creative product design of Chinese Museum for example. But the interaction of three levels, aesthetic level, behavior level and reflective level, is not considered [21]. In fact, all levels of user experience often do not exist in a single product. Meanwhile, from the perspective of user experience, the exploration of integrating culture into products is relatively small.

Based on previous studies, this paper offers a method of integrating culture with products. First, literature review is used to collect and sort out elements of a chosen culture. Secondly, brainstorm and visualize the results in the form of brainstorm map. Affinity Diagram [22] is used to summarize the results of brainstorming and sort out more useful information. Then, the product culture design is conceived based on the three levels of user experience. Finally, cultural products are designed according to the concept to create multi-dimensional cultural experience for users.

3 Motivation

3.1 Assisting Designers to Design Cultural Products

Effective creative design strategy of cultural products can help designers to make traditional cultural images and modern design complement each other. This research can assist the designer to choose the appropriate design means and methods in the more concrete process. By selecting cultural image elements at different levels and aiming at different types of consumers, it will help develop products with Chinese cultural characteristics in an all-round way and promote the vigorous development of related cultural industries.

3.2 Inheriting and Enriching the Forms and Connotations of Traditional Culture

The use of traditional cultural images in the design process can enable designers to understand and reflect on traditional culture. The cultural products designed in this way contain the designer's own thinking and act as a bridge between designers and users, helping them communicate with each other and giving users chances to experience the unique cultural image connotation. Through the continuous communication between people and products, the traditional culture can be well inherited.

4 Proposed Method

The paper presents a method that integrates culture identity into products design to provide effective references for designers. The flowchart is shown in Fig. 1. First, a specific culture is sorted out and interpreted. Second, based on the three levels of user experience, the conception of cultural design are sorted out, and then the conceptual design is carried out according to the conception of cultural design, which involves the cultural design of product form elements, the cultural design of product function interaction so as to evoke the emotional experience including memories or thoughts of users. Our approach consists of three main steps as follows.

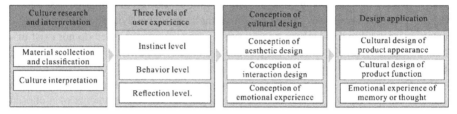

Fig. 1. Flowchart

4.1 Research and Interpretation of Culture

Cultural research needs to be based on huge data and information. First of all, the cultural materials are collected and classified through books, papers, journals, etc., then the specific culture is analyzed and interpreted by brainstorming, the brainstorm map is also drawn in a visual way to record brainstorming results in detail. By thoroughly interpreting culture connotation, we can get as many usable design elements as possible.

4.2 Conception of Cultural Design Based on User Experience

Before being applied to practice, the determined cultural elements are still scattered and unrelated to each other. The construction of design conception is an effective way to integrate and reorganize these cultural elements to the product design and provide users with cultural experience in an optimized way, which provides guiding information for the design and implementation of cultural and product integration. Conception of cultural design links culture and design practice.

Conception of Aesthetic Design. The conception of aesthetic design is the basis of the cultural design of product appearance, which creates the instinctive experience under particular cultural background for users. In other words, cultural elements are applied to product design by using the form elements of design, and aesthetic innovation is realized in the aspects of modeling, color and material quality by following the laws of formal beauty.

Conception of Interaction Design. It is cleared that how to let users experience culture when using products, get surprised by culture elements under the products meeting the availability, applicability and usability. In addition, to be based on ergonomics, it is supposed to pay attention to the cultural elements which can be used for enhancing users' experience. What's more, the details occurred in the communication and interaction between users and the product are also should be valued, so that users can feel the comfort and the cultural features of the product at the same time.

Conception of Emotional Experience. Conception of emotional experience provides the basis for the reflection layer design of product culture. Emotional experience is an abstract feeling, to some extent, it comes from user's perception of the appearance and using feelings of the product, which can enable users to reflect on the culture contained in the product after using the product, and arouse the deep memory or emotional resonance of users.

4.3 Design Application

Corresponding to the level of user experience, the application level of traditional culture in product design can also be divided into three levels.

Cultural Design of Product Appearance. The cultural design of the product's appearance refers to the application of visual elements peculiar to cultural elements, including color, texture, shape, decoration, line, detail processing, construction and composition, on the product, which contains both a detailed decoration on the surface of the product and an overall application of elements.

Cultural Design of Product Function. Take an in-depth understanding of the skill, function, operability, convenience of use, safety, combination relationship and other attributes contained in cultural elements, then realize them in the product application.

Emotional Experience of Memory and Thought. The traditional spiritual emotion is placed in the design expression and the spiritual connotation contained in the cultural elements is conveyed through the product design. In results, cultural creative products at the reflective level can easily arouse people's inner cultural identity.

5   Case Study and Evaluation

To verify the validity of the proposed method, the smart watch design project was introduced for verification. At present, the phenomenon of market homogeneity of wearable products is becoming common, especially in the field of smart watches. Through the exploration of Chinese traditional culture, the combination of cultural concept and wearable products endue products with cultural meaning and the competitiveness of differentiated products.

Five participants were recruited to design the wristwatch. The test results show that the method achieves a satisfactory interpretation. The evaluation suggests that the scheme designed by this method contains cultural connotation and can satisfy the users' requirements of culture.

5.1 Research and Interpretation of Kungfu

Literature review was used to obtain information about Kungfu through books, papers, journals, academic websites and other means. After the collection and arrangement of related materials of Kungfu, the author analyzed and interpreted Kungfu culture, and combined personal thoughts and opinions to the brainstorm map, visualizing the results in detail. Figure 2 shows the brainstorm map of one participant (participant number P3). See English translated version in Appendix A.

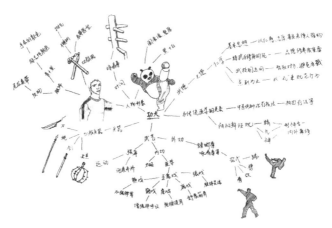

Fig. 2. Brainstorm map of Kungfu

5.2 Design Conception

Affinity Diagram was used to summarize and group the products of brainstorming according to the three aspects of aesthetic experience, interaction experience and cultural emotion experience, as shown in Fig. 3. Then, based on the three levels of user experience, the product culture design is conceived. See English translated version in Appendix B.

Based on the three levels of user experience, the cultural design conception referring to some ideas in the affinity graph was completed. Figure 4 and Fig. 5 show the conception of the two participants (participant number P3, P5).

5.3 Design Application

In this part, participants should pay attention to how to integrate the design elements of Kungfu culture into the product according to the design conception, when drawing sketches, and realize the user's three levels of experience on the basis of integration. Concrete sketches are able to iterate quickly, refine ideas and generate new ideas, and also help designers to refine the shape, color, material and pattern of design objects quickly.

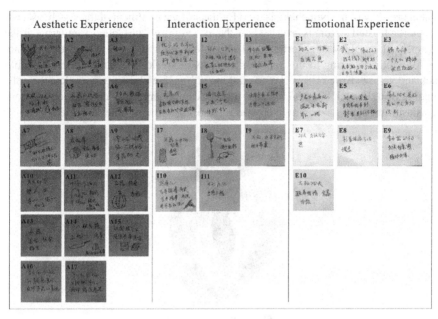

Fig. 3. Affinity diagram

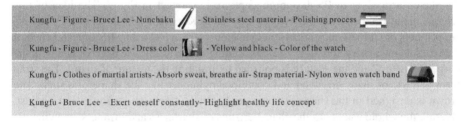

Fig. 4. Conception of P3

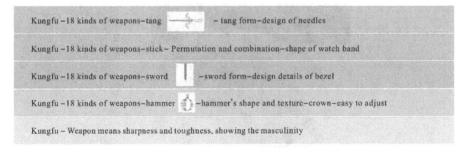

Fig. 5. Conception of P5

Sketch according to P3 design conception, is showed in Fig. 6. Nunchaku has become the symbol of Bruce Lee because of his excellent technique of it. In this scheme, the rough wooden property of the Nunchaku is used for the surface treatment of the watchcase, and the metallic frosted texture is used to show the masculinity. The clothes for martial artists are mostly made of sweat absorbing and ventilating materials. The watch strap is made of nylon cloth, referring to this feature, which is more skin friendly and comfortable. Black and yellow, Bruce Lee's classic clothing colors, are chosen to complete the color matching of the watch band. Generally, it satisfies the personal identification of sports enthusiasts and people who are interested in Kungfu, showing a healthy life style.

Fig. 6. Sketch of P3

Sketch according to P5 design conception, is showed in Fig. 7. The idea of this scheme mainly comes from the weapons related to Kungfu. Eighteen kinds of weapons are symbols of Chinese martial arts and weapons, which has a long history, and each of them has its own characteristics. The surrounding of the watchcase is decorated with the shape of "sword" and the surface is polished to imitating the texture of sword. The strap form is made of many arranged and combined cylinders like the shape of a stick. The crown is designed in the shape of "hammer", which is convenient for users to adjust. Overall, the whole watch has the characteristics of sharpness and firmness, showing the masculinity.

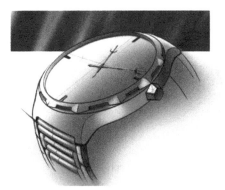

Fig. 7. Sketch of P5

5.4 Design Evaluation

8 users evaluated the schemes (23–34, 4 female). Liker scale method was used to score whether the scheme conforms to the cultural image of Kungfu. The 7-level attitude is: very consistent/3, consistent/2, fairly consistent/1, general/0, hardly consistent/- 1, inconsistent/- 2, very inconsistent/- 3. The evaluation results are shown in Figure n. More than 60% of participants thought that these two design schemes are fairly consistent, or consistent, or very consistent with cultural images (Table 1).

Table 1. Evaluation results

	U1	U2	U3	U4	U5	U6	U7	U8
	3	2	-1	3	2	1	0	2
	2	2	1	0	-1	0	3	1

6 Conclusions and Future Work

Taking user experience as the core, this paper proposes a method of cultural products design. It deeply excavates the needs of user's experience and satisfies their deeper needs for culture when defining the cultural connotation of products, Cultural design clues are constructed from aesthetics, using experience and feelings. It shows how to attract users by the external cultural features of products so that users desire further contact with the product. At the same time, it makes users feel the culture characteristics when interacting with the product and cause emotional resonance and value identification of users, creating a physiological, psychological and other multi-dimensional spiritual and cultural experience for users. The integration of culture and products is an effective method to enrich the cultural connotation and increase value to products, which realizes the organic unity of cultural value and practical value of products, providing valuable reference for designers to design successful cross-cultural products.

In the future research, in addition to extensive literature review, field investigation and interviews are suggested to accurately understand the Kungfu culture and avoid incorrect interpretation when transforming cultural characteristics into modern product design.

Appendix A

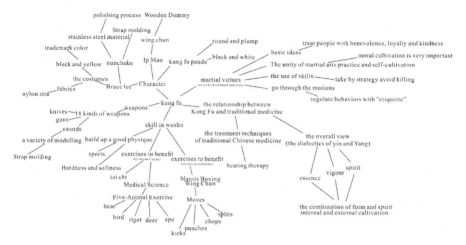

Brainstorm map of Kungfu

Appendix B

Aesthetic Experience	Interaction Experience	Emotional Experience
A1.Wushu movements, watch in dynamic style A2.Using the silhouette of martial arts practitioners to design hands A3.boring, pointer A4.The sweat of the martial artist, a drop of water, crown A5.The texture of the weapon, rough, Metal frosted A6.Kung Fu Panda, black and white, round and rolling A7.The cloth strip is tied to the arm, so the watchband can be designed this way A8.Tai Chi ,black and white A9.Bruce lee' nunchaku, stainless steel, metal polishing A10.Wushu Moves splits, chops, punches kicks A11.Ip Man' Wooden Dummy, rough, frosted, well-behaved A12.weapon--an ancient weapon, composed of two iron balls fixed on a long iron chain--crown A13.Weapon shape, sharp angle, spearhead A14.18 kinds of weapons A15.The shape of the double stick can be used for the shape of the watch band A16.The color of Bruce Lee's clothes can be used A17.The texture of Bruce Lee's clothes, woven watch band	I1.The strap can be made of new materials I2.Clothes of martial artists , absorb sweat, breathe air, and the strap material should be the same I3.Bruce Lee's clothes , fabric, skin friendly, Nylon Watch Band I4.Five-Animal Exercise, the buckle of watch in animals shape, interesting to tie I5.Nylon strap, avoid the fine hairs on the wrist being caught, good experience I6.in the use of watches, Personalization of hand gestures I7.Weapon handle, antiskid. watch crown I8.watch crown, Easy to rotate I9.Tai Chi, conquering the unyielding with the yielding I10.Palm and Fist Salute, posture, buckle design I11.frosted texture, good feel	E1.Kung Fu - exert oneself constantly E2.Martial arts - "stop fighting", Kungfu itself is not for attack, but for self-cultivation E3."Spirit", have a good mental state E4.Product differentiation, Satisfy the psychology of seeking novelty E5.Weapon, masculine , show masculinity E6.Personal identification for Kung Fu fans E7.National pride E8.Highlight healthy life concept E9.Bruce Lee, national pride, cultural confidence E10.Tai Chi, Keep in good temper, implicit and introverted

Affinity Diagram

References

1. Yu, J.: Product design in the era of experience economy (2011)
2. Cao, X.-D., Su, X.-P.: Analysis of product design in the age of experience economy. Packag. Eng. 242–244 (2018)
3. Klinger, K: Encyclopedia of multimedia technology and networking. Choice Rev Online **43**, 43:3770 (2006)

4. He, J., Wang, C.L.: How global brands incorporating local cultural elements increase consumer purchase likelihood. Int. Mark. Rev. **34**, 463–479 (2017)
5. Jiang, J., Kang, Y., Liu, C., Zhang, Y.: Elements analysis of Shaanxi' s cultural tourism products based on the perspective of the whole industry chain, pp. 970–976 (2016)
6. Hsu, C.-H., Lin, C.-L., Lin, R.: A study of framework and process development for cultural product design. In: Rau, P.L.Patrick (ed.) IDGD 2011. LNCS, vol. 6775, pp. 55–64. Springer, Heidelberg (2011). https://doi.org/10.1007/978-3-642-21660-2_7
7. Lin, R.T.: Transforming Taiwan aboriginal cultural features into modern product design: a case study of a cross- cultural product design model. Int. J. Des. **1**, 45–53 (2007)
8. Liu, Y., Zhang, C., Zhou, L.: Adoption of Chinese ink painting elements in modern poster design inspired by innovation research on physical and chemical reactions. Curr. Sci. **108**, 2017–2022 (2015)
9. He, J., Wang, C.L.: How global brands incorporating local cultural elements increase consumer purchase likelihood: an empirical study in China (2017). https://doi.org/10.1108/IMR-08-2014-0272
10. Lin, R., Sun, M.-X., Chang, Y.-P., Chan, Y.-C., Hsieh, Y.-C., Huang, Y.-C.: Designing "culture" into modern product: a case study of cultural product design. In: Aykin, N. (ed.) UI-HCII 2007. LNCS, vol. 4559, pp. 146–153. Springer, Heidelberg (2007). https://doi.org/10.1007/978-3-540-73287-7_19
11. The cultural product: integration and relational approach (2014). https://doi.org/10.4018/978-1-4666-5007-7.ch001
12. Dong, H.: Application of regional cultural elements in the design of wenbo culture product. In: Appl. Reg. Cult. Elem. Des. Wenbo Cult. Prod., 3rd International Conference on Arts, Design and Contemporary Education (ICADCE 2017) Application, pp. 395–397 (2017)
13. Yan, F., Wang, S.: Research on the heritage and development of traditional cultural elements in product design. Bol. Tec./Tech. Bull. **55**, 554–560 (2017)
14. Wu, X., Xie, J., Mao, Y.: The application of Chinese elements in the product design. Asian Soc. Sci. **4**, 109–111 (2008)
15. Wu, T.-Y., Huang, W.-H.: Appearance–behavior–culture in creating consumer products with cultural meaning meant to evoke emotion. In: Rau, P.-L.P. (ed.) CCD 2018. LNCS, vol. 10912, pp. 245–253. Springer, Cham (2018). https://doi.org/10.1007/978-3-319-92252-2_19
16. Rau, P.L.P. (ed.): CCD 2015. LNCS, vol. 9180. Springer, Cham (2015). https://doi.org/10.1007/978-3-319-20907-4
17. Song, Y., Li, M.: Research on cultural and creative product development based on museum resources. In: IOP Conference Series Materials Science Engineering (2018). https://doi.org/10.1088/1757-899X/452/2/022090
18. Lee, Y.: Exploration of local culture elements and design of cultural creativity products products (2013). https://doi.org/10.1080/09720510.2010.10701505
19. Yeh, M.-L., Lin, P.-H.: Applying local culture features into creative craft products design. In: Rau, P.L.Patrick (ed.) IDGD 2011. LNCS, vol. 6775, pp. 114–122. Springer, Heidelberg (2011). https://doi.org/10.1007/978-3-642-21660-2_13
20. Yang, L., Yu, W., Jiang, S., Jia, S.: The Application of "Emotion Retrospection" in the Design of Museum Cultural Creative Products. In: Marcus, A., Wang, W. (eds.) HCII 2019. LNCS, vol. 11583, pp. 547–556. Springer, Cham (2019). https://doi.org/10.1007/978-3-030-23570-3_41

21. Sun, X., Jin, W., Li, C.: Research on the design of Nanjing museum cultural and creative product from the perspective of experience. In: Marcus, A., Wang, W. (eds.) DUXU 2017. LNCS, vol. 10290, pp. 529–539. Springer, Cham (2017). https://doi.org/10.1007/978-3-319-58640-3_38

22. Walny, J., Carpendale, S., Riche, N.H., Venolia, G., Fawcett, P.: Visual thinking in action: Visualizations as used on whiteboards. IEEE Trans. Vis. Comput. Graph. **17**, 2508–2517 (2011)

Play It My Way: Participatory Mobile Game Design with Children in Rural Nepal

Dev Raj Lamichhane[(⊠)] and Janet C. Read

University of Central Lancashire, Preston PR1 2HE, UK
drlamichhane1@uclan.ac.uk

Abstract. This paper describes a participatory study with 35 children, aged 9–12, from a rural Nepalese village in the Himalayan Mountains where they were asked to draw a picture or pictures of a game that they would love to have in a digital pen pal app that is being developed. Altogether 58 game designs were collected. Seven of them were new game ideas with some reflecting the children's own culture and community. There were also inevitably some cases where the children appeared to have copied from each other.

This paper gives details on how the study was conducted, it gathers together the full results and then reflects on both the worth of the drawn images for the game design and the usefulness of the drawing-based design method. The paper also analyses and critiques the effectiveness of the participatory design method with children.

Keywords: Participatory design · Child computer interaction · Game design · HCI4D

1 Background and Related Research

With conflict in many parts of the world, helping children to learn about the world, by communicating with other children, is timely. One way of doing that is with the help of pen pal activities. Using pen pals in childhood used to be a common activity during the 1950s and 1960s where children used to make friends from different parts of the world, and learn about different cultures and languages, through letter writing. Shulman explains in his paper how the pen pal activity initiated in school managed to develop a safe, close and open relationship which continued to adolescence [1]. This really helps to bring the world together.

Times are quite different now and the writing of letters to people in other lands is much scarcer. Technology has taken a big leap and people can now talk to anyone they want with the help of it. There are several applications where people can make friends and talk to each other and while some are reported for adults [2], pen pal like systems for children are uncommon and are not seriously researched. In this work we are exploring how such a product, a digital pen pal app (hereinafter referred to as 'Digipal') can be developed and what functionalities it should need. The app is being iteratively developed whilst research is also taking place.

© Springer Nature Switzerland AG 2020
P.-L. P. Rau (Ed.): HCII 2020, LNCS 12192, pp. 325–336, 2020.
https://doi.org/10.1007/978-3-030-49788-0_24

One difficulty for this kind of application is that the children from different countries often do not have a shared language. One solution to this is to create applications that use English as this is a widely used second language – but this then creates an asymmetric conversation with the power being given to the person with the stronger ability in the English language. In cases where intended communication is between a child with English as a second language and with a child with English as a first language – the latter has the advantage; but also, this model diminishes local languages by relegating them to a secondary position. Our aim, in our work, is to create a method by which children can talk one with the other in their own languages with no preference being given to one language over another.

This Digipal app, which is being built at the moment (see Fig. 1), is supporting research to understand how children from different parts of the world can make friends with other children in different places, each using their first languages. In making friends in this way, the children will learn about each other's different cultures and lifestyles. The app will let the children send letters to other children in their own languages which will be translated with the help of the Google API which is integrated within the app.

Children are also participating in the design process, contributing to requirement gathering and critiquing the product and giving feedback. The inclusion of children in the design process is really important to ensure that the app is attractive and useful for the children who will use it.

Fig. 1. Digital pen pal app

Having children participating in the design process is advantageous as they think differently than adults. Read et al. [3], explained in their paper that participatory methods for children to contribute to design are beneficiary. This type of participatory design with children has increased in recent decades with an interest of taking ideas from the children as the social actors and listening to their experiences rather than treating them as the passive part of the society and as pre adults waiting to be competent. Researchers argue how children can aid understanding and generate data that cannot be achieved by working with or on adults [4, 5]. The premise is that children should decide on what they want in the system rather than the adults making the decisions for them.

When children act as co-researchers they help the principal researcher in the process to find a clear view of the design matters or to see the children's perspective. For example, Honkanen et al. [6], worked with children to research on subjective well-being in residential area. Children (aged 2–13) along with young people (aged 14–16) participated in the study where they were asked to take pictures of the places they spent time in as well as the situations that made them feel good. The children and young people were free to take any pictures they liked. This provided them with freedom for their actions. In this way, children were participating in research rather than being subjects or objects of research as used to be in traditional approaches. The researchers used a socio-cultural approach believing that the time, place and actors each affect the participants' experience and perspective.

For the initial requirement gathering and development purpose for this Digipal app, only children from Nepal and the United Kingdom are being designed for. For this inclusion, an interaction problem is encountered as due to the time difference between these two countries, the communication between the children doesn't happen at the same time. A letter sent today can only be read and replied to on another day. This makes the letter exchanging activity a bit boring because of the waiting.

According to Yarosh et al. [7], communication in different place and time is a complex phenomenon especially when there is a time zone difference. Their research included an experiment where children from different locations were enabled to video conference with one another to engage in social free play. Though the children hesitated and were self-conscious initially, they were later seen to be enjoying, interacting and reacting to each other's activities and being creative during the session. This gives an idea on how children can interact with each other and can have a playful experience even when they are not present in the same place. More research is needed to bridge different place/different time, however. One way to deal with waiting is to include games within the app. Such games will add the fun and the engaging aspect to the app to make the communication much more sustainable. As an example, children could play the game and challenge their partners, or share their scores, until they get their reply thus keeping the children within the environment.

As the app is developed for the children, it is very important and logical to include children to get the ideas for the games too. That is why a participatory design method is used to get ideas for games.

Sim et al. [8], explored how a group of children from England developed a fun game for Ugandan children. In the study, children developed a health-related game intended to elementary school children from Uganda. There was evidence that the designer children really thought about the target group and that this was a successful process. In another similar research, children participated in a game design session where they were told that the game was being designed for other children but actually, they were designing for themselves. The study showed that the children were able to participate in such activity and really come up with new and cool game ideas [9].

Gathering game design ideas with the help of drawings is encouraged because, according to Salmon and Lucas [10], art-based methods are very suitable for children because children find it hard to express things orally. Drawings are a way to express how the children feel about something, convey their point of view and ascertain their development. Drawings also richly explain how children see the world and they are a familiar task to them and they usually like to do it too [11]. Children naturally start to draw as soon as they can hold a drawing utensil. This is their first representational way to express and communicate with the world which is why drawing is a great way for children to express the views and interpretations of their experiences and also their hopes and fears [12].

For any drawing activity with children, it is important that they draw freely without constraints. On the study by Villarroel et al. [13], children were asked to draw pictures of plants. The children were given instructions before the drawing activity started about what was expected but once the activity started, they were left with full freedom on what they wanted to draw. The authors found remarkable results on how much children can express themselves. Kullman [14], also found, in his photo and video taking study, that if children are left free to do whatever they want, they will try new things, be more creative, and even find new ways of using the materials provided.

Using drawings to capture 'cultural expression' has been a common activity in the research [15]. Barraza [16], gathered 741 drawings from 247 children aged 7–9 where they were asked to draw about the perception, expectation and concerns about the environment. This work showed that more that 37% of the children expressed deep environmental concerns. That shows children are experts in their surroundings and the environment around them. Other researchers described this in their paper that the children seem to know a lot about their surrounding and their culture and they can be taken as the experts in those fields as they were clearly showing that in their drawings [17].

2 Research Design

The question in this paper was '*to what extent, and with what results, could a group of children in Nepal contribute ideas towards an app*?' The novelty was that very few studies have looked at participatory design with children from such communities. Children in Nepal typically have highly structured education and their experiences of working freely on creative activities was expected to be different from those of UK children. A design aim for the paper was '*to also uncover some ways to culturally align any included game with the Nepalese context*'.

Given that the work was also being situated in a rural part of Nepal, another question was '*to understand game preferences of children from rural Nepal.*' This paper describes design work from a big fieldwork study where the children from a small school in a rural village of Nepal carried out a number of activities. The study was conducted by the first author and the other author contributed to the analysis and paper writing.

2.1 Participants

Children were chosen by the head teacher of the school; he chose 35 children within the age group of 9–12. Each child and his/her parents consented to take part in this study. The study was clearly explained to the children before the activity started. The work was covered by ethical approval from the host university and approved by the school board. The children had limited English and the school was situated several hours into the hills surrounding Kathmandu. The researcher had to take a bus and walk several miles to get there.

2.2 Tools

As explained above, the app development process was incremental. At the time of this study, the app was in its second version (shown in Fig. 1). The children used this app, on an Android mobile device, for the first phase of the activity. In the second phase, they used a sketched prototype of a mobile phone (shown in the Fig. 2), and were given a pencil, some colours, an eraser, and a sharpener for the drawing activity.

Fig. 2. Mobile paper prototype

2.3 Procedure

The participating children were taken into a classroom which was set aside for the sole use of the researcher for the study. The participating children did a number of different activities in a full day session. This paper describes two of these activities associated with app design.

Activity 1. In this activity, the children used the second version of app (Fig. 1.) installed in an Android device. Due to the researcher only having a limited number of devices, only five participants took part in this activity at a time. Children were each given a mobile phone with the app running and then they would select their country flag which would take them to the second page where they would write a letter to an imaginary friend in the UK in their own language (Nepalese). They would then give the phone back to the researcher and go back to their seat to continue with next activity.

Activity 2. After everyone had used the app, the children were asked if they liked to play games on the mobile phone and everyone answered "Yes". Then they were asked if they wanted to have a game in the app they were using before, and the answer was again a unanimous "Yes". After that, each child was given the paper mobile prototype and all the drawing tools required as explained above. Then they were asked to draw a picture or pictures of a game that they would like to have in the digital pen pal app. They were encouraged to be creative and to think about a new game. They were also asked not to copy from the person sitting next to them. They were also asked to name their imagined games, describe in words how the games would be played, including how to score points if relevant, and how to win. No time limit was imposed, and no instructions were given. The children drew freely and handed the drawings to the researcher once they finished.

3 Results

All the children were able to draw some games. Altogether 58 game designs were collected from 35 children out of which 7 were new game ideas. That means some of the children gave more than one game design idea. The data shows that 18 of the children drew one game, 14 drew 2 games and 3 drew 4 games. The average number of games drawn by the children came to 1.65. This multiple game design might be because they were drawing freely without any time limit, so many had enough time to draw more than one. The results seen in [14] are similar as the children were drawing freely that encouraged them to be more creative. The results are further explained in following points.

3.1 Games Frequency

The games were examined and coded using thematic coding to bring similar games together. This coding resulted in 20 unique game design themes with 6 of these being drawn by more than one child. These six are shown in Table 1.

Table 1. Frequency of the type of games

Game	Number of children
Driving	19
Snake	14
Car race	4
Bubble Shooter	2
Shooting	2
Break Wall	2

Driving and Snake games were both very popular with 19 and 14 of the drawings being these respectively. Examining the similarities in drawings it did appear that this was sometimes because the children were sitting next to each other. This is one of the disadvantages of group work for research. It is noted that a similar observation was seen in [6]. However, [11], argues that the drawings still represent personal expressions and thoughts so should not be summarily dismissed. Even if they were influenced from another, the drawings looked a bit different anyway with some extra or less details on them. Other researchers add to this aspect mentioning that children's drawings' results are easily influenced by so many factors like what others draw or say in the group [18] (Fig. 3).

3.2 Instructions and Result

Children were specifically asked to name the game, write how to play, and detail how to score. Results gathered are represented in Table 2.

Seven of the designs contained all three of the descriptions in writing; these could be said to be complete. Naming the games appeared to be quite easy for the children – as evidenced by the large number of games (44) that had names. Much fewer children were able to describe how to play (12) and how to score (17). Regardless, only nine (c15%) of the drawings didn't have any detail on them. It could be that was because the children didn't know what to add but alternatively it could also be that these children thought the drawings are self-explanatory.

3.3 Game Designs Inspired by Culture and Daily Lives

Seven of the drawings seemed to be influenced by Nepalese culture. For example: one was a drawing of a card game that is played only in Nepal. Another example included a Nepalese flag on the finish line of a car race. The roads for the vehicles were curvy which is normal for the context of Nepal and one child drew a game that involved fighting a tiger which was possibly a result of the village they lived in being close to the forest where tigers were found. These cultural images confirm research that reports that children include social and cultural elements in their drawings [15] (Fig. 4).

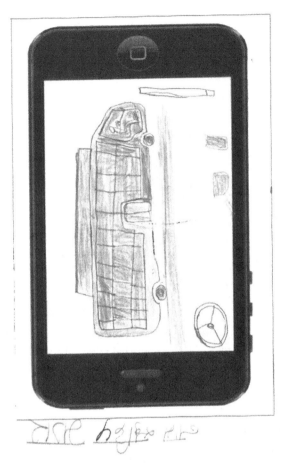

Fig. 3. Most popular driving game picture

Table 2. Number of games with the details

Instruction	Number of game design containing
Name	44
How to play	12
How to score	17
None	9

There were some cases where the children designed a game which seemed quite well related to their daily life or to something they love to do. Examples included games of cricket which is very popular in Nepal, Slingshot Catapult games, Carrom board, Snakes and Ladders board game, Cycling, Card games etc. These are all based on games or activities the children play or do in their daily lives.

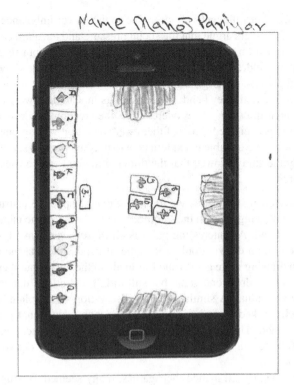

Fig. 4. Card game: played only in Nepal

4 Discussion

Returning to the research question '*to what extent, and with what results, could a group of children in Nepal contribute ideas towards an app?*' it was seen in the findings that children could in fact do this activity and they were able to provide some, if not very imaginative, results.

It was found that the activity did '*uncover some ways to culturally align any included game with the Nepalese context*', and some of the game preferences of children in Nepal could be said to have been uncovered although the issue of sharing and perhaps copying was possibly a confound to this.

4.1 Process

Children Could Do It. Sometimes participatory methods are criticized by researchers saying that they are largely managed by the researchers not the participants. Researchers are concerned that participatory methods are in danger of being seen as a 'fool-proof' technology, writing: 'we have not been arguing against participatory methods as such- we have no particular issue with researchers asking children to draw, dance or build – we are simply concerned that such methods are not used naively.' Methods used for child participatory design are equally problematic being ethically ambiguous to other research methods. The researcher influences children due to the power imbalance to participate in certain ways [19]. Others argue that the social pressure and the adult controlled study session has some effect on the responses of the children [11]. In our research we have tried to ensure that children were given the full power on what to draw and what to include in their drawings. After giving the mandatory instructions for the activity, the children were left free until they handed the drawings in. Children were able to express their desires through drawings. They were given the freedom to draw anything they liked, and everyone participated. Most of them were able to give a unique name to their game and some of them were able to explain how to play and how to score the game too. Most importantly the study showed that the drawing method can be used to get design ideas from children.

Game Details. Referring to results, we found out that, despite the instructions clearly asking, only seven drawings had all the 'required' details – viz. game name, how to play and how to score detail. An interesting point is all those seven drawings are from the children who focused on drawing only one game. It is possible that the other children were focusing on drawing more games and forgot about the details but there were eleven children who also only drew one game but still didn't add all the details, so, it really depended on the individual. A similar kind of observation was explained in paper [8], when the researcher asked Ugandan children to draw an educational game design for children from England. The researchers were expecting to get educational stuff, game design and interface design from the children but the children only delivered a subset of these requirements.

While children almost all named their games, many seemed to struggle to explain how to score and how to play. This could be because the games they know have English language interfaces which they found difficult to express in their own language. As an example of an under described game, example, one drawing had floating coins in front of a bus. We can imagine that the bus is collecting the coins which will add to the player's score but that is open to doubt. Some of the game designs had scores on the screen but didn't explain the scoring 'mechanism' like in the game by one child who wrote that in the driving game the score was by having passengers or by collecting coins.

In describing how to play there was also some uncertainty and some obvious things missing. For example, for a snake and ladder board game, the child forgot to write that there had to be a dice. There was also some confusion with some children thinking only about the technical and user interface part of the game and so writing and drawing about the buttons they needed to click to start the game.

Interpretation. Even if the children had the full power on what to draw, at the end the researcher is the one who does the analysis of the data and there were several cases where the researcher had to make a guess as to what was intended. When drawings are interpreted, do the researchers interpret them too little or too much? This is another critical question to consider [11] That is why many researchers have emphasized the importance of listening to children's perspectives [13, 20, 21]. One option is to capture both audio and drawings which is a much more effective way. [22] explained in his paper that creative visual methods do not really give the researcher access to what people really think or feel. Interpretation in this study was improved by having the drawings looked at by a native Nepalese researcher but still we conceded there were some uncertainties.

4.2 Product

Games that are Already Out There. Even though we encouraged the children to come up with new game ideas, most of the children drew games that are already out there. This could in part be considered a consequence of working in groups where it is known children 'borrow' from one another but, as pointed out in [11], the drawings still have to be considered to represent personal expressions and thoughts and cannot be completely ignored just because they are not new game ideas. On top of that, the app is being developed for the children, but we are asking for their ideas and therefore have to consider them. This is the whole point of participatory design.

Cultural Insights. The children included some cultural objects in the game designs for example a Nepalese flag at the finish line of the car race game. We refer to these as *low-cost cultural overlays* – namely things that could easily and cheaply be skinned onto a relatively complex game design in order to give it some cultural positioning. There is a much bigger challenge to design deeper culturally situated games especially when the app is intended to cross cultures. Skinning a UK or Nepalese flag is easy – making a game that is on the one hand Solitaire and the other Carrom, is much more challenging. The Nepalese children are already playing games that have been developed for children from western and developed countries and that was evident in the drawings they made, but the children in the UK were not accessing Nepalese traditions which provides a lop-sided ness to cultural balancing. The challenge is therefore how to enable this deeper cultural meaning in cross cultural design.

5 Conclusion

Children can take part in a game design session and can come up with nice game ideas if they are given the freedom on what to draw. The drawing method helps them to be creative and express freely. Some of the games were culturally influenced by things the children do in their daily lives, and by their surroundings. The responses from the children seem to be affected by the people they live with, and the environment where the study is done. Most of the children were able to name the game they designed some even explained how to play and score. In several cases the drawings were not clear, and no detail was written on how to play. That necessitated some interpretation which was challenging. An interview session with the children after the activity could have improved the results.

Future work aims to replicate the same process with children from United Kingdom which will allow for some cultural comparisons. The ideas described in this paper will be used to improve the sustainability of the Digipal app and bring the participants closer with game play and share experience.

Acknowledgement. We thank all the children for the participation, the Teachers for the assistance and support and the Head Teacher for allowing us to do the study in the School.

References

1. Shulman, S., Seiffge-Krenke, I., Dimitrovsky, L.: The functions of pen pals for adolescents. J. Psychol. **128**(1), 89–100 (1994)
2. Chauhan, R.: Social media and society. Our Heritage **68**(1), 1938–1951 (2020)
3. Read, J.C., et al.: An investigation of participatory design with children-informant, balanced and facilitated design. In: Presented at the Interaction Design and Children, pp. 53–64. Eindhoven (2002)
4. Holloway, S.L., Valentine, G.: Children's geographies and the new social studies of childhood. Child. Geogr. Playing Living Learn. 1–26 (2000)
5. Barker, J., Weller, S.: Emerald article:" Is it fun?" developing children centred research methods. Int. J. Sociol. Soc. Pol. **23**(1), 33–58 (2003)
6. Honkanen, K., Poikolainen, J., Karlsson, L.: Children and young people as co-researchers–researching subjective well-being in residential area with visual and verbal methods. Child. Geogr. **16**(2), 184–195 (2018)
7. Yarosh, S., Inkpen, K.M., Brush, A.B.: Video playdate: toward free play across distance. In: Presented at the Proceedings of the SIGCHI Conference on Human Factors in Computing Systems (2010)
8. Sim, G., Read, J.C., Gregory, P., Xu, D.: From England to Uganda: children designing and evaluating serious games. Hum. Comput. Interact. **30**(3–4), 263–293 (2015)
9. Mazzone, E., Read, J.C., Beale, R.: Design with and for disaffected teenagers. In: Presented at the Proceedings of the 5th Nordic Conference on Human-Computer Interaction: Building Bridges (2008)
10. Salmon, A.K., Lucas, T.: Exploring young children's conceptions about thinking. J. Res. Child. Educ. **25**(4), 364–375 (2011)
11. Punch, S.: Interviewing strategies with young people: the 'secret box', stimulus material and task-based activities. Child. Soc. **16**(1), 45–56 (2002)
12. Farokhi, M., Hashemi, M.: The analysis of children's drawings: social, emotional, physical, and psychological aspects. Procedia Soc. Behav. Sci. **30**, 2219–2224 (2011)
13. Villarroel, J.D., Antón, A., Zuazagoitia, D., Nuño, T.: Young children's understanding of plant life: a study exploring rural–urban differences in their drawings. J. Biol. Educ. **52**(3), 331–341 (2018)
14. Kullman, K.: Experiments with moving children and digital cameras. Child. Geogr. **10**(1), 1–16 (2012)
15. Cox, S.: Intention and meaning in young children's drawing. Int. J. Art Des. Educ. **24**(2), 115–125 (2005)
16. Barraza, L.: Children's drawings about the environment. Environ. Educ. Res. **5**(1), 49–66 (1999)
17. Farrugia, D.: Towards a spatialised youth sociology: the rural and the urban in times of change. J. Youth Stud. **17**(3), 293–307 (2014)
18. Richards, R., Richards, R.: My drawing sucks!': children's belief in themselves as artists. In: Presented at the NZARE/AARE Conference (2003)
19. Gallacher, L.-A., Gallagher, M.: Methodological immaturity in childhood research? Thinking through 'participatory methods'. Childhood **15**(4), 499–516 (2008)
20. Einarsdottir, J., Dockett, S., Perry, B.: Making meaning: children's perspectives expressed through drawings. Early Child Dev. Care **179**(2), 217–232 (2009)
21. Profice, C.: Nature as a living presence: drawings by Tupinambá and New York children. PLoS ONE **13**(10) (2018)
22. Buckingham, D.: Creative' visual methods in media research: possibilities, problems and proposals. Media Cult. Soc. **31**(4), 633–652 (2009)

Flow and Interflow: The Design Principles of Cooperative Mandala Coloring (CMC)

Yu-Chao Liang[1][✉], Mei-Ling Lin[1][✉], Ding-Hau Huang[2], and Wen-Ko Chiou[1][✉]

[1] Department of Industrial Design, Chang Gung University, Taoyuan City, Taiwan
ycliang465264@gmail.com, march19670324@gmail.com,
wkchiu@mail.cgu.edu.tw
[2] Institute of Creative Design and Management, National Taipei University of Business,
Taoyuan City, Taiwan
hauhuang@ntub.edu.tw

Abstract. Design is an activity with creativeness that can be deemed as a life-philosophy for exploring from flow of individuals to group interflow. This Study aims to combine Western psychology and Oriental dynamic meditation activities on basis of Flow Theory by Mihaly Csikszentmihalyi and Carl Gustav Jung's in depth psychological analysis and concept of Mandala, to develop an extended cooperative mandala coloring (CMC) and to attempt on proposing the design principle of CMC. Through researching the collective unconscious for specific groups at specific ages, the Bildung education for newbie designers.

The subjects of this study are 14 students from Taiwanese junior high and senior high schools who jointly participated the APDEC 2017 held in Japan in August 2017 with the researcher. The three approaches of action research are taken on basis of study objectives. First, "Pre-action Research", which involves redesign of the individual mandala coloring activity by Jung to Cooperative Mandala Coloring; then, "In-action Research", which involves interpretation of Jung phenomena by Mandala works by the adolescents; lastly, "Post-action Research" for introspection on overall effects for facilitation of flow and interflow.

Through CMC activities, this study discovers that adolescents can not only satisfy their own expressed needs but also situate themselves in spiritual safe space in order to facilitate themselves to obtain self-organization at the transient moment of conversion to complexity when at the status of being empathized, while opening up to reveal the self for broadening the experiences in interflow with others.

Keywords: Flow · Interflow · Jung (Carl gustav Jung) · Cooperative Mandala Coloring (CMC) · Design education

1 Introduction

1.1 Individuation and Collective Unconscious

Buddha said, "He who walks away from the brilliance of light sees only himself. He who walks toward the brilliance of light sees the whole world. Physical shape is of no

© Springer Nature Switzerland AG 2020
P.-L. P. Rau (Ed.): HCII 2020, LNCS 12192, pp. 337–355, 2020.
https://doi.org/10.1007/978-3-030-49788-0_25

importance, one can take in all matters regardless of its size. If one only sees oneself, then one is nothing more than a mustard seed in the vastness of Mount Sumeru. If one were to open oneself up and look afar, even things as enormous as Mount Sumeru can be contained in the mind." If we are to discuss individuation, we are in fact entering the holistic collective unconscious. Visualizing our mind being enlarged as a container able to contain the universe is like a grain of tiny mustard seed containing the energy and vigor of the entire universe, it involves the so-called "self-organization", just as how people accept to play the game of life: literally "playing a game is a voluntary attempt to overcome unnecessary obstacles." [1], even a participation to an external activity would contain changes because of self instead of non-external factors, which is the equilibrium and harmony adjusted by the system with reference to self-condition or correlatives. Human beings should realize their own unique life lessons and should sense that he and his partners are close-knit, therefore, should the intrinsic and extrinsic integration be attained throughout development of human history, we must enter the orientation of the personal flow and collective interflow by human beings.

Jung (1875–1961) also considers the original intention of human minds is the tendency of balance to self- improvement and self-balance, and the existence of unconscious is to pursue the realization of individuation process, expecting the various opposites in life to coordinate and unify deep in people's mind and form the character holistic but with unique personality. In other words, as an individual obtains the short balance of phasal consciousness and unconsciousness, the individual still needs to infuse the upcoming contradictions in binary opposites in order to allow the contents of unconscious to be understood by the consciousness and to perform correction to the biased to induce growth in character. The psychological experience of us projecting the personal growth to the crowds, also known as the process of individuation progressing toward totality, is even more a continual separation of self to the overall universe for dialectic. Such psychological experience with intrinsic and extrinsic cultivation is namely "individuation process" [2].

Buddhism from the East has a consistency with Jung's concept on human totality, "contains unity in intrinsic and extrinsic personalities enabling balance and harmony between personal consciousness and collective unconscious". As our psychic life cluster through correspondence to environmental and intrinsic hints, they will be revealed in the archetype, which is the process of individuation progressing toward totality. Between 1918 and 1920, Jung found Mandala as an instrument to demonstrate id. This type of painting featuring repeated circulation around the center of circle allows us to discover the form and color rooted from our subconsciousness; expressing the energy of subjectivity to furthermore assist us in linking all intrinsic and svabhava qualities. Jung indicated that Mandala shows the natural power enabling the development of personal potential and full realization of intrinsic essence inside human, it is the core in the depth of mind, is a piece of art of imagination, and is also the source of spirit and vitality proving the progression of individuation [3, 4]. In Tibetan Buddhism, Mandala does not only represent the diagram of the universe, it represents the life with cycles arising and passing away and is a crucial archetype image for collective unconscious of human beings. Mandala was first applied in the East as an auxiliary tool for meditation and reflection, also as one of the utterly important methodology for meditation through the process of Tibetan Lamas' cultivation; meanwhile, Mandala is also used as artistic approach to express specific mind reality, as Mandala Coloring can express the intrinsic mind through externalization with graphics. In art therapy, it is often manipulated as a method to assist cases at all ages to explore intrinsic mind and self-integration. Mandala requires the learner to conduct creations in a focused and calm status. Mandala also serves as mediation when a person's mental status is in chaos, and Mandala is therefore the archetype for mental unification and natural completion [5–7].

Jung discovered that when there are benign interactions among consciousness, unconsciousness, self and id, people feel their unique personality meanwhile channeled to the ocean of experience in the depth of human subsistence, making the lives of ourselves truly contain creativity, symbol and uniqueness. Jung referred to this process of reaching mind and spirit balance as individuation, a basic conception of Jung psychology, and considered this the principle and process all mind and spirit activities shall follow as he indicated that the mind contained tendency of self-improvement and balance. Jung frequently made repetitive quote as such: The gigantic catastrophes that threaten us today are not elemental happenings of a physical or biological order, but psychic events. To a quite terrifying degree we are threatened by wars and revolutions which are nothing other than psychic epidemics. At any moment several million human beings may be smitten with a new madness, and then we shall have another world war or devastating revolution. Instead of being at the mercy of wild beasts, earthquakes, landslides, and inundations, modern man is battered by the elemental forces of his own psyche [8–10].

1.2 Individuation and Collective Unconscious

Mihaly Csikszentmihalyi (1934~) began the study on concept of flow about 40 years ago, exploring the power of human optimism, creativity, intrinsic motivation and sense of responsibility, which not only pioneered the research orientation for creativity of cognitive science, it further influenced the study of global positive psychology [11].

Flow, also known as Shenchi (literally "running of mind" in Chinese) or Zone, is fully immersed in a feeling of energized focus, full involvement, and enjoyment in the process of the activity; When flow occurs, it will simultaneously generate a high degree of excitement and fulfillment. Famous cognitive scientist and creativity scholar Mihaly Csikszentmihalyi first introduced the concept of flow in 1975 and explored it in a scientific manner. However, the technique of flow is not developed by Csikszentmihalyi. The Eastern thoughts systems had used this technique as a tradition for more than 2,500 years, such how the Buddhists and Taoists used techniques of flow to develop their spiritual strength. The Japanese Zen used flow to determine its expression. In the field of education, the saying "being tireless in teaching, and never being wearied in learning" is also caused by free flow. Many modern athletes also experience flow during exercise, and they call this experience "in the zone".

Csikszentmihalyi classified flows as follows: "the body in flow": exploration on how to make good use of the body to improve the quality of experience, and the various ways in Eastern culture to control the mind by training the body. With mention of Oriental Yoga as one of the oldest and most systematic ways to create a flow experience, he believes that the "spiritual training" in a monk's life in the western world is still far behind the yoga in the Oriental. He then talked about the "flow of thought": it refers to the challenge of lifelong learning and the enjoyment gained from historical, scientific and philosophical psychology games. The last is "work as flow": training oneself to become a happy worker, and finding opportunities for learning and feasible goals [11].

Judging from such, we can conclude that the flow does not occur from the unconscious state, most of it can be cultivated consciously. The Mandala painting activity advocated by Jung is a probing of the psychoanalytic science, but also an artistic expression of human spirituality, and art is one of the ways to mutate mental images and express the cognitive world which not only sees the shapes of the external world but also depends on its own creativity via the operation of brain tissues to describe the essence of what humans are, and art, the active process of self-organization. Semir Zeki, neurologist from the University of London, is the pioneer of modern visual brain and neuro-esthetics research. He also believes that the process of mental imagery visualization represents that the brain is in motion. "Brain is like art, as both are pursuing the constancy and essence of objects, as the objects we see are not the images of the object projected on the retina. It is an explanation of the cognitive experience of the brain" [12–16].

Jung's individuation process is to form a solid and strong ego so that the individual can establish a sense of self in the world and then associate with the other and the collective culture in which he is placed.

Interflow represents any unification of heterogeneity, which facilitates greater harmony and fusion among self and others. Mysterium Coniunctionis is very important in Jung's ideology [17], as it is an image that can guide individual needs and external connections, and is very challenging. Jung acquired images of interflow from the medieval Eastern Taoist alchemy works. In these seemingly physical and chemical processes of alchemy, what Jung saw was to project the inner psychological activities on the outer matter, and his purpose was not just to transform base metal into gold, but an internal spiritual conversion of the alchemist. The combination of opposites (coniunctio oppositorum) is one of the last stages of alchemy. At this stage, heterogeneous objects separated from each other finally merge into a new invert-the ultimate goal of alchemy performance [2, 10].

Archetypal image is a deep foundation image formed by the accumulated experience of the individual mind, and is the basic content of the object mind. The archetype itself cannot be directly observed, instead, it resembles a magnetic field identifiable from the image content in the brain it affects [10, 18]. The process of driving the peripheral nerve through the hand to holding the colored pen for visualizing the images in the brain makes the Mandala Painting serving as the methodology for self-expression and self-healing.

Jung's Mandala painting can be deemed as the perfect match of Csikszentmihalyi theory of flow and neuroaesthetics of the brain. Through physical, mental, and work experiences, it presents the process of a personal flow reaching intrinsic and extrinsic interflow.

1.3 Self-organization and Synchronicity

From the classical physics concept of "sympathy of all things." Jung was inspired by Hippocrates, the father of ancient Greek medicine, Jung considered synchronicity to be a coincidence and is not limited to the realm of psychology. It can enter the state of consciousness from "the interior of the mind matrix" and "our external world" or even from both aspects. When the two occur simultaneously, it is called "synchronicity"1. There are three major elements of synchronic theory: "meaningful coincidence", "acausal connecting principle", and "mystery". Synchronicity refers to the phenomenon of the extra-connection between events in life and the state of mind [18].

"Synchronicity" was a concept that had been in development by Jung in 1929. In the Memorial of Richard Wilhelm in 1930, Jung publicly used the term "synchronicity" for the first time. It wasn't until 1951 that Jung made a keynote speech on "On Synchronicity" at the Eranos Conference, and in 1952, the more complete German dissertation of "Synchronicity: An Acausal Connecting Principle" was first published, which furthermore defined the concepts to be dealt with. Jung challenged the status quo, requiring us to go beyond the limits of the description of the world by causal induction, as believed that the casual personal experiences of life should be included and given appropriate status, and exploring the potential meaning of coincidence or random events with a completely open mind.

Self-organization theory is a systems theory established and developed in the late 1960s. Its research object is mainly the formation and development of complex self-organized systems (life systems, social systems), that is, under certain conditions, how the system evolve from disorder to order automatically, and from low-level order to high-level order. Without external prompts, as the system runs the mutual hidden rules with each part fulfilling their duties while coordinately and automatically form ordered structure, it becomes self-organization. Contemporary complexity theory and self-organization were introduced synchronically after Jung's synchronicity, and therefore a direct relationship between the two has been considered.

The researcher believe that the context between the two theories exist by the fact that the Mandala provided by Jung is an epitome of the essence of the self-spirit, which often appears under the condition of confusion and disorders, being clamped by the circle of protection. Mandala is a self-organizing tool to help restore mental order [19] Cooperative Mandala Coloring: Expressive Art Therapy.

The discourse "Art is the primitive activity of the mind" has always been the core point of aesthetic education. It explicitly demonstrates that art originates from the unity of being, which is what Jung said: "Einwärtsgewandtsein", once the entity of the ego can be transcended by artistic revelation, it may form a non-ego role replacing the original self.

The expressive art therapy developed by Paolo J. Knill originated from a phenomenological art therapy movement in 1970. Eventually, he broke away from the framework of psychology and identified and developed from the perspective of anthropology [20], Paulo Knill believes that expressive art focuses on "hope," and "we don't just need hope, we hope the needs to be created." What matters for the forthcoming ones is: "What is the best learning?" He also said, "Art is not a pathological phenomenon, not the creation of disease, but the realization of expression, learning, challenge, and achievement." It is a hopeful creation! [20].

The researcher carefully studied Jung's discourse on Mandala as well as researched and collected Jung's personal diary "Red Book", containing Mandala creations to discover that Jung himself made Mandala paintings between 1916 and 1928, but the concept was not mentioned until 1929 in the review of The Secret of the Golden Flower [21], but at the same time, Jung stopped the creation of the Mandala in the Red Book. Hence, did his cherished Mandala evolve in sync with his research? Is Jung's Mandala completed, or stopped growing? Jung claimed that he had spent 13 years remaining silent on the effects of the Mandala. Jung performed his own experiments from 1913 to 1930. He wrote down a series of illusions, fantasies and thoughts in a small black notebook, the "Black Book", which collected the first Mandala he painted and colored on January 16th, 1916. Since 1915, he has made reflections on the words and transcribed them by calligraphy to the red leather booklet, the Red Book. In Richard Wilhelm's memorial in 1930, he first used the term "synchronicity" publicly. It wasn't until 1951 that Jung made a speech "On Synchronicity" at the Eranos Conference. In 1952, a more complete German dissertation on "Synchronicity: An Acausal Connecting Principle" was first published.

Since 1952, in Jung's late life, with inspiration from the Red Book, he published many important theories about the symphony between nature and the others. Researchers believe that the individuation mandala images that stopped growing before 1930 could no longer fully respond or implement the important theories on complex and exploratory mental imitations proposed by Jung in his later years such as "synchronicity", "simulation theory", "symphony ability", "mirror theory" and "empathy". Although in 1959, Jung, who was about to reach the end of his life, published the article "On the Symbol of Mandala" [8, 9].

How would the researcher's design education objective based on facilitation of developing empathy and observation on adolescents' participation through this expressive art activity convey and co-create aesthetics experience between the group? The researcher redesigned Jung's personal Mandala painting activities and performed contingent combinations on the spot based on factors of number of participants, age, gender, topics to be discussed, available time, making various segmentations to a complete circle and furthermore developing a Mandala painting course with versatile cooperation model and named it Cooperative Mandala Coloring (hereinafter referred to as CMC).

A single CMC always starts from a circle. Participants draw circles with the compass at hand in a fixed and consistent radius, however, the form of the circle division is ever-changing and can be flexibly manipulated by teachers according to the various possibilities and restrictions on the scene. Basically, there are pie chart, a circle divided into several sectors, and the arc length (and the central angle and area) of each sector is the proportion of the quantity it represents. These sectors put together form a complete circle. The quarter round is the drawing of perpendicular cross lines through the center of the circle. A Venn diagram shows the overlapping area of a part of the set after two or more circles intersect, and a directed acyclic graph (DAG).

The practice of CMC involves the collaborative creation of two or more people, allowing adolescents not only express their inner self but also present the spiritual reality in the process of co-creating a Mandala with the other, assisting them to generate flow, and to create symphony with the team while appreciating the beauty of interflow.

1.4 Research Purpose

The researcher has continued to use the Mandala painting activities developed by Jung as an artistic experience course of expression and meditation since 2008, and have successively implemented on learners of different ages, thus accumulating tens of thousands of works and photos gained from learners' participations, which among them contain certain quantity of pictures showing looks of flows on the 14 to 17-year-olds who focus on Mandala painting. For many years, I have devoted myself to asking what makes them all able to heartfully [22] devote into this artistic creation? In addition to the perfect inner self, can the mandala flow this aesthetic experience between people and me with positive energy?

The adolescent stage usually begins in the second decade of life and can be roughly divided into three stages: early stage adolescents (ten to thirteen years old), mid-stage adolescents (fourteen to seventeen years old), late-stage adolescents (eighteen to twenty years old), the target group of this study are adolescents at early to middle stages (13 to 17 years old), which is a stage bridging childhood and adulthood [23]. From the perspective of the Jung, the stage of adolescence represents the formation of personal identity and the beginning of changes in all growth in human life.

We can understand that there are chaos in the pursuit of "becoming self", but Jung found: At any moment like this, some Mandala images emerge spontaneously from his mind. He further described these mandala paintings as: "Eternal creation of composition, transformation and eternal heart" [21]. The researcher believes that the image of Mandala not only embodies the current state of the mind, but also often implies a "balance effect" of self-healing. The growth metaphor of synchronicity between adolescents and Mandala is as Mandala came from different civilizations, different times under different moods but from the "collective unconscious" inherent in human beings (Hall, 1983), it can be said that the Mandala is also universal structure in human's mind becoming artsy.

Mandala developed by Jung was able to highlight the inherent uniqueness of the individual and create a personal symbol of the moment [6], but Dr. Murray Steine (1991), who had more than 40 years of practice as Jung's Analytical Psychologist, once proposed, "Do collectives and organizations also apply to the principle of individuation when they touch the other as time advances?" [24]; The researcher also wishes to further inquire: "Every Mandala created by each individual is unique; however, is this kind of personalized flow activity suitable for studying and observing how to build and develop one's body, mind and soul interflow with others?"

The development of CMC is a reflection on the important theories "synchronicity", "simulation theory", "symphony ability", "mirror theory" and "empathy", the researcher gradually builds the CMC design principles based on Jung's theory after years of action research in the teaching field, and explore how CMC's artistic creation activities evoke participants' interactive mutation concerning "psychology/mind" "Flow/Interflow".

The purpose of this study, based on the theoretical deduction and long-term field experience, has completed to redesign a CMC activity that can cross-reference and interactive games between specific teams/communities/peers, to assist learners/designers/users at growth stage to lifted the crisis of archetypal self-arrogance and dismantle its defense mechanism, and created a safe anchorage for collective unconscious with the magic circles of Mandala, introducing them to similar situations or initiate a spiritual dialogue with others who wish to develop empathy, in addition to making them the material of symphony in each other's individuation process, from which they can also obtain a positive life experience of interflow. Lastly, this study will make concrete suggestions for the oneness of design education theory through the fusion of Eastern aesthetic philosophy and Western scientific theory.

2 Methodology

2.1 Participants

The researcher explores the relationship between Jungian Mandala and individuation, collective unconscious, self-organization, synchronicity, and flow and interflow. These processes can be said to serve self-salvation and personality mutation, which is the struggle for the highest goal of individuation. The Mandala is not only a kind of surrounding regardless of time and space, it also serves as the center that defines the whole and the self that is defined by the whole.

Adolescents are in a period of internal and external pressure conflicts, as personal factors interact with environmental factors, they need a spiritual space to expand and integrate. Field Theory scholar Kurt Lewin believes that in order to fulfill the need of adolescents for "movement" away from the goal, it requires parents and teachers to provide ample sense of freedom and security, and reducing guidance and restrictions is necessary, in addition, avoid dependence on the family to conflict with the independence required by the culture [23, 25].

Through expressive art such as Mandala creation, one can fully talk to him/herself, so that sensory perception, moody sorrow, value beliefs, and unclear self-confidence can be interwoven through colors, shapes and lines to express that the Mandala can contain all positive and negative emotions, and let the adolescents in the creation fully control themselves, so that they can leave deep or shallow, fast or slow images in the circle of the Mandala, and that is what the learners' mind image revealed from within. The researcher have found that when the Mandala creation is completed and the learner gazes upon his/her work from the trance of creation, most learners cheer in joy and stretches for relaxation, while a mentor with a chromatics background or relevant experience can make the learner see what issues of life and hidden positive or negative emotions the learners are facing [26].

Fourteen adolescents were invited to, on a blank workbook, draw a Venn diagram of three centimeters in diameter with their compasses. Using their own pencils in more than twelve colors, they first doodled one of them, and then they exchange creations with the other two partners in turn, until the three circles of Venn diagram for CMC is completed and returned to the original owner. During the period, only the meditation music Sa Ta Na Ma [27] that has been proven to be beneficial to the hippocampal enhancement by the neuroscience of the University of Pennsylvania is played, and the learners are asked to reduce unnecessary conversations, while the mentors go around the learners to provide timely helps (including sharpening color pencils or fetching water) and merely waiting patiently for all learners to put down the pen with satisfaction, and then carefully arrange all the works on an open table or ground or wall, asking learners to leave the seat to appreciate the Mandala that evaluates the others. Each person is provided with three post-it notes for them choose three of the CMC works from others and give them a "content-filled and factual basis" comments for praise and post next to the works. Then the mentor publicly express, observe, appreciate, reflect and encourage each learner on their mandala painting process, followed by the learners bringing their works back to their, and the mentor read aloud the poem of the mystic Persian poet Rumi from the 13th century with the participants' random transcription or echo allowed. Finally, the adolescents are invited to three-minute free-writing to describe the beauty in the experience of cooperation with others. The overall CMC, peer feedback and free writing takes about two hours.

The perception and learning of expressive art must be obtained through actual experience. The researcher believes that it can be said to be both psychotherapy and an art aesthetic training, and to further examine its ultimate goal is to achieve the participants' mind exercise from flow to interflow. We should allow adolescents to wander freely between the process and chaos occurring in sculpting and shaping of art, and the guide should keenly capture the self-organizing phenomenon of chaos and order interaction.

2.2 Procedure and Design

As we prepare to lead adolescents into expressive art courses, the first issue we may encounter is: if individuals who feel that they have no artistic talent, can they also be satisfied in investing in expressive art activities? If the cooperation model is changed, can this psychological resistance be lifted? Most of the art course activities tend to express individual originality, and art courses with a collaborative process that satisfies individual expression are not common. It will be difficult for adolescents to recognize their existence value from the inside out in the aesthetic experience, and it will be more difficult for the researcher to observe the projection and transformation of mind energy in the process of co-creation by adolescents.

Mandala painting only requires a coloring pen, no expensive or special tools and skills are needed, and it can be started anytime, anywhere. According to the researcher's years of teaching practice experience, participants can feel self-satisfaction, happiness, focus and creativity in the activity of Mandala painting. Furthermore, after a period of peer training, participants can continue to improve in producing creative and design mandala paintings. Mandala thinking is one of the creative thinking methods. We can achieve personal growth and self-realization through mandala [28].

There is no rigid method to paint the Mandala. When the learners immerse into the scribbling status where the boundaries of time and space disappears and the state of being in the flow, they can create art freely with the pen they hold, while the mentors can adjust ways of conduct following the atmosphere. Generally, the beginning of Mandala painting starts with a blank to-be-painted pattern with a circular shape and a geometric pattern inside. It is recommended to use a color pen to paint from the outside to the inside. Most of the internal patterns have a dual relationship. The time is about 30 min. At the beginning, most painters will be immersed in the focus of painting without going beyond the line, but the subsequent courses will reduce the instruction, no longer restricted to directionality or established frames, participants can show more focus and autonomy, creativity will gradually show up in an unsatisfactory state of mind. Mandala is not only a way of meditation, but the creative process is also regarded as a hypnotic state. It absorbs the images of the mind and surrounds the learner with the essence of life philosophy or self-spirit, thereby demonstrating the spiritual value of art transformation.

2.3 Analysis and Interpretation

In many arts-oriented courses, many related scientific studies have shown that when people come into contact with Mandala painting activities, they can often achieve emotional relief, focus and relax, and they can fulfill the sense of security on expression, compated to being required to speak or write directly in the first place, and can better meet expressive needs [5, 7, 8, 29, 31]. Mandala painting is a creative activity that involves the reorganization of painting, design and self-life experience. It is the presence of creativity and meditation. Learners can use various tools such as color pencils, watercolors, crayons and even plants and stones in nature by arranging various shapes and color blocks for liberal creation, the "Cooperative Mandala Coloring" activity that I redesigned based on Jung's theory from his late years can not only help learners indulge in artistic activities to achieve a joyous meditation state, but also beneficial in observe the cooperative aesthetic image between the self and the other [26].

This study will eventually produce three parts of data available for analysis: (1) cross-disciplinary literature from educational aesthetics, neuroscience, expressive art, and Jung's psychological analysis; (2) CMC Works form the 14 adolescents; (3) Manuscript of adolescents' free writing after Mandala painting.

This study boldly revokes the researcher's role as conductor first, and, replace with a girl at senior year at high school who had participated in the CMC for many years but happened to be a new participant to the camp to anonymously cooperate to paint the Mandala to complete the diagram, and intuitively interpret the various phenomena relationships and symbolic meanings she sees. This is an important process of adolescents' mutual text. It inevitably covers cognitive structures including the archetype experience as an interpreter in her mind, the self-conscious aesthetic experience and the subconscious shadow and other identification structures as well as the special mind map for relationship structure of persona, anima, animus, etc. The process is highly exciting and touching.

Hermeneutics is the most uncontrollable subject of art philosophy or literature to many philosophers. It has been plaguing the philosophical community for thousands of years (it had been in discussion since the Stoic school in ancient Greece), whose biggest argument lies in the relationship between author's intention and the meaning of the text. The purpose of hermeneutics, as far as its negativity is concerned, is that hermeneutics stands an objection to the methodological monolithic view claimed by positivism, that is, the application of scientific experience analysis methods and causal interpretation rules to social science inquiry, ignoring subjective considerations and emphasis on instrumental rationality and technical tendencies. On the positive side, hermeneutics takes human subjectivity as the basis to construct the meaning of the living world, emphasizing the need for self-examination, criticism, and interpersonal relationships as the subjective communication of each other. That is, hermeneutics aims to form an interactive dialogue situation through the development of human subjectivity to construct a reasonable and harmonious life-world relationship.

Roland Barthes said: "The Text is plural. Which is not simply to say that it has several meanings, but that it accomplishes the very plural of meaning: an irreducible (and not merely an acceptable) plural. The Text is not a co-existence of meanings but a passage, an overcrossing; thus it answers not to an interpretation, even a liberal one, but to an explosion, a dissemination." Therefore, this study takes the young reader's "commentary" to discover the enlightening meaning of their own work and form a new creative text. l'autonymine flows between the research data of the fusion of horizons between teachers and students [31.32]. The critical interpretation of pedagogy can be considered more a method but as a practical wisdom. This study expects to capture the possibility of youth self-cultivation through the aesthetic experience of artsy education. As far as self-cultivation is concerned, the right of interpretation will be returned to the youth as much as possible. In this chapter, researchers will only submit the phenomenon of Cooperative Mandala explained by the youth, the phenomenon of hanging, the phenomenon of intuition, and the phenomenon of shelving.

The meditation effect of Mandala is equivalent to the method of mindful meditation practice used in the research of cerebral neuroscience. In Jung's psychoanalysis, he also had an important discourse on the inner spiritual perfection of mandala; however, various studies in the international stage on Mandala still falls in the experimental topic of creativity and emotional sense with hollow argument about the change of human spirit. Jung's important theories such as "synchronicity", "simulation theory", "Symphony ability", "mirror theory" and "empathy" will be captured and interpreted in Jungian phenomenology in the CMC course of this case.

3 Results

3.1 Field, Border and Uncertainty

The CMC painting in Fig. 1 is considered completed by the three girls who just met in this camp, who were not yet familiar with each other and cannot figure out the ways to interact with one another. The circle on the lower right corner is considered to be made by the one who drew first, the main painter, while the second person drew the polkadots in the lower left corner, and the third one couldn't handle the unexpected ideas brought by polkadots, in order to maintain the relationship she just established with the master, she chose to follow the style of the master's painting, without concerns on the purpose of the polkadots. The third painter chose to be a dependent of the main painter in the new group of friends. In fact, she may not yet have a sense of identity with this group relationship of the three people, and therefore she would not help the master to care for and organize the final picture, and she would not conceive the method to make the polkadots into one.

For the second girl who painted the polkadots which were different from everyone else's works, why wasn't her behavior questioned? It was because she just wanted to create a different style, we shall first assume that she is well-intentioned, however, if she was the third person taking over the painting, she would feel very bad and considered the embellished part self-centered. As for the master, she should have mixed feelings when she saw the result of the cooperation, therefore, she drew three patterns with water

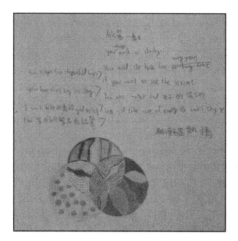

Fig. 1. The polkadot new relationship.

ripples like a canal or decorative carpets in the center, indicating that there may be a long way before the two can fraternalize.

3.2 Repetition in Form, Imitation and Crystal

The CMC painting in Fig. 2 seemed to be completed by the three girls. The overall picture was bright and harmonious, but I don't think that friendship can last forever, as this kind of painting technique involving direct cooperations with others can be developed through cooperation by acquaintances. The three black frames were eye-catching, but the three yielded similar touches making Mandala contained with similar style, but not a single painter had lost sight in this Mandala as they also chose their own colors. This relationship

Fig. 2. Rationality over friendship

was more rational, and this was likely that the Mandala was done by individuals who developed temporary friendship under the circumstances, once faded out, one might be in less intimacy with others. Even if it was drawn in complete harmony, it did not necessarily mean that there was true friendship, it was all about helping others achieve others' expectations. If they were true friends, they would try their best to paint better, no matter how hard it was, and the level of efforts would be different. At this point, the interpreter cried, saying it might be her obstacle. She didn't want to easily define friendship, but the friendship might fade away after she was in isolated environment and then re-associated. She had a shy personality with high self-demands. She could not open her heart to others, with only few friends.

3.3 Breaking the Symmetricity, Singularity and Implicate Order

The CMC painting in Fig. 3 seemed to have been completed by three girls. I always thought girls make works in nice and neat lines and patterns while boys make works in bad images with cluttered lines. With a touch of cuteness, the painters could form an everlasting friendship. As it needed something with special care to develop to show cuteness on the picture, and the overall sense of Fig. 2 can be achieved only by rational cooperation. And usually some rationality were required to blend in emotional friendship, but a relationship with more rationality contained less of emotional content. From this CMC painting, I could see that this friendship was with firm foundation, unlike the friends one just met, but also, the painters were very homogeneous with each other, and they liked things similar, resulting in good moods when in cooperation. The interpreter suddenly mentioned that he was unable to get back to the kind of person in the friendship after being separated from the environment, and it was always difficult to maintain, and he would get stuck for his less proactive acts. The one looked forward to a lifelong friendship, probably like the ones who hang out to talk about their thoughts, just like how the American TV series went. In terms of replying message s from friends, ones he replied instantly were the friends in frequent interactions recently, or it was about seizing the moment, or it's about saving face. If the message were from old friends, he/she would stall the replies or even more leave the message not replied when expired.

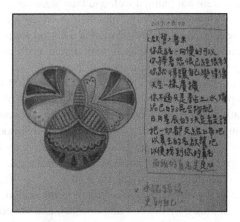

Fig. 3. Rationality over friendship

3.4   Group Logic, Symphony and Self-organization

The CMC painting in Fig. 4 yielded a feeling that it was a cooperation by one girl and two boys, and it is an unconventional girl with a clumsy hand and unconventional boys with thoughts. Although the colors were not neatly painted, the overall look was great. They were not good friend yet, but they were all good people. I guess the first painter worked on the pink background, followed by the second one taking over to paint the green background, and the last one must paint the black. Black showed high confidence, had a sense of superiority and knew his strengths, wanting to show his integrity and thinking that he would be the best in this cooperation process. The second one working on green background showed more confidence than the first one adding pink background. The picture had the symbolic look of a compass also induced the combination in association of car brand logo, neutrality, and good mathematical ability. Figure 4 (Left) was a commonly seen group of friends. Some of them were very confident, others were less confident, and each knew others' status of accepting others which will contribute to decent overall cooperation.

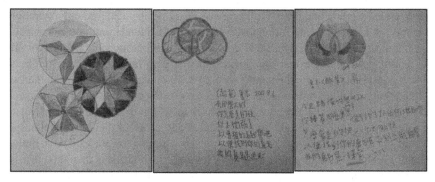

Fig. 4. Unconventional boys and girl (Left), buddies (Middle), pubescent graffiti (Right)

The CMC painting in Fig. 4 (Middle) seemed to have been completed by the three boys, because the edges were not painted well and they were not being careful enough, plus, there were issues with the three circles interlacing. Weren't they feeling strange at all? The one painting red in the middle had completed all the wrong shapes by himself, and the one painting red was too self-exposed, very domineering and not careful. However, men's friendship was inconceivable. Even if they were very heterogeneous, they could still be friends in close relationship suddenly, and it would last for a long time. This was the friendship between men, which I was quite envious of. Although the three of them were not painting delicately, they all painted things crossing the entire picture which looked like channels or routes: red object looking like neurons was the sun; blue object was a serpent, with a blue sky, and green was the path in the meadows. The sun, the blue sky, and the green meadows made a sense of wholeness. The latter two were more like underlings contrasting and obeying their boss.

The CMC painting in Fig. 4 (Right) looked like a combination of male-and-female group. Graph below was drawn by the boy and one girl, and the other girl finished the diagram on top. They were pursuing beauty and harmony, and they were drawing curves, which indicated the boy in the group was in prepubescent with less aggression. Overall, it was a landscape painting with mountains and rivers. One boy and one supporting each other to maintain symmetry. The sun above was a result of making the best out of a bad bargain, yielding a feeling which the painters wished to explore adolescence together.

This study first invites adolescents to perform an intuitive interpretation of inter-textuality, which is another method of shelving the phenomenon. This is to highlight the unreadable nature of the researcher at the first time, and to encourage readers and interpreters to develop the meaning of the work, rather than catering to intentions of the collaborators. In terms of design pedagogy, the discussion of the nature of education, the exploration of methodology, the enlightenment of the subjectivity of education objects, and the cultivation of self-reflection ability have always been regarded as the main areas of discussion for pedagogy. More critical thinking is needed. In order to introduce the force of Jung's "synchronic" field, internal and external events can be resonantly reflected, presenting a non-linear force, so that the object itself can affect the space; therefore, the teenager's telepathy served as a scalpel of the mind, through simulation theory, without the need to reproduce explicit behaviors, adolescents can use their own minds to simulate the minds of others, and to observe the goals and intentions of them, which is what Jung called Symphony. This process of interpretation is like "weaving a net crossing the sky." although it is not attached to the phenomenon, it brings a great deal of inspiration and cognitive impact to the cooperative relationship and self-organized object minds of adolescents in expressive artistic activities, and the core values of artsy education are in emergence from chaos one by one.

4 Discussion

Jung said that when the archetype is summoned, it is "realization" transcending the awareness of matter in space and time. This chapter represents the mutation process of self-analysis by adolescents after the CMC painting, through their own reflexes and reflections, they are no longer relying on the traditional role of teachers but are achieving the following possibilities for self-cultivation, which embodies the realm of symbiosis of "Morals and Aesthetics in Consistency".

4.1 Field of Consciousness

In this study, through flow process of the CMC, it can be said that the mandala is connected from one circle to another circle, so that the archetype itself which could not be directly observed become a safe channel formed after artsy formation summoning the collective unconscious of the adolescent, or object mind. Using psychology to become the guardian of art and science to master the mysteries of cosmology and ontology was what Jung dedicated himself in development and guarding [18] in his later years, and this study wishes to maintain was the aesthetic meaning of interflow inspired by synchronicity.

When adolescents are placed in unfamiliar fields, even though they are full of uncertain relationships, and they also faced the challenge of new intervention forces in the process of encountering various other personalities, however, through the channels of aesthetic experience, the adolescents' emotional rollercoaster would become the character of being willing to try to understand accidents, to accept conflicts and overcome internal noise, and gradually settling in this relationship. When own-being emerged, higher-order phenomena seeking meaningful coincidence also comes, appearing in the circles of borders to chaos and order.

"Me in Mandala, at the beginning of painting the mandala, it is always difficult to initiate a stroke. The way to combine the Mandala with the mood has always been difficult. But beginning is always the hardest! It's nothing. When I started painting, I slowly calmed down.

"We in Mandala find it easier to paint in collaboration than painting individually, but it is a hassle to find the common grounds of each Mandala, because everyone's painting style is different, and it is impossible to fully cooperate, but I think this is an opportunity to practice observation, observe the common grounds of things, and handle it in the same way, I find it quite fun. However, with many people and Mandalas, and sometimes it is painful to draw."

This free writing was done by a sophomore student at high school. In his experience, Mandala was an artistic activity with hard-to-complete strokes, however, once he entered the state and he started painting, he was in the flow. The same was true for the Mandala painting which require the painter to considerately cooperates with others, fun could not be sensed easily. Through the process of continuous mental repulse and practice, willpower is awakened, of course, when there are many Mandala for painters to collaborate on, it was actually the handling of interpersonal relationship processing more complicated conditions, and it was complaining in full candor. Being able to feel the process of CMC painting, the mind of the adolescents are playing a game of role exchange, which increases the intimacy of resonating with the other, resulting in attempts to imitate and develop empathy, which is a prerequisite for human consciousness, and it is also the self-cultivation experience hard to be intrigued by informative lectures.

4.2 Energetics

Since the complex in the individual mind is based on the archetype in the mind of the object, as long as one understands any complex in depth, one can find associations about its archetype. The art of Jung format analysis mostly relied on extending the meaning of the image, allowing the self to feel its connection with the prototype world in a therapeutic way without drowning in the scattered sea of archetypes. For example, if the ego feels its connection with the id, then the ego-id axis emerges as the times require, and therefore the ego can feel the core of the self and the soul more. However, if this connection is experienced by a weak, immature self, it may be assimilated by the ego, showing an expansion of the mind, and losing a clear foothold in consciousness [33].

From the adolescents' CMC works, we can sense the motions to own being after self recognizes collective unconscious from the chaotic and complex system. Their works created order through the repetition of patterns, making people feel the wholesome aesthetics, soothing minds, and generating the so-called tranquility as referred to by the Jungian school through the symphony of symmetric harmony.

"In the mandala, I feel different from my appearance. It makes me emptied in body, mind and spirit and relaxed. I draw timeless masterpieces of my heart box by box. When entering that world of color and shadows, it felt like having fun in my dreams.

"We in the mandala, when I am no longer alone, but with a large group of people, our people are brought together by this painting, and also move towards the same goal together. When I get other people's paintings, I always feel pressure, fearing that I would leave some defects over someone else's masterpiece, but I am step by step towards the door of courage and confidence. I need to step out to complete a real Mandala."

This is a free writing by a 9th grader. It seems to have experienced a warm, brave and positive peer relationship. Although I felt the social pressure to pursue perfection, I finally greeted it with a complete replacement for perfection. By challenging yourself. It seems that at the end of each of the children's journeys of cooperatively Mandala painting, I heard cheers from the heroes, full of positive energy, and thanks for practicing altruistic and altruistic minds.

4.3 Alchemy

The premise of self-organization is to promote the mental and class flow among adolescents, and allow children with different learning qualities to perform artsy self-expressions. CMC painting not only can express self, but also can help each other, imitate, create, and autonomously in the process of learning, the motivation of teamwork can be enhanced, the social fun and the development of interpersonal relationships that are "altruistic beauty". Spiritual growth, no matter through school, religion, art activities, etc., the factors that can finally achieve successful mutation and return to the starting point of "self-cultivation". Sufficient safe space, aesthetic experience of expressive art, mutual help story of team learning, and full immersion in the space and time of meditation and meditation, the CMC painting adopted by this study has indeed discovered the implementation of moral aesthetics of self-cultivation from the paint works and free writing of adolescents. In this process, the mentor did no unnecessary reminders and educated instructions, but only yielded the aforementioned liberal time and space conditions, letting the beautiful own-being occur, and, from the beginning of the splitting to the final self-integration, the adolescents have completed the collective alchemy of the spirit. It is an education game with artistic traits, but never a daydream.

"Me in Mandala, infinite tiny universes are contained in every circle, and I am immersed in the world of painting. I like painting very much, and I enjoy the process of creating the mandala very much. I can still maintain my passion even with 12 circles."

"We in mandala, I like to know who does the original work, and then refer to his/her ideas before add my own ideas. Cooperating with others while escaping from the general ideas, it made the painting less lonely with a rich sense of surprise and a way of thinking as well as mood in order to let each other stay in each other's stories.

This is a free writing by a junior student at high school vividly describing the satisfaction brought by the individualized Mandala, the thoughtfulness he wished to provide when interacting with others and surprise after collaboration, with the willingness to contribute to others' stories, his own story became wonderful and meaningful. With the youth spark of self-cultivation started, one can wholeheartedly feedback to own-beings of children.

5   Conclusion and Future Research

In the process of artistic creation, the precious inspiration obtained from the free writing by the adolescents in this study we can conclude that the, artistic activities are not only the creation of self flow, but also the creation of stories of interflow with others, and the entire process makes design principles of CMC fall into the altruistic aesthetic act.

Jung said that everything started from his heart are artistic. In this study, the artistic activity of adolescents collaborating on CMC painting also showed us the complex dynamics of the real world, although the implementation process has many details that can be improved, such as requesting participants to code the drawing order so as to make the process of mutual influence clearer and more convincing for advanced psychological analysis and comparison, also, the time for free writing can also be increased to make the self-organized experience more complete. In addition, in the future, the research object of this CMC painting shall be rooted upwards or downwards to explore the possibility and related restrictions of artistic education at all ages.

References

1. Suits, B.: The Grasshopper-: Games. Life and Utopia. Broadview Press, Petervorough (2014)
2. Wilhelm, R.: The Secret of the Golden Flower: A Chinese book of life. Routledge, London (2013)
3. Fincher, S.F., Johnson, R.A.: Creating Mandalas: For Insight, Healing, and Self-Expression. Shambhala, Boston (1991)
4. Malchiodi, C.A.: The Soul's Palette: Drawing on Art's Transformative Powers. Shambhala Publications, Boulder (2002)
5. Curry, N.A., Kasser, T.: Can coloring mandalas reduce anxiety? Art Ther. **22**(2), 81–85 (2005)
6. Fincher, S.F.: Creating Mandalas: For Insight, Healing, and Self-Expression. Shambhala, Boston (1991)
7. Jung, C.: Mandala Symbolism from The Archetypes and the Collective Unconscious in the Collected Works of CG Jung, ed. William McGuire et al. trans. RFC Hull, Bollingen Series XX, Vol. 9/1. Princeton University Press, Princeton (1959)
8. Hopcke, R.H.: A Guided Tour of the Collected Works of CG Jung. Shambhala Publications, Boulder (2013)

9. Jung, C.G.: The Development of Personality, vol. 17. Collected works, Pantheon, New York (1954)
10. Jung, C.G.: Collected Works of CG Jung: Symbols of Transformation, vol. 5. Routledge, London (2014)
11. Csikszentmihalyi, M.: Flow, The Psychology of Optimal Experience. Haper and Row, New York (1990)
12. Zeki, S.: Inner vision: An exploration of art and the brain. J. Aesthetic Art Crit. **60**(4), 365–366 (2002)
13. Zeki, S.: Splendors and Miseries of the Brain: Love, Creativity, and the Quest for Human Happiness. Wiley, Hoboken (2011)
14. Zeki, S., Marini, L.: Three cortical stages of colour processing in the human brain, Brain J. Neurol. **121**(9), 1669–1685 (1998)
15. Zeki, S., Nash, J.: Inner Vision: An Exploration of Art and the Brain. Oxford University Press, Oxford (1999)
16. Zeki, S., Watson, J.D., Frackowiak, R.S.: Going beyond the information given: the relation of illusory visual motion to brain activity. Proc. Roy. Soc. London B Biol. Sci. **252**(1335), 215–222 (1993)
17. Stein, M.: Transformation: Emergence of the Self. Texas A&M University Press, College Station (2004)
18. Cambray, J.: Synchronicity: Nature and Psyche in an Interconnected Universe, vol. 15. Texas A&M University Press, college Station (2009)
19. He, C.C., et al.: Expressive Art Therapy Lecture 15: The Good Medicine for Grief Consultation. Wu-Nan, Taipei (2017). (in Chinese)
20. Knill, P., Barba, H., Fuchs, M.: Minstrels of soul: intermodal expressive therapy. Arts Psychother. **3**(23), 275–276 (1996)
21. Jung, C., Ed Shamdasani, S., Kyburz, M., Peck, J.: The Red Book: Liber Novus. WW Norton, New York (2009)
22. Kabat-Zinn, J., Davidson, R.: The Mind's Own Physician: A Scientific Dialogue with the Dalai Lama on the Healing Power of Meditation. New Harbinger Publications, Oakland (2012)
23. Huang, D.X.: Youth Development and Counselling. Wu-Nan, Taipei (2015). (in Chinese)
24. Stein, M.: The Principle of Individuation: Toward the Development of Human Consciousness. Chiron Publications, Asheville (2006)
25. Lewin, K.: Field theory and experiment in social psychology: concepts and methods. Am. J. Sociol. **44**(6), 868–896 (1939)
26. Lin, M.L.: Research on the consistency of postmodern educational philosophy and brain plasticity-Human images of learners of Mandala through cooperative painting. In: Paper presented at the forth conference of educational aesthetics, National Dong Hwa University. (2017). (in Chinese)
27. Newberg, A., Waldman, M.R.: How God Changes your Brain: Breakthrough Findings from a Leading Neuroscientist. Ballantine Books, New York (2009)
28. Hsu, C.C., Hsieh, C.T., Shih, Y.L.: Application of Mandala Drawing for Meditation. Taiwan Counseling Quarterly **1**(1), 8–17 (2009). (In Chinese)
29. Henderson, P., Rosen, D., Mascaro, N.: Empirical study on the healing nature of mandalas. Psychol. Aesthetics Creativity Arts **1**(3), 148 (2007)
30. Jung, C.: Mandala Symbolism (RFC Hull, Trans.), Princeton University Press, Princeton (1973). 3rd printing
31. Barthes, R.: Introduction to the Structural Analysis of the Narrative. The Johns Hopkins University Press, Baltimore (1966)
32. Barthes, R.: Roland Barthes by Roland Barthes. Macmillan, London (2010)
33. Hall, J.A.: Jungian Dream Interpretation: A handbook of Theory and Practice, vol. 13. Inner City Books, Ontario (1983)

An Exploration of the Development of Visual Design in Taiwan - A Case Study of the Cover Design of Industrial Design Magazine

Po-Hsien Lin[1]([⊠]), Jianping Huang[1], Rungtai Lin[1], and Mo-Li Yeh[2]

[1] Graduate School of Creative Industry Design, National Taiwan University of Arts, Daguan Rd., Banqiao Dist, New Taipei City 22058, Taiwan
{t0131,rtlin}@mail.ntua.edu.tw, 50516059@qq.com
[2] Department of Cultural Creativity and Digital Media Design, Lunghwa University of Science and Technology, No. 300, Sec. 1, Wanshou Rd., Guishan District, Taoyuan City 33306, Taiwan, Republic of China
1101moli@gmail.com

Abstract. This study took the cover design of the first ten issues of the Taiwanese design magazine founded in the 1960s, "Industrial Design", as the research target. The purpose of the study is to examine the similarities and differences between the two sets of publications of Industrial Design and Bauhaus, through which to explore the continuation and blending of Bauhaus visual design in the cultural context of Taiwan after half a century of communication and development. The research results show that Industrial Design has continued Bauhaus's visual design language in the color application, graphics composition and layout design, while the font style is full of distinct characteristics of the times and region. Through the analysis, it is found that the subjects are more interested in visual design with strong color contrast, rich graphics composition, multiple design techniques and spatial representation. This study aims to explore the influence and spread of the Bauhaus movement in Taiwan's visual design, hoping to contribute to the construction of Taiwan's graphic design history and provide it to the design field.

Keywords: Industrial Design magazine · Bauhaus · Visual design

1 Introduction

"Industrial Design", which was the first design journal in Taiwan, has been published for over fifty years. The magazine contains the most complete history of the design industry in Taiwan, from the development of the industry and technology to how they affect the field worldwide. It also covers the transformation of the design industry from mechanized to Human Factors Engineering.

Walter Gropius established the first design school, Bauhaus, in Germany in 1919. The school was founded based on the educational theory "the fusion of arts and techniques"; it was a crucial starting point for Taiwanese modern design in 1960s [9] when

© Springer Nature Switzerland AG 2020
P.-L. P. Rau (Ed.): HCII 2020, LNCS 12192, pp. 356–366, 2020.
https://doi.org/10.1007/978-3-030-49788-0_26

government set policies and brought back design courses and ideas from abroad. Non-governmental group also started to publish reading material to express their ideas and values. Based on this background, scholars from America, Japan and Germany ran five "Industrial Design Training" workshops in Taiwan between 1963 and 1967. At the same time, our government sent skilled students abroad for further studies and gave grants to schools which set up related programmes; later on, these became the solid foundation of Taiwanese Industrial Design. [19]. This research investigates the magazine which recorded the Taiwanese Industrial Design field for over half of a century and focused on the comparison between development and the cultural content of Taiwanese Design Industry under the influence of culture, research, education and industry. Figure 1 is the structure of this research.

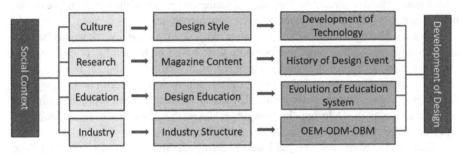

Fig. 1. Framework of the study

2   Research Purpose

Visual Design is the face of a product. It shapes and creates the expression of the product [10]. This research sampling analyzed the visual design of "Industrial Design" journal. The design elements of the magazine are quite similar to Bauhaus. The research used the four basic visual design aspects: font, image content, color choice and format design as the criteria for the analysis and compared them with classic Bauhaus posters. We aimed to understand the influence of Bauhaus on Taiwanese visual design in the 1960s, and in addition to contribute to the development of the historical narrative of Taiwanese modern design.

The purposes of research are as followed:

1. To understand the influence of Bauhaus on Taiwanese visual design in 1960s from the general public's point of view.
2. To understand the different versions and understandings of the public toward modern visual design in order to provide a reference for the design field.

3 Literature Review

3.1 "Industrial Design" Magazine and the Development of Taiwanese Industrial Design Education

It has been over half a century since the first issue of "Industrial Design" was published. Reading the magazine is like walking through the history of the development of Taiwanese Industrial Design Education. The cover for each volume was carefully chosen to represent the evolution of Taiwanese industrial design themes [10]. From the works on the covers, we can gauge the changing tastes in product design, which include the focus on operational functions; user friendly design; the combination of creativities and technology development; beautifying the exteriors etc. On the earlier volumes, they used dots, lines, surfaces to create aesthetically pleasing images on white covers while on the later magazines, photographical techniques were largely used.

After the photographical techniques were introduced, a design work would be used as a cover image. Within these works, some were based on human factor engineering, some were modeling-focused and some were highly culture oriented. For the more recent volumes, computer aided drawing once again took over from photography and 3D images have become the mainstream. To sum up, from 2D images to photography and from photography to 3D design, the changing styles on the magazine covers undoubtedly represent the transformation of techniques and design in Taiwan.

"Industrial Design" was first published by Ming Zhi Technical College in 1968. The government invited foreign scholars to the college to give speeches and teach course, for example the German scholar Frank Sander, who visited Ming Zhi in 1968; and Japanese scholar, Hiroshi Ohchi who held the workshop "Basic Design" in the same year [5, 18]. The covers for the first ten volumes of "Industrial Design" were colorful and focused on geometric design, which indicated the influence of German and Japanese scholars. In addition, some of these Japanese scholars studied in Germany and were introduced to Bauhaus during that time: we can say that the design education in the early Taiwan was highly influenced by Japan and Germany. This is also the reason why this research used the first ten issues of the journal: to study how Bauhaus extended its influence over Taiwanese visual design.

3.2 How Did the Bauhaus Effects Taiwanese Design Industry

Early in the 20th century, Taiwan was still a colony of Japan. Most Taiwanese artist were educated by the Japanese system; therefore, they imported the most advance craft art education and resources from Japan [20].

After the Second World War, financial and design industries developed quickly in Taiwan. Bauhaus came to the country at the same time, which started the modernization of Taiwanese design [6]. Bauhaus changed the way Taiwanese designers think, the way they understand their works and the way they present their designs and all of these changing processes were recorded in journals of the time.

In the mid 20th, there were many design journals published in Taiwan, for instance: "Designer" by National Taiwanese College of Art in 1964, "People of Design" by Guo Cheng Feng in 1967, Ming Zhi Technology College's "Industrial Design" in 1967 etc.

[18]. Not only did these journals introduce and comment on works from worldwide, but they also brought back ideas from foreign designers, which became a bridge between Bauhaus and the Taiwanese design industry [9].

3.3 Visual Design in the Bauhaus

In the early 20th Century, visual design in Europe was still highly influenced by Decorativism. Curving lines and bright cheerful colors created a novelty art style (Art Nouveau, 1890–1910) which was widely used in different forms of arts, such as architecture, painting, furniture design etc. However, after the cultural industry was introduced, product design no longer only targeted the top rank consumers but the general public.

The requirements and expectations of products were also becoming more practical and functional. In order to accommodate this change, Bauhaus accepted ideas from Russian Constructivism and De Stijl, highly focused on rationality, functions and geometry [17]. These features also matched perfectly with Expressionism and Suprematism in the early 20th century which were pointing in the direction of Abstract Expressionism. Under the influence of Abstract Expressionism, Bauhaus started a series of creative experiments in visual design, including font, image content, color choice and format design

Font. The most common font in German visual design in the early 20th century was "Garamond. This font was trivial and difficult to use and some designers noticed the problem, among them Herbert Bayer (1900–1985). Bayer was a former student of the Bauhaus who inherited the functional, practical and rational design from his school. Bayer stated in his book "On Typography": "from the history of font design, we found that font design is lacking standards, structure, accuracy and effectiveness. The new font design should focus on the original design of letters. We need to create a font that can actually let the letters speak. This is what we need in a scientific era. [1, 13]" Based on this theory, Bayer created a font of exclusively lower-case letters in 1925, the Universal. Ever since the Universal was released, Bayer used it to design the posters for Bauhaus exhibitions. It was clear and with no complicated lines, which made communication much smoother. These were also the features of the Bauhaus' font. The Universal was highly regarded by Walter Gropius and used widely in the Bauhaus' publications [3, 4]. It was the most advanced design in font and graphic design.

Image Content. Effected by the new arts movement in 1920s, Bauhaus gave up on Expressionism and turned to the Abstractism theory which Wassily Kandinsky (1866–1944) brought out, and which matched the need of Bauhaus perfectly. Kandinsky specifically defined "form" in his publication "Über das Geistige in der Kunst": "strictly speaking, forms are the boundary between two surfaces, however, the surfaces should always contain the inner essence. In other words, all forms have their meaning." Kandinsky focused on the integration of form and inner essence. He believed the abstract format is the best way to present deeply hidden meanings [2, 16]. Another Bauhaus master, Johannes Itten (1888–1967) started his research for Abstract Geometry in 1915. He was fascinated by the beauty of geometric shapes [7]. and according to his research, he believed only though prosodies, geometry and abstracts forms can designers represent the inner essence [11]. Therefore, under the leadership of Kandinsky and Itten, the foundation education of Bauhaus highlighted basic geometry combinations to train the students design ability. They believed squares, circles and triangles are the foundations of all other forms.

Color Choice. On the basis of geometry, Kandinsky believed that if color elements can be added into the design, this would bring different feelings for the audiences. He stated in the "Über das Geistige in der Kunst": "the thickness of the lines, position of the imagines, the shape of their forms are all extensions of paintings. Only the yellow triangle, red square and blue circle are the most beautiful combinations [8]." Itten believed colors should not only effect the audiences visually but also mentally [7]. His color theory made a solid foundation for modern color education.

Format Design. In the early 20th century, Bauhaus started to use asymmetry to create unbalance and dynamic elements in visual design, and to create a contradictory sensation. At the forefront of the movement was Moholy-Nagy (1895–1946), who was strongly influenced by Russian Suprematism and Constructivism [12]. Nagy believed in rationality and focused on the tension and spaces in the second dimension, trying to make geometry more stereoscopic through perspective transformation. It was a real departure from the original second dimension and highlighted the sense of space in geometry design [14]. Gropius observed how Nagy used light as a method to understand the space in the preface of Nagy's publication "New Vision". Nagy raised the ideas of the fourth dimension in space and Synchronicity. He also used new techniques to strengthen the light inside the space, which makes it closer to natural light [14]. Nagy tried to explore the limits of space through the use of lights and tried to transform his own impression of light into a whole new dimension relationship in his works [15]. This series of experiments pushed dimension design to a new level.

To sum up, the experiments for font, image content, color choice and format design led by the Bauhaus created a new language in the field of visual design. This new method of displaying became the foundation of design in several different industries. It also allowed Bauhaus to overcome the barriers of nationality, language and time, becoming a long running movement around the world.

4 Research Methodology

This research cross-compares the classic early 20th century Bauhaus poster to the first ten "Industrial Design" magazine's covers from the aspect of visual design in order to understand the development history of the two. The research started with literature reviews, and then used an online survey to gathered views from the subjects on perceptions of the difference between the images. Lastly, a series of interviews were held with magazine-related individuals. We didn't only receive firsthand oral statements from the interviewees but also photos, documents etc., which provided us with some valuable documents for qualitative research.

4.1 Research Instrument

The subjects of the internet survey are drawn from the Taiwanese public. We targeted four different demographic groups and surveyed their understanding of four basic visual design aspects. The survey included sets of visual design: two-dimension focused (Table 1) and three-dimension focused (Table 2). Both sets contained five classic Bauhaus

Table 1. Stimulus group 1: visual design with 2D feature

Table 2. Stimulus group 2: visual design with 3D feature

posters and five "Industrial Design" covers. The subjects completed the survey based on their feeling for the four basic visual design aspects.

4.2 Research Participants

The four targeted demographic groups are as follows: arts industry related, graphic design related, other design field related and others. 579 surveys were received and 579 were valid. The semi-structured interviews were conducted with cover designers, authors and editors of the "Industrial Design" magazine from various eras. We interviewed the subjects either once or twice. The interview time was between 20 min and an hour. The interviews were recorded and transcribed. We then analyzed them from the perspectives of formats, eras and region.

5   Research Results and Data Analysis

5.1   Visual Cognition Analysis of the Cover Design of Industrial Design Magazines

The research started by surveying the demographic groups' understanding toward four basic visual design aspects. We tested 579 subjects. According to the result showing in Table 3 and Table 4, the "color choices" in the Bauhaus poster and the Industrial Design magazine were very similar. However, the "font" and "format design" in the two objects didn't show the same similarity.

Table 3. Visual similarity between Bauhaus posters and Industrial Design magazines for group 1

Total (N=579)	Group 1				
	Item	Font Style	Color Application	Graphics Composition	Layout Design
	M	2.6*	3.2*	2.9	2.6*
	SD	1.1	1.0	1.1	1.1

*p <.05

Table 4. Visual similarity between Bauhaus posters and Industrial Design magazines for group 2

Total (N=579)	Group 2				
	Item	Font Style	Color Application	Graphics Composition	Layout Design
	M	2.7*	3.2*	3.0	3.0
	SD	1.1	1.0	1.0	1.0

*p <.05

Furthermore, we analyzed the relationship between the subjects' background and their responses. Among the 579 subjects, 85 of them were art industry related, 145 were graphic design related, 196 were from other design fields and 153 were from other backgrounds.

The result of group one: 2D focus, One-way ANOVA shows the F number didn't reach a significant level, which indicates that subjects' background doesn't affect their choices. However, in the results of the second group "3D focus" we see the F number reach a significant level in both Font Style and Layout design using One-way ANOVA (Table 5). The result also shows that the value for those who come from "other design field" is higher than for the other backgrounds, which means they think the similarity is greater between the groups.

Table 5. Analysis of variance for visual cognition by learning backgrounds

	Item	Source of Variance	SS	DF	MS	F	Dependent Variables	M	SD	Scheffe
Group 2	Font Style	Between Groups	12.11	3	4.04	3.398*	1.Art Major（N=85）	2.55	1.14	3>4
		Within Group	683.05	575	1.19		2.Graphic Design (N=145)	2.67	1.01	
							3.Design Major（N=196）	2.87	1.05	
		Total	695.16	578			4.Others（N=153）	2.53	1.18	
	Layout Design	Between Groups	12.75	3	4.25	4.041**	1.Art Major（N=85）	2.88	1.06	3>4
		Within Group	604.99	575	1.05		2.Graphic Design (N=145)	2.90	0.99	
							3.Design Major（N=196）	3.15	.97	
		Total	617.74	578			4.Others（N=153）	2.79	1.10	

*p <.05 **p <.01

We analyzed the results above and came to the following conclusion:

1) For the "font style" the two groups share less similarity. The Bauhaus emphasized simple, sans-serif font to make the communication clearer. On the other hand, the font "Industrial Design" magazine chose to use despite its boldface feature, is obviously squashed to make the font more consistent. We also see an inclined angle at the end of the character, which shows the lively characteristics of the industry. These are the main reasons for the difference between the Bauhaus and the magazine.

2) As for the "color application", a high similarity was detected. The Bauhaus tended to use original or secondary colors in design. These features were inherited by the "Industrial Design" Magazine which led to this result.

3) Graphics Design between the two also shows significant similarity. Emphasizing rationality and practical functions, the Bauhaus used abstractive and geographical design in visual design. Geographical design was also used in the cover for the magazine, which made for similarity between the two.

4) The only similarity seems to show in the second group "3D focus". From the prospect of professional background, subjects from "other design field" saw more similarity than others. However, in the group of "2D focus" none of the subject groups think they look alike.

5.2   A Verification of Formats in Visual Design

The semi-structured interviews were taken with staff from the "Industrial Design" magazine from various eras: we interviewed four cover designers, one authors and one editor who worked between 1967 and 1979; one author, one main editor and the president from 1980 to 1998; we also interviewed one current editor, one current main editor and the current president. We used these interviews to verify the results we obtained from the studies above. At the same time, we used these opportunities to understand the influence from the Bauhaus on Taiwanese visual design between 1920s and 1960s. We used the same two groups of subjects to test the twelve objects. We noticed that their reactions were very consistent. They all thought both groups show high similarity in "color application" but that "font style" were not so similar. In the following sections we will show further explanation for the interview results.

Font Style. Subjects thought similarity between the Bauhaus and the magazine in the prospect of Font Style is low. Subject A2, a cover designer from the first era, believed the difference came from the different focus of the two institutions. The magazine is still a commercial organization which will highlight this characteristic in its design. In order to match the lively and strong personality of "Industrial Design", the magazine not only used boldface but also cut an inclined angle at the end of the character to create a sensation of sharpness for the readers. Yet not everyone agreed with this opinion, some of the subjects thought font design from the Bauhaus was more vivid while "Industrial Design" shows more consistency. Nevertheless, both font styles used boldface as their basis; they are both clear, easy to read, recognizable and show modernity and practicality, which made for a certain level of similarity between the two.

Color Application. Subjects all think the two groups of work showed high similarity in color application. The main difference lies in the Bauhaus using big black blocks to show visual contrast, a technique which was rarely used in the magazine. The reason for this may come from the training for designers in the early days which did not usually encourage them to use black in their works. This made the color choices from the magazine gentler than the Bauhaus. Overall, the Bauhaus used color more boldly and without restriction while "Industrial Design" was more reserved. The magazine used secondary colors more often in their designs, which is also commonly seen in Japanese design.

Graphic Composition. Subjects thought the Graphic Composition between the two showed some similarity. Though they both used abstractive and geometry elements in their design, the styles were not the same and this led to different visual affect. In the works of the Bauhaus, we see they not only used surfaces but also used dots and lines to create multiple layers. This resulted in a more accurate and rational visual style and contained more depth and complexity. Most subjects believe this difference came from the cultural background. The Bauhaus focused on a lively style while "Industrial Design" tried to use geometry design to create harmony which is closer to the values in Taiwan.

Layout Design. Subjects believed some similarity showed in the two works. In group one, Graphic Design was thought to have significant difference. This came from the

Bauhaus showing more rationality and order in their font and graphics while "Industrial Design" had more sensitivity and balance. However, in the group two works, Layout Design seems to share more similarity. Lines and surfaces were both used to create perspective in order to make three-dimensional effects inside two-dimensional images.

Apart from the above four basic visual design aspects, the subjects also mention how texture, collage, images and size showed their importance in visual display. They used wooden texture geometry with black, white and gray backgrounds to highlight contrast and several green color blocks to lighten the picture on the cover of their first issue. Some subjects also pointed out most magazines were 21 * 28.5 cm in 1960s while "Industrial Design" used 21 * 21 cm. This novelty size also made the magazine more modern than others available at the same time.

The designer for the third issue was Japanese designer Hiroshi Ohchi who kept the original colorful feature from "Industrial Design" but also added more brand-new features: firstly, he extended the size of color block on the cover to full size and made even greater contrast in choice color, making the magazine more lively; secondly, he used new texture in the design. Inspired by collage methods, Hiroshi used the special serrated effect from torn paper to break the usual geometrical rules. This created a new visual effect for the cover and a more varied way of displaying.

"Industrial Design" used new elements and methods to experiment in their design. They inherit some from the Bauhaus but at the same time melded western ideals with Taiwanese strengths to create a modern magazine.

6   Conclusion and Recommendations

This research cross-compares the classic early 20th century Bauhaus's poster to the first ten "Industrial Design" magazine's covers from the aspect of visual design, gathering opinions from contemporary audiences in order to understand the similarity between the two. While surveying the understanding of our subjects, we also tried to learn how the Bauhaus rooted itself and grew in the Taiwanese design industry.

After more than half a century's development, the Bauhaus came to Taiwan in the 60 s and melded with local design features. This distinctive generational and regional visual design not only showed the development path but also the growth of Taiwanese society and economy.

The history of Taiwanese visual design is the foundation of modern design. For the aspect of great history, we can see a "routine cycle" feature. Therefore, we not only can study the core of original modern design and analyze it in a more complete manner but also use this to predict future trends.

This research also used group psychological cognition to understand the audiences feeling towards different designs. Hopefully, this data will be beneficial in allowing designers to understand the needs and preferences of their audiences. We also hope this research will contribute to the development of modern design history in Taiwan.

References

1. Bondy, J.A.: The "graph theory" of the Greek alphabet. In: Alavi, Y., Lick, D.R., White, A.T. (eds.) Graph Theory and Applications. LNM, vol. 303, pp. 43–54. Springer, Heidelberg (1972). https://doi.org/10.1007/BFb0067356
2. Düchting, H., Kandinsky, W.: Wassily Kandinsky, 1866–1944: A Revolution in Painting. Taschen America LLC, New York (2000)
3. Gropius, W.: The New Architecture and the Bauhaus. The M.I.T. Press, Cambridge (1965)
4. Gropius, W.: Scope of Total Architecture. G. Allen & Unwin, London (1956)
5. Hsiao, L., White, M.: The Bauhaus and China: present, past, and future. West 86th J. Decorative Arts Des. Hist. Mater. Cult. **22**(2), 176–189 (2015)
6. Huang, S.H., Lee, Y.H.: Counteraction against the industrial revolution: the pioneer of arts and crafts—Williams Morris and Yen Shui-Long. In: Lin, D.T. (ed.) Crafts pioneer—Yen Shui-Long, pp. 6–15. National Taiwan Crafts Research and Development Institute, Nantou County (2006)
7. Itten, J.: Design and Form: the Basic Course at the Bauhaus. Van Nostrand Reinhold, New York (1975)
8. Kandinsky, W.: Concerning the Spiritual in Art. Courier Corporation, North Chelmsford (1910)
9. Lin, H.L.: An identification and validation of competencies for industrial designers in Taiwan (Doctoral dissertation, The Ohio State University) (1991)
10. Lin, M.T., Lin, R.T.: The analysis of the changes to Taiwan's designs from industrial design magazine cover. Taiwan J. Arts **93**(10), 153–184 (2013)
11. Lu, M.: Johannes Itten's ideas on design education. Art Des. **3**, 85 (2006)
12. Malevich, K.: The Non-Objective World. Paul Theobald & Co, Chicago (1959)
13. McCoy, K.: Graphic design in a multicultural world. Design Studies—Theory and Research in Graphic Design, a Reader, pp. 200–205 (2006)
14. Moholy-Nagy, L.: Vision in Motion. Paul Theobald & Co, Chicago (1947)
15. Moholy-Nagy, L.: The New Vision: Fundamentals of Bauhaus Design, Painting, Sculpture, and Architecture (Hoffmann, D.M.: Trans.). Dover Publications, New York (2005)
16. Roskill, M.W.: Klee, Kandinsky, and the Thought of their Time: a Critical Perspective, p. xix. University of Illinois Press, Urbana (1992)
17. Wolfe, T.: From Bauhaus to Our House. St. Martin's Press, New York (1981)
18. Yang, C.: The process and influence of the "Industrial Design Training Programs" in Taiwan supervised by foreign experts in the 1960s. J. Des. **15**(4), 81–101 (2010)
19. Yang, C., Hasumi, T.: The analysis on the formulation and effect of the industrial design development strategy in the early stage of Taiwan. J. Sci. Technol. Humanity Sci. **18**(1), 1–14 (2009)
20. Yao, T.H., Sun, C.Y., Lin, P.C.: Modern design in Taiwan: the Japanese period, 1895–1945. Des. Issues **29**(3), 38–51 (2013)

Innovative Application of the Research on Material Properties of Chinese Lacquer in the Design of Pile Lacquer Implement

Yonghui Lin[✉] and Hailin Liu

Beijing Institute of Technology, Zhuhai, China
53907024@qq.com, puccayama@qq.com

Abstract. Lacquer art is an excellent traditional culture of China. As early as the Neolithic period, lacquer has been used to paint patterns on pottery spittoon or woodware. The wooden-body lacquer bowl unearthed from the third floor of the Hemudu primitive social and cultural site is the earliest lacquer ware known to date, with a history of more than 6,000 years. For us, lacquer art is not only a perfect expression of visual art, but also a symbol of national culture and spiritual symbols, as well as a materialized form of the spiritual life of the Chinese people.

Pile lacquer art is a kind of skill created by Taiwan's famous lacquer artist Yang Fengcheng. While inheriting the traditional lacquer culture and skills, he has also made creative transformation. He has created a large number of high-quality lacquer artworks with tubeless Pile lacquer skill, breaking the traditional pattern of using inner tube to put on lacquer and solving the problem that thousand years of burial and soaking would decay the wooden body in the lacquer ware, which enables the lacquer ware to go further in preservation and cultural heritage in the future.

In the age of chemical lacquer, especially today when the lacquer design gradually returns to human life, we pay attention to the style in terms of art. However, we should focus more on the purpose that design should serve people. That is, design should inherit the true traditional culture for everyone, and serve the quality of life and health for people. It is well known that chemical substances in chemical lacquer are harmful to human body and the environment, and also violate the research and inheritance of the use of Chinese lacquers in traditional lacquer art. Chinese lacquer not only has such physical properties as acid resistance, heat resistance, moisture resistance, anti-corrosion, hard film, high abrasion resistance and colorfast, but also has unique aesthetic characters.

Through in-depth research and experiment on the characteristics of Chinese lacquer materials, this topic understands changes of its color, texture, shape, and drying time under different environments, in different production processes, with different tools and different techniques. On this basis, apply the characteristics of Chinese lacquer to the design of Pile lacquer ware through the tubeless lacquer skill, which can help achieve practical and beautiful design of living lacquers. Inherit traditions in materials, draw on traditions in skills, break traditions in terms of shapes, and integrate lacquer art into human life, so as to inherit Chinese traditional culture.

© Springer Nature Switzerland AG 2020
P.-L. P. Rau (Ed.): HCII 2020, LNCS 12192, pp. 367–382, 2020.
https://doi.org/10.1007/978-3-030-49788-0_27

Keywords: Chinese traditional lacquer culture · Natural lacquer · Study on lacquer property · Pile of lacquer art · Implements the design · Innovative applications

1 Introduction

Lacquer art appeared as a practical utensil. The world has witnessed a splendid history of lacquer art since the wooden-body lacquer bowl of the Hemudu culture. The emergence of lacquerware not only brought convenience to human life but also delighted our spirit. It was major progress in the history of human civilization. Raw Chinese lacquer is the main material of lacquer art. It combines with wood, bamboo, rattan, leather, pottery, porcelain, metal, bone, fibre, etc. to form different lacquerware. From the first day when lacquer was used, its beautiful lustre has given it both practical and aesthetic functions, giving lacquer art both material and spiritual cultural values. The unity of practicality and aesthetics also gives lacquer art a wider living space, which makes lacquer art develop into two aspects, practicality and appreciation. The Chinese nation is the first nation to discover and use lacquer liquid and the Chinese history of lacquer art has experienced the wooden-body lacquer bowl of the Neolithic Age, the prosperity of the Chu and Han lacquer culture, and the flourishing period of the Ming and Qing dynasty. The fascinated charm of lacquer has gone on for thousands of years. Speaker of the World Lacquer Culture Conference, Mr. Nagatoshi Onishi said that lacquerware has its own deep, steady and graceful style. Her intoxicating and fascinating charm has lasted for thousands of years. Now it has attracted worldwide attention and is regarded as "the mystery of the oriental world". Chinese lacquer art has a splendid history of more than 7,000 years. From devices used in tribal ritual activities to the exclusive supplies of the upper class, from the art appreciated by literati and poet, and the daily necessities in the homes of ordinary people, lacquerware has occupied a certain position in the life of Chinese people.

Until now, lacquer art, a traditional craft born on the ancient land of the oriental world has been regarded as a form of artistic expression after undergoing various historical changes. Modern lacquer art has developed from a practical form of arts and crafts into an independent art category. The lacquer art discussed here contains two meanings, lacquer craftsmanship and lacquer art. Its artistry has been strengthened like never before. The practical functions of lacquerware are constantly evolving towards aesthetic functions. At this transition stage, it is difficult to distinguish the practical and aesthetic functions of lacquer. Through the study of this ancient but young field, we can have a clear picture of the status and development of this oriental traditional handicraft after the baptism of the western industrial revolution. The problems we face are how to continue to spread the tradition under the modern history and culture, how to have a new development at the same time, and how to find the way for lacquer art to survive and grow.

In the development of lacquer art today, the organic integration of aesthetics and functions, and the effective inheritance of Chinese lacquer material will become the focus of modern lacquer art creation. However, the variable property of Chinese lacquer has made it difficult to control and to our creation. Therefore, through the research and understanding of Chinese lacquer, this essay aims to awaken people who use chemical lacquer in the creation and draw their attentions of Chinese lacquer.

2   Research on the Characteristics of Chinese Lacquer Materials

2.1   Growth Environment and Morphological Characteristic of Lacquer Tree

China is a country rich in lacquer trees and has a wide variety of lacquer trees. The centralized growth of lacquer trees are in Yunnan, Hubei and Guizhou and mainly in Shaanxi Province. In August 2018, under the leadership of the famous lacquer artist Yang Fengcheng, the author and other members of Yang school came to Shaanxi and Hubei province to inspect lacquer trees (as shown in Fig. 2). The lacquer trees are suitable to grow in the humid climate and sunny hillside, in temperature 8–10 °C. The minimum temperature must be higher than −20 °C. At the same time, the annual sunshine duration for lacquer tree should be between 1,400 and 2,500 h, and the average annual precipitation should reach 500 mm. Lacquer trees are more suitable to grow in an environment where the soil is acidic, and the relative humidity is higher than 70%. The population density is higher in such an environment. Most of them are distributed at lower altitudes, such as at the foot of the mountain, on the mountainside or on the banks of farmland.

Lacquer trees are deciduous trees and grow up to 20 m tall with large leaves in ellipse, oval and ovate-lanceolate shapes (as shown in Fig. 1). The bark is greyish-white and smooth at first. When it grows older, it shows longitudinal cracks. There is milky white paint under the bark. In the early summer, it comes into yellow and green flowers and in autumn, they become morphological oblate with smooth and yellow appearance. Lacquer trees that have grown to eight to ten years can be used to extract lacquer liquid. As early as the Shang Dynasty in China (about the early 17th century BC to the 11th century BC), exquisite lacquerware had been made from lacquer. Laccase in the lacquer can accelerate the urushiol to oxidize to form a film, and it is a natural drier for cured forming a film of raw lacquer coating.

Fig. 1. Morphological characteristics of lacquer tree (Color figure online)

Fig. 2. Inspecting lacquer trees in Shaanxi and Hubei province with master Yang Fengcheng

2.2 Characteristics and Tapping Technology of Chinese Lacquer

Lacquer-reaping is related to the farming season, but the climate varies from place to place. On middle and low mountains with high temperature, the lacquer-reaping starts from summer, lasts for 120 days and stops in the frost descent. In the high mountains with low temperature, the reaping period is shorter, from minor heat to frost descent, lasting for 90 days and sometimes even shorter. Because the temperature plays a decisive role in lacquer-reaping, the farmers usually refer to the solar term for relevant activities. There is a rule 'Inspecting the lacquer at grain rain, setting up the shelf at the start of summer, peeling and releasing the liquid at the summer solstice, and start busy with reaping when seeing leafy trees.'

The age at which lacquer trees are harvested varies greatly from species to species. The earliest is 6-year-old while the latest is 12-year-old. Normally the trees are reaped at 8 to 10 years old and the quality is better if the tree is older than 10 years. When to cut the tree is mainly decided by the diameter at breast height (DBH) of the trunk. Large trunks with a DBH larger than 16 cm and small trunks with DBH larger than 13 cm are suitable for cutting. The test shows that if small lacquer trees are not cut in time, the thick lacquer liquid will lead to the leaves' drying and dying within 2–3 years, which is called bloat death. However, the large lacquer tree will not be affected even not cut in time.

Lacquer-reaping time refers to the best time to reap the tree in a day. Tests have shown that the amount of lacquer liquid is related to atmospheric temperature. When the temperature is low and the humidity is high, the amount of liquid is large; when the temperature is high and the humidity is low, the amount of liquid is small. Therefore, the best time to reap in a day is before sunrise, the sooner the better. Before sunrise, there is less transpiration in the tree canopy, the lacquer liquid comes out faster. The cut will not get dry easily since the temperature is high, so the secretion time is longer, and production is higher. After sunrise, the canopy is exposed to the sun, the transpiration is enhanced and the air humidity is low, the cut is easy to dry, so the amount of lacquer liquid is reduced. In the cloudy day, the tree can be reaped for the whole day. It is necessary to stop reaping on a rainy day because if the rain goes into the lacquer liquid, the cut will get rotten easily and the quality of the liquid will be affected.

Before reaping, tools including lacquer knives, scrapers, lacquer buckets and clamshells should be prepared. The cut has several shapes, such as thrush eye, willow leaf, fishtail, bull-nosed shape and so on (as shown in Fig. 3). The practice in recent years has proved that the bull-nosed shape is a better form. The bull-nosed form does less harm to the trees and they would heals quickly, which can shorten the recovery period of the tree and improve the resource utilization, and increase the lacquer production. The way of using the lacquer knife in reaping has an effect on the yield of lacquer. There are upward cutting and downward cutting. Upward cutting goes first and then downward cutting. When cutting the lacquer, one should take the knife firmly, cut and lift the knife quickly to ensure the one-shot, with no residue around the cut. The cut depth should be appropriate. If the cut is too shallow, the lacquer liquid flows slowly and produce less lacquer; while if the cut is too deep, it will hurt the wood and affect the normal growth of the lacquer tree; the exuding water will also affect the quality of the lacquer liquid. The width of the bark cut is usually within 3 mm. Each cut should ensure a sharp top and flat bottom because this shape is easier for liquid flows down. Flat or concave place on the trunk is better than a raised area for a cut. Cuts should be kept away from insect pests. Clamshells are used to store the lacquer. After the cutting, insert the clamshell into the crevice on the bark. The crevice is deep enough if the clamshell can be inserted firmly.

Fig. 3. Reaping tools and methods

2.3   Character and Compositions of Chinese Lacquer

The main ingredients of Chinese lacquer are urushiol, nitrogen, gum, and water, with a small amount of other organic substance. The content of various components varies with the different species, growth environment and harvesting time of lacquer trees. In China, raw lacquer roughly contains 40–80% urushiol, lower than 10% nitrogenous, lower than 10% tree resin, 20–30% water and a small amount of other organic substance. The content of each composition is shown in Fig. 5. Urushiol is the main component of raw lacquer. It is insoluble in water, but soluble in various organic solvents such as ethanol, ether, acetone, xylene and vegetable oil. Laccase exists in nitrogenous substance and also known as raw lacquer protein and oxidative enzymes, which account for about several ten-thousandths of the entire lacquer solution. It is insoluble in water and organic solvents, but soluble in urushiol. It can promote the oxidation of urushiol and accelerate its drying and film-forming process. So, laccase is an indispensable natural organic drier when raw lacquer is drying at room temperature. The activity of laccase is closely related to temperature and humidity. Practice shows that when the raw lacquer is at a temperature of 20–35 °C and relative humidity of 70–80%, it is easier to dry and form a film. Otherwise, it is not suitable for drying. When the temperature is 35 °C and the relative humidity is 80%, the laccase has the highest activity. When the temperature is raised to 75 °C, its activity will be completely destroyed within one hour. Laccase activity is related to the condition of raw lacquer. The laccase separated for new lacquer is blue and has high activity, while that separated from aged or partially oxidized raw lacquer is white and has low activity. Tree resin is soluble in water but insoluble in organic solvents. It belongs to polysaccharides and contains a trace amount of calcium, potassium, lead, sodium, silicon and other elements. Tree resin is transparent and has a faint scent. The content of tree resin in raw lacquer often varies with the type and origin of the lacquer. Generally, large lacquer contains more tree resin, and small lacquer has lower content. Tree resin enables the main components including water in raw lacquer to be a more uniform colloid, making it more stable and not easy to deteriorate. Water in Chinese lacquer is one of the main ingredients for forming emulsoid, and it is also a necessary component for laccase to play its role in the drying process of raw lacquer. Even refined raw lacquer contains 4–6% water, otherwise, it is extremely difficult to get dry by the lacquer itself. Generally speaking, in raw lacquers, the less water, the better the quality (Fig. 4).

Fig. 4. Filtration of Chinese lacquer

	China sumac	Vietnam sumac	molecular weight	Polar groups
urushiol	40-80%	52%	320	-OH
Gum tree	5-7%	17%	2200	-COO-, Metal ions
polysaccharide			2700,8400	-OH,-O⁻
glycoprotein	2-5%	2%	(8,000)6-15	Protein+10% sugar
laccase	<1.0%	<1.0%	120,000	Protein+45% sugar
moisture	15-40%	30%	18	

Fig. 5. Compositions of Chinese lacquer

2.4 Difference Between Chemical Lacquer and Chinese Lacquer

The modern chemical industry is well developed. There are many types of synthetic lacquers in many colours. Lacquerware and souvenirs made from chemical lacquer are exquisite and beautiful. Solidification of chemical lacquer is faster and the production cycle is shorter. While Chinese lacquer features a longer production cycle and takes more time and labour. So, what are the difference between chemical and Chinese lacquer? Types of chemical lacquers include polyurethane resin lacquer, nitrocellulose varnish and spray lacquer. Chemical lacquer is easy to buy and use, not requiring humidity conditions, fast-drying, transparent, suitable for various light-colour modulations, especially suitable for gold and silver topcoat. But after long-term use, chemical lacquer is easy to get yellow, fading, cracking, not abrasion resistant and too shiny. From an aesthetical perspective, it is not as soft, subtle and restrained as Chinese lacquer, and what is worse, it contains more harmful chemical ingredients. Chinese lacquer has strong corrosion resistance, durability, heat insulation, waterproofness and antiseptic property. It can be used in medicine or in food. The lacquer tree buds can also be cooked, and the fruits can be processed to make lamp oil and candles. In addition, the film of Chinese lacquer has strong corrosion resistance, it will not deform or deteriorate even contacting with acidic materials such as hydrochloric acid and sulfuric acid. Chinese lacquer offers a long-lasting gloss, which cannot be provided by the chemical one. The strong adhesion of Chinese lacquer plays a significant role in repairing pottery, stoneware, metal and other implements. It only takes a day for the chemical lacquer to come off when it gets on hands, but it takes a week or even longer for Chinese lacquer to come off. The surface hardness of the lacquer film is extremely strong, and it will not be easily damaged a few days after applying layers. With prolonged use, the lacquer will show softer and more beautiful and gloss and becomes more transparent. Such changes are natural and cannot be achieved by chemical lacquer. It is a kind of animate changes. Nowadays, in an industrial society full of synthetics and substitutes and a modern society where industrial pollution is getting worse and the ecological environment is being destroyed, human beings are paying more attention to natural things because of their wishes to protect themselves and the living environment of the planet. Therefore, the value of Chinese lacquer and its art will also receive more and more attention. For example, in Japan, for the same two pieces of lacquerware, the price of natural lacquer is tens, hundreds or even thousands of times that of chemical lacquer. Tourists would rather spend more money on lacquerware made from Chinese lacquer because they know that such lacquerware is the most representative of Japanese craftsmanship, which fully reflects the national characteristics and beauty of Japanese art. Though modern material production in Japan is well-developed, Japanese people still insist on creating crafts with natural taste using natural materials. In terms of materials, if chemical lacquer is used, real gold and silver decoration will not be used. Instead, it is a Japanese convention to use fake gold and silver decoration in this case. The Japanese lacquer art world is divided into traditional school and modern school. Though the two schools have different opinions in their ideology, they agree on the use of Chinese lacquer. Therefore, in lacquer art, chemical lacquer is only regarded as a supplement to Chinese lacquer, rather than a substitute.

As shown in Fig. 6, the colours of work made of chemical lacquer are dull, jumpy and bright. During the production process, the colour of the lacquer almost remains unchanged. In the pile lacquer works of Chinese lacquer by Mr. Yang Fengcheng, the colour of the lacquerware is almost the same as the colour of the dried lacquer. The pattern in Chinese lacquer is soft, subtle, restrained and has subtle gradation changes. It looks totally different with naked eyes and under strong light. Colours of layers painted at different times will change with time and the environment, from dimness to transparentness, from depth to lightness. Such change is animate under the action of nature. As Mr. Yang says, "Lacquer is able to breathe. It absorbs water and spray water, which an organic life form. Under the action of human craftsmanship, it will give you natural surprises".

Chinese lacquer and chemical lacquer are compared in terms of colour, taste, quality, drying time, impacts on the environment and other factors, and they have totally different reactions. They have different colour and colour changes. Fresh raw lacquer is milky white at first. After being exposed to air, it gradually becomes golden yellow, reddish-yellow, red or purplish-red (also called liver-coloured), and finally, it turns to dark brown or pure black. The colour changes of raw lacquer are called as "white like snow, red like blood and black like iron". While the colour of chemical lacquer will not change from beginning to end (as shown in Fig. 6). Chinese lacquer has a strong lacquer acid scent. For example, large wood lacquer features acid scent, small wood lacquer features light acid scent and Maoba lacquer has a soft aroma. On the contrary, chemical lacquer has a harsh chemical odour. The texture of the two lacquers is different as well. For Chinese lacquer, if lifting the lacquer with a stick, the liquid flows down in strips. When the flow breaks, the upper and lower end retracts strongly, and then the liquid drips down. The drops fall down elastically when dripping into the bucket, small vortexes appear in the

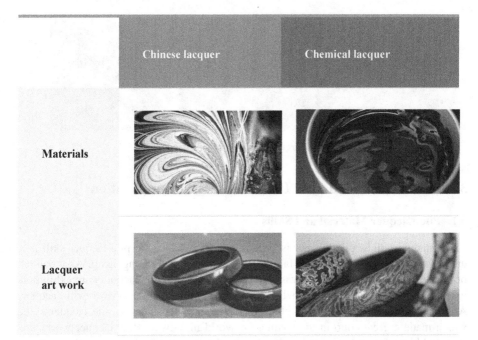

Fig. 6. Comparison of Chinese lacquer and chemical lacquer on materials and lacquer art works (Color figure online)

surface and then disappear quickly. For chemical lacquer, it does not flow down in strips. It is thick and drips into piles quickly. Chinese lacquer needs a certain temperature and humidity to get dried. To get the ideal condition in the warm and humid room, one can use artificial heating, water splashing or putting wet sacks and straw bags on the ground. The warm and humid room is an essential place for Chinese lacquer production. On the contrary, chemical lacquer can be dried regardless of climatic conditions.

2.5 Lacquer Character and Symptoms of Allergies in Humans

Chinese lacquer is non-toxic, but some people will have skin irritation when they contact with lacquer. Some even suffer from skin rash when smelling lacquer. The saying goes that people are bitten by lacquer, but in fact, it is a kind of lacquer allergy (as shown in Fig. 7). Degree of allergy varies depending on the person's physical fitness and health status. After 2–3 days after contacting Chinese lacquer, one may feel itching in some spots and get patches of red spots. The sports may start to swell and gradually form a blister. Small blisters will grow into scabs. After the scabs have fallen off, the skin will recover, leaving no scars. The allergy results from the urushiol contained in Chinese lacquer, but urushiol will solidify after oxidization, which is an essential characteristic in the production of lacquer artwork. For people who work with lacquer for a long time, most of them will develop antibodies if they don't use the medicine after allergies, so they will no longer be allergic.

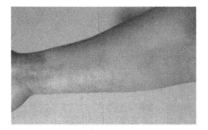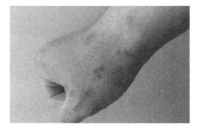

Fig. 7. Allergic symptoms of Chinese lacquer

3 Pile Lacquer Skills and Innovative Design Application

3.1 Pile Lacquer Material and Skills

As its name implies, pile lacquer is a stack of Chinese lacquer. pile lacquer skill is a unique skill created by Taiwan's famous lacquer artist Yang Fengcheng. As a painter of high-end furniture, Mr. Yang was shocked when visiting the lacquerware exhibition in the National Palace Museum in Taiwan and made up his mind to work with lacquer. After years of experiment and exploration, he created the tubeless pile lacquer skill, which made a great coup in the lacquer art world in Taiwan. Raw lacquer works are usually black after drying, but Mr. Yang's pile lacquer works show multiple layers of brown light. Such creation is kind of accidental. Once, Mr. Yang jumped into the bathtub for a few hours, holding the work he had just finished. But later he found that the pile lacquer slowly changed its colour in the water. So, he did more experiments and finally mastered how would lacquerware change colour with different time in the water for the implement's color.

Pile lacquerware starts from materials. Experiments and exploration of lacquer properties from different aspects in the early stage play a decisive role in the production of pile lacquer implement. Only by understanding the property of lacquer, can one control the changes of lacquer and better integrate his artistic ideas into the design of work. Chinese lacquer itself is a kind of slightly thick emulsion. How does pile lacquer production stack lacquer layer by layer to achieve the solidification?

First, ensure the quality of Chinese lacquer. Barrels of lacquer is divided into upper, middle and lower layers. The upper layer is a surface film. The film crusts and becomes shiny black. It is in good quality if it shows wrinkles and as light as chicken skin. The middle layer should be yellow with red heads. The lower layer should be milky white, with high density and thickness. As shown in Fig. 8, if the colours of the three layers show little or no difference, the lacquer has low quality or is mixed with impurities. Only with a deep understanding of lacquer quality, can we guarantee the solid and bright texture of the pile lacquerware.

Second, learn the drying time of the lacquer in different environments. Put on lacquers on several small wooden boards, store them in different seasons, climates, spaces and locations, making corresponding records of the drying time and colour changes. And finally, learn about how the property of lacquer changes in different environments. Applying these changes in the design of lacquerware, and we can conclude that the best

Fig. 8. Colours of different layers of lacquer (Color figure online)

drying environment for lacquer is relatively high apparent temperature and high humidity. The layer that dries faster is darker, while the layer that dries slowly is brown and more transparent.

Third, pile lacquer needs to add a small amount of special edible stone powder to Chinese lacquer. Stir the two evenly to allow the powder to completely penetrate the Chinese lacquer, without damaging the quality and property of lacquer. On this basis, put the stirred lacquer in the bottom corner of the fresh-keeping bag, seal the knot and make a small opening in the corner of the bag with scissors. When painting, put on the pile lacquer little by little on the corresponding position of the lacquerware (as shown in Fig. 9). Apply the lacquer slowly from the bottom to the top. The height should be determined by climate condition or environments. In higher temperature and humidity, the lacquer dries faster, so it can be stacked up to four layers. Under such condition, piling can be repeated every three hours. In the case of low temperature and dry weather, the number of layers of the pile lacquer should be reduced accordingly, and the time between two operations should be longer to ensure the lacquer gets dried entirely. Piling skill is also a basic requirement in the pile lacquer skill. Different people have different strength, different direction and position when putting on pile lacquer, which will lead to different changes in the shapes of lacquerware. Therefore, one should not only know well about properties of lacquer but also know how to apply the changes of its properties in the design of pile lacquer implements. In the piling process, the shape of the lacquerware cannot be repaired if the lacquer has not got dried. Otherwise, the overall shape will collapse due to the effect of lacquer property. We should wait for the paint to get dry before repairing.

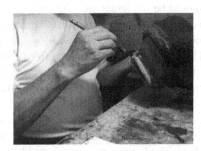

Fig. 9. Paint mixing, piling and detailed painting in the pile lacquer skill (Author: Yang Fengcheng)

Fourth, after the lacquer base is finished and dried, detailed painting is required to ensure the overall smooth surface of the implement. As shown in Fig. 9, use a painting brush to put on lacquer into the gaps, then use a horn scraper to smooth the lacquer to make the surface flat. When the lacquerware is 80% dry, it can be put into water for two weeks. The colour of the implement will become lighter during the immersion. But the immersion time should not be too long, otherwise, it will affect the quality of the lacquer. The colour changes after the immersion provide a foundation for rich changes of layers in later lacquer application and polishing.

Fifth, polish the lacquer base with good-quality water-abrasive paper. The first pass is using thicker sandpaper (180–300 g). During the polishing, some holes left during the piling will be found. After the primary polishing and drying, use lacquer paste to fill the holes. After the repeated operations, the lacquer left at different times will have different colour changes, thereby enriching the layer. Then use 600–800 g sandpaper, and then 1200–1500 g sandpaper for polishing. At the last stage, use 2,000 g sandpaper. The well-polished lacquerware surface feels like human skin, delicate and smooth (as shown in Fig. 10).

Fig. 10. Polishing and painting of pile lacquer implement (Author: Lin Yonghui)

Sixth, put on lacquer on the polished lacquerware. Put the lacquer on the work and then spread. The layer of lacquer put on the lacquerware can be thick, but when spreading the lacquer, the strength should be moderate, and the speed should be slow. The brush maintains a slight slope to ensure uniform paint surface. The brush should be made from horse mane, the length should be within 2–3 cm. Polishing and lacquer applying should be repeated for 3 times. The selection of sandpaper is from coarse to fine. This step is finished until the lacquer surface shows uniform gloss without slight unevenness.

Seventh, lacquer rubbing. Use a cotton ball with good texture (especially used for lacquer rubbing) to wipe the surface of implement with Chinese lacquer. The strength should be moderate to ensure a flat and even surface, and that the lacquer will remain on the surface. Excessive strength will erase the lacquer layer, while too light strength will lead to an uneven surface. The lacquer rubbing should be repeated for three times, and the surface should be dried completely before each rubbing, as shown in Fig. 11.

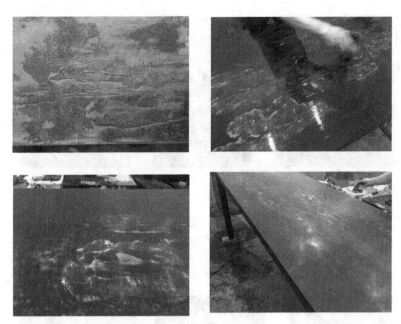

Fig. 11. Design and production of lacquer art furniture (polishing and lacquer rubbing) (Author: Lin Yonghui, Huang Guangjie, Wu xueyin, Xu Ming)

3.2 Design and Application of Pile Lacquer Implement

The traditional lacquerware design elements are mainly divided into two types, function and decoration. Functions of traditional lacquerware refer to the structure of lacquerware, which is generated by a certain function. In other words, the function is displayed in the form of physical objects. The specific division of functional forms of tradition lacquerware can be divided into the followings: 1) Practical design. The ancient food utensils such as ear cups, spoons, tripods and boxes were forms of practical art from the perspective of harmony and integration. 2) Bionic design. The bionic design generally refers to the artistic design according to animal images in nature. In bionic design, the morphological characteristics of animals are perfectly combined with the material form of lacquerware. The pig-shaped box, the tiger bird, the phoenix cup, and the lying deer are representatives in traditional lacquerware. 3) Lacquer musical instruments and weapons. In ancient life, musical instruments and weapons were used frequently, therefore, they were produced in large quantities, such as harps and lyres, lacquered bows, patent leather armour, and lacquer shields. The decorative form of traditional lacquerware refers to all the external characteristics of lacquerware. This external feature can trigger people's association and make them realize its function. For example, the abstract geometric ornaments, animal patterns, monster patterns, semi-abstract decorative ornaments, and theme ornaments reflecting social life on lacquerware are reflections of traditional lacquerware.

Fig. 12. Pile lacquer implement works created by Mr. Yang Fengcheng

In the pile lacquerware design of Mr. Yang Fengcheng, we can see the clam, restrained and elegant characters of traditional lacquerware. Such character is the nature of lacquer. Based on his own understanding of beauty and the refinement of realistic form, Mr. Yang has done innovative designs. He broke the relatively simple painting and flat application of traditional lacquer. Mr. Yang takes advantages of the characteristics of lacquer to do experiments by immersion, cooking, light drying, lamp drying and other methods. He learns about the characters of lacquer from the perspective of artistic production, which helps him apply the properties of lacquer in the production freely and appropriately. The natural character of lacquer has become his expression of beauty. The gradation of lacquer colour is integrated with the shape and colour of the pile lacquerware. Mr. Cai Kezhen, former director of the Department of Crafts of Guangzhou Academy of Fine Arts, comments on Mr. Yang that he integrates innovation into traditions, keeping both of them during his production.

In the design of implements, Mr. Yang has inherited and borrowed from the traditional classics, he has changed from decoration intention to practical design. The simple and straight pen pot, brush washer, the teapot, and the fruit plate all reflect the aesthetics and wisdom of the artist (as shown in Fig. 12).

4 Conclusion

The beauty of lacquer is reflected in its properties and appropriate operating techniques. The higher the technical difficulty, the more challenging and higher aesthetical value it is and has. Chinese lacquer has many limitations. For example, it is expensive, requires complicated techniques, easy to cause skin allergy and has many limitations in colours. But without these limitations, the soul and beauty of lacquer will disappear. The value of lacquer art lies in its irreplaceability. Lacquer creation should first apply the rule of lacquer itself. We should realize and accept its limitations before any creation.

Modern lacquer art integrating into modern space requires us to use traditional techniques and combine new design concepts. Due to change of concepts, there are certain differences in aesthetic standards and production techniques between modern and traditional lacquer art. Traditional lacquerware pays more attention to the neatness, preciseness, and symmetry of shapes, as well as the practicality and decorations. While modern lacquerware design emphasizes the aesthetic value of lacquerware. The shape and decoration of modern lacquerware are integrated, in other words, combined perfectly. The modern lacquer art breaks the boundaries between painting and craftsmanship, absorbs modern design concepts in the design, attach significance to the texture of the material itself and applies the characteristics of the lacquer to combine various techniques to express the unique charm of lacquer. Hegel once said that implication is more profound than visible presence. A work of art should have implication. We should create lacquer works with a sense of the times, and the unique oriental charm. Innovating and expanding on the basis of elegance and quietness and ancient charm will better meet the aesthetic characteristics and development path of lacquer itself.

References

1. Ji, N.: King of raw lacquer. Mt. Area Dev. (1989)
2. Chang, F., Zhang, W., Wei, S.: Study on lacquer resources in China and its fine application. China Lacquer **26**(02), 36–60 (2007)
3. Du, Y.: Survey of natural lacquer chemistry in recent years. Coat. Ind. (01), 51–59 (1980)
4. Wang, Q., Liu, D.: Identification of raw lacquer. Shaanxi For. Sci. Technol. (05), 20–29 (1980)
5. Xiao, M., Hu, S.: Harvesting technology of raw lacquer. Spec. Econ. Anim. Plants (04) (2010)
6. Speech by Mr. Nagatoshi Onishi, speaker of World Lacquer Culture Conference and Professor of Tokyo University of the Arts, at the International Symposium on Prospects of Lacquer Culture in the 21st Century
7. He, H.: Walking from Hemudu Culture- Selected Papers of China Modern Lacquer Art Exhibition 2005, Brief Introduction of Lacuquer Renaissance, China National Photographic Art Publishing House, p 196 (2005)
8. Qiao, S.: Qiao Shiguang's Lacquer Works. People's Fine Arts Publishing House, P68
9. Si, L.: The current status and industrial development prospects of lacquer resources and varieties in China. Green Technol. (01) (2016)
10. Tu, G.: Research and application of traditional lacquerware design elements. Northern Lights (10) (2015)

Enlightenments to the Beijing Winter Olympic Games: A Review of Main Color Schemes for Olympic Games

Meiyu Lv[1,2] and Huijun Qin[1(✉)] (iD)

[1] School of Digital Media and Design Arts, Beijing University of Posts and Telecommunications, Beijing 100876, China
usaginomifi@bupt.edu.cn
[2] Beijing Key Laboratory of Network and Network Culture, Beijing University of Posts and Telecommunications, Beijing 100876, China

Abstract. The upcoming Beijing 2022 Winter Olympic Game is the largest sports event for Beijing since the 2008 Olympic Games. Both the Chinese government and Chinese people are giving great efforts presenting the perfect Winter Olympics to the world. From the perspective of visual design scheme for the Beijing Winter Olympic Games, color scheme is the foundation of the whole visual design system, thus the importance of color scheme is self-evident. This paper discusses the main color schemes of the Summer Olympic Games and the Winter Olympic Games during 1980–2020 through analyzing the reasons of color selection, color image scales, word image scales, and the properties of hue, blackness and chromaticness based on the NCS color space. According to the results of analysis, the color schemes for the Summer Olympics focus more on appearing the local landscape or culture of the hosting city/country, while the color schemes for the Winter are often developed with the season and Olympic elements. Hard colors should be avoided, soft and warm colors or soft and cool colors are more popular choices to choose. colors which combined by green and blue are seldom used, medium or high blackness colors are rarely used, and low chromaticness colors are rarely used. But the proportion of using low chromaticness colors is bigger in the color schemes of the Winter Games. These findings give instructive advices to the color scheme for Beijing 2022 Winter Olympic Games.

Keywords: Olympic games · Color · Color scheme · Beijing 2022

1 Introduction

The Olympic Games itself resembles the image of peace, vitality and development. Hosting the Summer Olympics or the Winter Olympics has been an excellent chance to present the culture and achievements in infrastructure building and economy strength of a country to the world, which attracts more opportunities for development.

Visual design for Olympic Games is a subtle but powerful tool to facilitate the success of the grand event. The first image people around the world perceive of an Olympic Game

© Springer Nature Switzerland AG 2020
P.-L. P. Rau (Ed.): HCII 2020, LNCS 12192, pp. 383–395, 2020.
https://doi.org/10.1007/978-3-030-49788-0_28

would definitely be the logo, then the mascots. Hence visual design serves for the event since the very early stage of it and exerts strong influences on the satisfaction level of the audience. While color scheme is the foundation of the whole visual design system, thus the importance of color scheme is self-evident. Therefore, studying the color schemes of previous Olympic Games and learn from them can give valuable instructions to the color scheme for the Beijing 2022 Olympic Games. Through improving the quality of visual design for the Beijing 2022 Olympic Games, a positive effect can do good to the image of our country as well as the design industry of our country [1].

Based on the literature review, though many research on the Olympics have been done, the investigations on color schemes are much less than other subjects, reviews of visual design are mostly focusing on 2 or 3 certain events. Considering of this, a preliminary study which only focus on the main color schemes for the Olympics is necessary for further study. Thus the goal of this paper is to discuss a collection of the main color schemes of the Summer Olympic Games and the Winter Olympic Games during 1980–2020, and refining instructive advices for Beijing Winter Olympics through analyzing the reasons of color selection, color image scales, word image scales, and the properties of hue, blackness and chromaticness based on the NCS color space.

2 Literature Review

2.1 Research on Olympic Color Schemes

Zhang [2] analyzed the main color scheme of Athens, Beijing and London by the NCS color space as a means of quantitative research, calculating the whiteness, blackness and chromaticness and give visualized charts to show the characteristics of the schemes. Wang [3] studied the main color scheme of the Beijing 2022 Olympic Games by analyzing the reasons of color selection and application. Shi [4] also studied the reasons of color selection and application in the Turin, Vancouver, Sochi Winter Olympic Games, as a part of studying the communication value of the Olympic color schemes. While Zhang [5] analyzed the reasons of color selection and application in his study on the visual image design of the Rio Olympic Games. Dong [6] took the Rio Olympic Games as an example, too. He investigated the color scheme from a semiotic perspective.

In general, the number of qualitative researches are much more than quantitative researches on the Olympic color schemes, and more researches focus on the reasons of color selection. Therefore, this paper choose to use quantitative way to compare the main color schemes of Olympic Games to find the common characteristics of the reasons of color selection and the colors, exploring a new strategy to analyze the Olympic color schemes and try to discover instructive advices for Beijing 2022 Winter Olympics.

2.2 Color Image Scale and Word Image Scale

The Image Scale System is invented by the Nippon Color and Design research Institute based on wide researches, which contains of color image scale, color combination image scale, word image scale, and image scale for interior designs (see Fig. 1 [8]). This system can effectively transform the colors into their corresponding Kansei adjectives in the form of scale maps [7, 8]. Among all the scales, color image scale and word image scale are suitable to describe the images of Olympic colors.

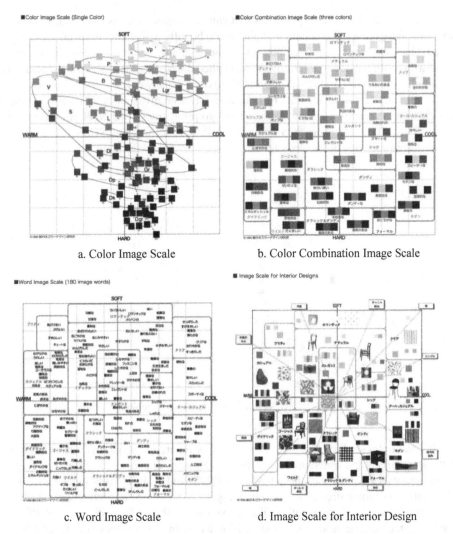

a. Color Image Scale

b. Color Combination Image Scale

c. Word Image Scale

d. Image Scale for Interior Design

Fig. 1. Image scale system, retrieved from *Kobayashi, S.: Color image scale* [8]

2.3 NCS Color Space

The Natural Color System (NCS) is a proprietary perceptual color model. It is based on the color opponency hypothesis of color vision, first proposed by German physiologist Ewald Hering [9]. The current version of the NCS was developed by the Swedish Color Centre Foundation, from 1964 onwards. The NCS states that there are six elementary color percepts of human vision—which might coincide with the psychological primaries—as proposed by the hypothesis of color opponency: white, black, red, yellow, green, and blue. The last four are also called unique hues. In the NCS all six are defined as elementary colors, irreducible qualia, each of which would be impossible to define in terms of the other elementary colors. All other experienced colors are considered composite perceptions, i.e. experiences that can be defined in terms of similarity to the six elementary colors. E.g. a saturated pink could be fully defined by its visual similarity to red, blue, black and white. Beginning with the elementary colors, it is possible to construct a three-dimensional descriptive model called the NCS color space, which includes the whole color world and makes it possible to describe any conceivable color percept. All imaginable surface colors can be placed and thus be given an exact NCS-notation. For the sake of clarity, it is usually shown in two projections – the color circle and the color triangle [10, 11]. The NCS color circle shows the hue, the NCS color triangle shows the nuance of a color (see Fig. 2).

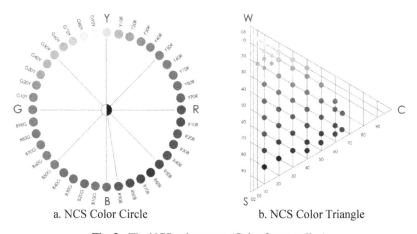

| a. NCS Color Circle | b. NCS Color Triangle |

Fig. 2. The NCS color space (Color figure online)

For example (see Fig. 3), S means that the color is part of the visual selection of NCS 1950 standard colors, which illustrates the NCS System. The first part of the code (e.g.) 1050 describes the nuance of the actual color R90B [11]. This has 10% blackness, 50% chromaticness (color strength). The remaining 40% out of 100% is whiteness which is not printed out in the NCS notation. The second part of the NCS notation, e.g. R90B, is the hue, which can be described as the position within the NCS color circle. Code R90B is a reddish blue color described as: A red (R) with 10% resemblance to red and 90% resemblance to blue (B). Another example: B40G is a blue color with 40% green, and so on [11].

Fig. 3. Example of a NCS color (Color figure online)

By transforming the main Olympic colors into their NCS forms, the distribution of the three properties (hue, chromaticness, and blackness) would be revealed.

3   Methodology

The goal of this study is to compare the main color schemes of Olympic Games to find the common characteristics of the reasons of color selection and the colors, exploring a new strategy to analyze the Olympic color schemes and try to discover instructive advices for Beijing 2022 Winter Olympics. To achieve this goal, main color schemes of both the Summer and Winter Olympic Games need to be collected, as well as the reasons of the color selections. Then the color image scale, word image scale and NCS color space would be applied to characterize the images and properties of the colors. Next, comparing the two groups (Winter Olympics and Summer Olympics) and refining valuable characteristics for the Beijing 2022 Winter Olympics.

3.1   Analysis of Main Color Schemes for the Summer and Winter Olympic Games

The main color schemes and reasons of color selections of the Summer and Winter Olympic Games can be collected from the respective official websites or other websites (for the sake of insufficient literature materials), the links would be affiliated in the references of this paper. Then separately build the color image scales, word image scales and NCS color spaces for the two kinds of Olympic Games, and conclude their images and characteristics.

The sample of color image scale and word image scale (translated by authors) (see Fig. 4) are below.

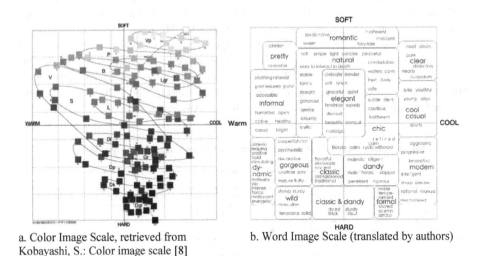

a. Color Image Scale, retrieved from   b. Word Image Scale (translated by authors)
Kobayashi, S.: Color image scale [8]

Fig. 4. The word image scale and word image scale

The NCS color spaces are built basing on the original system as well, and the rules of characterizing the hue, blackness and chromaticness properties are presented in Fig. 5, indicating the eight kinds of hue, three degrees (0–35, 36–65, 66–100) of blackness, and three degrees of chromaticness (0–35, 36–65, 66–100) [1].

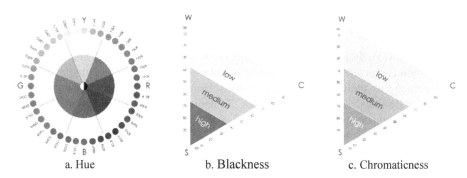

a. Hue   b. Blackness   c. Chromaticness

Fig. 5. Rules of characterizing the hue, blackness and chromaticness properties

3.2 Refining Enlightenments to the Beijing Winter Olympics by Comparison

After characterizing the reasons of selecting the colors, the images and properties of the colors, the advantages of the main color schemes of Summer and Winter Olympics are discussed, and instructive advices for the Beijing Winter Olympics are refined.

4 Analysis of Main Color Schemes for the Summer Olympic Games

4.1 Collection

The collection of main color schemes for the Summer Olympic Games from 1980 to 2020 are arranged below in Table 1. A main color scheme includes the representative colors used in logo design, mascots design and actual application.

The reasons of color selections are arranged below in Table 2. The reasons of color selecting can be summarized to three categories: characteristic natural landscape (1988-Seoul-KR, 1992-Barcelona-ES, 1996-Atlanta-US, 2000-Sydney-AU, 2016-Rio-BR), cultural characteristics (2004-Athens-GR, 2008-Beijing-CN, 2020-Tokyo-JP), and national identity (1980-Moscow- USSR, 1984-Los Angeles-USA).

Table 1. Main color schemes for the Summer Olympic Games during 1980–2020

Year-City-Country	Main color schemes with NCS properties						
1980-Moscow- USSR[12]	S 3560-Y70R	S 1085-Y90R	S 3050-R90B	S 9000-N	S 1080-Y	S 2570-G30Y	S 7020-Y30R
1984-Los Angeles-USA[12]	S 1085-Y90R	S 0300-N	S 1565-B	S 1565-B			
1988-Seoul-KR[12]	S 1070-R	S 1070-Y	S 4050-R80B	S 0300-N			
1992-Barcelona-ES[12]	S 0580-Y70R	S 0580-Y70R	S 0580-Y70R	S 0580-Y70R			
1996-Atlanta-US[12]	S 1080-Y70R	S 0550-G90Y	S 1050-R90B	S 1060-R20B	S 2050-R50B	S 5040-G20Y	
2000-Sydney-AU[12]	S 1080-Y70R	S 1060-G90Y	S 3060-R70B	S 1070-G20Y			
2004-Athens-GR[1]	S 1060-B	S 0585-Y40R	S 2070-G50Y	S 2060-R30B			
2008-Beijing-CN[1]	S 1085-Y80R	S 0550-Y20R	S 4050-G	S 3060-R80B	S 3502-B	S 0300-N	
2012-London-UK[1]	S 1070-R20B	S 1080-Y80R	S 1050-B	S 1050-B70G	S 0570-G90Y		
2016-Rio-BR[12]	S 0580-Y	S 1080-Y50R	S 2060-R90B	S 1050-B40G	S 1070-R		
2020-Tokyo-JP[13]	S 2070-R	S 4040-Y	S 5540-R70B	S 1060-R30B	S 0580-Y	S 3060-G	S 1060-B

Table 2. Reasons of color selections for the Summer Olympic games during 1980–2020

Year-City-Country	Reasons
1980-Moscow- USSR [12]	Social system: socialism; visual impact
1984-Los Angeles-USA [12]	Colors of national flag
1988-Seoul-KR [12]	Nature and traditional philosophy: sky, earth and life
1992-Barcelona-ES [12]	Landscape and culture: sun, sea and life
1996-Atlanta-US [12]	Medals, athletes and laurel branch (city tree)
2000-Sydney-AU [12]	Landscape and culture: sun, beach, sea and red land
2004-Athens-GR [1]	Traditional architectures and myths
2008-Beijing-CN [1]	Traditional architectures and crafts
2012-London-UK [1]	Highness; the Olympic rings
2016-Rio-BR [12]	Plants and animals
2020-Tokyo-JP [13]	Traditional color

As the 2008 Beijing Olympics had chosen color from Chinese traditional architectures and crafts, it's worthy to try seeking some characteristic natural landscape for color scheme refining.

4.2 Color Image Scale and Word Image Scale

The color image scale and word image scale of the Summer Olympics are presented in Fig. 6. As it is shown in the figure, soft and warm colors are most widely used, soft and cool colors are the second, hard colors are rarely used in the Summer Olympic Games.

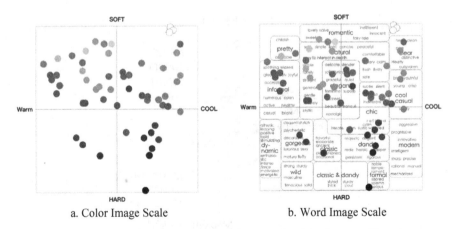

a. Color Image Scale   b. Word Image Scale

Fig. 6. The color image scale and word image scale of the Summer Olympics

4.3 NCS Color Analysis

The NCS Color Spaces of the Summer Olympics are presented in Fig. 7.

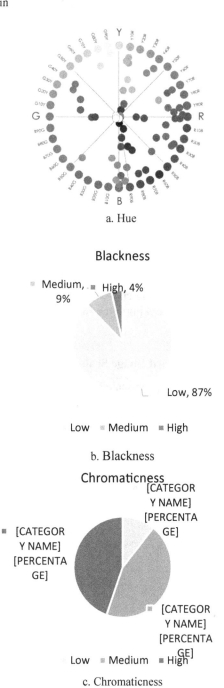

a. Hue

Blackness

Medium, High, 4%
9%

Low, 87%

Low ▪ Medium ▪ High

b. Blackness

Chromaticness
[CATEGOR
Y NAME]
[PERCENTA
GE]

[CATEGOR
Y NAME]
[PERCENTA
GE]

[CATEGOR
Y NAME]
[PERCENTA
GE]

Low ▪ Medium ▪ High

c. Chromaticness

Fig. 7. The NCS color spaces of the Summer Olympics

The NCS color paces analysis shows that among the color schemes for Summer Olympics, colors which combined by green and blue are rarely used, medium or high blackness colors are rarely used, and low chromaticness colors are rarely used.

5 Analysis of Main Color Schemes for the Winter Olympic Games

5.1 Collection

The collection of main color schemes for the Winter Olympic Games from 1980 to 2020 are arranged below in Table 3. Compared to the color schemes of the Summer Olympics, the application of white and blue is obviously increased in the color schemes of Winter Olympics.

Table 3. Main color schemes for the Winter Olympic Games during 1980–2020

Year-City-Country	Main color schemes with NCS properties				
1980-Lake Placid- USA[14]	S 1565-B	S 1080-Y70R	S 0300-N	S 4550-R70B	
1980-Lake Placid- USA[14]	S 1080-Y80R	S 0300-N	S 1050-R90B		
1988- Calgary -CA[14]	S 1070-Y80R	S 0570-G80Y	S 5540-R70B	S 0300-N	
1992- Albertville -FR[14]	S 1580-Y90R	S 2065-R90B	S 0580-Y		
1994-Lillehammer-NO[14]	S 3560-R80B	S 0300-N	S 1070-Y	S 2070-Y60R	S 5040-G20Y
1998- Nagano -JP[14]	S 1085-Y80R	S 0550-Y20R	S 2060-R70B	S 1070-G20Y	S 1070-Y40R	S 1560-R90B
2002- Salt Lake City -USA[14]	S 3050-Y80R	S 2050-Y50R	S 2050-Y10R		
2006-Turin-IT[14]	S 1060-B	S 2070-R	S 0300-N		
2010- Vancouver -CA[14]	S 1070-G20Y	S 5540-R70B	S 1050-R90B	S 1085-Y80R	S 0560-Y20R
2014-Sochi-RU[14]	S 2070-R	S 2065-R90B	S 0300-N		
2018- Pyeongchang -KR[14]	S 1070-R	S 8502-G	S 1070-Y	S 2070-G10Y	S 2065-R90B

The reasons of color selections are arranged below in Table 4. The most common principle for making the color schemes is employing the image of winter, ice and snow (1980-Lake Placid-USA, 1984-Sarajevo-YU, 1994-Lillehammer-NO, 2006-Turin-IT), while perfectly combing them with the characteristics of the host city/country requires sophisticated design, just like the way Calgary and Lillehammer did.

For the Beijing 2022 Winter Olympics, color refining can not only conducted from the perspective of season, weather, landscape, traditional color and Olympics spirits, but also from the view of traditional crafts and local winter culture, since seldom of the colors schemes presented above has taken this perspective.

5.2 Color Image Scale and Word Image Scale

The color image scale and word image scale of the Winter Olympics are presented in Fig. 8. As it is shown in the figure, both warm and cool colors are widely used, however soft colors and hard colors are relatively less used, hard colors were least used. This result reveals the season characteristic of the Winter Olympics, while at the same time, indicates the hot spirits of sports.

Table 4. Reasons of color selections for the Summer Olympic Games during 1980–2020

Year-City-Country	Reasons
1980-Lake Placid-USA [14]	Winter; ice; national flag
1984-Sarajevo-YU [14]	Enthusiasm; winter
1988-Calgary-CA [14]	National identity: maple leaf; local characteristic: cowboy
1992-Albertville-FR [14]	Olympic flame; skater
1994-Lillehammer-NO [14]	Landscape and season: sky, aurora and snowflake
1998-Nagano-JP [14]	Traditional color; enthusiasm
2002-Salt Lake City-USA [14]	Landscape
2006-Turin-IT [14]	Winter
2010-Vancouver-CA [14]	Landscape; the Olympic rings
2014-Sochi-RU [14]	National flag; skater
2018-Pyeongchang-KR [14]	Traditional color

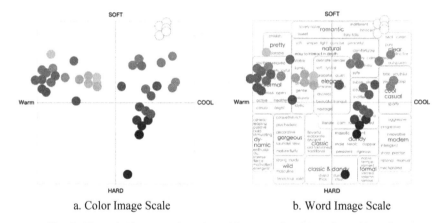

a. Color Image Scale   b. Word Image Scale

Fig. 8. The color image scale and word image scale of the Winter Olympics

5.3 NCS Color Analysis

The NCS color paces analysis shows that among the color schemes for Winter Olympics, colors which combined by green and blue are rarely used, medium or high blackness colors are rarely used, and low chromaticness colors are rarely used. Though the results are in line with the results of the Summer Olympics, the proportion of using low chromaticness colors is bigger in the color schemes of the Winter Games (Fig. 9).

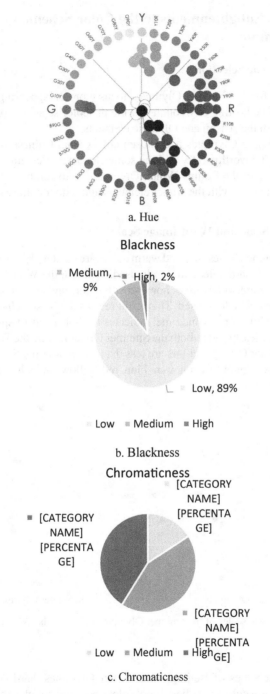

a. Hue

b. Blackness

c. Chromaticness

Fig. 9. The NCS color spaces of the Winter Olympics (Color figure online)

6 Conclusion: Enlightenments to the Color Scheme for the Beijing Winter Olympics

6.1 Reasons of Color Selection

The color schemes for the Summer Olympics focus more on appearing the local landscape or culture of the hosting city/country, while the color schemes for the Winter are often developed with the season and Olympic elements.

As the 2008 Beijing Olympics had chosen colors from Chinese traditional architectures and crafts, it's worthy to try seeking some characteristic natural landscapes for color scheme refining of the Beijing Winter Olympics, or to combine the characteristic colors of traditional crafts with the winter elements and winter culture of Beijing.

6.2 Color Image Scale and Word Image Scale

In the Summer Olympic Games, soft and warm colors are most widely used, soft and cool colors are the second, hard colors are rarely used. While in the Winter Olympics, both warm and cool colors are widely used, however soft colors and hard colors are relatively less used, hard colors were least used. This result reveals the season characteristic of the Winter Olympics, while at the same time, indicates the hot spirits of sports.

Hard colors were least used in both the Summer Olympics and the Winter Olympics. The hues of the Winter Olympic colors are less than the ones in the Summer Olympics, popular colors in the Winter Olympics are blue, red, yellow, and white (Fig. 10).

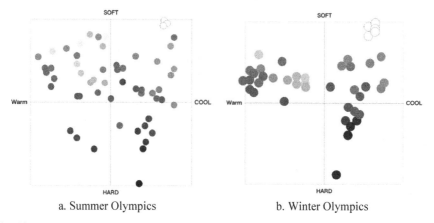

a. Summer Olympics   b. Winter Olympics

Fig. 10. The color image scales of the Summer Olympics (left) and the Winter Olympics (right) (Color figure online)

For the color schemes of Beijing 2022 Winter Olympics, hard colors should be avoided, soft and warm colors or soft and cool colors are more popular choices to choose, and importing more kinds of hues into the color schemes of the Winter Olympics might be a good choice.

6.3 NCS Color Analysis

The NCS color spaces analysis shows that among the color schemes for both summer and winter Olympics, colors which combined by green and blue are seldom used, medium or high blackness colors are rarely used, and low chromaticness colors are rarely used. But the proportion of using low chromaticness colors is bigger in the color schemes of the Winter Games.

The results implies that when making the color scheme for the Beijing Winter Olympics, no matter how many kinds and what kind of hues are in the color scheme, low blackness and medium-high chromaticness colors are favorable, which can be explained by the vital and energetic nature of the Olympic Games.

6.4 Limitations

The main color schemes of the Olympics are collected from online websites, which are not comprehensive enough for analyzing. Also, the study only collected the main Olympic color schemes during 1980–2020. The analysis is not deep enough and requires further research, such as evaluating the cross-cultural effects of the color schemes.

References

1. Wang, J.: Historical research on olympic visual image. Sports Sci. **12**(1000–677X), 28–36+73 (2004)
2. Zhang, W.: Athens, Beijing and London olympic color vision applications system research (Master). Beijing Institute of Fashion Technology (2013)
3. Wang, J.: Research on the color system design of Beijing 2022 winter olympics (Master). Beijing Institute of Graphic Communication (2017)
4. Shi, J.: Research on the transmission value of visual culture in winter olympic games—— taking Turin, Vancouver, Sochi as examples. China Winter Sports **03**(1002–3488), 22–26 (2015)
5. Zhang, J.: A study on the visual image design of the Rio Olympic Games 2016. In: 2017 International Conference on Applied System Innovation (ICASI), Sapporo, pp. 970–972. IEEE (2017)
6. Dong, Y.: Research on visual symbols of the 31st Rio olympic games. Sports Cult. Guide **05**(1671–1572), 192–197 (2017)
7. Kobayashi, S.: The aim and method of the color image scale. Color Res. Appl. **6**(2), 93–107 (2009)
8. Kobayashi, S.: Color Image Scale. Kodansha International, Tokyo (1992)
9. Ewald, H.: Outlines of a Theory of the Light Sense, 1st edn. Harvard University Press, Cambridge (1964)
10. Hård, A., Sivik, L., Tonnquist, G.: NCS natural color system - from concepts to research and applications. Part I and II. Color Res. Appl. **21**, 180–220 (1996)
11. How the NCS system works – NCS. https://ncscolour.com/about-us/how-the-ncs-system-works. Accessed 31 Jan 2020
12. List of previous Olympic Games - Olympic Database - NetEase Olympics. http://info.2012.163.com/match/olympic. Accessed 01 July 2019
13. Tokyo 2020|The Tokyo Organising Committee of the Olympic and Paralympic Games. https://tokyo2020.org/en. Accessed 01 July 2019
14. History Beijing Organising Committee for the 2022 Olympic and Paralympic Winter Games. https://www.beijing2022.cn/en/olympics_/history.html. 01 July 2019

Research on the Development Path of "New Technology" and "Traditionalization" of Chinese Embroidery

Shuang Ou[1]([⊠]), Minghong Shi[2,3]([⊠]), Wei Deng[2]([⊠]), and Rungtai Lin[3]([⊠])

[1] Industrial Culture Development Center of MIIT, Beijing, People's Republic of China
oushuang@miit-icdc.org
[2] Shenzhen Technology University, Shenzhen, Guangdong, People's Republic of China
jun101786@126.com, david_deng@foxmail.com
[3] Graduate School of Creative Industry Design, National Taiwan University of Arts, New Taipei City, Taiwan
rtlin@mail.ntua.edu.tw

Abstract. The contemporary development of Chinese traditional embroidery encounters both an external crisis brought by the development of science and technology as well as an internal crisis caused by the changes of consumption structure. Through the research conducted on contemporary environment and consumption trend in China, the article explicates a specific way of embroidery development, and puts forward a model of parallel development of traditionalization and new technology as a way to cope with specific problems met in embroidery development. And two different development paths, the innovative development and creative transformation, are analyzed in some typical cases. By means of discussion of innovative design concepts, integration of emerging materials as well as the assistant of science and technology, the embroidery would go deep in populace's life to convey its culture value and to make traditional embroidery skills inherited. In this way, philosophical thinking and lifestyle of contemporary people will be reflected.

Keywords: Chinese embroidery · New technology · Traditional

1 Introduction

Chinese traditional embroidery is inseparable culture from human being since Chinese ancient time. The embroidery was originated from the tattoo custom. When the tattoo was covered by clothing, the embellishment gradually transferred to clothing. In slave society, clothing materials were possessed by ruling class, therefore an ancient uniform system gradually came into being in order to stabilize internal order of class. In Shang and Zhou period, delicate embroidery combined with painting appeared. In Zhou Dynasty, there were records of "colored threads being embroidered to form unique emblazon". In Tang Dynasty, the embroidery was not only used in clothing and accessories, but also employed in calligraphy and painting. Such major shift make embroidery gradually

© Springer Nature Switzerland AG 2020
P.-L. P. Rau (Ed.): HCII 2020, LNCS 12192, pp. 396–407, 2020.
https://doi.org/10.1007/978-3-030-49788-0_29

separate from textile decoration and became a relatively independent art category. In Song dynasty, the embroidery was combined with the flourishing art of calligraphy and painting. The consequent form of unique ornamental work has carry the development of embroidery art ramped to a peak. In Ming and Qing Dynasties, the embroiderers working for the courts were in a large scale, and the folk embroidery also developed at that time. Suzhou Embroidery, Hunan Embroidery, Guangdong Embroidery and Sichuan Embroidery and other styles of embroidery boomed [4]. After a long period of development, there are more than 300 embroidery techniques and unique classification methods that are differentiated based on various stitching skills. The traditional embroidery plays a role both in decoration of clothing and in society function. A large number of auspicious patterns convey their connotation meaning through embroidery. Through successive generations of constant improvement and extension, the principle of embroidery inheritance and creation eventually formed [5]. All of the above can be summarized as the development path of traditional embroidery.

However the traditional embroidery is facing aesthetics crise, inheritance crise and development crise due to the rapid development of modern science and technology and social economy. According to the official data of the National Bureau of Statistics, the constantly increased residents' income and disposable income per capita has accelerated the change of consumption habits and concepts. The rising residents in urban areas are massive consumer force that can't be ignored. The young people are gradually becoming the main body of consumption. They pay more attention to featured handicrafts that convey history and culture and are related to life [23]. The consumers' demands for embroidery products is increasingly diverse as day passes [22]. Impacted by the development of science and technology, embroidery has continued its fusion development with other industries and went into people's life from many aspects and dimensions of mass consumption. The limitations of time and space have been broken. The interaction among various cultures enables a variety of innovative embroidery products to gradually construct a contemporary development path of embroidery (Figs. 1 and 2).

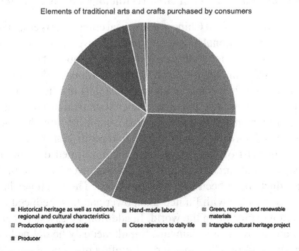

Fig. 1. Elements of traditional arts and crafts purchased by consumers

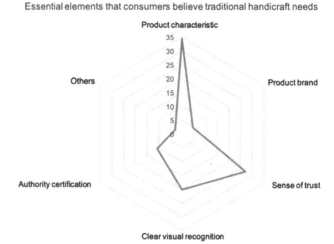

Fig. 2. Essential elements that consumers believe traditional handicraft needs

There are two purposes of this research. On one hand, we want to realize internal innovation of embroidery by exploring how to achieve the fusion development of various industries. On the other hand, we want to figure out how to realize the creative transformation of embroidery with the assistance of science and technology. By clarifying the relationship between tradition and science in the embroidery development, we demonstrate how embroidery construct contemporary aesthetic with consumption culture, show contemporary lifestyle and convey contemporary social values.

2 Literature Review

2.1 External Factor: The Impact of New Technology on Embroidery Production

The contemporary development of Chinese traditional embroider is essentially a question of relationship between traditional handicrafts and the development of modern mechanical technology and meanwhile it's a question set for us about how to coordinate the development of art and technology. Currently, the coordinated development of handicrafts and contemporary science and technology just gets off the ground and is still in the initial stage of exploration and practice. Some theoretical discussions on the development of fusing craftsmanship and science and technology together were conducted by scholars at home and abroad, most of whom hold positive attitude.

Zongyue, L., a theorist of folk art in Japan, demonstrated that the machines possess amazing speed and accuracy and powerful horsepower by which a large number of inexpensive products have been tirelessly produced. The machines have completed the tasks that were incapable with limited manpower and have reduced financial stress caused by population explosion in the world. In addition, he asserted that the invention of machines was an endowment to humans. Simultaneously he defined the boundaries of mechanical crafts, and rectified mechanical crafts [14]. Lin, R., a Taiwan scholar, put forward that "technology always comes from human nature". Namely the rational "Hi-tech" and the emotional "Hi-touch" should be both emphasized in product design, which pursued a balance between science and human nature [20]. He also pointed out the "function" provided by technological products enabled people to live comfortable lives. What he emphasized was the universality [9]. Bhandari, V., an Indian scholar, believed that both handicraft and machine could realize creative production of embroidery. And both of them should have a place in contemporary textile culture. Embroidery produced by machines also belonged to human's creative labor, which contributed to wider understanding of constitution of hand-made handicrafts. And the chronergy of such understanding was becoming stronger and stronger [18]. Juchnevičienė et al. studied the opportunities and challenges of the fusion production of embroidery and wearable antenna. They hold the view that, in the case of machine production, the colored threads of embroidery could be applied to create beautiful shapes. Specialized conductive threads could be embroidered to the foundation of the fabric to accurately reflect the size and shape and meanwhile to make it more washable [19].

Some Chinese scholars believe that the development of handicrafts and industrial production are not contradictory. They served as supplements to each other and only existed in form of service no matter how far the development of science and technology have come [13]. And some other scholars focused on the impact of science and technology on Chinese traditional embroidery from perspectives like technological aesthetics, economic value and industrial ethics. They believed that 3D printing technology possessed the practical value in solution optimization and challenge, art performance and visual presentation in the field of creation design. The new technology would boost the development of aesthetics and culture [2].

2.2   Internal Factor: The Development Path of Traditional Embroidery Itself

If traditional handicraft continues to develop, it must advance with the time. Anying, C., a professor in Tsinghua University, proposed that traditional handicrafts should keep pace with the time to constantly adapt to modern life. Through innovative development, the fine works of times were remained. Under the new circumstance of culture and economy, traditional handicraft should provide product and service that consumers really need with the help of designers [6].

For questions of contemporary development of embroidery, experts and scholars also provided their own ideas from various aspects. After analyzing the development of Yangzhou embroidery, Jingjing, Y. proposed some targeted measurements, such as cost reduction, senior artist protection, training innovative design talent and integration of modern handicraft to promote the sustainable development of Yangzhou embroidery [15]. Taiwan scholar Yan et al. demonstrated that cultural creativity was a lifestyle and a taste for life. The cultural and creative products made creative life full of emotional fields and moving experiences and further showed the charm of sensual goods [17]. Apart from that, Lin, R. proposed that cultural and creative products were originated from cultural creativity. The diversity of products was the prerequisite to gain the favors of consumers [11]. Jie, L., Kuang, R. conducted a research on the development of the cultural and creative products of embroidery of Yi nationality, targeting on their function and ductility. He pointed out the embroidery products should base on traditional

Yi embroidery culture. But it was also important to industrialize the products to the market. It showed what role consumer market played in the contemporary development of embroidery [3]. Haomin, C., Jinlian, L. carried out a research on the innovative application of folk traditional embroidery of Xibe nationality. He indicated that innovative tourism products with national characteristics were developed depending on traditional culture elements of Xibe, creative color matching and combination of new materials. Embroidery, an important mean to record history, entered the contemporary consumer market through design innovation. It not only preserved history and embroidery skills, but also effectively expand the cultural capacity of embroidery [7].

Embroidery is integrated with design power in different fields to seek innovation, and there are practical cases all over the world. In terms of the fusion development of embroidery with other industries, Mecnika et al. did some research on the characteristics of embroidery. Embroidery had three - dimensional and stable characteristics. And it has been successfully used in tissue development of wound dressing and other innovative solutions. Thus it effectively expanded the application attributes of embroidery [21]. Arora, A., a textile and clothing manufacturer in India, embroidered on the surface of soluble plastic with sewing machine. When the background layer, soluble plastic, was resolved, lacy form of embroidery was formed to create a large tent-like structure. It expanded the fusion development of embroidery and other materials. Shaikh, A., a handicraftsman and designer, focused on the innovative work of embroidery with delicate silk and metal. What he created made Indian embroidery extremely modern [18]. Wenting, M conducted an applicable research on the fusion development of embroidery art and packaging design, integrated the embroidery elements and packaging design, and analyzed the technical problems of embroidery stitches, packaging structure and materials combination to bring new visual and emotional experiences to consumers. The quintessence of embroidery culture was passed on through modern packaging design. It further expanded the application value of embroidery [8].

3 Methodology

Firstly, literature research method is adopted in this article. By sorting out the viewpoints of the coordinated development of art and technology from scholars at home and abroad, it analyzes both the external factor of the impact of new technology on embroidery production, and the internal factor of contemporary development path of traditional embroidery. In addition, issues that should be paid heed to in the path of its exploration and development are mentioned as well. Secondly, data analysis method is adopted. According to official data released by the National Bureau of Statistics of China and authoritative department data measurement center, we better understand the social environment and the trend of the times, developing condition of national economy, different levels of demand in consumption structure, as well as development trend that traditional handicraft demand for. By this mean, it points out the direction of development of contemporary embroidery. Lastly, we adopt a method of studying specific cases. Through the analysis of two successful typical cases of contemporary embroidery in China, it discusses how embroidery conducts innovative transformation and creative development, and further to discuss more possibilities and new historical development dimension of embroidery development.

4 Research on Specific Cases: Innovative Development and Creative Transformation

4.1 Transmission of Values of Handicraft: Innovative Development of Embroidery

The innovative development of embroidery, giving priority to traditional hand-made methods and based on regional culture, achieves the inherent values of embroidery handcraft, which takes traditional embroidery as its carrier. The key path of developing embroidery is respecting traditional embroidery culture, attaching importance to injecting cultural and creative factors into the embroidery industry, promoting the fusion development of embroidery and other industries based on the former and developing emerging embroidery industries with culture and creation.

In 2019, the dazzling and shining collector's edition of BMW 7 Series was an excellent example of embroidery entering the design field in automobile industry. The interior space of the car uses center-shaft layout for reference. Traditional embroidery technique was reflected on the center-shaft armrest and seat belts, creating a peaceful atmosphere. It was a masterpiece created with masters' own understanding of traditional embroidery and their unique aesthetic concept as well as the superb stitching skills. Embroider selected 4 gray silk threads whose tinct differences are very subtle from over 4000 different silks. The intensity and shades of the color were embroidered with different stitching skills. The seat belts were embroidered with overlapping threads in different directions, called irregular embroidery technique, to achieve a delicate effect, which reflected the highest handicraft skills and mild temperament. It is not only a fusion and innovation of Chinese and western design concepts, but the perfect fusion of traditional Chinese culture and handicrafts with the western automobile industry spirit. It is a tribute to traditional Chinese culture, and also to convey the zeitgeist of embroidery.

Apart form the basic function of the cars, the innovative application of embroidery on automobile accessories stimulate the imagination of consumers to enable them indulge in a certain vibe and keeps advancing towards arts [10]. Combined with cool-natured automobile styling created by craftsman, embroidery when in use attaches strong empathy to reviewers. The case above breaks the consumers' stereotypical understanding of the relation between embroidery and automobile and further extends dimension of innovation and development of embroidery and automobile. The embroiderer inherits tradition handicraft, combines the characteristics of vehicles and the aesthetic needs of consumers and innovates automobile interior accessories. The machinery production allows embroidery accessories to be widely applied in the interior of all vehicles. Thus it expands the application value of embroidery, improves the art value of cars and meets consumers' spiritual demand (Fig. 3).

4.2 Transmission of Value of Culture: Creative Transformation of Embroidery

The creative transformation of embroidery, based on modern industries except embroidery, grafts itself, an intangible cultural resource, into the modern industry to enhance the cultural connotation of embroidery products and services, and then to form derivative values of embroidery handicraft. In the filed of embroidery production today, there are

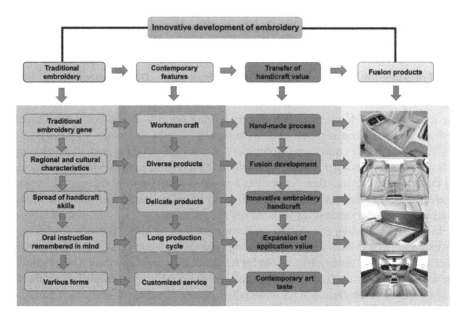

Fig. 3. Innovative development path of embroidery (compiled by the research)

traditional hand-made embroidery and more machine-manufactured embroidery. The development of science and technology enable newly-researched materials and updated new technologies to constantly participate in innovation of embroidery products, and serve as a huge force in promoting the contemporary development of embroidery.

In 2018, the Beijing Palace Museum introduced popular lipsticks, getting design inspiration from the collection of Qing Dynasty concubine clothing. The lipsticks were provided with six colors. Their external tubes respectively imitated six collections of concubine clothing, reflecting the court system, costume culture, aesthetic connotation and former fashion trends [1]. There is an old saying "review the past to understand the present". The memory of the palace life combined with exquisite skills provided with a moving experience and constructed a sensuous life. Contemporary life ceremony was also constructed through the innovation of embroidery culture. Designing cultural and innovation products should take not only consumers' demand into consideration, but also their feelings while using in daily life. From the perspective of cultural innovation, that's the purpose of cultural innovation design. With the assistance of 3D printing technology, the outer tubes perfectly reappeared the thread and embroidery textures of those six clothing, and also showed their arrangement texture and concavo-convex characteristics. Embroidery, which were exclusive for court, now can be afforded by general public thanks to mass production. When putting lipstick on lip, people can also experience the exquisite lifestyle of the court in Qing Dynasty. Such communication of consumption culture established on the basis of mass tastes and living habits can fully express consumers' freedom of behavior, expression and emotion.

The Palace lipsticks are different from general ones using flat stickers as outer tubes. They adopted 3D printing technology, based on digital model files, and used powdery metal or plastic and other adhesive materials to print metal outer tubes layer by layer with overlapping oil ink. The texture of textile was reappeared on the metal tubes. The concavo-convex characteristics consequently formed to simulate the characteristics of embroidery. The core of this unique and innovative way to convey embroidery skills is not to duplicate original embroidery works, but to promote embroidery to form a new aesthetic perspectives and cultural values. Production on demand bursts the restraints of historical stereotypical images and social function of embroidery, and makes the value of new materials fully exploited. This production method adds a new dimension to the historical development tradition of embroidery. In feudal society, princes and nobles occupied top resources of the society. And a large number of lower class people were always in passive position in the resource distribution. The hierarchy system prevents many excellent cultural resources from being shared, but the development of technology makes those resources available for common people. Culture shows its fairness with help of technology. 3D printing technology used in the field of embroidery production has a clear cultural consumerism, which conforms to cultural logic of consumption in current age. And fair ethical attributes are further reflected in material culture, which opens a door for ordinary people to equally enjoy the culture. Such fusion development is a creative transformation of embroidery and also an important means to promote cultural fairness. Modern technology is adopted in traditional embroidery, the embroidery culture is inherited in new forms of interaction of tradition and modernity (Fig. 4).

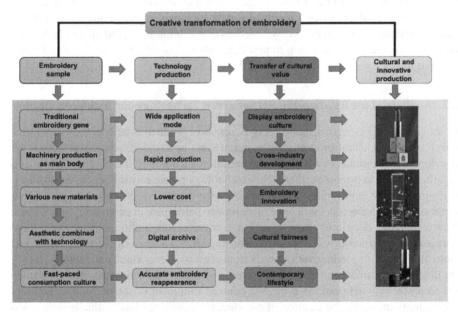

Fig. 4. Creative transformation path of embroidery (compiled by the research)

5 Research Results and Discussion

5.1 The Impact of New Technology on the Development of Embroidery

From external circumstance, new technology has significant advantages in improving working environment, alleviating labor intensity and enhancing artistic expression of embroidery. From internal circumstance, new technology is conducive to enhancing the application values of embroidery skills and further to develop new forms of embroidery and special products.

Exemplified as 3D printing technology, for enterprises, it reduces early cost and investment in the research and development stage, accurately edit the existing electronic templates to lower the cost in production stage, explore and apply new materials to realize the diversified production and to enhance productivity. In addition, the products can be printed on site, which reduce risks of surplus products in stock. Advantages above bring considerable benefits to enterprises, which can effectively push the rapid expansion of embroidery industry. And for embroiderers, new technology allows them to design regardless of production cost and techniques so it expands their freedom to create. Understanding of the properties of new materials contributes to generating and developing a new design trend. Furthermore, the new technology is able to establish a true and accurate 3D digital archives [12]. It allows digital models to be highly restored without limitation of time and space, which is conducive to exploring new handicraft, researching and restoring lost handicraft. People's expectations and demands are met by machinery production, as a result of the development of technology. The field of traditional handicraft introduces new materials and technologies. And the newly-formed order and force are an adaptation to new demands for social innovation. It also further expands the cultural capacity of handicraft.

5.2 Analysis of the Relationship Between New Technology and Traditionalization

From a material perspective, embroidery is a creative production activity that transforms materials. The inherent properties of material can't be avoided, which determines the development of embroidery is always impacted by technology. Technology serves as a significant force to promote culture. New technology potentially influences embroidery. It also internally guides embroiderers to improve their work efficiency in production process, so that embroiders will have more time to participate in embroidery innovation. New technology does have advantages, but, compared with traditional handicraft, there are also problems needed to be optimized and improved. Facing a situation of science and technology involving in innovation production of embroidery, which undoubtedly satisfy the diverse needs of consumers, we must have a boundary consciousness. At a time when the technical rational thinking brought by the development of science and technology increasingly influence the creation of embroidery, technology has gradually dominated the direction of embroidery development instead of potentially influence it. We should be aware of the excessive involvement of science and technology in embroidery creation and production, as well as the negative effects it consequently brings.

On the one hand, when it comes to embroidery innovation, it's necessary to distinguish the difference between traditional embroidery innovation and cultural and creative products. Innovation of traditional embroidery, whether in forms of involvement of new materials and technology or cross-industries development, should take embroidery as main carrier to strike a balance between authentic embroidery handicraft and innovation gene. Cultural and creative products have added multiple dimensions of product application to embroidery innovation. But many products simply use embroidery as an element reflected on the products. In this way, traditional embroidery handicraft can not be inherited, letting alone mention innovation. On the other hand, in the process of machinery production, blindly worshiping machines and technologies is heavily domained. If it continues, embroidery will face a dilemma of losing regional cultural characteristics and the crisis of losing the traditional embroidery handicraft. However, if overemphasizing hand-made production, embroidery will face a desperate situation of losing its market and can not be inherited and developed.

Based on the arguments above, it's of vital significance to establish a correct role for machines and embroiderers. They are not opponents in labor relationship. And new technology and traditionalization are not entirely opposed. For embroiderers, they can absorb ideas of modern culture, scientific and technological inventions, and artistic trends to create embroidery, improving skills and enriching creative themes. Embroidery carrying contemporary characteristics can be created with needles. For machines, they aren't capable of innovating so that embroiderers or programmers are needed for cooperation. The key points and concepts of embroidery are transferred into computer and saved as electronic template. In this way, the machinery products can satisfy consumers' spiritual needs [16]. That process is a significant component of hand-made handicraft, an important part for embroiders to inherit tradition and convey the contemporary value of embroidery, and an embodiment of the creative labor of embroiders. From this point of view, relationship between new technology and traditionalization draws on mutual strength and promotes each other. Their development relation are mutually complement.

6 Conclusion and Suggestions

Contemporary development of Chinese traditional embroidery is facing challenges and difficulties, simultaneously undergoing various development opportunities as well. Only embroidery that adapts themself to contemporary consumer market can be inherited. Innovative development and creative transformation are the two main directions for contemporary embroidery development. One effective means of retaining traditional embroidery handicraft is combining other industries with embroidery as the carrier. Embroidery, as culture asset, is grafted in modern industries. Thanks to technology, the needs of mass consumption can be satisfied through increased productivity and lower costs. Eventually cultural fairness is embodied in mass consumption. Consequently contemporary culture value is passed on, which makes up the characteristics of the times.

Traditionalization and involvement of technology are contemporary characteristics for development of embroidery. And it's extremely crucial to build up a coordination system between tradition and technology. Embroidery production should be innovated as contemporary social environment changes, actively exploring fusion development of embroidery and new materials and technology, and continuously expanding the application value and cultural capacity of contemporary embroidery. Only in this way, embroidery can be endowed with new life to create products that have national features and adapt to modern consumer market, further reflecting the spirit of the new era.

References

1. Wei, F., Kejun, Z.: Possibility of embroidery digital transformation based on Miao embroidery cases. Silk **9**, 42–47 (2016)
2. Zeng, W., Xuyi, L.: Artistic aesthetic and cultural value orientation of 3D printing technology. Jiangxi Soc. Sci. **8**, 239–244 (2016)
3. Jie, L., Kuang, R.: Research on designs of cultural and innovative products to protect intangible cultural heritage of Yi nationality: take Yi traditional embroidery products in Ganluo county, Liangshan Yi autonomous prefecture as example. Heritage Conserv. Stud. **4**, 91–94 (2019)
4. Pingshan, X., Hui, L., Qin, L.: Investigation and analysis of the development of Chinese traditional embroidery. J. Zhengzhou Univ. Aeronaut. **35**, 149–152 (2016)
5. Congwen, S., Xu, W.T.: Chinese Costume History. Citic Press, Beijing (2018)
6. Anying, C.T.: Handicraft Ought to Keep Up with Contemporary. Research on a Case of Revitalization of Traditional Craft. China Light Industry Press, Beijing (2019)
7. Haomin, C., Jinlian, L.: Research on innovative application in tourism products of Xibe folk traditional embroidery. Art Sci. Technol. **11**, 1–3 (2017)
8. Wenting, M.: Comprehensive study on embroidery used in modern packaging design. Art Panorama **6**, 136–137 (2017)
9. Lin, R.: Techniques and arts ought to be inherited-conversation between craft and technology. Artistic Appreciation **9**(3), 49–66 (2013)
10. Lin, R.: Value-added design in cultural innovation. Artistic Appreciation **7**, 1–9 (2005)
11. Lin, R.: Preface: study on and essence of cultural innovative industries. J. Des. **16**(4) (2011)
12. Jiamin, Z.: Guard and cross borders: discussion and analysis of the value construction of 3D printing technology to contemporary handicraft. Beauty Times **06**, 31–33 (2019)
13. Yuchen, L., Wu, J., Xiuwei, L.: 3D printing technology can endow intelligence to creative design and cultural and innovative products. Beauty Times **06**, 28–30 (2019)
14. Zongyue, L.T.: Craft Culture. Guangxi Normal University Press, Nanning (2011)
15. Jingjing, Y.: Research and development strategy of Yangzhou embroidery. Art Sci. Technol. **8**, 119 (2018)
16. Ming, N., Fei, B., Xiaohui, L.: Comparison of stitching skills between traditional embroidery and computer aided embroidery and instance analysis. Shanghai Text. Sci. Technol. **09**(45) (2017)
17. Huiyun, Y., Jingrong, L., Lin, R.: Discussion of research process form designer factors to cultural factors in Taiwan: opinions from design educator. J. Commercial Des. **22**, 39–58 (2018)
18. Bhandari, V.: Trends in Embroidery. Marg-A Mag. Arts **67**(4), 88–95 (2016)
19. Juchnevičienė, Ž., Jucienė, M., Dobilaitė, V., Sacevičienė, V., Gulbinienė, A.: The Research on the accuracy of the geometrical parameters of the closed-circuit embroidery element. Mater. Sci. **24**(4) (2018). https://doi.org/10.5755/j01.ms.24.4.18843
20. Lin, R., Kreifeldt, J., Hung, P.-H., Chen, J.-L.: From technology to humart – a case study of Taiwan design development. In: Rau, P.L.P. (ed.) CCD 2015. LNCS, vol. 9181, pp. 263–273. Springer, Cham (2015). https://doi.org/10.1007/978-3-319-20934-0_25
21. Mecnika, V., Hoerr, M., Krievins, I., Jockenhoevel, S., Gries, T.: Technical embroidery for smart textiles: review. Mater. Sci. Text. Cloth. Technol. **9**, 56 (2015). https://doi.org/10.7250/mstct.2014.009

22. Disposable Income of Urban Residents in China. Official Websites of National Statistics Bureau in China (2015–2019). http://www.stats.gov.cn/
23. Valid Questionnaires Collected from Micro Survey of Handicraft Ecology in China, Chinese Handicraft (2019)

Research on Development of Guangdong Porcelain Design Driven by Export Trade in the 16th–19th Century

Research on Development of Guangdong
Porcelain Design Driven by Export Trade
in the 16th–19th Century

Xiao Song[1,2]([⊠])

[1] Beijing Technology Institute, Zhuhai, People's Republic of China
842766039@qq.com
[2] Bangkok Thonburi University, Bangkok, Thailand

Abstract. As the starting point of the Maritime Silk Road, Guangdong has been a major foreign trade province famous at home and abroad since ancient times. It is not only an important trading hub of Chinese porcelain export, but also the location of kilns for porcelain manufacturing. Compared with kilns in other provinces, the porcelain design in Guangdong has caught less attention in relevant academic history. Guangdong porcelain has a long history of design, manufacturing, processing and export trades. The discovery and collection of Guangdong porcelain in Japan, the United States, Europe, Africa and other places demonstrate the acceptance of them all over the world. The export trade of porcelain becomes a medium of design aesthetics exchange between China and the Western countries with profound effects on the world's creation culture.

At present the research achievements on Guangdong porcelain design mostly focus on the perspective of trading, relics, archaeology or art appreciation, while few of them are on the design development sequences of porcelain produced in Guangdong kilns. Taking the relationship between Guangdong traditional ceramic and porcelain design as the basis of research, this paper will focus on the driving effects of Guangdong porcelain export during the 16th–19th centuries on the improvement of local porcelain design techniques, modeling, decorative styles and Sino-Western exchanges from several aspects including folk kiln firing, famous kiln imitation, faience custom-making and Sino-Western interconnections.

Keywords: Export driven · Guangdong · Porcelain design

1 Introduction

The development of Guangdong porcelain design is a kind of interaction between traditional culture and modern culture, local culture and world culture. Since the overseas demand for porcelain fueled the rapid development of Guangdong porcelain, even until now new ways of porcelain manufacturing and design have a strong impact on traditional handicraft manufacturing and design approaches. This paper takes a historic and local, traditional and innovative view of Guangdong and starts from the perspective of traditional market demands and modern people's demands for porcelain design to illustrate the continuity of porcelain export in Guangdong and analyze the development of Guangdong's porcelain export in the history of design.

© Springer Nature Switzerland AG 2020
P.-L. P. Rau (Ed.): HCII 2020, LNCS 12192, pp. 408–417, 2020.
https://doi.org/10.1007/978-3-030-49788-0_30

2 Ceramics and Modern Porcelain Design

As we all know, the word "design" is a loanword that has been used in China since the early 20th century. The word "craft" originated from "百工(*baigong*)" in Chinese history which was first used in the pre-Qin period (before 221BC), referring to all kinds of skills for living. Ceramics, as a representative pearl in the long history of Chinese creation art, had been praised by the world for its original technology and material properties suitable for human life. In the 20th century, along with the wide use of the new term "design", porcelain design began to be used more extensively. From then on there have always been arguments and debates in various academic areas on the similarities and differences between "ceramics" used for years and "porcelain design" in all kinds of academic field. The debates on the terms just reflect changes of the times and the development of Chinese design art education disciplines.

As for the definition of porcelain design, most scholars discuss the relationship between traditional ceramics and porcelain design from the perspective of the macro-history. Mr. Zhang Daoyi pointed out in his paper "Design Concept: A Key Problem in the Teaching of Arts and Crafts" that "How did the design concept come into being? It originated in practice and was based on the direct needs of daily life and the needs of production [1]." "A Brief History of Chinese and Foreign Design" edited by Li Yanzu and Zhang Fuye states from a historical perspective that "creation" is always followed by "design", and "creation" without "design" is unimaginable. Fang Xiaofeng (2018) also pointed out that the term "arts and crafts" expressed the meaning of "design" in the context of that time, and also reflected the design concept at that time. In this context, it just means "design" [2]. By these ideas we can tell that there has been a long history of porcelain design in our country, and from the ancient times "design" existed when people making ceramic according to their own needs. The understanding of porcelain design in this paper quotes the definition by Li Zheng'an (2002): it is the way to express and produce ceramic products [3].

Based on the change of terms, this paper discusses the development of porcelain design based on the ceramic techniques applied in Guangdong, mainly concentrating on the development of Guangdong's porcelain design during the 16th–19th centuries, and demonstrates the driving effects of porcelain export on local porcelain design from several aspects including folk kiln firing, famous kiln imitation, faience custom-making and Sino-Western exchanges.

3 Porcelain Export: The Driving Force for the Development of Guangdong's Ceramics

The rise of the ceramic industry in Guangdong is closely related to overseas trade. Guangdong locates right at the shipping hub of the South China Sea. Ancient Yue people living in Lingnan region sailed on the vast ocean very early. With technological advances in navigation and shipbuilding, they went abroad by sea to communicate with countries all over the world in order to increase their openness, and eventually brought Guangdong infinite open advantages and vitality through the sea. According to archaeological records, in the ancient times stamped pottery of Lingnan region spread to Indochina and the Malay Peninsula via Vietnam, and then to the south Pacific islands. Stamped pottery has been found in Vietnam, Malaysia, the Philippines, Indonesia and the South Pacific islands, while the ceramic, techniques quality and patterns are similar to those of Lingnan region [4]. By the time of Sui and Tang Dynasties (581–907AD) and the Five Dynasties Period (907–960AD), the porcelain making technology in Guangdong was quite mature. Merchants in Lingnan travelled far and did maritime trade all over the world for a long time.

Taking advantage of the unique natural and ecological environment, the ceramic industry in Guangdong has formed the export-oriented business model. Driven by porcelain export and great economic profits, porcelain production and export sites have been set up including Chaozhou Kiln, Bijiashan Kiln, Raoping Kiln, Mei county Waterwheel Kiln, Dapu Kiln and Huizhou Kiln in east Guangdong, Xicun Kiln, Shiwan Kiln, the South China Sea Rare Stone Kiln and Xinhui Guanchong Kiln in central Guangdong, Leizhou Kiln, Lianjiang Kiln and Suixi Kiln in west Guangdong. From the Neolithic Age to Ming and Qing Dynasties (1369–1840), there were 1382 kilns found in Guangdong province [5] (see Fig. 1).

Fig. 1. Distribution map of kilns in Lingnan in history

4 Development of Porcelain Design in Guangdong Driven by Export

The rise of ceramic industry in Guangdong was closely related to overseas trade. In the Tang Dynasty (618–907AD), Guangdong ranked top among the four ports, and the ceramic industry began to rise though it was at the primary phase and the techniques were relatively rough. But just as the magnificent sea gave birth to brave, pioneering and enterprising Guangdong residents who have the courage to face the challenges of the natural environment and opportunities, driven by foreign trade, ceramic practitioners in Guangdong improved the manufacturing technology of ceramic by practicability, imitation of ancient kilns and learning from Western countries, and creatively developed the modeling and decoration design methods of Guangdong ceramics.

4.1 Using Folk Kilns to Benefit People

Professor Li Lixin defines the characteristics of the study of design art as below: the subject of study is something useful rather than pure art; the approach of study is systematic rather than empirical; the problem of study is realistic rather than abstract [6]. The production objects of Guangdong ceramics have always focused on folk kilns, mainly for practical purposes, and were produced as daily necessities of the common people as well as for commercial trade. In addition to Guangzhou, in nearby areas imperial kilns were used to make porcelain for the imperial court, but they didn't last long. Therefore, the porcelain of Lingnan was not as gorgeous, noble or elegant as imperial kiln products in northern China, but was more economical, more practical, simpler and fresher. The style was a unity of functions and forms while the appearance was originated from the usage. The decorative patterns reflected the aesthetic tastes of common people and citizens and were popular with them [7].

Although most of the design of the porcelain in Guangdong originally derived from the imitation of other kilns with rough techniques. Just as Gombrich mentioned that the history of art development is not about technical proficiency development but about the change of concept and demands [8], people's unconscious appreciation and admiration only associate with the means and level of performance, and it is necessary to have an introspection and review [9]. Because of the changes of concept and demands, the ancestors of Guangdong laid the foundation for the future porcelain design and manufacturing in Guangdong.

4.2 Imitating Famous Kilns to Be All-Inclusive

Although the history of ceramic manufacturing in Guangdong is quite long, in terms of techniques and craftsmanship there was a lack of representative category in history. In order to meet the great demand for porcelain products at home and abroad, ancient people in Guangdong continuously absorbed manufacturing experience from other famous kilns and formed a set of design and manufacturing models to adapt to the mode of export sales. As far as export demand is concerned, in Yuan and Ming Dynasties the export volume of Longquan Kiln was the largest. Therefore, a large number of kilns imitating Longquan Kiln mushroomed in Lingnan, and most of them located in Huiyang near Guangzhou. Celadon was quite comparable with that of Longquan Kiln in Zhejiang from the late Yuan Dynasty (1271–1368) to the Ming Dynasty (1368–1644) in terms of the characteristics, decorative styles, loading and firing technologies and kilning technology [10]. Thus it can be seen that by the Ming Dynasty Guangdong had mastered the manufacturing technology of celadon (see Fig. 2 and Fig. 3).

Fig. 2. Green glaze vase with two ears produced in Chaozhou Kiln (Color figure online)

Fig. 3. Green glaze vial produced in Chaozhou Kiln (Color figure online)

The bluish white porcelain, egg white porcelain and blue and white porcelain produced in Jingdezhen Kiln were the second only to Longquan celadon in terms of the quantity of export. Figure 4 and Fig. 5 show the blue and white porcelain made in Haikang Kiln in Leizhou, Guangdong province, imitating Jingdezhen kilns. Thanks to the convenience of transportation on Leizhou Peninsula, people could sail from Leizhou port to Guangzhou, Fujian and other important ports, and also to some countries in Southeast Asia. The emergence and setup of Haikang Kiln were to meet the great demands of overseas trade. Take Shiwan kiln as another example. In the Qing Dynasty there was the saying that "no kiln is not imitated", with Tang tri-color glazed ceramics, porcelain of Jun Kiln and Ge Kiln all imitated (see Fig. 6). Among these, Shiwan Kiln was the most famous for imitating the flambe-glazed porcelain of Jun Kiln, whose imitation-based creation and creative development were known all over the world, for instance, the blue glaze caused by transmutation during firing as shown in Fig. 7. By imitating famous kilns, it provides corresponding technical support for the future development of porcelain design in Guangdong.

4.3   Custom-Making Faience with Creation and Innovation

Canton glazed porcelain, a representative type of Guangdong overglazed porcelain, originated during the reign of Emperor Kangxi of the Qing Dynasty, developed during the reign of Emperor Yongzheng and matured during the reign of Emperor Qianlong. There are documentary records on the origin of Cantonese glazed porcelain. Liu Zifen mentioned in *Records on Porcelain in Bamboo Garden* that by the middle of Qing Dynasty the ships gathered and the business flourished. Europeans highly appreciated Chinese porcelain, so Chinese businessmen catered to their pleasure by firing white porcelain in Jingdezhen, shipped porcelain products to Guangdong, employed craftsmen to paint in western style, set up kilns on the southern bank of the Pearl River to dry the paint,

Fig. 4. Porcelain jar with lotus leaf-shaped lid produced in Chaozhou Kiln

Fig. 5. Blue and white porcelain jar with lid produced in Haikang Kiln (Color figure online)

Fig. 6. Vial of Shiwan Kiln by imitating Ge kiln

Fig. 7. Ewer in Shiwan Kiln by imitating Jun Kiln

made them into faience and then sold them to the Western countries, that's to say, "the glaze produced at the southern bank of the river was called Cantonese glaze " [11], "the southern bank of the river" here refers to the southern bank of the Pearl River.

In *Ancient Chinese Porcelainware Exported to Southeast Asia* edited by Han Huaizhun it records that at that time, trading company in Guangzhou also made over-glazed porcelain including five-color or tri-color glazed ones for European merchants. The bisque products were mostly made in Jingdezhen and transported to Guangzhou. In accordance with Europeans' habits and preferences, sometimes the patterns were drafts or army emblems drawn by Europeans totally in European style, sometimes being a mixture of Chinese and western style, and sometimes copying Chinese painting. The porcelain painters in Guangzhou painted on bisque with color enamels and golden paint, and then the porcelain was fired, became five-color or tri-color glazed and was exported to Europe [12]".

Since the late 17th century, the Qing government began to accept custom-made Chinese porcelain with shapes and styles influenced by European counterparts. Among these orders there was a type of porcelain decorated with the armorial bearings of European aristocracy, royalty and corporations, also known as heraldic porcelain (see Figs. 8 and 9). Due to the need of export, the European style of ware and decorative patterns expanded the field of vision of Guangdong porcelain design development, and enriched the raw materials of Guangdong porcelain production. Moreover Cantonese glazed porcelain was deeply affected by European drawings and printmaking works. The porcelain tableware in Fig. 10 was an imitation of George Carter's work "The Injured Hussar". This plays a very positive role in divererifying Guangdong-style paintings and fostering the tendency of design combing painting and ceramic modeling (Fig. 11).

Fig. 8. Plate decorated with armorial bearings of French family De Neuverville

Fig. 9. Plate decorated with the English armorial bearings and inscriptions

In the process of porcelain making, porcelain makers in Guangdong would try their best to produce porcelain according to the models and patterns provided by western customers. In the generation of export orders, the special cultural phenomenon of the

Fig. 10. A copy of George Carter's etchings by Valentine Green

Fig. 11. Cup and plate decorated with a picture of hussar sitting beside the gun carriage

integration of Chinese and western cultures appeared in the communication and mutual assistance of porcelain design, for example, the modeling of the Western style. As shown in Fig. 12, the style of the kettle obviously originated from the European modeling of gold and silver, while the decoration was of typically Chinese style. For the need of the orders, Chinese and western cultures could be integrated with porcelain as a carrier, and created more possibilities for the development of porcelain design. In Figs. 13 and 14, the shapes of the porcelain are not of traditional Chinese shapes. Through the exchange

Fig. 12. Kettle

Fig. 13. Kettle for washing

Fig. 14. Dessert plate decorated with the armorial bearings of the Dutch family Kamps

of trade, the design forms were transferred, and economy promoted the communication on design and the development of Guangdong's ceramic industry.

By the 18th century, in Europe and the Middle East flourished another popular production process in which metal accessories or painted decoration were inlaid on finished or semi-finished porcelain made in China. Thus it can be seen that the foreign customized Chinese porcelain had become more diversified and attractive, reflecting the tastes and preferences of Western customers (see Figs. 15, 16 and 17). The innovative technique of combining various materials such as gold-plated copper, silver inlay, brass and bronze accessories extended the design form of Guangdong porcelain, and the concept of design and development for needs gradually came into being.

Fig. 15. China's porcelain cup inlaid with English silver base

Fig. 16. Five-color kettleinlaid with copper-clad fittings in Kangxi period

Fig. 17. Five-color goblet with lid inlaid with copper-clad fittings in Kangxi period

5   Conclusion

To sum up, under the strong overseas demands from the 17th century through the 19th century, through the commercial and cultural exchanges between China and Western countries, Guangdong ceramics kept its own characteristics on the one hand, but on the other hand it constantly stimulated, inspired and enriched its own design forms. The constant reference, absorption and utilization of various designs are also changing the design features of the world in this historical period [13]. These decorative methods and technical characteristics driven by export have laid a solid foundation for Guangdong to become a modern porcelain production site in China. All kinds of ceramic production techniques can be realized in Guangdong, making Guangdong the largest porcelain manufacturing and export province in China for many years in a row. From the tableware design in Fig. 18, we can feel the combination of Chinese and Western elements in terms of functions and shapes, and the design concept of decoration with various crafts such as Cantonese glazing and partial gold plating.

Communication is essential for cultural and ideological exchanges, and the demand of human life is the driving force of design development. This paper discusses the development of Guangdong porcelain design driven by the eastward spread of Western

Fig. 18. Auratic porcelain, Shenzhen, Guangdong

designs, culture, lifestyles and design concepts. From the perspective of design culture, the paper is aimed to make full use of the predecessors' research results during the 16th— 19th centuries to demonstrate the basic context and draw out some beneficial ideas of Chinese and Western cultural exchanges on porcelain design in order to help with the development of modern Chinese porcelain design.

References

1. Xi, C.: Selected Reading on Art Design, p. 150. Southeast University Press, Nanjing (2002)
2. Fang, X.: Practice guide, research drive: how design study establishes its own disciplinary paradigm. Ornament **12**(9), 14 (2018)
3. Li, Z.: Porcelain Design, p. 84. China Academy of Art Press, Hangzhou (2002)
4. Shen, J.: History of Ceramics in Lingnan, p. 52. Guangdong Higher Education Press, Beijing (2003)
5. Zeng, G.: Summary of archaeology in Ci Kiln site in Guangdong. Cult. Relics Jiangxi **32**(4), 73 (1991)
6. Li, L.: Research Methods of Design Art, p. 3. Jiangsu Art Press, Nanjing (2010)
7. Shen, J.: History of Ceramics in Lingnan, p. 215. Guangdong Higher Education Press, Beijing (2003)
8. Gombrich, E.H.: The Story of Art, p. 18. Tianjin People's Fine Arts Press, Tianjin (2006)
9. Li, G.: History of Art in Guangdong, p. 115. Guangdong People's Publishing House, Guangzhou (2008)
10. Xu, Z.: Discussing ceramics in Yinliu House edited by Zheng Tinggui, A Compilation of Chinese Classics on Ceramic, Chinese Library, 144(1991)
11. Liu, Z.: Records on Porcelain in Bamboo Garden, proofread by Sun Yan, Ancient Porcelain Identification Guide, pp. 93–94. Yanshan Press, Beijing (1993)
12. Han, H.: Ancient Chinese Porcelainware Exported to Southeast Asia, p. 68. The Youth Book Co., Singapore (1960)
13. Jiang, Q.: Research on design exchanges in the context of Sino-Britishtrade, p. 1. Doctoral dissertation of Nanjing University of the Arts (2017)

A Study of the Qualia Characteristics
of Taiwanese Food

Cheng Hsiang Yang[✉] and Po-Hsien Lin

Graduate School of Creative Industry Design, National Taiwan University of Arts,
New Taipei City, Taiwan
yjs.amo@gmail.com, t0131@ntua.edu.tw

Abstract. Food experience is crucial for international tourists to understand local cultures. Studies have revealed that qualia are critical for product design. Qualia are closely related to consumer satisfaction with food experience. However, most studies have only mentioned factors that influence qualia, and no quantitative studies on food qualia have been conducted. Taiwan is an Asian hub with a history incorporating diverse cultures, thus creating a unique and diverse food culture. Taiwanese food has become a key medium for foreigners to experience Taiwanese culture. Therefore, integrating food into qualia is an essential topic for international food marketing. In this study, we selected eight typical Taiwanese foods and 12 qualia to conduct a quantitative questionnaire survey. Overall, 225 valid responses were collected. By conducting variance, correlation, cluster, and multidimensional analyses, consumers' perception and classification of the qualia of Taiwanese foods were explored. The results were as follows. (1) People with food experience can be oriented toward affordability, locality, gourmet, and popularity. (2) Food preferences were significantly correlated with diverse flavors, rich taste, professional cooking, healthy nutrition, convenience, locality, and delicateness. (3) Food representativeness was significantly correlated with rich taste and locality.

Keywords: Qualia · Experience · Taiwanese food

1 Introduction

Food experience is a primary marketing policy in various countries. In recent years, Asian countries have actively promoted themselves through food provision. For example, in Singapore, an Asian food capital project was developed incorporating Chinese, Indian, and Malay cuisines; in 2009, a Korean food globalization plan was initiated to place Korean food among the top five dishes globally [8]. Among Taiwanese food, brands such as Din Tai Fung have been popular for many years, and bubble tea has recently been prevalent in Japan and major international cities. In 2018, Michelin Guide Taipei was published, and Taiwanese foods have been approved by the international media (Taiwan was voted among the world's top 10 food locations by CNN [2], and CNN [3] also broadcasted a selection of the top 40 foods in Taiwan); thus, the image of Taiwanese food has entered the international market, indicating that food is a crucial

© Springer Nature Switzerland AG 2020
P.-L. P. Rau (Ed.): HCII 2020, LNCS 12192, pp. 418–430, 2020.
https://doi.org/10.1007/978-3-030-49788-0_31

medium for internationally promoting Taiwanese culture [4]. In addition, the Taiwanese government has emphasized the food economy and developed related policies. In 2010, the Ministry of Economic Affairs in Taiwan designated food internationalization as an ongoing policy and developed a Taiwanese food internationalization project as well as promoting technological service provision. In addition, the Ministry of Economic Affairs used two strategies (i.e., local internationalization and international localization) to promote Taiwanese food internationalization, indicating that food tourism is a crucial method to provide the international community with an understanding of Taiwanese culture and to create economic benefits for Taiwan [15].

Food culture can attract tourists because food makes an excellent present [9], and food-related tourist experience is connected to memories [11]. Local food is considered an authentic product from the local culture of a tourist place [17]. Food is a part of culture and can reflect cultural experience and identity. In addition, typical local food reflects the characteristics and personalities of local residents [6]. In terms of the economy, local food can manifest a region's unique character, thereby facilitating sustainable tourism development [8]. If a region lacks a famous tourist attraction, food experience can be used as a measure to attract tourists, thereby stimulating local economic development [5, 12, 22].

Food experience can be diverse and complex, contains tangible and intangible elements [15], and provides participants with five sensations through sight, smell, touch, hearing, and taste. In addition, participants can be influenced by physiological, psychological, cultural, and social factors during the experience of food [13]. The modern product design should satisfy users' functional needs and consider their feelings when using the product. The former chief executive officer of SONY Corporation, Nobuyuki Idei, used the term qualia to describe a delighted feeling obtained through sight and touch. Studies have shown that consumers have begun to emphasize the importance of feeling in addition to product functionality and related services (e.g., novelty, local culture, and environmental friendliness) [1, 4, 10, 11]. Therefore, in the future, food experience should provide an affectional communication function with consumers through food design. The aim of this study was to explore tourists' perceptions of Taiwanese foods from the perspective of qualia and to investigate the relationship between qualia and perceived preferences.

2 Food Experience

Food provides tourists with a playful means for discovering the culture of a destination. In particular, the gastronomy of a locality, as a component of the intangible cultural heritage of a locality, reflects the local character and creates a sense of place [16]. However, only until recent decades that the relationship between food and tourism has received attention by academics [7]. The current study on food and tourism focused on five broad topics, including. The paper discovers that the literature on food tourism is dominated by five themes: motivation, culture, authenticity, management and marketing, and destination orientation [4].

Food and experience cannot be separated. Reaching customers through their experience is crucial for food experience. In Welcome to the experience economy, Pine II and Gilmore indicated that value creation includes material extraction from primary products, product production, service provision, and display experience. The tourist search for food-related information before their trips and value originality, newness and locality, as well as authenticity and uniqueness in local food, which eventually have an impact on travel satisfaction [20]. The researcher proposed five positive qualia, including attractiveness, beauty, creativity, delicacy, and engineering, indicating that aesthetics and attraction must be considered in addition to product-related techniques. The results indicated that the more qualia traits a product features, the more likely consumers desire the product. Thus, quality traits should be considered in the design of contemporary cultural and creative products [7].

The researcher also indicated that, in addition to product functionality, the quality, innovation, practicality, and emblem levels must be considered, as well as cultural, tradition-related, and story-related factors [19]. Björk and Kauppinen-Räisänen proposed 17 factors that influence local food experience, including allied production and unique dining manner, and claimed that experience processes have a crucial influence on perception and feeling [1]. In the present study, qualia were considered discovery, perception, and touching experience. Food experience was explored from three levels (i.e., sensory, utility, and symbolic) [10].

3   Methodology

3.1   Food Samples

In the present study, eight typical foods were selected from among 40 popular foods chosen by Taipei city governmental agencies and international media according to the assessment of experts in design and tourism (Table 1) [3, 18]. The typical foods include Minced Pork Rice (p1), Beef Noodle (p2), Soup Dumplings (p3), Bubble Tea (p4), Stinky Tofu (p5), Intestine and Oyster Vermicelli (p6), Oyster Omelet (p7) and Pineapple Cake (p8).

3.2   Research Framework

To explore consumers perceptions of and preferences for Taiwanese foods, according to aforementioned description, this study divided qualia into three levels (i.e., sensory, utility, and symbolic) [18]. A questionnaire was developed for this study according to the proposals of Björk and Lin [16, 18]. On the basis of experts' opinions, reliability analysis was conducted for the questionnaire; the results revealed a Cronbach's α of .801, indicating satisfactory reliability. The qualia indicators used in the questionnaire are described briefly as follows.

1. Sensory indicator: Sensory stimulation through five senses was used to assess visual aesthetics and design. This dimension included four items, namely single flavor

Table 1. Eight foods for this study.

No.	Name	Picture	Description
p1	Minced Pork Rice		A bowl of regular-looking minced pork rice will surprise your taste buds with the tenderness of the pork without greasiness. You can find minced pork rice everywhere from a street vendor to a five-star hotel. Stewed pork rice commonly refers to the minced pork rice served in Northern and Central Taiwan of which the pork is minced by a knife and stewed with soy sauce.
p2	Beef Noodles		Delicious beef noodles require three essential elements which are noodles, broth and beef. Broth is especially crucial because it is the soul of beef noodles. There are two kinds of beef noodles, braised and clear stewed that come with stewed sirloin and cow tendon.
p3	Soup Dumplings		The famous food originated from the south of Changjiang. It features small size, considerable amount of fillings, juicy and fresh taste, thin wrapper and exquisite shape. Soup is the soul of traditional soup dumplings. In pursuant of paper-thin wrappers and nice presentation, chefs spread flour on the rolling pin before rolling the dough. Traditional soup dumplings have at least 14 folds each, but in some stores that are particularly dainty about soup dumplings like Din Tai Fung, a soup dumpling might have over 18 folds.
p4	Bubble Tea		Bubble tea looks awesome with different layers, the chewy tapioca balls (pearls) immersed in the mellow milk tea are the spirit of the drink that takes you on a journey of revolutionary tastes.
p5	Stinky Tofu		Stinky tofu was named after the special aroma of fermented tofu. Whether it is deep fried or steamed and stewed, that indescribable flavor is what makes it an iconic local food. Most stinky tofu vendors in Taiwan offer deep fried ones with non-spicy Taiwanese kimchi or sichuan kimchi as a side dish to balance off the greasiness of the tofu.

(*continued*)

Table 1. (*continued*)

No.	Name	Picture	Description
p6	Intestine and Oyster Vermicelli		The main ingredients of the dish are oysters and vermicelli. Oysters are coated with cornstarch. Sometimes stewed intestines are used as well. Whether a dish of oyster vermicelli is successful depends on the freshness and size of the oysters. The stewing process of the intestines and the ratio over vermicelli are crucial, too. When enjoying a nice bowl of intestine and oyster vermicelli, you can add some of the store's signature chili sauce, black vinegar, minced garlic and coriander to elevate the flavor!
p7	Oyster Omelet		A perfectly cooked oyster omelet has a crispy texture from the coating made of yam flour and cornstarch. Each store boasts their signature chili sauce that adds to the smooth and chewy mouthfeel of oyster omelets. The rich scent of eggs explode in your mouth as you first take a bite of the omelet with refreshing vegetables including garland chrysanthemum and bok choy.
p8	Pineapple Cake		Pineapple cake is a renowned Taiwanese food made of flour, butter, sugar, eggs, Chinese squash and pineapple jam. The tender fillings inside the shortcrust coating is full of local flavors that are simple yet authentic. The Taiwanese pronunciation of pineapple sounds like "ong lai" which signifies auspiciousness and prosperity. It is also one of the most popular souvenirs in the minds of foreign tourists in Taiwan.

versus multiple flavors (f1), rich and mellow versus fresh and elegant (f2), simple taste versus rich taste (f3), and simple appearance versus novel appearance (f4).

2. Utility indicator: Food was used to benefit health. This dimension included four items, namely simple and fast cooking versus professional cooking (f5), artificial seasoning versus natural and original taste (f6), casual foods versus nutrition and health (f7), and eating on site versus convenience (f8).

3. Symbolic indicator: Food was used to express local culture and personal identity by focusing on local cultural experience and popularity pursuit. This dimension included four items, namely tradition and nostalgia versus fashion and innovation (f9), exotic derivation versus locality (f10), simplicity versus delicateness (f11), and inexpensive versus luxury (f12) (Fig. 1).

3.3 Research Samples

An online questionnaire survey was conducted for this study. Overall, 228 responses were returned. After two invalid responses were removed, a total of 225 valid responses were collected. Therefore, the valid return rate was 98.7% (Table 2). Regarding demographic

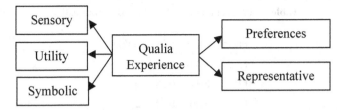

Fig. 1. The conceptual framework

variables, most participants were women (165 women; 73.3%), and most were young (115 participants aged 30 years or younger [51.1%], 43 participants aged 31–40 years [19.1%], 34 participants aged 41–50 years [15.1%], 20 participants aged 51–60 [8.9%], and 13 participants aged 61 years or older [5.9%]).

Table 2. Sample statistics

Variable	Category	Number	Percent (%)
Gender	Male	60	26.7
	Female	165	73.3
Age	~30	115	51.1
	31–40	43	19.1
	41–50	34	15.1
	51–60	20	8.9
	61~	13	5.9

4   Research Results and Discussion

4.1   Descriptive Statistics for Qualia

Table 3 presents the statistics for participants' assessments of the various attributes of qualia of Taiwanese foods. According to their mean values, f10(M = 4.19, SD = 0.62) was the attribute with the highest score, followed by f3(M = 3.60, SD = 0.62), f1(M = 3.34, SD = 0.74), f5(M = 3.28, SD = 0.68), f8(M = 2.82, SD = 0.63), f11(M = 2.82, SD = 0.63), f6(M = 2.75, SD = 0.74), f7(M = 2.66, SD = 0.60), f4(M = 2.46, SD = 0.67), f2(M = 2.34, SD = 0.58), f12(M = 2.27, SD = 0.61) and f9(M = 2.22, SD = 0.64). Thus, according to consumers' qualia of Taiwanese foods, Taiwanese foods tend to be multiple flavors, rich and mellow, rich taste, simple appearance, professional cooking, artificial seasoning, casual foods, eating on site, namely tradition and nostalgia, locality, simplicity and inexpensive.

Table 3. Product ranking of original data for qualia

Factor	Mean	SD	Ranking
f1	3.43	0.74	3
f2	2.34	0.58	10
f3	3.60	0.62	2
f4	2.46	0.67	9
f5	3.28	0.68	4
f6	2.75	0.74	7
f7	2.66	0.60	8
f8	2.82	0.63	5
f9	2.22	0.64	12
f10	4.19	0.62	1
f11	2.82	0.61	5
f12	2.27	0.61	11

4.2 Preference for and Representativeness of Foods

According to participants' preferences, eight foods exhibited high preference ($M = 3.93$, $SD = 0.47$) and representative ($M = 4.29$, $SD = 0.47$) levels. Particularly, p4($M = 4.07$, $SD = 0.88$) exhibited the highest preference level, followed by p3($M = 4.00$, $SD = 0.76$); p6($M = 3.80$, $SD = 0.92$) and p8($M = 3.79$, $SD = 0.89$) exhibited the lowest preference level. Moreover, p4($M = 4.58$, $SD = 0.70$) exhibited the highest representativeness level, followed by p1($M = 4.48$, $SD = 0.60$); p6($M = 4.14$, $SD = 0.80$) and p3($M = 4.06$, $SD = 0.82$) exhibited the lowest representativeness levels (Table 4).

Table 4. Product statistics of original data with their preferences and representative

Product	Preferences			Representative		
	Mean	SD	Ranking	Mean	SD	Ranking
p1	3.86	0.83	6	4.48	0.60	2
p2	3.97	0.81	4	4.16	0.75	6
p3	4.00	0.76	2	4.06	0.82	8
p4	4.07	0.88	1	4.58	0.70	1
p5	3.99	0.93	3	4.37	0.75	3
p6	3.80	0.92	7	4.14	0.80	7
p7	3.94	0.92	5	4.25	0.76	4
p8	3.79	0.89	8	4.25	0.71	4
AVG	3.93	0.47		4.29	0.47	

4.3 Correlation Analysis of Preferences and Representative

According to correlation analysis of preferences and representative, food preferences were significantly correlated with diverse flavors, rich taste, professional cooking, healthy nutrition, convenience, locality, and delicateness. Particularly, f3($r = .303$, $p < .001$) exhibited the highest preference level, followed by f11($r = .275$, $p < .001$); f5($r = .260$, $p < .001$); f10($r = .169$, $p < .05$); f8($r = .151$, $p < .05$) and f7($r = .135$, $p < .05$). Moreover, food representativeness was significantly correlated with rich taste and locality. f11($r = .385$, $p < .001$) exhibited the highest representativeness level, followed by f3($r = .174.$, $p < .01$) (Table 5).

Table 5. The correlation matrix of attributes with their preferences and representative

Factor	Preferences	Representative
	Correlation coefficient	
f1	.243***	.076
f2	.043	−.034
f3	.303***	.174**
f4	.116	.000
f5	.260***	.130
f6	.071	.130
f7	.135*	−.021
f8	.151*	.015
f9	.073	.017
f10	.169*	−.104
f11	.275***	.385***
f12	.058	−.041

4.4 Multidimensional Analysis of Qualia

This study used a multidimensional scale to handle the related data. A spatial distribution diagram was obtained for eight typical Taiwanese foods and four attributes. The stress index and coefficient of determinant were as follows: Kruskal's stress $= .141$ and R squared $= .940$. This result confirmed the suitability of the diagram for describing the spatial relationship between the eight foods in this study and 12 qualia attributes (Fig. 2). Three clusters were formed. For the first cluster, the similar attributes of bubble tea and pineapple cake caused a cluster to form at the right side of the center of the coordinate space. Both bubble tea and pineapple cake were sweet and suitable for taking away. A second cluster for Soup Dumplings and beef noodles was formed at the lower left of the center of the coordinate space. Soup Dumplings and beef noodles were considered as delicate foods. The third cluster included minced pork rice, oyster noodles, oyster pancake, and stinky tofu; this was formed at the upper left of the center of the coordinate space and was considered as affordable traditional street food (Fig. 2).

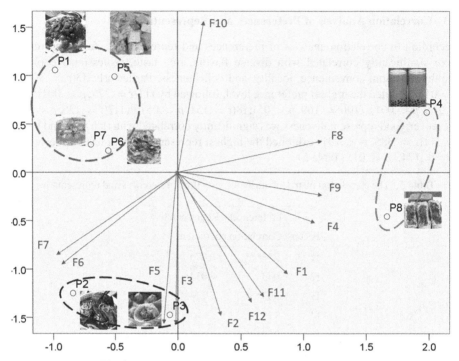

Fig. 2. The spatial representation of MDPREF analysis

When the mean value of a qualia attribute was closer to 5, the attribute tended to be at the right side. By contrast, when the mean value of an attribute was closer to 1, the attribute tended to be at the left side. The ranking for the foods is as follows: p2 > p4 > p6 > p7 > p5 > p8 > p3 > p7 > p1. Bubble tea (p4), which exhibited the highest preference and representativeness levels, had the attributes of multiple flavors and rich and novel taste. Pineapple cake (p8), which exhibited the lowest preference level, had the attributes of being suitable for taking away and having local features. Finally, Soup Dumplings (p3) exhibited the lowest representativeness level and was considered a type of food with exotic culture but without a connection to local culture; accordingly, consumers considered this food less representative (Table 6).

4.5 Cluster Analysis on Qualia

This study first adopted a hierarchical clustering method to determine the most suitable number of clusters; then, k-means clustering was used to perform one-way analysis of the variances of the clusters. Table 7 presents the results. Overall, Factor 10 was the most crucial factor. Therefore, local culture is an important source of qualia for foods. The mean value for Cluster 1 was lower than those for the other clusters. Qualia in this cluster were inexpensiveness, simple appearance, and fast cooking. Therefore, this cluster was termed the "affordable cluster." Cluster 3 emphasized the importance of flavors, cooking skills, and health; therefore, this cluster was called the "delicious

Table 6. Mean value and ranking of 8 works in 12 factors

	p1	p2	p3	p4	p5	p6	p7	p8
f1	2.44	3.80	3.20	3.77	3.48	3.75	3.56	3.42
Ranking	8	1	7	2	5	3	4	6
f2	1.92	2.12	3.28	2.19	1.56	2.10	2.49	3.08
Ranking	7	4	1	5	8	6	3	2
f3	3.10	3.96	3.60	3.73	3.59	3.71	3.67	3.43
Ranking	8	1	5	2	6	3	4	7
f4	1.65	2.60	2.58	3.47	2.20	2.26	2.56	2.33
Ranking	8	2	3	1	7	6	4	5
f5	2.82	4.10	4.07	2.44	2.92	3.22	2.76	3.93
Ranking	6	1	2	8	5	4	7	3
f6	3.12	3.29	3.11	1.95	2.46	2.70	2.78	2.60
Ranking	2	1	3	8	7	5	4	6
f7	3.10	3.71	3.36	1.56	2.06	2.88	2.84	1.79
Ranking	3	1	2	8	6	4	5	7
f8	2.82	1.71	2.47	4.36	2.24	2.39	1.99	4.56
Ranking	3	8	4	2	6	5	7	1
f9	1.40	2.12	2.32	3.67	1.83	1.80	1.89	2.77
Ranking	8	4	3	1	6	7	5	2
f10	4.66	3.94	3.59	4.42	4.24	4.20	4.20	4.24
Ranking	1	7	8	2	3	5	5	3
f11	2.07	3.16	3.81	3.25	2.00	2.41	2.28	3.57
Ranking	7	4	1	3	8	5	6	2
f12	1.54	3.03	3.00	2.23	1.72	1.88	1.94	2.83
Ranking	8	1	2	4	7	6	5	3

Table 7. Analysis of evaluation attributes of foods by different cluster (ANOVA)

Factor	Cluster 1 Affordable (N = 70)		Cluster 2 Local (N = 43)		Cluster 3 Delicious (N = 8)		Cluster 4 Popular (N = 101)		F	Scheffe'
	Mean	SD	Mean	SD	Mean	SD	Mean	SD		
F1	2.84	.62	3.82	.76	4.44	.47	3.60	.53	37.80***	3, 4 > 2 > 1
F2	2.23	.51	2.19	.63	3.83	.70	2.36	.41	26.04***	3 > 1, 2, 4
F3	3.13	.51	4.17	.49	4.41	.45	3.61	.47	49.50***	2, 3 > 4 > 1
F4	2.06	.46	2.01	.46	4.14	.56	2.79	.47	81.45***	3 > 4 > 1, 2
F5	2.73	.59	3.61	.61	4.34	.53	3.46	.49	41.81***	3 > 2, 4 > 1
F6	2.36	.62	3.28	.75	3.97	.75	2.68	.58	29.21***	2, 3 > 4 > 1
F7	2.44	.56	2.63	.59	4.00	.58	2.73	.48	21.53***	3 > 4 > 1&3 > 2
F8	2.62	.52	2.91	.65	3.97	.68	2.83	.58	13.56***	3 > 1, 2, 4
F9	1.85	.42	1.88	.42	3.89	.74	2.50	.48	69.96***	3 > 4 > 1, 2
F10	4.39	.59	4.47	.50	4.39	.47	3.92	.58	14.57***	1, 2 > 4
F11	2.43	.47	2.71	.66	4.28	.50	3.01	.41	44.69***	3 > 4>2 > 1
F12	1.90	.41	1.98	.50	3.83	.76	2.51	.41	65.10***	3 > 4 > 1, 2

cluster." Cluster 2 emphasized the importance of local and exotic attributes, including the importance of the origins of foods and related stories; therefore, this cluster was called the "local cluster." Finally, Cluster 4 emphasized novelty, innovation, and luxury; therefore, this cluster was called the "popular cluster."

5 Discussion and Suggestions

The participants in this study indicated high preference and representativeness levels for Taiwanese foods. According to the participants' reported qualia, the image of Taiwanese foods tended to be unnatural, unhealthy, and fast cooking. Because the international market emphasizes the importance of health and nature, this unnatural and unhealthy image may be disadvantageous to the promotion and marketing of Taiwanese foods. According to the multidimensional analysis, foods can be divided into three types as follows: desserts (P4 and P8), exotic delicate foods (P2 and P3), and local traditional foods (P1, P5, P6, and P7). Exotic delicate foods were highly preferred but less representative. For such foods, consumers considered that they were not local foods (i.e., they were exotic). Minced pork rice and pineapple cake were less preferred but highly representative. These foods were considered as local food, indicating that the representative level of foods was associated with the connection between a food and the local culture. In addition, bubble tea and stinky tofu exhibited high preference and representativeness levels because they have gained publicity through brand marketing. Although stinky tofu is frequently rated among the most unacceptable Taiwanese foods by international tourists, it was considered as a challenging local food. Therefore, for foods, brand promotion may influence the representativeness level of an attribute. Both active and passive promotion can motivate consumers to try a food.

In the results of multidimensional analysis, bubble tea and pineapple cake belonged to the same cluster. However, bubble tea was the most preferred whereas pineapple cake was the least preferred. Because bubble tea is a soft drink that is easy to purchase and can be drunk daily, it was preferred by consumers. Pineapple cake is a traditional dessert; most consumers purchased pineapple cake as a gift but seldom purchased it for themselves. Therefore, the preference level for a food varied according to consumers' motivation. In addition, according to the cluster analysis, consumers can be categorized as affordability-oriented, locality-oriented, gourmet-oriented, and popularity-oriented. A large proportion of consumers were popularity-oriented, and they preferred exotic foods. However, the local features of a food were a crucial factor affecting the food's representativeness. Moreover, consumers' preferences were influenced by flavor diversity, taste richness, cooking professionalism, healthiness, and delicateness. Local culture was another essential factor influencing the representativeness level of a food. Nevertheless, most Taiwanese foods lack relevant qualia attributes, indicating that food business operators should improve product design to enhance Taiwanese foods to diversify their qualia.

Finally, stinky tofu, Soup Dumplings, and beef noodles are exotic foods that consumers highly prefer. However, consumers gave different ratings to the representativeness levels of these foods according to their perceptions of locality. This indicated that consumers' perceptions of local food culture were influenced by diverse and complex factors. Therefore, future studies should explore the actual qualia of food. In addition, the participants in this study were Taiwanese consumers, and foreign consumers were not included. However, eating habits can be influenced by cultures. Therefore, we suggest that future studies should compare Taiwanese and foreign consumers to explore their differences.

References

1. Björk, P., Kauppinen-Räisänen, H.: Exploring the multi-dimensionality of travellers' culinary-gastronomic experiences. Curr. Issues Tour. **19**(12), 1260–1280 (2016)
2. CNN: 40 of the best Taiwanese foods and drinks. http://edition.cnn.com/travel/article/40-taiwan-food/index.html. Accessed 25 Jan 2020
3. CNN: You voted: World's best food is in. http://edition.cnn.com/travel/article/world-best-food-culinary-journeys/index.html. Accessed 25 Jan 2020
4. Ellis, A., Park, E., Kim, S., Yeoman, I.: What is food tourism? Tour. Manag. **68**, 250–263 (2018). https://doi.org/10.1016/j.tourman.2018.03.025
5. Florek, M., Conejo, F.: Export flagships in branding small developing countries: the cases of Costa Rica and Moldova. Place Brand. Public Dipl. **3**(1), 53–72 (2007). https://doi.org/10.1057/palgrave.pb.6000048
6. Forristal, L.J., Lehto, X.Y.: Place branding with native species: personality as a criterion. Place Brand. Public Dipl. **5**(3), 213–225 (2009). https://doi.org/10.1057/pb.2009.16
7. Gordin, V., Trabskaya, J.: The role of gastronomic brands in tourist destination promotion: The case of St. Petersburg. Place Brand. Public Diplomacy **9**(3), 189–201 (2013). https://doi.org/10.1057/pb.2013.23
8. Ho, S.-N.: The globalization of Korean food: Korean cuisine and the construction and imagination of a Korean national brand. J. Chin. Diet. Cult. **13**(1), 165–203 (2017)
9. Lee, K.-H., Scott, N.: Food tourism reviewed using the paradigm funnel approach. J. Culin. Sci. Tech. **13**(2), 95–115 (2015). https://doi.org/10.1080/15428052.2014.952480
10. Lin, L., Mao, P.C.: Food for memories and culture–a content analysis study of food specialties and souvenirs. J. Hospitality Tour. Manag. **22**, 19–29 (2015). https://doi.org/10.1016/j.jhtm.2014.12.001
11. Lin, L.: Food souvenirs as gifts: tourist perspectives and their motivational basis in Chinese culture. J. Tour. Cult. Change **15**(5), 439–454 (2017). https://doi.org/10.1080/14766825.2016.1170841
12. Lin, Y.C., Pearson, T.E., Cai, L.A.: Food as a form of destination identity: a tourism destination brand perspective. Tour. Hospitality Res. **11**(1), 30–48 (2011). https://doi.org/10.1057/thr.2010.22
13. Mak, A.H., Lumbers, M., Eves, A., Chang, R.C.: Factors influencing tourist food consumption. Int. J. Hospitality Manag. **31**(3), 928–936 (2012)
14. Ministry of Economic Affairs: Gourmet TAIWAN. https://ws.ndc.gov.tw/Download.ashx?u=LzAwMS9hZG1pbmlzdHJhdG9yLzEwL1JlbEZpbGUvNTU2Ni81MjMzLzAwMTM5OTEucGRm&n=OTkwNTI25Y%2bw54Gj576O6aOf5ZyL6Zqb5YyW6KGM5YuV6KiI55WrKOS%2fruato%2baguOWumuacrCkucGRm&icon=..pdf. Accessed 25 Jan 2020
15. Ron, A.S., Timothy, D.J.: The land of milk and honey: biblical foods, heritage and holy land tourism. J. Herit. Tour. **8**(2–3), 234–247 (2013). https://doi.org/10.1080/1743873X.2013.767817
16. Sanchez-Cañizares, S., Castillo-Canalejo, A.: A comparative study of tourist attitudes towards culinary tourism in Spain and Slovenia. Br. Food J. **117**(9), 2387–2411 (2015). https://doi.org/10.1108/BFJ-01-2015-0008

17. Sims, R.: Food, place and authenticity: local food and the sustainable tourism experience. J. Sustain. Tour. **17**(3), 321–336 (2009). https://doi.org/10.1080/09669580802359293

18. Taipei City Government.: Must Eat in Taiwan. https://www.travel.taipei/zh-tw/must-visit/snacks-top10. Accessed 25 Jan 2020

19. Yen, H.Y., Lin, P.H., Lin, R.: Qualia characteristics of cultural and creative products. J. Kansei **2**(1), 34–61 (2014)

20. Yen, H.Y., Lin, R., Lin, R.: A study of value-added from qualia to business model of cultural and creative industries. J. Natl. Taiwan Univ. Arts **91**, 127–152 (2012). https://doi.org/10.6793/JNTCA.201210.0127

21. Chou, Y.-H., Ou-Yang, H.-Y., Chen, K.-C.: Taiwan tourism 2020: a sustainable tourism development strategy. Taiwan Modern Tour. **1**(1), 1–20 (2018). https://doi.org/10.6348/TMT.201809_1(1).0001

22. Yurtseven, H.R., Kaya, O.: Local food in local menus: the case of Gokceada. Tourismos **6**(2), 263–275 (2011)

Aesthetic Contemplation of the Tang Dynasty Dunhuang Frescoes Elements on Contemporary Costume Design

Hong Zhang[✉]

Beijing Institute of Technology, Zhuhai, Zhuhai, China
693108832@qq.com

Abstract. With a history of over one thousand years, the art of Dunhuang has developed into a series and diversified schools of design, highlighting the profound connotation of the Chinese culture. Influenced by social politics, economy, culture and aesthetic orientation at that time, Dunhuang fresco in Dang dynasty reached its peak in decorative forms and decorative techniques, obtaining high historical literature value and aesthetic value. To apply the aesthetic elements of Dunhuang murals in Dang dynasty to modern fashion design, the designer is supposed to be innovative, and endeavor to analyze, research, extract, reconstruct the national culture embodied in Dunhuang murals in Dang dynasty, so that they can present the ethnic culture characteristics in the modern visual design language, which may conform to modern aesthetic orientation.

This paper applies methods of literature research, case analysis, and comparative study to summarize and classify the aesthetic literature of Dunhuang murals in Dang dynasty in an effort to study the material, design, image, detail and cultural connotation of Dunhuang murals in Dang dynasty to understand its aesthetic value and its development progress. The paper also gives an analysis on some representative cases of the application of the aesthetic elements of Dunhuang murals in Dang dynasty in the fashion design. Meanwhile, the paper is also an investigation and analysis about the trend of contemporary fashion design as to the style, color, craft and so on, to see possibility to incorporate the aesthetic elements of Dunhuang Dang fresco into the concept of contemporary fashion design and its practice.

On the basis of the study, this paper points out that the Dunhuang fresco in Dang dynasty may offer enlightenment to the contemporary fashion design in fields of material, color, composition, delicacy and cultural connotation: pure natural mineral toner and other materials used in the Dunhuang fresco can be applied to the corresponding clothing fabrics, trying to produce special clothing materials; and the color composition of Dunhuang Dang fresco can also be a reference for the clothing color collocation; the design of clothing style and pattern can be analyzed, extracted and reconstructed from the pattern of Dunhuang fresco in Dang dynasty; and the delicacy of Dunhuang frescoes in Dang dynasty may help the designer in the detail-designing. The Buddhist culture in Dang dynasty was unprecedentedly active, showing the magnificence of the time. It is also necessary to implant such culture into contemporary fashion design.

Clothing is a walking culture. It provides a more intuitive window for the dissemination and exchange of Chinese culture to apply aesthetic elements of

© Springer Nature Switzerland AG 2020
P.-L. P. Rau (Ed.): HCII 2020, LNCS 12192, pp. 431–440, 2020.
https://doi.org/10.1007/978-3-030-49788-0_32

Dang dynasty frescoes in Dunhuang into contemporary fashion design. Therefore, the study is of a high research value and importance.

Keywords: Dunhuang · Dang dynasty fresco · Aesthetics · Fashion design

1 Decorative Significance of the Tang Dynasty Dunhuang Frescoes

Dunhuang art can date back to 366 AD. With a history of over a thousand years, it implies the profound cultural connotations of China. It has gone through several periods and there are a variety of masterpieces of design art in different forms. During the Tang Dynasty, politics, economy and culture were more flourishing than ever before in China. At that time, Dunhuang frescoes reached the peak in decorative forms and techniques of great value in historical literature and artistic aesthetics as well.

1.1 Analysis of Decorative Significance of Caisson Patterns in the Tang Dynasty Dunhuang Frescoes

The decorative function of patterns was greatly developed in the Tang Dynasty Dunhuang frescoes; especially the layouts of caissons on the top of the grottoes were in good order, complicated and diverse. In early Tang Dynasty, the central parts of Dunhuang caissons were designed relatively wide in structure, and the patterns in the center were highly prioritized with little special design on other patterns around. The three main decorative patterns are grape & pomegranate pattern, pomegranate & lotus pattern, and among them lotus pattern was the mainstream. The patterns are in concentric circles structure and almost conform to the golden proportion. The aesthetic feeling of design is illustrated by the perfect proportion. Taking Cave 209 in Dunhuang Mogao Grottoes as an example (as shown in Fig. 1), the caisson patterns are quite symmetrical on the whole. In the center four pomegranates make up a cross, and around them four pieces of grape leaves and branches make up another cross. Two crosses build the frame with 8 bunches of grapes around pomegranate which is the exotic round and square structure. The center of the frame extends outwards to a symmetrical square frame, in which the decoration is composed of points, lines and surfaces formed by simplified figure of grapes. The center of the caisson is mainly in realistic style, while the decoration outside the center is mainly abstract. On the whole, the design is in contrast and unity as well.

In the flourishing period of the Tang Dynasty, the overall structure of the patterns was designed in a more harmonious proportion. The center of the caisson gradually became smaller and protruding upward. The edge outside the center was decorated with more layers, so the overall patterns were luxuriant and complicated in a flourishing scene but solemn and well-ordered as well.

In this period, the decorative patterns in the center of the caisson were mostly lotus and could be divided into four types: peach-shaped lotus, leaf-shaped lotus, round lotus and hybrid lotus. The difference of lotus patterns presents the attributes and characteristics of caisson decoration in different periods. For example, the caisson of Cave 320 in Mogao Grottoes (as shown in Fig. 2) was with group lotus pattern in the flourishing period of

Fig. 1. Cave 209 in Dunhuang Mogao Grottoes

Fig. 2. The caisson of cave 320 in Mogao Grottoes

Tang. The center was relatively small and filled with a large flower. During this period, large flowers basically lost the characteristics of lotus; they had luxuriant layers of petal and were round in shape. The outer patterns were half round lotus pattern, diamond pattern, one and two halves round lotus pattern, scale pattern, triangle pattern and bell pattern. Compared with the caissons in the early Tang Dynasty, caissons painted in this period were more patterning with less realistic elements and more abstract contents.

1.2 Analysis of Decorative Significance of Curling Grass Pattern in the Tang Dynasty Dunhuang Frescoes

The curling grass pattern is a typical decorative pattern of the Tang Dynasty Dunhuang frescoes and mainly used in halo, caisson and background. The theory of harmony advocated by Buddhism has a certain representation in this structure. The effective combination of various elements of curling grass pattern adds poetic charm to Dunhuang. In addition, as people in the Tang Dynasty loved flowers, the plant decoration was a leading decorative pattern at that time.

The curling grass pattern in the Tang Dynasty Dunhuang frescoes was in a dynamic line with an elastic curve, which was flexible and decorated the edge of frescoes. The plants and flowers depicted on the curve were mostly symbolic, conveying good intentions to people. For example, the lotus patterns commonly seen in the Tang Dynasty

Dunhuang frescoes symbolized the pure land of Buddhism, while phoenixes and birds symbolized nobility and elegance.

Most of the curling patterns in the Tang Dynasty Dunhuang frescoes adopted the design method of combining lotus pattern elements with realistic techniques, which can effectively improve the authenticity of the lotus patterns on the curling grass patterns. The curling grass pattern was set in the fresco reasonably as the curve of the main stem changed. The lotus leaves and branches were intertwined with each other, which not only highlighted the sense of depth of the pattern, but also gave people a vibrant sensory experience.

1.3   Analysis of Decorative Significance of Oval or Circular Designs in the Tang Dynasty Dunhuang Frescoes

Oval or circular designs were familiar patterns in the Tang Dynasty Dunhuang frescoes. Such patterns are in a round shape with petals prioritized. Petals are around different flowers like lotus, camellia, peony and pomegranate and form a relatively circular pattern. Oval or circular designs are usually made in areas such as caisson, edge and pedestal of Buddha statue, etc., with a wide range of applications. Because of the oval or circular shapes, such designs can well fit with round or square frames.

In early and flourishing periods of Tang, oval or circular designs diverged outward from the center of the circle, and the patterns of different types were arranged from the inside out. The mix of realistic and abstract styles made the decoration full and fruity. Also the use of various patterns had a strong sense of rhythm and made the caves gorgeous.

The most famous oval or circular design in the Tang Dynasty Dunhuang frescoes was Baoxiang design. "Baoxiang (宝相)" means the solemnness of Buddha statues. In the Tang Dynasty, the open policy enabled the peak of cultural absorption and communication, and Baoxiang design was a combination of different ethnic cultures. During the middle Tang Dynasty, various plant patterns such as peony and lotus were integrated into Baoxiang design. Peony implies wealth and honor, and fits the aesthetic concept of elegance and magnificence of the Tang Dynasty. Lotus stands for Buddhism, pure land and holiness. The round lotus pattern in the caisson and the lotus base of the Buddha statue form a perfect pair. The cave was like a dome with mysterious artistic conception. Also the pattern was often used in the clothing accessories of donors and statues.

1.4   Analysis of Decorative Significance of Color in the Tang Dynasty Dunhuang Frescoes

Color is an indispensable painting language in the field of art. Through reasonable design and application of color elements, the distance between art works and common people can be shortened, and it can help people deeply experience the emotions of works. Most Tang Dynasty Dunhuang frescoes were painted with mineral pigments, which are characterized by high purity, exquisite colors and long preservation. During the Tang Dynasty, the most common pigments in Dunhuang were cyan, green, vermilion, reddish brown, black and white. Different ores were ground and bleached to get different colors. In order to ensure the direct use of pigments, the modulation was difficult. The study of color composition of Dunhuang frescoes has a high artistic value. After the baptism of a thousand years, the colors are still bright now. At the same time, through the application of formal beauty principles such as contrast and unity, and through the design techniques such as overlapping and foil, the contents of Dunhuang frescoes were more life-like. The dominant hues used in Dunhuang frescoes in different periods were different. When viewing and analyzing the frescoes in various caves, the researchers have found that they can clearly reflect the characteristics of the times and ensure the high artistic value of the frescoes under harmonious and unified tones.

2   Status Analysis of Dunhuang Culture in Costume Design

The application of Dunhuang elements in modern and contemporary fashion design is a hot topic recently. But it can be found that in many designs the elements were applied quite mechanically. First, the combination with clothing characteristics was not considered. For example, some designs only applied certain elements of Dunhuang, such as oval or circular design, in modern clothing by apparel technology without carrying out appropriate designs for specific clothing styles. Secondly, the specific craft was not taken into consideration. For example, some patterns could have been better represented by embroidery, but in consideration of cost some designers just turned to other relatively low-cost technologies, so the effect would be greatly diminished. Thirdly, the fabric was not considered. Famous fashion designer Lawrence Xu once hosted a fashion show themed on "Dunhuang". For one of the clothes he "visited a number of silk markets in China and finally chose brocade from Zhejiang to make it more original". How many designers are so persistent on the effect of clothes now?

3   Innovative Application of Dunhuang Culture in Costume Design

3.1   Innovative Application in Modeling and Decorative Parts

In regards to structure modeling, the three-dimensional tailoring modeling mode can be added. Through the analysis of the structural characteristics of the objects selected in the Tang Dynasty Dunhuang frescoes, the structural forms were integrated into the structural modeling of clothing, and the unity and modeling sense of clothing were formed. In the 2015 Spring and Summer Paris Fashion Week, Lawrence Xu designed the Dunhuang series of clothes, some of which had excellent applications in structural modeling. In the design of waist part, he used the modeling characteristics of lotus flower petals for reference, and extended them in a three-dimensional modeling way, making the apparel modeling multilayered (as shown in Fig. 3).

3.2   Innovative Application in Pattern and Texture

The fabric is remanufactured and innovated. As the basic means of modeling, fabric is of great significance to enrich the layers of clothing, strengthen the visual effect, and further expand the expressive force of modeling. Using fabrics to assist modeling is a new expressive force pursued on the basis of making full use of modeling elements -

Fig. 3. Lawrence Xu designed the Dunhuang series of clothes

shape, space and structure. By molding fabrics or using color blocks, the layers of the pattern can be enriched by stitching. Through hollowing out, piling up, inlay and other basic modeling techniques, new fabrics is created with decorative elements of patterns integrated.

Hollowing-Out. Hollowing-out is a relatively commonly used technique in modern women's dress, that is, to strengthen females' sexy charm by hollowing out part of their clothes. At present, the most representative hollowing-out technique is applied in lace fabrics. Many designers generally like the design pattern and style which is light but not over-revealing. The flying Apsaras and other traditional decorative patterns in Tang Dynasty Dunhuang frescoes are very suitable to be presented by the technique of hollowing-out.

Piling-Up. In contrast to hollowing-out, piling-up is the modeling technique of adding materials to the overall design of clothing in order to make the contour of clothing fuller and three-dimensional, and enhance the concave and convex shapes. Piling-up, if not well designed, can easily lead to a mess. Therefore, it needs to be designed as a whole to make sure the structure is well organized and balanced. In terms of specific materials, suitable ones can be used in combination with the characteristics of the selected objects of the Tang Dynasty Dunhuang frescoes. For example, piling-up was applied to the clothes designed by Lawrence Xu (as shown in Fig. 4, Fig. 5) for several times. In Fig. 4 he used three-dimensional patches to pile up a circular design. In Fig. 5 he applied woven ropes for piling-up. Other designers often use this technique as well, such as famous Chinese designer Vivienne Hu who basically applied piling-up to the dress shown in Fig. 6.

Fig. 4. The clothes designed by Lawrence Xu

Fig. 5. The clothes designed by Lawrence Xu

Fig. 6. The clothes designed by Vivienne Hu

Inlay. Inlay can enhance the sense of three dimensions and layering of clothing. After learning the decorative patterns in Dunhuang in the Tang Dynasty, different gemstones in different shapes and colors can be arrayed to form decorative patterns of the embroidery (as shown in Fig. 7). Other materials such as inlaid pearls can be used as well. The technique of inlaying is not too complicated, but attention should be paid to the aesthetic feeling of the array of inlaid materials.

Fig. 7. Inlay can enhance the sense of three dimensions and layering of clothing

3.3 Combination of Colors

In the process of designing clothes, one important factor is colors. It is very important to deeply analyze and understand the application and the using principles of colors, and how to harmoniously apply it to costume design. First, the basic tone of clothes should be determined.

The color application of Dunhuang frescoes in Tang Dynasty was very mature. During the learning, we should strictly follow color extraction, recombination and innovation. In the frescoes, contrasty colors like cinnabar, stone green and earth red were used, but it seems that there was no confusion or uneasiness. One of the important methods was to choose one color as the main tone and the others as secondary colors. The second method was to reduce the purity between contrasting colors, such as adding a color to red and green to increase or decrease their lightness, and reduce the strong contrast between colors. The third method, which emphasizes the importance of white and black in the picture, separates the contrasting colors with colorless shades of black, white or gray, also to weaken the contrast between them. Taking the Dunhuang series designed by Xiong Ying as an example (as shown in Fig. 8), she took white as background color. Although cinnabar and indigo were used on it, the two colors were separated with white, and there was no stark contrast of red and indigo thereby, which made the dress delicate and elegant.

Fig. 8. The Dunhuang series designed by Xiong Ying (Color figure online)

3.4   Embodiment of Apparel Technology

Due to the restriction of style, fabric, tailoring, craft and so on, the traditional clothing usually applies graphic decorative pattern. Thanks to the rapid progress of science and technology in modern times, clothing production technology and methods have entered a new period of development. Coupled with the high speed of information dissemination of fashion trends, pattern is not only a decorative existence, but a personalized application as well. Clothing is more about personal emotions and personality. Therefore, now fashion designers put more emphasis on creativity in the design of patterns.

Digital Printing. Digital printing first appeared in the 1990s, the use of digital technology was the fastest way to achieve personalized clothing. Customers only need to select fabrics and patterns, and in 1 to 2 h the clothes are ready. This is the best way to meet personalized demands and improve people's quality of life. Digital printing is rich in colors. By using four colors plus spot colors, 16.7 million colors can be printed. For fashion designers, by using digital printing they can show their personal creativity and talent effectively and can turn the ideal works into practical clothes. Moreover, when applying digital printing, dye is directly added to a special box and sprayed on fabrics as needed without any waste or water pollution. This can skip the discharge of used dye in traditional process, ensure no pollution in the printing process and reduce the wastage of material like films, silver tubes and silk screens. It not only reduces the burden of the company, but also contributes to environmental protection. If this technology can be combined with the Dunhuang frescoes of the Tang Dynasty and applied in clothing, the gorgeous Chinese traditional decorative patterns can be better inherited and developed.

Laser Engraving. Laser engraving is built on the level of numerical control technology through the laser processing method and the medium. The physical changes of instantaneous gasification and melting of materials in laser make the process happen. It is with higher processing accuracy and faster in speed. In the process of clothing design, laser engraving can accurately produce multiform complicated patterns, and the blank material for cutting does not need to be continuously disposed. In the Tang Dynasty Dunhuang frescoes there were a variety of patterns, and the use of traditional types of technology could hardly achieve good results when producing complicated ones. Laser engraving can be used to solve this problem by hollowing out or collaging and cutting the fabric according to different styles without the need of subsequent modification. At present, the technology can still be used in special types of textiles.

Printing and Dyeing. Printing and dyeing is also called dyeing and finishing. It is the general term of processing methods including dyeing, printing, washing water and so on. In the Neolithic Age about 6,000–7,000 years ago, the ancestors used hematite powder to make sackcloth into red. By the Shang and Zhou Dynasties, the skills of dyeing improved continuously. In the imperial manual workshop, there were special officials in charge of dyeing management, and the number of available colors increased continuously. By the Han dynasty, dyeing techniques had reached a relatively high level. Printing and dyeing can be divided into four types: tie-dyeing, batik, clip-dyeing and gray dyeing. Dye sources are divided into mineral pigments and vegetable dyes. Basically they are both from the nature; therefore it is a kind of craft being environment-friendly. For example, tie-dyeing is to use a line to tie the printing and dyeing fabrics, and then put them into the dye vat for printing and dyeing, then remove the line. The line can be arranged in over 100 ways with different characteristics respectively. The dyed patterns are with rich color shading and very natural. Its unique artistic effect is not easily achieved by mechanical printing and dyeing technology. In the advocacy of sustainable development of materials today, the tie-dyeing technology will arouse more and more attention.

4 Conclusion

The artistic and cultural heritage of Dunhuang is a precious treasure handed down from the ancestors, and the Tang Dynasty frescoes are an indispensable part of the artistic and cultural heritage of Dunhuang. We should not only carry out in-depth research and serious study of this culture, but also comprehend and feel the sincerity of the lonely craftsmen who painted in the caves thousands of years ago devotionally.

At present, modern costume design taking Chinese traditional cultural elements as models has begun to get more attention. For Chinese fashion designers, it is a compelling obligation to study the application of traditional culture. Clothing is a kind of walking culture. To apply the aesthetic elements of Tang Dynasty Dunhuang frescoes to contemporary costume design, we should not only express the culture through appropriate techniques and inherit it, but also provide a more intuitive display platform for the dissemination and exchange of Chinese culture.

References

Yi, C.: Dunhuang Art Aesthetics. Shanghai People's Publishing House, Shanghai (2005)

Zhao, S.: A Brief History of Dunhuang Grottoes Art. China Youth Press, Beijing (2015)

Chang, S.: Dunhuang Costume Patterns in History. China Light Industry Press, Beijing (2001)

Fan, J.: Dunhuang Decorative Patterns. East China Normal University Press, Shanghai (2010)

Chang, S.: Fifty Years in Dunhuang. Peking University Press, Beijing (2011)

Ling, L.: The Application of Tang dynasty Dunhuang Frescoes Caisson patterns in modern clothing, Tianjin Polytechnic University, Tianjin (2015)(12)

Li, X.: Study on the decorative characteristics of the Tang dynasty Dunhuang curling grass patterns and the application in costume design, Beijing Institute of Fashion Technology, Beijing (2013)(12)

Chen, Y.: A study on the modeling use and design experience of edge patterns of Dunhuang Frescoes. Ornament (2006)(9)

Zhang, J.: A study on the application of round flower pattern in the Tang Dynasty Dunhuang Frescoes in costume design. West Leather (2018)(10)

Yang, M.: The cultural function and commercial value of curling grass pattern. Art Obser. (2011)(4)

Zhou, M., Zhou, J.: Design practice on digital jacquard fabric with Dunhuang art style. Silk (2012)(11)

Cross-Cultural Behavior and Attitude

Evaluating Trust, Trustworthiness and Bullwhip Effect: A Three-Echelon Supply Chain Interactive Experiment

Pin-Hsuan Chen and Pei-Luen Patrick Rau$^{(\boxtimes)}$

Department of Industrial Engineering, Tsinghua University, Beijing, China
rpl@mail.tsinghua.edu.cn

Abstract. Trust is essential in supply chain management, and it maintains relationships among members in a supply chain. Moreover, trustworthiness influences an individual's behavior, which is related to trust-building. Therefore, this study aims to investigate trust and trustworthiness in a three-echelon supply chain based on the bullwhip effect, trust diffusion, and suppliers' production adjustment. This study conducts experiment involving 36 participants. Two tasks are performed with unknown and known (high/low) partners' trustworthiness levels. The results present three findings: a) The bullwhip effect could be controlled to a lower level when partners in a supply chain were with high trustworthiness but not those with low trustworthiness. b) Trust diffusion is observed in both high and low trustworthiness cases, which could be useful for developing long-term trust relationships. c) Suppliers are more likely to distrust low trustworthiness partners and adjust their production strategies relative to the high trustworthiness partners. To sum up, this study establishes a connection between trust and the bullwhip effect through experimental results. Additionally, this suggests that trustworthiness is vital information in a supply chain for enhancing performances, especially for upstream members, which is one of the critical elements for information sharing in a supply chain.

Keywords: Supply chain management · Trust · Trustworthiness · Bullwhip effect · Trust diffusion

1 Introduction

Supply chain management is essential for successful business models. Previous studies have proposed several methods to improve supply chain performance. For example, Bowersox et al. [1] presented seven ways to maintain supply chain operating, including integrated behaviors, mutually sharing of information, mutually sharing of risks and rewards, cooperation, same goal and target customers, integrating processes and building of long-term relationships.

Among the mentioned methods, trust is critical to the stability and performance of a supply chain. With trust developing among supply chain partnerships, mutual communications between members could be much easier and more efficient. In this way, it

© Springer Nature Switzerland AG 2020
P.-L. P. Rau (Ed.): HCII 2020, LNCS 12192, pp. 443–453, 2020.
https://doi.org/10.1007/978-3-030-49788-0_33

increases the transparency of a supply chain and strengthens the relationships among members. Therefore, many business models are designed based on trust, particularly for the supply chain management. The Toyota Production System (TPS) is one of the best practice examples for lean production as it integrates production systems, and maintains relationships.

In addition to the production processes, the TPS model highlights the importance of integration. In other words, the model could ensure the accessibility of relevant information and develop the trust by the understanding of members. As trust is one of the keys to the performance of supply chain management, effective mutual communications would be a vital approach to establish.

The purpose of this study is to investigate members' trust and trustworthiness through experiment in a three-echelon supply chain, which includes a supplier, a wholesaler, and a retailer. The bullwhip effect, trust diffusion, and suppliers' production adjustment are considered in this study to be the performance of the supply chain.

2 Literature Review

2.1 Information Sharing and Trust

Companies invested considerable resources in improving supply chain visibility and collaborating with partners to be beneficial in supply chain management. While focusing on the bullwhip effect, studies showed that developing favorable relationships through partnership could mitigate its adverse impact for all members in the supply chain, such as suppliers and retailers [2, 3].

Comparing forecast information sharing and non-sharing operational systems, Ali et al. [4] found that sharing information was much more effective. Moreover, the trust could be established as a result. They further pointed out that partners' credit was related to the level that they trust each other, which meant that high information sharing quality guaranteed trust level and strengthened members' collaboration within the supply chain. Moreover, Simatupang et al. [5] reported that it would be hard to collaborate without a trust-based relationship in the supply chain. Thus, to build trust relationships, minimizing the uncertainty on demand forecasting was helpful, which kept the information quality by information sharing [6].

Because information sharing would be suitable for trust development and collaboration in a supply chain, the reliable information about demand forecasting was associated with both processes and performances. Besides, trustworthiness indicated information sharing quality. Members would distinguish whether the information could be trusted or not based on trustworthiness. Hence, trustworthiness and trust were to be considered in supply chain management.

2.2 Information Sharing and the Bullwhip Effect

A supply chain became unstable and inefficient under the bullwhip effect. Lee et al. [2] proposed four causes of the bullwhip effect, demand forecast updating, order batching, price fluctuation, and rationing and shortage gaming. Accurately forecasting demand by updating information was the most critical for reducing the bullwhip effect among four causes [8].

Therefore, researchers put their effort into lowering the uncertainty via the latest information and then enhancing the operation of information sharing to ensure the demand forecasting quality. Seferlis et al. [9] developed a demand prediction model to eliminate the uncertainty in inventory control strategies with the autoregressive integrated moving average, which increased information quality. Moreover, Li et al. [10] reduced uncertainty probability from the aspect of operational causes, including lead time delays, demand, production processes, and inventory processes. In terms of the simple two-echelon supply chain, Tanweer et al. [11] improved effectiveness by increasing supply chain transparency and demand forecasting accuracy.

Other techniques such as collaborative planning, vendor-managed inventory, sales and operations planning were introduced based on the concept of transparency improvement on information sharing and coordination in a supply chain. Also, immediately updating the information about product availability was useful to correct and adjust demand and ordering policy, reliving the impact due to the bullwhip effect and other adverse outcomes [7].

In addition to determining the reasons for the bullwhip effect, evaluating the bullwhip effect was the way to observe the performance of a supply chain. Dejonckheere et al. [12] measured the bullwhip effect in an order-up-to policy by the variance amplification of orders from a control-theoretic approach. Another evaluation method was to calculate with a coefficient of variance with input (order placing) and output (order receiving) quantity, which provided an early indication of the bullwhip effect [13].

2.3 Experiments About Trust and Supply Chain

Trust game [14] and investment game [15] were simulated to measure the spontaneous trust in several fields, especially in financial and supply chain management. The repeated interaction in games was to examine the pattern of trust increasing and decreasing during the focusing period. Consequently, a long-term relationship could be noticed as the period went on.

Since the trust was the key to maintain relationships and performances [16], conducting experiments with trust and supply chain could well simulate the real situation. Özer et al. [17] investigated the information sharing of demand forecasting in a two-echelon supply chain, which involved a retailer and a supplier. Their experiments showed the demand forecast and production decision and measured trust and trustworthiness according to subjects' decision behaviors.

The beer game was introduced to demonstrate the bullwhip effect by considering market demand, ordering policy, and cost in trials. Except for downstream retailers, other members in the supply chain had little market knowledge. Nevertheless, the real supply chain was more dynamic and complex, making the simulation's performance weak and unrealistic; other implicit issues were adopted to improve the investigation [18]. For example, increasing supply chain members' motivations and incentives could facilitate excellent performance and make simulation realistic.

In light of the previous studies, this study combined trust and supply chain management in the experiment to learn trust between members and the bullwhip effect in a supply chain.

2.4 Hypotheses

This study aimed to observe trust and trustworthiness in a supply chain from three aspects. First, in terms of information sharing, the comprehension of other members could be an effective way of sharing information and enhancing performance by mitigating the bullwhip effect. Second, for trust diffusion, the trust relationships between partners could indicate a change of trust in different situations. Third, this study also interested in the suppliers' behaviors for production adjustment since they lacked market information. Experiments would be designed to highlight people's behaviors, such as trust and trustworthiness in a supply chain. And there were three hypotheses proposed below.

Hypothesis 1. With adequate information sharing on the trustworthiness of partners, the bullwhip effect could be reduced in both high and low trustworthiness cases.
Hypothesis 2. Trust diffusion is significant when partners in a supply chain have high trustworthiness.
Hypothesis 3. Suppliers tend to adjust their production strategy when cooperating with low trustworthiness partners to minimize risk.

3 Methodology

3.1 Experimental Design

The experiment in this study was based on researches about information sharing in the supply chain [17, 19] and the beer game. This study designed a three-echelon supply chain, which included a retailer, a wholesaler, and a supplier. There were two transactions in the experiment: one deal occurred between retailer and wholesaler, and the other was between wholesaler and supplier. Participants were assigned as retailers and suppliers randomly, while the computer program was set to be wholesalers.

This study divided the experiment into two tasks. The first one was the baseline of the initial trustworthiness and trust without the understanding of others, showing behaviors when there was no information sharing in a supply chain. In the second task, information sharing existed for learning about partners. Wholesalers with high or low trustworthiness scenarios were given in this case to control the uncertainty in the experiment.

For the high trustworthiness, the wholesaler in the experiment always delivered an unmodified retailer's demand forecast report to the supplier. For the low trustworthiness levels, the wholesaler would inflate the number of reports received from the retailer, which was set as twenty in the experiment.

In terms of numerical setting, a retailer could forecast demand with firsthand market demand information ξ, while a wholesaler and supplier could merely know the demand distribution followed by uniform distribution on the interval [100, 400] and the market uncertainty also followed by uniform distribution on the interval [−75, 75].

Moreover, the retailer had to report a forecast demand $\widehat{\xi_1}$ to the wholesaler, and then the wholesaler would place the demand forecasting report $\widehat{\xi_2}$ to the supplier after receiving the retailer's report. For the number of physical goods, Q_1 units meant the wholesaler's selling quantity, and Q_2 units denoted the suppliers' production quantity. Market demand was formulated as $D = \xi + \epsilon$, and ϵ symbolized the market uncertainty. The actual amount of retailer sales was indicated with $\min(D, Q_1)$, and the actual amount of supplier sales was shown with $\min(Q_1, Q_2)$.

3.2 Experiment

Four participants were in a group for an experiment, and they were paired randomly to avoid partners having preconceived impressions on others. The experiment was realized with z-tree and z-leaf [20]. Additionally, two tasks were involved in an experiment.

Researchers would explain the experiment in the beginning, and then have participants sign consent forms. In the first part of the experiment, participants were asked to complete task 1 by being suppliers and entering production quantity according to the given report from wholesalers for ten trials. The second part of the experiment was for task 2, involving thirty trials. Participants were asked to be either suppliers or retailers. As suppliers, they had to design strategy and decide on production quantity. For retailers, they had to input the demand forecasting amount to the upstream member, the wholesaler, by considering their firsthand market demand information and wholesalers' behaviors.

3.3 Measurement

Due to retailers' proximity to the market, they could evaluate market demand more precisely based on their understanding of customer and market trends. With the information in mind, retailers then created a forecast report for wholesalers. To ensure sufficient supply for customers, retailers tended to inflate the data on their forecast reports. The retailers' inflation depended on two main factors: trust between retailers and wholesalers and the trustworthiness of wholesalers. In terms of suppliers, since they had minimal information about market demand and retailers' forecast reports, they could only manage production with the wholesalers' forecast report. Thus, trust and trustworthiness mattered for suppliers. Accordingly, adverse effects, such as the bullwhip effect, would occur under different levels of trust and trustworthiness.

The information sharing between downstream retailers and upstream suppliers was related to the level of trustworthiness and trust [7, 19]. For retailers, low trustworthiness referred to the amount of inflation in the forecast report. On the other hand, retailers had high trustworthiness when they reported the same amount of their firsthand forecast demand. Thus, retailers' trustworthiness was measured by $\widehat{\xi_1} - \xi$. The trust between retailers and wholesalers was determined by $Q_1 - \widehat{\xi_1}$. The value would approach zero when the high trust relationship existed and would be more than zero otherwise.

For suppliers, the trustworthiness was examined with the difference in production and the forecast demand from wholesalers, $Q_2 - \widehat{\xi_2}$. Because suppliers had limited information on the downstream, their production strategies relied on the forecast demand report from wholesalers. High trustworthiness meant that their behaviors were consistent with the forecast demand, although some adjustments might be necessary, while low trustworthiness was the opposite. As for the trust between suppliers and wholesalers, the differences between actual supplier sales and forecast demand reports from wholesalers were calculated. A high-trust relationship was shown by the value equal to zero. In contrast, the low-trust relationship was represented by the positive value, and trust would become lower as the value higher.

As suppliers could only receive little market information, observing impacts of the information sharing of partners' trustworthiness provided methods to improve the performance of a supply chain. Therefore, this study focused on production adjustment in two tasks, which was evaluated with the differences in trust values in both tasks.

Besides, this study investigated how the bullwhip effect was influenced by trustworthiness and trust in a supply chain. In order to measure the bullwhip effect, ω_i was the notation of the bullwhip effect ratio. The mean and standard deviation of outgoing orders $(\widehat{\xi_i})$, and incoming demands (Q_i) for $i = 1, 2$, in a specific period (from time t to t $+$ T) were used to calculate the coefficient of variation. Then the coefficient of variation of outgoing orders was divided by the coefficient of variation of incoming demands to get ω_i. The formula was shown below:

$$\omega_i = \frac{c_{out}}{c_{in}} \text{ where } c_{in} = \frac{\sigma\left(\widehat{\xi_i}(t, t+T)\right)}{\mu\left(\widehat{\xi_i}(t, t+T)\right)} \text{ and } c_{out} = \frac{\sigma(Q_i(t, t+T))}{\mu(Q_i(t, t+T))} \quad (1)$$

3.4 Participants

There were 36 participants aged from 20 to 26 involved in this study (mean $= 22.89$, standard deviation $= 1.26$). All of them were either undergraduate or graduated students. They were paired in the high or low trustworthiness group and randomly assigned as retailers or suppliers by the researchers.

4 Results

4.1 Trustworthiness and the Bullwhip Effect

Hypothesis 1 stated that the trustworthiness of partners could reduce the bullwhip effect whether their trustworthiness level was high or low. This study calculated the bullwhip effect based on the input forecast report and output sales to display the fluctuation of information sharing biases in a supply chain, which could lead to the bullwhip effect and other negative influences in a supply chain.

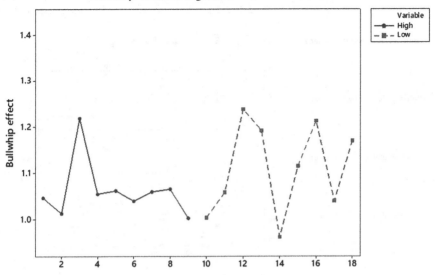

Fig. 1. The bullwhip effect in high and low trustworthiness

According to the bullwhip effect indicator in this study, the value 1 was the threshold and implied that the bullwhip effect was slight when the value was approaching 1. Comparing both trustworthiness levels, the bullwhip effect was lower in a high-level case (mean = 1.06, standard deviation = 0.06) than in a low-level case (mean = 1.11, standard deviation = 0.10). In other words, the results pointed out that the bullwhip effect was worse in low trustworthiness case despite members being given their partners' trustworthiness levels. Thus, Hypothesis 1 was not supported by the results.

Despite that, results indicated the importance of the information sharing quality. High trustworthiness could lead to a lower bullwhip effect since the provided information was reliable. On the other hand, it inferred that information delivered to or provided by low trustworthiness partners was suspected by the majority of participants, which caused the unstable information sharing quality, and then worsen the bullwhip effect (see Fig. 1).

4.2 Trust Diffusion

Hypothesis 2 could be observed by comparing trust in high and low trustworthiness cases. This study calculated the average trust of each period to examine trust diffusion. The trust decreased as the value lower. The pattern in Fig. 2 showed how trust varied from period to period.

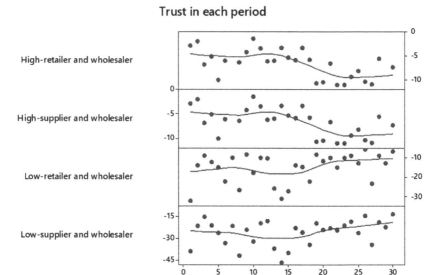

Fig. 2. Trust in each period

The results had three implications. First, it indicated that both patterns in the high trustworthiness case were consistent. Namely, the trust between one of them was related to the trust of the other. Second, Fig. 2 presented that the average trust between retailers and wholesalers (mean $= -15.22$, standard deviation $= 7.54$) was higher than suppliers and wholesalers (mean $= -25.93$, standard deviation $= 8.70$) in a low trustworthiness case. This result proved that trust levels were lower while interacting with upstream, who were suppliers and wholesalers. Third, results implied similar trends among periods in both cases, which might imply that trust between supplier and wholesaler changed with the retailer and wholesaler.

Based on three implications, trust diffusion was in both high and low trustworthiness cases, especially for high trustworthiness. Therefore, Hypothesis 2 was supported. However, the results further indicated that trust in upstream was smaller and more varied, making long-term and trust relationships challenging to develop and maintain even though trust diffusion might help increase trust level. Thus, this study reported that trust diffusion in a supply chain with low trustworthiness partners might lead to trust fluctuation, particularly for upstream partners who lack market knowledge.

4.3 Suppliers' Production Adjustment

Hypothesis 3 examined the suppliers' production adjustment before and after understanding the trustworthiness level of partners. Since production strategies were made with a wholesalers' forecast report, trust should be taken into consideration. Changes in suppliers' production strategies between task 1 and task 2 would measure with trust adjustment in the present study.

Production adjustment of supplier and wholesaler

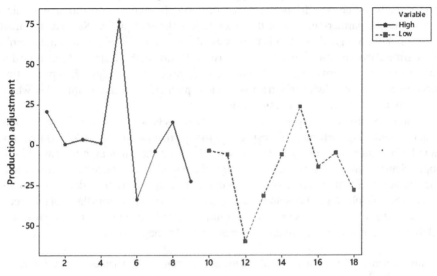

Fig. 3. Production adjustment of supplier and wholesaler

Because suppliers could receive less market information, making an accurate demand forecast was difficult. More details from downstream partners would be helpful for scheduling production strategies and enhancing overall performances. Figure 3 displayed the trust adjustment of supplier and wholesaler. The higher the production adjustment, the more trust was between suppliers and wholesalers.

In terms of high trustworthiness, 6 out of 9 groups of suppliers increased their trust in wholesalers (mean = 19.30, standard deviation = 29.1). As for low trustworthiness, 8 out of 9 groups of suppliers decreased trust in wholesalers (mean = −19.22, standard deviation = 19.73). Their differences reported that participants would distrust partners with low trustworthiness and be more trust in those with high trustworthiness. It indicated from two points, one was risk tolerance, and the other was the development of trust relationship.

For the risk tolerance, participants, who were suppliers, tended to be risk-averse while cooperating with low trustworthiness partners. Since they believed that wholesalers would modify the demand report to a higher number and ensure an ample supply of goods, they could protect themselves from the loss of uncertainty in this way. For the trust relationship, participants were likely to turn to distrust partners in low trustworthiness that trust in partners with high trustworthiness. Namely, trust became fragile and hard to be mended in a short time in supply chain management with low trustworthiness.

5   Conclusions

This study investigates three aspects related to trust and trustworthiness in supply chain management, including the bullwhip effect, trust diffusion, and suppliers' production adjustment. There are three findings: first, the bullwhip effect would be mitigated to a lower level when partners in a supply chain had high trustworthiness. Since the bullwhip effect directly influences the performance of a supply chain, cooperating with high trustworthiness partners enhances the information sharing quality. Second, the trust diffuses in both high and low levels of trustworthiness supply chain, which helps build long-term and trust relationships. However, trust diffusion in the upstream takes time due to a lack of market information. Third, suppliers are more likely to adjust their production strategies to cooperate with low trustworthiness partners with minimum risks, which potentially leads to a break in trust relationships.

To sum up, this study suggests that trustworthiness be vital information in a supply chain for enhancing performances, especially for upstream members. Additionally, suppliers behave more cautiously in low trustworthiness rather than high trustworthiness supply chains, which presents risk aversion and the tendency of uncertainty avoidance. Since participants in this study have similar cultural backgrounds, their decision-making procedures, and tolerance of risk and uncertainty may be different from those of other cultures. This study recommends that cross-cultural studies be conducted in future research and that international supply chain management will benefit from it.

Acknowledgement. The authors would like to thank Ta-Ping Lu (Sichuan University, China) for assistance with this paper.

References

1. Bowersox, D.J., Closs, D.J., Cooper, M.B.: Supply Chain Logistics Management, 4th edn. McGraw-Hill Education, New York (1996)
2. Lee, H.L., Padmanabhan, V., Whang, S.: The bullwhip effect in supply chains. Sloan Manag. Rev. **38**, 93–102 (1997)
3. Lummus, R.R., Duclos, L.K., Vokurka, R.J.: The impact of marketing initiatives on the supply chain. Supply Chain Manag. **8**(4), 317–323 (2003)
4. Ali, M.M., Babai, M.Z., Boylan, J.E., Syntetos, A.A.: Supply chain information forecasting when information is not shared. Eur. J. Oper. Res. **260**(3), 984–994 (2017)
5. Simatupang, T.M., Sridharan, R.: The collaborative supply chain. Int. J. Logist. Manage. **13**(1), 15–30 (2002)
6. Laeequddin, M., Sahay, B.S., Sahay, V., Abdul Waheed, K.: Measuring trust in supply chain partners' relationships. Meas. Bus. Excell. **14**(3), 53–69 (2010)
7. de Almeida, M.M.K., Marins, F.A.S., Salgado, A.M.P., Santos, F.C.A., da Silva, S.L.: Mitigation of the bullwhip effect considering trust and collaboration in supply chain management: a literature review. Int. J. Adv. Manuf. Technol. **77**(1), 495–513 (2014). https://doi.org/10.1007/s00170-014-6444-9
8. Jaipuria, S., Mahapatra, S.S.: An improved demand forecasting method to reduce bullwhip effect in supply chains. Expert Syst. Appl. **41**(5), 2395–2408 (2014)
9. Seferlis, P., Giannelos, N.F.: A two-layered optimization-based control strategy for multi-echelon supply chain networks. Comput. Aided Chem. Eng. **14**, 509–514 (2004)
10. Li, C., Liu, S.: A robust optimization approach to reduce the bullwhip effect of supply chains with vendor order placement lead time delays in an uncertain environment. Appl. Math. Model. **37**(3), 707–718 (2013)
11. Tanweer, A., Li, Y.Z., Duan, G., Song, J.Y.: An optimization model for mitigating bullwhip effect in a two-echelon supply chain. Procedia – Soc. Behav. Sci. **138**, 289–297 (2014)

12. Dejonckheere, J., Disney, S.M., Lambrecht, M.R., Towill, D.R.: Measuring and avoiding the bullwhip effect: a control theoretic approach. Eur. J. Oper. Res. **147**(3), 567–590 (2003)
13. Fansoo, J.C., Wouters, M.J.F.: Measuring the bullwhip effect in the supply chain. Supply Chain Manag. **5**(2), 78–89 (2000)
14. Kreps, D.M.: Markets and hierarchies and (mathematical) economic theory. Ind. Corp. Change **5**(2), 561–595 (1996)
15. Berg, J., Dickhaut, J., McCabe, K.: Trust, reciprocity, and social history. Games Econ. Behav. **10**(1), 122–142 (1995)
16. Barratt, M.: Understanding the meaning of collaboration in the supply chain. Supply Chain Manag. **9**(1), 30–42 (2004)
17. Özer, Ö., Zheng, Y., Chen, K.Y.: Trust in forecast information sharing. Manage. Sci. **57**(6), 1111–1137 (2011)
18. Sterman, J.D.: Modeling managerial behavior: misperceptions of feedback in a dynamic decision making experiment. Manage. Sci. **35**(3), 321–339 (1989)
19. Özer, Ö., Zheng, Y., Ren, Y.: Trust, trustworthiness, and information sharing in supply chains bridging China and the United States. Manage. Sci. **60**(10), 2435–2460 (2014)
20. Fischbacher, U.: z-Tree: Zurich toolbox for ready-made economic experiments. Exp. Econ. **10**, 171–178 (2007)

The Effect of a Long Simulated Voyage on Sailors' Alertness

Jin Liang[1], Xin Wang[1(✉)], Liang Zhang[1], Ye Deng[1], Yingwei Zhou[1],
Yuqian Zhang[1,3], Yang Yu[1], Zhen Liao[1,2], Zhiqiang Tian[3], Zhanshuo Zhang[1],
and Yongjiang Fu[3]

[1] China Institute of Marine Technology and Economy, Beijing 100081, China
xinwang_thu@126.com
[2] Tsinghua University, Beijing 100084, China
[3] China Institute of Marine Human Factors Engineering, Qingdao 266400, China

Abstract. Aiming at the long-term voyage demand of ships, a long voyage simulated experiment was carried out in the self-developed experimental chamber in this study, to probe the crews' alertness changing and the optimal performing opportunity of alertness maintenance measures during the simulated long voyage. The results showed that the fatigue level of the subjects increased and the alertness decreased during the simulated long voyage. Participants' alertness had begun to decay from the second stage of the simulated voyage, and they were unable to maintain their alertness as same as the general people's level after the first two stages of the simulated voyage. It means that alertness maintenance measures need to be carried out from the start of the voyage and execute intensive intervention before (preventive intervention) or on (immediate intervention) the second stage end to reduce safety problems caused by increased fatigue and decreased alertness.

Keywords: Long voyage · Alertness · Fatigue · Optimal opportunity of intervention

1 Introduction

The long-term maritime navigation task requires sailors to pass through multiple sea areas and experience complex sea conditions and changeable climate. How to ensure the safety of ships and complete the voyage tasks under the ever-changing sea conditions has become one of the most important challenges in the field of maritime transportation. A large number of maritime accidents showed that, in addition to the necessary detection equipment and safety regulations, the working state (including attention, alertness, etc.) of the duty sailors during the navigation have an important impact on the maritime navigation safety. However, during the long voyage, the crew's physical and mental state would gradually decline, fatigue would accumulate, and alertness would decay during voyage under the influence of lots of adverse factors, such as jolting, vibration, social isolation, strong noise, high humidity, and taking turns on duty, which would seriously threaten the safety of ship navigation [1–4]. Hence, it's necessary to understand the

© Springer Nature Switzerland AG 2020
P.-L. P. Rau (Ed.): HCII 2020, LNCS 12192, pp. 454–462, 2020.
https://doi.org/10.1007/978-3-030-49788-0_34

changing patterns of the crew's alertness during the long ocean voyage. The accumulation of the related data and the solution of the problems are of great significance for the establishment of the relevant models, such as alertness changing models, and thus provide support for the fatigue reducing during the long voyage. Therefore, aiming at the long-term voyage demand of ships, a long voyage simulated experiments were carried out in the self-developed experimental chamber in this study, to probe the crews' alertness changing and the optimal performing opportunity of alertness maintenance measures during the simulated long voyage.

2 Methods

2.1 Participants

12 male subjects were recruited through the society in this experiment. All of the subjects had high school education or above, physically and mentally healthy, no drug dependence, no alcohol dependence, no smoking addiction, no Internet addiction, no history of inherited diseases, hepatitis b, hepatitis c, AIDS or other infectious diseases, and no history of severe allergies. They had no mental disorders, no psychological diseases, no organic and functional mental and neurological lesions, no sleep disorders or abnormalities in the two and three generations. All participants possessed normal visual acuity (or corrected visual acuity), hearing, smell and language expression ability, without color blindness, color weakness. All subjects passed the interview, systematically physical examination and psychological evaluation.

2.2 Procedure

The study passed an ethical review (Certificate No: 2019013) before the experiment was conducted. The experiment included training period, experimental period, recovery period and return visit period.

(1) Training period: 12 subjects were fully informed of the experimental procedure and content, signed the informed consent after passing the physical examination, interview, psychological evaluation. They were assembled for experimental training before the experiment begins. The training includes test tasks, simulated ship tasks, emergency handling, and other matters needing attention in experiments, in order to guarantee that every subject was familiar with all test tasks, simulated ship tasks and matters needing attention before the experiment, and can effectively respond to emergency situations according to the contingency plan.

(2) Experiment period: the subjects entered the experimental period after all them achieved the training standards. 12 subjects carried out the shifting, work and life styles as the actual crew during the simulated long voyage, and implemented the ship simulation task when they were on duty. Before and after their duty, they were asked to accomplish alertness tasks and tests, including psychomotor vigilance task (PVT task), and Karolinska Sleepiness Scale (KSS). The participants were also asked to fulfilled Subjective Fatigue Symptoms Check List (SFSCL) every day during the

simulated voyage. The project team had signed the cooperation agreement with a mental hospital and a general hospital, to safeguard the experiment volunteers could get timely and effective assistance when they met with abnormal physical and mental state during the whole experiment. Their mental health status were assessed by psychologist regularly and their body symptoms were evaluated by the clinical doctors during the experiment.

(3) Recovery period: the recovery period shall be entered after the end of the experiment. During the recovery period, the physical and mental functions of the subjects were evaluated again by the same hospital that which conducted a systematically checkup on the physical and mental functions of the subjects before the experiment. The physical and mental assessment results of the subjects after the experiment were compared with their results before the experiment. The subjects could leave by themselves after they were confirmed no abnormalities in the physical and mental functions though a systematically assessment.

(4) Return visit period: the physical and mental status of the subjects was re-evaluated and followed up to confirm whether the simulation experiment had a long-term impact on their physical and mental status during the return visit period.

2.3 Test Tasks

Psychomotor Vigilance Task (PVT Task). The subjects were asked to look at the fixation point in the center of the screen after the start of the psychomotor vigilance task (PVT task), and press the "J" button immediately when they saw a number. The number would continue to increase until the right key was pressed, and the response time of the key was recorded. Participants were asked to respond as soon as possible, but not before the number appears, otherwise an error message would appear. The experimental process is shown in Fig. 1.

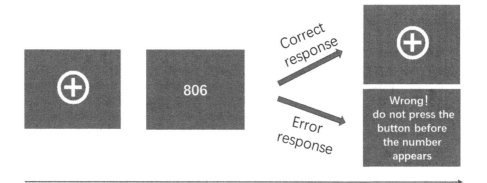

Fig. 1. Psychomotor vigilance task (PVT task)

Karolinska Sleepiness Scale (KSS). Karolinska Sleepiness Scale (KSS) only had one project with the trial 1–9 figures representing one's current awake status. 1 and 9 delegated extremely sober and sleepiness respectively, and scores among 1–9 expressed the subjects' state between extreme sober and sleepiness. The larger KSS score was, the higher alert participants had.

Subjective Fatigue Symptoms Check List (SFSCL). Subjects' subjective fatigue of the daily were assessed by checklist of subjective fatigue symptoms (SFSCL), which include "body fatigue", "psychological fatigue", "fatigue symptoms" three factors, each factor within ten items, participants could use the 0–4 to state their fatigue degree. 0 represented no this situation; 1 represented this rarely happened; 2 represented this sometimes happened; 3 represented this happened a lot; 4 represented this happened all the time. The lowest factor score is 0, the highest factor score is 40, the lowest total score is 0, the highest total score is 120. The higher the score was, the higher fatigue degree participants had.

3 Results

3.1 The Trend of Subjects' Alertness During Long Simulated Voyage

PVT Task Results. The average reaction time of PVT task was calculated after eliminate errors trials, antedating trials, and trials beyond three standard deviations. Then the correlation between average reaction time of PVT task and voyage duration was analyzed. The results showed that PVT task reaction time had significantly positive correlation with simulated voyage duration, $r = 0.73$, $p < 0.01$, that is, with the increase of voyage duration, participants' of alertness gradually decline during the simulated voyage.

KSS Scale Results. The KSS scale scores of the subjects had positively correlated with the voyage duration, $r = 0.54$, $p < 0.01$, that is, with the increase of voyage duration, the subjects' subjective fatigue increased and their alertness decreased, which was consistent with the results of the PVT task.

SFSCL Results. The voyage duration and subjects fatigue rating scale total score, body fatigue factor, psychological fatigue factor, fatigue factor correlation were analyzed. The result showed that there was positive correlation between the subjects SFSCL score and voyage duration, $r = 0.61$, $p < 0.01$, between the mental fatigue factor score and voyage duration, $r = 0.73$, $p < 0.01$, between the subjects fatigue symptoms score and voyage duration, $r = 0.69$, $p < 0.01$). The body fatigue score of subjects was no significantly related to voyage duration. These results indicated that with the passage of the simulated voyage duration, the fatigue state of the subjects increased gradually, and mainly in mental fatigue and fatigue symptoms, while physical fatigue did not increase with the voyage duration.

3.2 The Participants' Alertness Changes in the 24-h Dimension

In consideration of the factors such as shifting, eating, sleep, etc., the subjects' alertness was evaluated at different time in the forenoon, afternoon and evening to understand the change characteristics of the subjects' alertness in the 24-h dimension during the simulated long voyage. Due to the influence of the subjects' shift, eating, sleep and other factors, a subjects' alertness test was not completed within 24 h in a row. The study balanced the test order of different subjects at the specific time.

PVT Task Results. ANOVA was completed by taking participants' state (After work vs. After rest) and test time as independent variable and participants' PVT reaction time as the dependent variable. The results showed that the participants' state has significantly main effect, $F(1, 11) = 13.53$, $p < 0.01$, $\eta_p^2 = 0.55$, the test time had no significant main effects, and there was significantly interaction between participants' state and test time, $F(6,6) = 10.86$, $p < 0.01$, $\eta_p^2 = 0.92$. Further analysis showed that after rest, the subjects had the fastest PVT response time and the best alertness at night, the slowest reaction time and the worst alertness in the morning. There was significantly different among test time in 24 h dimension, characterized by participants had the fastest PVT reaction time and best alertness at dusk, the slowest PVT reaction time and worst alertness at midday.

KSS Scale Results. In order to understand the participants' subjective state of alertness and its changes in the 24 h of dimension, the study carried out an ANOVA with taking participants' state and test time for independent variables, subjects' KSS scale score as the dependent variable. The results showed that the participants' state had significantly main effect, $F(1, 11) = 6.39$, $p < 0.05$, $\eta_p^2 = 0.37$, test time had significantly main effect, $F(6, 6) = 6.62$, $p < 0.05$, $\eta_p^2 = 0.87$, and there was significant interactions between participants' state and test time, $F(6, 6) = 22.08$, $p < 0.01$, $\eta_p^2 = 0.96$. Further analysis found that, the subjects owned the highest KSS scale score and most drowsiness in the morning, the participants had the lowest KSS scale score and the most sober at night after rest; the subjects owned the highest KSS scale score and most drowsiness in the morning, the participants had the lowest KSS scale score and the most sober at dusk after work.

3.3 Key Time of Participants' Alertness Changes During Long Simulated Voyage

PVT Task Results. In order to explore critical timing of subjects' alertness transformation during the long voyage and guide alertness intervention strategy design, the navigation phase was analyzed, the results showed that there was significant difference in different periods of navigation, $F(4, 8) = 5.98$, $p < 0.05$, $\eta_p^2 = 0.75$. The participants' PVT reaction time of the stage 1 was the fastest, and significantly faster than the rest of the four stages; PVT task response time of stage 2 significantly slower than the first stage, but significantly faster than the 3rd, 4th, 5th stage; the 3rd and 4th stage reaction time had no significant difference, stage 5 volunteers with the slowest response time, significantly slower than the other four stages.

In order to further explore the subjects PVT task performance in different segment features, this research took sailing stage, test time, participants' state as independent variables, PVT reaction time as the dependent variable to carry on the ANOVA, the results showed that the main effect of participants' PVT task reaction time on the sailing stage (F (4, 8) = 18.05, p < 0.001, η_p^2 = 0.90) and participants' state (F (1, 11) = 13.83, p < 0.01, η_p^2 = 0.56) was significantly, and there was significantly interactions between participants' state and test time (F (4, 8) = 8.39, p < 0.05, η_p^2 = 0.89), between sailing stage and test time (F (6.68, 73.48) = 2.39, p < 0.05, η_p^2 = 0.18). The main effect on the test time, the interaction between the participants' state and the sailing stage as well as the interactions among test time, sailing stage and participants' state were not significant. To further understand the interaction between variables, the response time of PVT after rest and after work was analyzed.

PVT Task Response Analysis After Rest. The sailing stage and the test time were taken as independent variables, and the response time of the PVT task after the rest was taken as the dependent variable for variance analysis. The results showed that the main effect of sailing stage was significant, F (4, 8) = 28.45, p < 0.001, η_p^2 = 0.93, while the main effect of test time was not significant, and the interaction between the test time and the sailing stage was significant, F (18.03, 198.33) = 1.99, p < 0.05, η_p^2 = 0.15. Further analysis showed that in the first stage of the voyage, the response time of the subjects was the slowest at morning and the fastest at midnight. In the second stage of the voyage, the volunteers had the slowest response time at morning and the fastest response time at night. In the third stage of the voyage, the volunteers had the slowest reaction time at morning and the fastest PVT reaction time at midday. In the fourth stage of the voyage, the volunteers had the slowest response time at morning and the fastest at midnight. In the fifth stage of the voyage, the volunteers had the slowest response time at morning and the fastest at night.

PVT Task Response Time Analysis After Work. The results showed that the main effect of the sailing stage was significant, F (4, 8) = 13.19, p < 0.001, η_p^2 = 0.87, while the main effect of test time was not significant, and the interaction between the test time and the sailing stage was significant, F (11.13, 122.41) = 2.06, p < 0.05, η_p^2 = 0.16. Further analysis showed that during the first stage of the voyage, the response time of volunteers was the slowest at morning, while the fastest at dusk. In the second stage of the voyage, the volunteers had the slowest reaction time at morning and the fastest reaction time at dusk. In the third stage of the voyage, the volunteers had the slowest response time at midday, and he fastest at night. In the fourth stage of the voyage, the subjects had the slowest reaction time at midday and the fastest reaction time at morning. In the fifth stage of the voyage, the subjects had the slowest reaction time at morning and the fastest at dusk.

KSS Scale Results. In order to explore the key time of the change of the participants' alertness during the long voyage and guide the design of the alertness intervention strategy, the study took the sailing stage and test time as independent variables, and the volunteers' KSS scale score as the dependent variable for the variance analysis. The results showed that there was no significant difference on sailing stage for the KSS scale scores. In order to further explore the characteristics of the subjects' subjective alert on different sailing stage, the research had taken sailing stage, test time, participants' state as independent variables, taken participants KSS scale score as the dependent variable to carry on the analysis of variance. The results showed that the test time had significantly main effect, $F(6, 6) = 6.17$, $p < 0.05$, $\eta_p^2 = 0.86$, the main effect of the participants' state was significant, $F(1, 11) = 6.41$, $p < 0.05$, $\eta_p^2 = 0.37$, and the interaction between participants' state and test time was significant, $F(6, 6) = 17.84$, $p < 0.001$, $\eta_p^2 = 0.95$, the interaction between sailing stage and test time was significant, $F(5.98, 65.79) = 2.84$, $p < 0.05$, $\eta_p^2 = 0.21$, the interactions among sailing stage, test time, participants' state was significant, $F(14.11, 155.20) = 2.15$, $p < 0.05$, $\eta_p^2 = 0.16$, but main effects of sailing stage as well as the interactions between participants status and sailing stage was not significant. To further understand the interaction between the variables, the study analyzed the KSS scale scores of the subjects after rest and after work.

KSS Scale Score Analysis After Rest. Taking sailing stage and test time as independent variables and KSS score of subjects as dependent variables, ANOVA was executed. The results showed that the main effect of the volunteer KSS scale was not significant in the sailing stage, but significant in the test time, $F(6, 6) = 10.51$, $p < 0.01$, $\eta_p^2 = 0.91$, and the interaction between the test time and the sailing stage was significant $F(7.41, 81.46) = 2.06$, $p < 0.05$, $\eta_p^2 = 0.17$. Further analysis showed that in the first stage of the voyage, the subjects' highest score at morning the lowest at night. In the second stage of the voyage, the subjects had the highest score at morning and the lowest score at night. In the third stage of the voyage, the subjects had the highest score at morning and the lowest score at afternoon. In the fourth stage of the voyage, the subjects the highest score at morning and the lowest at afternoon. In the fifth stage of the voyage, the subjects had the highest score at morning and the lowest score at night.

KSS Scale Score Analysis of Subjects After Work. The variance analysis was conducted with the KSS scale score as the dependent variable and the sailing stage and the test time as the independent variables after work. The results showed that the main effect of sailing stage and test time was not significant and the test period, but the interaction between the test time and the sailing stage was significant, $F(5.46, 60.02) = 2.66$, $p < 0.05$, $\eta_p^2 = 0.20$. Further analysis showed that in the first stage of the voyage, the subjects had the highest score at morning and the lowest at dusk. In the second stage of the voyage, the subjects had the highest score at morning and the lowest score at dusk. In the third stage of the voyage, the subjects had the highest score at midday and the lowest at dusk. In the fourth stage of the voyage, the subjects had the highest score at morning and the lowest score at night. In the fifth stage of the voyage, the subjects had the highest score at midday and the lowest score at midnight.

SFSCL Results. In order to explore the key time of the change of alertness during long voyages and guide the design of the alertness intervention strategy, the study took the sailing stage as the independent variable and the fatigue scale analysis of volunteers as the dependent variable for the variance analysis. And the result has shown that there was no significantly effect on sailing stage.

4 Discussion

The results which was consistent with previous studies [4, 5] had shown that there was significant positive correlation between voyage duration and participants' PVT task response time, (r = 0.73, p < 0.01), alertness score(r = 0.54, p < 0.01), as well as the fatigue scores (r = 0.61, p < 0.01) during the simulated voyage. PVT task response time showed as the first stage < second stage < third stage/fourth stage < fifth stage of the simulated voyage, and there was no significant difference between the third and fourth stage, which meant the subjects' alertness had decreased significantly from the second stage of the simulated voyage, and continued to decrease significantly during the rest stages of simulated voyage.

In the 24 h dimensions, the subjects' PVT task response time was slowest at morning and fastest at midnight and midday, alertness scores was highest(tiredest) in the morning and lowest(soberest) at dusk before duty (after break); the subjects' PVT task response time was slowest in the morning and midday and fastest in the dusk, alertness scores was highest at morning and lowest at dusk after duty (after work). These results were different from the alertness trend of the ordinary people [6, 7], which meant that participants' alertness rhythm was changed by the simulated voyage in 24 h dimensions [8–10]. The more precise pattern of alertness in the 24-h dimension needed further research.

There was different change for subjects' alertness between the first two stages and the last three stages of the simulated voyage. In the first two stages, participants possessed best alertness at dusk and night and worst alertness at morning and midday before duty, best alertness at afternoon and worst alertness at midnight and morning after duty. But in the last three stages, participants possessed best alertness at midnight and midday and worst alertness at morning before duty, best alertness at afternoon and dusk and worst alertness at midday after duty. These results meant the end of the second stage might be the turning point of participants' alertness during the simulated voyage, so in order to reduce safety problems caused by increased fatigue and decreased alertness, the alertness maintenance measures could be carried out on the second stage end, which was the best opportunity for immediate intervention. And the alertness maintenance measures could be carried out before the second stage end, which was the best opportunity for preventive intervention. The alertness maintenance measures which was executed from the start of the voyage might be effective to prevent the decreased alertness.

5 Conclusion

All above results implied the participants' fatigue increased steadily and their alertness continuously declined during the simulated voyage. Participants' alertness had begun to decay from the second stage of the simulated voyage, and they were unable to maintain their alertness as same as the general people's level after the first two stages of the simulated voyage. It meant that alertness maintenance measures needed to be carried out from the start of the voyage and execute intensive intervention on before (preventive intervention) or on (immediate intervention) the second stage end to reduce safety problems caused by increased fatigue and decreased alertness.

References

1. Barnett, M., Pekcan, C., Gatfield, D.: The use of linked simulators in project "Horizon": research into seafarer fatigue (2012)
2. Goel, N., et al.: Circadian rhythms, sleep deprivation, and human performance. Prog. molecular Biol. Transl. Sci. **119**, 155–190 (2013)
3. Vidacek, S., Sarić, M.: Circadian variation in alertness, readiness for work and work efficiency. Arch. Ind. Hyg. Toxicol. **42**(1), 13–25 (1991)
4. Dinges, D.F., et al.: Summary of the key features of seven biomathematical models of human fatigue and performance. Aviat. Space Environ. Med. **75**(3), A4–A14 (2004)
5. Zulley, J.: The influence of isolation on psychological and physiological variables. Aviat. Space Environ. Med. **71**(9 Suppl), A44–A47 (2000)
6. Parkes, K.R.: Sleep patterns, shiftwork, and individual differences: a comparison of onshore and offshore control-room operators. Ergonomics **37**(5), 827–844 (1994)
7. Rutenfranz, J., et al.: Work at sea: a study of sleep, and of circadian rhythms in physiological and psychological functions, in watchkeepers on merchant vessels. Int. Arch. Occup. Environ. Heath **60**(5), 331–339 (1988). https://doi.org/10.1007/BF00405666
8. J., et al.: Work at sea: a study of sleep, and of circadian rhythms in physiological and psychological functions, in watchkeepers on merchant vessels. Int. Arch. Occup. Environ Health (1988)
9. Dijk, D.J., Duffy, J.F., Czeisler, C.A.: Circadian and sleep/wake dependent aspects of subjective alertness and cognitive performance. J. Sleep Res. **1**(2), 112–117 (1992)
10. Matchock, R.L.: Circadian and sleep episode duration influences on cognitive performance following the process of awakening. Int. Rev. Neurobiol. **93**, 129–151 (2010)

The Effect of Long Time Simulated Voyage on Sailors' Athletic Ability

Zhen Liao[1,2,3], Chi Zhang[2]([⊠]), Yuqian Zhang[2,3], Zhanshuo Zhang[2], Ye Deng[2],
Yingwei Zhou[2], Yang Yu[2], Jin Liang[2], Zhiqiang Tian[3], Xin Wang[2], and Hao Meng[3]

[1] Tsinghua University, Beijing 100084, China
[2] China Institute of Marine Technology and Economy, Beijing 100081, China
zc810@126.com
[3] China Institute of Marine Human Factors Engineering, Qingdao 266400, China

Abstract. Due to the special task condition, sailors of oceangoing vessels have to conduct monotonous tasks in environment isolated from outside information. During a long voyage, such isolated environment may lead to the decline of the sailors' physical and mental state and decrease in task motivation, which may directly influence the health and capability of sailors, and even result in human error and task failure. The purpose of this study was to investigate how sailors' athletic ability changes under the long-time isolated condition. In the experiment, participants worked in a lab module isolated from outside for a period of time. The participants should not only finish the tasks on the watch, but also conduct regular physical fitness tests with professional equipment, including hand muscle strength, upper limb muscle strength, lung capacity, body flexibility, body balance and coordination. By analyzing the data collected at different time points during the experiment, comparison and significance analysis were conducted to evaluate the changes in the athletic ability of the subjects in the long-time isolated environment. Generally, the strength of upper limb muscle, body flexibility, and body balance and coordination slightly decreased. However, the fine muscle strength of the hands increased significantly. This phenomenon may relate to the handicap of wide body movement in isolated narrow space. More influence factors should be considered in future studies.

Keywords: Long time voyage · Physical ability · Isolated environment · Experimental study

1 Introduction

The working environment of ocean vessel sailors is special, which is narrow, isolated from outside. The daily task of them is special, which is filled with monotonous task units and change shifts. Meanwhile, the living condition is also special, lack of adequate sunshine, ozone and supplies for daily life. Because of the above reasons, sports conditions and opportunities of sailors are limited, and motivation of physical practice may decrease.

© Springer Nature Switzerland AG 2020
P.-L. P. Rau (Ed.): HCII 2020, LNCS 12192, pp. 463–473, 2020.
https://doi.org/10.1007/978-3-030-49788-0_35

During a long voyage, many factors, such as narrow space and dead air, extremely limit the mode and amount of sailors' exercise. At the same time, long-time isolated environment and irregular sleep patterns probably lead to sailors' low desire of sports. Multifaceted factors significantly impact the athletic ability of sailors. Sailors of ocean-going vessels live in marine environment for long time. Due to complex and multiple factors, the influence of marine environment on human body is manifold. Zhang et al. [1] and Fu et al. [2] summarized the effects of ship environment on the physiological and psychological functions and biochemical indexes of seamen. They believed that the noise in the cabin, high temperature and high humidity, and the irregular shaking of the hull had great influence on the human body. Shen et al. [3] and Ou et al. [4] investigated the effects of long voyages on the pulmonary ventilation function of sailors. According to previous studies, the time of standing with closed eyes decreases with age, which is related to muscle strength and reaction time [5].

However, according to the literature research, there are many studies focusing on the changes in psychological state during long voyage [6–9]. Ren et al. [10] researched the effect of long flight on the health of mission personnel and corresponding interventions. There are relatively few studies on the change of personnel capability under such special conditions. It is necessary to study how the ability will change with time under such special task conditions, and explore the mechanism of change in ability, which can provide strategy for personnel capacity maintenance and improvement.

This paper aimed to investigate the effect of long time voyage on sailors' athletic ability. We proposed a hypothesis that long time voyage has significant effect on sailors' athletic ability, and as time goes on, sailors' athletic ability gradually reduces. We designed a long-time simulated voyage experiment to verify the hypothesis. Through the study of long-time simulated voyage tasks, we explored whether such task model would affect the sailors' athletic ability, including hand muscle strength, upper limb muscle strength, lung capacity, body flexibility, body balance and coordination. The results could provide some suggestions for sailors' personnel capacity improvement and long- time voyage environment design, to maintain better sailors' status and give full play to their personnel capabilities.

2 Method

2.1 Experiment Design

We performed a long time isolated experiment to simulate a long time voyage and verify the hypothesis that long time voyage has significant effect on sailors' athletic ability. The independent variable was the sampling points of time and different groups of participants. There were six sampling points from baseline period to the end, and there were three different groups completing experiment with different schedules. The dependent variables were grip strength (left and right hand), number of push-ups in one minute, lung capacity, time of standing on one foot with eyes closed (hereafter referred to as eye-closed standing), sit-and-reach, which were considered as the athletic variables influenced by sampling time points.

2.2 Task Design

In the experiment, participants worked in a lab module isolated from outside for a period of time. Those participants could not get out of the narrow space and outside staff could not get in. Communication between inside and outside was also forbidden. Divided into three groups, the participants should work and rest inside according to established arrangement at different times, and complete the experimental tasks and various physical fitness tests. During the experiment, the participants mainly engaged in three aspects:

1. took turns on duty according to the schedule and carried out simulated voyage tasks.
2. conducted regular physical fitness tests with professional equipment and record data every few days, including hand muscle strength, upper limb muscle strength, lung capacity, body flexibility, body balance and coordination.
3. committed limited physical activities, daily living care and recreational activities.

2.3 Participants

12 male participants aged 24–37 years took part in this experiment. They were divided into three groups, marked as A, B, C, and all of them completed the experiment and no one dropped out. This study was approved by the Ethics Committee of Beijing University of Aeronautics and Astronautics, and each participant signed on the written informed consent.

2.4 Experimental Facility

Grip dynamometer: We chose grip dynamometer to measure grip strength of participants. The unit of measurement is kilograms (kg), and the measurement was accurate to 0.1 kg.

Second chronograph: Push-ups is manual muscle test, and no special equipment is required. Second chronograph was used to record the number of push-ups in one minute and the duration of eye-closed standing. The unit of measurement is second (s), and the measurement was accurate to 0.01 s.

Spirometer: Spirometer was used to measure the lung capacity of participants. The unit of measurement is milliliter (mL), and the measurement was accurate to 0.1 mL.

Sit and reach tester: Sit and reach tester was used to measure the result of sit and reach. The unit of measurement is centimeter (cm), and the measurement was accurate to 0.2 cm.

2.5 Data Analysis

Firstly, we performed descriptive analysis on the dependent variables, summarized a large amount of data preliminarily, and found out inherent laws. Secondly, we performed paired-samples t test to analyze the difference of participants' athletic data during the experiment.

3 Results

The descriptive analysis data of dependent variables is shown in Table 1. The data of T1 (baseline) and T6 are compared as below. (The data of one time point in experiment was labeled with 'T' and the test number)

Table 1. Descriptive analysis data of dependent variables

		T1		T6		
		Mean value	Standard deviation	Mean value	Standard deviation	
Grip strength (kg)	Left hand	44.3250	5.22548	46.7167	6.13512	
	Right hand	47.3667	7.78651	50.3083	7.26779	
Push-ups (number per minute)		31.7000	11.53786	29.7000	8.98208	.493
Lung capacity (mL)		4591.5833	831.14723	4546.5833	753.72547	.767
Eye-closed standing (s)		39.2500	31.55118	32.2133	23.00831	.226
Sit-and-reach (cm)		18.0667	8.97191	17.1750	9.19111	.437

Obviously, comparing the data of T1 and T6, the results indicated that the grip strength increased (see Fig. 1 and Fig. 2), while the other dependent variables decreased. During the whole experiment, the grip strength of left hand right hand increased from 47.4 (kg) to 50.3 (kg) by 6.12%. This trend is significant (Sig < 0.05). (The data of T0 represents the initial level measured before the experiment).

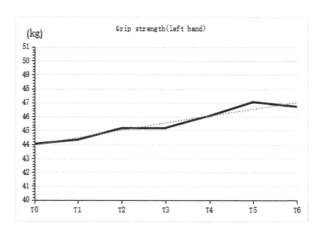

Fig. 1. The variation trend of grip strength (left hand)

Moreover, both hands' strength of participants in different groups all improved to some extent. Left hands' grip strength of different groups are shown in Fig. 3, 4 and 5,

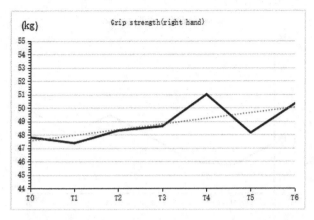

Fig. 2. The variation trend of grip strength (right hand)

and right ones of different groups are shown in Fig. 6, 7 and 8. Narrow space greatly limited wide body movement and the amount of exercise. In terms of daily activities, limited by the space of exercise, it was not good for large muscle exercises and long aerobic exercises. In spare time, participants engaged in local muscle exercises, such as push-ups and punch. As the experiment progressed, isolated closed environment would bring boredom and anxiety. Participants tended to improve their mood by small range of exercises, could lead to an increase of grip strength.

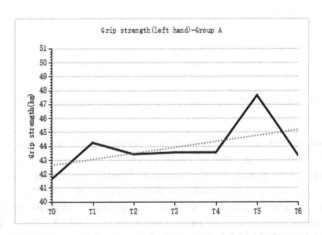

Fig. 3. The variation trend of grip strength (left hand)-Group A

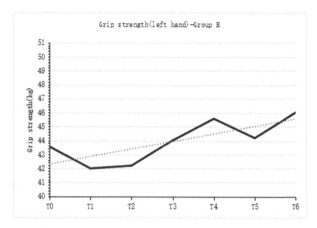

Fig. 4. The variation trend of grip strength (left hand)-Group B

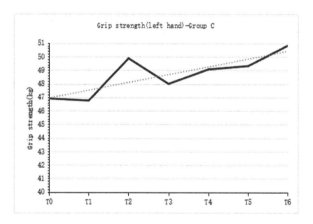

Fig. 5. The variation trend of grip strength (left hand)-Group C

Further analysis revealed that there was no growth trend in the number of push-ups. On the contrary, the number of push-ups in one minute didn't increase, which decreased from 31.7 per minute to 29.7 per minute. Even though the data is not significant (Sig > 0.05), it presents a slight downward trend basically (see Fig. 9).

This might occur because of the following reasons. 1) Although the participants had some upper limb muscle exercises, the effect was more on mood adjustment. The amount of exercise was not enough to stimulate the muscle strength to rise, thus improving the score of push-ups in one minute. 2) In contrast, the muscle size of the upper limbs was large, making it difficult to improve significantly in a shorter experimental period. 3) Affected by the rest of isolated closed environment, the body fatigue increased, the desire for short-term anaerobic exercise in decline, to some extent limited the results of push-ups. In general, the upper limb strength of the participants was not greatly affected during the test period for the long time isolated and closed experiment.

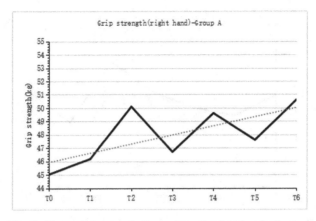

Fig. 6. The variation trend of grip strength (right hand)-Group A

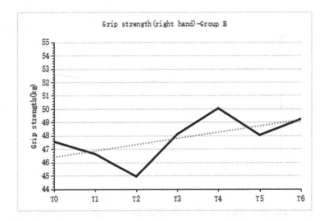

Fig. 7. The variation trend of grip strength (right hand)-Group B

As can be seen from the overall results of lung capacity, the data of lung capacity presents a relatively stable cyclical fluctuation, with no significant rise or fall (see Fig. 10). This might be because the experiment period was relatively short, and the decline in personnel activity did not directly affect cardiopulmonary function, so there was no significant decline in lung capacity.

The eye-closed standing means that the body relies on the balance regulation system to maintain the center of gravity on the supporting surface of the single foot when the eye is closed (i.e. when the reference is lost). Its duration is a measure of balance. According to previous studies, the time of eye-closed standing decreases with age, which is related to muscle strength and reaction time. As can be seen from the results of this experiment, there was no significant change in the time period of the experiment (see Fig. 11). The experiment duration was relatively short, and the isolated closed environment had no significant effect on the participants' ability to balance.

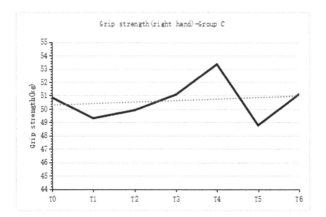

Fig. 8. The variation trend of grip strength (right hand)-Group C

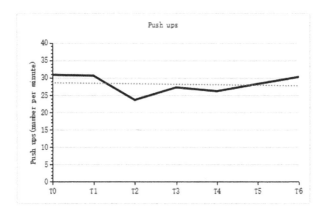

Fig. 9. The variation trend of push-ups per minute

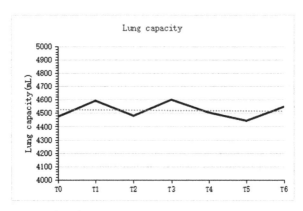

Fig. 10. The variation trend of lung capacity

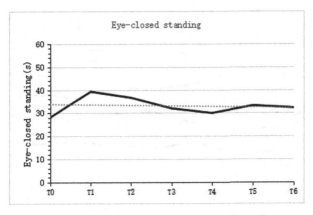

Fig. 11. The variation trend of eye-closed standing

Sit-and-reach is an important test to measure the flexibility of human body. According to the experimental results, sit-and-reach did not change significantly with the advance of the experimental time, showing no significant upward or downward trend (see Fig. 12).

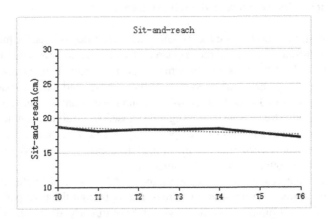

Fig. 12. The variation trend of sit-and-reach

According to the data of each group, the results of sit-and-reach were the best in group C, showing a certain downward trend (see Fig. 13). However, group A, whose scores were generally poor, showed a slight upward trend. The results of participants with different flexibility gradually converged with the progress of the isolated closed experiment.

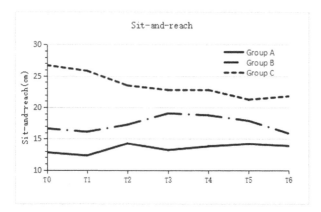

Fig. 13. The variation trend of sit-and-reach (each group)

4 Conclusion

In this experiment, from the beginning to the end of the experiment, each athletic ability index was tested 7 times, and the corresponding sample point data was collected. The following conclusions were obtained through data analysis.

(1) In the long-time experiment under isolated and narrow conditions, the hand strength showed an upward trend, and the grip strength of the left and right hands increased by 5.42% and 6.12% respectively, which was significant. This phenomenon may relate to the handicap of wide body movement in isolated narrow space. Narrow space greatly limited wide body movement and the amount of exercise.

(2) The upper body strength showed a slight decrease, but no significant decrease. The reason may be that muscles of the upper limbs act as larger muscle groups, there was no significant degeneration in the limited time span of the experiment, although there was not enough movement stimulation.

(3) The data of lung capacity showed a relatively stable cyclical fluctuation, with no significant increase or decrease. The reason may be that the time period was relatively short and the decrease in activity did not directly affect cardiopulmonary function, so there was no significant decrease in lung capacity.

(4) In the time period of the experiment, there was no significant change in the time of eye-closed standing. The experiment time is relatively short, and the isolated environment had no significant effect on the participants' ability to balance.

(5) There was no significant change in sit-and-reach. However, gradual decrease was found in participants with better performance, and there was a slight increase in the participants with lower performance.

References

1. Zhang, R.P., Sun, X.C., Zhang, B., et al.: Advance in research of effects of ship environment on seamen. J. Prev. Med. Chin. People's **24**(2), 149–151 (2006)

2. Fu, M.Z., Wang, M.K., Ba, J.B., et al.: Research progress on the influence of warship environment on the physical ability of shipboard personnel. Chin. J. Naut. Med. Hyperb. Med. **23**(6), 483–486 (2016)
3. Shen, W.M., Chen, H.S., Li, A.F., et al.: Effects of prolonged navigation on the pulmonary functions of sailors. Chin. J. Nauti. Med. Hyperb. Med. **11**(2), 68–70 (2004)
4. Ou, M., Xu, H.T., Lai, L.F., et al.: Influence of oceangoing voyage on the respiratory function of naval medical staff. Med. Pharm. J. Chin. People's Liber. Army **4**, 89–91 (2015)
5. Yuan, J.F., Zhang, Q.X., Lu, A.M., et al.: One-legged standing with eyes closed in physical fitness testing. J. Clin. Rehabil. Tissue Eng. Res. **33**, 6049–6054 (2013)
6. Yu, H., Xu, L.H., Hu, P.K., et al.: Influence of long-term navigation on mental health in crews. Med. J. Chin. People's Liber. Army **37**(7), 745–748 (2012)
7. Yu, H.Z., Yu, H.B., Wang, Y.X., et al.: Analysis of mental health status and its influencing factors of navy crews during long-term sailing. Acad. J. Second Mil. Med. Univ. **38**(4), 506–510 (2017)
8. Hu, A.X., Wu, X.S., Li, C.X., et al.: Investigation and analysis of psychological health state among the ship crew deployed at sea for a prolonged time. J. Navy Med. **34**(2), 86–88 (2013)
9. Bao, J.: Psychological intervention on navy officers and soldiers on a long sea voyage. J. Naval Univ. Eng. **07**(3), 49–53 (2010)
10. Ren, J.L., Wang, H.W., Ou, M., et al.: Influence on health condition of voyage on board staff. Chin. J. Health Care Med. **18**(6), 462–464 (2016)

How the Linguistic Context Influences the Decision-Making Process of Bilingual Individuals with a Comparison Between Eastern and Western Languages

Giorgio Manenti[✉], Jean-Raphaël Eid, Abdul Qadeer Khoso, Marius Julian Vogel, Md Redwan Ahmed Reyadh, and Victor Lopez Perez

Department of Industrial Engineering, Tsinghua University, Beijing, China
manenti.giorgio@outlook.com, {zhugt18,ks18,
rdw18}@mails.tsinghua.edu.cn, marius.v@hotmail.com,
a01336434@itesm.mx

Abstract. The decision-making process of bilingual individuals differ based on the language used, this phenomenon is called "Foreign Language Effect", and in this paper, the authors will demonstrate just how pervasive and ubiquitous this effect is. To extend the current understanding of this phenomenon, three different games were designed in order to measure risk propensity, preference towards foreign products and empathy. Eight different first languages were tested against two main second languages, English and Chinese, resulting in the collection of 650 data samples. The research was conducted through questionnaires and the results obtained showed that different first languages reacted in different ways to foreign language effects, people are less inclined towards risky behaviour when using a second language, and they generally prefer to buy products when those are presented in their native language. Age and proficiency level do not significantly affect the effect discovered, and there is a reduction in the empathy level when individuals need to take delicate decisions within second language contexts. Some cultural causes have been identified and provide explanations for some of the anomalous behaviour observed in certain first languages, opening future research questions on the role that culture plays in bilingual decision making.

Keywords: Bilingualism · Decision making · Foreign language · Linguistic context · Culture · Foreign language effect

1 Literature Review

More than half of the world's population can use at least two different languages, and not only will the number of bilingual people likely grow, but also over half of the total world population is already a bilingual decision-maker [1]. Therefore, grasping the implications of bilingualism in human's decision-making processes means having a better understanding of the modern world itself.

© Springer Nature Switzerland AG 2020
P.-L. P. Rau (Ed.): HCII 2020, LNCS 12192, pp. 474–492, 2020.
https://doi.org/10.1007/978-3-030-49788-0_36

Foreign language usage affects people's judgments of risks and benefits, reducing the perception of risks and increasing the perception of benefits. Keysar et al. [2] explains in his paper: "The Foreign-Language Effect: Thinking in a foreign tongue reduces decision biases" how using and thinking in a foreign language reduces emotional attachment and increases people's rational thinking during decision making.

More than half of the world's population uses words in two or more languages daily. In the work done by Costa et al. [3], it is mentioned that emotion is one of the factors associated with foreign language processing and that it can alter the interaction between intuition and deliberation. Messages processed in a foreign language usually elicit a milder emotional response compared to those processed in a native language. Furthermore, foreign-language users present lower emotional reactivity, as, for example, shown by reduced electrodermal responses to swearing words and childhood reprimands [5].

1.1 Culture, Context and Ethnicity

One of the variables which have to be taken into account when talking about interactions and taking decisions in a foreign language is the culture of both speakers. Briley et al. [6] demonstrated that prompting decision-makers to provide reasons before making selections from choices set with a "compromise" and two "extreme" options produces culturally dependent choice patterns. This observed effect can be seen as a precedent to posterior research that shows how bicultural individuals shift the values that they manifest depending on the language that they use. Briley et al. [7] conducted further researches on bilingualism by looking at how the strategies used to solve problems would be affected by the language used. When asked to solve a problem in one of the languages that they could speak, subjects tended to choose strategies that reflected values usually associated with the culture of the language. This area of research has been further explored in a marketing, human resources, and negotiating context, as the implications of cultural interaction can be widely exploited in these contexts.

1.2 Customer-Seller Relationship and Advertisement in Bilingual Contexts

Cayla et al. [8] described service encounters as "moments of truths" in which customers decide if they will either continue doing business with the service providers or exit and deprive the service provider of all future revenues. Therefore, the seller needs to maintain a good relationship and always work on meeting his/her buyer's or consumer's expectations.

While the tangible aspects of customer-seller interactions after a service failure have been studied extensively, the intangible verbal cues (language used, tenor) and nonverbal cues (displayed emotions and ethnicity) have not [9]. Cayla et al. [8] states that sellers have to understand the impact of language and ethnicity on the customer experience in order to increase marketing effectiveness, customer retention and to apply appropriate recovery strategies when customers are disappointed with the service provided.

2 Research Questions and Hypotheses

This paper aims at highlighting the ramifications of the foreign language effect and its implications by analysing multiple languages and cultures. In order to explore such effects, four hypotheses have been formulated and tested. In particular, this research focus on rationality, susceptibility and empathy with a complete comparison between such effects in eastern and western languages.

The hypotheses tested are the followings:

- **Hypothesis no. 1:** "People are more rational when using their second language."
- **Hypothesis no. 2:** "People prefer products described in their native language when choosing between two identical items."
- **Hypothesis no. 3:** "The usage of a foreign language influences the subject's level of empathy."
- **Hypothesis no. 4:** "Different first languages reacts in different proportions to foreign language effects."

In order to test these four hypotheses, three games were designed in a questionnaire form, and a total of 650 data samples were collected. Three types of questionnaires were developed: one questionnaire, the native questionnaire, entirely written into the subject's native language (Bangla, Chinese, French, German, Italian, Spanish, Swiss and Urdu), the foreign questionnaire, written in the subject's second language (English or Chinese) with parts of the games written in the subject's native language and one control questionnaire, entirely written in English.

In this research's figures and tables, native and control questionnaires will be indicated as "Native Q" and "Control Q", whereas for second language questionnaires, "Foreign Q" will indicate that the data include both English and Chinese second language questionnaire results. "Foreign questionnaire – English" and "Foreign questionnaire – Chinese" will indicate the results of English second language questionnaires and Chinese second language questionnaires.

2.1 Experiment's Variables

Variables are those factors which limit and define the boundaries of the experiment. In the proposed case, variables have been classified into three categories: Independent, control and outputs variables.

Independent Variables. Independent variables are those variables which cannot be controlled. In this research, the independent variables are the following: nationality, language proficiency level, gender and age.

Control Variables. Control variables are those variables under the control of the researchers during the experiments and can be categorised, expanded, diminished, or changed whenever required during the project execution. For this research, the control variables are the followings:

- **Native language:** Since the questionnaires are designed in multiple languages, this is a control variable. The authors have chosen eight native languages.
- **Second language:** In this research, two languages English and Chinese, have been chosen for the roles of second languages.

Output Variables. Output variables are the free dummy variables that are the outcomes of the questionnaires.

- **Risk propensity:** This variable regards the propensity of the subjects towards risk, and it is the output variable of the first game in the questionnaires.
- **Susceptibility:** Susceptibility measures the propensity of individuals to make purchasing decisions based on emotions rather than considering the actual value of the products.
- **Empathy:** Empathy measures the level of emotional attachment to a specific cause, even when not rationally justified.

2.2 Experiment Procedure

Three games were designed, each game focusing on one of the previously stated output variables.

Every questionnaire contains the same information, but each time translated in the subject's native language of or in his/her second language. English and Chinese were chosen as second languages.

Control Questionnaire. The control questionnaire aims at providing reference thresholds for the outcomes of the three experiments. It acts as a filter to remove possible unwanted biases by identifying the objectively best option without any bilingualism effects. Control questionnaire has been designed entirely in English and does not present any trace of a second language. Thus, those questionnaires' answers were used as unbiased reference points.

3 The Experiment

3.1 Experiment 1 – Coin Toss Game

The first experiment is a coin toss game. As the name suggests, the participants are given the chance to participate in a hypothetical game, where they have to choose either head or tail of a possible coin toss. In case the participants win the toss, they are given a fixed amount of monetary reward, if they lose the toss, they do not get anything. However, the participants can also decide to not participate in the game but to take home a secure amount which is 51% of the monetary reward for a winning toss. In this game, the participants are tested on their risk behaviour regarding a possible reward. The standard amount of the reward is around 80$ and the secure amount around 41$. The reference currency was converted to the specific one associated with the questionnaire's language. In this game, the researchers aimed at identifying the risk propensity of individuals when the game is presented in their native or foreign language. In this game, the more rational choice is the risk-averse one. The output variable for game 1 is risk propensity, which indicates whether people are risk-averse or not and in which language this effect is stronger.

3.2   Experiment 2 – E-Commerce Game

This experiment focuses on the influence of the language used in a product description on the decision-maker. The participants will read the descriptions of two daily used products, in this specific game, two USB-chargers. The chargers will have the same price point, same description and specifications; however, the product description is given in either the subject's native or second language. USB-chargers give an excellent basis to test certain tendencies since many people use them, and they can be categorised as commodities.

This game, along with the control questionnaire helps to identify the impact of the subject's native or foreign language on his/her purchasing decisions. The output variable for this game is susceptibility, and an example of the game is shown in Fig. 1. In the specific case of game 2, there are no differences between the native language questionnaires and the second language questionnaire because in both cases, there were two products with the same bilingual descriptions.

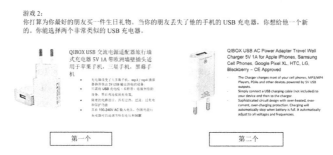

Fig. 1.   Game 2, Chinese first language - English second language.

3.3   Experiment 3 – Charity Game

The participants are given a description explaining a scenario in which they are passing away soon, but their last wish is to help their grandchildren who are likely to have heart problems in the future. The story narrates that the participant is wealthy but does, however, only carry a small amount of cash with himself/herself. The participants are then advised to donate a rather small amount to two charities which help battling heart diseases.

One of the charities is called "Saviors of Heart" in the corresponding native language of the participant. For instance, in German, this charity is called "Herzensretter", in Italian "Salvatori del Cuore" and so forth. The charity is active around the world, including in the participants' origin country. The second charity is called "Heart Without Borders" which is a smaller worldwide organisation operating actively in the participants' region. The participants then have to decide how to split the small amount of cash between the two charities. This experiment focuses on the possible biases of the subjects towards the local charity with the name in the participants' native language. The currency, like in the other experiments, is converted according to the questionnaire's language.

This game is designed to measure the output variable "empathy". It helps to identify the empathetic factor of the language; whether or not people are more emphatic/emotional when the name and the context of the charity are presented in the subject's native language.

3.4 Description of the Participants

In this section, the participants that took part in the research are described in detail. In total, 650 subjects participated. An overview of the participants for the different questionnaires is found in the following tables. The participants were contacted either online or through printouts of the questionnaire. The authors of this paper shared the questionnaire to close friends, family and university associates. Most of the questionnaires (588) have been filled online, whereas 62 were collected on paper. The participants have been selected according to their native language. For instance, the native French questionnaire was distributed to French native speakers and so forth. There were no constraints in age, gender and religion for the participants.

Table 1. Participants of the native language questionnaire.

Participant language	Sample size	Average age	Percentage of males	Percentage of females	English proficiency level
French	80	35.83	28.8%	71.3%	4.23
Italian	59	43.73	32.2%	67.8%	3.03
Swiss	29	52.93	44.8%	55.2%	3.54
Urdu	96	23.57	21.9%	78.1%	4.91
Chinese	28	21.04	39.3%	60.7%	4.07
Mexican	11	49.27	72.7%	27.3%	4.27
Bangla	26	23.34	87.0%	13.0%	5.41
Total	**329**	**33.38**	**35.8%**	**64.2%**	**4.23**

In Table 1, the participants who took part in the native questionnaire are shown. The English proficiency level of the participants ranges from a value of 1, which indicates a low English level and 7, for a high proficiency level. In total, 329 participants with an average age of 33.38 took part in this questionnaire, 35.8% male and 64.2% female.

The participants of the native questionnaire have an average English proficiency level of 4.23. With an average age of 52.93, the Swiss participants are the oldest, whereas the Chinese are the youngest averaging at 21.04. In general, the average age of the French, Italian, Swiss and Mexican participants of the native questionnaire is relatively higher than the Urdu, Bangla and Chinese one. The French, Italian, Urdu and Chinese participants are mainly females, whereas a large portion of the Mexican and Bangla participants are males.

Table 2. Participants of the foreign language questionnaire – English.

Participant language	Sample size	Average age	Percentage of males	Percentage of females	English proficiency level
French	101	24.62	49.5%	50.5%	4.86
Italian	40	25.3	57.5%	42.5%	4.67
Swiss	15	26.67	80.0%	20.0%	5.33
Urdu	52	24.6	71.2%	28.8%	4.82
Chinese	25	24.04	28.0%	72.0%	4.71
Mexican	14	22.69	57.1%	42.9%	6.07
Bangla	28	22.89	75.0%	25.0%	5.59
Total	**275**	**24.50**	**57.5%**	**42.5%**	**4.98**

Table 2 is built in the same manner as Table 1. In total, 275 participants took part in the foreign language questionnaire. The average age of the participants is 24.50 years, and the male/female distribution is quite even at 57.5% males and 42.5% females. The average English proficiency level of the participants is at 4.98. The average age of the participants is quite similar between the different languages and sways between 22.69 for the Mexicans to a 26.67 for the Swiss. The Swiss and Bangla participants of the second questionnaire are mainly males, while the Chinese participants are mainly female.

In Table 3, the native and second language questionnaire are summarised, native indicates that the subject with the first language indicated took part in the native questionnaire, whereas second means that the person with the first language indicated took part in a questionnaire written in his/her second language. In total, 604 participants took part in either of these two questionnaires. The average age of the participants is 29.3 years with a male/female distribution of 45.6% to 54.4%. The average English proficiency level of the participants is 4.57 from a scale from one to seven and the average age ranges from 22.45 of the Chinese participants to 43.98 of the Swiss participants. The average age of the Italian, French, Swiss and Mexican subjects is generally higher than those of the Urdu, Bangla and Chinese participants. The distribution of Male and Female is close to 50% for the Italian, French and Swiss participants.

The participants estimated their average English proficiency level rather high at an average of 4.57. The level is ranging from a low of 3.85 for Italians to a high of 5.50 for Bangla participants although the rating is solely dependent on the self-evaluation of the participants, which might differ from one culture to another.

Table 3. Summary of the participants for the native and foreign language questionnaire.

Type of questionnaire	Language	Sample size	Average age	Percentage of males	Percentage of females	English proficiency level
Native	Italian	59	43.73	32.2%	67.8%	3.03
Second	Italian	40	25.30	57.5%	42.5%	4.67
Total/Average	–	99	36.28	42.4%	57.6%	3.85
Native	French	80	35.83	28.8%	71.3%	4.23
Second	French	101	24.62	49.5%	50.5%	4.86
Total/Average	–	181	29.57	40.3%	59.7%	4.55
Native	Swiss	29	52.93	44.8%	55.2%	3.54
Second	Swiss	15	26.67	80.0%	20.0%	5.33
Total/Average	–	44	43.98	56.8%	43.2%	4.44
Native	Urdu	96	23.57	21.9%	78.1%	4.91
Second	Urdu	52	24.6	71.2%	28.8%	4.82
Total/Average	–	148	23.93	39.2%	60.8%	4.87
Native	Chinese	28	21.04	39.3%	60.7%	4.07
Second	Chinese	25	24.04	28.0%	72.0%	4.71
Total/Average	–	53	22.45	34.0%	66.0%	4.39
Native	Mexican	11	49.27	72.7%	27.3%	4.27
Second	Mexican	14	22.69	57.1%	42.9%	6.07
Total/Average	–	29	38.66	64%	36%	4.92
Native	Bangla	26	23.34	87.0%	13.0%	5.41
Second	Bangla	28	22.89	75.0%	25.0%	5.59
Total/Average	–	54	23.1	80.8%	19.2%	5.50
Total Native	–	**329**	**33.38**	**35.8%**	**64.2%**	**4.23**
Total Second	–	**275**	**24.5**	**57.5%**	**42.5%**	**4.98**
Total/Average	–	**604**	**29.3**	**45.6%**	**54.4%**	**4.57**

Table 4. Participants of the control group questionnaire.

Number of countries	Sample size	Average age	Percentage of males	Percentage of females	Average english proficiency level
25 Countries	33	23.12	48.5%	51.5%	5.61

For the control questionnaire, as shown in Table 4, there were a total of 33 participants from 25 different countries. The average age is 23.12 years, and the male/female distribution is 48.5% male and 51.5% female.

Table 5. Participants of the foreign language questionnaire – Chinese.

Number of countries	Sample size	Average age	Percentage of males	Percentage of females	Average english proficiency level
10 Countries	13	23.31	61.5%	38.5%	5.50

Table 5 shows the participants of the foreign language questionnaire in Chinese. The small sample size of 13 participants shows the difficulty of gathering people for this questionnaire. The average age of the participants is 23.31 years, 61.5% of the participants are male, and 38.5% are female. The average English proficiency level is 5.50. Figure 2 shows the age distribution of the participants in a graphical way. Clearly visible is the age difference from the native to second language questionnaires for French, Italian, Swiss, and Mexican participants.

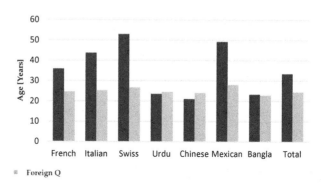

Fig. 2. Average age of participants by questionnaire type.

In Fig. 3, the average English proficiency level of the participants is displayed.

The gender distribution of the participants is seen in Fig. 4, where native indicates that the subject with the first language indicated took part in the native questionnaire, whereas second means that the person with the first language indicated took part in a questionnaire written in his/her second language. 45.6% of the participants are male, and 54.4% female.

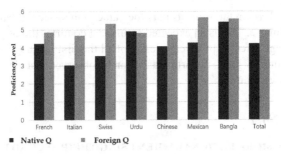

Fig. 3. Average English proficiency level of participants by questionnaire language.

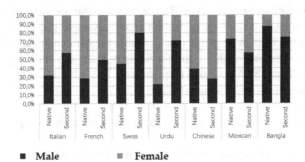

Fig. 4. Gender distribution by questionnaire type.

4 Data Analysis

4.1 Game 1 – Results

Game 1 was a coin toss game testing the risk propensity of the participants. The results are seen in Fig. 5. The more rational choice was the risk-averse one, therefore the less the value of the risk propensity variable, the higher the level of rationality of the subject. Figure 5 shows how the risk propensity decreases, indicating an increment in

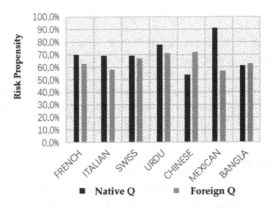

Fig. 5. Game 1: risk propensity by questionnaire type.

the rationality level, when the subjects play the game in their second language, whether it is English or Chinese. This is an expected result as it confirms the original findings of the foreign language effect [2]. Nevertheless, not all the first languages benefitted from a boost in the rationality level, in particular, it can be observed that Chinese and Bangla went against the overall trend.

4.2 Hypothesis No. 1

"PEOPLE ARE MORE RATIONAL WHEN USING THEIR SECOND LANGUAGE."

To test this hypothesis, a T-Test was applied. This is because the Standard Deviation (SD) of the population was unknown. The output variable measured here is the risk propensity. For the native language questionnaires, the sample size consists of 329 data points while for the foreign language questionnaire the data points are 288. The mean of Sample 1 is represented with $\mu 1$ and for Sample 2 $\mu 2$. The Null Hypothesis is that H0: $\mu 1 - \mu 2 = 0$ and the Alternative Hypothesis is H1: $\mu 1 + \mu 2 = 0$

Table 6 highlights the following information:

1. 70.7% of the sample subjects are risk propense when the game is presented in their native language.
2. 64.4% of the sample subject are risk propense when the game is presented in their second language.

Table 6. Quantitative analysis hypothesis no. 1.

Independent variable	Dependent variable	Sample size	Sample mean	Sample std	SE Mean	T-value	DF	P-value
Native language	Risk_Propensity	329	0.707	0.025	0.001378	26.39	530	<0.0001
Second language	Risk_Propensity	288	0.644	0.033	0.001945			

With a confidence level of more than 99.99%, it can be inferred that the decision-making process varies when the subject's first or second language is used.

What emerges is that people are more risk-averse while using their second language because they try to think more rationally and are emotionally detached, so they choose the second option which is reasonably more convenient than the first one and requires a moment of detached reflection to understand its value.

4.3 Game 2 – Results

Game 2 tests the susceptibility of the participants, depending on the language context in which the game is presented. The games describe two products, one in the subjects' native language, the other one described their second language. Results are summarized in Fig. 6. In this game, the control questionnaire, with its 33 data points, shows that the objectively best option is the second charger presented.

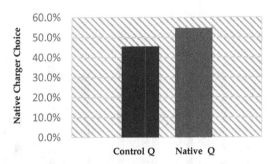

Fig. 6. Game 2: Comparison of native charger choice by questionnaire type.

In game 2, there are no differences between the native language questionnaires and the second language ones because in both cases, there were two products with the same bilingual descriptions. Even with two identical games, people when presented with a questionnaire, and therefore a linguistic context, in their first language, preferred to buy the charger which has the description in their first language, whereas when the questionnaire is presented in their second language, there was a slight but noticeable trend towards buying the second charger, deemed as objectively better by the control subjects. This change in the decision-making process can be explained by a higher level of rationality when approaching a problem in a subject's second language context, already demonstrated by the foreign language effect [2].

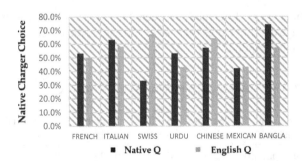

Fig. 7. Game 2: Comparison of native charger choice by subject's language (Native Q – English Q).

As Fig. 7 shows, Swiss and Chinese participants preferred to buy the product described in their native language when the problem and the context were framed in English, their second language.

When the questionnaire was presented in Chinese, for people with Chinese as their second language, there was a quasi-absolute preference towards the charger described using the participant's native language, as shown in Fig. 8. More than a language decision, it is most likely a mistrust towards Chinese products.

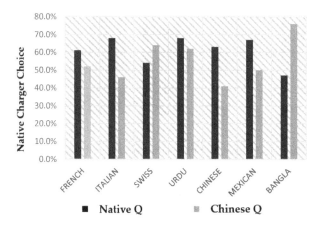

Fig. 8. Game 2: Comparison of native charger choice by subject's language (Native Q – Chinese Q).

4.4 Hypothesis No. 2

"PEOPLE PREFER PRODUCTS DESCRIBED IN THEIR NATIVE LANGUAGE WHEN CHOOSING BETWEEN TWO IDENTICAL ITEMS."

In game 2, there are no differences between the native language questionnaires and the second language questionnaires because in both cases, there were two products with the same bilingual descriptions.

Table 7. Quantitative analysis hypothesis no. 2 – Native Q vs Control Q.

Independent variable	Dependent variable	Sample size	Sample mean	Sample std	T-value	DF	P-value
Native language	Susceptibility	329	0.547	0.025	2.85	32	0.0075
Control sample	Susceptibility	33	0.455	0.185			

Table 7 shows that with a sample size of 329, 54.7% of the subjects preferred to buy the product presented in their native language, while the threshold was 45.5% from the control questionnaire. With a confidence level of more than 99.25%, it can be inferred that people preferred products in their native language.

Table 8. Quantitative analysis hypothesis no. 2 – Foreign Q vs Control Q.

Independent variable	Dependent variable	Sample size	Sample mean	Sample std	T-value	DF	P-value
Second language	Susceptibility	288	0.524	0.033	2.14	32	0.0402
Control sample	Susceptibility	33	0.455	0.185			

Similarly, Table 8 shows that with a sample size of 288, 52.4% of the subjects preferred to buy the product presented in their native language, while the threshold was 45.5% from the control questionnaire. In this second case, the confidence level decreases from 99.25% to 95.98%, but the values of the susceptibility variables from the two questionnaires types remain very close. This similarity in the outputs is due to game 2 being identical regardless of the questionnaire type. Only the control questionnaire had two chargers both described in the same language, English, in order to gain a reference threshold regarding the objectively best product. From the results, it can be inferred that there is a difference in the susceptibility level based on the language context and that people are more inclined to buy the product presented in their native language. So, a native language context has a higher persuasion, and it can be exploited as a marketing tactic.

4.5 Game 3 – Results

In the third game, participants read a small case that described two charities. The first one is a local charity, with a named in the subject's native language, for instance, written in Italian if the questionnaire target Italians. The second one is an international charity, with an English name, always the same. Both operate at least in the subject's own country. The participants have a standard budget of 999$ (or 9999¥ for Chinese subjects) that they need to split between the local and international charities. This amount has been chosen as such in order to prevent the subjects to always split in half the money.

This game has been designed to test the subject's empathy, which has been defined as the attitude to act in favour of a cause even when it is the less rational choice. In this game, the more rational choice was the second charity because the money spent there have a vastly higher impact on society compared to the same amount spent on the local charity. The results of game three are displayed in Fig. 9.

From the control questionnaire, the international charity emerges as the objectively best one, the charity associated with the highest rationality and the least emotional and therefore, empathy level.

When the subjects were presented with the game written in their native language, they favoured more than 60% of the times the local charity. This choice shows an increment in the emotional attachment level as the more rational choice is the international

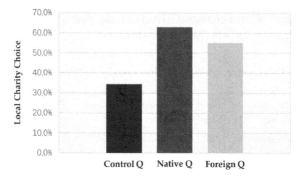

Fig. 9. Game 3: Comparison of local charity choice by questionnaire type.

charity. Such international charity is described as being able to save many more children compared to the local one. Therefore, the choice of the local charity was driven by emotions and empathy rather than by rationality.

Figure 9 shows an increase in the percentage of subjects who favoured the international charity, compared to the native language questionnaire. This implies a lower level of empathy, explained by a higher level of rationality triggered by the second language frame of the game.

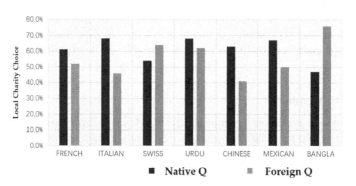

Fig. 10. Game 3: Comparison of local charity choice by subject's language (Native Q – Foreign Q).

Figure 10 shows a uniform distribution in the empathy level variation across all languages except for Swiss and Bangla. These two languages seem to prefer the local charity when the game is presented in English, showing more empathy in a second language context.

4.6 Hypothesis No. 3

"THE USAGE OF A FOREIGN LANGUAGE INFLUENCES THE SUBJECT'S LEVEL OF EMPATHY."

The data analysis in Table 9 shows how around 63% of the subjects selected the native charity when the scenario was presented in their native language, while this percentage decreased to 55% when the scenario was presented in the subjects' second language.

With a confidence level of more than 99.99%, it can be inferred that the usage of a foreign language influences the level of empathy while making decisions.

In a second language context, people are less empathetic because they become more rational and less emotional while using their second language. The mean of Sample 1 is represented with $\mu 1$ and for Sample 2 $\mu 2$.

Table 9. Quantitative analysis hypothesis no. 3.

Independent variable	Dependent variable	Sample size	Sample mean	Sample std	T-value	DF	P-value
Native language	Empathy	329	0.629	0.025	19.88	330	<0.0001
Second language	Empathy	288	0.550	0.033			

4.7 Hypothesis No. 4

"DIFFERENT FIRST LANGUAGES REACTS IN DIFFERENT PROPORTIONS TO FOREIGN LANGUAGE EFFECTS."

For this hypothesis, a test could not be developed since multiple factors needed to be considered. However, the data were compared graphically. Figure 11 describes the behaviour of peoples of different countries against the output variables risk propensity, susceptibility and empathy.

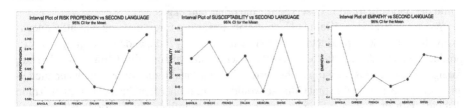

Fig. 11. Quantitative analysis hypothesis no. 4.

From Fig. 11 it can be inferred that the hypothesis is validated. Different first languages reacted in different ways to the three games when played in a second language context, and the strength of the foreign language effect was different for every language. This implies that there are some additional factors which influence the decision-making process of bilingual individuals, like culture and traditions.

As far as risk propensity in a second language context, Chinese tend to be more risk propense compared to first language contexts, whereas all the others tend to be more rational and risk-averse.

For susceptibility, only Mexicans and Pakistani preferred the English charger in a second language context, whereas all the others preferred their native language option in either language contexts.

For empathy, only Chinese and Italians had an increased emotional detachment when in a second language context, with an empathy level close to the control subjects'. Swiss tend to have a reduction in rationality when tested in their second language compared to a usually high level of rationality when tested in their first language.

5   Peculiar Behaviours

5.1   Chinese Participants

In game 1, Chinese participants went completely off-trend with their choices, becoming more risk-seeking, and in this research context, less rational, when speaking their second language, while subjects from all the other countries (except Bangla) had the opposite behaviour demonstrating a boost in rationality when in a second language context.

According to Lau et al. [10], native Chinese speakers and native English speakers do not have the same betting behaviour. Chinese subjects exhibit higher levels of risk-taking behaviours than the English subjects, partly because of different probabilistic thinking.

The majority of the subjects were more risk-averse and in the case of game 1, more rational while reading questionnaires in their second language. Chinese participants though, went completely off-trend with their choices, becoming more risk-seeking when using English, and in this case, less rational, against what stated by the foreign language effect. Chinese culture generally favours moderation and caution, and it could be assumed that speaking in English releases some cultural constraints, resulting in riskier behaviours.

In game 2, when the game is written in English rather than in Mandarin, Chinese subjects are more inclined towards choosing the charger written in their native language, while the opposite trend is observed for the other languages.

As demonstrated by Klein [11], animosity towards a foreign nation will affect negatively the purchase of products produced by that country independently of judgments of product quality. While it has been widely observed that historically, Chinese liked to buy foreign quality goods, looking at the current international trade context, between the US and China, it is reasonable that Chinese would favour more often the local product. Trade disputes with English speaking countries could be an influencing factor, thus leading to a stronger attraction towards buying local products.

5.2   Swiss Participants

The Swiss participants have in comparison to the other languages chosen differently in the e-commerce game. In the Swiss native language questionnaire, 19 of the 29 Swiss participants chose English charger, which is anomalous compared to the participants of the other languages. With an average age of almost 53 years, the participants of the native questionnaire are amongst the oldest of all participants.

The English language plays an essential role in many areas of public and private life in Switzerland. English words and expressions, such as "sorry", "cool" or "oh my God", are used in all the national languages of Switzerland, large population groups can speak English and use it regularly in business, education or travel contexts [12]. The pervasiveness of the English language in Swiss culture could be the reason why most Swiss participants choose the second charger. The subjects probably are so used to buy English products that it only seemed natural to them to choose the English charger.

6 Conclusions

Taking decisions while using a foreign language or while immersed in a foreign language context has already been proven to influence how people make decisions and the results of this study broaden the knowledge on the foreign language effect's ramifications.

The results obtained showed that different first languages reacted in different ways to foreign language Effects, people are less inclined towards risky behaviour when using a second language, and they generally prefer to buy products when those are presented in their native language. Age and proficiency level do not significantly affect the effect discovered, and there is a reduction in the empathy level when individuals need to take delicate decisions within second language contexts.

Since this research has shown that different first languages react in different ways to foreign language effects, it can be implied that there are some additional factors which influence the decision-making process of bilingual individuals, like culture and traditions. This finding open future research opportunities on the role that culture plays in bilingual decision making.

References

1. Benesch, S.: Emotions as agency: feeling rules emotion labor, and English language teachers' decision making. System **79**, 60–69 (2018)
2. Keysar, B., Hayakawa, S.L., An, S.G.: The foreign-language effect: thinking in a foreign tongue reduces decision biases. Psychol. Sci. **23**(6), 661–668 (2012)
3. Costa, A., Vives, M., Corey, J.D.: On language processing shaping decision making. Curr. Dir. Psychol. Sci. **26**(2), 146–151 (2017)
4. Colish, M.: The mime of God: vives on the nature of man. J. Hist. Ideas **23**(1), 3–20 (1962)
5. Harris, C.L., Ayçiçeĝi, A., Gleason, J.B.: Taboo words and reprimands elicit greater autonomic reactivity in a first language than in a second language. Appl. Psycholinguist. **24**(4), 561–579 (2003)
6. Briley, D.A., Morris, M.W., Simonson, I.: Reasons as carriers of culture: dynamic versus dispositional models of cultural influence on decision making. J. Consum. Res. **27**(2), 157–178 (2000)
7. Briley, D.A., Morris, M.W., Simonson, I.: Cultural chameleons: biculturals, conformity, motives and decision making. J. Consum. Psychol. **15**(4), 351–362 (2005)
8. Cayla, J., Bhatnagar, K.: Language and power in India's "new services". J. Bus. Res. **72**, 189–198 (2017)

9. Holmqvist, J.: Consumer language preferences in service encounters: a cross-cultural perspective. Manag. Serv. Qual.: Int. J. **21**(2), 178–191 (2011)
10. Lau, L.-Y., Ranyard, R.: Chinese and English probabilistic thinking and risk taking in gambling. J. Cross Cult. Psychol. **36**(5), 621–627 (2005)
11. Klein, A.: Firm performance and board committee structure. J. Law Econ. **41**(1), 275–304 (1998)
12. Ronan, P.: Perspectives on English in Switzerland. Universite de Lausanne (2016)

The Influence of Chinese-English Bilingualism on Rationality in Decision Making Behaviors

Mostafa Moazami, Nan Qie, and Pei-Luen Patrick Rau(⊠)

Institute of Human Factors and Ergonomics, Department of Industrial Engineering,
Tsinghua University, Beijing 100084, China
rpl@mail.tsinghua.edu.cn

Abstract. This research examined the effect of bilingualism on Chinese-English bilinguals through three decision-making scenarios, including risk framing, loss aversion, and myopic loss aversion. Twenty-four Chinese-English bilinguals who were raised acquiring both Chinese and English with an almost equal proficiency were recruited. The bilinguals demonstrated less asymmetry between gain and loss frames, behaved less loss-averse and myopic loss-averse in the English contexts than in Chinese contexts. The results indicated that Chinese-English bilinguals were more rational in their decisions using the English language than the Chinese language. Finally, the discrepancies in the decision making behaviors of bilinguals using Chinese and English in different linguistic contexts have been discussed from the psycholinguistic perspective.

Keywords: Chinese–English bilingualism · Psycholinguistics · Risk framing · Loss aversion · Myopic loss aversion

1 Introduction

As a result of globalization and immigration, hundreds of millions of people all over the world use more than one language to communicate with their coworkers at their workplaces or to talk with their family members at home. Therefore, it is very common that bilinguals are confronted with decision-making problems in different linguistic contexts. It remains a mystery for a long time whether bilinguals would make the same decisions when using different languages.

Cumulative Chinese investment in the United States has seen a significant growth from US$280 million in 2004 to approximately US$36 billion in 2014 [1]. This trend will continue to grow in the coming years. It is predicted that if China FDI follows the current growing rate, Chinese firms that have international trades with the United States will need to hire 200,000 to 400,000 American employees by 2020 [2]. Not only for the U.S., Chinese OFDI in Europe hits another record high in 2015, indicating that China's role as a source of productive capital [3]. Due to OBOR strategies and discussion on G20 in Hangzhou, China's impact in international economics will continue increasing. As the number of Chinese who were born and raised overseas is increasing significantly, more and more Chinese-English bilinguals with native-like fluency in both

© Springer Nature Switzerland AG 2020
P.-L. P. Rau (Ed.): HCII 2020, LNCS 12192, pp. 493–504, 2020.
https://doi.org/10.1007/978-3-030-49788-0_37

Chinese and English will be available in the job markets in and out of China in the foreseeable future. They will play a critical role in many aspects of China's international relationships. Therefore, understanding the influences of bilingualism on their decision-making behaviors is very important.

Researchers believe that a bilingual's behaviors, thoughts, actions and responses are consistent with the cultural values associated with the language that the circumstance presented. Bond and Yang call this phenomenon cultural accommodation [4]. In their study, Chinese native speakers with great fluency in English were given a questionnaire randomly, and the language of it was either Chinese or English. Tendency towards Chinese cultural values were found in the participants' responses when they answered the questionnaire in Chinese, while their responses contained more western cultural values and attributes when they answered the same questions in English.

One of the main concepts used to explain the behavior of bilinguals under different circumstances is cultural frame switching. Most researchers focusing on bilingualism and cross-cultural studies believe that bilinguals are bicultural [5]. Two cultures exist simultaneously inside bilinguals' minds and accompany them in their daily life. Cultures mainly control bilinguals' behaviors in their corresponding linguistic contexts. Hong, Morris, Chiu, and Benet-Martinez studied the cultural frame switching behavior of Chinese-American bicultural people [6]. They found that participants displayed more Western characteristics when presented with American symbols such as American flag, Marilyn Monroe and Superman, while they tended to demonstrate more Chinese cultural features in their behaviors when presented with Chinese symbols such as Chinese dragon, the Great Wall and the Peking Opera. It concluded that the presence of elements of a certain culture can make bicultural people display behaviors and values associated with that specific culture.

The cultural framing of bilinguals can also influence their decision-making behaviors. Briley, Morris, and Simonson believe that, in the decision-making process, bilinguals try to make their decisions in favor of the cultural framing of the presented linguistic context [7]. They studied Hong Kong Chinese decision-making behaviors in their use of Cantonese and English proverbs. They found that using proverbs in a particular language triggered the related cultural frame of that language. Hong Kong Chinese tried to make their decisions under the Chinese cultural frame when using Cantonese, while they moderated their decision-making behaviors using the English proverbs.

Besides cultural framing, there exists another bilingual decision-making pattern which is called the second language effect. Keysar, Hayakawa and An studied the decision-making behaviors of bilinguals in their use of both the native languages and foreign languages [8]. They found that bilinguals behaved more rationally using their foreign languages than their native languages. They argued that using foreign languages makes bilinguals rely more on their systematic decision-making process which results in less bias in decision making. Since their six experiments produced the same results, they argued that this language effect was common regardless of what the native language and the foreign language were.

The second language effect can also be explained by emotional reactions. Bilinguals experience reduction in emotional reaction when using their second acquired language or the language which they are less proficient in [9]. Most bilingualism and emotional reaction studies used self-report measures and physiological measures to examine emotional reaction. Researchers have examined bilinguals' emotional reactions to the curse words and taboo words, and the results of self-rating as well as skin conductance responses demonstrated that bilinguals display higher emotional reaction to such words in their native language than their second language [10, 11].

Harris and his colleges measured the skin conductance of Spanish-English bilinguals who were born in the United States or immigrated in their early childhood [12]. The presented data in her research indicated that there was higher emotional resonance to endearments, insults and reprimands stimuli in Spanish, which is the participants' first acquired language, than in English, which is their second acquired language. This result can be explained from the angle of language acquisition. Although bilinguals had almost the same proficiency in their first and second language, they were more emotionally attached to their mother language that is acquired from the family than their second language that is acquired from peers and the environment.

The language effect on rationality of decision-making behaviors of bilinguals has also become a hot topic in China. Peng focused on the gain-loss framing effect on Chinese university students when using the foreign language and native language [13]. He recruited Chinese bilingual students majoring in English or in finance. Each sample members were randomly assigned to Asian disease scenario or monetary scenario. The results indicated that regardless of professional experience of bilinguals and whether the risk context is monetary or non-monetary, the gain-loss asymmetry in risk preference of Chinese bilinguals diminished in the use of English language. The author pointed out that the reason behind the rational decision-making behaviors of Chinese bilinguals using the foreign language mainly came from the reduction in the emotional resonance which is associated with the use of foreign language. The result of this study was consistent with the finding of Keysar [8].

Each researcher has defined bilingualism from the perspectives which are in service of his/her research. De Mejía defined bilingualism based on the privilege and needs of acquiring the second language [14]. Schrauf used the age of acquiring the language as a criterion for defining bilingualism. People who acquire their second language in early childhood can switch between languages very quickly with negligible grammatical and lexical errors, and this group of bilinguals is called simultaneous bilinguals [15]. The other group of bilinguals is called sequential bilinguals, which refers to bilinguals who acquire their second language after childhood. Sequential bilinguals are less proficient than simultaneous bilinguals. Furthermore, bilingualism can be defined from the angle of proficiency level. Based on the proficiency level of the second language, bilinguals can be classified into four groups including limited, passive, dominant and balanced bilinguals [16]. Pavlenko argued that maintaining the same level of proficiency between two languages is not possible and not necessary [17]. For example, a balanced bilingual may be better in speaking and writing about his/her professional job in the language that the job training adopts than his/her other language that is not adopted in job training process. In contrast, some researchers pointed out that balanced bilinguals have equal proficiency level in his/her languages; otherwise they are not classified as balanced bilingual [18].

Based on the purpose of our study, we focus on Chinese-English bilinguals in two aspects, including the manner of acquiring the languages and the proficiency level of languages. Our research involves participants who acquire both Chinese and English in a natural setting since early childhood, and their proficiency levels of Chinese and English are equal or almost equal.

2   General Methods

In this study, we aim to investigate the decision-making behaviors of Chinese-English bilinguals under risk-related circumstances in Chinese or English language contexts. Three decision-making scenarios including risk framing, loss aversion and myopic loss aversion were selected based on the research by Keysar et al. [8]. We hope to gain a deep understanding of the decision-making behaviors of Chinese-English bilinguals through three studies.

In order to determine whether potential participants are eligible to participate in the experiments or not, we designed a self-rating questionnaire on language learning history to gather their personal, linguistic and cultural background information before conducting the experiments. Questions included country of birth, growing up country, native language(s), parents' native language(s), the age when being exposed to Chinese and English for the first time, and self-rating on proficiency of the Chinese language and English language. Only candidates who acquire both Chinese and English in a natural setting since early childhood and are equally or almost equally proficient in Chinese and English were recruited. WeChat, the most popular instant messaging App in China, and SoJump questionnaire, the most professional online questionnaire platform in China, were used as main channels for recruitment.

Twenty-four Chinese-English bilinguals participated in this study and every participant went through all three experiments. All participants were living in Beijing, China at the time of the experiment. Three out of four participants in this study were born in China with their parents being Chinese, and moved with their family to North America in the early childhood. The rest were born in the United States whose parents were Chinese immigrants. Among all the participants, 12% grew up in Canada and 63% in the United States, while 25% of them claimed both China and United States as the places where they grew up. Participants began to acquire both Chinese and English since early childhood. Based on our questionnaire, participants started to acquire Chinese at birth and English at the mean age of 3.60 years old (SD = 2.44). Their Chinese parents and experience of growing up in North America provided the participants with an opportunity to acquire both Chinese and English in a natural setting since early childhood. All participants reported Chinese and English as their native languages. All participants reported themselves as highly proficient in both Chinese and English on a 1–7 scale (1-very poor, 7-excellent). The mean score of Chinese proficiency was 6.58 (SD = 0.58), and that of English proficiency was 6.33 (SD = 0.63).

Participants in this study were randomly divided into Chinese language context group and English language context group. The whole experiment procedure was designed in English, and was translated into Chinese by a native Chinese-English speaker. To ensure that the Chinese and English versions are equivalent, the Chinese translation was checked by two other Chinese-English bilinguals and was revised several times until both versions were the same in meaning.

3 Study I: Risk Framing

Tversky and Kahneman points out a significant effect of gain-loss framing on human risk decision making [19]. When an option is presented in terms of gains, people choose the sure option and behave risk averse, but when the same option is framed in terms of losses, people choose the risky option and act more risk seeking. They point out this phenomenon through an "Asian disease problem". People prefer to save the lives of 200 out of 600 people for sure rather than take a chance of saving all of them or none. Whereas, if the choice is framed as losing 400 lives out of 600, people become risk seeking. The results point out that the willingness to take or avoid risk is related to how the situation is described.

3.1 Methods

The first experiment of current study was designed based on "Asian disease problem". The participants were randomly arranged into four groups, with six participants in each group. The four groups were Chinese-gain, Chinese-loss, English-gain, English-loss framing respectively. The problem was described as below:

Currently, a fatal disease has been spreading around. Six Hundred patients will die without using any medicinal treatment. Therefore, two types of medicine have been developed to save these patients' lives which we call Medicine A and Medicine B.
The Gain Frame

If Medicine A is used for treatment, two hundred patients will be saved.

If Medicine B is used for treatment, there is a one-third chance that all Six Hundred patients will be saved and a two-third chance that they will all die.
The Loss Frame

If Medicine A is used for treatment, Four Hundred patients will die.

If Medicine B is used for treatment, there is a one-third chance that none of the patients will die and a two-third chance that all Six Hundred patients will die.

Which medicine would you choose to use? Medicine A or Medicine B?

3.2 Results and Discussion

In order to understand the distribution of bilingual participants' responses under the gain and loss frame in both English and Chinese language contexts, we gathered the frequencies of choosing the sure option (Medicine A) by participants under all four groups. As the sample size in this experiment was small, Fisher's exact test was adopted to analyze the data (Table 1).

In English language context, eight participants preferred the sure option (Medicine A) when presented with the gain frame problem, while six participants preferred sure option when presented with the lose frame problem. In the English language context,

Table 1. Frequencies of sure option (Medicine A) in each language context and Fisher's exact test result in Study I

Language	Frame		P value
	Gain	Loss	
English	8	6	0.233
Chinese	12	2	0.015

there was no significant difference in participants' responses under gain and loss frames (p = 0.233).

In the Chinese language context, all 12 participants preferred the sure option (Medicine A) when presented with the gain frame problem, while only two participants preferred sure option when presented with the loss frame problem. There was significant difference in participants' responses under gain and loss frames when decision was made in the Chinese language context (p = 0.015).

The use of English language reduced the asymmetry between gain and loss frames. It can be concluded that participant's response was frame-independent in the English language context, whereas the use of Chinese language demonstrated significant asymmetry between gain and loss frames. The use of English language as one native language of Chinese-English bilinguals affected their risk preference reversals and made their decision-making behaviors more rational, but the use of Chinese language did not demonstrate the same effect.

4   Study II: Loss Aversion

Loss aversion is one of the assumptions underlying Prospect Theory [20]. It implies that people are loss averse in the sense that they anticipate the negative impact of a potential loss would outweigh the positive impact of an identical potential gain. For instance, most people would decline a positive expected value bet that offers a chance of winning $50 or losing $50. People typically require a potential gain of at least $100 to make up for a potential loss of $50 because the subjective impact of losses is roughly twice as much as that of gains [21]. People in general behave more loss averse when dealing with risky options. Furthermore, loss aversion is more significant when people face bets with larger sizes.

4.1   Methods

In this experiment, participants were randomly assigned into two sample groups in either the Chinese language context (n = 12) or the English language context (n = 12). Participants were presented with 10 rounds of positive expected value bets. Half of the bets had a large size (amount of winning = $250/1500 RMB) and half of the bets had a small size (amount of winning = $6/36 RMB). Within each bet size, the loss-to-gain ratio ranged from a highly attractive bet with one to six to an unattractive bet with five to six. Different bet sizes and bet attractiveness were presented to the participants in random order.

4.2 Results and Discussion

The results showed that Chinese-English bilinguals were willing to take more bets in the English language context than in the Chinese language context. In the English language context, bilinguals took 9.17 out of 10 rounds of bets on average (including both small and large size bets), while in the Chinese language context the average number of accepted rounds significantly dropped to 5.3, which demonstrates that bilingual participants were more loss-averse in the Chinese language context than in the English language context.

In the English language context, bilingual participants had higher tendency to accept bets with both small and large sizes, and the number was 4.75 rounds and 4.42 rounds respectively on average. By contrast, bilingual participants in the Chinese language context decreased the tendency to take positive expected value bets. They took 3.58 rounds out of 5 rounds of bets with the small size, and the reduction in number of accepted bets was more significant in bets with the large size with only 1.75 accepted rounds on average.

Two-way ANOVA was adopted to study the effects of the language context and the bet size on bilinguals' decision making (See results in Table 2). There was a significant difference between the numbers of accepted bets in the Chinese and English language contexts. The willingness to accept the bets by Chinese-English bilinguals was significantly higher in the English language context than in the Chinese language context (F = 28.39, p < 0.001).

Table 2. Results of two-way ANOVA in Study II

Source of variation	df	F	P-value
Language context	1	28.39	<0.001
Bet size	1	9.07	<0.001
Language context * Bet size	1	4.35	0.04
Within	44		
Total	47		

The English language context increased rationality in the decision making of Chinese-English bilinguals and made them less loss-averse in both large and small size bet scenarios than the Chinese language context. The results of two-way ANOVA demonstrated that there was a significant interaction between the language used in the experiment and the bet size (F = 4.35, p = 0.04). The effect of language on participants' willingness to take bets was more significant in the large bet size scenario than in the small bet size scenario.

In conclusion, the English language context, compared with the Chinese language context, can help Chinese-English bilinguals reduce the negative feeling of a potential loss. Therefore, they are willing to take more bets in English language context. This experiment demonstrated that switching from the Chinese language context to the English language context can increase the rationality in Chinese-English bilinguals' decision-making behavior.

5 Study III: Myopic Loss Aversion

Most people not only emphasize more on the negative feeling of losses than on the positive feeling of gains, they also hold the view that the short-term outcomes outweigh the long-term ones. Benartzi and Thaler introduced myopic loss aversion [22]. Myopic loss aversion relies on two behavioral hypotheses, namely loss aversion and mental accounting, with the latter implying that long-term investments, in financial decision making, are evaluated according to their short-term returns.

Suppose that a bet with positive expected value is repeated several times in the long run. When an individual just lost one bet, the emotional reaction to the loss will lead him/her to hesitate about whether to accept more bets. In other words, the emotional impact of short-term loss causes people to focus on recent loss so that they will miss a big picture of gaining in the long run.

5.1 Methods

In this experiment, we aim to investigate the myopic loss aversion behavior of Chinese-English bilinguals. At the beginning of the experiment, each participant received 10 dollars in the form of 1dollar bill. Participants took part in 10 rounds of bets by using a coin flip procedure. In each round, the participant got 1-dollar bill out of his/her remaining bills and made a decision to place a bet or not. If participant decided to bet, the experimenter would flip a coin while the participant made his/her desired choice by calling heads or tails. If the participant was correct, he/she kept that 1-dollar bill and received an additional 1.5 dollar. If participant was wrong, he/she lost that 1-dollar bill and should return it to the examiner. At the end of each round, the participant had a chance to stop betting and keep the rest of the money. In the Chinese language context, the dollar was converted into RMB was based on a rough exchange rate of 1 Dollar = 6 RMB. At the beginning of this experiment, participants were told that the money earned in this experiment was their incentive for their participation in this study.

5.2 Results and Discussion

Chinese-English bilinguals were less affected by the short-term outcome in the English language context than in the Chinese language context. Bilingual participants showed rationality in their decision-making behavior by taking more bets in the English language context than in the Chinese language context.

Bilingual participants in the English language context displayed more willingness to take bets with accepting 8.9 rounds of bets on average. Whereas, the number of accepted rounds declined to 6.5 rounds on average in Chinese context. Independent sample t-test showed difference in the numbers of accepted bets in the English language context and in the Chinese language context samples was significant (t = 2.42, p = 0.024). The higher accepted number of rounds in the English language context indicates that Chinese-English bilinguals behave less myopic loss-averse and more risk seeking in the English language context than in the Chinese language context.

6 General Discussion

The results of all three experiments showed that Chinese-English bilinguals are more rational in the decision-making process in the English language context than in the Chinese language context. The result of this study shows that Chinese-English bilinguals can overcome the bias in their decision-making behaviors by thinking in English rather than in Chinese.

As we mentioned before, participants in this study acquired both Chinese and English since early childhood in a natural setting with approximately equal proficiency in both languages. Now the key question is that why Chinese-English bilinguals have different risk preferences and decision-making behaviors when using English and Chinese languages while both languages are the participants' native languages. The results can be explained by psycho-linguistic theories.

Psycho-linguistics or psychology of language refers to the effects of psychological factors on acquiring, producing and using a language. It covers cognitive processing of a language which makes a human talk, think and make decisions. We consider emotion toward a language and grammatical structure of a language as two main psycho-linguistic factors that influence bilinguals' behaviors in the risk-related, decision-making process.

Emotional process in decision making differs between languages. Loewenstein et al. and Quartz presented that the decision-making process is not only affected by cognitive process but also by emotional process [23, 24]. Based on their studies, individual's emotional reactions to the risk play a very important role in the decision-making process under uncertain circumstances. Stanovich et al. and Kahneman argued that decision-making process involves two types of processes, namely systematic process and heuristic process. Systematic decision-making process is more cognitive, analytic, and follows specific rules [25, 26]. When individuals go through the systematic process, they make risk-related decisions based on the likelihood of possible outcomes through calculating cost and benefit of their choices. On the other hand, heuristic decision-making process is more intuitive and affective.

Keysar et al. pointed out that the emotional reaction can lead to decisions less systematic and analytic and make individuals rely more on heuristic decisions which lead to bias and less rationality in decision making [8]. De Martino et al. found that individuals who were presented with the framing effect experiment showed higher activation in the amygdale part of their brains [27]. Amygdala plays an important role in the emotional decision-making process. Higher activation in amygdale part in the framing effect experiment reflected individual's additional emotional tendency to choose sure options in the gain frame and strong negative emotional reaction to sure losses in the loss frame.

Furthermore, research has shown that the recall of autobiographical memory is language dependent [28]. Bilinguals have the ability to re-experience their past memories such as personal events and life experiences using the language context in which those events had originally been encoded and experienced. Therefore, the use of first language by bilinguals who were brought up in an immigrant family makes them associate the context related to family and close relatives with higher emotional resonance, while the use of second acquired language makes them associate the context related to school peers and workmates with lower emotional resonance.

This serves as a good evidence to believe that Chinese-English bilingual participants in the current study have higher emotional resonance in the Chinese language context than in the English language context. All of the recruited participants in the current study were born with Chinese mothers and Chinese fathers Although they reported both Chinese and English as their native languages with high proficiency, they acquired their Chinese language at home by interacting with their parents but acquired English at kindergartens or schools. Based on previous research findings, we believe that Chinese-English bilinguals demonstrate higher emotional resonance in the Chinese language context than in the English language context.

Therefore, it can be concluded that higher emotional resonance in using the Chinese language can lead decision-making process to be less systematic and more intuitive, which can result in bias in decision. Nevertheless, the lower emotional resonance in using the English language make bilinguals less affected by their emotion and rely more on systematic decision- making process which result in decisions more rational.

Besides emotions towards a language, the grammatical characteristics of a language can also influence such speakers' decision making. M. K. Chen adopted future-time reference (FTR) criteria which divided languages into Weak FTR and Strong FTR [29]. Weak FTR language refers to a language that does not require grammatical marking when predicting future events such as most Germanic languages, while Strong FTR language refers to a language which requires mandatory use of grammatical marking when predicting future events such as English. Strong FTR language speakers significantly distinguish the present events from the future events in their language which leads them to put more emphasis on the present than on the future in their decision-making process. In contrast, Weak FTR language speakers treat the present and the future almost in the same way which results in taking the future into account at the present time. Based on his study, the differences in the grammatical structures of languages in expressing the present and future events can affect risk preferences of speakers of such languages.

Jakobson and Halle pointed out that the core difference between languages is not in what they may convey but is in what they must convey [30]. In English grammar, the time reference in sentences is expressed by the tense of the sentences. Tense in English is carried by the use of auxiliary and/or verb inflection. In future simple tense, the modal verbs "will" and "be going to" is mainly adopted as future event prediction marker to express prediction of future events. In contrast, Chinese language is a tense free language with no inflection. If you check out Chinese grammar books, there is tense categorization such as past, present and future. In the aspect of grammar, the time reference in Chinese is mainly carried by aspectual morphemes or by the use of temporal adverbs before verbs. For an English language speaker, it is mandatory to use future-time reference marker to convey future information grammatically, but for a Chinese language speaker it is very natural to neglect the use of future-time reference grammatical markers and it won't impact the meaning expressed. Chinese can also convey future information by relying on non-grammatical markers such as the whole environment of communication or the presence of temporal nouns in the sentences.

Based on the above discussion, we can conclude that Chinese speakers put extra weight on the future at the present time which makes them avoid any risk that can influence their future as a consequence of their present decisions. Therefore, Chinese language speakers are more loss-averse when making risk-related decisions. On the other hand, there is a very clear distinction between the present and the future in the minds of English speakers as they speak a Strong FTR language. When they make a risk-related decision, they don't put an emphasis on considering whether their present decisions will bring about positive or negative results in the future. Therefore, they are more open to take risks when making risk-related decisions.

This conclusion can be extended to the decision-making behaviors of Chinese-English bilinguals in the risk-related contexts. The use of Chinese language can make bilinguals follow associated risk-related decision-making behaviors due to Chinese language being a Weak FTR language. Bilinguals behave less rational by outweighing the influences of their present decisions on their future. Therefore, they prefer to save more for security at the present time and put their present and future at a lower risk. In contrast, the use of English language makes Chinese- English bilinguals not exaggerate the consequences of their current decisions on their future lives; they are less influenced by the negative feeling of loss at the present time in a risky context, since they are willing to take more risks.

References

1. Lu, H.Q.: China top spot for FDI in world in 2014 (2015). http://usa.chinadaily.com.cn/bus iness/2015-02/02/content_19471524.htm
2. Hanemann, T., Lysenko, A.: Employment impacts of Chinese investment in the United States. In: East Asia Forum, October 2012
3. Hanemann, T., Huotari, M.: A new record year for Chinese outbound investment in Europe. Mercator Institute for China Studies, February 2016
4. Bond, M.H., Yang, K.-S.: Ethnic affirmation versus cross-cultural accommodation the variable impact of questionnaire language on Chinese bilinguals from Hong Kong. J. Cross-Cult. Psychol. **13**, 169–185 (1982)
5. LaFromboise, T., Coleman, H.L., Gerton, J.: Psychological impact of biculturalism: evidence and theory. Psychol. Bull. **114**, 395 (1993)
6. Hong, Y., Morris, M.W., Chiu, C., Benet-Martinez, V.: Multicultural minds: a dynamic constructivist approach to culture and cognition. Am. Psychol. **55**, 709 (2000)
7. Briley, D.A., Morris, M.W., Simonson, I.: Cultural chameleons: biculturals, conformity motives, and decision making. J. Consum. Psychol. **15**, 351–362 (2005)
8. Keysar, B., Hayakawa, S.L., An, S.G.: The foreign-language effect thinking in a foreign tongue reduces decision biases. Psychol. Sci. **23**, 661–668 (2012)
9. Altarriba, J., Santiago-Rivera, A.L.: Current perspectives on using linguistic and cultural factors in counseling the Hispanic client. Prof. Psychol.: Res. Pract. **25**, 388 (1994)
10. Dewaele, J.-M.: The emotional force of swearwords and taboo words in the speech of multilinguals. J. Multil. Multicult. Dev. **25**, 204–222 (2004)
11. Harris, C.L., Aycicegi, A., Gleason, J.B.: Taboo words and reprimands elicit greater autonomic reactivity in a first language than in a second language. Appl. Psychol. **24**, 561–579 (2003)
12. Harris, C.L., Gleason, J.B., Aycicegi, A.: When is a first language more emotional? Psychophysiological evidence from bilingual speakers. Biling. Educ. Biling. **56**, 257 (2006)
13. Zhang, H., Shaheen, S.A., Chen, X.: Bicycle evolution in China: from the 1900s to the present. Int. J. Sustain. Transp. **8**, 317–335 (2014)
14. De Mejía, A.-M.: Power, prestige, and bilingualism: International perspectives on elite bilingual education. Multilingual Matters (2002)

15. Schrauf, R.W.: Bilingual autobiographical memory: experimental studies and clinical cases. Cult. Psychol. **6**, 387–417 (2000)
16. Ng, B.C., Wigglesworth, G.: Bilingualism: An Advanced Resource Book. Taylor & Francis, Milton Park (2007)
17. Pavlenko, A.: Emotions and Multilingualism. Cambridge University Press, Cambridge (2007)
18. Hamers, J., Blanc, M.: Bilinguality and Bilingualism. Cambridge University Press, Cambridge (1989)
19. Tversky, A., Kahneman, D.: The framing of decisions and the rationality of choice. DTIC Document (1980)
20. Kahneman, D., Tversky, A.: Prospect theory: an analysis of decision under risk. Econ.: J. Econ. Soc. **47**, 263–291 (1979)
21. Tom, S.M., Fox, C.R., Trepel, C., Poldrack, R.A.: The neural basis of loss aversion in decision-making under risk. Science **315**, 515–518 (2007)
22. Benartzi, S., Thaler, R.H.: Myopic loss aversion and the equity premium puzzle. National Bureau of Economic Research (1993)
23. Loewenstein, G.F., Weber, E.U., Hsee, C.K., Welch, N.: Risk as feelings. Psychol. Bull. **127**, 267 (2001)
24. Quartz, S.R.: Reason, emotion and decision-making: risk and reward computation with feeling. Trends Cogn. Sci. **13**, 209–215 (2009)
25. Stanovich, K.E., West, R.F., Hertwig, R.: Individual differences in reasoning: Implications for the rationality debate?-Open Peer Commentary-The questionable utility of cognitive ability in explaining cognitive illusions (2000)
26. Kahneman, D.: A perspective on judgment and choice: mapping bounded rationality. Am. Psychol. **58**, 697 (2003)
27. De Martino, B., Kumaran, D., Seymour, B., Dolan, R.J.: Frames, biases, and rational decision-making in the human brain. Science **313**, 684–687 (2006)
28. Marian, V., Neisser, U.: Language-dependent recall of autobiographical memories. J. Exp. Psychol.: Gener. **129**, 361 (2000)
29. Chen, M.K.: The effect of language on economic behavior: evidence from savings rates, health behaviors, and retirement assets. Am. Econ. Rev. **103**, 690–731 (2013)
30. Jakobson, R., Halle, M.: Phonology and phonetics. Fundam. Lang. 4–51 (1956)

Research on the Path Integration Behavior
of Firefighters in the Dark

Hua Qin[1,2], Xiao-Tong Gao[1,2], Wei Zhao[1,2], and Yi-Jing Zhang[1,2(✉)]

[1] Department of Industrial Engineering, Beijing University of Civil, Engineering and
Architecture, Beijing 100044, People's Republic of China
ldlzyj@163.com
[2] Department of Industrial Engineering, Tsinghua University,
Beijing 100084, People's Republic of China

Abstract. Firefighters often need to complete combat missions in dark environments, such as finding trapped people, exploring unknown rooms and forming corresponding internal fire maps. The judgment of spatial orientation is especially important in these tasks because it can be effectively optimized. Rescue strategy to improve the personal protection level of firefighters. Therefore. this study conducted a $2 \times 2 \times 2 \times 3$ inter-group experiment on the spatial orientation of firefighters in a dark environment based on posture, personnel composition, difficulty level of the scene and memory demand level. The subject needs to pass through the 5×14 rectangular space clockwise along the wall and stop at the measurement points on the other three sides except the side of the departure position. The direction judgment is performed by measuring the oral report angle of the subject and the actual arm pointing angle. The posture is divided into upright and half-turn. The composition of the personnel is divided into one person and two people. Scene difficulty is divided into simple and complex. A complicated and simple way to distinguish is to increase the edge of the scene by placing obstacles on the corners of the rectangle. The memory requirement level is divided into three levels: low, medium, and high. The experimental results show that the complexity of the scene significantly affects the azimuth backlash error of the subject, the complex scene error is higher, and the simple scene error is lower. The composition of the personnel greatly affected the azimuth adaptation error. In the case of a group of 2 people, the orientation judgment error is lower. The level of posture and memory requirements did not significantly affect the azimuth anaphora error. The results of this study can provide a reference for firefighters in space cognitive design.

Keywords: Firefighter · Way-finding · Path integration · Dark environment

1 Introduction

Navigation is an important activity related to human survival and reproduction. Cruisers rely on different spatial cues to update the spatial relationship between themselves and

---

Y. Zhang—Associate researcher and doctor, mainly focuses on human reliability, human-computer interaction and astronaut selection and training.

the environment, which is called spatial updating. According to the types of spatial cues, cruising can be divided into piloting relying on environmental cues such as road signs and path integration relying on its own motion information. When the surrounding environment provides rich visual clues, people rely on navigation. When visual cues are lost, people rely on path integration. Path integration is an important strategy of non vision navigation for blind people. Non visual navigation can not only help the road finding subject in the field environment space such as navigation, mountaineering, desert travel, etc., or in the environment space with narrow road, limited vision and lack of striking road signs, but also protect the road finding subject losing visual clues from the danger of unknown environment in the operation of strange space or high-risk space.

Firefighters often fight in the dark and dangerous environment. Compared with the congenital blind, because they do not have the ability to find the way in the dark environment, it will be more difficult to determine the reference, estimate the distance, angle, direction and memory route [1, 2]. Because the difference between the dark space environment and the visual environment will make them face the adverse effects of the path direction effect, which makes it more difficult to construct the surrounding environment to layout the mental space. In the dark environment, they can only rely on their own movement elements to construct direction and integrate memory, but unfortunately, this ability seems to be untrained [3]. Although the update of fire-fighting facilities and equipment, the restriction of intact rules and regulations, and fire emergency assessment can ensure the safety of firefighters [4]. However, firefighters have to work in the above high-risk cognitive environment, which is often accompanied by many tasks, responsibilities, orders, and overworked physical and mental labor [5]. In our interview with firefighters, we found that although there was no corresponding training, the elderly fire chief developed the ability of path integration in the dark space. To improve fire-fighter's cognitive ability of dangerous environment can improve the ability of controlling the crisis scene [6], so this study focuses on the path integration process in the dark environment.

2 Method

Reviewing the research on human path integration, we can see that the vast majority of research paradigms are path completion tasks. This task can better reflect the ability of human path integration, and is also suitable for laboratory operation, so this study uses path to complete the task. In order to investigate the influencing factors of fire-fighter's ability of path integration in the dark environment, an inter group experiment was designed based on the changes of environment, posture, personnel composition and task difficulty.

2.1 Experimental Variables

Posture: combat power, half squat
Personnel composition: single or multiple
Location of measuring points: simple, medium and difficult
Complexity of environment: simple and difficult.

2.2 Dependent Variables

Angle of error of limb pointing; angle of error of oral report1. Actual

2.3 Experimental Site

Point O is set as the starting point, and all test points are based on this point. Point a-the angle between test point 1 and point a is 40°, and the distance from point O is 1241.47 cm. Point B-the angle between test point 2 and point a is 15°, and the distance from point O is 1447.32 cm. Point C-the angle between test point 3 and point a is 5°, and the distance from point O is 628.39 cm (Figs. 1, 2 and 3).

Fig. 1. Fire education and training center

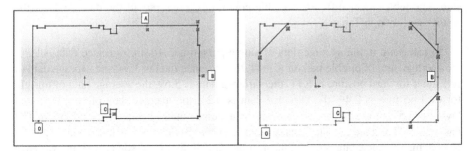

Fig. 2. Simple experimental environment   **Fig. 3.** Difficult experimental environment

2.4 Obstacle

There are three exhibition boards in the obstacle use site, each of which has the same size. Specification: 321 L × 17 W × 22 h (Fig. 4).

Fig. 4. Obstacles

3 Result

After the experiment, 144 data of distance, azimuth, position and time were collected, and the data were analyzed by data collation, calculation mean, variance analysis and error analysis.

3.1 Basic Statistics

Expression Distance of Participants. Take the number of group members, forward posture and the complexity of indoor space layout as a single variable, get the basic statistical table under different variable conditions, combine all the data, and get the results as shown in Table 1.

In test point a, the average data of single person is 5.15, the variance data is 9.20, the average data of double person is 3.44, the variance data is 3.54; the average data of full squat forward posture is 6.11, the variance data is 5.69, the average data of upright forward posture is 2.49, the variance data is 0.24; the average data of simple spatial layout is 3.57, the variance data is 3.31, the average data of complex spatial layout The data was 5.03 and the variance was 9.96. In test point B, the average data of single person is 3.68, the variance data is 6.338, the average data of double person is 1.63, the variance data is 2.28; the average data of full squat forward posture is 3.64, the variance data is 7.55, the average data of upright forward posture is 1.67, the variance data is 1.29; the average data of simple spatial layout is 2.61, the variance data is 9.26, the average data of complex spatial layout The data was 2.70 and the variance was 2.17. At test point C, the mean data of single person is 1.87, the variance data is 0.91, the mean data of double person is 1.27, the variance data is 0.39; the mean data of full squat forward posture is 1.61, the variance data is 0.77, the mean data of upright forward posture is 1.52, the variance data is 0.77; the mean data of simple spatial layout is 1.56, the variance data is 0.64, and the mean data of complex spatial layout is 1.56. The data was 1.57 and the variance was 0.89.

Table 1. Basic statistics of distance

Variable test			Point A	Point B	Point C
Number of teams	Mean value	Single	5.15	3.68	1.87
		Double	3.44	1.63	1.27
	Variance	Single	9.20	6.33	0.91
		Double	3.54	2.28	0.39
Forward position	Mean value	Squat	6.11	3.64	1.61
		Up-right	2.49	1.67	1.52
	Variance	Squat	5.69	7.55	0.77
		Up-right	0.24	1.29	0.77
Complexity of spatial layout	Mean value	Simple	3.57	2.61	1.56
		Complex	5.03	2.70	1.57
	Variance	Simple	3.31	9.26	0.64
		Complex	9.96	2.17	0.89

Orientation of the Participants. The basic statistical tables under different variable conditions are obtained by using the group number, the forward position and the complexity of the indoor space layout as a single variable, and all the data is combined to obtain the results as shown in Table 2 below.

Table 2. Shows the basic statistical table of orientation

Variable test			Point A	Point B	Point C
Number of teams	Mean value	Single	14.38	33.13	45.25
		Double	12.29	21.46	17.92
	Variance	Single	118.69	33.51	224.32
		Double	4.80	39.53	113.66
Forward position	Mean value	Squat	9.17	22.92	35.67
		Up-right	17.50	31.67	27.50
	Variance	Squat	28.24	71.53	587.43
		Up-right	51.85	41.20	204.17
Complexity of spatial layout	Mean value	Simple	15.83	26.25	25.42
		Complex	10.83	28.33	37.75
	Variance	Simple	62.96	38.19	240.97
		Complex	46.76	122.69	493.68

At the test point A, the mean data of the single person is 14.38, the variance data is 118.69, the mean data of the double person is 12.29, the variance data is 4.80, the mean data of the forward position of the squatting position is 9.17, the variance data is 28.24, the average data of the straight forward posture is 17.50, and the variance data is 51.85; The simple spatial layout mean data is 15.83,the variance data is 62.96, the complex spatial layout mean data is 10.83, and the variance data is 46.76. At test point B, the single person's mean data is 33.13 and the variance data is 3 3.51. The mean data of the two persons is 21.46, the variance data is 39.53, the mean data of the forward position of the squat is 22.92, the variance data is 71.53, the average data of the upright forward position is 31.67, the variance data is 41.20, the simple spatial layout mean data is 26.25, the variance data is 38.19, The complex spatial layout mean data is 28.33 and the variance data is 122.69. At the test point C, the mean data of the single person is 45.25, the variance data is 224.32, the average data of the double person is 17.92, and the variance data is 1.13.66. The mean data of forward posture were 40.67, the variance data were 587.43, the vertical forward posture mean data were 35.67, the variance data were 204.17, the simple spatial layout mean data were 25.42, the variance data were 240.97, the complex spatial layout mean data were 37.75,and the variance data were 493.68.

Position and Orientation of the Participant. Taking the number of groups, forward posture and the complexity of indoor space layout as a single variable, the basic statistical tables under different variables are obtained, and all the data are combined, and the results are shown in Table 3 below.

Table 3. Basic statistical table of position orientation

Variable test			Point A	Point B	Point C
Number of teams	Mean value	Single	47.21	10.13	22.13
		Double	31.54	3.92	39.00
	Variance	Single	244.04	66.93	17.06
		Double	518.64	3.27	115.64
Forward position	Mean value	Squat	23.33	8.37	43.38
		Up-right	55.42	5.67	17.75
	Variance	Squat	179.15	71.88	655.23
		up-right	60.94	19.13	115.64
Complexity of spatial layout	Mean value	Simple	39.00	3.75	33.54
		Complex	39.75	10.29	27.58
	Variance	Simple	590.06	2.31	342.56
		Complex	335.88	65.06	842.40

At the test point A, the mean data of the single person is 47.21,the variance data is 244.04, the average data of the double person is 31.54, the variance data is 518.64, the average data of the forward position of the squatting position is 23.33, the variance data is 179.15, the average data of the upright advancing posture is 55.42, and the variance data is 60.94; The simple spatial layout mean data is 39.00, the variance data is 590.06, the complex spatial layout mean data is 39.75, and the variance data is 335.88. At test point B, the single person's mean data is 10.13, The variance data is 66.93, the average data of the two persons is 3.92, the variance data is 3.27, the mean data of the forward position of the squatting position is 8.37, the variance data is 71.88, the average data of the straight forward posture is 5.67, the variance data is 19.13, the simple spatial layout mean data is 3.75, the variance data is 2.31, The complex spatial layout mean data is 10.29 and the variance data is 65.06. At the test point C, the mean data of the single person is 22.13, the variance data is 17.06, the average data of the double person is 39.00, and the variance data is 11.5.64. The mean data of the forward position of the squatting position is 43.38, the variance data is 655.23, the average data of the upright forward position is 17.75, the variance data is 115.64, the simple spatial layout mean data is 33.54, the variance data is 342.56, the complex spatial layout mean data is 27.58, and the variance data is 842.40.

Time. The time basic statistic table obtained by the experiment is obtained under different variable conditions by using the group number, the forward position and the complexity of the indoor space layout as a single variable, and all the data is combined, and the result is shown in the following Table 4.

Table 4. Basic time statistics

Variable test		Mean value	Variance
Number of teams	Single	227.42	419.94
	Double	266.75	1127.64
Forward position	Squat	228.13	707.62
	Up-right	266.04	912.91
Complexity of spatial layout	Simple	255.79	1614.58
	Complex	238.38	762.17

It can be seen from the table that the average time of completing an experiment is 227.42 s for a single person and 266.75 s for a double. The average time for completing an experiment in a squatting forward posture is 228.13 s, and the average completion time in an upright forward position is 266.04 s. The average time for a simple spatial layout to complete an experiment is 255.79 s, and the average completion time for a complex spatial layout is 238.38 s. The time variance of single person completing an experiment is 419.94, that of double is 1127.64, and that of squatting forward posture is 70.7.62, the time variance of upright forward posture is 912.91, the time variance of simple spatial layout is 1614.58, and the time variance of complex spatial layout is 762.17.

3.2 Analysis of Variance

The Participant's Stated Distance. Taking the number of groups, forward posture and the complexity of indoor spatial layout as a single variable, the variance analysis table under different variables is obtained, and the data of the three test points are combined, and the results are shown in Table 5.

Table 5. Analysis of distance variance.

Variable	Test-point		Quadratic sum	Free degree	Mean square	F	P-value
Number of teams	A	Inter group	5.85	1.00	5.85	0.92	0.38
		In group	38.19	6.00	6.37		
	B	Inter group	8.47	1.00	8.47	1.97	0.21
		In group	25.84	6.00	4.31		
	C	Inter group	0.72	1.00	0.72	1.12	0.33
		In group	3.88	6.00	0.65		
Forward position	A	Inter group	26.28	1.00	26.28	8.88	0.03
		In group	17.76	6.00	2.96		
	B	Inter group	7.78	1.00	7.78	1.76	0.23
		In group	26.52	6.00	4.42		
	C	Inter group	0.02	1.00	0.02	0.02	0.89
		In group	4.58	6.00	0.76		
Complexity of spatial layout	A	Inter group	4.23	1.00	4.23	0.64	0.46
		In group	39.81	6.00	6.64		
	B	Inter group	0.02	1.00	0.02	0.00	0.96
		In group	34.29	6.00	5.72		
	C	Inter group	0.00	1.00	0.00	0.00	0.99
		In group	4.60	6.00	0.77		

The significance of group A, B and C was 0.38, 0.21 and 0.33, respectively. The significance of the advanced posture in A, B and C was 0.03, 0.23 and 0.89, respectively. The significance of space layout complexity in A, B and C was 0.46, 0.96, and 0.99, respectively.

Orientation of the Participants. Taking the number of groups, forward posture and the complexity of indoor space layout as a single variable, the variance analysis table under different variables is obtained, and the data of the three test points are combined, and the results are shown in Table 6.

The significant number of group members in A, B, C was 0.72, 0.03, 0.03, the significant of forward posture in A, B, C was 0.11, 0.15, 0.58, and the significant of spatial layout complexity in A, B, C was 0.38, 0.75, 0.40, respectively.

Table 6. Represents the azimuth variance analysis table

Variable	Test-point		Quadratic sum	Free degree	Mean square	F	P-value
Number of teams	A	Inter group	8.67	1.00	8.67	0.14	0.72
		In group	370.46	6.00	61.74		
	B	Inter group	272.14	1.00	272.14	7.45	0.03
		In group	219.09	6.00	36.51		
	C	Inter group	1494.13	1.00	1494.13	8.84	0.03
		In group	1013.81	6.00	168.97		
Forward position	A	Inter group	138.86	1.00	138.86	3.47	0.11
		In group	240.27	6.00	40.05		
	B	Inter group	153.13	1.00	153.13	2.72	0.15
		In group	338.11	6.00	56.35		
	C	Inter group	133.25	1.00	133.25	0.34	0.58
		In group	2374.69	6.00	395.78		
Complexity of spatial layout	A	Inter group	49.95	1.00	49.95	0.91	0.38
		In group	329.18	6.00	54.86		
	B	Inter group	8.69	1.00	8.69	0.11	0.75
		In group	482.54	6.00	80.42		
	C	Inter group	304.18	1.00	304.18	0.83	0.40
		In group	2203.76	6.00	367.29		

Position and Orientation of the Participants. Taking the number of groups, the forward posture and the complexity of indoor space layout as single variables, the variance analysis table under different variable conditions is obtained, and the three test point data are combined, and the results are shown in Table 7.

The significance of group A, B and C was 0.30, 0.19 and 0.33, respectively. The significance of the advanced posture in A,B and C was 0.01, 0.59 and 0.11, respectively. The significance of the spatial layout was 0.96, 0.16 and 0.74, respectively.

Table 7. Position azimuth variance analysis table

Variable	Test-point		Quadratic sum	Free degree	Mean square	F	P-value
Number of teams	A	Inter group	490.94	1.00	490.94	1.29	0.30
		In group	2287.94	6.00	381.32		
	B	Inter group	77.07	1.00	77.07	2.20	0.19
		In group	210.41	6.00	35.07		
	C	Inter group	569.53	1.00	569.53	1.12	0.33
		In group	3056.15	6.00	509.36		
Forward position	A	Inter group	2058.57	1.00	2058.57	17.15	0.01
		In group	720.31	6.00	120.05		
	B	Inter group	14.66	1.00	14.66	0.32	0.59
		In group	272.82	6.00	45.47		
	C	Inter group	1313.28	1.00	1313.28	3.41	0.11
		In group	2312.40	6.00	385.40		
Complexity of spatial layout	A	Inter group	1.12	1.00	1.12	0.00	0.96
		In group	2777.76	6.00	462.96		
	B	Inter group	85.48	1.00	85.48	2.54	0.16
		In group	202.00	6.00	33.67		
	C	Inter group	71.04	1.00	71.04	0.12	0.74
		In group	3554.64	6.00	592.44		

Time. Taking the number of groups, forward posture and the complexity of indoor spatial layout as a single variable, the variance analysis table under different variables is obtained, and the data of the three test points are combined, and the results are shown in Table 8.

The significant values of C and C in A, B and C were 0.88, 0.20 and 0.09, respectively, and the forward posture in A, B and C were 0.01, 0.02 and 0.11, respectively. The complexity of spatial layout was 0.41, 0.66 and 0.50, respectively.

Table 8. Time-of-variance table

Variable	Test-point		Quadratic sum	Free degree	Mean square	F	P-value
Number of teams	A	Inter group	1.62	1.00	1.62	0.03	0.88
		In group	378.30	6.00	63.05		
	B	Inter group	840.50	1.00	840.50	2.04	0.20
		In group	2468.24	6.00	411.37		
	C	Inter group	3096.85	1.00	3096.85	4.01	0.09
		In group	4639.12	6.00	773.19		
Forward position	A	Inter group	262.21	1.00	262.21	13.37	0.01
		In group	117.72	6.00	19.62		
	B	Inter group	2112.50	1.00	2112.50	10.60	0.02
		In group	1196.24	6.00	199.37		
	C	Inter group	2872.82	1.00	2872.82	3.54	0.11
		In group	4863.14	6.00	810.52		
Complexity of spatial layout	A	Inter group	43.25	1.00	43.25	0.77	0.41
		In group	336.68	6.00	56.11		
	B	Inter group	115.52	1.00	115.52	0.22	0.66
		In group	3193.22	6.00	532.20		
	C	Inter group	605.52	1.00	605.52	0.51	0.50
		In group	7130.44	6.00	1188.41		

4 Data Analysis

4.1 Basic Statistics Analysis Results

The data analysis is carried out according to the basic statistics of the distance expressed by the participants, the number of the groups is a single variable, and the mean data of the test points A, B and C are more close to the basic value when the number of the teams is a double experiment, and the two people are more accurate through the communication and the exchange of the data, With the order of the test points, the accuracy of the recording data is also becoming higher and higher. The mean data of the test points A, B and C are the single variable in the forward posture, and the distance is closer to the base value when the forward attitude is the vertical forward experiment, and the accuracy of the record is higher in the form of a space The complexity of the office is a single variable, and the mean data of the test points A, B and C are more close to the basic value when the space layout is a simple experiment, and the accuracy of the recorded data is higher as the test points are in sequence.

According to the data analysis of the basic statistics of the participants, taking the number of groups as a single variable, the mean data of test points A, B and C are closer to the basic values when the number of groups is a double experiment. Taking forward posture as a single variable, the mean data of test point A and B are closer to the basic value when the forward posture is squatting forward experiment, and the data of test point C are closer to the basic value when the forward posture is upright forward experiment. Taking the complexity of spatial layout as a single variable, the mean data of test point A is closer to each other when the spatial layout is complex. Base value, test point B, C is in the space layout as a simple experiment, the expression azimuth is closer to the base value.

According to the basic statistics of the position and azimuth of the participants, taking the number of groups as a single variable, the data of test point A and B are closer to the basic value when the number of groups is a double experiment, and the mean data of test point C are closer to the basic value when the number of groups is a single experiment. In the past, the forward posture was a single variable, and the mean data of C and C were closer to the basic value and more accurate to record the azimuth information when the forward posture was vertical forward experiment, while the position azimuth of test point A was closer to the basic value when the forward posture was squatting forward experiment. In space The complexity of the layout is a single variable. The mean data of the test point A and B are closer to the basic value when the spatial layout is a simple experiment, and the position azimuth of the test point C is closer to the basic value when the spatial layout is a complex experiment.

Taking the number of groups as a single variable, the time for the participants to complete the experiment is faster than the time for the two to complete the experiment together. Taking the forward posture as a single variable, the time to complete the experiment in the squatting posture is faster than that in the upright posture. When the complexity of spatial layout is taken as a single variable, the time to complete the experiment when the spatial layout is complex is faster than that when the spatial layout is simple. Single experiment does not need to communicate with teammates, so it saves more time, double experiment needs team communication and cooperation, so it is more time-consuming than single experiment. When squatting forward, the center of gravity of people is reduced, and there is more sense of security. In 2005,the participants were able to move forward more safely, which may be the reason for the faster completion of the experiment. The obstacles in the experiment are placed at the corner, and the distance of the participants becomes shorter, which may be the reason why the participants complete the experiment faster under the complex spatial layout.

4.2 Analysis of Variance Results

Through the analysis of variance, there was no significant difference in the number of participants in the test points A,B,C and C in the distance expressed by the participants. There was no significant difference in forward posture at test point B and C, but there was significant difference between squatting forward and vertical forward at test point A. because of the different forward posture, the accuracy of recording information was very low at first, but then the difference of accuracy of recording distance information disappeared gradually by adapting to the environment, and the complexity of spatial layout was not significantly different in test point A, B and C. In terms of the orientation of the participants, there was no significant difference in the data of the group at the test point A, but there was no significant difference in the test. There was significant difference between single experiment and double experiment at point B. with the order of test points, the accuracy of azimuth information recording was lower; there was no significant difference in forward posture at test point A,B and C, and the complexity of spatial layout was not significantly different in test point A, B and C. In the position and orientation of the participants, there was no significant difference in the number of groups at the test points A, B and C, and there was no significant difference in the forward posture at the test points B and C.at the test point A, the difference between squatting forward and upright forward was very significant. The difference of the starting position had an impact on the adaptive space of the participants, but with the order of the test points and the influence of the posture. There is no significant difference in the complexity of spatial layout between test points A, B and C. In the completion time of the participants, there was no significant difference in the data of the group at the test points A, B, C, but there were significant differences in the forward posture at the test points A, B and C, and there was no significant difference in the complexity of the spatial layout between the test points A, B and C, and there was no significant difference in the complexity of the spatial layout between the test points A, B and C. Combined with the results of four data analysis, it is found that the number of groups and forward posture have the greatest influence on the recording distance, orientation and time. The more backward the test point, the greater the influence of the number of groups, the closer the test point is, the greater the influence of forward posture is.

5   Conclusion

In order to explore the influencing factors of firefighters' path integration ability in the process of indoor fire search and rescue in non visual environment, and to explore the firefighters' spatial perception ability in different conditions, the researchers designed experiments to simulate the indoor environment of smoke building where fire occurred, and analyzed the influence of the number of group members, forward posture and complexity of spatial layout on firefighters' recording distance and orientation information In order to improve the efficiency and safety of firefighters in the process of indoor search and rescue, the best indoor rescue method is proposed. The research results show that when people search and rescue in the smoke room, firefighters can complete the indoor search and rescue task more efficiently through team cooperation and using the upright forward posture. This is preliminary data processing, and there will be complete research results in the future.

Acknowledgment. The research presented here is part of the "The influence of remote interactive mode of dynamic spatial information on path-finding decision-making in high-rise building fire" project, which is supported by the Beijing Natural Science Foundation. The authors would like to acknowledge the support of the Beijing Natural Science Foundation for this project(9172008) and the National Natural Science Foundation of China (Project No. 71671167).

References

1. Golledge, R.G., Klatzky, R.L., Loomis, J.M.: Cognitive mapping and wayfinding by adults without vision. The Construction of Cognitive Maps (1996)
2. Denef, S., Ramirez, L., Dyrks, T., et al.: Handy navigation in ever-changing spaces-an ethnographic study of firefighting practices (2008)
3. Klann, M., Geissler, M.: Experience prototyping: a new approach to designing firefighter navigation support. IEEE Pervasive Comput. 11(4), 68–77 (2012)
4. Tracey, E.A.: Firefighter workplace learning: an exploratory case study. Proquest Llc 189 (2014)
5. McWhorter, C.: Bachelor of arts. "That all seems more like common sense": an examination of firefighter cognitive maps (2014)
6. Guo, R.-Y., Huang, H.-J., Wong, S.C.: Route choice in pedestrian evacuation under conditions of good and zero visibility: experimental and simulation results. Transp. Res. Part B: Methodol. 46(6), 669–686 (2012)

Risk-Taking Propensity During a Prolonged Voyage at Sea: A Simulator Experiment Study

Xin Wang[1], Liang Zhang[1(✉)], Tuoyang Zhou[1], Zhen Liao[1,2], Zhanshuo Zhang[1], Ning Li[1], Qiang Yao[1], Jin Liang[1], Yang Yu[1], Zhiqiang Tian[3], and Tianqi Chen[3]

[1] China Institute of Marine Technology & Economy, Beijing 100081, China
zhl_thu@163.com
[2] Tsinghua University, Beijing 100084, China
[3] China Institute of Marine Human Factors Engineering, Qingdao 266400, China

Abstract. Long-term isolation and rotating watch-keeping schedule in extended voyage might impact seafarers' risk-taking propensity. The present study aimed to explore how seafarer's risk-taking propensity varies with the prolonged voyage. 12 subjects were recruited to commit a prolonged "voyage" in a self-developed maritime chamber simulator. They lived and worked together following a rotating watch-standing schedule. Social isolation was realized by cutting off the communication between inside and outside world. Balloon Analogue Risk Task (BART) and Sensation Seeking Scale V (SSSV) were adopted as behavioral and self-assessed measures of risk-taking. Participants completed these two assessments before the "voyage" (baseline), at five different time points during the "voyage" and after the "voyage" (post-voyage). Number of explosions and adjusted number of pumps, which represent the riskiness on BART, decreased with the ongoing of the voyage, though the magnitude of the declination is small in size. Accordingly, the scores on SSS and its four subscales, especially the thrill and adventure seeking (TAS) subscale, also declined with the voyage. The results implicated that the sleep deprivation caused by the rotating shift and the stimulation deprivation caused by isolation together would reduce people's riskiness in decision making and the willingness to engage in high-risk and sensational activities as the prolonged voyage going on.

Keywords: Risk-taking propensity · Prolonged voyage · Sleep loss · Social isolation

1 Introduction

1.1 Risk-Taking at Sea

When at sea, the environment can be very complex and change rapidly. The ship should be ready to cope with very unpredictable situations or even some critical disasters, such as fire, environmental pollution, ship hull damage, ship grounding, and some unlawful acts that would threaten the safety of the members on the ship. Therefore, the capacity of seafarers, especially the officer of the watch (OOW) and the master, to make reasonable

© Springer Nature Switzerland AG 2020
P.-L. P. Rau (Ed.): HCII 2020, LNCS 12192, pp. 519–529, 2020.
https://doi.org/10.1007/978-3-030-49788-0_39

decisions and take reasonable actions at risk is crucial for the safety of the ship and all members on it. The trait of risk-taking propensity plays an important role in risk decision making.

Risk-taking was constituted by two components: decision making in a high-level cognitive task (e.g., Balloon Analogue Risk Task, Iowa Gambling Task) and self-assessment of risk-taking propensity (e.g. sensation seeking). Balloon Analogue Risk Task (BART), developed by Lejuez et al. (2002), measures impulsive decision-making as risk-taking. It involves decisions based on evaluation of potential outcomes (i.e., risks and rewards) and choices to risk rewards already gained for the potential of higher rewards. Sensation seeking personality is characterized by the need for varied, novel, complex, and intense sensations and experiences, and the willingness to take physical, social, legal and financial risks for the sake of such experiences (Zuckerman 1994). The experience is needed for subjects to reach their optimal level of stimulation required to maintain the optimal level of arousal. Sensation seeking is related to risk taking in all kinds of risk areas, including driving, crime, financial, social violations, sports and so on (Jonah 1997; Dahlen et al. 2005; Horvath and Zuckerman 1993).

Working at sea, such as on large container ships, is regarded as stressful due to monotony, social isolation, fatigue, stress and rotating shift during the long voyage. Seafarers' mental health, cognitive function, as well as risk-taking propensity might be impaired.

1.2   Effect of Rotating Watch-Keeping Schedule on Risk-Taking

Ships are operated under a rapidly rotating watch-keeping schedule. This irregular working time arrangements will bring problems of fatigue, circadian rhythms, and sleep disruption to seafarers, which may affect their physiological and psychological functions (Eriksen et al. 2006; Colquhoun et al. 1988).

Sleep loss has been proved to induce sleepiness, fatigue and mood changes, and cause impairments in cognitive performance according to a review by Reynolds and Banks (2010). However, there is still no clear consensus about whether sleep deprivation impairs risk-taking. A review by Womack et al. (2013) concluded that sleep loss was overall positively associated with risk-taking behavior. Telzer et al. (2013) even revealed by functional magnetic imaging (fMRI) that poor sleep disrupts brain function related to cognitive control and reward processing. However, studies with different research findings cannot be neglected. Killgore (2007) showed that sleep deprivation decreased risk-taking in both decision making with the balloon analogue risk task and risk propensity with the EVAR scales. In the study of Chaumet et al. (2009), risk-taking propensity decreases during the night of sleep deprivation, and remains stable the following day. Moreover, Demos et al. (2016) showed partial deprivation did not alter the risky decision making assessed with the balloon analogue risk task.

1.3   Effect of Long-Term Social Isolation and Monotony on Risk-Taking

The monotony in the prolonged voyage, restricted possibilities for social contacts, sexual problems, insufficient leisure-time activities have adverse effects on seafarers and have been recognized as risk factors for seafarers' mental health and welfare (Lileikis 2014;

Iversen 2012; Sampson and Thomas 2003). However, knowledge about how social isolation, confinement and monotony affect cognitive functions is still limited. Few studies, especially experimental studies, have investigated the long-term effect of isolation (i.e., social isolation, information isolation) on decision making and risk-taking propensity.

A longitudinal study (Gow et al. 2013) with older adults found that perceived isolation was associated with deficits in cognitive functions, as well as slower processing speed. In the study of Duclos et al. (2013), feeling of isolation and exclusion engendered by a Cyberball game increased financial risk-taking. The authors state that interpersonal rejection exacerbates financial risk-taking by heightening the instrumentality of money (as a substitute for popularity) to obtain benefits in life. An experiment study by Chaumet et al. (2009) showed that impulsiveness increases in a confined environment under a normal sleep-wake schedule, as combined to the baseline.

The present study aimed to explore how seafarer's risk-taking propensity change with the extended voyage under sleep deprivation and isolation conditions (i.e., social isolation, confinement).

2 Method

2.1 Subjects

12 male subjects aged 24−37 years took part in this experiment. This study passed the ethics review [Certificate No: 2019013], and each subject signed on the written informed consent.

2.2 Measures

The Balloon Analogue Risk Task (BART) was applied in this experiment as the objective and behavioral measure of risk-taking. At the start of the BART, the computer screen displayed four items: a small balloon accompanied by a balloon pump, a reset button labeled Collect, a Total Earned display, and a second display labeled Last Balloon that listed the money earned on the last balloon. The subjects can choose between "collect" and "pump" as long as the balloon does not explode. Each click on the pump inflated the balloon a little, and money was accumulated. When a balloon was pumped past its individual explosion point and exploded, all money for this trial was lost. The number of pumps at which the balloon explodes obeys a uniform distribution from 1 to 128. There are a total of 30 balloons (i.e., trials) in an assessment (Lejuez et al. 2002).

Two highly correlated indices, the adjusted number of pumps across balloons (defined as the average number of pumps on balloons that did not explode) and the number of explosions, were adopted to represent participants' riskiness on BART. According to White et al. (White et al. 2008), risky behavior on the BART (adjusted average pumps) has been proved to have acceptable test−retest reliability across days ($r = +.77, p < .001$). Test−retest reliabilities for the present study are higher than $+.77$, as shown in Table 1.

The 40-item self-reported Sensation Seeking Scale V (SSSV) was selected as a subjective measure of risk propensity. The SSS form V has the following subscales: (1) Thrill and Adventure Seeking (TAS), which expresses a desire to engage in sports or other physical activities involving speed or danger; (2) Experience Seeking (ES), represents the seeking of experience through the mind and senses, travel, and a nonconformist lifestyle; (3) Disinhibition (DIS) represents the desire for social and sexual disinhibition as expressed in social drinking, partying, and variety in sexual partners; and (4) Boredom Susceptibility (BS), represents an aversion to repetition, routine, and restlessness when things are not changing (Zuckerman 1994).

Zuckerman et al. (1978) have provided data supporting the internal consistency of the measure, with alpha coefficients ranging from .83 to .86. The alpha coefficient for the current sample was mostly higher than .76, except for the alpha (.64) for the first-time assessment. Test–retest reliabilities shows in Table 1. The correlation coefficients of the total score, TAS and ES were all greater than .69. The test-retest reliability was especially high between the five-time points within the voyage. The reliability coefficient for Disinhibition on the SSS remained above .88 (all significant) across the five assessments within the voyage, but reduced to .42 (not significant) between the last assessment within the voyage and the assessment just after the voyage. The test-retest reliability for Boredom Susceptibility (BS) is poor, with the lowest correlation coefficient as $-.03$.

Table 1. Test-retest reliability: correlations between successive time points

Test	Measures	BL-T1	T1-T2	T2-T3	T3-T4	T4-T5	T5-Post
BART	Explosions	.76**	.85**	.83**	.92**	.83**	.77**
	Pumps	.92**	.98**	.95**	.94**	.96**	.87**
SSS	SSS total score	.87**	.87**	.91**	.90**	.91**	.80**
	Thrill and Adventure Seeking	.73**	.96**	.97**	.94**	.95**	.97**
	Disinhibition	.67*	.88**	.92**	.91**	.91**	.42
	Experience Seeking	.92**	.69*	.73**	.73**	.86**	.71*
	Boredom Susceptibility[b]	.43	.66*	.61*	.52	.12	−.03

[b]Spearman rho correlation coefficient
**significance level < .01
*significance level < .05

2.3 Procedures

The 12 subjects committed a prolonged "voyage" in a self-developed maritime chamber simulator. They lived and worked in the chamber following a rotating watch-keeping schedule. The communication and network connection with the outside was cut off, and subjects were provided no chances to meet anyone else except the other subjects in the chamber. These manipulations are to simulate the social isolation and monotony on real oceangoing ships. Subjects completed BART and SSS right before the "voyage" (baseline), at 5 evenly distributed time points during the "voyage", and just after the "voyage" (post-voyage).

2.4 Statistics

Repeated measures ANOVA or Friedman ANOVA with SPSS 26.0 was adopted to test
the time effect based on the normality of the data at all the timepoints. Since there are
7 levels of treatment, post-hoc with Bonfferoni correction will inflate the type 2 error.
Therefore, post-hoc with Tuckey's HSD was used. Studentized Range Statistic (q) for
Tuckey test is directly tied to the t statistic. Since there is no Tuckey test in SPSS for
within subject measures. we converted the paired t test on the means to a q statistic, by
multiplying t by the square root of 2.

3 Results

3.1 Effect of Time on Riskiness on Balloon Analogue Risk Task

As illustrated in Fig. 1 and Fig. 2, the overall trend for both the number of explosions
and the adjusted number of pumps for BART is downward.

With a relatively large number of explosions from baseline to the 2nd time point,
the declination in explosions mainly occurred from the 2nd time block to the 4th time
block within the voyage. According to the repeated measures ANOVA, the differences
between the time blocks did not reach significance. Post-hoc analysis with Tuckey's HSD
correction revealed that the reduction in number of explosions from the 2nd time block
to the 4th time block reached marginal significance, $(14.58 \pm 5.68$ vs. 12.13 ± 5.50,
respectively, $q(7, 11) = 4.872$, $p = .062$). No other significance was found between the
other time blocks.

The number of pumps exhibited a similar but more modest downward trend as
compared with the number of explosions. Repeated measures ANOVA and post hoc
analysis with Tuckey's HSD correction revealed no significant differences between the
time points.

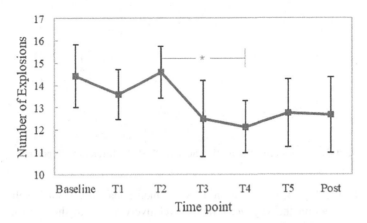

Fig. 1. Number of explosions in BART at different time points (*significance level < .05)

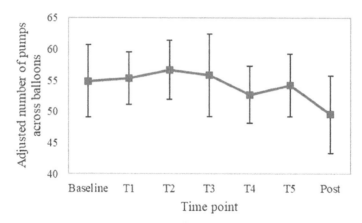

Fig. 2. Adjusted number of pumps across balloons in BART at different time points

3.2 Effect of Time on Sensation Seeking

The means and standard errors for total score on SSS at each time point are shown in Fig. 3, from which an overall declining trend can be found. A repeated measures ANOVA with a Greenhouse-Geisser correction determined that the differences in the total score on SSS between the 7 time points reached marginal statistical significance ($F(1.335, 14.684) = 3.650$, $p = .066$). However, post-hoc analysis corrected by Tuckey's HSD did not reveal any significance on the paired differences.

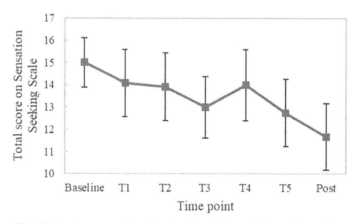

Fig. 3. Total score on Sensation Seeking Scale at different time points

According to Fig. 4, the score on TAS declined apparently from Baseline to the 1st time point within the voyage, remained relatively stable to the 3rd time block, then declined steadily afterwards until the post-voyage time point. A repeated measures ANOVA with a Greenhouse-Geisser correction determined a significant effect of time on the TAS ($F(1.268, 12.947) = 4.776$, $P < .039$). Post hoc tests using the Tuckey's

HSD correction revealed that the TAS reduced significantly from 6.00 (S.D. = 2.92) in the 3rd time block to 5.42 (S.D. = 3.03) in the 5th time block within the voyage (q(7, 11) = 5.549, p = .029), and further reduced significantly to 4.92 (S.D. = 2.94) after the voyage (q(7, 11) = 5.327, p = .038).

The difference between the 1st voyage time point and the baseline did not reach significance. After checking the data separately for all subjects, the declination in TAS was found to be mainly formed by the sharp drop of two subjects. Change in the TAS score for the other subjects was relatively slight and the direction of change (upward, downward or unchanged) varied among subjects.

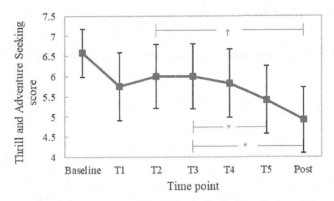

Fig. 4. Score on Thrill and Adventure Seeking in Sensation Seeking Scale at different time points (*significance level < .05; ✝ significance level < .1)

After a quick increase from the baseline to the 1st time point, the score on disinhibition exhibited an overall declining trend (see Fig. 5). However, examination of time effect with a repeated measures ANOVA corrected by Greenhouse-Geisser determined that the differences on Disinhibition (DIS) between time points did not reach statistical significance.

A repeated measures ANOVA with a Greenhouse-Geisser correction determined no significant overall effect of time on mean ES. Post hoc tests using Tuckey's HSD correction revealed that ES of the first time block reduced to 2.58 ± 1.51, which was statistically significantly different from baseline (3.36 ± 1.69, q(7, 11) = 5.911, p = .020). However, the ES score in the remaining assessments remains at the low level of the first time block as shown in Fig. 6, and no significance was reached between any pair of the remaining assessments.

The means and standard errors for total score on SSS at each time point are shown in Fig. 7. Repeated measures Friedman ANOVA determined that no significant overall effect of time on mean BS ($\chi^2(6) = 12.159$, p = .059).

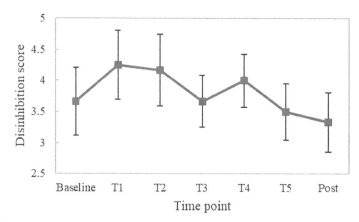

Fig. 5. Score on Disinhibition in Sensation Seeking Scale at different time points

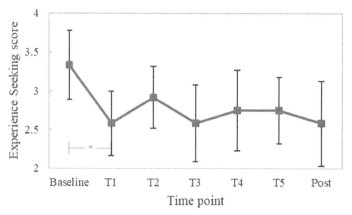

Fig. 6. Score on Experience Seeking in Sensation Seeking Scale at different time points (*significance level < .05)

4 Discussion

An overall descending trend was observed on both the behavioral and self-assessed risk-taking propensity in the subjects with the proceeding of the simulated "voyage", characterized by rotating watch-keeping schedule and isolation. This declination might be results of the combined effect of both sleep loss and isolation. However, it is difficult to tell whether they are joint effects or counteracting ones, since little or inconsistent knowledge is known about the effect of each factor. The authors try to elaborate on these effects based on related research findings to investigate the implications of this declination for further research.

The risk-taking propensity, as depicted by sensation seeking, represents the optimal level of stimulation the subjects need to maintain their optimal level of arousal. The information isolation in the prolonged voyage is essentially the deprivation of stimulation. As compensation for the lack of stimulation, people would seek for new sensations,

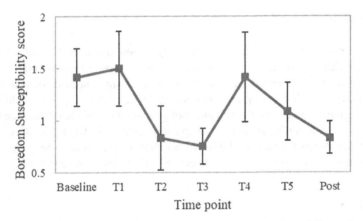

Fig. 7. Score on Boredom Susceptibility in Sensation Seeking Scale at different time points

as shown in the studies of Zuckerman et al. (1996) and Ha and Jang (2015). However, the isolated condition in the present experiment make opportunities of new sensations very limited, and the repeated assessments of the test and scale could not serve as new stimulus at all. Hence, a compromise, which means reducing the need for stimulation, might be adopted by the subjects. Reduction in the scores of Sensation Seeking Scale might be the reflection of this compromise. This could be regarded as an adaptation to the conditions.

Contradicted with the overall positive effect of sleep loss on risk-taking according to the review by Reynolds and Banks (2010), sleep loss caused by the rotating shift in the present study was correlated with less risk propensity. One possible causation of this contradiction might the differences in the form and duration of sleep loss between the present study and many other studies. The majority of studies focused on the total sleep deprivation, which means elimination of sleep for a period of time and no chances to catch up on sleep. The rotating watch-keeping schedule in the present study, on the other hand, provided people with reasonable duration of the sleep time window, but the time window changed every day. Besides, the sleep loss in previous experimental studies only lasted for several days, which could not reveal the long-term effect as shown in the present study. The other two possible reasons for this contradiction might be that the negative effect of isolation suppressed that of sleep loss, and that the declination in the number of pumps is due to loss of interest or a sense of futility rather than risky behavior in itself.

The other thing to be addressed is the low significance of the time effect on the risk-taking propensity. One straight forward reason for this might be that the magnitude of the time effect itself is small in size. The other reason is that, even though the overall trend might be downward for the majority of subjects, the descending process is fluctuated, other than straight and steady. This would result in great individual differences in the change directions between two specific time points, which impact the significance.

5   Conclusion

An experimental study was conducted to investigate how seafarers' risk-taking propensity change with the proceeding of extended voyage characterized by rotating watch-keeping schedule and isolation (i.e. social isolation, information isolation and information isolation). Both behavioral and self-assessed risk-taking propensity exhibited an overall descending trend with the voyage, though magnitude of the declination seemed to be small in size. It implicated that the sleep deprivation caused by the rotating shift and the stimulation deprivation caused by isolation together would reduce people's riskiness in decision making and the willingness to engage in high-risk and sensational activities.

References

Chaumet, G., et al.: Confinement and sleep deprivation effects on propensity to take risks. Aviat. Space Environ. Med. **80**, 73–80 (2009)

Colquhoun, W.P., et al.: Work at sea: a study of sleep, and of circadian rhythms in physiological and psychological functions, in watchkeepers on merchant vessels. Int. Arch. Occup. Environ. Health **60**, 321–329 (1988)

Dahlen, E.R., Martin, R.C., Ragan, K., Kuhlman, M.M.: Driving anger, sensation seeking, impulsiveness, and boredom proneness in the prediction of unsafe driving. Accid. Anal. Prev. **37**, 341–348 (2005)

Demos, K.E., et al.: Partial sleep deprivation impacts impulsive action but not impulsive decision-making. Physiol. Behav. **164**, 214–219 (2016)

Duclos, R., Wan, E.W., Jiang, Y.: Show me the honey! Effects of social exclusion on financial risk-taking. J. Consum. Res. **40**, 122–135 (2013)

Eriksen, C.A., Gillberg, M., Vestergren, P.: Sleepiness and sleep in a simulated "six hours on/six hours off" sea watch system. Chronobiol. Int. **23**, 1193–1202 (2006)

Gow, A.J., Corley, J., Starr, J.M., Deary, I.J.: Which social network or support factors are associated with cognitive abilities in old age? Gerontology **59**, 454–463 (2013)

Ha, J., Jang, S.(Shawn): Boredom and moderating variables for customers' novelty seeking. J. Foodserv. Bus. Res. **18**, 404–422 (2015)

Horvath, P., Zuckerman, M.: Sensation seeking, risk appraisal, and risky behavior. Personality Individ. Differ. **14**, 41–52 (1993)

Iversen, R.T.B.: The mental health of seafarers. Int. Marit. Health **63**, 78–89 (2012)

Killgore, W.D.S.: Effects of sleep deprivation and morningness-eveningness traits on risk-taking. Psychol. Rep. **100**, 613–626 (2007)

Lejuez, C.W., et al.: Evaluation of a behavioral measure of risk taking: the Balloon Analogue Risk Task (BART). J. Exp. Psychol. Appl. **8**(2), 75 (2002)

Lileikis, S.: What kind of leadership do seafarers need in regard to their main emotional states caused by the physical and psychosocial maritime work environment. J. Marit. Trans. Eng. **3**(2), 24–33 (2014)

Reynolds, A.C., Banks, S.: Total sleep deprivation, chronic sleep restriction and sleep disruption. In: Kerkhof, G.A., van Dongen, H.P.A. (eds.) Progress in Brain Research, vol. 185, pp. 91–103. Elsevier, New York (2010)

Sampson, H., Thomas, M.: The social isolation of seafarers: causes, effects, and remedies. Int. Marit. Health **54**, 58–67 (2003)

Telzer, E.H., Fuligni, A.J., Lieberman, M.D., Galván, A.: The effects of poor quality sleep on brain function and risk taking in adolescence. NeuroImage **71**, 275–283 (2013)

White, T.L., Lejuez, C.W., de Wit, H.: Test-retest characteristics of the Balloon Analogue Risk Task (BART). Exp. Clin. Psychopharmacol. **16**(6), 565 (2008)

Womack, S.D., Hook, J.N., Reyna, S.H., Ramos, M.: Sleep loss and risk-taking behavior: a review of the literature. Behav. Sleep Med. **11**, 343–359 (2013)

Zuckerman, M.: Behavioral Expressions and Biosocial Bases of Sensation Seeking. Cambridge University Press, New York (1994)

Jonah, B.A.: Sensation seeking and risky driving: A review and synthesis of the literature. Accid. Anal. Prev. **29**, 651–665 (1997)

Zuckerman, M., Eysenck, S.B., Eysenck, H.J.: Sensation seeking in England and America: cross-cultural, age, and sex comparisons. J. Consult. Clin. Psychol. **46**, 139–149 (1978)

Zuckerman, M., Persky, H., Hopkins, T.R., Murtaugh, T., Basu, G.K., Schilling, M.: Comparison of stress effects of perceptual and social isolation. Arch. Gen. Psychiatry **14**(4), 356–365 (1966)

The Influence of a Long Voyage on Mental Status: An Experimental Study

Yang Yu[1], Zhanshuo Zhang[1], Jin Liang[1], Zhiqiang Tian[3(✉)], Chi Zhang[1], Qiang Yao[1], Ning Li[1], Tuoyang Zhou[1], Xin Wang[1], Zhen Liao[1,2], and Baochao Zong[3]

[1] China Institute of Marine Technology and Economy, Beijing 100081, China
[2] Tsinghua University, Beijing 100084, China
[3] China Institute of Marine Human Factors Engineering, Qingdao 266400, China
tianzhiqiang2000@163.com

Abstract. Objective: The current study aimed to explore how did the long voyage affect the mental status in the experimental environment and construct a quantitative relation model of the mental status during a long voyage. **Method:** To investigate participants' mental health state during a long simulated long voyage, 12 participates were equally divided into three groups and were instructed to live in a simulated sailing cabin. The cabin provided a narrow and closed space and a social isolation environment. Symptom checklist-90 (SCL-90), Minnesota Multiphasic Personality Inventory-2 (MMPI-2) were adopted to evaluate participates' anxiety, depression, somatic symptoms and sleep, fatigue and stress status respectively before and after the long voyage. The positive effect and negative affect schedule (PANAS) was used to investigate the daily change of positive and negative emotions in the whole simulated voyage period. These three measurements reveal the change of mental status from different perspectives before, during, after a long voyage. **Results:** The pretest scores of somatization were lower than the post-test scores of somatization on SCL-90. The post-test T-scores of depression and paranoia were significantly higher than those of pretest(P < 0.05) on MMPI. The change of the positive emotion could be described as a quadratic polynomial model, and the change of the negative emotion could be described as a cubic polynomial model. **Conclusions:** The long voyage had a significant influence on participants' state of mind and mood, which resulted in increased negative emotion and declined positive emotion, and brought out depression and Paranoia. The positive and negative emotions showed different trends during the long term simulated voyage and different starting times of recovery, but the time of positive emotions recovering to the level of baseline after the voyage was very close to the recovery time of negative emotions after the voyage.

Keywords: A long voyage · Mental status · Before-after experimental study

1 Introduction

The health of the crews had a great influence on the safety of a long term voyage, so it was necessary to investigate the physiological and psychological change during

© Springer Nature Switzerland AG 2020
P.-L. P. Rau (Ed.): HCII 2020, LNCS 12192, pp. 530–539, 2020.
https://doi.org/10.1007/978-3-030-49788-0_40

the whole voyage. **Lu et al.** [1] adopted a data classification approach to present a computational study, in which they used 41 chemistry indicators to measure seamen' blood samples collected before and after the sailing, they found that nine of 41 indicators were significantly affected during the sailing.

Some researchers had demonstrated that people show physiological adaptation to the environment. For example, seasickness was a common phenomenon during a long voyage. **LI et al.** [2] investigated that the habituation of the seasickness occurred in a long voyage, the prevalence and degree of seasickness started to decrease after two weeks. The result inevitably raised the question of whether it was possible that the habituation occurred in some aspects of psychological functioning. **Xia et al.** [3] examined the psychological and physiological effects on Chinese female seamen before, during, and after long navigation, they found that the scores of SCL-90 psychological evaluation of somatization, anxiety, paranoia were significantly higher following navigation than before the long voyage. This voyage was the first time for female seamen to go to sea that lasted for 4 months.

During the long-term voyage, the sailors underwent enormous mental load because of their special working environment and tense tasks. They were liable to undertake huge psychological stress under such special environmental conditions at sea. **Zhang** [4] **and Zhang et al.** [5] reported the general state of Chinese seamen's mental health was worse than the general population. A self-administered questionnaire was adopted by **Kamada et al.** [6] seven times to collect the data before and during the voyage, it reported that complaints of neurotic symptoms and fatigue were reported to be much greater during the 40-day voyage than those in a general occupational setting, and "quality of sleep" was significantly correlated with neurotic symptoms for the crews. **Jirásek and Hurych** [7] proposed a five-factor model to exam spiritual health in a proper framework. These five factors were the authentic mode of existence, relationship with other people, understanding of the world of nature, question of the purpose and meaning of life, transition and transcendence. **Yu et al.** [8] proposed that long sailing has a bad influence on the mental health of crews, and they found the total score of SCL-90 was mainly related to sleep, fatigue, working pressure, mental quality and ship adaptability based on multivariate linear regression analysis. However, the psychological change of crews didn't get the attention as it should deserve. This research mainly explored the correlation between the mental health of ocean seamen and the voyage to investigate the effects of a long voyage on their mental health.

Effective measures could be taken to intervene in the crew's emotional changes. The controlled study of **Norris and Weinman** [9] has shown statistically significant improvements in self-esteem and coping in the long sail training voyage. In their study, volunteers completed various psychometric questionnaires before and after the transatlantic voyage, the whole voyage takes 3 months and the longest time between ports was 26 days in this case. There was a trend toward reduction in psychological distress after the long sail training voyage.

There were too many disadvantageous environmental factors which would influence mental health, for example, the narrow in closed space, noise, radiation, air pollution, dissimilar cycle of day and night et al. And there are physiological factors mariners need to face, for instance, disturbances in biological rhythm, sleep disorder, social isolation and barrier of personal communications et al. These factors would cause mental stress to mariners during a long voyage, severely influenced their psychological state, increased anxiety and depression, induced cognitive competence and operational performance, increase their error rates of decision-making and operations, and even cause more serious consequences. However, the previous studies rarely concern and control these environmental and social factors. Therefore, it was necessary to explore how did the long voyage affect the mental status in the experimental environment and construct a time prediction model of the mental status during a long voyage.

2 Methods

2.1 Experimental Design

A before-after experimental study and comparisons among homogeneous groups were used to investigate their mental health status before and after a long simulated voyage. 12 people were randomly enrolled to participate in this experiment. Participants filled in the SCL-90, the MMPI-2, and the PANAS in accordance with a procedure that has been approved by the local Ethics Committee [Certificate No: 2019013]. They were equally divided into three groups to live in a simulated ship. This ship provided narrow and closed space, and social isolation to simulate the environment during a long voyage.

2.2 Participants

The study recruited 12 Chinese male seamen that live in a simulated ship with several functional cabins. The 12 male crewmembers were selected in compliance with the standards of the study subjects. Participants were between 24 and 37 years-old and no drug dependence, no infection history, no mental illness, no hereditary diseases, no history of severe allergies, no chronic diseases that require sustained medication, no neurogenic or functional lesions. They also did not have any sleep disturbance nor receive psychotherapy.

2.3 Measures

Such as Symptomchecklist-90 (SCL-90), Minnesota Multiphasic Personality Inventory-2 (MMPI-2), were adopted to evaluate their subjective experience in anxiety and depression, self-rating somatic symptoms and sleep, fatigue and stress status respectively before and after a long voyage. The positive effect and negative affect schedule (PANAS) were used to investigate the change of positive and negative emotions during the whole experiment period. All surveys had no time limit.

SCL-90. The Symptom Checklist (SCL-90) [10] was a go-item self-report inventory. It was composed of 90 questions, including 10 sub-scales such as somatization, obsessive-compulsive, symptom, depression and paranoia. Because of its convenient checklist format and its multidimensionality, it had been adopted in many scientific studies since it was originated in the U.S. in the early 1970s. **Hafkenscheid** [10] believed that SCL-90 was a valuable instrument to detect mental states for the general population. An online test was conducted to the volunteers before and after the simulated voyage.

Minnesota Multiphasic Personality Inventory-2 (MMPI-2). The MMPI-2 [11] is the widely researched instrument as a multidimensional self-report inventory of psychopathology. **Williams et al.** [11] reported that one of the most valuable effects of the MMPI-2 was to provide information about whether the examinee cooperated with the test administration. The mainstay of the MMPI-2 were the empirically derived Clinical Scales: Scale 1 Hypochondriasis (Hs); Scale 2, Depression (D); Scale 3, Hysteria (Hy); Scale 4, Psychasthenia (Pt); Scale 5, Psychopathic Deviation (Pd); Scale 6, Masculinity/Femininity (Mf); Scale 7, Paranoia (Pa); Scale 8, Schizophrenia (Sc); and Scale 9, Mania (Ma); Scale 10, Social Introversion (Si).

PANAS. Because of its state-like and trait-like formats, the Positive and Negative Affect Schedule (PANAS) [12] became a widely used adjective-based questionnaire to measure positive affect (PA) and negative affect (NA) in diverse research projects. The PANAS comprises 10 items measuring PA and 10 items measuring NA. The whole simulated voyage was divided into 6 stages according to the watch system. The subjects needed to complete the PANAS several times in every stage. The average scores of all subjects were used to evaluate positive emotion and negative emotion.

2.4 Statistical Analyses and Fitting Analysis

The collected data were processed and analyzed by the software of Statistic Package for Social Science (SPSS) 21.0. SCL-90 Questionnaire test and MMPI-2 Questionnaire test were expressed as mean ± SD. The descriptive statistics, one-way ANOVA and the paired-samples T-tests were adopted for this study in statistical approaches, where $P < 0.05$ was considered statistically significant. As for PANAS tests, two fitting analysis models were applied to explore the rule of PA and NA change, respectively.

3 Results and Discussion

3.1 Mental Health Assessment Based on SCL-90

Table 1 presents SCL-90 information about the sample as a whole (N = 12). The SCL-90 had 10 dimensions, the total score of the pretest was less than that of the posttest, but there was no significant difference between pretest and posttest ($P > 0.05$). The average score of the posttest had improved 0.02 comparing to the pretest, and the standard deviation of the posttest decreased, which implied the average level improved. The pretest scores of the dimensions in somatization, depression, terror, paranoia, sleeping and eating were less than those of posttest, the pretest scores of the dimensions in compulsion, interpersonal relationship, anxiety, hostility, psychosis were higher than those of posttest, but there was no statistically significant difference between pretest and posttest ($P > 0.05$).

Any score greater than 2 was considered abnormal. The detection rate of psychological abnormalities in the dimension of sleeping and eating after the long term simulated voyage was 0.42. The detection rate before the long term simulated voyage was 0.08. It indicated that volunteers had a higher rate in the dimension of sleeping and eating after the long term simulated voyage than before. The detection rates of psychological abnormalities in other dimensions of SCL-90 were less than 0.17. Therefore, sleeping and eating was a sensitive factor reflecting the mental health of the crews on a long voyage.

The pretest scores in all dimensions of SCL-90 were not significantly different from the posttest. The reasons were manifold: Firstly, the duration of the long voyage might be too short, the mental health had no obvious change, or the duration of the long voyage might be too long, the volunteers had adapted the simulated environment, the mental health had already recovered to a normal level after the voyage. Secondly, the simulated environment was a lack of emergency conditions and life-threatening situations. However, it was very difficult to clarify which one was dominant.

Table 1. SCL-90 information about the sample (N = 12).

SCL-90 factors	Before a long voyage		After a long voyage		t value	p-value
	Average score	Standard deviation	Average score	Standard deviation		
Somatization	1.16	0.23	1.29	0.38	−1.31	0.22
Compulsion	1.43	0.45	1.43	0.36	0.00	1.00
Interpersonal relationship	1.43	0.38	1.26	0.33	2.01	0.07
Depression	1.15	0.32	1.17	0.27	−0.51	0.62
Anxiety	1.23	0.45	1.29	0.31	−0.94	0.37
Hostility	1.31	0.33	1.24	0.24	0.79	0.45
Terror	1.17	0.38	1.20	0.38	−1.40	0.19
Paranoia	1.14	0.15	1.21	0.38	−0.69	0.50
Psychosis	1.20	0.27	1.10	0.16	1.69	0.12
Sleeping and eating	1.31	0.52	1.60	0.58	−1.65	0.13
Total score	112.08	27.27	113.75	25.85	−0.29	0.78
Average score	1.25	0.30	1.26	0.29	−0.31	0.77

3.2   Mental Health Assessment Based on MMPI-2

Table 2 presented MMPI-2 information about the sample as a whole (N = 12). The MMPI-2 could be divided into two parts: 4 validity scales and 10 clinical scales. Based on items included in the test, scores were calculated for both validity scales and clinical scales. The results revealed significant differences on one validity scale and two clinical scales. The results were also presented below in graphical form (Fig. 1).

The Minnesota Multiphasic Personality Inventory-2 (MMPI-2) has been shown as a useful measure of psychopathology in extensive projects. With reference to the recommended criteria of **Graham (2006)** [13], MMPI-2 profiles were validated. The posttest score of the dimension in lying was significantly less than that of pretest (P < 0.05). Compared with the pretest, the participants had higher F scores on the validity scales in the posttest, while the participants had lower L scores. F scores indicated that there were more manifestations of psychosis after the long term simulated voyage than before. L scores indicated that the volunteers had more psychologically defensive in the pretest than in the post-test in order to participate in the experiment successfully.

The posttest T-scores of the dimensions in depression and Paranoia were significantly higher than those of pretest (P < 0.05). The other posttest T-scores of the clinical scales were higher than those of pretest, but there was no statistically significant difference between pretest and posttest (P > 0.05). In the clinical scale, the participants agreed to more items in all subscales of MMPI-2, indicating that the participants had more neurosis and psychotic symptoms after the long term simulated voyage than before the voyage.

High scorers of the dimensions in depression were often associated with depressed mood, low self-esteem, lethargy, and feelings of guilt. The posttest T-scores of depression scale were significantly higher than pretest (P = 0.03), which demonstrated that the long term voyage adversely affected the mental health of crew members and increased the risk of depression among crew members. It's worth noting that D scale was also associated with both psychotic disorders and neurotic disorders. Based on the perspective of clinical practice, depression was a clinical symptom. It could occur not only in severe mood, depression and bipolar depression, but also in schizophrenia, and even obsessive-compulsive disorder and anxiety disorder could be accompanied by depression.

Paranoia Scale elevations were usually associated with being suspicious, aloof, guarded, and overly sensitive. The posttest T-scores of the paranoia scale were significantly higher than pretest (P = 0.04), which indicated that the participants may be more likely to externalize blame and hold grudges after the long voyage.

3.3 Mental Health Assessment Based on PANAS

The whole simulated voyage had 6 stages. The subjects needed to complete the PANAS several times in every stage. The average scores of all subjects in every stage were used to evaluate positive emotion and negative emotion during the long term voyage. The results of positive emotion change and negative emotion change during the long voyage were shown in Fig. 2 and Fig. 3, respectively. According to the recorded data, the minimum value of the positive emotions was in the 5th stage during the whole voyage, the minimum value was 18.25. On the basis of the recorded data, the maximum value of the negative emotions was in the 6th stage during the whole voyage, the minimum value is 12.27.

Based on all recorded experimental data of PANAS, the fitting of a polynomial was used to analyze the change of positive emotion and negative emotion during the long simulated voyage. The red lines in Fig. 2 and Fig. 3 were the trend lines of positive and negative emotions, they were obtained by fitting different models. The trend lines projected forward for one stage were figured to show the recovery of positive and negative emotions.

Table 2. MMPI-2 information about the sample (N = 12).

MMPI-2 factors	Before a long voyage		After a long voyage		t value	p-value
	Average score	Standard deviation	Average score	Standard deviation		
Q score	4.25	8.68	0.58	1.16	1.53	0.15
L-Lying*	54.16	9.73	49.20	9.19	3.19	0.01*
F-infrequency	39.65	6.16	43.05	5.91	−1.70	0.11
K-correction	58.40	10.76	58.58	10.19	−.067	0.94
Hypochondriasis	61.06	4.10	64.84	10.59	−1.25	0.23
Depression*	39.11	5.56	42.30	5.09	−2.60	0.03*
Hysteria	50.49	5.05	51.26	8.64	−0.39	0.70
Psychasthenia	59.25	3.80	61.84	5.02	−1.93	0.07
Psychopathic deviation	61.55	7.98	66.21	9.09	−1.71	0.11
Masculinity/ femininity	49.02	8.62	51.29	9.04	−1.01	0.33
Paranoia*	41.47	4.55	44.88	6.51	−2.34	0.04*
Schizophrenia	53.27	3.36	55.74	4.23	−1.58	0.14
Mania	49.49	8.152	51.59	6.07	−1.21	0.24
Social introversion	37.51	10.32	40.66	10.70	−1.12	0.28

* indicates that there is a significant difference before and after test.

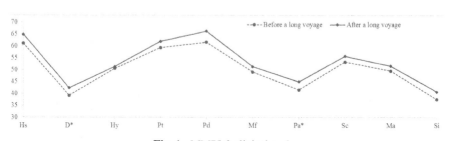

Fig. 1. MMPI-2 clinical scales

As for positive emotion, the trend line was calculated by the quadratic polynomial interpolation and minimum mean square error fitting. The time prediction model was described by

$$y = 0.02019t^2 - 0.8399t + 28.33 + \varepsilon \ \left(R^2 = 0.6745\right) \tag{1}$$

Based on this fitting model, the score of positive emotion gradually declined in the first 4 stages, gradually increased in the remaining 2 stages, and the minimum score of positive emotion was 19.60. The model predicted that participants still needed almost 2 stages after the long simulated voyage to adjust their positive emotions to the level of baseline.

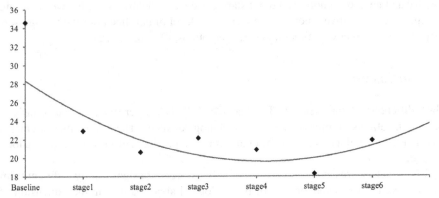

Fig. 2. Positive emotion change during the long voyage (The red line is the trend line) (Color figure online)

In regard to negative emotion, the trend line was calculated by the cubic polynomial interpolation and minimum mean square error fitting. The time prediction model was described by

$$y = -0.0001581t^3 + 0.006199t^2 + 0.009890t + 10.44 + \varepsilon \left(R^2 = 0.9204 \right) \quad (2)$$

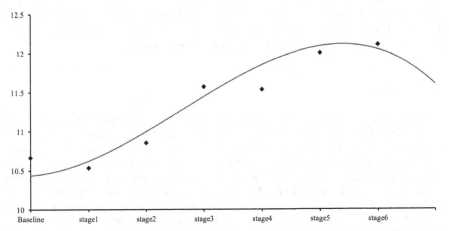

Fig. 3. Negative emotion change during the long voyage (The red line is the trend line) (Color figure online)

According to this prediction model, the score of negative emotion gradually declined in the whole voyage, and the maximum score of negative emotion was 12.10. The model predicted that participants still also needed 2 stages after the simulated long voyage to adjust their negative emotions to the level of baseline.

Comparing the changes of positive and negative emotions, this study found that the positive and negative emotions did not start to recover synchronously. It's worth noting that the time of positive emotions recover to the level of baseline after the voyage was very close to the recovery time of negative emotions after the voyage.

4 Conclusion

Three different measurements, SCL-90, MMPI-2, PANAS, were used simultaneously to explore the effects of long voyages on psychological states. These three measurements reveal the change of mental status from different perspectives before, during, after a long voyage.

The pretest scores in all dimensions of SCL-90 were not significantly different from the posttest, but the detection rate of psychological abnormalities in the dimension of sleeping and eating after the long term simulated voyage was very high. Therefore, the sleeping and eating scale in dimensions of SCL-90 could be regarded as a sensitive factor reflecting the mental health of the crews on a long voyage.

Based on MMPI-2, this research caught sight of that the mental behavioral disorders of mariners were mainly expressed as depression and Paranoia, and that long voyages influence their mental health. Because the participants agreed to more items in all clinical subscales of MMPI-2, it indicated that the participants had more neurosis and psychotic symptoms after the long term simulated voyage than before the voyage.

PANAS could be used as an effective tool to measure the mental status of a long voyage. The change of the positive emotion could be described as a quadratic polynomial model, and the change of the negative emotion could be described as a cubic polynomial model. Though the positive and negative emotions did not start to recover synchronously, the time of positive emotions recovered to the level of baseline after the voyage was very close to the recovery time of negative emotions after the voyage.

References

1. Lu, Y., et al.: A study of health effects of long-distance ocean voyages on seamen using a data classification approach. BMC Med. Informat. Decis. Mak. **10**(1), 1–7 (2010)
2. Li, J.R., Zhu, L., Yuan, W., et al.: Habituation of seasickness in adult during a long voyage. Chinese J. Otorhinolaryngol Head Neck Surg. **47**(8), 642–645 (2012)
3. Xia, S.Y., Zhang, K., Shang, W., et al.: Psychological and physiological effects of a long voyage on female seamen in China. Int. J. Clin. Exp. Med. **9**(4), 7314–7321 (2016)
4. Zhang, M.: An investigation on personality and mental health of round-the-globe sailors. Acta Academiae Medic Inae Qingdao Unive Rs Itatis **40**(3), 265–266 (2004)
5. Zhang, X.L., Gao, Y.D., Zhou, Y.L., et al.: Study of the general state of Chinese Seamen' s Mental Health. Navig. China **64**(3), 72–77 (2005)
6. Kamada, T., Iwata, N., Kojima, Y.: Analyses of neurotic symptoms and subjective symptoms of fatigue in seamen during a long voyage. Sangyo igaku. Japanese J. Ind. Health **32**(6), 461–469 (1990)

7. Jirásek, I., Hurych, E.: Experience of long-term transoceanic sailing: Cape Horn example. J. Outdoor Recreat. Tour. **28**, 1–9 (2019)
8. Yu, H.Z., Yu, H.B., Wang, Y.X.: Analysis of mental health status and its influencing factors of navy crews during long-term sailing. Acad. J. Second Mil. Med. Univ. **38**(4), 506–510 (2017)
9. Norris, R.M., Weinman, J.A.: Psychological change following a long sail training voyage. Personality Individ. Differ. **21**(2), 189–194 (1996)
10. Hafkenscheid, A.: Psychometric evaluation of the symptom checklist (SCL-90) in psychiatric inpatients. Personality Individ. Differ. **14**(6), 751–756 (1993)
11. Williams, C.L., Butcher, J.N., Paulsen, J.A.: Overview of multidimensional inventories of psychopathology with a focus on the MMPI-2. In: Handbook of Psychological Assessment, pp. 397–417. Academic Press (2019)
12. Seib-Pfeifer, L.E., Pugnaghi, G., Beauducel, A., et al.: On the replication of factor structures of the Positive and Negative Affect Schedule (PANAS). Personality Individ. Differ. **107**, 201–207 (2017)
13. Graham, J.R.: MMPI-2: Assessing Personality and Psychopathology. Oxford University Press, Oxford (2006)

Experimental Analysis of Cultural Factors on Trust in Global Supply Chain Management

Fansheng Zhou[1], Pin-Hsuan Chen[2(✉)], Ta-Ping Lu[1], and Pei-Luen Patrick Rau[2]

[1] Sichuan University-Pittsburgh Institute, Sichuan University, Chengdu, China
2019225025130@stu.scu.edu.cn, robertlu@scupi.cn
[2] Department of Industrial Engineering, Tsinghua University, Beijing, China
cpx19@mails.tsinghua.edu.cn, rpl@mail.tsinghua.edu.cn

Abstract. Trust influences the stability and development of the global supply chain and is a unique phenomenon across different cultures. This study aimed to investigate how culture influences trust and the consequential global supply chain through the computerized supply chain management experiment, which was conducted with a two-echelon supply chain involving retailers and wholesalers. In the experiment, forty-two participants with supply chain management background from seventeen countries were asked to take the role of wholesalers in the supply chain and to develop ordering strategies in response to the quantity forecast generated by a computer. The uncertainty avoidance index proposed by Hofstede was applied to quantify the definition of culture in this study. Results indicated that participants from countries with low uncertainty avoidance index present a higher trust level than those with a high uncertainty avoidance index. This study can offer managers a theoretical backup on how to best utilize cultural information to develop and maintain successful global supply chain relationships.

Keywords: Uncertainty avoidance · Cross-cultural · Supply chain management · Decision making · Trust

1 Introduction

Cross-cultural study in supply chain management has gained more scholarly interests as international business grows. Establishing and maintaining strong relationships would lead to a successful global supply chain, and it depends on the right selection of channel members [1]. Therefore, building a trustworthy relationship with partners is the prerequisite for a robust supply chain. While focusing on intercultural relations, the trust would be a vital element to its development [2].

However, cultural differences are regarded as significant obstacles to building trust with other parties in the global supply chain [3], which would lead to inefficiency and other undesirable consequences in supply chain managerial activities. Despite the negative effects, working with individuals and cooperation from other countries is still a necessary practice to reduce costs and expand the business scale for many enterprises. Therefore, investigating a way to enhance supply chain performances with trust from the cultural perspective is an important strategy in supply chain management.

© Springer Nature Switzerland AG 2020
P.-L. P. Rau (Ed.): HCII 2020, LNCS 12192, pp. 540–550, 2020.
https://doi.org/10.1007/978-3-030-49788-0_41

Additionally, decision making is essential for supply chain. For upstream members, the suppliers for example, a forward-looking schedule and decision on their production policies are required. These decisions are based on two sources of information: the demand forecast reports from downstream and the exterior market and order history. For downstream members, the retailers for example, a decision must be made regarding their demand so that they are able to generate forecasting reports for upstream members to prepare for orders. Improper decisions will result in backlog and high inventory costs, which would consequently lead to reduced profit in the supply chain. For this reason, it is advisable to have strategic decision-making support to supply chain management.

Meanwhile, uncertainty avoidance proposed by Hofstede [4] was adopted as the qualifying measurement in the experiment to differentiate trust in different cultures, since it is one of the five dimensions of culture, as well as a moderator to trust [5]. However, it has not been studied how uncertainty avoidance specifically affects trust, hence, the objective of this study is to investigate how uncertainty avoidance impacts trust in global supply chain management through a computerized supply chain management (SCM) experiment.

2 Literature Review

2.1 Trust Within the Supply Chain

Trust was defined as a willingness to take risks frequently [6]. The prerequisite of trust is that a party shows certainty over the credibility and integrity of an exchange partner [7]. Therefore, as a beneficial and valuable contributor, trust influences several forms of exchange and network, including the supply chain [8]. Trust in the supply chain could lead to the "firm's belief that another company will perform actions that will result in positive outcomes for the firm as well as not take unexpected actions that result in negative outcomes" [9]. Beccerra [10] reported that both negative and positive consequences would occur in low and high trust through field data analysis.

Trust is especially significant for the supply chain, as it could help people to determine the reliability of words and other given information [11]. Trust deficiency made it hard for advanced supply chain collaboration due to the higher costs of transactions and agency [1]; as a result, managers spend more time and energy dealing with low-trust relationships than high-trust relationships. Besides, it was pointed out that people with high trust were much more willing to sharing known information and believing received knowledge [10]. In summary, a lack of trust among supply chain partners often resulted in inefficient and poor performance as the transaction costs (verification, inspections, and certifications of their trading partners) mounting.

High trust is the intangible attribute for the success of supply chain [1], which reduces cycle time and then improves supply chain responsiveness [12, 13]. In other words, it indicated that trust increased the agility in the supply chain [14]. Mutually trusted partners had a positive supply chain relationship and established cooperation objectives to enhance product quality and add value to consumers [15]. With the increasing flexibility of supply chain and low level of out-of-date inventories, trust improved efficiency and constructed long-term relationship thorough supply chain members [16]. In a continuous supply chain relationship, the exchange partners would be willing to accept the short-term imbalance of the results to ensure the overall supply chain performances [17].

2.2 Impact of Cultures in Trust

Several factors have impacts on developing the trust-building process, and they could be divided into two aspects: subjective and objective. In terms of subjective, people tend to build trust based on emotions, including empathy, politeness, and attitudes (kind and friendly) [18]. As for objective measurements, people's past experiences and numerical data would determine whether they should trust others or not. Confidentiality, credit, reliability, experiences from regular interaction, customization, information sharing, and even shared value were involved in objective consideration [19–21]. By considering the reciprocity, information sharing was the dominant factor in supply chain management. Trust could moderate people's information sharing quality and willingness. Besides, cultural dimensions, including individualism, uncertainty avoidance was related to building trust [2].

The definition of culture varies according to previous researches. One defined culture as "the sum of learned beliefs, values, and customs that create behavioral norms for a given society" [22]. In addition, the culture was viewed as the "collective programming of the mind" [4], it revealed what people have, think, and do as members of their society [23]. Furthermore, Rokeach [24] pointed out that cultural values were the fundamental driving force of life and the prescription of behavior.

The concept of national culture was noted as a mode of thinking, emotion, and action, which was rooted in shared values and social customs [25]. National culture was classified in many ways, high and low uncertainty avoidance proposed by Hofstede [4] were widely adopted in international marketing [26]. Besides, the importance of national culture was shown in numbers of studies since it has influences on the followings: personal values and lifestyles [27], perceptions of service quality and satisfaction [28], negotiations [29], perceptions about ethical behavior [30]. Hence, it was expected that culture would affect the trust in supply chain management.

Despite that culture was defined by many researchers, it was hard to interpret it. In global business activities, members must take their impact into account carefully to achieve the benefit of global works [31]. Johnson [6] reported that the main reason for those failing international businesses was the cultural challenges of overseas operation. Because managers would be unwilling to customize their management strategies and fit into local cultures. Cultural differences were the main obstacle to building trust among members in the global supply chain [3, 32].

According to Hofstede's cultural dimensions theory, cultures can be specified with six dimensions, including power distance, individualism, masculinity, uncertainty avoidance, long term orientation, and indulgence while uncertainty avoidance indicates the degree of social members to the tolerance of uncertainty, ambiguity, and risk [21]. Moreover, most opinions about trust recognized that it was relevant to risk, which influenced choices and behaviors [33]. Therefore, people's uncertainty avoidance could be an indicator of trust while conducting a study of cross-cultural decision making. Meanwhile, uncertainty avoidance index (range from 0 to 100) was applied to quantify uncertainty avoidance [4].

2.3 Scope and Limitation of Research

Using the measurement of culture at the individual level would be useful to implement further interpretation. More specifically, uncertainty avoidance was posited as a moderator to trust [5]. Their research assessed cultural characteristics through personality tests and found the moderating effect of uncertainty avoidance in the relationship between subjective norms and behavioral intentions.

Accordingly, uncertainty avoidance was related to prediction, intentionality, capability, and transference; an individual's intent to avoid uncertainty or risk remained a significant predictor of trust between upstream and downstream in the supply chain.

3 Research Framework and Methodology

3.1 Objectives and Hypotheses

Previous studies have proven that trust is one of the major factors to be considered in the development of supply chain management strategies, and culture will affect trust among different countries. However, few studies focused on how cultural factors affect trust in the global supply chain. Therefore, this study provided a quantitative analysis of trust and measured cultural factors with uncertainty avoidance index (UAI). The UAI's range between 0 and 50 is low UAI, and between 50 to 100 is high UAI. This study will start with the following research hypotheses:

- Hypothesis 1: In a cross-cultural supply chain, there is a correlation between trust level and UAI of retailers and wholesalers.
- Hypothesis 2: For high UAI countries, there is a low trust level with high integrity partners in the cross-cultural supply chain.
- Hypothesis 3: For low UAI countries, there is a high trust level with low integrity partners in the cross-cultural supply chain.
- Hypothesis 4: For high UAI countries, building trust with partners is easier and faster than those with low UAI in the cross-cultural supply chain.

The above hypotheses were principally proposed based on studies done on trust and culture [7, 8, 34, 35]. Several works have been conducted on the establishment of trust, but mainly within the western countries, few Asian countries are involved. Hence, this study will identify and fill gaps between western and Asian countries as reported in previous works by applying an experiment and the measurement of uncertainty avoidance in sampled countries. There are two independent variables, the UAI and the trust level.

3.2 Task and Setting

Participants are required to play the role of a wholesaler and computers to play the role of a retailer. In the simple supply chain, the retailer will give the demand forecast to wholesalers for them to prepare goods or inventory for the retailers in advance. However, in a real-world situation of the supply chain, retailers tend to inflate their forecasts to wholesalers to guarantee sufficient goods to sell. As for wholesalers, knowing the forecast is inflated, they tend to prepare fewer goods or inventory for retailers to avoid high inventory cost. Therefore, it is fair to assume, the smaller the gap between demand forecast received and inventory prepared, the higher the trust level is seen in the supply chain. Otherwise, it is a low trust supply chain. In this experiment, the computer is set to high integrity, which means that the computer would tend to provide accurate quantities of demand forecast to participants.

Therefore, to test the trust level of participants in this supply chain, we set the quantities of demand forecast given by retailer (computer) following uniform distribution [100,400]. The wholesalers (participants) would give feedback on quantities of goods or inventory prepared according to the demand forecast.

A total of 40 rounds were given to each participant in this experiment since time was needed for participants to familiarize themselves with the flow of the experiment. Therefore, it is fair to assume that after the participants got familiar with the rule, the result would be meaningful and useful. In sum, the trust level for every participant could be defined by

$$T = \frac{\sum_{i=stable}^{40}|m_i - n_i|}{40 - stable} \tag{1}$$

In **Formula** (1). i represents the rounds, m_i represents the quantities of demand forecast by a retailer (computer played), and n_i represents the quantities of prepared goods or inventory by a wholesaler (participants played). Since participants might be unfamiliar with the experiment flow, an unnecessary fluctuation might occur in the data collected from them. To address this issue, *stable* was defined as the number of the last round, when the deviation between demand forecast and prepared goods of a round remains less than ten for continuous five rounds. According to **Formula** (1), great T means participants have low trust level relationships in this supply chain, and small T means participants have a high trust level relationship in this supply chain.

To test hypothesis 4 based on the definition of *stable* by **Formula** (1), the speed of building trust can be defined as

$$V = stable \tag{2}$$

3.3 Procedure

There are two main parts of the supply chain management experiment. The first part is a paper-based experiment aiming to help participants get familiar with the business processes in the supply chain and to attain their permission to participate. The second part is the computerized experiment aiming to let the participants play the role of wholesalers, and the computer plays the role of the retailers. This part was conducted with Z-tree software [36] with 40 rounds for participants to complete. A monthly trade between a wholesaler and a retailer is regarded as a completed round.

3.4 Participants

Forty-two participants from 17 different countries with supply chain management related background were invited to participate in the experiment: 10 of them were Chinese (UAI = 30), 3 of them were Vietnamese (UAI = 30), 3 of them were Thai (UAI = 64), 2 of them were Polish (UAI = 93), 3 of them were Pakistanis (UAI = 70), 2 of them were Nepalese (UAI = 40), 2 of them were Malaysian (UAI = 36), 6 Indian (UAI = 40), 2 Egyptian (UAI = 80), 2 Bangladeshi (UAI = 60), 1 Yemen (UAI $\approx$ 75), 1 Russians (UAI = 95), 1 Rwandan (UAI $\approx$ 80), 1 Liberians (UAI = 68), 1 Iranian (UAI = 59), 1 Guinean (UAI $\approx$ 80), and 1 Belgian (UAI = 94). Thus, these subjects could highlight the influence of culture on the level of trust within global supply chain relationships.

According to different index of uncertainty avoidance, this study divides subjects into two categories: 19 of them are with high uncertainty avoidance (UAI $\geq$ 50), and 23 of them are with low uncertainty avoidance (UAI < 50).

4   Results

4.1   The Relationships Between UAI and Trust Level

To Explore whether there is a correlation between the UAI and trust level, the mean values of the trust level of different people are divided into two kinds of groups- low uncertainty avoidance groups (0 $\leq$ UAI < 50) and high uncertainty avoidance groups (50 $\leq$ UAI $\leq$ 100). According to the category of UAI, a two-sample T-test was used to test if there was a statistically significant difference between low UAI and high UAI with the trust level. An outlier test was conducted before T-test (as Table 1), and five outlier trust level was eliminated.

Table 1. The comparison of the low and high UAI towards different trust levels.

UAI level	N	Mean	STD
Low	22	3.16	1.16
High	16	5.53	1.52

Table 2. T-test result for H1, H2, and H3

T-value	Degree of freedom	P-value
−5.51	36	0.00

According to Table 2, the p-value is smaller than 0.05, which meant that the trust level is significantly different between low UAI and high UAI group, and there existed a correlation between trust level and UAI. The results of Hypothesis 1 indicated that there should exist a correlation between trust level and UAI in cross-cultural business. Hence,

UAI had to be considered by the business before designing cooperative conditions and terms across countries. Although the trust relationship would reach a balance point in their future cooperation, it might take a long time to make it stable, especially in the supply chain. Since the competition in the global market would concentrate more on the advanced and reliable supply chain, establishing a stable and long-term relationship is foremost in supply chain management.

Based on Table 1, the mean value was 3.16, and the value of high uncertainty avoidance groups was 5.53, which means participants from countries with high UAI showed a low trust level in the supply chain collaboration scenario. In contrast, those from countries with low UAI behaved oppositely.

The results of Hypothesis 2 and 3 indicated that people in countries with lower UAI have a higher level of trust in cross-cultural business trade. For these countries, the results are practical and efficient before cross-cultural business trade. For countries with a low UAI, participants tended to show more trust in their partners, and reward the trust back from their partners. Even if the quantities of the order fluctuate, this trust relationship will exist and remain reliable. However, for countries with a high UAI, the trust had been built in the long run, the trust level of them remained at a low level. Since they would be vigilant and protect themselves from loss in trade and would feel threatened by uncertain or unknown situations. Thus, their partners tended to be facing more risks from them.

4.2   The Relationship Between UAI and the Speed of Building Trust

Table 3 displays the descriptive statistics of the low and high UAI towards the different stable groups. According to Table 4, the p-value is greater than 0.05, the difference is not statistically significant. Therefore, higher stability of trust was observed in the cooperation between computer and participants of low uncertainty avoidance level, although fluctuations existed.

Table 3. The comparison of the low and high UAI towards the different stable groups

UAI level	N	Mean	STD
Low	22	3.82	3.58
High	16	5.63	5.86

Table 4. T-test result for H4

T-value	Degree of freedom	P-value
−1.22	36	0.23

Although the research is devoted to studying the varying speeds of building trust between countries with high and low UAI, as the results from testing of the fourth research hypothesis (H4) indicated that there is no significant statistical difference in the speed of building trust for a different culture. However, which it does not necessarily mean that there are is no correlation because since the sample was limited by 42 and this conclusion cannot be indicated drawn with by limitation of sample numbers by in a statistic method. Therefore, future related researches can collect more sample data that will need to be collected in future researches to study and prove this hypothesis. If this hypothesis can be proven true, more strategies need to be considered beforehand when trading with high uncertainty avoidance culture countries.

Our results show that not all the above hypotheses on the moderating effects of UAI on the trust-building processes in business trading can be could be proven true. The only significant effect was shown for the level of trust, which is conceptually very close to the definition of uncertainty avoidance. The subjects come from 17 countries, but the total samples are only 42. Moreover, the distribution of people in these countries is not uniform even. Fortunately, the UAI categorization category method did not rely on the countries but relied on the UAI. Therefore, for future researches, more country-based data need to be collected because there are fewer so as to avoid system errors and help prove the last hypothesis.

5 Discussions

It is critical for managers to assess possible differences between domestic and foreign markets, especially when it comes to choosing the appropriate strategies for the foreign markets [37]. Though there is a limitation due to the scope of data, this study can offer managers an insight into the impact and importance of culture on supply chain relationships.

5.1 Trust Relationship Will Eventually Be Established

There are findings from our experience that trust is significant to whatever culture-based countries. The Brazilian distributor lends further credence to the work of Hutchings and Weir [38], who found through their work in China and the Middle East that no matter high or low UAI culture one deals with, relationships and trust are of the utmost importance. The experiment's result shows that whatever culture they belong to and how high the UAI is, if their partner keeps high integrity consistently, they will trust their partner in the end even if someone needs more time to reach the state of trust. After all, trust-building relies on long-term cooperation. Hence, the point that trust can always be built in the long run for everyone can be identified by this experiment. However, even if people can build trust with each other in the long run, the trust level is different among them, and there is a correlation between trust level and culture.

5.2 Suggestions for Trading with High and Low UAI Countries' Company

High UAI countries tend to have a low trust level, so companies have to prepare their strategies accordingly before trading with high UAI countries. First, the company may need to sign more terms and conditions with these countries' companies to ensure a stable and persistent relationship. Secondly, more time and effort are needed to maintain a transparent and reliable communication process, especially when a company is stepping into a new market of high UAI countries.

Furthermore, the company should pay more attention to definite word-of-mouth improvement, such as getting customer trust through recommendation when available, been proven to have a powerful indirect impact on trust. On the other hand, a transparent customer relationship with integrity is even more important for companies from lower UAI background, since it is the way for them to establish trust. Therefore, the possible higher cost incurred during cooperation with high UAI can be regarded as a reasonable opportunity cost.

6 Conclusions

To sum up, this study investigates the effect of cultural factors on trust in the global supply chain and reveals the hardship of building a trust relationship when collaborating with people from countries with high UAI. Thus, it indicated that a more considerate and well-round plan should be adopted in the collaboration to develop a long-term trust relationship since it may take time and effort to achieve stable trust between partners. Also, trust was found as one of the crucial factors for supply chain management, particularly for a cross-cultural supply chain. Therefore, a thorough understanding of partners' cultural backgrounds, such as their decision-making processes and business models, would be conducive to developing and maintaining a long-term cooperative relationship.

As economic globalization proceeds, the most critical challenge will be to understand and tackle the often-subtle cultural differences. The findings of this study have demonstrated how culture affects the level of trust among different countries in business trading and how to use the difference of culture to prompt the development of the global supply chain in the future. It is recommended that companies prioritize trust as a critical factor to consider and explore how to establish and maintain it while entering a new market and then narrow down the trust level gap between two companies. Before the global business trading session starts, it is advisable that the companies bear in mind how UAI is conducive to making decisions for long-term cooperation with partners. When it comes to the stage of maintaining the business relationship, more personal contacts, such as telephone calls and visits, are suggested to be made from time to time.

The results of this study will provide useful information for all cross-cultural companies to help establish and maintain successful global supply chain relationships and for current supply management professionals have to develop cross-cultural skills. Effective management of the global supply chain and the realization of the primary role of trust in this relationship will have a direct and positive impact on the company's competitive advantages in the market.

References

1. Fawcett, S.E., Magnan, G.M., Williams, A.J.: Supply chain trust is within your grasp. Supply Chain Manag. Rev. **8**(2), 20–26 (2004)

2. Medlin, C.J.: Interaction in business relationships: a time perspective. Ind. Mark. Manag. **33**(3), 185–193 (2004)
3. Smyrlis, L.: Cultural differences can trump the most logical of supply chain planning. Can. Transp. Logist. **107**(9), 4–5 (2004)
4. Hofstede, G., Hofstede, G.J., Minkov, M.: Cultures and Organizations: Software of the Mind, 3rd edn. McGraw-Hill, New York (2010). (Revised and expanded)
5. Karahanna, E., Evaristo, J.R., Srite, M.: Levels of culture and individual behavior: an investigative perspective. J. Glob. Inf. Manag. (JGIM) **13**(2), 1–20 (2005)
6. Johnson, J.P., Lenartowicz, T., Apud, S.: Cross-cultural competence in international business: toward a definition and a model. J. Int. Bus. Stud. **37**(4), 525–543 (2006). https://doi.org/10.1057/palgrave.jibs.8400205
7. Morgan, R.M., Hunt, S.D.: The commitment-trust theory of relationship marketing. J. Mark. **58**(3), 20–38 (1994)
8. Doney, P.M., Cannon, J.P., Mullen, M.R.: Understanding the influence of national culture on the development of trust. Acad. Manag. Rev. **23**(3), 601–620 (1998)
9. Anderson, J.C., Narus, J.A.: A model of distributor firm and manufacturer firm working partnerships. J. Mark. **54**(1), 42–58 (1990)
10. Beccerra, M., Gupta, A.K.: Trust within the organization: integrating the trust literature with agency theory and transaction costs economics. Public Adm. Q. **23**, 177–203 (1999)
11. Cavusgil, S.T., Deligonul, S., Zhang, C.: Curbing foreign distributor opportunism: an examination of trust, contracts, and the legal environment in international channel relationships. J. Int. Mark. **12**(2), 7–27 (2004)
12. Monczka, R.M., Petersen, K.J., Handfield, R.B., Ragatz, G.L.: Success factors in strategic supplier alliances: the buying company perspective. Decis. Sci. **29**(3), 553–577 (1998)
13. Handfield, R.B., Bechtel, C.: The role of trust and relationship structure in improving supply chain responsiveness. Ind. Mark. Manag. **31**(4), 367–382 (2002)
14. Hult, G.T.M., Nichols Jr., E.L., Giunipero, L.C., Hurley, R.F.: Global organizational learning in the supply chain: a low versus high learning study. J. Int. Mark. **8**(3), 61–83 (2000)
15. Wong, A., Tjosvold, D., Zhang, P.: Supply chain relationships for customer satisfaction in China: interdependence and cooperative goals. Asia Pacific J. Manag. **22**(2), 179–199 (2005). https://doi.org/10.1007/s10490-005-1254-0
16. Johnston, D.A., McCutcheon, D.M., Stuart, F.I., Kerwood, H.: Effects of supplier trust on performance of cooperative supplier relationships. J. Oper. Manag. **22**(1), 23–38 (2004)
17. Griffith, D.A., Harvey, M.G., Lusch, R.F.: Social exchange in supply chain relationships: the resulting benefits of procedural and distributive justice. J. Oper. Manag. **24**(2), 85–98 (2006)
18. Coulter, K.S., Coulter, R.A.: Determinants of trust in a service provider: the moderating role of length of relationship. J. Serv. Mark. **16**, 35–50 (2002)
19. Heffernan, T.: Trust formation in cross-cultural business-to-business relationships. Qual. Mark. Res.: Int. J. **7**, 114–125 (2004)
20. Hausman, A., Myhr, N., Spekman, R.E.: Collaborative supply-chain partnerships built upon trust and electronically mediated exchange. J. Bus. Ind. Mark. **20**, 179–186 (2005)
21. Batt, P.J.: Building trust between growers and market agents. Supply Chain Manag.: Int. J. **8**(1), 65–78 (2003)
22. Yau, O.H., You, H.: Consumer behaviour in China: customer satisfaction and cultural values. Taylor & Francis, Milton Park (1994)
23. Ferraro, G.P.: The Cultural Dimension of International Business. Prentice Hall, Englewood Cliffs (1994)
24. Rokeach, M.: The Nature of Human Values. Free Press, New York (1973)
25. Nakata, C., Sivakumar, K.: Instituting the marketing concept in a multinational setting: the role of national culture. J. Acad. Mark. Sci. **29**(3), 255–276 (2001)

26. Kim, D., Pan, Y., Park, H.S.: High-versus low-context culture: a comparison of Chinese, Korean, and American cultures. Psychol. Mark. **15**(6), 507–521 (1998)

27. Sun, T., Horn, M., Merritt, D.: Values and lifestyles of individualists and collectivists: a study on Chinese, Japanese, British and US consumers. J. Consum. Mark. **21**(5), 318–331 (2004)

28. Ueltschy, L.C., Ueltschy, M.L., Fachinelli, A.C.: The impact of culture on the generation of trust in global supply chain relationships. Mark. Manag. J. **17**(1), 15–26 (2007)

29. Adair, W.L., Brett, J.M.: The negotiation dance: time, culture, and behavioral sequences in negotiation. Organ. Sci. **16**(1), 33–51 (2005)

30. Elahee, M.N., Kirby, S.L., Nasif, E.: National culture, trust, and perceptions about ethical behavior in intra-and cross-cultural negotiations: an analysis of NAFTA countries. Thunderbird Int. Bus. Rev. **44**(6), 799–818 (2002)

31. Daniels, J.D., Radebaugh, L.H., Sullivan, D.P.: Globalization and Business. Prentice Hall, New York (2002)

32. Smagalla, D.: Supply-chain culture clash. MIT Sloan Manag. Rev. **46**(1), 6 (2004)

33. Lewis, J.D., Weigert, A.: Trust as a social reality. Soc. Forces **63**(4), 967–985 (1985)

34. Gulati, R.: Does familiarity breed trust? The implications of repeated ties for contractual choice in alliances. Acad. Manag. J. **38**(1), 85–112 (1995)

35. Ha, J., Karande, K., Singhapakdi, A.: Importers' relationships with exporters: does culture matter. Int. Mark. Rev. **21**, 447–461 (2004)

36. Fischbacher, U.: z-Tree: Zurich toolbox for ready-made economic experiments. Exp. Econ. **10**(2), 171–178 (2007). https://doi.org/10.1007/s10683-006-9159-4

37. Sousa, C.M., Bradley, F.: Cultural distance and psychic distance: two peas in a pod. J. Int. Mark. **14**(1), 49–70 (2006)

38. Hutchings, K., Weir, D.: Understanding networking in China and the Arab world. J. Eur. Ind. Train. **30**, 272–290 (2006)

Cultural Facets of Interactions with Autonomous Agents and Intelligent Environments

Fostering Cross-Cultural Research by Cross-Cultural Student Teams: A Case Study Related to Kawaii (Cute) Robot Design

Dave Berque[1]($\boxtimes$), Hiroko Chiba[1], Michiko Ohkura[2], Peeraya Sripian[2], and Midori Sugaya[2]

[1] DePauw University, Greencastle, USA
{dberque,hchiba}@depauw.edu
[2] Shibaura Institute of Technology, Tokyo, Japan
ohkura@sic.shibaura-it.ac.jp, {peeraya,doly}@shibaura-it.ac.jp

Abstract. As robotic gadgets, and eventually robots, become increasingly common in daily life, it is critical that roboticists design devices that are accepted across cultures. Previous studies have examined cross-cultural differences in robot acceptance based on various design characteristics. Similarly, prior studies have examined cross-cultural perceptions of kawaii (Japanese cuteness). Building on these two prior research strands, this paper reports on our developing approach, with support from a United States National Science Foundation (NSF) International Research Experiences for Undergraduates (IRES) grant, to use a cross-cultural, faculty-student design team to gain a deeper understanding of the role that kawaii (Japanese cuteness) plays in fostering positive human response to, and acceptance of, robotic gadgets across cultures.

After explaining the motivation for the work, we outline our approach from both a technical and educational perspective. In doing so, we provide a case study that demonstrates how a cross-cultural design team involving students can simultaneously generate new knowledge and provide research training for future Human Computer Interaction professionals.

Keywords: Human-robot interaction · Cross-cultural design · Kawaii

1 Introduction and Motivation

1.1 Kawaii

The word kawaii stemmed from the word *kawayushi* that appeared in *Konjaku Momogatarishu* (Tales of Times Now Past) in classical Japanese literature [1]. At this time, the word kawaii meant pitiful, shameful, or too sad to see [1]. Over the course of Japanese history, the meaning of the word evolved to describe the small, weak, and someone or something that invokes the feeling of "wanting to protect" [1]. More recently, the meaning of kawaii has been extended to the concept of "Japanese cuteness". Indeed, in their book "Cuteness Engineering: Designing Adorable Products and Services" Marcus,

© Springer Nature Switzerland AG 2020
P.-L. P. Rau (Ed.): HCII 2020, LNCS 12192, pp. 553–563, 2020.
https://doi.org/10.1007/978-3-030-49788-0_42

Kurosu, Ma and Hashizume confirm that kawaii is the closest Japanese word to the English word cute [2]. In this modern context of cuteness, the notion of kawaii is now pervasive in Japan and can be seen in designs ranging from Hello Kitty products to road signs to robotic gadgets, to name just a few examples.

Increasingly, kawaii design principles are incorporated into successful products that are used globally including robotic gadgets [3, 4]. An object's kawaii-ness is an important part of its affect, and may influence a user's acceptance of the object. Therefore, when designing products, such as robotic gadgets, for global use, it is important for designers to understand how specific groups of target users perceive and relate to kawaii.

1.2   Related Work

Previous studies have examined cross-cultural differences in robot acceptance based on design characteristics. For example, researchers have documented the impact of localizing a robot's greeting style (gestures and language) on acceptance by Japanese versus Egyptian users [5] and other studies have obtained similar results varying only robot gesture styles across users from Japan, China, Thailand and Bangladesh [6]. In another study, differences in likeability of robots were found between Chinese, Korean and German users based on variations in the robot's appearance [7].

Similarly, a number of studies, including studies by the authors, have examined perceptions of kawaii including differences in perceptions across cultures and genders. For example, third author Ohkura and her collaborators have found gender differences in preferences for various kawaii spoon designs based on shape, color and geometric pattern [8]. In a broader study, Hashizume and Kurosu examined the extent to which perceptions of kawaii in 225 photographs differ between male and female Japanese college students [9]. The photographs were divided into subgroups including products, objects, foods (see Fig. 1), geometric shapes, animals, characters and people.

Fig. 1. Japanese sweets

Gender differences were established, depending on the subgroup of object studied. For example, male subjects found spherical geometric objects to be more kawaii than female subjects [9]. In collaboration with Hashizume and Kurosu, first author Berque and second author Chiba extended the original study by presenting 217 of the original 225 images to American college students and gathering data about their perceptions of kawaii-ness in each image. Differences were found in a variety of images, particularly between Japanese males and American males as well as between Japanese females and all other groups [10].

Prior work that investigates the role of kawaii in user perceptions and user acceptance of robots or robotic gadgets is limited. One pair of papers reports on studies of kawaii-ness in the motion of robotic vacuum cleaners [11, 12]. The authors programmed a visually plain Create robot (a programmable version of a Roomba vacuum cleaner) to move according to 24 different patterns, including patterns that the authors describe with terms such as bounce, spiral, attack, spin and dizzy [11, 12]. While subjects did not judge the static robotic vacuums to exhibit kawaii-ness, most subjects did consider the vacuums to exhibit kawaii-ness when they made specific rotational motions such as a spiral [11]. These studies demonstrate that kawaii-ness can be expressed through motion even in the absence of more traditional visual kawaii-ness; however, the studies did not consider cultural or gender differences.

Increasingly, Japanese robots are being designed with kawaii-ness principles in mind. As an example, consider Sony's Aibo robotic companion dog (see Fig. 2, which depicts an Aibo cuddled with a ball). Many people find Aibo to be kawaii; however, it is not clear which features (size, shape, sound, motion) of Aibo evoke this feeling and it is not clear if the key features vary across genders or cultures.

Fig. 2. Sony Aibo cuddled with a ball

1.3   The NSF International Research Experiences for Students Program

The International Research Experiences for Students (IRES) program is one of many programs supported by the United States National Science Foundation (NSF). As described by the NSF, the program supports "…international research and research-related activities for U.S. science and engineering students". [13] The NSF further explains that the program "contributes to development of a diverse, globally-engaged workforce with world-class skills. IRES focuses on active research participation by undergraduate or graduate students in high quality international research, education and professional development experiences in NSF-funded research areas" [13].

The IRES program consists of several tracks and this paper describes a proposal that falls under a track which must provide a cohort international research experience for a group of students. Although each student should have a specific role on the research team, the research experiences must have a single intellectual theme that is aligned with an area that the NSF supports. We organized this project, which we describe in the remainder of this paper, around developing a better understanding the role that kawaii (Japanese cuteness) plays in robot design.

1.4   Project Overview

In the remainder of this paper, parts of which are adapted from our IRES proposal, we describe a collaboration between DePauw University in the United States and Shibaura Institute of Technology (SIT) in Japan. The collaboration is supported by a three-year $286,761 National Science Foundation IRES grant entitled "Involving Undergraduates in Research on Design and Cross-Cultural Perceptions of Cuteness in Robotic Gadgets". The NSF grant provides funding for 12 DePauw students (four per summer in 2020, 2021 and 2022) to spend seven weeks in Japan collaborating with faculty and student researchers at SIT.

From a technical viewpoint, the grant is enabling us to investigate cross-cultural and gender differences related to perceptions of kawaii in robotic gadgets, with the aim of informing the design of robotic gadgets that are accepted across cultures. In parallel, this work provides a case study that demonstrates how a cross-cultural design team involving students can generate new knowledge while simultaneously providing research training for future Human Computer Interaction professionals. We believe a similar two-pronged approach can be taken by other HCI researches and educators by engaging cross-cultural student teams in studying cross-cultural research questions and we intend for this paper to provide a case study that can be adapted by others.

2   Research Methodology

2.1   Example Project Domains

We present two example research project domains that can provide context for the IRES robotic gadget prototypes, although additional domains will be explored. The first example domain centers on the design and evaluation of prototype kawaii robotic vacuum cleaners for use in a home environment. As mentioned previously, prior work has investigated user perceptions of kawaii-ness based on movement patterns of robotic vacuums [11, 12]. While this work established that users can perceive kawaii-ness based solely on motion, the results are limited in a number of ways. First, the work does not consider differences across culture or gender. Second, the work does not incorporate other common kawaii attributes such as colors, geometric shapes, child-like or animal-like qualities, and sounds. Using the Wizard of Oz technique, IRES students will design and prototype robotic vacuums that incorporate kawaii-ness through their appearance, or by making sounds or changing expression upon bumping into an obstacle. These kawaii visual, motion-based, and auditory attributes can also be combined synergistically.

The second example domain centers on the design and evaluation of prototype robotic gadget companions/assistants for college students. As early as 2003, Fogg proposed what he then considered to be a futuristic "study buddy" that would assist students by prompting them to adopt good living and study habits and by providing other helpful information [14]. What seemed futuristic in 2003 is now realistic. In the past year, several schools have begun providing all students with Amazon Echo Dots that use Alexa to equip every student dorm room with a student assistant service, customized with campus-specific information [15]. The service is helpful in some ways; however, user reception has been mixed due to privacy concerns [15]. We will build prototypes of novel robotic gadget student assistants that exhibit various forms of kawaii-ness to understand the role that kawaii can play in user-acceptance of these gadgets. IRES students will be particularly well-equipped to the creative design of these prototypes because they will have a good understanding of how this type of technology can be matched to the needs of a college student.

2.2 Cross-Cultural Evaluation Approaches

There are well-established approaches to determining a user's emotional reaction to an object, including self-reports, external measures such as analyzing facial expressions, and internal measures such as using biosensors to measure brainwaves (EEG) and heart rate (ECG). Advantages and disadvantages of these techniques are delineated in the literature (see, for example, [16] as well as prior work by fifth author Sugaya [17]). As explained below, we will use both self-reports and biosensors in the IRES project.

One challenge in running a cross-cultural, multi-site user study involving prototypes of robotic gadgets involves replicability. It is difficult to build several identical proto-types for use at different sites and even more difficult to build several versions of each prototype that incorporate different kawaii properties. Therefore, we will explore the use of traditional video, as well as stereoscopic 3D video, to record each prototype and to present the prototype consistently to users at different sites.

Presenting the robotic gadget prototypes via video brings the added benefit of allow-ing us to use the Wizard of Oz Technique to run our studies. Specifically, rather than building fully functioning robotic gadget prototypes, the prototypes only need to have enough functionality to make a convincing video. For example, a robotic gadget that looks like it is taking an action in response to a human stimulus, or in response to bumping into another object, may actually be responding to a command given by an off-camera human operator.

Each student will work on a project that involves prototyping several versions (with varying degrees and types of kawaii-ness) of a domain specific robotic gadget, and then conducting one or more experiments from our user-study suite.

Across domains, research will follow the following general methodology:

1. Consider the relevant kawaii literature and discuss with SIT faculty mentors.
2. Design and prototype multiple versions of robotic gadgets in the given domain with varying degree and types of kawaii. Feedback will be provided by the SIT faculty mentors throughout the design and prototyping stage.
3. Create traditional and, or stereoscopic 3D videos of the prototypes in action.
4. Conduct one or more of the user studies using appropriate measures (e.g., Self-Reported Kawaii, Measured Internal Joy, Self-Reported Acceptance) under the guidance of the SIT faculty mentors. For example, comparing Measured Internal Joy to Self-Reported Acceptance across cultures.
5. Analyze results and iterate as appropriate.

When cross-cultural user studies are required, data will be gathered from Japanese participants at SIT and from American participants at DePauw University. User studies will first be run at SIT under the guidance of the SIT faculty mentors and then will be replicated at DePauw.

3 Student Preparation for Cross-Cultural Experience

3.1 Student Recruitment

Students will be recruited from DePauw University Computer Science Majors. We expect students to participate in the IRES program during the summer following their sophomore or junior year.

Students must have completed, or have a plan to complete, core computer science course-work including courses in Data Structures, Object Oriented Software Development, and Computer Systems prior to departure for Japan. Coursework related to Japanese language and/or culture will strengthen a candidate's application.

3.2 Pre-departure Technical and Cultural Preparation at DePauw University

Prior to departure, students will be encouraged to enroll in relevant computer science and Japan-related courses such as Human Computer Interaction, Japanese Language, and Japanese Art History. Table 1 shows the relevant courses that will have been completed by each of the students who are participating in this project during the summer of 2020. We believe that the balance of technical, language and cultural preparation will prepare the participants to engage in a cross-cultural design experience.

The first and second authors will meet regularly with the IRES students in the months prior to departure in a workshop series that will provide an introduction to: Japanese culture, Japanese customs and etiquette, Japanese design principles including traditional aesthetic concepts as well as kawaii, the role of robotics in Japan, the benefits and challenges of working on cross-cultural teams, and research ethics. As part of the workshop

Table 1. Technical, Japanese language and Japanese culture background of participants

Student	Computer Science Courses	Language and Culture Courses
Student 1	Computer Science I; Data Structures; Object Oriented Software Development; Computer Systems; Foundations of Computation; Data Mining; Artificial Intelligence; Database and File-systems	Three Years of Japanese Language; Introduction to China and Japan Early Modern and Modern Japanese Art; Readings in Asian Studies; Japanese Culture, Technology and Design; Warrior Art of Japan; Advanced Readings and Projects in Japanese
Student 2	Data Structures; Computer Systems; Object Oriented Software Development; Writing in Computer Science	Three Years of Japanese Language; Introduction to China and Japan; Japanese Culture, Technology and Design; Martial Arts Intensive; Supernatural in Japanese Art
Student 3	Computer Science One; Data Structures; Object Oriented Software Development; Foundations of Computation; Computer Systems	One Year of Japanese Language
Student 4	Computer Science One; Data Structures; Object Oriented Software Development; Computer Systems; Foundations of Computation; Artificial Intelligence; Computer Networking	One Year of Japanese Language; Japanese Culture, Technology and Design

series, the students will also use DePauw's Tenzer Technology Center to learn how to make stereoscopic 3D videos and to learn the basics of the programming environments they will use at SIT.

The students will also be introduced to the LINE social media system. This is the most popular messaging and social media system in Japan [18] and will be indispensable to the students in Japan. LINE's is itself an example of kawaii design and the first two authors have published about the use of LINE to introduce kawaii to American students and to enable student communication during an American travel course to Japan [19]. The cohort will use LINE to communicate with each other and with the faculty leaders so they can experience kawaii design and to prepare them to use the tool, including its language translation features, in Japan.

DePauw enrolls more than twenty Japanese students and a few of them will meet with the IRES students, as well as participate in the LINE discussions, to offer their perspective on Japanese culture. The SIT faculty mentors will also join a few workshops (virtually, via LINE) to introduce themselves and to provide information about specific research projects. In addition, in years two and three of the IRES program, students who participated in the program in the prior year will attend a workshop to share their experiences.

IRES students will attend a session that DePauw's off-campus study offers students who are preparing to study abroad. These sessions cover Identity Abroad, Integrating International Experiences into Career Planning, and Health and Safety Abroad.

As shown in Table 1, each IRES student brings different experiences to this program. For example, some students know more Japanese language than others and some students have deeper technical knowledge. Each student will be encouraged to use his or her strengths to help other students. In our experience, this approach is as helpful for team-building as it is for building technical and cultural competencies. IRES students will travel to Japan as a team. To assist with travel and acclimation for the first week, the students will be accompanied by DePauw faculty members.

3.3 Technical and Cultural Support at Shibaura Institute of Technology

Shibaura Institute of Technology (SIT) is the ideal site to host the proposed IRES project from both a technical and global learning perspective. SIT is a private technology university with three greater Tokyo area campuses that collectively enroll more than 7,000 undergraduates and more than 900 graduate students. The Institute holds three colleges that each house a cluster of engineering and science departments. SIT is the only private technology university in Japan to be selected for the Japanese Government's Top Global University Project. This project seeks to recognize and enhance Japanese universities that are innovators in internationalizing their educational offerings and reputations. As such, SIT has significant experience in hosting student and faculty researchers for short-term, medium-term and long-term research collaborations. At any given time, SIT hosts approximately 500 international exchange students and the IRES students will benefit from this environment.

The IRES project will be housed in the Department of Information Science and Engineering in the College of Engineering, which is part of SIT's main campus, the Toyosu campus, located in the Tokyo Bay area. This campus is home to junior and senior undergraduate and graduate students in the College of Engineering and is also home to the SIT's Robotics Research Square, which is the base of the Shibaura Institute of Technology Robotics Consortium.

IRES students will have access to the research facilities that are overseen by the SIT faculty research mentors as well as to appropriate departmental resources. These facilities will provide the hardware materials and software tools for building, customizing and programming prototypes of physical-electronic gadgets, such as robot components, that will be central to this research. The labs also have several Neurosky EEG and ECG biosensors that can be used to gather biological readings from human subjects.

The grant will provide funding (admission tickets, travel expenses) for the IRES students and a roughly equal number of Japanese student lab-mates to take advantage of a number of relevant cultural activities while they are on site in Japan. These group activities will strengthen the local research community. Students will visit Japan's National Museum of Emerging Science and Innovation, which is located in Tokyo. Among the interactive robot displays are the Paro therapeutic robot and Honda's Asimo robot. Students will also visit the Chiba Institute of Technology Tokyo Sky Tree Town Campus which houses a public display of robots including rescue robots. Students will also visit the International Tokyo Toy Show, which takes place each June and showcases thousands of toys, including robotic gadgets.

Because contemporary Japanese design aesthetics are rooted in tradition, students will visit the Edo-Tokyo museum, which showcases Japanese design in the Edo period from 1603–1868. Students will also take a weekend excursion to Kyoto, Japan's old capital which is famous for preserving traditional designs, to see historical sites such as castles, temples, traditional handcrafts and the masterful Zen gardens that greatly influenced Steve Jobs's design aesthetic.

4 Conclusion

This IRES project will benefit greatly from a cross-cultural collaboration and is ideally suited to be conducted in Japan due to the strong integration of kawaii in Japanese culture. The synergistic contributions of a research team that includes faculty mentors who have a deep and broad understanding of the Japanese concept of kawaii, working alongside students who understand American culture are nicely situated to tease out subtle differences that may influence differences in perceptions of these two populations. In addition, the cross-cultural team will be well positioned to run user studies with both Japanese and American subjects. The Shibaura Institute of Technology faculty mentors are excited about applying their ongoing investigation of kawaii in a new cultural context, with the help of their American student and faculty collaborators.

This collaboration will advance knowledge related to several scientific and method-ological challenges. Robots and robotic gadgets can induce feelings of both fascination and fear, especially when operating in personal spaces such as homes [20]. Researchers have shown that a robot's appearance can impact the degree to which users accept the robot, at least in some contexts [21]. The IRES research will apply previous work that examines perceptions of kawaii-ness in simple static objects to the new context of dynamic robotic gadgets. This work will result in a set of kawaii design guidelines that help designers build robotic gadgets with a kawaii appearance that makes them more likely to be accepted by specific populations.

The research will also explore correlations between several measures that can be used to evaluate a range of robotic gadgets. This work will advance our understanding of the most appropriate measures to use when evaluating robotic gadgets in specific contexts, and for use with specific populations.

Additionally, this research will deepen our understanding of the benefits and limi-tations of using traditional video and stereoscopic 3D video for studying robotic gadget prototypes in a cross-cultural, multi-site setting. If effective, the use of video will enable other researchers to replicate our studies in different cultural contexts, which will result in a robust dataset.

While each IRES student will be responsible for designing, building and evaluating specific prototypes, the students will employ the same research methodology, which will result in a shared knowledge base. This will enable the twelve student participants over the three-year project to contribute to a research result that is significantly greater than the sum of the contributions made by each student. This approach has the potential to impact the way other researchers design multi-student undergraduate research experiences in computer science.

Acknowledgements. This material is based upon work supported by the National Science Foundation under Grant No. OISE-1854255. Any opinions, findings, and conclusions or recommendations expressed in this material are those of the author(s) and do not necessarily reflect the views of the National Science Foundation.

References

1. Yomota, I.: Kawaii Ron (The Theory of Kawaii). Chikuma Shobō, Tokyo (2006)
2. Marcus, A., Kurosu, M., Ma, X., Hashizume, A.: Cuteness Engineering: Designing Adorable Products and Services. Springer, Cham (2017). https://doi.org/10.1007/978-3-319-61961-3
3. Cole, S.: The Most Kawaii Robots of 2016 (2016). https://motherboard.vice.com/en_us/article/xygky3/the-most-kawaii-robots-of-2016-5886b75a358cef455d864759. Accessed 08 Sept 2018
4. Prosser, M.: Why Japan's cute robot's could be coming for you (2017). www.redbull.com/us-en/japan-cute-robot-obsession. Accessed 8 Sept 2018
5. Trovato, G., et al.: Cross-cultural study on human-robot greeting interaction: acceptance and discomfort by Egyptians and Japanese. J. Behav. Robot. **4**(2), 83–93 (2013)
6. Shidujaman, M., Mi, H.: Which country are you from? A cross-cultural study on greeting interaction design for social robots. In: Rau, P.L. (ed.) CCD 2018. LNCS, vol. 10911, pp. 362–374. Springer, Heidelberg (2018). https://doi.org/10.1007/978-3-319-92141-9_28
7. Li, D., Rau, P., Li, Y.: A cross-cultural study: effect of robot appearance and task. Int. J. Social Robot. **2**, 175 (2010)
8. Tipporn L., Ohkura, M.: Comparison of spoon designs based on kawaiiness between genders and nationalities. In: A4-3 Proceedings of ISASE2017, March 2017
9. Hashizume, A., Kurosu, M.: The gender difference of impression evaluation of visual images among young people. In: Kurosu, M. (ed.) HCI 2017, Part II. LNCS, vol. 10272, pp. 664–677. Springer, Cham (2017). https://doi.org/10.1007/978-3-319-58077-7_51
10. Berque, D., Chiba, H., Hashizume, A., Kurosu, M., Showalter, S.: Cuteness in Japanese design: investigating perceptions of kawaii among American college students. In: Fukuda, S. (ed.) AHFE 2018, vol. 774, pp. 392–402. Springer, Cham (2018). https://doi.org/10.1007/978-3-319-94944-4_43
11. Sugano, S., Miyaji, Y., Tomiyama, K.: Study of kawaii-ness in motion – physical properties of kawaii motion of roomba. In: Kurosu, M. (ed.) HCII 2013, Part I. LNCS, vol. 8004, pp. 620–629. Springer, Heidelberg (2013). https://doi.org/10.1007/978-3-642-39232-0_67
12. Sugano, S., Morita, H., Tomiyama, K.: Study on kawaii-ness in motion–classifying kawaii motion using Roomba. In: International Conference on Applied Human Factors and Ergonomics, San Francisco, California, USA (2012)
13. National Science Foundation International Research Experiences for Students. www.nsf.gov/publications/pub_summ.jsp?WT.z_pims_id=505656&ods_key=nsf19585. Accessed 24 Feb 2020
14. Fogg, B.J.: Persuasive Technology: Using Computers to Change What We Think and Do. Morgan Kaufmann, San Francisco (2003)
15. McKenzie, L.: Alexa, what's the deal with you, anyway? Inside Higher Ed (2018). www.insidehighered.com/news/2018/08/22/meet-new-kid-campus-alexa. Accessed 8 Sept 2018
16. Zhao, M., Adib, F., Katabi, D.: Emotion recognition using wireless signals. Commun. ACM **61**(9), 91–100 (2018)
17. Ikeda, Y., Horie, R., Sugaya M.: Estimating emotion with biological information for robot interaction. In: 21st International Conference on Knowledge-Based and Intelligent Information & Engineering Systems (KES-2017), pp. 6–8 (2017). Proc. Comput. Sci. **112**, 1589-1600

18. Bloomberg: Has Japan's most popular messaging app peaked? (2016). www.bloomberg.com/graphics/2016-line-ipo/. Accessed 08 Sept 2018
19. Berque, D., Chiba, H.: Evaluating the use of LINE software to support interaction during an American travel course in Japan. In: Rau, P.L. (ed.) CCD 2017. LNCS, vol. 10281, pp. 614–623. Springer, Cham (2017). https://doi.org/10.1007/978-3-319-57931-3_49
20. Korn, O., Bieber, G., Fron, C.: Perspectives on social robots. From the historic background to an experts' view on future developments. In: PETRA 2018 Proceedings of the 11th International Conference on PErvasive Technologies Related to Assistive Environments. ACM, New York (2018)
21. Sonya, S., Kwak, J., Jung, J.: Can robots be sold? The effects of robot designs on the consumers' acceptance of robots. In: Proceedings of the 2014 ACM/IEEE International Conference on Human-Robot Interaction, 03–06 March 2014, Bielefeld, Germany (2014)

Can Older Adults' Acceptance Toward Robots Be Enhanced by Observational Learning?

Sung-En Chien[1]([✉]), Ching-Ju Yu[1], Yueh-Yi Lai[2], Jen-Chi Liu[2], Li-Chen Fu[3,4,5], and Su-Ling Yeh[1,5,6,7,8]

[1] Department of Psychology, National Taiwan University, Taipei, Taiwan
chiensungen@gmail.com, suling@g.ntu.edu.tw
[2] Industrial Technology Research Institute, Hsin-Chu, Taiwan
[3] Department of Electrical Engineering, National Taiwan University, Taipei, Taiwan
[4] Department of Computer Science and Information Engineering, National Taiwan University, Taipei, Taiwan
[5] Center for Artificial Intelligence and Advanced Robotics, National Taiwan University, Taipei, Taiwan
[6] Graduate Institute of Brain and Mind Sciences, National Taiwan University, Taipei, Taiwan
[7] Neurobiology and Cognitive Science Center, National Taiwan University, Taipei, Taiwan
[8] Center for Advanced Study in the Behavioral Sciences, Stanford University, Stanford, CA, USA

Abstract. It has been shown that older adults' negative attitudes toward robots are stronger than those of younger adults, which presumably causes older adults to lack motivation in interacting with robots. We examined the possibility to improve older adults' attitudes toward robots through observational learning. 40 younger and 40 older adults watched one of two video clips introducing features of an assistive robot. Within each age group, half watched a video clip containing human-robot interaction scenarios (the observational-learning group) and the other half watched a video clip without human-robot interactions (the control group). After watching the video, participants decided if they would like to interact with the actual robot. Participants' explicit attitudes were measured by questionnaires, and implicit attitudes were measured by the implicit association test (IAT) and the name-shape association task (NSAT). Results showed that (1) for those who chose to interact with the robot, the observational-learning group reported higher perceived safety of the robot; (2) implicit negative attitudes toward the robot as measured by the IAT and the NAST was updated after watching video clips. Our results suggest that direct human-robot interaction cannot be overlooked. Furthermore, contrary to conventional assumptions, implicit attitudes toward robots can be rapidly shaped.

Keywords: Older adults · Observational learning · Robot · Acceptance · Implicit attitude

© Springer Nature Switzerland AG 2020
P.-L. P. Rau (Ed.): HCII 2020, LNCS 12192, pp. 564–576, 2020.
https://doi.org/10.1007/978-3-030-49788-0_43

1   Introduction

1.1   Aging Society and Robots

Population aging and low birth rates have become global trends in many developed and developing countries, and thus efforts made to remedy such potential labor shortage are desperately needed. For example, Taiwan became an aged society (14% over 65 years old) in 2018 and will become a super-aged society by 2026 when at least 20% of the population will be 65 or older [1]. With population aging, demand for older adult caregivers are expected to rise, and yet labor supply is evidently short to meet such demands due to low birth rates. Nevertheless, thanks to advances in robotic technology nowadays, novel robotic products have been invented to potentially provide daily assistance to older adults [2, 3].

Two factors, however, may prevent older adults from using robots. First, compared to younger adults, older adults tend to have stronger *negative* attitudes toward robots [4, 5] and are *less* willing to use robots [6, 7]. Second, learning and adapting to new technology could be a tedious process, which may negatively affect users' attitudes toward the products (i.e., robots), especially for older adults [8].

Previous studies also suggested that introducing interactive experience of human-robot interaction plays a critical role in forming positive attitudes toward robots. For instance, older adults tended to perceive robots as easier to use after having an experience with service functions of an assistive robot [4] or performing a partnered stepping task with a dancing robot [9]. Since attitude is a key component of intention to use a targeted technology, it is a precondition for eventual production adoption [10], and enhancing positive attitudes toward robots is an important issue if AI and robot technology are intended to be useful for aging societies. However, older adults' negative attitudes toward robots and the tedious technology adopting process as mentioned above would prevent them from interacting with robots in the first place. If this is the case, older adults might even be less willing to let others (e.g., their family) buy a robot for them to assist their daily lives. Therefore, an approach other than direct interaction should be tested to see whether there is a plausible alternative to improve older adults' attitude and acceptance of robots without direct human-robot interaction.

1.2   Observational Learning

Learning can occur through observing others' behaviors and thereafter acquiring attitudes, values, thinking styles, and behavior styles, and subsequently retaining the information and replicating the behaviors that were observed [11]. Such observational learning is a form of social learning [11, 12], which is not simply an imitation of others' behaviors but also involves changes in motivations, attitudes, and even beliefs by understanding behavioral consequences, rules, and principles embodied in actions.

Observational learning has been shown effective when applied to learning about technology products. Yi and Davis [13] developed a training intervention for computer task performance based on the observational learning processes. They found that observational learning yielded better training outcomes than lecture-based training. Beer et al. found that older adults' attitudes toward robots could be improved after observing a mobile manipulator robot demonstrating functions [14]. However, in Beer et al.'s work, the robot either directly interacted with the participants (i.e., delivering medication to participants) or simply demonstrated functions that did not involve any human-robot interaction (i.e., table clean-up and 3D-location detection). Therefore, it is still unclear whether observational learning could further enhance older adults' positive attitudes and acceptance toward robots.

1.3 Explicit and Implicit Attitudes

We investigated whether observational learning plays an important role in enhancing older adults' acceptance and attitudes toward robots with both explicit and implicit measures. On the one hand, explicit attitude captures deliberate and conscious thinking styles. On the other hand, implicit attitude functions without one's full awareness and has its own manifestation in human behavior. Typically, implicit attitude is thought to be more reflective of spontaneous behaviors and explicit attitude is more predictive of deliberative behaviors [15].

Among various models which have been developed to theoretically predict acceptance of new technology, the Unified Theory of Acceptance and Use of Technology (UTAUT) [8] and the Subjective Technology Adaptivity Inventory (STAI) [16] are shown to be able to predict older adults' actual usage and perceived competence of robots [6, 8, 16]. Therefore, we adopted UTAUT and STAI to measure older adults' acceptance of robots.

Moreover, curiosity, which reflects a person's willingness and desire for information seeking, can also be helpful to predict future initiation of learning, engagement, and acceptance of a particular object, such as a robot [4]. Curiosity can be measured at trait and state levels [17]. Trait curiosity measures individuals' overall tendency for seeking novel information, while state curiosity measures individuals' momentary desire for information related to a particular object. Since we were interested in investigating attitudes toward robots, we focused on state curiosity toward robots while the trait curiosity was measured as a control variable of personality traits. We also measured participants' personality traits with the Big Five personality scale [18] to further explore correlations between robot acceptance and personality traits.

Additionally, implicit attitudes toward robots were measured by the name-shape association task (NSAT) [19] and the implicit association test (IAT). The NSAT was designed to measure the processing priority of a stimulus which could be modulated when attached to identities. Neutral objects (e.g., geometric shapes) associated with the names of significant others (e.g., best friends) are processed faster and more accurately than when objects are associated with strangers [25]. This kind of identity referential advantage could be used as an index of implicit association toward specific identities (i.e., faster responses represent stronger implicit association). Since the implicit association toward a specific robot is rarely examined in previous studies, we examined with the NSAT whether observation learning could affect participants' implicit associations. Self- and other-advantages refer to faster reaction times (RTs) when shapes are associated with the self and significant others, respectively, as compared to shapes associated with strangers. As opposed to self-advantage, which has absolute priority of processing, other-advantage is thought to be weaker and more plastic so that it could be easily built or shaped

[19]. Hence, the present study compared the relative implicit associations between the best friend, the assistive robot, and a stranger. We expected to observe partner-advantage of the friend and the robot (i.e., faster RTs for friend- and robot-related associations than those of stranger-related associations).

Previous studies using IAT already showed that older adults have stronger implicit negative attitudes toward robots than that of younger adults [4, 20]. Although changes in implicit attitudes are slower than explicit attitudes [4, 20], recent studies have identified that one's implicit attitudes could be rapidly updated if he or she is provided with sufficient information to dispel existing impressions. New experiences could also help the individual reinterpret previous implicit evaluations [21]. Based on these studies, we examined whether observational learning could help older adults update their implicit negative attitudes toward robots.

1.4 Hypotheses

We hypothesized that (1) attitude and acceptance of robots could be enhanced by observational learning even without direct human-robot interaction; (2) observational learning could enhance both implicit and explicit attitudes toward robots; and (3) it is possible to rapidly update participants' general implicit impression toward robots. To these goals, we used video clips as materials for participants to observe how an assistive robot could help them do daily chores because a previous study showed that video-model presentation could lead to better improvement of early acquisition of a complex judo movement [22].

2 Experiment

2.1 Overview of the Design

Two different video clips were filmed. One (video-with-interaction) contained scenarios of human-robot interactions in which one actor used five different selected functions of the assistive robot and the actor received positive outcomes after using these functions (e.g., the robot successfully helped her take a selfie). In the other video clip (video-without-interaction), the assistive robot introduced the same five functions by itself with no human-robot interaction.

Specifically, we examined if observational learning could be applied to reach the goal of improving attitudes toward robots by observing others' human-robot interactions. Alternatively, observational learning may not work in certain situations where social interaction is critical. Kuhl and colleagues showed that for American infants, interacting with native Mandarin Chinese speakers could enhance phonetic learning of a foreign language, but merely exposing them to audiovisual recordings of Mandarin materials could not [23]. Since observational learning is a form of social learning, it is also possible that the effect of observational learning cannot be revealed without direct interaction.

Participants were assigned to watch one of the two video clips in a counterbalanced manner. Before watching the video, they filled in questionnaires of trait curiosity and the Big Five personality scale. Then, participants performed the IAT as the pretest. After watching the video, we measured both the participants' explicit attitudes and implicit attitudes toward the assistive robot with the UTAUT, state curiosity questionnaire, and the NSAT to examine whether attitudes toward robots could be enhanced without direct human-robot interaction. Participants also performed the IAT again at the posttest.

Since the processes in observational learning also include reproduction of observed behaviors, participants could choose whether or not they would like to interact with the assistive robot. If they did, their adaptivity of the robot would be measured by the STAI after the interaction (Fig. 1).

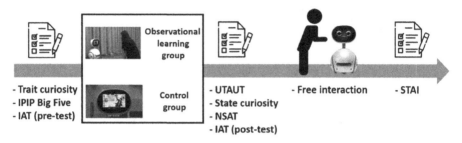

Fig. 1. The overview of the experimental design.

We predicted that participants in the observational learning group would show higher acceptance of robots in both explicit (high scores on questionnaires) and implicit measures (stronger partner-advantage of the robot), and the implicit negative attitudes toward robots would decrease more compared to those of the control group. Finally, previous studies related to acceptance of robots usually reported age-related differences. Therefore, we also recruited younger adults as the reference group to examine age-related differences in the present study.

2.2 Method

Participants. Younger participants were college students. For older participants, we recruited people who were 60 years or older. Forty younger adults (mean age: 20.84, range = 18 to 26 years; 18 males) and 40 older adults (mean age: 67.27, range = 60 to 75 years; 6 males) were recruited. They received either financial reimbursement in cash or course credits for their participation. All participants were healthy (i.e., without neuropsychological disorder) and had normal or corrected-to-normal vision. They all gave informed consent before their participation. This study was approved by the research ethics committee of National Taiwan University (REC code: 201706HS062).

3   Materials

The Assistive Robot and the Video Clips. We introduced Zenbo (an assistive robot developed by ASUSTek Computer. Inc.) to participants. Zenbo will be addressed as the assistive robot from this point onward.

We filmed two video clips: (1) video-with-interaction and (2) video-without-interaction for introducing five main functions and features (camera functions, cooking recipes finder, storytelling, schedule reminder, and remote control of household appliances) of the assistive robot. The scenarios, contents, and lengths (around 3 min) of the two video clips were almost identical. The only difference is that the video-with-interaction clip contained scenarios of human-robot interactions in which operating procedures of the selected functions were displayed and introduced to participants via a series of scenarios where one actor interacted with the assistive robot. Since the assistive robot received commands via voice control, the video-with-interaction contained dialogue between the assistive robot and the actor (note: another actor was shown as the audience in the scenarios of storytelling, but she did not interact with the assistive robot). In the video-without interaction clip, operating procedures were introduced by sequential displays of the standard operating procedures by the voice of the assistive robot for instructions.

Self-reported Questionnaires. UTAUT contains a total of 13 constructs (e.g., anxiety, perceived ease of use, attitude, etc.). Each construct included 2–5 items on a 5-point Likert scale ranging from 1 (strongly disagree) to 5 (strongly agree). The internal reliability of UTAUT for the assistive robot was checked, and only constructs with satisfactory reliability were included for further analysis (i.e., Cronbach's $\alpha > 0.70$).

STAI included three constructs (4 items in each construct) on a 5-point Likert scale for investigating different aspects of adaptivity: (1) perceived adaptivity utility (PAU), (2) technology-related goal-engagement (TGE), and (3) perceived safety of technology (PST). All constructs of the STAI passed the reliability test (PAU = .92, TGE = .94, PST = .74).

State curiosity toward robots [4] is a 3-item questionnaire on a 7-point Likert scale. The reliability was good ($\alpha = .93$). The trait curiosity questionnaire is a 7-item measure on a 7-point Likert scale. The reliability was below standard ($\alpha = .65$), and therefore the measure was removed for further analysis.

The Chinese version of the IPIP-50 Big Five personality scale, which has an across age-level measurement invariance [18, 24], was used to measure five different aspects of personality traits: openness, conscientiousness, extraversion, agreeableness, and neuroticism. as control variables of participants' personality traits.

Implicit Association Task (IAT). We used the robot impression IAT [4] to estimate the implicit negative attitudes toward robots. Participants' automatic implicit associations of two target concepts (e.g., robot vs. human) along an attribute dimension (e.g., negative vs. positive emotions) was compared based on participants' RTs in the combined categorization task. In the robot impression IAT, five positive words (ecstasy, surprise, satisfaction, happiness, and trust) and five negative words (abuse, resentment, hate, abandoned, and fear) were used for the attribute dimension of attitude toward robots. Five silhouettes of humanoid robots were used to represent the target concept "robot" and five silhouettes of humans were used to represent the target concept "human". If participants' implicit association of robot-negative pairs was stronger than that of human-negative pairs, we would expect to observe faster RTs when the robot-negative pairs were presented than when the human-negative pairs were presented. The IAT-D1 score was then calculated as an index of the effect size of the implicit negative attitudes toward robots to exclude the influences of individual differences in RT. Positive D1 scores in the robot impression IAT indicate that participants have stronger associations of robot-negative pairs than those of human-negative pairs, and larger D1 scores represent stronger implicit negative attitudes toward robots. See [4] for details and procedures of the robot impression IAT.

Name-Shape Association Task (NSAT). The procedure of the NSAT were adopted from [25]. It included two phases. The first phase was a learning phase in which participants were required to associate geometric shapes (i.e., circle, rectangle, triangle) with three targets: the name of the participant's best friend, the name of the assistive robot (i.e., Zenbo), and the name of a neutral stranger who did not exist (i.e., Ching-Ming in Mandarin). Participants were asked to memorize the target-shape associations. In the second phase, participants performed the NSAT. Using two buttons on the keyboard (i.e., "v" for matching and "n" for mismatching), they were asked to report whether a series of shape-name pairings (e.g., circle and the best friend, rectangle and the assistive robot, and triangle and stranger) was correct or incorrect based on the name-shape associations learned previously. Each trial began with a central fixation presented for 500 ms (ms). Then a shape (i.e., circle, rectangle, and triangle) and a name (i.e., best friend, the assistive robot, and stranger) were presented above and below the fixation for 100 ms, respectively. Then, there was a blank display presented for 1000 ms. Participants had to judge whether the pairings matched or mismatched the previously learned associations via pressing the corresponding key as fast as possible within the time frame of the blank display. A visual feedback ("Correct" or "Incorrect") was given on the display for 500 ms after key press. If the participant did not respond within 1000 ms, the program displayed the word "Slow" for 500 ms. Participants initially performed 30 trials for practice. Then, there were 2 blocks, each block consisted of 75 trials in which matched-pairings and mismatched-pairings were presented in a random sequence. In each block, there were 45 matching trials for three possible matching pairs (i.e., friend-matching, robot matching, and stranger matching, 15 repetitions for each pairing) and 30 mismatching trials (i.e., friend mismatching, robot mismatching, and stranger mismatching, 10 repetitions for each pairing). There were 150 trials in total.

4   Results

4.1   UTAUT

We performed a two-way between-subjects (younger vs. older and video-with-interaction vs. video-without-interaction) ANOVA to compare the UTAUT acceptance between groups for all constructs with satisfying reliability. For the "use intention" construct, there was a significant effect of age ($F(1, 76) = 10.686, p < .001, \eta p^2 = .123$). However, The main effect of video type was not significant, $F(1, 76) = 1.138, p = .290, \eta p^2 = .015$. The interaction between age and video type was not significant, either, $F(1, 76) = .096, p = .758, \eta p^2 = .001$. In terms of age differences, older adults showed higher use intention of the assistive robot ($M = 3.66$, SD = 1.11) than younger adults ($M = 2.88$, SD = 1.06) after watching the video clip, regardless of whether or not the video included human-robot interactions. As for other constructs, there were no significant main effects of age and video type and there were no interactions.

4.2 State Curiosity Toward Robots

A two-way between-subjects ANOVA was performed to test the effects of age group and video type on state curiosity toward robots after watching the video clip. Both the main effects of age ($F(1, 76) = 2.216, p = .141, \eta p^2 = .028$) and video type ($F(1, 76) = .267, p = .607, \eta p^2 = .004$) were not significant, and neither was the interaction of age and video type ($F(1, 76) = 1.131, p = .291, \eta p2 = .015$). The results suggested that there were no differences in state curiosity toward robots between groups.

4.3 STAI

Participants could choose whether or not they would like to interact with the assistive robot after watching the video clips. Fifty-five out of 70 participants chose to interact with the assistive robot, after excluding 10 participants who had to leave due to prior engagements. For older adults, 72% (13/18) and 78% (14/18) of participants in the observational learning group and the control group respectively chose to interact with the robot; for younger adults, 80% (12/15) and 84% (16/19) of participants in the observational learning group and the control group respectively chose to interact with the robot. Chi-square tests showed that neither older ($\chi^2(1) = 0.000$, p = 1) nor younger adults' ($\chi^2(1) = 0.000$, p = 1) choices were affected by video types.

We performed a two-way between-subjects ANOVA on the three constructs of STAI. We found significant main effects of video type ($F(1, 51) = 5.793, p < .05, \eta p^2 = .10$) and age ($F(1, 51) = 4.080, p < .05, \eta p^2 = .074$) in the PST construct. The interaction between age and video types ($F(1, 51) = 1.335, p = .253, \eta p^2 = .026$) was not significant. These findings indicate that younger adults had higher PST scores ($M = 3.88$, SD = .48) than older adults ($M = 3.58$, SD = .62). More importantly, participants in the observational group had stronger beliefs that using the assistive robot was safe ($M = 3.92$, SD = .47) after interacting with it than those in the control group ($M = 3.59$, SD = .60). As for other constructs, there were no significant main effects of age and video type and there were no interactions (Fig. 2).

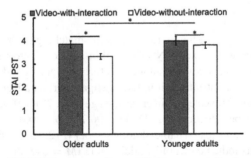

Fig. 2. Results of the STAI PST construct (*$p < .05$). Younger adults showed higher PST scores than older adults and participants who watched the video-with-interaction clip had higher PST scores than those who watched the video-without-interaction clip. Error bars represent one standard error from the mean.

4.4 IAT

A three-way ANOVA with one within-subjects factor (pretest vs. posttest) and two between-subject factors (younger vs. older; video-with-interaction vs. video-without-interaction) was performed on the average D1 scores. A significant pre-post effect was observed ($F(1, 75) = 24.179$, $p < .001$, $\eta p^2 = .244$), showing that D1 scores in the posttest ($M = .38$, $SD = .30$) were significantly lower than those of the pretest ($M = .59$, $SD = .30$). However, the main effects of age ($F(1, 75) = .364$, $p = .548$, $\eta p^2 = .005$) and video-type ($F(1, 75) = .175$, $p = .677$, $\eta p^2 = .002$) were not significant, and there was no significant two-way interaction (age × video-type: $F(1, 75) = 2.434$, $p = .123$, $\eta p^2 = .031$; age × pre-post: $F(1, 75) = 1.045$, $p = .310$, $\eta p^2 = .014$; video-type × pre-post: $F(1, 75) = 1.435$, $p = .235$, $\eta p^2 = .019$) or three-way interaction ($F(1, 75) = .013$, $p = .907$, $\eta p^2 = .000$). These results implied that participants' implicit negative attitudes were decreased after watching the video clips regardless of their types (Fig. 3).

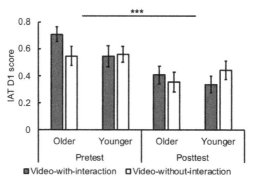

Fig. 3. The results of IAT (***p < .001). D1 scores in posttest were significantly lower than those of the pretest. Error bars represent one standard error from the mean.

4.5 NAST

We compared the RTs in both the matching and mismatching trials related to different identities with a three-way mixed ANOVA (within-subjects factor: best friend vs. the assistive robot vs. stranger; between-subjects factor: younger vs. older and video-with-interaction vs. video-without-interaction) to investigate partner advantages of the friend and the robot. The results revealed main effects of age ($F(1, 76) = 109.921$, $p < .001$, $\eta p^2 = .591$) and name-shape category ($F(2, 152) = 106.721$, $p < .001$, $\eta p^2 = .584$), as well as an interaction between age and name-shape category ($F(2, 152) = 6.559$, $p < .01$, $\eta p^2 = .079$). These results indicate that older adults' RTs ($M = 647.36$, $SD = 107.29$) were slower than those of younger adults ($M = 511.99$, $SD = 59.34$). For younger adults, post-hoc tests showed that friend-related ($M = 471.16$ ms, $SD = 45.26$ ms) and robot-related trials ($M = 521.35$ ms, $SD = 54.22$ ms) exhibited faster RTs than those of stranger-related trials ($M = 542.46$ ms, $SD = 55.05$ ms, $p < .001$, with Shaffer's modified sequentially rejective Bonferroni procedure). For older adults, friend-related ($M = 608.24$ ms, SD

= 83.38 ms) and robot-related trials (M = 685.69 ms, SD = 101.55 ms) also exhibited faster RTs than those of stranger-related trials (M = 729.14 ms, SD = 102.29 ms, $p <$ .001, with Shaffer's modified sequentially rejective Bonferroni procedure). The results demonstrated partner-advantages of both the friend and the robot regardless of video types, indicating that the identity-associated label of the robot could be established and distinguished from strangers by watching video clips even without viewing human-robot interactions (Fig. 4).

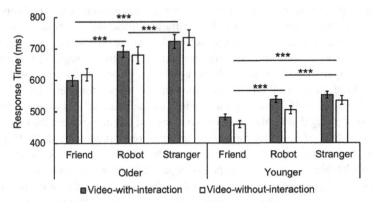

Fig. 4. RTs for different name-shape categories (***p < .001). RTs were faster in friend-related and robot-related trials regardless of video types. Error bars represent one standard error from the mean.

5 Discussion

5.1 Effects of Observational Learning on Explicit Attitudes

Regarding the question of whether or not explicit attitudes could be improved by observational learning, we first examined participants' explicit attitudes after watching videos introducing functions of the robot using questionnaires of UTAUT and state curiosity. The results showed that video type had no effect on participants' scores on both UTAUT and state curiosity questionnaires in both older-adult and younger-adult groups, indicating that observation learning could not enhance participants' acceptance of robots through explicit attitude change without interaction.

Notably, after direct interaction with the assistive robot, participants in the observational learning group had higher STAI PST scores than those of the control group. This implies that observational learning could enhance participants' sense of safety during their interactions with the robot as a result of observing a behavioral model of a safe direct interaction experience. More specifically, participants in the observational learning group had observed behavior models in which the actor interacted with the robot without any issues, which might bring the participants a sense of safety to a certain extent, and such a sense of safety could then be further verified via direct human-robot interaction. In other words, participants' explicit attitudes could be enhanced by observational learning, but only for those who actually stayed afterwards and experienced for themselves the human-robot interaction.

5.2 Rapid Changes in Implicit Attitudes Toward Robots

Although the effects of video type and age were both insignificant, implicit attitudes toward robots could be rapidly updated. The decreased IAT-D1 scores after watching a video clip implied that implicit association between the general concept of robots and negative words was attenuated by the demonstration of the assistive robot's functions.

It should be noted that IAT-D1 scores were still positive even in the posttest across all experimental groups, indicating that the implicit association of negative-robot pairing was still stronger than the negative-human pairings. However, the pre-post effect of video watching on the IAT showed that the implicit association between the general concept of "robots" and the negative attribution was attenuated. It is in line with recent evidence supporting that implicit first impressions and attitudes toward the object could be rapidly updated if new and convincing information was provided [21]. Hence, the results imply that the implicit negative attitudes toward the concept of "robots" might not be unshakeable, and could be shaped or adjusted by new information.

On top of that, the partner-advantage of the robot (i.e., faster RTs in the robot-shape than the stranger-shape pairing) observed in the NSAT implied that a brief introduction of the assistive robot could enhance the human-robot association. It might not be surprising that the partner-advantage of the robot was built after video-watching because both video clips demonstrated its functions to assist users. Participants could easily understand that the assistive robot was designed to provide daily assistance, and hence the association between the participants and the assistive robot was enhanced. Recent evidence in human-human interaction experiments showed results similar to the present study: a newly met human partner in a joint NSAT could form the partner-advantage [25], implying that forming partnerships in human-robot and human-human interactions may be similar to each other.

6 Summary

Our results showed that both older and younger adults' explicit attitudes toward the assistive robot could be enhanced via observational learning, but direct human-robot interaction was crucial for this modulating effect. Furthermore, implicit attitudes toward the robot could be rapidly improved by introducing information in the utility aspect of the robot to users, suggesting that negative implicit attitudes toward robots are not unchangeable.

References

1. Lin, Y.-Y., Huang, C.-S.: Aging in Taiwan: building a society for active aging and aging in place. Gerontologist **56**, 176–183 (2015). https://doi.org/10.1093/geront/gnv107
2. Roy, N., et al.: In: Workshop on Interactive Robots and Entertainment (WIRE 2000), pp. 184

3. Smarr, C.-A., et al.: Domestic robots for older adults: attitudes, preferences, and potential. Int. J. Social Robot. **6**, 229–247 (2014)
4. Chien, S.E., et al.: Age difference in perceived ease of use, curiosity, and implicit negative attitude toward robots. ACM Trans. Hum. Robot Interact. **8**, 1–19 (2019). https://doi.org/10.1145/3311788
5. Scopelliti, M., Giuliani, M.V., Fornara, F.: Robots in a domestic setting: a psychological approach. Univ. Access Inf. Soc. **4**, 146–155 (2005)
6. Heerink, M.: In Proceedings of the 6th International Conference on Human-Robot Interaction, pp. 147–148 (ACM)
7. Mast, M., et al.: User-centered design of a dynamic-autonomy remote interaction concept for manipulation-capable robots to assist elderly people in the home. J. Hum. Robot Interact. **1**, 96–118 (2012). https://doi.org/10.5898/jhri.1.1.mast
8. Heerink, M., Kröse, B., Evers, V., Wielinga, B.: Assessing acceptance of assistive social agent technology by older adults: the almere model. Int. J. Social Robot. **2**, 361–375 (2010)
9. Chen, T.L., et al.: Older adults' acceptance of a robot for partner dance-based exercise. PLoS ONE **12**, e0182736 (2017)
10. Barnard, Y., Bradley, M.D., Hodgson, F., Lloyd, A.D.: Learning to use new technologies by older adults: Perceived difficulties, experimentation behaviour and usability. Comput. Hum. Behav. **29**, 1715–1724 (2013)
11. Bandura, A.: Observational learning. The international encyclopedia of communication (2008)
12. Bandura, A., Ross, D., Ross, S.A.: Transmission of aggression through imitation of aggressive models. J. Abnormal Soc. Psychol. **63**, 575–582 (1961). https://doi.org/10.1037/h0045925
13. Yi, M.Y., Davis, F.D.: Developing and validating an observational learning model of computer software training and skill acquisition. Inf. Syst. Res. **14**, 146–169 (2003)
14. Beer, J.M., et al.: Older users' acceptance of an assistive robot: Attitudinal changes following brief exposure. Gerontechnol. Int. J. Fundam. Aspects technol. Serve Ageing Soc. **16**, 21 (2017)
15. Dovidio, J.F., Kawakami, K., Gaertner, S.L.: Implicit and explicit prejudice and interracial interaction. J. Pers. Soc. Psychol. **82**, 62 (2002)
16. Kamin, S.T., Lang, F.R.: The Subjective Technology Adaptivity Inventory (STAI): a motivational measure of technology usage in old age. Gerontechnology **12**, 16–25 (2013). https://doi.org/10.4017/gt.2013.12.1.008.00
17. Kashdan, T.B., Rose, P., Fincham, F.D.: Curiosity and exploration: facilitating positive subjective experiences and personal growth opportunities. J. Pers. Assess. **82**, 291–305 (2004)
18. Goldberg, L.R.: The development of markers for the big-five factor structure. Psychol. Assess. **4**, 26 (1992)
19. Sui, J., He, X., Humphreys, G.W.: Perceptual effects of social salience: evidence from self-prioritization effects on perceptual matching. J. Exp. Psychol. Hum. Percept. Perform. **38**, 1105 (2012)
20. MacDorman, K.F., Vasudevan, S.K., Ho, C.-C.: Does Japan really have robot mania? Comparing attitudes by implicit and explicit measures. AI Soc. **23**, 485–510 (2008). https://doi.org/10.1007/s00146-008-0181-2
21. Ferguson, M.J., Mann, T.C., Cone, J., Shen, X.: When and how implicit first impressions can be updated. Curr. Dir. Psychol. Sci. **28**, 331–336 (2019). https://doi.org/10.1177/0963721419835206
22. Lhuisset, L., Margnes, E.: The influence of live- vs. video-model presentation on the early acquisition of a new complex coordination. Phys. Educ. Sport Pedagogy **20**, 490–502 (2015). https://doi.org/10.1080/17408989.2014.923989

23. Kuhl, P.K., Tsao, F.-M., Liu, H.-M.: Foreign-language experience in infancy: effects of short-term exposure and social interaction on phonetic learning. Proc. Natl. Acad. Sci. **100**, 9096 (2003). https://doi.org/10.1073/pnas.1532872100

24. Li, R.-H., Chen, Y.-C.: The development of a shortened version of ipip big five personality scale and the testing of its measurement invariance between middle-aged and older people. J. Educ. Res. Dev. **12**, 87–119 (2016). https://doi.org/10.3966/181665042016121204004

25. Cheng, M., Tseng, C.-H.: Saliency at first sight: instant identity referential advantage toward a newly met partner. Cogn. Res. Principles Implications **4**, 42 (2019). https://doi.org/10.1186/s41235-019-0186-z

Design Smart Living Scenario Through Future Research Tools

Songling Gao, Jeongeun Song, and Zhiyong Fu[✉]

Tsinghua University, Beijing, China
{gs119,songzy19}@mails.tsinghua.edu.cn,
fuzhiyong@tsinghua.edu.cn

Abstract. The paper presents a design method in progress that will develop future scenario within the context of the smart living. First, we collated the literature on futurology to study the logical relationships and the effects of interactions between theories. Through these studies, we try to explore the layers between theories to help us better explore future scenarios. There are three stages in our design model, namely the Discovery stage, the Deepen stage and the Transformation stage. Each stage we will use different tools, and the tools will help to predict and design the future situation in our lives. In the second study, we take a smart living scenario as an example. we categorized smart living into two sections of health and security, and apply the different stages of the model we designed in the previous part to practical examples. Finally, we wrote a service blueprint on how those technologies would be developed in the future and be used for smart devices or application in smart living, displaying its service and the underlying entire process of smart system.

Keywords: AI cities · Smart cities · Future study · Design research

1 Introduction

Based on theoretical research, this article applies theoretical research on futurology to the practice of designing the future. Integrate existing theoretical tools on futurology through a modelled thinking process. We are trying to build a design model, which is a tool for the transformation between futuristic theoretical research and designing actual future scenarios. Through the integration of theoretical methods and traditional design tools, it can help us create more possibilities for future scenarios.

2 Background

Futurology explores the development prospects of future society. Through research, everyone pays more attention to the prospects and conflicts of possible future development. Design involves intelligence, economics, culture and other aspects, and constantly enters the core of society. A preliminary exploration of the futurology of design can allow

© Springer Nature Switzerland AG 2020
P.-L. P. Rau (Ed.): HCII 2020, LNCS 12192, pp. 577–590, 2020.
https://doi.org/10.1007/978-3-030-49788-0_44

more people to stand in the perspective of futurology, look back to the present, and identify design development opportunities. First, we sorted out previous scholars' research on futurology and smart cities.

Futurology was first proposed and used by the German scholar Ossip Flechtheim in the 1940s [1]. It explores the prospects for the future development of science and technology and society, and reveals the possibility of moving towards the future in accordance with various choices made by humans.

Zhang Shaohong proposed the connection between three worlds and three dialectics [2]. The three worlds are the unknown world, the essential world, and the future world. They correspond to three dialectics, namely, dialectics of discovery, dialectics of reproduction, and dialectics of practice. Together, they constitute the methodology of futurology. In the future, we must study the past and the present. Only by knowing the past can we predict the future.

Han Jingxian and Xu Yunpeng introduced the main ideas of the three books of Toffler's "Future Shock", "The Third Wave" and "Transfer of Rights" [1]. The three books analyzed the thoughts of the information society on people's understanding of the world in a progressive narrative mode Impact. He proposed that while exploring the future, research should be deeply rooted in history, and the future should be considered based on history. Therefore, we must follow a certain theoretical method for future research and can't be described in fantasy.

Sohail Inayatullah proposed the Causal layered analysis (CLA) in 1993. CLA is not in predicting the future but in creating transformative spaces for the creation of alternative futures [3]. He mainly focuses on the vertical level of future research. It consists of four levels: the litany, systemic causes, discourse/worldview and myth/metaphor.

Chris Riedy explored the theoretical and conceptual differences between Integral Futures work and CLA in 2007 [4]. His purpose is to determine whether the theory behind CLA is consistent with Integral Theory and how CLA relates to Integral Futures work. He compared the two theories. The starting point of these two theories is the same. They are both understand development as a process that brings greater depth and both use depth as the basis for judging which perspectives are more valid. But when establishing the level of stratification, they made a difference.

Integration theory is more comprehensive. It has four quadrants: behavioral quadrant, systemic quadrant, cultural quadrant, and psychological quadrant. But the four levels of CLA are mainly focused on the cultural quadrant. the psychological perspective is ignored. Therefore, CLA as an integration theory is not very complete. In the future, we can develop a more complete CLA and go deeper into each quadrant level to deepen the understanding of reality at each level.

Sohail Inayatullah further considered and summarized futuristic research methods based on the previously studied Causal layered analysis (CLA) in 2008 [5]. He proposed six basic concepts of futures thinking, the six basic futures questions, and the six pillars of futures studies. Through a more systematic and comprehensive summary, he presented the futuristic research with more possibilities. As he said, "Futures thinking does not wish to condemn us to hope alone."

These theories provide important theoretical support for our research. In addition to, we have also compiled some researches by other scholars on the vision of future cities;

Wang Ding points out while technology proliferates dramatically in short time, the 'smart city policy' is not only superficial, but it also accommodates contesting priorities in each different city faced by very divergent problems [6]. In fact, even though the current development of technology is grown rapidly, the smart city has to demonstrate its own financial value. This is what designers have to consider when designing smart city based on ICT development which requires financial needs. To such a complicated issue, this is really good example to address the 'real smart city' how it should be more sensibly designed and what might be a point to be considered.

Louise Mullagh seeks to address how human values can be made manifest within a physical smart city and identify potential effectiveness a visual approach to 'the value reflection map'(VRM) tool [7]. He fundamentally focused on technology as a component of developing user-centric design methodology into the urban environment. He stresses a co-dependence of relationship between human values and design as it is vital to seek to investigate how consideration of human values could enable designers to question whether introducing new technologies into a space was appropriate and if so, how values might be designed into such technologies.

3   Research Process

Based on Sohail Inayatullah's research methods and other scholars' theoretical research on futurology, we have conducted a lot of testing and discussion. According to the characteristics of the design discipline, we try to set how to transform futuristic theory into design practice. We designed a set of tool models to design the future, and tested through workshop and other forms.

The design model is divided into three stages:

3.1   The First Stage: Discovery

At this stage we will use three theories about futurology:

The Future Triangle
The Futures Triangle refers to today's view of the future through three dimensions. The three sides of this triangle represent *Pull of the Future, Push of the Present*, and *Weight of Historical.*

Pull of the Future pulls us forward. While there are many images of the future, it can bring social progress through science and technology, or imagine the limits that humans may reach in the future, and whether history will return to its original state and so on. Push of the Present refers to a current trend and driving force. This current trend will affect our thinking about the future. The last one is Weight of Historical, which is an obstacle we couldn't change. By using this tool, we can have a preliminary understanding of the relationship between the present and the future.

For example, in the future, autonomous driving technology will bring convenience to our lives and increase our efficiency. Now we can find that today many manufacturers of autonomous driving technology are increasing their R&D efforts, and their technology is constantly being improved. But in history, people have doubts about the safety and

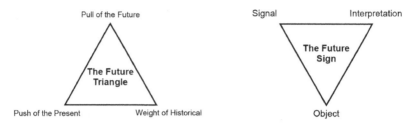

Fig. 1. The Future Triangle and the Future Sign

reliability of autonomous driving technology, and their acceptance is not very high (Fig. 1).

The Future Sign

The second tool is *the Sign of the Future*, which has some similarities with the future triangle tool. It also has three dimensions, which are signal, object, and interpretation. Signals are messages from newspapers or news. These signals have a certain amount and visibility. The object indicates that what we see and feel is really happening around us and is an objective fact. Interpretation is an understanding of future trends, biased towards subjective reasoning.

Through the three dimensions of the content, you can analyze these behaviors may have an impact on future development trends. For example, in terms of signal, there is increasing publicity about autonomous driving technology in the news. In terms of object, we can see the increase in private cars causing traffic jams. At the same time, Baidu's Apollo self-driving cars started testing on the road. Therefore, in terms of interpretation, we can infer that future autonomous driving technology will improve our transportation environment, and sharing autonomous driving will be a development trend (Fig. 2).

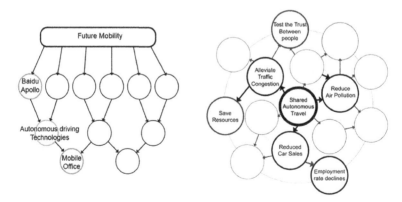

Fig. 2. The Future State and the Future Wheel

The Future State

The Future State is a transformation tool that infers the possibility of the future from the current products and technologies. we looked for current apps and products related to smart living, analyzed their technologies and systems, and listed the possibilities that may arise in the future. For example, for future mobility, we first organized existing products and technologies, such as Baidu Apollo. Behind the product is autonomous driving technology. With the continuous development of these technologies, mobile office may be generated, or autonomous driving technology will gradually mature.

The three tools we use to explore the future at this stage are to understand the current state. This is a relatively shallow research.

3.2 The Second Stage: Deepen

At this stage we will use three tools about future design: *The Future Wheel, Causal layered analysis (CLA)* and F*our-Quadrant Mapping.* Through the use of these three tools, we will deepen the future based on the previous stage.

The Future Wheel

The Future Wheel seeks to find out the consequences of the current problem from a longer-term perspective. It intends to explore and deduce unexpected results. The circle in the center is the space to write down with the theme of future possibility, and the surrounding circles of the center are to list the effects on it and infer the causes and consequences from it. In this way, we can move from seeing the world as a simple, separated level to seeing the world as a complex and interconnected level. For example, we use this method to discuss the impact of shared autonomous travel in the future, which may reduce air pollution and alleviate traffic congestion. Going further out may save resources, but at the same time, it will have an impact on car sales and test the trust between people.

Causal layered analysis (CLA)

Causal layered analysis (CLA) seeks to unpack, to deepen the future. It includes four levels in the vertical level: *the litany, systemic causes, discourse/worldview, and myth/metaphor.* It is a research method that precisely defines and deconstructs problems.

i.   Litany: It is the official public or media description of an issue. The problem can be exaggerated to form a fear, which can be avoided through proper action, but the solutions used at this level are short-term.
ii. Systemic Causes: It focuses on the systemic causes of the problem itself, including political, economic, environmental, historical, social, and scientific aspects. Stakeholders will stand on each side to explain the incident.
iii. Discourse/Worldview: It is a holistic view of the overall situation, how we use our cognition to understand or shape the world. Go beyond the deeper social, historical, and cultural reasons of the parties, redefine the problem, and restructure our perception of the problem.

iv. Myth/Metaphor:It is the deepest layer, focusing on the deep stories and collective archetypes. It is a subconscious and emotional aspect. As this requires telling a new story, rewiring the brain and building new memories and the individual and collective body.

For example, using CLA for analysis of urban traffic. In the level of litany, we see the problem of traffic congestion and shortage of parking spaces. The possible solution is just to stagger the time to get off work and avoid peak periods.

In the level of systemic causes, we can see that the improvement of living standards has increased the number of private cars, which has caused problems such as environmental pollution and waste of resources. The possible solutions include the development of clean energy and autonomous driving technologies.

At the third level, we look for deeper issues. People's values may be inherent in thinking that when they buy items, they are considered to own them. At this time, we can promote the sharing economy and appeal users to focus on the service attributes of the product rather than owning the product itself.

At the last level, people always think that as many things and humans as possible are the smartest beings. Then we can create value: robots can do things for humans, goods can be either their own or the best, and the sharing economy can maximize the use of resources (Fig. 3).

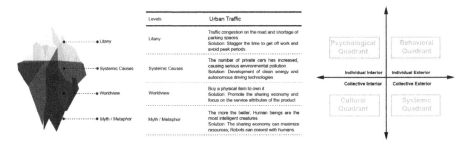

Fig. 3. Causal layered analysis (CLA) and four-quadrant mapping

Four-Quadrant Mapping
This tool can be considered as a supplement to CLA. It involves a wider range of aspects. We need to consider the relationship between internal, external, individual and collective.

There will be a total of four quadrants here. The first quadrant is the individual interior perspective, it is also a psychological quadrant, which is the realm of the self, consciousness, personal experiences and values. The second is the individual exterior perspective, which is the behavior we do, similar to the first level of CLA. The third is the collective exterior perspective, which is the system strategy adopted by the organization. It is similar to the second level of the CLA. The last quadrant is the collective interior perspective, which is a cultural-oriented thinking, similar to the last two levels of CLA.

Comparing the CLA and four-quadrant mapping, we can find that the four-quadrant mapping considers the psychological quadrant, but CLA does not consider it. The combination of these two methods will make our second deepening stage more comprehensive.

For example, when we mentioned shared autonomous driving as an example. In addition to the three aspects covered by CLA, the interior level of the individual is equal to the level of the psychological. We also need to think about the user's acceptance of autonomous driving, including issues such as safety and accuracy, as well as issues such as trust between people in a shared environment.

The successive use of these three tools makes our exploration of the future more comprehensive, and the aspects and areas involved also have a certain depth and credibility. The analysis from the surface to the deeper layer laid a solid foundation for our next stage of design.

3.3 The Third Stage: Transform

The third stage is the *Transform Stage*. At this stage, we will use design tools we are familiar with, such as *User Journey Maps*, *Service Blueprints*, or *Storyboards*. These tools are used to show our vision of future scenarios.

User Journey Map

User Journey Map is a common design tool. Through observation and research, draw the user's existing journey. However, at this stage, we use the tool to draw what the future journey will look like. Through the first two stages, we have sorted out the current problems and propose solutions. At this stage, we use this method to express.

As shown in the figure, in the future journey, our lives will be smarter. Shared autonomous driving makes people travel more efficiently. Of course, the journey in the future will not be entirely happy. When the intelligence of everything becomes higher, it will also make people feel lonely (Fig. 4).

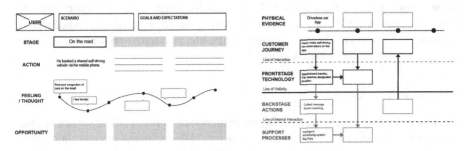

Fig. 4. User Journey Map and Service Blueprint

Service Blueprint

The Service Blueprint is also a familiar design expression tool. We use the service blueprint to describe the entire system of future scenarios. Including *customer journey, frontstage technology, backstage actions, support processes, etc.*

Storyboard

The tool of the *Storyboard* is a choice of presentation method. After we have drawn the future service blueprint, we can express it in the form of a story, so that readers can better understand it. Of course, in addition to using the storyboard, you can also use tools such as role-playing to demonstrate.

By now, we have completed the tasks of the transformation stage, and the three phases of the entire design model have been completed. Through the above three stages, from the initial discovery stage to the deepening stage, and then the theory to the design stage, showing our vision of the future. This model can help us to translate theoretical research into design, and create more possibilities for future scenarios. We drew the whole process of design research in the diagram below to make it easier for everyone to understand (Fig. 5).

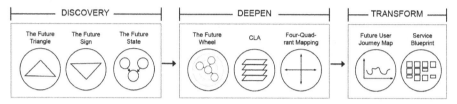

Fig. 5. The whole flow of our tool model

In the next part, we will introduce the results of each step of our entire experimental process, so that readers can clearly feel the role of our tools.

4 Validation Examples: Smart Living

Smart cities are defined along with six dimensions. Smart economy, smart mobility, smart governance, smart environment, smart living and smart people. To validate these research methods and predict our plausible future life, inspired by present user scenario, smart living is chosen among six dimensions.

4.1 Future Triangle

Pull of future: At home, you can enjoy convenient one-to-one medical service without going to the hospital to register and see a doctor. 24-h intelligent security monitoring, control every corner of the home anytime and anywhere and could ensure the safety of family and pets from hidden dangers. In addition to all linked home furnishings, the increase in individual computing performance can also perform more tasks with energy saving. There are a large number of applications and highspeed networks, making humans more and more capable of acquiring information.

Push of the present: Medical resources are tense, medical trends are digitized and online appointments and consultations are made on the Internet. For this, the network speed is improved, and real-time remote security monitoring is gradually mature. The birth of AI speech recognition technology had begun with machine learning which was successfully applied to issues in industry and the application of powerful computer algorithms through acoustic and linguistic modeling. It can be used in homes and businesses, allowing users to speak on the computer and convert words into text through word processing and speech recognition.

Weight of history: Fundamentally, there are four steps when patients get diagnosis of a professional in china. "Look, smell, ask, cut" which represents that we hope to observe the patient's physical condition, such as face color, patient's voice, cough, smell in one's mouth or body for any odor and ask patient about their symptoms in person. Therefore, face to face diagnosis and treatment make patients feel reassured. People prefer emotional and humanized equipment and products to cold impersonal technology. High-tech networks and smart device systems are packed with electronic sensors that track and respond to the movements of residents and are artificially changing the social habits and movements that have been rooted for hundreds of years, creating a uniform pattern.

4.2 Future Sign

Signal: Businesses take smart life as their selling point, and put the advertising of smart technology plus home in a prominent position. They, for example, began to invest a lot of cost to produce and promote a variety of intelligent furniture products. On the other hand, in the medical fields, people can get a medical checkup themselves by using medical smart kits instantly at home.

Issue: "Google Ngram Viewer" shows how often words related to 'smart living' are mentioned in English books, and the graph for 'smart living' is rising. Therefore, we can see how dominant the word 'smart living' is in the recent trend. Home products with intelligent technology can be easily seen in the store and physical E-commerce. More and more people choose to buy smart furniture such as AI speaker that play songs and tell user the expected weather.

Interpretation: All household devices are connected and have self-judgement abilities. The reasons for purchasing smart furniture may change from display trend and function trend to emotion trend. Thus, people seek more emotional and humanized interaction with household equipment. Thanks to smart self - medical checkup equipment which is simply used, we can save our time and check our body data instead of getting medical information from doctors. Therefore, we might don't need any medical expert in the future. However, people are too dependent on smart devices and systems for living space, which arouses criticism that what human do is no longer. Furthermore, along with the development of smart cities filled with complex systems, political and policy aspects such as the new social regulation policy should be developed as well.

4.3 Future State

Let us start by smart living's future state diagram in our discovery stages. Here we categorized living into two parts: health/balance and security. Based on researching smart products and applications which are now in its stages of commercialization, we inspected in detail that which technologies are used in current application and make us place into smart living (third layer). We found that, for example, voice recognition and differentiation via biometric technology is applied to 'Pria' product; smart personal medication assistant. An important part of this diagram is the fourth layer. That is, biometric technology can have an effect on possibility of voice-based personal predictive health analysis product in the future living (Fig. 6).

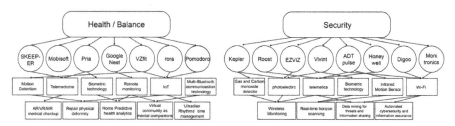

Fig. 6. Two parts [health and security] of future state diagram

4.4 Future Wheel

These future wheels map the consequences of medical issue if voice is used as diagnostic and medical tools and implication of upgraded IoT security (Fig. 7).

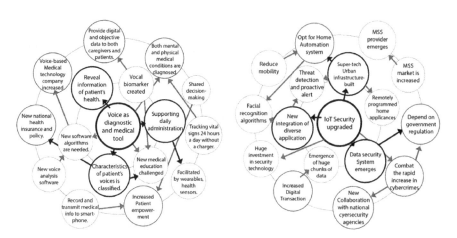

Fig. 7. Future wheel of [healthy] and [security]

For example, for using personal smart medication assistant, characteristics of user's voice should be classified. As voice can reveal information of patient's health, vocal biomarker is created… As well, track vital signs both mental and physical conditions are diagnosed 24 h a day. After recording medical information, provide digitalized and objective those data and transmit to both caregivers and patients by smart phone. Voice-based medical technology companies increase and they develop new voice software algorithms. At the same time, new national health insurance and policy are needed.

Another example, we anticipated that upgraded IoT security might influence data security systems and super-tech urban infrastructure-built. Moreover, security is ensured so that multiple applications can be converged, and this new form of integration forms the results of the face recognition algorithm for a strong security system. Still, though security system is upgraded and "Managed Security Services (MSS)" market is increased, as digital automatic transaction between parties is continue to increase, in fact, it undoubtedly makes to increase the likelihood of exposure to cybercrime. Hence new collaboration with national cyber security agencies can combat the rapid increase in cybercrimes .

4.5   Causal Layered Analysis (CLA)

Litany problems are often loneliness, crime in life and AI technology still has a long way to go. Solutions focus on revolutions of home security technology and are to form both virtual and reality community. Systemic solutions are to provide new future career and extend the terms of employment. In addition, ensuring legal and ethical technology frameworks could be long-term strategy. Policy strategy can be deeper. For example, New national technology policy by government need to be redefined (Table 1).

At the worldview level, we see the issue of smart living may be: sustainable development, instead of undermining stability of the natural system and resources, can enhance energy efficiency. It can include pursuing eco-friendly residential construction design, using green materials. Create equal opportunity to ensure healthy society without affecting quality of smart living in negative aspects. At the myth and metaphor level, the pursuit of value of residents in smart cities can be metaphorized as 'high quality snail life'. In place of inhuman contained in productivity, efficiency, competition, speed and affluence, which are the ideological characteristics of modern society, it is a way of smart life that seeks a harmonious life between nature and human beings. At the same time, quality life that meets human needs.

4.6   A Four-Quadrant Map

Four-quadrant mapping is important as CLA. The upper-left is the inner-individual – this is psychological quadrant which is what we find people's own interior emotions and thoughts. For instance, people are content with good quality of living that reduces energy and maximizes the use of reusable and recycled materials. Similarly, people feel convenient when they use smart appliance at home. In contrast, they feel depressed by solitary pastime and chronic diseases. The upper-right is the external-individual - the behavior people engage in as they are blessed with smart living situation. For example, they get medical checkup conveniently themselves and health care by smart medical devices at home. Also, they seek green roof for sustainable smart living. Purchasing AI robot as a friend helps not to feel depressed.

Table 1. Applying causal layered analysis to smart living.

Litany	The average usage rate of voice assistant has been higher
	The way of voice interaction is very mechanical and still clumsy
	Single girls become targets of crime
	The number of people living alone is increasing
	Solution: Revolution of home security technology and form both virtual and reality community
Systemic causes	The employment rate is low
	All internet enterprises are chasing the technological commanding point
	IoT technology makes everything interconnected
	The problem of aging is serious and the marriage rate is declining
	Solution:
	Provide new future career and extend the terms of employment
	Form health care cities and redefine national new technology policy
	Ensure legal and ethical technology frameworks
Worldview	The intelligent of human life style
	The diversity of residential choice
	Pursuing eco-friendly of residential design
	Effort to expand the field of intelligent technology
	Enhance the concept of sustainable development
	Solution: Cost-saving construction and home automation
Myth/ Metaphor	High quality snail life
	Solution: Balance with super technology and slow tempo life

The lower-left is that developing intelligent of human life style and let resident be able to choose diversity of residential style. Additionally, pursue eco-friendly of residential design. The last – enhance the concept of sustainable development and develop networking technology are the strategy official organization have to take into account (Table 2).

Table 2. Four-quadrant map of the smart living. This map is related to causal layered analysis.

Feeling good living in house that reduces energy and maximizes the use of reusable and recycled materials	Home self-medical checkup
	Seek green roof living construction
	Purchase AI robot as a friend
Feel depressed by solitary pastime and chronic diseases such as depression after retirement	
The intelligent of human life style	Enhance the concept of sustainable developments
The diversity of residential choice	
Pursuing eco-friendly of residential design	Develop networking technology

4.7 Blueprint

See Fig. 8

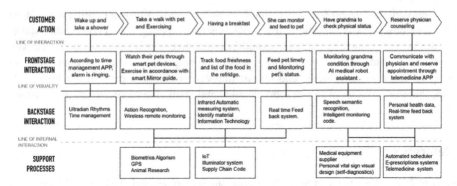

Fig. 8. Future service blueprint in smart living.

5 Related Works

Stanford d. school 'human cities expo' is program in which Stanford students and Beijing Tsinghua University students present innovative approaches to urban planning and solve problems in the modern world. We were grateful to be involved this program. The expo, which took place in December 2019, was announced by students to plan a human-centered and sustainable city, with an interdisciplinary program that not only allows future cities to work and live comfortably for people, but also to become smart places where they can thrive and connect with each other. To this end, first, students conducted interviews and analyzed the data with deep insight to identify the serious urban problems currently underway. Then, they suggested, for instance, urban regeneration through community engagement and design of community landscape furniture as solutions. In the next breath, we then shared our research in 'smart cities, vision future' chapter. We have successfully introduced our scenarios of future smart living and smart people, including those on technical solutions (service blueprint), as whole of academic studies.

6 Conclusion

In summary, through the research and arrangement of the process, we have established the use of the tool model. There are three stages, the first stage is observing, the second stage is deepening, and the third stage is transforming. Through the use of different tools at each stage, progressing from the external surface to the interior, we could help us suggest the possibility of future smart living scenarios.

Of course, these tools not only could be used to comprehend current smart living scenario and to how technology affects future state of smart living, but also be applied to six dimensions such as smart education and smart governance, as mentioned in section [validation example]. Therefore, in the course and other environments, we tested the tool model in different scenarios, and verified the feasibility and effectiveness of the tool through tests.

Although the current speed of technology development is grown rapidly, since smart city is so obscure and superficial that when designers, policy makers, and other stakeholders arrange smart city plan, they could face very divergent problems. This tool model would help them identify opportunities and vulnerabilities of smart city plans, discovering potential improvements and eliminating each redundant part. In the future, we hope that this tool model can be more detailed and applied to more aspects about future scenarios.

References

1. Han, J., Xu, Y.: Review of Toffler's "Future Studies" trilogy. Sci. Technol. Inf. Dev. Econ. **19**(32), 124–126 (2009)
2. Shaohong, Z.: Three worlds and three dialectics——on the methodology of futurology. J. Xiangtan Univ. (Philos. Soc. Sci.) **01**, 44–51 (2004)
3. Inayatullah, S.: Causal layered analysis. Futures **30**(8), 815–829 (1998)
4. Riedy, C.: An integral extension of causal layered analysis. Futures **40**(2), 150–159 (2007)
5. Inayatullah, S.: Causal layered analysis defined. The Futurist **48**(1), 26 (2014)
6. Wang, D.: HCI, Policy and the Smart City (2016). https://doi.org/10.14236/ewic/HCI2016.35
7. Mullagh, L., Lynne, B., Nick, D.: Beyond the 'Smart' City: Reflecting Human Values in the Urban Environment (2014)

Cross-Cultural Design of Facial Expressions of Robots

Ichi Kanaya[1]([⊠]) [iD], Meina Tawaki[1], and Keiko Yamamoto[2]

[1] The University of Nagasaki, Manabino 1-1-1, Nagayo, Nagasaki, Japan
kanaya@pineapple.cc
[2] Kyoto Institute of Technology, Matsugasaki, Sakyo-ku, Kyoto, Japan
https://pineapple.cc

Abstract. In this research, the authors succeeded in creating facial expressions made with the minimum necessary elements for recognizing a face. The elements are two eyes and a mouth made using precise circles, which are transformed to make facial expressions geometrically, through rotation and vertically scaling transformation. The facial expression patterns made by the geometric elements and transformations were composed employing three dimensions of visual information that had been suggested by many previous researches, slantedness of the mouth, openness of the face, and slantedness of the eyes. In addition, the relationships between the affective meanings of the visual information also corresponded to the results of the previous researches.

The authors found that facial expressions can be classified into 10 emotions: happy, angry, sad, disgust, fear, surprised, angry*, fear*, neutral (pleasant) indicating positive emotion, and neutral (unpleasant) indicating negative emotion. These emotions were portrayed by different geometric transformations. Furthermore, the authors discovered the "Tetrahedral model," which can express most clearly the geometric relationships between facial expressions. In this model, each side connecting the face is an axis that controlled the rotational and vertically scaling transformations of the eyes and mouth.

Keywords: Facial expression · Emotion · Design · Human factors

1 Introduction

When human beings communicate with each other, their faces convey the most important and richest information, and their facial expressions are by far the most essential element for understanding the other's emotion. Various elements such as the eyes, mouth, and nose are contained in the face, and these are transformed to form facial expressions. Transformation is the method we use to make expressions. For many primates including human beings, the eyes are one of the most important factor used to recognize a face as a face [1–3]. In addition, visual preference for face figures (patterns in which three figures are

© Springer Nature Switzerland AG 2020
P.-L. P. Rau (Ed.): HCII 2020, LNCS 12192, pp. 591–605, 2020.
https://doi.org/10.1007/978-3-030-49788-0_45

arranged at the top of an upside-down triangle) [4,5] and research on perceptions of upside-down faces [6,7] are well known regarding facial recognition. These research results have suggested that primates recognize faces as sets of configural information comprised of facial elements, and in particular, the eyes and the mouth are the most important elements of all.

Meanwhile, these days there are two theories regarding the recognition of facial expressions: the "category perception theory" and the "dimension theory" [8–11].

The category perception theory states that human beings judge the meaning of facial expressions through 7 ± 2 universal categories common to all human beings. This theory is based on a basic theory of cognition and emotion [12] stemming from evolutionary theory [4]. Those who advocate this theory insist that facial expression is not a continuous variate but a discrete variate, and deny the presence of the psychological dimension described later in the cognitive process of expression [10]. In terms of emotional categories, the six basic emotions (happy, angry, sad, disgust, fear, and surprise) advocated by Ekman and many other researchers are the most typical. Basic emotions synchronize with physiological responses and signals to the body such as facial expressions [14], and it is proposed that facial expressions can be classified under one of the six basic emotions without exception irrespective of culture [15–21].

The dimension theory proposes that affective category judgment is conducted following previous judgment that the facial expression is located as one point in two or three dimensional space [8,10], and that facial expression is a continuous variate. In addition, those who advocate this theory insist that the universal factor for human beings is not category but dimension [13]. Dimension theory begins with Schlosberg's theory of the dimension of emotion, for example the circular ring model [22] comprising two dimensions, namely pleasant vs. unpleasant and attention vs. rejection, and the circular cone model [23] comprising the previous two dimensions plus tension vs. sleep [8,10,24]. Since this research, many researchers have discussed such affective meaning dimensions and have repeatedly encountered three dimensions: the pleasantness dimension (pleasant vs. unpleasant), the attention vs. rejection dimension, and the activeness (awareness) dimension (aware vs. asleep) [8–10,24]. Especially in recent times, the circumplex model comprising a pleasantness dimension and an awareness dimension suggested by Russell [13] has been validated in terms of its universality and robustness by many previous researches [24,25].

Thus, since Schlosberg's research [22,23], many researches have been conducted attempting to find an affective meaning dimension (psychological variable) relating to the recognition of facial expressions. However the visual information related to the recognition of facial expressions has been little researched. This means that the relationship between the psychological variable and the physical variable has not yet been identified regarding human cognizance of emotion by facial expression.

Yamada [26] conducted a study to clarify visual information (the physical variable) related to the cognition of facial expressions using a line-drawing figure in which eight points of the eyebrows, eyes and mouth are manipulated. From the results, two physical variables have been found: slantedness, meaning the curve and indication of face elements; and openness and curvedness, meaning the level of curve and openness of facial elements. In addition, they have suggested that there are strong relationships between slantedness and the pleasantness dimension, and between openness and activeness [27,28]. Based on this knowledge, they proposed that there are three processes used to cognize facial expressions: (1) acknowledgement of visual information (the physical variable) of the face, (2) evaluation of affective meaning based on the physical variable, and (3) judgment of the emotional category based on affective meaning [10]. Moreover, from similar research using actual human faces, it has been suggested that there are three physical variables of visual information—openness, slantedness of mouth, slantedness of eyebrows and eyes [29]—and that the same results as those from in a three-dimensional structure would be produced in the case of using a line-drawing face [30]. Furthermore, it has been determined that there is a strong relationship between the slantedness of the mouth and the pleasantness dimension, and between the slantedness of the eyebrows/eyes and activeness, as well as a moderate relationship between openness and both forms of slantedness [31]. Through competing research using line- drawing faces and actual human faces, a strong relationship was also found between the openness of the eyes and mouth and pleasantness [32]. In this respect, there is a close relationship between visual information (the physical variable) and emotional information (the affective variable).

The objective of this thesis is to validate the relationship between the physical variable and the affective variable discovered by previous researches using geometrical faces, and to apply the existing knowledge to robot facial design. Clarification of this relationship is one of the most important aspects for research on cognition of facial expression. The geometrical face used for this thesis is comprised of the minimum necessary elements for recognizing the face, and the elements are transformed geometrically to form the various facial expression patterns. The author introduces the physical variables, slantedness and openness, to the geometrical transformation. These facial patterns are classified by basic emotions, and evaluation of the relationship between the physical and affective variables is conducted by applying principal component analysis to the facial expression space centered on physical variables. Based on the results of the research, the author finally developed a model to create facial expressions.

I addition, in this thesis the author researched the neutral face, a concept that has rarely been mentioned by previous researches. In recent research, it has been suggested that the neutral face does not strictly speaking show a neutral emotion, but rather conveys some meaning concerning an actual circumstance of life.

By not using a realistic-looking human face, but rather a face made with limited elements and employing geometrical transformation, the relationship between the elements and factors used in making facial expression is shown more clearly. In addition, the result of this relationship expands in application to not only robot facial design, but also to medical fields such as curing cognitive impairment of facial expression. Thus, it is aimed to apply the results of this research to other fields [33–35].

2   Experiment Concerning Cognition and Structure of Facial Expressions

In this chapter, an expression pattern is generated using the geometrical transformation of a face made with the minimum necessary elements, the eyes and the mouth. Elements are transformed by the parameters of slantedness and openness. The category of basic emotions is used to classify the facial patterns. The classified facial expression patterns are evaluated by analyzing the spatial distribution of the facial expression patterns generated based on the values of the transformation parameters.

These processes are aimed at discovering the geometrical, spatial and quantitative relations between the elements and the factors used to make facial expressions.

2.1   Making Expression Patterns

The elements composing the face are limited to just three: the two eyes and the mouth, composed of precise circles and placed on the top of the upside-down triangle, the face figure (Fig. 1). Facial expression patterns are made by adding transformations to the three elements within the two parameters of slantedness and openness, comprising the physical variable for cognition of facial expressions. Slantedness of the eyes is a parameter that expresses the curve according to the opening of the eyes by their rotational deformation, while slantedness of the mouth expresses the rise or fall of the corners of the mouth. Openness is a parameter that expresses the change in the opening state of the eyes and mouth by the change in the vertically scaling transformation of the precise circle. Figure 2 and Fig. 3 show the changes to the eyes and mouth according to the value of the parameters. The eyes make 19 patterns and mouth makes 7 patterns as shown in Fig. 2, 3, making a total of 133 expression patterns. The facial expression is assumed to be completely facing the observer, and all faces are symmetrical.

Thus, each facial expression is defined by four values: two parameters of the eyes and two of the mouth, namely slantedness and openness respectively. So, the coordinates of one face can be shown as in the following mathematical expression.

$$f_i = \begin{pmatrix} E_{si} \\ E_{oi} \\ M_{si} \\ M_{oi} \end{pmatrix}, i \in \{1, 2, \ldots, n\}$$

The above mathematical expression shows the coordinates of the ith facial expression, where E_s and E_o represent the eyes' slantedness and openness respectively, and M_s and M_o represent the mouth's slantedness and openness respectively.

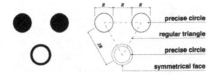

Fig. 1. Basic face before transformation

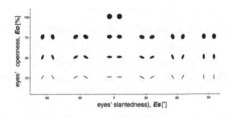

Fig. 2. Transformation of eyes by each parameter

2.2 Classification of Facial Expression Patterns

The obtained expression patterns are classified by emotional category. For the emotional categories, the author uses the six basic emotions (happy, angry, sad, disgust, fear, and surprised) advocated by Ekman and many other researchers. In this thesis, the author additionally utilizes a neutral (pleasant) emotion, displaying no emotion but showing a pleasant expression, and a neutral (unpleasant) emotion, showing no emotion but displaying an unpleasant expression, in order to research the facial expressions of neutral emotions. Clarification of the neutral face is important for robot faces, especially for humanoid robots without the function of forming facial expressions. In total, the eight emotional categories are defined.

Classification of facial expression patterns is conducted through an identification task. There are three rules for this task as follows: (1) the answerer must choose one emotion category for one face; (2) the answerer may adopt the same emotion category for more than one face; (3) the answerer does not need to select an emotion category if no category fits the face.

2.3 Distributing Facial Expression Patterns in Mathematical Space

As described in Sect. 2.1, each expression is defined by four-dimensional coordinates. Based on these coordinates, the author distributes the classified facial expression patterns in four-dimensional space. The facial patterns distributed in space are the faces selected by more than $p/2$ answerers (p is the maximum number of answerers selecting the face as the emotion category).

Fig. 3. Transformation of mouth by each parameter

After creating the facial expression space, analysis is applied to the space and the dimensions of the space are reduced to render it perceptible to the human eye. Through this analysis and visualization, the author can observe the difference in spatial distribution and parametric values of each facial pattern.

3   Result of Classification and Spatial Distribution of Facial Expressions

From the results of the principal component analysis, the face distribution space is composed of three variables: openness of the mouth, slantedness of the eyes, and openness of the face. Moreover, this result almost completely concurs with expression distributions in the space composed of the emotional meaning dimension in previous research, and shows similarity to the results of expression cognition research using faces of actual human beings.

In addition, the research of the past is consolidated, and 10 basic facial expressions—happy, angry, angry*, sad, disgust, fear, fear*, surprised, neutral (pleasant), and neutral (unpleasant)—are advocated.

3.1   Three-Dimensional Space for Distribution of Face

The identification task was undertaken by 140 men and women ranging in age from their teens to their 60s (average age in their 20s). The results of the principal component analysis of four-dimensional space are shown in Table 1. Four-dimensional space can be reduced to three-dimensional space from the result of the cumulative proportion from principal component-1 to principal component-2, shown to be 0.807. In addition, the value of each principal component is shown in Table 2. Principal component-1 was be judged to mean the slantedness of the mouth; principal component-2 shows the slantedness of mouth; and principal component-3 shows the openness of the face, meaning openness of both eyes and mouth, according to this value. This result corresponded to the three types of visual information (the visual variable) that had been obtained by previous researches [12–16].

To assess the relationship between the emotional meaning dimension obtained by this research and the visual information dimension obtained by previous researches, the author compared two planes, a projection plane of three-dimensional space obtained by this research comprising slantedness of mouth and openness of face, and a plane comprising the pleasantness dimension and the activeness dimension [10], which have strong relationships with the slantedness of the mouth and the openness of the face [17, 18]. The former plane is shown in Fig. 4, and the latter in Fig. 5. Each facial expression is assigned a weight according to the number of selections. An important point in Fig. 5 and Fig. 6 is that the average coordinates of each emotion category took these weights into consideration.

Comparing Fig. 4 and Fig. 5, the distribution of happy, surprised, and fear can be seen to almost correspond, and in addition, angry, sad, and disgust were also seen to correspond in terms of closeness of distribution. Moreover, fear could be seen to separate into two clusters in terms of distribution, which could be read in Fig. 4. As mentioned above, the author determined that the distribution of the facial expressions in three-dimensional space obtained through this thesis was valid.

In addition, angry, sad and disgust were separately observed by constructing and observing a projection plane of three-dimensional space comprising slantedness of the eyes and openness of the face (Fig. 6). From this result, by considering the third dimension, the slantedness of the eyes, the distribution of each facial expression was easy to separate and read. Thus the author discovered that the visual information dimension (physical variable) comprises three variables for cognizance of facial expressions.

Table 1. Result of principal component analysis

Principal component (PC)	PC1	PC2	PC3	PC4
Standard deviation	1.119	1.034	0.951	0.873
Comulative proportion	0.313	0.581	0.807	1.00

Table 2. Meaning of each principal component

Principal component (PC)		PC1	PC2	PC3
Eyes	Slantedness	0.280	−0.773	−0.298
	Openness	0.557	0.125	0.749
Mounth	Slantedness	0.674	−0.137	−0.160
	Openness	−0.396	−0.609	0.570
Meaning of PC		Mouth's slantedness	Eye's slantedness	Face's openness

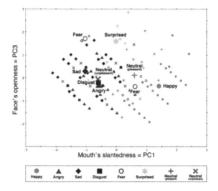

Fig. 4. Projection plane of three-dimensional space for facial expression composed of slantedness of the mouth and openness of the face

3.2 Eight Facial Expressions and Two Neutral Facial Expressions

In addition to the observation of face distributions in space, the author found 10 basic facial expressions, eight basic facial expressions: happy, angry, angry*, sad, disgust, fear, fear*, and surprised; and two neutral facial expressions: neutral (pleasant) and neutral (unpleasant), according to the parametric values and actual facial patterns. Figure 7 shows each typical facial expression. The difference between angry and angry* is especially apparent in terms of the openness of the eyes and mouth, so angry can be separated in terms of the facial expression showing the anger emotion. Moreover, it can be seen that fear and fear* can be separated because the slantedness of the eyes and the mouth indicate an opposite value.

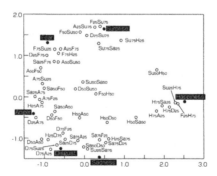

Fig. 5. Circumplex mode constructed by Takehara & Suzuki (2001)

Furthermore these facial expressions are distributed in three-dimensional space based on continuous geometrical transformation of the eyes and mouth, and it is understood that each facial expression generates a network comprising a visual and a physical variable (geometrical transformation).

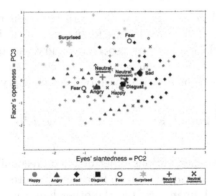

Fig. 6. Projection plane of three-dimensional space for facial expressions composed of the slantedness of the eyes and the openness of the face

Fig. 7. Typical facial patterns of 10 basic facial expressions

4   Discussion

In this chapter, a tetrahedral model for making facial expressions is advocated. This model comprises the geometrical and spatial relationships between the 10 facial expressions, and the actual facial patterns and the parametric values of each element.

4.1   Relationship Between the Visual Variable Dimension and the Affective Meaning Dimension

Through the facial expression pattern presented in this thesis, the visual variable dimension was obtained, namely, the slantedness of the eyes, the slantedness of the mouth, and the openness of eyes and mouth. Moreover, by observation and comparing with three-dimensional space and the results of previous research, it was found that there is a strong relationship between the slantedness of the mouth and the pleasantness dimension, and the openness of the eyes and mouth and activeness, as indicated in previous research.

In addition, based on observation of the distribution in three-dimensional space, the slantedness of the eyes is an effective means of discerning the distribution areas of the facial expressions, especially angry, sad and disgust. Thus the third effective variable's dimension serves to aid cognition of facial expression, and bears a strong relationship with the judgment of angry, sad and disgust.

4.2 Limitation of Facial Elements and Cognition of Facial Expressions

In many previous researches, various experiments on recognition of facial expressions was conducted using actual human faces and pictures or line-drawing faces closely resembling the human face. On the other hand, in this thesis, the minimum elements needed to recognize the face as a face were selected, and the face used for the experiment consisted of only two eyes and a mouth. The eyes and the mouth were used to make facial patterns through geometrical transformations based on precise circles.

From the results, it was found that human beings could recognize faces and judge facial expressions and emotions even when observing a face composed of limited elements subjected to geometrical transformations, as well as being able to recognize the expressions of actual human beings.

Thus, it is possible to read emotion sufficiently, not only from the facial expression of an actual human being, but also from the expression made by a geometrical transformation (rotation and vertically scaling transformation) of the minimum geometrical elements (two eyes and a mouth) needed to recognize a face.

4.3 Tetrahedral Model to Make Facial Expressions

According to the results and discussion, a tetrahedral model used to make facial expressions is advocated using the 10 basic facial expressions. The model's structure is shown in Fig. 8. This facial model's target is a face made with two eyes

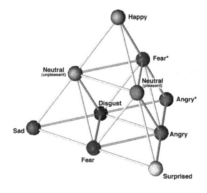

Fig. 8. Tetrahedral-model on structure of facial expressions

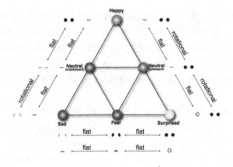

Fig. 9. Happy-sad-surprised surface of tetrahedral model

and a mouth, and facial patterns made by rotation and vertically scaling transformation of the elements. The 10 emotions are placed in four locations, on top of the body and at a middle point on each side. The position in which each expression is produced is expressed with the ball. The sides between the balls are the axes of transformation of the eyes and the mouth in a constant direction. The actual transformation of the elements is shown in Figs. 9, 10 and 11.

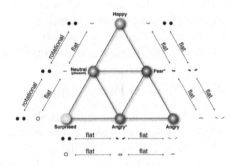

Fig. 10. Happy-surprised-angry surface of tetrahedral model

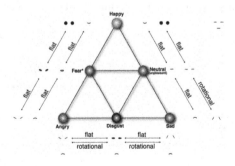

Fig. 11. Happy-angry-sad surface of tetrahedral model

These figure show the equilateral triangle plane made from each of the three tops of the tetrahedral model. Rotation and vertically scaling transformation adds the eyes, while vertically scaling transformation alone adds the mouth, along the axis (side) connecting each expression. By these transformations and the relationships between each parametric value, the facial expressions can be made to change.

To take the triangle in Fig. 9 as an example, on the side connecting happy and sad, first of all the eye flatness rate is changed to 10% (this forms the eye for neutral (unpleasant)), and secondly, rotation is added and the angle of the eye is changed to −60°, creating the eye for sad. Meanwhile, the mouth's form is changed to flat, (neutral (unpleasant)), after that it is changed to form the sad face. The happy face's mouth and the sad face's mouth indicate fully opposite values. As shown in Figs. 9, 10 and 11, the topological distance between each emotion is equal, and each facial expression changes to the adjacent facial expression along the side connecting the emotions.

The tetrahedral model can be applied to all face making facial expressions by using geometrical elements and their transformation. For instance, facial expression is assumed to become very important in future communications between humanoid robots, as well as for communication with human beings. In such situations, making facial expressions of humanoid robots is easier using the tetrahedral model, because this model has succeeded in creating facial expressions by geometrical elements and transformation. As well as in the robot field, the application range of the tetrahedral model in the future is thought to be vast, encompassing the realm of healing, curing cognitive impairment, and so on.

4.4   Facial Expression of Neutral Facial Expression

In this thesis, the neutral (pleasant) and neutral (unpleasant) expressions were added to the classification of facial expressions. These expressions were found to exist even though they do not show a specific emotion, and these neutral facial expressions can be generated through transformation of facial elements. Moreover, according to the facial pattern, the neutral face can be divided into two facial expressions, showing pleasant and unpleasant emotions respectively. From the results of observing the spatial and geometrical differences between faces, and

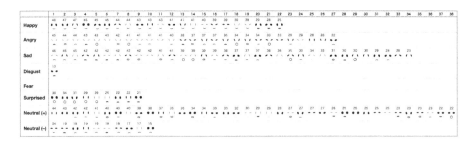

Fig. 12. Survey in Nagasaki, Japan.

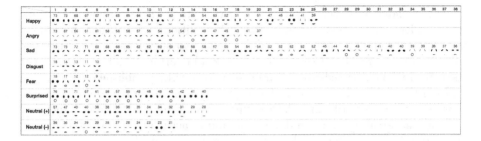

Fig. 13. Survey in Aalborg, Denmark.

the tetrahedral model, the author defined the neutral (pleasant) facial expression as between happy and surprised, and the neutral (unpleasant) expression as between happy and sad.

5 Conclusion

In this research, the authors succeeded in creating facial expressions made with the minimum necessary elements for recognizing a face, and enabling human recognition of the expressions created. The elements used for the minimal face were two eyes and a mouth made by precise circles and transformed to make facial expressions geometrically, through rotation and vertically scaling transformation. The facial expression patterns made by the geometric elements and transformations comprised three dimensions of visual information (visual variables) that had been suggested by many previous researches: slantedness of the mouth, openness of face, and slantedness of the eyes. Thus the results of this research indicate that human beings can classify expression patterns of minimal faces to particular emotional categories just as they would with an actual human face. In addition, the relationships between visual affection dimension and the affective meaning dimension also corresponded to the results of the previous researches. These relationships were strong between the slantedness of the mouth and pleasantness; and between the openness of the face and activeness; and the existence of a third affective variable strongly related to the slantedness of the eyes was also suggested.

The authors found that facial expressions could be classified into 10 different facial expressions: happy, angry, sad, disgust, fear, surprised, and angry*, fear*, neutral (pleasant) showing a positive emotion, and neutral (unpleasant) showing a negative emotion. These facial expressions are composed by different geometric transformations and combinations of the eyes and the mouth. Furthermore, the authors discovered a tetrahedral model that could express most clearly the geometric relationships between the facial expressions. This model is structured in the form of a tetrahedron, with each facial expression located on the top or the middle of the sides of the tetrahedron. Each side connecting faces is an axis determining the rotation and vertically scaling transformation of the eyes and the mouth.

The authors have began survey to investigate if those findings were universal under *inter-, cross-,* or *trans-cultural* conditions. The authors have done the initial survey in Denmark and also a *follow-up* survey in Japan. Figure 12 and 13 show the results of survey in Nagasaki (Japan) and Aalborg (Denmark) respectively. The detailed results and discussions will be published shortly.

In future, it hoped to research the affective dimension for recognition of facial expressions and to clarify the relationship between the visual variable dimension and the emotional information dimension, and to actually apply the tetrahedral model to humanoid robots and to other wide- ranging fields.

Acknowledgement. This work was supported by Grant-in-Aid for Scientific Research on Innovative Areas No. 18H04203 "Construction of the Face-Body Studies in Transcultural Conditions".

References

1. Ellis, H.D., Shepherd, J.W., Davis, G.M.: Identification of familiar and unfamiliar face from internal and external features: some implications for theories of face recognition. Perception **8**, 431–439 (1979)
2. Shepherd, J.W., Davis, G.M., Ellis, H.D.: "Studies of cue salienc" Perceiving and Remembering Faces, pp. 105–131. Academic Press, Cambridge (1981)
3. Young, A.W., Hay, D.C., McWeeny, K.H., Fluid, B.M., Ellis, A.W.: Matching familiar and unfamiliar faces on internal and external features. Perception **14**, 737–746 (1985)
4. Yin, R.K.: Looking at upside-down faces. J. Exp. Psychol. **81**, 141–145 (1969)
5. Diamond, R., Carey, S.: Why faces are and are not special: an effect of expertise. J. Exp. Psychol. **115**, 107–117 (1986)
6. Goren, C., Sartty, M., Wu, P.: Visual scan patterns of rhesus monkeys viewing faces. Perception **11**, 133–140 (1975)
7. Johnson, M.H., Morton, J.: Biology and Cognitive Development: The Case of Face Recognition. Blackwell, Cambridge (1991)
8. Ekman, P., Friesen, W.V., Ellsworth, P.: What emotion categories or dimensions can observers judge from facial behaviour? In: Goldstein, A.P., Krasner, L. (eds.) Emotion in the Human Face, pp. 39–55. Cambridge University Press, Cambridge (1982)
9. Smith, C.A., Ellsworth, P.C.: Patterns of cognitive appraisal in emotion. J. Pers. Soc. Psychol. **48**(4), 813–838 (1985)
10. Yamada, H.: Expression model for process of perceptive estimation of facial expression. Jpn. Psychol. Rev. **43**(2), 225–245 (2000). (in Japanese)
11. Darwin, C.: The Expression of the Emotions in Man and Animals. Oxford University Press, Oxford (1998). (original work published 1872)
12. Ekman, P.: Basic Emotions, Handbook of Cognition and Emotion, pp. 45–60. Wiley, New York (1999)
13. Russell, J.A.: Reading Emotions from and into Faces: Resurrecting a Dimensional-Contextual Perspective, The Psychology of Facial Expression, pp. 295–320. Cambridge University Press, Cambridge (1997)
14. Ekman, P.: Basic Emotions. In: Daglesh, T., Power, M. (eds.) Handbook of Cognition and Emotion, pp. 45–60. Wiley, New York (1999)

15. Ekman, P.: Universals and cultural differences in facial expressions of emotion. Nebr. Symp. Motiv. **19**, 207–283 (1972)
16. Ekman, P.: Cross-cultural studies of facial expression. In: Ekman, P. (ed.) Darwin and Facial Expression: A Century of Research in Review, pp. 169–222. Academic Press, Cambridge (1973)
17. Ekman, P.: An argument for basic emotions. Cogn. Emot. **6**(3), 169–200 (1992)
18. Ekman, P.: Constants across cultures in the face and emotion. J. Pers. Soc. Psychol. **17**(2), 124–129 (1971)
19. Ekman, P.: Pan-cultural elements in facial displays of emotions. Science **169**, 86–88 (1969)
20. Ekman, P.: Conceptual ambiguities. In: Goldstein, A.P., Krasner, L. (eds.) Emotion in the Human Face. Cambridge University Press, Cambridge (1982)
21. Ekman, P.: Universals and cultural differences in the judgments of facial expressions of emotion. J. Pers. Soc. Psychol. **53**(4), 712–717 (1987)
22. Schlosberg, H.: The description of facial expressions in terms of two dimensions. J. Exp. Psychol. **44**(4), 229–237 (1952)
23. Schlosberg, H.: Three dimensions of emotion. Psychol. Rev. **10**(5), 81–88 (1954)
24. Shah, R., Lewis, M.B.: Locating the neutral expression in the facial-emotion space. Vis. Cogn. **1**(5), 549–566 (2003)
25. Takehara, T., Suzuki, N.: Robustness of the two- dimensional structure of recognition of emotionality. Percept. Mot. Skills **93**(3), 739–753 (2001)
26. Yamada, H.: Visual information for categorizing facial expression of emotions. Appl. Cogn. Psychol. **3**(7), 252–270 (1993)
27. Yamada, H., Shibui, S.: The relationship between visual information and affective meanings from facial expressions of emotion. Perception **27**, 133 (1998)
28. Watanabe, N.: Research of distribution of facial expressions in emotional space by using affect grid method. Jpn. Psychol. Res. **65**, 274 (2001). (in Japanese)
29. Yamada, H., Matsuda, T., Watari, C., Suenaga, T.: Dimenson of visual information for categorizing facial expressions of emotion. Jpn. Psychol. Res. **35**(4), 172–181 (1993)
30. Watanabe, N., Suzuki, R., Yamada, H.: Visual information related to facial expression recognition - Research of 3 dimensional structure, The Technical report of The Proceeding of The Institute of Electronics, Information and communication Engineers, vol. 102, no. 598, HCS2002-35, pp. 43–48 (2003)
31. Watanabe, N., Suzuki, R., Yamada, H.: Re-consideration about relationship between mental and physical information for facial expression recognition. Jpn. Soc. Cogn. Psychol. **3**(2), 167–179 (2006). (in Japanese)
32. Russell, J.A., Weiss, A., Medelsohn, G.A.: Affect grid: a single-item scale of pleasure and arousal. J. Pers. Soc. Psychol. **57**(3), 493–502 (1989)
33. Kawahara, H., Matsumoto, N.: A cognitive analysis to the variety in facial features of prehistoric figurines: an approach from cognitive and esthetic archaeology, Technical report of IEICE. HIP, vol. 107, no. 369, pp. 79-84 (2007)
34. Carrera, P., Fernandez-Dos, J.M.: Neutral faces in context, Their emotional and their function. J. Nonverbal Behav. **18**, 281–299 (1994)
35. Okamoto, T.: Numerical experiment for assessment of reproducibility of each method of multiple classification analysis. Behav. Soc. Japan **18**(2), 47–56 (1991). (in Japanese)

How Drivers Categorize ADAS Functions
–Insights from a Card Sorting Study

Liping Li[⊠], Hsinwen Chang, Weihan Sun, Jin Guo, and Jianchao Gao

Baidu Intelligent Driving Experience Center, Shenzhen 518000, China
liliping01@baidu.com

Abstract. Advanced driver assistance systems (ADAS) are intelligent systems that can help vehicle drivers to drive with ease and safety. A growing body of ADAS technologies bring benefits to vehicle drivers. However, understanding increasing ADAS terminologies becomes a potential problem for ordinary drivers. An individual open card sorting study was conducted to test Chinese driver's mental model of ADAS categories.

14 private car drivers were invited to do card sorting, visualizing their mental model of ADAS function categorization. 18 ADAS-related terminologies and their brief explanations were printed on separate paper cards. Participants are told to sort these cards into groups and then name their card groups as what they would want to see in their own cars.

To analyze card sorting result, 2 researchers merged similar group names into 7 groups: Driving Assist, Collision Prevention, Safety Setting, Parking Assist, Start-Stop Assist, Information, and General. 15 out of 18 cards can be clearly grouped, being put under the same group by more than (or equals) 40% participants. 3 out of 7 group get cards clearly followed. They are Driving Assist, Collision Prevention, and Parking Assist. We also compare the result with Troppmann's Model and United States Department of Transportation (NHTSA, 2020) Model and find Chinese drivers tend to use a different driving status clue.

The result suggests ADAS technology categories maybe different between industry practice and Chinese private drivers' mental model. Culture influence and ADAS knowledge can be possible reasons for the difference.

Keywords: Advanced Driver Assistance Systems (ADAS) · Mental model · Card sorting

1 Introduction

1.1 Adoption of ADAS

Road safety is believed to be mainly impaired by 'human factor' [1]. 90% of traffic accidents are caused by human failure, including fatigue, inattention, alcohol and etc.

Therefore, researchers, professionals and governments try to countermeasure 'human factor' with driving assistant systems. Effective ADAS can reduce 70% crash rate. In real world, according to the 2005–2008 U.S. GES Crash Records, ADAS technologies

© Springer Nature Switzerland AG 2020
P.-L. P. Rau (Ed.): HCII 2020, LNCS 12192, pp. 606–615, 2020.
https://doi.org/10.1007/978-3-030-49788-0_46

successfully reduced light vehicles' crashes by 32.99% and heavy trucks' crashes by 40.88% [2].

In 2003, European Union opened 'eSafety' program. In 2005, America started assessment of 'Integrated Vehicle-Based Safety System'. In 1991, Japan started 'Advanced Safety Vehicle Development Program'. These programs have facilitated ADAS adoption in major vehicle markets [3]. From policy perspective, China acts slow in this area. Till 'Made in Chine 2025' (2015), China officially encourages ADAS technologies' development and adoption [4].

1.2 Understanding of ADAS Scope

Troppmann's Model
Troppmann (2006) depicts ADAS technologies in **Active-Safety dimensions** [5]. In Active dimension, technologies are allocated to demonstrate the degree of machine control: they are just text reminders or can take over the vehicle control from its driver. In Safety dimension, technologies are arranged to show their Safety impact: they are crash-avoiding or making driving easier (Fig. 1).

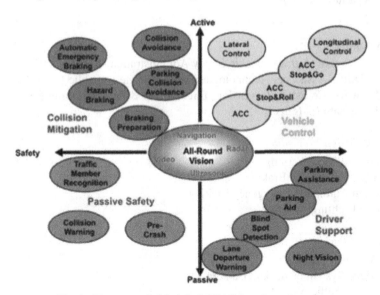

Fig. 1. Troppmann's Model of ADAS function categories

According to this framework, ADAS technologies can be categorized as:

Collision Mitigation (Active-Safety)
– Collision Avoidance
– Automatic Emergency Braking
– Hazard Braking
– Braking Preparation

Vehicle Control (Active-no-Safety).
- Longitudinal Control
- Lateral Control
- Adaptive Cruise Control Stop &Go
- Adaptive Cruise Control Stop &Roll
- Adaptive Cruise Control

Passive Safety (Passive-Safety)
- Traffic Member Recognition
- Collision Warning
- Pre-Crash

Driver Support (Passive-no-Safety)
- Parking Assistance
- Parking Aid
- Blind Spot Detection
- Night Vision
- Lane Departure warning

NHTSA Model

America's National Highway Traffic Safety Administration (NHTSA, 2020) describes ADAS as Driver Assistance Technologies on its website to public [6]. It has 4 categories: **Forward Collision Prevention, Backing Up &Parking, Lane &Side Assist, Maintaining Safe Distance**.

Forward Collision Prevention
- Forward Collision Warning
- Automatic Emergency Braking
- Pedestrian Automatic Emergency Braking
- Adaptive Lighting

Backing up &Parking
- Rear Automatic Braking
- Rear Video System or Backup Camera
- Rear Cross Traffic Alert

Lane & Side Assist
- Lane Departure Warning
- Lane Keeping Assist
- Blind Spot Detection
- Lane Centering Assist

Maintaining Safe Distance
- Traffic Jam Assist
- Highway Pilot
- Adaptive Cruise Control

ADAS is a collection of systems and subsystems that transform manual driving to autonomous driving. Though it has a relative clear main purpose, its sub-categories can be slightly various in different researches.

1.3 Challenge of ADAS Category Design

All category models above could be references for in-car HMI designers to structure ADAS manual/setting system. And we can learn the categorizing strategy from both models. In Troppmann's Model, drivers need understand the vehicle's active control level and safety impact. In NHTSA's category, drivers should tell the technological configuration difference of ADAS functions.

However, all models above cannot directly represent Chinese drivers' understanding of ADAS Category. Although China's vehicle ownership grows rapidly, it is far behind major developed countries, like United States. Chinese drivers probably have less knowledge about ADAS. Also, China has a large population. People's concern about ADAS may vary from other countries.

Based on these references and manufacturer information, we design a card sorting study to examine Chinese customers' mental model of ADAS categories. In this study, we will compare drivers' mental model with these 2 existed models to reveal ordinary drivers' grouping strategy.

2 Method

2.1 Material

Based on collaborating manufacturer information, we prepare 18 ADAS terminology paper cards. Every card is about 5 cm × 4 cm with the terminology name and a brief explanation on it (Table 1).

2.2 Participants

14 private car drivers participate this study. Their driving experience range from 5 months to 8 years. All of them have ADAS function using experience (Table 2).

2.3 Procedures

It is an individual open card-sorting research design, each participant sorts cards individually. There is no number limit on group quantities and participants could name their sorted groups by their own language. In this study, each participant trial takes about a half hour in conduct.

Table 1. ADAS terminologies and brief explanations used in the study

Terminology	Brief explanation
Cruise control	Keep the vehicle running at the set speed without pressing the accelerator pedal
Speed limit	When the vehicle reaches the set speed, you will see a warning on the instrument cluster and hear a warning tone
Hill descent control	Prevent the downhill speed from being too fast, and active braking will intervene
Lane keeping assist	Include Lane Keeping Assist switch, warning intensity and sensitivity adjustment
Traction control system	Reduce wheel slip
Hill start assist	The vehicle can remain on a slope for 2–3 s after the brake pedal is released
Auto stop	After starting, the engine will shut down automatically when the vehicle stops. When the brake pedal is released, the engine will restart automatically
Auto hold	After opening, the vehicle will brake automatically when it is stationary, and the brake will be cancelled automatically when the accelerator is pressed
Blind spot detection	If a vehicle enters your blind area during driving, the system will give an alarm
Traffic sign recognition	Identify the traffic signs and display traffic signs in the instrument
Rear park assist	When the vehicle is in reverse gear (R), there will be a warning tone if approaching an obstacle
Wrong Way Warning	Identify the no-entry sign and one-way street, and warm the driver when the vehicle goes wrong way
Pedestrian detection	Activated when the vehicle speed reaches a set high speed. If the sensor detects a pedestrian and the driver does not respond, the system will give visual and audible warning
Rear automatic braking	Automatic brake in case of obstacles, too fast reversing speed, and wrong accelerator in rear parking
Automatic post-collision braking	Automatic brake after a serious collision to reduce the potential risk of a second collision
Collision warning	Prompt collision risk
Fatigue driving alarm	The system alerts the driver that he needs to rest at set conditions
Rear cross traffic alert	When reverse gear (R) is engaged, there will be warnings if other vehicles coming cross from sides

Table 2. Participants information

No.	Driving experience	Sex	Subjective reported ADAS using experience
01	3 Years	Male	Cruise control, parking assist
02	7 Years	Male	Speed control, parking assist
03	5 Months	Female	Lane keeping assist, parking assist
04	6 Months	Male	Cruise control, lane keeping assist, collision warning
05	3 Years	Male	Cruise control, lane keeping assist, collision warning
06	4 Years	Male	Cruise control, lane keeping assist, parking assist
07	5 Years	Male	Cruise control
08	2 Years	Male	Lane keeping assist, collision warning
09	1.5 Years	Male	Parking assist
10	6 Yeas	Male	Cruise control, collision warning, blind spot detection, parking assist
11	2 Yeas	Male	Cruise control, collision warning, parking assist
12	5.5 Years	Female	Cruise control, lane keeping assist, collision warning, blind spot detection, parking assist
13	8 Years	Male	Cruise control
14	4 Years	Male	Cruise control, parking assist

Fig. 2. Picture examples of the study conducting process

Step 1 - Understand 18 ADAS terminologies. 18 ADAS-related terminologies and their brief explanations were printed on separate paper cards. A participant has enough time to read these cards and ask questions until he/she is well prepared to do the card sorting.

Since all participants have ADAS feature using experience, with brief explanation along with the terminology, they can understand these 18 ADAS terminologies easily. No one ask for further terminological explanation, but some inquiry about the technological feasibility in real products. For example, some participants show interests on Traffic Sign Recognition technological feasibility.

Step 2 - Sort 18 ADAS terminologies into groups. Participants are told to sort these cards into groups that could show the relationship of cards in their mental model. Participants are encouraged to do a rough categorization first, and then refine it on the second round.

Step 3 Name the groups. Participants are told to name their card groups as what they would want to see in their own car. They were given blank cards to write down those names and put them on the top of related card stacks. They can organize these ADAS terminologies in their own language, not limited to existed models (Fig. 2).

3   Result

3.1   Category Distribution

To analyze card sorting result, 2 researchers merged similar group names into 7 groups based on participant interview: Driving Assist, Collision Prevention, Parking Assist, Safety Setting, General, Information, and Start-Stop Assist, showing in Table 3.

The percentage represents the ratio of participants who group a terminology under this category. The background color goes deeper as the number increases.

12 out of 18 cards can be clearly grouped, being put under the same group by more than (or equals) 50% participants. 6 terminologies that cannot be clearly grouped are Auto Hold, Blind Spot Detection, Traffic Sign Recognition, Pedestrian Detection, Automatic Post-Collision Braking, and Rear Cross Traffic Alert. If we set 40% as the group reference, Auto Hold can be grouped into Driving Assist, and Pedestrian Detection, Rear Cross Traffic Alert into Collision Prevention. 3 out of 7 group get cards clearly followed. They are Driving Assist, Collision Prevention, and Parking Assist.

3.2   Model Comparison

In Table 4, we compare mental model extracted in this study with Troppmann's Model and NHTSA Model. We can find that participants tend to organize terminologies by driving status: driving, colliding, and parking.

Table 3. ADAS category distribution in this study

	Driv- ing Assist	Colli- sion Pre- vention	Park- ing Assist	Safety Setting	Gen- eral	Infor- mation	Start- Stop Assist
Cruise Control	93%	0%	0%	0%	7%	0%	0%
Speed Limit	86%	0%	7%	7%	0%	0%	0%
Hill Descent Control	79%	0%	21%	0%	0%	0%	0%
Lane Keeping Assist	71%	7%	7%	14%	0%	0%	0%
Traction Control System	57%	0%	14%	7%	21%	0%	0%
Hill Start Assist	57%	7%	29%	0%	0%	0%	7%
Auto Stop	50%	7%	14%	7%	7%	0%	14%
Auto Hold	43%	7%	14%	7%	14%	0%	14%
Blind Spot Detection	36%	36%	0%	21%	0%	7%	0%
Traffic Sign Recognition	29%	21%	7%	7%	14%	21%	0%
Rear Park Assist	29%	7%	50%	7%	0%	0%	7%
Wrong Way Warning	21%	50%	0%	21%	0%	7%	0%
Pedestrian Detection	21%	43%	0%	29%	0%	7%	0%
Rear Automatic Braking	21%	7%	50%	14%	0%	0%	7%
Automatic Post-Collision Braking	14%	36%	14%	36%	0%	0%	0%
Collision Warning	7%	57%	7%	21%	0%	7%	0%
Fatigue Driving Alarm	7%	50%	7%	14%	7%	14%	0%
Rear Cross Traffic Alert	7%	43%	21%	14%	0%	7%	7%

Unlike the Active concern in Troppmann's model, participants are not so aware of ADAS's vehicle control extent. Also, participants would not differentiate assists from which sides of the vehicle. Therefore, some category scope differs from what in NHTSA Model. For example, Lane Keeping Assist is categorized into Lane &Side Assist in NHTSA Model, while it's labeled as Driving Assist with Cruise Control in this study (Maintaining Safe Distance in NHTSA).

Table 4. ADAS category model comparison

	Troppmann's model	NHTSA model	This study
Cruise control	**Vehicle control**	**Maintaining safe distance**	**Driving assist**
Speed limit			**Driving assist**
Hill descent control			**Driving assist**
Lane keeping assist	**Driver support**	**Lane &side assist**	**Driving assist**
Traction control system			**Driving assist**
Hill start assist			**Driving assist**
Auto stop			**Driving assist**
Auto hold			**Driving assist**
Blind spot detection	**Driver support**	**Lane &side assist**	*
Traffic sign recognition			*
Rear park assist	**Driver support**	**Backing up &parking**	**Parking assist**
Wrong way warning			**Collision prevention**
Pedestrian detection	**Passive safety**	**Forward collision prevention**	**Collision prevention**
Rear automatic braking	**Collision mitigation**	**Backing up &parking**	**Parking assist**
Automatic post-collision braking	**Collision mitigation**		*
Collision warning	**Passive safety**	**Forward collision prevention**	**Collision prevention**
Fatigue driving alarm			**Collision prevention**
Rear cross traffic alert	**Passive safety**	**Backing up &parking**	**Collision prevention**

*Represents no clear category, blank means not mentioned.

4   Discussion

4.1   Inspiration for in-Car HMI Design

This study reveals normal drivers' mental model of ADAS functions that in-car HMI designer can exploit. Optimizing the manual/setting information architecture based on the study model, drivers can have a higher probability to find the target function items easily.

The results demonstrate that drivers are incline to group these functions according to driving status. They are not so aware of vehicle's active control level and which direction side the function is tackling.

This phenomenon suggests liability between ADAS and drivers is ambiguous in driver's mental model. In current situation, all liabilities are on drivers' side. So, they do not argue for machine's liability. However, once cars become more autonomous, technology's liability may be need to clearly informed to drivers and considered by lawmakers.

This phenomenon also suggests drivers are not familiar with technology configuration. No matter collision detection in the front or in the back, they tend to group all these function as 'Collision Prevention'. Once they get more technology familiarity, they may incline to group as NHTSA Model.

4.2 Limitation of Card Sorting

The card sorting method makes it more reasonable to retrieve information from the structure when users browse information, but it is difficult to solve personalized settings, such as shortening the information level of common settings. Designers should consider shortcuts, i.e. 'my collection', to empower user personalize their in-car HMI, reducing the search cost of users.

4.3 Limitation of Sample Representation

This study suggests Chinese drivers using driving status as a clue to group ADAS technologies. However, whether all ordinary drivers tend to use this strategy need to test in other country and culture as well. It is possible in other culture, with more technological knowledge, people will agree to Troppmann's Model or NHTSA Model. And we can dig deep into the influence of clear vehicle active control level, safety impact and technological configuration.

References

1. Brookhuis, K.A., Waard, D.D., Janssen, W.H.: Behavioural impacts of advanced driver assistance systems-an overview. Eur. J. Transp. Inf. Res. 1(3), 245–253 (2001). https://doi.org/10.18757/ejtir.2001.1.3.3667
2. Yue, L., Abdel-Aty, M., Wu, Y., Wang, L.: Assessment of the safety benefits of vehicles' advanced driver assistance, connectivity and low level automation systems. Accid. Anal. Prev. 117, 55–64 (2018)
3. Ma, J., Cao, J.: Research on developing trend of automotive advanced driver assistant system based on specific market demand in China. Agric. Equip. Veh. Eng. 50(3), 5–10 (2012)
4. Made in China 2025. http://www.miit.gov.cn/n973401/n1234620/. Accessed 22 Jan 2020
5. Troppmann, R.: Tech tutorial: driver assistance systems, an introduction to adaptive cruise control: part 1(2006). http://www.automotivedesignline.com/howto/189600772
6. NHTSA's Driver Assistance Technologies webpage. https://www.nhtsa.gov/equipment/driver-assistance-technologies. Accessed 22 Jan 2020

Can Smart Voice Assistant Induce Social Facilitation Effect? A Preliminary Study

Na Liu[(✉)] and Quanlin Pu

School of Economics and Management, Beijing University of Posts and Telecommunications, Beijing, China
{liuna18,ql2018}@bupt.edu.cn

Abstract. The present study investigated whether a smart voice assistant can induce social facilitation effect. We designed a 3 (social presence: alone, smart voice assistant, human assistant) × 2 (task difficulty: easy, difficult) within-subject experiment. Sixteen university students participated in the experiment. The results showed that a smart voice assistant could induce social facilitation effect when participants performed modular arithmetic tasks, i.e., the response time of easy tasks was shortened while that of difficult tasks was lengthened. No significant difference in accuracy rates was found among three social presence conditions. Intensities of social facilitation effect between smart voice assistant present and human assistant present showed similar strength. These findings provide practical implications in designing work organization modes involving smart voice assistants.

Keywords: Smart voice assistant · Social facilitation effect · Social presence · Work organization

1 Introduction

Smart voice assistant is an emerging device and is becoming widely used in people's daily life. It is a platform composed of hardware and software which has the capability of understanding voice commands of a user and executes corresponding tasks. Current smart voice assistants can provide a wide variety of services such as reporting weather information and making calls. Amazon Alexa, Google Assistant, Apple's Siri, Microsoft's Cortana, and Samsung's Bixby are leading smart voice assistants on the current market, and they are integrated into various smart devices, such as smartphones, smart speakers, smart wearable devices, and smart TVs [1]. The smart voice assistant market continues to grow. Accenture surveyed 22,500 online consumers in 21 countries to reveal key trends in consumer preferences. 50% of respondents use smart voice assistant, up from 42% one year ago [2].

Smart voice assistants can do the following tasks: (a) make phone calls and send and read text messages and emails; (b) set timers and reminders, organize calendar schedules, and make lists; (c) answer basic informational questions (e.g., reporting weather information); (d) play media, such as music and video files, requested by the user;

© Springer Nature Switzerland AG 2020
P.-L. P. Rau (Ed.): HCII 2020, LNCS 12192, pp. 616–624, 2020.
https://doi.org/10.1007/978-3-030-49788-0_47

(e) chat and tell jokes and stories; (f) control Internet-of-Things (IoT) devices such as locks, lights, home security cameras, and thermostats; and (g) provide online shopping [3]. With the support of natural language processing (NLP) and automatic speech recognition (ASR), smart voice assistant provides services just like human assistants in such contexts as online shopping and information querying. The Social Response Theory states that people interact with computers and media as they do with other humans [4, 5]. Regardless of their embodiment, smart voice assistant's ability to have language-based communications with people serves as a critical human-like cue that evokes a sense of social presence in people's minds, making them treat the smart voice assistant as they do other humans and respond to them socially [6].

Previous studies have showed that social presence can induce social facilitation effect. Social facilitation effect is a classical concept in social psychology and it describes a phenomenon that people's task performance differ between working alone condition and working in the presence of others [7]. Recent research revealed that humans are not the only entities that can produce social facilitation effect; virtual humans and robots can also influence people's task performance and induce social facilitation effect [8–11]. However, no study investigated effects of smart voice assistant on people's task performance.

The theories explaining social facilitation effect can be grouped into three categories: drive, social conformity, and cognitive process [12]. According to drive theory [13], social presence increases people's arousal, and the increased arousal can enhance the dominant responses. In general, the dominant response is a correct and/or quick solution in easy or learned tasks and a false and/or slow solution in difficult or unlearned tasks. Hence, the performance of easy or learned tasks is improved and performance of difficult or unlearned tasks is impaired by the presence of others [13]. Theories in the second category is based on social conformity. In the evaluation apprehension theory, evaluation apprehension results in increased arousal level, causing performance enhancement or impairment in the presence of other people [14]. As people become afraid of being evaluated by other people who are present, their arousal level increases. In other theories, increase in arousal level is attributed to comparison between people and social standards (norms) or ideal expectancy [15, 16]. The third category comprises cognitive process theories focusing on variations in the manner by which people process information in the presence of others. Distraction-conflict theory regards distraction as the mediating mechanism that can explain social facilitation effect [17–19]. This theory suggests that the presence of others creates attentional conflict and distracts individuals when they perform tasks. Attentional conflict arises from individuals' intention to simultaneously concentrate on tasks and others and thus stimulates them to increase their focus to avoid distraction. People's arousal level increases, and social facilitation effect is induced [20].

In summary, this study aimed to determine whether smart voice assistant can evoke social facilitation effect and compare the intensities of social facilitation effect induced by smart voice assistant and human assistant. To address such issues, the present study used typical cognitive task to investigate the effects of smart voice assistant presence on people's task performance. Task performance in three conditions was measured: working alone, working in the presence of a smart voice assistant, and working in the presence of a human assistant. The findings in this study can provide practical implications for designing work organization modes involving smart voice assistants.

2 Method

2.1 Participants

Sixteen university students (8 males and 8 females) were recruited in the experiment. The average age of the participants was 21.2 ($SD = 1.2$) years old. Two additional students (one male and one female) were recruited to act as the human assistant respectively. None of the participants reported any experience with similar modular arithmetic tasks.

2.2 Experimental Design

The present study used a 2 (task difficulty: difficult, easy) × 3 (social presence: alone, smart voice assistant, human assistant) within-subject design. Six combinations were produced in this experiment. The dependent variable was task performance measured by response time (RT) and accuracy rates (ARs).

2.3 Apparatus and Software

The software used in this experiment was developed by C# language on Microsoft Windows platform. This software was used to present stimuli and to record performance. The software was run on a computer with a 23 in. liquid crystal display monitor with a resolution of 1920 × 1080 pixels and refresh frequency of 60 Hz. From a viewing distance of 60 cm, participants sat on a chair with adjustable height. Participants conducted experimental tasks using a keyboard.

The tasks adopted in the present study were modular arithmetic tasks. The software had one interface to present the problem statement at the center of the screen, e.g., $51 \equiv 19 \mod 8$ (Fig. 1). Participants were asked to decide whether the statement was true or false by pressing the corresponding key (Q for true and P for false) on the keyboard. The problem statement consisted of three numbers, and the computation rule was as follows. Taking $51 \equiv 19 \mod 8$ as an example, the statement's middle number is subtracted from the first number (i.e., $51 - 19$) at first, and the result (i.e., 32) is divided by the last number (i.e., $32 \div 8$). If the quotient is a whole number (in this case, 4), then the problem statement is true. If the quotient is not a whole number, then the problem statement is false. The task difficulty was manipulated by controlling the number of digits presented to participants for the first two numbers of a given problem; one for an easy task ($7 \equiv 2$) and two for a difficult task ($51 \equiv 19$), which was similar to previous study [9].

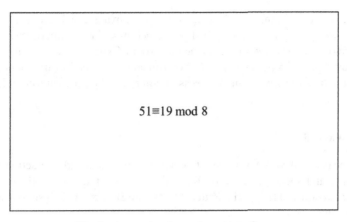

Fig. 1. A sample of experiment interface.

2.4 Procedure

Upon arrival at the laboratory, participants were informed of the purpose and experimental procedure and signed informed consent forms. Participants were then provided with instructions on how to conduct the experiment and use the software by smart voice assistant or human assistant in corresponding social presence conditions. Experiments consisted of seven sessions: one practice and six actual experimental sessions. Tasks used in practice session were the same as those in actual experimental sessions. After the practice session, participants took a 15 min rest and then proceeded to actual experimental sessions.

In actual experiments, participants were required to finish six sessions, which corresponded to six combinations of task difficulty and social presence: performing easy modular arithmetic tasks alone, performing difficult modular arithmetic tasks alone, performing easy modular arithmetic tasks with the presence of a smart voice assistant, performing difficult modular arithmetic tasks with a smart voice assistant, performing easy modular arithmetic tasks with a human assistant, and performing difficult modular arithmetic tasks with a human assistant. Orders of six combinations were randomly assigned and counterbalanced across participants. Each session included 30 modular arithmetic trials. Participants completed 180 trials (30 trials × 6 sessions) in actual experiments. Participants took a 10 min rest after each session.

In actual sessions, participants were instructed by smart voice assistant and human assistant about the experimental procedure and steps about how to perform modular arithmetic tasks. Participants were informed by smart voice assistant and human assistant that they should give the response as quickly and as accurately as possible. Through that, people could perceive the presence of human assistant and the presence of smart voice assistant. When working alone, participants finished modular arithmetic tasks on their own. In the condition of working in the presence of a human assistant, a same-gender student was seated near the actual participant at a distance of 90 cm and served as an audience. In the condition of working in the presence of a smart voice assistant, participants finished modular arithmetic tasks alone in a room. A smart speaker integrated with a smart voice assistant was placed near the participants at a distance of 90 cm. All participants were informed that they just performed modular arithmetic tasks with one assistive person or an assistive device in the same room. Experimenters left the laboratory to allow participants to perform tasks. Experimenters returned to the laboratory and helped set up the software for the next session after actual participants completed each session.

2.5 Data Analysis

Repeated-measures ANOVA was used to determine the main and interaction effects of task difficulty and social presence on RT and ARs. Effect size was calculated with a generalized eta-squared (η_G^2) [21]. Paired t tests with Bonferroni adjustment were used for post hoc analysis.

3 Results

3.1 Response Time

RT refers to the time interval between the appearance of experiment interface and the clicking of Q or P key in a trial. Results showed that the main effects of task difficulty ($F (1, 15) = 50.589; p < 0.0001; \eta_G^2 = 0.287$) on RT. The effect of social presence on RT was not significant ($F (1, 15) = 1.058; p = 0.34$).

A significant interaction was found between task difficulty and social presence on RT ($F (2, 30) = 62.173, p < 0.0001, \eta_G^2 = 0.102$). In the conditions of working in the presence of a smart voice assistant (SVA) or a human assistant (HA), participants responded faster in performing easy modular arithmetic tasks in contrast to working alone, whereas they responded slower in performing difficult modular arithmetic tasks in contrast to working alone (for easy tasks, $M_{alone} = 2687, M_{SVA} = 2115, t (15) = 5.789, p < 0.0001; M_{alone} = 2687, M_{HA} = 2087, t (15) = 5.988, p < 0.0001$; for difficult tasks, $M_{alone} = 5120, M_{SVA} = 6157, t (15) = 7.058, p < 0.0001; M_{alone} = 5120, M_{HA} = 6221, t (15) = 8.211, p < 0.0001$), revealing that both social presence forms (working in the presence of a smart voice assistant and working in the presence of a human assistant) can induce social facilitation effect, as illustrated in Fig. 2. Differences in RT between working in the presence of a smart voice assistant and working in the presence of a human assistant were not significant in both easy and difficult tasks (in easy tasks, $M_{SVA} = 2115, M_{HA} = 2087, t (15) = 1.254, p = 0.23$; in difficult tasks, $M_{SVA} = 6157, M_{HA} = 6221, t (15) = 1.347, p = 0.20$). Results suggested that the intensities of social facilitation effects induced by a smart voice assistant and a human assistant was similar.

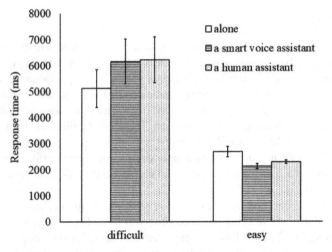

Fig. 2. RT in all conditions of task difficulty and social presence.

3.2 Accuracy Rates

Results showed that the main effect of task difficulty was significant (F (1, 30) = 17.542, $p < 0.01$, $\eta_G^2 = 0.171$). The accuracy rates of easy tasks were larger than those of difficult tasks ($M_{easy} = 0.952$, $M_{difficult} = 0.895$, t (15) = 5.021, $p < 0.0001$). However, the remaining main effect and interactions were insignificant ($ps > 0.05$) (Fig. 3). The results showed that the social facilitation effect induced by the presence of smart voice assistant and human assistant did not act on accuracy rates.

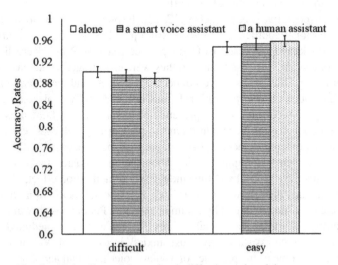

Fig. 3. AR in all conditions of task difficulty and social presence.

4 Discussion

The present study was a preliminary investigation that examined whether social facilitation effect could be induced by smart voice assistant. The results suggested that the presence of smart voice assistant could influence people's task performance just like the presence of human assistant. Our findings indicated that performance of easy modular arithmetic tasks was improved by the presence of smart voice assistant while that of difficult modular arithmetic tasks was impaired. The social facilitation effect induced by smart voice assistant showed similar intensity to the effect induced by human assistant. The presence of a smart voice assistant only exerted significant effects on RT rather than ARs when participants performed modular arithmetic tasks, as similarly shown in previous studies [22].

One important condition for occurrence of social facilitation effect is awareness of individuals that they are working in social situations [23]. There is evidence that social presence, defined by Lee [24] as "the experience of artificial objects as social actors that manifest humanness", is an important factor in social interaction [25, 26]. People tend to look at technological artefacts as social actors [5, 27]. Thus, consistent with previous studies, this study confirmed that smart voice assistant could possibly make people perceive social presence and thus induce social facilitation effect.

The aforementioned social facilitation theories suggest that the existence of social facilitation effect is due to the increase of arousal, which can enhance the dominant response when people perform tasks. The dominant response is one action that can be easily and quickly triggered by stimulus, i.e., quick response in easy tasks and slow response in difficult tasks. Previous studies indicated that the actual or anticipated use of information technology can increase an individual's arousal level [28, 29]. Thus, we infer that the presence of smart voice assistant can possibly elevate arousal level which possibly contributed to social facilitation effect.

According to cognitive process theory [17, 18], the presence of a smart voice assistant may cause attentional conflict, which arises from people's intention to simultaneously concentrate on tasks and the presence of a smart voice assistant. People were distracted by the presence of a smart voice assistant, and they were encouraged to increase their focus to avoid distraction. Thus, arousal levels of people increased, and social facilitation effect was induced. Besides, another person's role may influence individuals' expectations for their behavior [7]. According to self-presentation theory [15], a human assistant can influence people's behavior as people intend to present themselves as capable. Based on evaluation apprehension theory [14], people may be concerned with how human assistant evaluate them. The apprehension about being evaluated by human assistant could increase people's arousal and thus induce social facilitation effect.

The present study also showed similar intensity of social facilitation effect induced by a smart voice assistant with that by a human assistant. People possibly unconsciously regard technological artefacts as social entities [4]. The computer-mediated communication with smart voice assistant may cause anxiety in users [30]. Visual and auditory cues may remind people of the presence of a smart voice assistant and thus emphasize it social presence. Results in the present study implied that the increased arousal level may be similar in the presence of human assistant and smart voice assistant, which we will verify that in future investigation. In this study, only sixteen participants were recruited and only same-gender smart voice assistant was used to eliminate the possible effect of gender on social facilitation effect [22]. The influence of smart voice assistant gender in social facilitation effect could be examined in future research.

Results provide implications for the design of work organization modes for working with smart voice assistant. A smart voice assistant, similar to a human assistant, could elicit social facilitation effect, and such effect was mediated by task difficulty. Selection of work organization modes for working with smart voice assistant should vary with task difficulties. For easy tasks, the presence of smart voice assistant is advantageous to improve people's performance by applying social facilitation effect. For difficult tasks, the presence of smart voice assistant and human assistant could impair task performance. Hence, working alone mode should be mostly utilized.

In conclusion, the present study was a preliminary investigation to examine the social facilitation effect induced by smart voice assistant by applying modular arithmetic tasks. Results showed that the presence of smart voice assistant can shorten RT when people performed easy modular arithmetic tasks but lengthen it when they performed difficult modular arithmetic tasks. Intensity of social facilitation effect induced by a smart voice assistant was as strong as that by a human assistant. Both the presence of smart voice assistant and human assistant showed nonsignificant influence on accuracy rates. These findings provided practical implications in the design and improvement of work organization modes for people perform tasks in the presence of smart voice assistant.

Acknowledgements. This work was supported by grants from Natural Science Foundation of China (Project No. 71901033), Beijing Natural Science Foundation (Project No. 9204029) and Chinese Fundamental Research Funds for the Central Universities (Project No. 2482019RC43).

References

1. Dousay, T.A., Hall, C.: Alexa, tell me about using a virtual assistant in the classroom. In: EdMedia + Innovate Learning, pp. 1413–1419. Association for the Advancement of Computing in Education (AACE) (2018)
2. Accenture: Reshape Relevance (2019). https://www.accenture.com/us-en/insights/high-tech/reshape-relevance
3. Hoy, M.B.: Alexa, Siri, Cortana, and more: an introduction to voice assistants. Med. Ref. Serv. Q. **37**(1), 81–88 (2018)
4. Nass, C., Moon, Y.: Machines and mindlessness: social responses to computers. J. Soc. Issues **56**(1), 81–103 (2000)
5. Reeves, B., Nass, C.: The Media Equation: How People Treat Computers, Television, and New Media Like Real People and Places. CSLI Publications and Cambridge University Press, Cambridge (1996)
6. Chattaraman, V., Kwon, W.S., Gilbert, J.E., Ross, K.: Should AI-Based, conversational digital assistants employ social- or task-oriented interaction style? A task-competency and reciprocity perspective for older adults. Comput. Hum. Behav. **90**, 315–330 (2019)
7. Aiello, J.R., Douthitt, E.A.: Social facilitation from Triplett to electronic performance monitoring. Group Dyn. Theory Res. Pract. **5**(3), 163–180 (2001)
8. Kiesler, S., Hinds, P.: Introduction to this special issue on human-robot interaction. Hum. Comput. Interact. **19**(1–2), 1–8 (2004)

9. Park, S., Catrambone, R.: Social facilitation effects of virtual humans. Hum. Factors **49**(6), 1054–1060 (2007)
10. Riether, N., Hegel, F., Wrede, B., Horstmann, G.: Social facilitation with social robots? In: Proceedings of the Seventh Annual ACM/IEEE International Conference on Human-Robot Interaction, pp. 41–48. ACM (2012)
11. Zanbaka, C.A., Ulinski, A.C., Goolkasian, P., Hodges, L.F.: Social responses to virtual humans: Implications for future interface design. In: Proceedings of the SIGCHI Conference on Human factors in Computing Systems, pp. 1561–1570. ACM (2007)
12. Guerin, B.: Social Facilitation. Wiley Online Library (2010)
13. Zajonc, R.B.: Social facilitation. Science **149**(3681), 269–274 (1965)
14. Cottrell, N.B.: Social facilitation. In: McClintock, C.G. (ed.) Experimental Social Psychology, pp. 185–236. Holt, Rinehart & Winston, New York (1972)
15. Bond, C.F.: Social facilitation: a self-presentational view. J. Pers. Soc. Psychol. **42**(6), 1042–1050 (1982)
16. Carver, C.S., Scheier, M.F.: The self-attention-induced feedback loop and social facilitation. J. Exp. Soc. Psychol. **17**(6), 545–568 (1981)
17. Baron, R.S.: Distraction-conflict theory: Progress and problems. Adv. Exp. Soc. Psychol. **19**, 1–40 (1986)
18. Baron, R.S., Moore, D., Sanders, G.S.: Distraction as a source of drive in social facilitation research. J. Pers. Soc. Psychol. **36**(8), 816–824 (1978)
19. Groff, B.D., Baron, R.S., Moore, D.L.: Distraction, attentional conflict, and drivelike behavior. J. Exp. Soc. Psychol. **19**(4), 359–380 (1983)
20. Sanders, G.S., Baron, R.S., Moore, D.L.: Distraction and social comparison as mediators of social facilitation effects. J. Exp. Soc. Psychol. **14**(3), 291–303 (1978)
21. Olejnik, S., Algina, J.: Generalized eta and omega squared statistics: measures of effect size for some common research designs. Psychol. Methods **8**(4), 434–447 (2003)
22. Liu, N., Yu, R., Yang, L., Lin, X.: Gender composition mediates social facilitation effect in co-action condition. Sci. Rep. **7**(1), 15073 (2017)
23. Hayashi, Y.: Social facilitation effects by pedagogical conversational agent: lexical network analysis in an online explanation task. In: Proceedings of the 18th International Conference on Educational Data Mining, pp. 484–487. Madrid, Spain (2015)
24. Lee, K.M.: Presence, explicated. Commun. Theory **14**(1), 27–50 (2004)
25. Pereira, A., Martinho, C., Leite, I., Paiva, A.: iCat, the chess player: the influence of embodiment in the enjoyment of a game. In: 7th International Conference on Autonomous. Agents and Multiagent System, IFAAMAS, pp. 1253–1256 (2008)
26. Kennedy, J., Baxter, P., Belpaeme, T.: Comparing robot embodiments in a guided discovery learning interaction with children. Int. J. Soc. Robots **7**(2), 293–308 (2015)
27. Min Lee, K., Jung, Y., Kim, J., Kim, S.R.: Are physically embodied social agents better than disembodied social agents?: The effects of physical embodiment, tactile interaction, and people's loneliness in human–robot interaction. Int. J. Hum Comput Stud. **64**(10), 962–973 (2006)
28. Antheunis, M.L., Schouten, A.P., Valkenburg, P.M., Peter, J.: Interactive uncertainty reduction strategies and verbal affection in computer-mediated communication. Commun. Res. **39**(6), 757–780 (2012)
29. Brown, S.A., Fuller, R.M., Vician, C.: Who's afraid of the virtual world? Anxiety and computer-mediated communication. J. Assoc. Inf. Syst. **5**(2), 79–107 (2004)
30. Rickenberg, R., Reeves, B.: The effects of animated characters on anxiety, task performance, and evaluations of user interfaces. In: Proceedings of the SIGCHI conference on Human Factors in Computing Systems, pp. 49–56. ACM (2000)

Focus on Automotive User Interfaces Research: A Bibliometric Analysis and Social Network Analysis During 1994–2019

Chen Lu[1][(✉)] and Hao Tan[2]

[1] School of Design, Hunan University, Changsha, China
luchen@hnu.edu.cn
[2] State Key Laboratory of Advanced Design and Manufacturing for Vehicle Body, Changsha, China

Abstract. This paper aims to analyze 424 papers obtained by using "automotive user interfaces" as the keywords in the Web of Science database and discusses the core themes and cooperation in the research field of automotive user interfaces to understand the development and future trend of automotive user interfaces research. Employing bibliometrics and social network analysis, the study analyzed keywords, core themes, co-word networks, author's influence and collaborations patterns. The results identified "user interface", "Human-machine interfaces", "user-centred design", "sensors" and "usability" are the key research domains in automotive user interfaces research. From the social network analysis, we found that the overall density of the whole network is very low and academic cooperation among developed countries seems be closer in the field of automotive user interface research.

This research has revealed some important findings that identify the theme and key research areas, and reveals the current status of automotive user interfaces research in a quantitative way.

Keywords: Bibliometrics · Social network analysis · Automotive user interfaces

1 Introduction

With the rapid development of intelligent automobile and the new "style" of the new vehicles, automobile human-computer interaction design has become an important opportunity and challenge for automobile design theory and practice. Although bibliometrics has been applied in many other fields, few researchers have systematically reviewed the field of human-computer interfaces design of automobiles with quantitative methods. For this reason, we know little about the distribution and structure of the field of human-computer interfaces design of automobiles. Therefore, it is necessary to analyze the field of automobile human-machine interfaces design from the quantitative and empirical perspectives. In the paper, we adopted bibliometrics as quantitative research tools and review the current researches of this domain.

© Springer Nature Switzerland AG 2020
P.-L. P. Rau (Ed.): HCII 2020, LNCS 12192, pp. 625–640, 2020.
https://doi.org/10.1007/978-3-030-49788-0_48

The main purpose of this paper is to conduct bibliometrics research in the field of automobile human-machine interfaces design since 1994. This research will supplement the current vacant field of automobile human-computer interfaces design through empirical analysis of research structure, correlation between research topics, country of author, influence of author and qualitative work of cooperation mode.

2 Methodology

Bibliometrics analysis has gained considerable attention in the scientific community over the past few years, with the primary objective of reviewing or summarizing major trends in journal publication over a given period of time [6], This method is based on the quantitative and qualitative analysis of scientific literature and can be used to evaluate and compare the research performance of researchers, journals, institutions, countries or disciplines [7].

Traditional scientometrics research methods mainly include publication statistics [3], author statistics, citation analysis and word frequency analysis [22], and so on. With the deepening of scientometrics research, some new methods have appeared in this discipline, and many methods have been introduced by other disciplines. In recent years, common methods mainly include co-occurrence analysis [8], social network analysis [20], multivariate statistical analysis, and information visualization. This paper combines bibliometrics and social network analysis technology to analyze the field of automotive user interface research.

BibExcel is used in the bibliometrics section to provide data statistics including keywords, authors, and national and regional statistics. Some of the most popular tools are Citespace, HistCite, and BibExcel [3]. BibExcel was selected as the research tool of this study because BibExcel has a high degree of flexibility in modifying and/or adjusting input data imported from different databases such as Scopus and is compatible with various computer applications such as SPSS, Excel, Unicet and Gephi. BibExcel is also used to prepare input data for detailed network analysis to provide comprehensive data analysis for a range of network analysis tools including Gephi, VOSviewer, and Pajek [19].

In the part of social network analysis, Gephi is used to conduct keyword clustering analysis and cooperative mode analysis on the literature in the field of automotive user interface research. Gephi can efficiently process large data sets and different data formats, and at the same time provide a series of innovative visualization, analysis and investigation options. Therefore, in this study, we chose Gephi over existing network analysis software, such as Pajek [2] and VOSviewer [24].

2.1 Initial Data Statistics

Mongeon et al. [21] noted that most bibliometrics have a common data source: Thomson Reuters' Web of Science (WOS) and Elsevier's Scopus. The core set of Web of Science is a world-renowned citation index database, which is widely used for scientific research and evaluation because of its groundbreaking content, high-quality data and long history. It consists of three independent databases, namely Science Citation Index Expanded (SCI

Expanded), Social Sciences Citation Index (SSCI) and Arts & Humanities Citation Index (A&HCI). Through the platform, researchers can find bibliographic data of more than 12,000 scientific journals in the natural sciences, social sciences, arts and humanities fields for nearly a century. Therefore, Web of Science is selected as the database to obtain initial data.

Use the keywords "Automotive user interfaces" and limit search to headings, abstracts, authors, and keywords. Since Web of science only has data from 1994, the analysis was limited to the period 1994–2019. Export the full record of the data in ".txt" format as the input source for BibExcel.

2.2 Bibliometric Analysis

Bibliometrics techniques have evolved over time and continue to evolve: attributing papers by keywords, terms, institutions, and authors [21]. All these techniques combine to provide more detailed and efficient measurements. In this study, we use word frequency analysis, co-word analysis and author analysis to show the current situation of automotive user interfaces research.

Word frequency analysis is based on statistical linguistics to study the frequency distribution of words in scientific literature. The purpose of word frequency analysis is to establish word frequency dictionary according to the subject domain, and to conduct quantitative analysis and comparison of scientists' creative activities. The basis of word frequency analysis is the internal relationship between social phenomenon, literature content and word frequency. It involves the most basic subject word unit set, class title, author, etc. of knowledge and its disseminator in science communication, as well as the statistics of specific category words. Through analysis, we can find out the topic content, research hotspot and topic change of the literature. Word frequency analysis has been widely used in quantitative analysis. They include environmental sustainability [3], non-biomedical modality [13] and social entrepreneurship [14].

Co-word analysis is an important content in bibliometrics, which can express a topic field of research topic or research direction of professional terms together in literary works, analyze phenomena, explore the relationship between these words, and then analyze the structure changes of the sentence representing the theme and the theme. Co-word network analysis is to use statistical method to calculate the frequency of co-word pairs appearing in the same paper, obtain the co-word matrix, and then convert the co-word matrix into the co-word network analysis method. Co-word analysis is widely used in scientometrics research, scientology knowledge mapping research, subject hot spot research and subject structure research. Researchers have successfully used co-word analysis to identify patterns and trends in specific research areas of research, including recommendation systems [15], and pollution bioremediation [23], Internet of things (IoT) [16].

2.3 Social Network Analysis

Social Network Analysis (SNA) is one of the most prevalent methods of analyzing co-authorship [17] and constitutes a growing body of analytical approaches that aims to mathematically analyze the structure of networks [18]. A network composed of a large number of nodes would cause confusion, so SNA complements visualizations of networks with mathematical models that formally quantify network properties and the inter-actor relationships. The word co-occurrence matrix may be essentially considered a network, and social network analysis is considered a well-established and effective approach. In this study, several measures of social network analysis were used, such as network centrality, density and strategic diagram.

(1) In social network literature, centrality is an application measuring the structurally advantageous positions of nodes in the social network. For example, Keywords or themes with high centrality could be defined as lying on a central and important position in the cluster, or in the entire research field. Authors with high centrality could identifying the central actors in our network. Therefore, this value is often used to measure the importance of a node, or a sub-network of a research field [5].

(2) Density indicates the strength of correlation between nodes (keywords or authors) in network, within its corresponding research themes in a network. Network density reflects the internal coherence or cohesiveness among the nodes in a network. The density of the network, varies between 0 and 1, is a measure of the connectivity of a network. For example, if all nodes in the network is connected in pairs, the network density would be 1. If there are no connections in the network, the network density is 0. Low density means the network is not very well connected and low network density is typical in co-authorship networks with large numbers of authors [1].

(3) Based on centrality and density, a scatter graph known as a strategic diagram [4] was built to describe the internal relations within a cluster, as well as the interactions among different fields. In general, the strategic diagram uses a two-dimensional space to plot clusters according to their centrality and density. In a strategic diagram, the x-axis stands for centrality, while the y-axis stands for density and the origin of the strategic diagrams is the median of the respective axis values.

Thus, because of different centrality and density, the theme clusters are located in four quadrants, which could indicate the developing status of research themes. In quadrant I, with high centrality and density, research themes are more mature and central in the overall research field. In quadrant II, research themes are not central but are well developed. In quadrant III, research themes are peripheral and underdeveloped. And in quadrant IV, with low density and centrality, research themes are central in the field but are undeveloped or immature, or may possibly tend toward maturity [4, 12].

3   Results

A total of 424 articles, which constitute the work sample of the study. The track of total number of articles per year from 1994 to 2019 is shown in Fig. 1.

3.1 Keyword Analysis

The keywords for an article can express its main content, and the keyword frequency of academic journal articles can measure the importance of its related themes in a specific

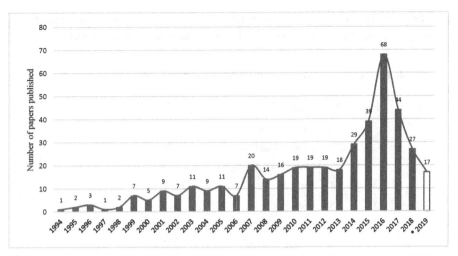

Fig. 1. Number of papers published in each year. (when data was collected, the year 2019 had not yet finished.)

field. In this study, records of 424 articles were selected for subsequent analysis. With the aid of BibExcel, 1559 keywords were extracted from these article records (3.68 keywords per article).

It was noted that not all the keywords provided by authors were normalized; thus, the extracted keywords were normalized using Affinity propagation according to the method mentioned by Brendan J in 2007. This algorithm takes the similarity matrix of the data set as input. At the initial stage of the algorithm, all samples are regarded as potential clustering center points. Meanwhile, each sample point is regarded as a node in the network. The normalization of these keywords was carried out to ensure consistent treatment of the singular and plural forms of words, unifying synonyms, and clarity of homonyms. For example, "HMI" was replaced by "Human-machine interfaces"; "Mixed-reality" was replaced by "Mixed reality"; "Computer-aided design" was replaced by "Computer aided design"; etc. General terms such as "automotive", "model", "assembly", "generation" were removed. The entire process of normalization helped to determine the final keywords for the analysis, with the aid of two professors specializing in this field, and discussion within the research team. Select keywords with frequency more than 3 and finally, 97 related keywords with a total frequency of 432 (about 27.8% of the frequency of the entire keywords) were determined to represent the main contents of Automotive user interfaces research. By BibExcel's frequency statistics of DE field in the original text, Table 1 shows the top 20 keywords. According to the word frequency and co-word frequency, the overall attributes of keywords or themes could be made more explicit. First, the top ten keywords with high co-word frequency is identified and shown in Table 2. Each of the top 10 keyword lists could be considered as a major focus in the field of automotive user interfaces research. In particular, "user interface", "Human-machine interfaces", and "user-centred design", occurring in the top total co-word frequency list and the top total co-word correlation coefficient list, could be considered as the representative of the subfield of automotive user interfaces research. These keywords have the most direct connections to other themes.

Table 1. The top 20 keywords in the field of automotive user interfaces research.

No.	Keyword	Frequency
1	User interface	30
2	Human-machine interfaces	26
3	User-centred design	8
4	Sensors	8
5	Usability	7
6	Driver distraction	7
7	Design method	7
8	Modelling	7
9	Human factors	6
10	Haptic feedback	6
11	Head-up displays	6
12	Automotive electronics	6
13	User experience	6
14	Interface design	6
15	Evaluation	6
16	Virtual reality	6
17	Automotive interface	5
18	ADAS	5
19	Multimodal interfaces	5
20	Human-computer interaction	5

3.2   Core Themes

Using the modularity clustering method of Gephi, with the clustering step being set up at one level, 9 clusters (named Cluster 1 to Cluster 9) were obtained. Each of these clusters could be regarded as a research theme. The 87 keywords from the normalization process were divided into 9 theme-clusters. If keywords are grouped into a cluster, they are more likely to have an identical research theme. In this study, 3 to 30 keywords with top frequencies and co-word data were selected to represent each theme-cluster, because these keywords were most likely to be chosen and used by researchers in each cluster. Table 3 shows the representative keywords within each theme-cluster.

Overall, there were 9 research clusters or themes in the field of automotive user interface research. In consultation with the experts in the automotive user interface field,

Table 2. The top 10 keywords with high co-word data in the field of automotive user interfaces research.

No.	Keyword	Total co-word frequency
1	Human-machine interfaces	103
2	User interface	81
3	Automotive electronics	31
4	Controller area network	27
5	User-centred design	26
6	Human factors	26
7	Design method	25
8	Interface design	23
9	Multimodal interfaces	21
10	Design thinking	21

Table 3. 9 clusters in the field of automotive user interfaces research.

Cluster	Number of keywords	Keywords
1	8	Modelling, haptic feedback, Simulation, robotics, Automotive design
2	6	Android, CAN, cloud computing, GPS
3	8	ADAS, Driving, Artificial intelligence, collaborative virtual environment, mixed reality
4	12	Human-machine interfaces, sensors, Usability, user experience, User Interface Design, user-centred design
5	20	User interface, driver distraction, interface design, Personalization, Multimodal Interfaces, Multi-Tasking, Human factors, head-up displays
6	6	Autonomous Driving, driving simulator, base-station, infotainment
7	2	Decision support system, programming
8	12	Automotive electronics, Embedded Systems, control wires, FPGA, Fuzzy logic
9	13	Design method, Design thinking, Inclusive design, case study, logistics, Java

we believe that this clustering result (9 theme-clusters) is a good representation of the development status of automotive user interfaces research.

In this study, co-word data includes the total frequency, total co-word frequency and their average values (shown in Table 4), which could more explicitly demonstrate the characteristics of each research theme-cluster. The comprehensive and proportional average values of co-word data can be designated as very specific indicators to distinguish each research theme. In order to obtain an accurate and explicit comparison among these 9 theme-clusters, rankings in their respective average frequency and average co-word frequency, are listed in Table 4. The ranking shows the relative position and status of related themes in the field of automotive user interfaces research.

Table 4. The frequency and co-word data of each theme-cluster.

Cluster	Total frequency	Total co-word frequency	Average frequency	Ranking of average frequency	Average co-word frequency	Ranking of average co-word frequency
1	35	83	4.36	3	10.36	8
2	20	67	3.33	7	11.17	7
3	32	107	4.00	4	13.36	5
4	85	237	7.08	1	19.75	1
5	110	361	5.5	2	18.05	2
6	20	79	3.33	8	13.17	6
7	6	10	3.00	9	5.00	9
8	42	190	3.50	6	15.83	3
9	49	176	3.77	5	13.54	4

When combining results in Table 3 with Table 4, it was found that several larger research themes or sub-directions exist within the field of automotive user interfaces during the period under study, such as Cluster 4 relating to "User-centered design", Cluster 5 relating to "principles of user interface design", Cluster 8 relating to "electronic circuit", and Cluster 9 relating to "Design thinking and methods". On the other hand, Cluster 1, Cluster 2, Cluster 3, Cluster 6, which are related to "Virtual simulation technology", "computer control system", "driving technical support", "automatic driving", and have relatively low indicators, were not found to be the major focus of automotive user interfaces research. With high-level total frequency and total co-word frequency, Clusters 4, 5, 8 and 9 appeared to have received more attention from researchers, and it can be considered as the focus of automotive user interfaces research on a global scale. It was found that Cluster 5 relating to "principles of user interface design", it had high-level co-word data, indicating that this theme played a central role in the research field. Meanwhile, Cluster 7 is related to "information technology", it had the lowest-level indicators, and themes in these clusters were unlikely to be a research focus. They might be isolated research themes in automotive user interfaces research.

Table 5 lists the top 10 keywords displaying a high degree centrality which helped identify the position of each keyword in the overall research structure. According to the high degree centrality and from the perspective of the whole network structure, these keywords were noted as the main focuses of automotive user interfaces research: "Human-machine interfaces"; "User interface"; "automotive electronics"; "Design method"; "Human factors". They also have high frequencies and occurring in the top total co-word frequency list. Therefore, this conclusion is further proved. Meanwhile, because of the high betweenness centrality, these keywords were defined as the important bridges connecting other research themes or subfields, such as "Human-machine interfaces"; "automotive electronics"; "User interface"; "evaluation"; "mixed reality".

Table 5. The top 10 keywords with high centrality.

No.	Keyword	Nrm degree centrality	No.	Keyword	Nrm betweenness centrality
1	Human-machine interfaces	0.184	1	Human-machine interfaces	0.281
2	User interface	0.149	2	Automotive electronics	0.173
3	Automotive electronics	0.149	3	User interface	0.133
4	Evaluation	0.103	4	Evaluation	0.131
5	Design thinking	0.092	5	Mixed reality	0.088
6	Visual demand	0.092	6	Augmented reality	0.083
7	Design method	0.080	7	Inclusive design	0.082
8	Driver distraction	0.069	8	User interface design	0.081
9	Human factors	0.069	9	Design thinking	0.076
10	ADAS	0.069	10	Human-machine-interaction	0.072

Based on the co-word analysis, the density of the overall research network was calculated. The density of the network with all keywords is 0.041, which is a relatively low level and indicates that automotive user interfaces research is decentralized [9, 11]. Further, the centrality and density of each theme cluster in one-step clustering were calculated and presented in Table 6, and the strategic diagram was generated as shown in Fig. 2. The averages of centrality and density were chosen as the origin (2.247, 0.359) of the strategic diagram. The strategic diagram reveals the status and trends of current researches of automotive user interfaces research by dividing these 9 clusters into four quadrants.

As shown in Fig. 2, there are no clusters in quadrant I, which represents a high centrality and high density. The high density indicates that this cluster has high internal correlation, and the research topic of this cluster tends to be mature. High centrality indicates that this cluster has extensive connections with other clusters. Therefore, in

Table 6. Density and centrality of each cluster.

Cluster	Density	Centrality
1	0.286	2.000
2	0.400	2.000
3	0.321	2.250
4	0.212	2.333
5	0.153	2.900
6	0.333	1.667
7	1.000	1.000
8	0.273	3.000
9	0.256	3.077

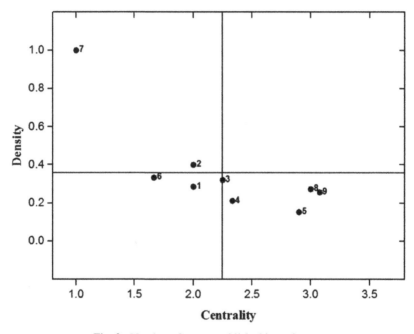

Fig. 2. Number of papers published in each year.

the field of automotive user interfaces research there is no topic to become the core. The main reason may be that the research field of automotive user interfaces is not very old, so there is still a lot of research and application space.

Displaying high density and low centrality, Cluster 2 and cluster 7 are located in quadrant II, showing high density and low centrality. That is to say, many researchers have paid attention to the research of these two clusters, but because of their low centrality, they have not become the core of automotive user interfaces research field. Especially suitable for cluster 7, which represents the research of decision support system and programming. Finally, we observed that the themes in Cluster 2 (research on Android, CAN, cloud computing, GPS, etc.) were close to quadrant I, indicating that they were more inclined to be well developed and become the core of the field.

Cluster 1 and cluster 6 are located in quadrant III. Their low density and low centrality indicate that the research topic of these clusters is marginal and immature. Subjects in the cluster include Modelling, haptic feedback, Simulation, robotics, Autonomous Driving, Driving simulator, etc. It is found that cluster 6 is close to quadrant II, indicating that subjects in the cluster are more and more closely related to other clusters and tend to be mature.

Quadrant IV shows several different types of clusters. These clusters have a lower density but a relatively higher centrality. This phenomenon indicates that the research topic in these clusters is the core of the research field of automotive user interfaces, but it is not mature yet. Cluster 3 mainly represents the research of ADAS, Driving, artificial intelligence, collaborative virtual environment, mixed reality and other fields. It is in the center, but it is not mature yet. Clusters 8 and 9 represent research on automotive electronics, embedded systems, control wires, design method, design thinking, inclusive design and case study. These two clusters have similar densities and centrality, which means that the research topics in the clusters are the core of the automotive user interfaces research field, but they are not yet well developed. We can find that cluster 8 and 9 are very close to quadrant I, so the themes in these two clusters have great development potential and tend to mature. Cluster 4 and 5 have relatively high centrality, indicating that the research topic in these clusters is the core of the research field of automotive user interfaces, but it is still in the immature stage.

3.3 Author's Influence

According to previous research by Behnam et al. [3], an author's influence in a field is measured by the number of papers he/she publishes. Based on that, BibExcel calculates how often authors appear to represent the number of articles published by authors in order to study their influence.

According to the Table 7, Langdon, P is the most prolific researcher with 8 papers published, ranking first among all the authors. Rigoll, G, Tscheligi, M, Meschtscherjakov, A, Pfleging, B and Skrypchuk, L have all published 7 papers, which is second only to Langdon, P. Five of the top ten are from Germany, three from the UK and two from Austria. This shows that Germany, Britain and Austria have made a lot of academic achievements in the field of automotive user interfaces, and have exerted a great influence on the field of automotive user interfaces research.

3.4 Collaborations Patterns

We found that the overall densities of the whole network (0.004) is very low. As mentioned in the method of social network analysis section above, the density of the network, varies between 0 and 1, is a measure of the connectivity of a network. Low density means the co-authorship network of automotive user interfaces research is not very well connected and low network density is typical in co-authorship networks with large numbers

Table 7. The top 10 author's publications.

No.	Name	Publications
1	Langdon, P	8
2	Rigoll, G	7
3	Tscheligi, M	7
4	Meschtscherjakov, A	7
5	Pfleging, B	7
6	Skrypchuk, L	7
7	Boll, S	6
8	Williams, M	6
9	Braun, M	5
10	Alt, F	5

of authors [1]. Nevertheless, the mean degree centrality of the whole network is 3.958, which means that, on average, every author has around four co-authors.

Besides the global network statistics, we then switch to Node-level centrality measures to identifying the central actors in our network with high levels of social and intellectual capital. Typically, central nodes in a network are important in so far as they are able to influence the rest of the network [10]. These central actors may also become gatekeepers of the research community and thus sought-after collaborators. It follows that identifying these actors is important to understand the overall network structure. In this study we employ two distinct centrality measures which are degree and betweenness centrality.

Table 8 shows degree centrality and normalized betweenness centrality measures for top 10 authors for the whole network with the mean values for these measures. It can be seen that Skrypchuk, L, Langdon, P and Shriberg, E are the top three in the list of degree centrality. We can notice that Langdon, P rank the first in the list of the top 10 author's publications at the same time. It indicates that Langdon, P is not only the most Influential author, but also are generous in collaboration in the field of automotive user interfaces.

In addition to the collaborations among individual scholars, it is also necessary to understand the collaborations state of automotive user interfaces research at the national level. The author's affiliations were extracted among all file data and calculate the co-occurrence of author's countries to analyze the collaborations between the countries and regions.

Table 8. The top 10 authors with high degree centrality and betweenness centrality.

No.	Name	Degree centrality	No.	Name	Betweenness centrality
1	Skrypchuk, L	26	1	Pfleging, B	0.001041
2	Langdon, P	19	2	Kun, A	0.000971
3	Shriberg, E	18	3	Skrypchuk, L	0.000883
4	Xu, K	18	4	Tscheligi, M	0.000787
5	Zhang, Q	18	5	Schmidt, A	0.000666
6	Mishra, R	18	6	Dey, A	0.000585
7	Varges, S	18	7	Boll, S	0.000580
8	Weng, F	18	8	Meschtscherjakov, A	0.000536
9	Scheideck, T	18	9	Jeon, J	0.000459
10	Schmidt, H	18	10	Braun, M	0.000424

Table 9 is the result of the frequency analysis of co- occurrence of author's countries and regions. It can be seen that Germany researchers collaborated considerable closely with scholars from the USA for they published 9 international collaborated papers. German and Austrian scholars jointly published 4 international papers, American and British scholars jointly published 4 international papers too. Most of the countries these authors belong to are developed countries. This shows that academic cooperation among scholars in developed countries is closer in the field of automotive user interface research.

Table 9. Collaborations between countries and regions.

No.	Countries and regions collaborated	Frequency
1	Germany and USA	9
2	Germany and Austria	4
3	USA and UK	4
4	USA and Austria	3
5	UK and India	2
6	UK and Sweden	2
7	UK and Germany	2
8	USA and South Korea	2
9	UK and Italy	2
10	Spain and Italy	2

4   Conclusion

In total, this paper analyzed keyword and authors of 424 publications published from 1994 to 2019 by using a bibliometrics and social network analysis. It is the first time to study the state of automotive user interfaces research using quantitative approach in international journal. This research has revealed some important findings that identify the theme and key research areas, and reveals the current status of automotive user interface research in a quantitative way.

Firstly, 97 keywords with a threshold value greater than 3 are selected by bibliometrics analysis technology, accounting for 27.8% of the total frequency of all keywords. Among them, the most popular keywords are "user interface", "Human-machine interfaces", "user-centred design" and "sensors", which represents the research theme and key field. Automotive user interfaces research is relatively immature on a global scale. According to the keyword analysis and co-word analysis, the main research areas focus on "user interface", "Human-machine interfaces", and "user-centred design". It proves that the research in these fields is relatively strong, which can be considered as the representative of the sub-field of automotive user interfaces research, while other topics and research emphases are relatively weak.

Secondly, by using the social network analysis method, 9 topic clusters are determined, which represent the sub-fields of automotive user interfaces research in recent 25 years. According to their metrics, these clusters are classified into high-level clusters and low-level clusters. For example, high clusters are mainly in the fields of "User-centered design", "principles of user interface design", "electronic circuit" and "Design thinking and methods" while low clusters are mainly in the fields of "Virtual simulation technology", "Computer control system", "Driving technical support", "Automatic driving" and "Information technology". By using strategic diagram analysis, 9 clusters are divided into 4 quadrants, revealing the current situation and trend of automotive user interfaces research. "User-centered design", "Principles of user interface design", "Electronic circuit" and "Design thinking and methods" are the cores of automotive user interfaces research, but these research topics is not mature yet. "Virtual simulation technology", "Driving technical support", "Automatic driving" and "Information technology", these research topics are at the edge and are in an immature state. Although "Computer control system" and "Information technology" are relatively mature, but they have not become the core of the field.

Thirdly, we analyze author's influence and collaboration pattern in the field of automotive user interfaces research, Langdon, P is the most prolific researcher with 8 papers published, ranking first among all the authors. Rigoll, G, Tscheligi, M, Meschtscherjakov, A, Pfleging, B and Skrypchuk, L all published seven papers. Through the country where the author belongs, we find that Germany, Britain and Austria have many academic achievements and have exerted a great influence in the field of automotive user interfaces research. Although the average author has four co-authors, but the co-author network is not well connected. Meanwhile, the central authors in network were identified. Langdon, P is not only the most Influential author, but also are generous in collaboration in the field of automotive user interfaces research. In addition, at national level of collaborations, academic cooperation among developed countries is closer in the field of automotive user interface research.

Nevertheless, this analysis is far from being perfect, and results from this study need to be further verified and recognized using other methods. Due to some constraints in this study, future improvements are recommended to address these limitations. The first limitation is that the accuracy of this study is affected by the integrity of the literature included in the Web of Science database. Because the literature included in the database may be incomplete. The second limitation is that the diversity of keywords and the keyword processing algorithm. For example, we use the affinity propagation method for keyword normalization, and the modularity clustering method provided by Gephi for keyword clustering. Future study could design and develop more specialized algorithms for keyword normalization and clustering to improve results.

References

1. Ilhan, A.O., Oguz, M.C.: Collaboration in design research: an analysis of co-authorship in 13 design research journals, 2000–2015. Des. J. **22**(1), 5–27 (2019)
2. Batagelj, V., Mrvar, A.: Pajek: program for analysis and visualization of large networks. Timeshift-The World in Twenty-Five Years: Ars Electronica, 242–251 (2004)
3. Fahimnia, B., et al.: Quantitative models for managing supply chain risks: a review. Eur. J. Oper. Res. **247**(1), 1–15 (2015)
4. Callon, M., Courtial, J.P., Laville, F.: Co-word analysis as a tool for describing the network of interactions between basic and technological research: the case of polymer chemsitry. Scientometrics **22**(1), 155–205 (1991)
5. Cambrosio, A., et al.: Historical scientometrics? Mapping over 70 years of biological safety research with coword analysis. Scientometrics **27**(2), 119–143 (1993)
6. Corrales, I.E., Reyes, J.J., Fornaris, Y.: Bibliometric analysis of the journal of oral research: period 2012-2015. J. Oral Res. **5**(5), 188–193 (2016)
7. Díaz-Faes, A.A., et al.: Unravelling the performance of individual scholars: use of Canonical Biplot analysis to explore the performance of scientists by academic rank and scientific field. J. Inf. **9**(4), 722–733 (2015)
8. Gan, C., Wang, W.: Research characteristics and status on social media in China: a bibliometric and co-word analysis. Scientometrics **105**(2), 1167–1182 (2015)
9. Hu, C.-P., et al.: A co-word analysis of library and information science in China. Scientometrics **97**(2), 369–382 (2013)
10. Burt, R.S.: Applied Network Analysis. Sage Publications (1978)
11. Liu, G.-Y., Hu, J.-M., Wang, H.-L.: A co-word analysis of digital library field in China. Scientometrics **91**(1), 203–217 (2012)
12. Muñoz-Leiva, F., et al.: An application of co-word analysis and bibliometric maps for detecting the most highlighting themes in the consumer behaviour research from a longitudinal perspective. Qual. Quant. **46**(4), 1077–1095 (2012)
13. Nguyen, D.: Mapping knowledge domains of non-biomedical modalities: a large-scale co-word analysis of literature 1987–2017. Soc. Sci. Med. **233**, 1–12 (2019)
14. Rey-Martí, A., Ribeiro-Soriano, D., Palacios-Marqués, D.: A bibliometric analysis of social entrepreneurship. J. Bus. Res. **69**(5), 1651–1655 (2016)
15. Hu, J., Zhang, Y.: Research patterns and trends of recommendation system in China using co-word analysis. Inf. Process. Manag. **51**(4), 329–339 (2015)
16. Yan, B.-N., Lee, T.-S., Lee, T.-P.: Mapping the intellectual structure of the Internet of Things (IoT) field (2000–2014): a co-word analysis. Scientometrics **105**(2), 1285–1300 (2015)
17. De Stefano, D., Giordano, G., Vitale, M.P.: Issues in the analysis of co-authorship networks. Qual. Quant. **45**(5), 1091–1107 (2011)

18. Wasserman, S., Katherine, F.: Social Network Analysis: Methods and Applications, vol. 8. Cambridge University Press, Cambridge (1994)
19. Okubo, Y.: Bibliometric indicators and analysis of research systems (1997)
20. Persson, O., Danell, R., Schneider, J.W.: How to use Bibexcel for various types of bibliometric analysis. In: Celebrating Scholarly Communication Studies: A Festschrift for Olle Persson at his 60th Birthday, vol. 5, pp. 9–24 (2009)
21. Mongeon, P., Paul-Hus, A.: The journal coverage of web of science and scopus: a comparative analysis. Scientometrics **106**(1), 213–228 (2016)
22. Rohani, V.A., et al.: An effective recommender algorithm for cold-start problem in academic social networks. Math. Probl. Eng. **2014**(pt.5), 123726.1–123726.11 (2014)
23. Luo, R., et al.: A critical review on the research topic system of soil heavy metal pollution bioremediation based on dynamic co-words network measures. Geoderma **305**, 281–292 (2017)
24. Van Eck, N.J., Waltman, L.: VOSviewer manual. Leiden: Univeristeit Leiden **1**(1), 1–53 (2013)

Exploring Universal and Cultural Preferences for Different Concepts of Autonomous Vehicles' External Communication in China, USA and Germany

Anne-Marie Julie Barthe Wesseling[1]($\boxtimes$), Ruth Mugge[2] ⓘ, Elmer van Grondelle[2], and Ina Othersen[1]

[1] Group Innovation, Volkswagen Aktiengesellschaft, 38436 Wolfsburg, Germany
anne-marie.wesseling@volkswagen.de

[2] Industrial Design Engineering, Delft University of Technology, 2628 CE Delft, The Netherlands

Abstract. External communication constitutes a research area that has emerged from the need to develop an effective and intuitive system that enables autonomous vehicles (AVs) to communicate with all human road users. Considering that AVs generally target different countries, it is important to understand potential cultural differences in user preferences to various external communication concepts. This research investigates user preferences for seven external communication concepts using an online questionnaire (N = 710) in three diverse cultures (i.e. China, Germany and the United States of America). Specifically, the study explores the potential of the following concepts: Display, LED, emphasized inclination caused by vehicle behavior ("Incline"), Directed Sound Beam ("Sound"), Smart Assistant in Wearables, Smart infrastructure, and Augmented Reality. After seeing short movies of these concepts, participants rated the concepts on usability and trust. The findings demonstrate that opportunities exist for both universalization and localization. In addition, these results describe an opportunity for a modular external communication concept that consist of a baseline concept and cultural-specific modules. The baseline concept can be derived from universal preferences for external communication concepts. This baseline concept is to be extended with cultural-specific modules that aim to optimize usability and human understanding in specific cultures based on information preferences.

Keywords: External communication · Human-Computer Interaction · Autonomous vehicles · Culture · Universal access · Design

1 Introduction

How would you know whether you can safely cross the road if the driver of the approaching vehicle is reading a newspaper? Although such a scenario does not seem likely yet, the rise of Autonomous Vehicles (AVs) could make such a scenario increasingly feasible in the near future. Here, AVs refer to vehicles that are capable to safely and efficiently

© Springer Nature Switzerland AG 2020
P.-L. P. Rau (Ed.): HCII 2020, LNCS 12192, pp. 641–660, 2020.
https://doi.org/10.1007/978-3-030-49788-0_49

navigate to their destination with little to no input from the driver [1]. By taking over the active driving task from the driver, AVs could benefit society through reduced emissions, enhanced traffic flow, increased productivity, higher convenience, mobility access-for-all and radically improved traffic safety [2–5]. However, next to taking over the active driving task, the AV should also take over all other responsibilities of the driver.

Communicating with other Human Road Users (HRUs) is one of the driver's responsibilities that should be taken over by the AV. The communication between HRUs allows humans to safely navigate and coexist in traffic as it enables us to interpret traffic scenarios, communicate our intentions, and negotiate, decide on and coordinate actions [2, 6–9]. For communication, HRUs (incl. drivers) rely on vehicle-centric cues (e.g., trajectory and velocity changes) and human-centric cues (e.g., eye-contact, posture and gestures) [2, 9]. In the case of AVs, the human-centric cues of the driver will be lost due to the removal of the driver [10]. Simply stated, AVs not having a driver presents a dilemma for HRUs, because HRUs do not know who controls any of the vehicles they see, "How can HRUs communicate and with whom?" With the vehicle-centric cues as the only interaction source left, bimodal communication becomes impossible [11]. Furthermore, vehicle-centric cues may also no longer be reliable: AVs are not operated by human drivers resulting in anomalous driving behavior [12–14]. This statement is supported by results of AV-testing in traffic, which indicate that minor accidents occur due to the AVs' anomalous behavior together with the loss of bimodal interaction [15]. As such, a need develops for an external communication system that allows the AV to facilitate communication with surrounding HRUs within the current traffic system. Thinking of the diversity in communication styles between people in different cultures, designing an effective and intuitive system is a challenging task that warrants more research attention. The present study contributes by exploring universal and cultural preferences for different external communication concepts of AVs.

1.1 External Communication

External communication constitutes a research area that has emerged from the need to develop a system that enables AVs to communicate with all HRUs effectively and intuitively [11, 16]. Past research in the field of external communication focuses on two topics: (1) communication of information and (2) developing external communication systems.

Communication of Information. Research on information needs for external communication points to three main areas: communication of intent, awareness and status (or mode). Communicating the intent of the AV refers to the trajectory and planned or executed maneuvers. Communicating awareness refers to the AV's understanding of the surrounding objects and regulations. The information related to the reason for the AV's behavior, current speed and whether the autonomous driving mode is activated or not, refers to communicating the status (or mode). By communicating the awareness, intent and status of the AV, the loss of bimodal interaction and the AV's anomalous behavior could be overcome leading to enhanced comfort and perceived safety for HRUs [17]. In addition to enhanced comfort and perceived safety, communicating awareness and intent could also aid in crossing decisions [10]. While communication of intent, awareness and status (or mode) are all considered important, these can be communicated using different perspectives. For example, in an "egocentric" perspective, the AV communicates its own actions, whereas in an "allocentric" perspective the AV gives you advice concerning your actions. At present, research in external communication has yet to explore HRUs' expectations and preferences for different perspectives of communication of external communication systems. The present research aims to explore the perspective and information expectations of HRUs to provide recommendations for the development of future AVs.

Developing External Communication Systems. Research on the development of external communication systems focuses on two approaches: (1) emphasizing vehicle-cues and (2) simulating driver-cues through additional modalities. The raison d'être of the research direction of emphasizing vehicle-cues is based on current situations (e.g., during the night) where HRUs can successfully deduce intent and awareness from the vehicle's natural inclination behavior and currently available systems, such as blinkers. In addition, the added benefit of vehicle-cues is that they are visible from multiple angles increasing visibility [7, 8]. Different studies concluded that additional modalities could aid AVs in addressing the missing driver-cues and enhance the interaction experience [10, 18]. Potential benefits of developing positive interaction experiences are improved trust and acceptance towards AVs [19–22]. External communication systems that have been researched focus on lighting technologies [17, 23–25]; vehicle behavior [7, 8]; displays [2, 26, 27]; acoustics [28, 29] and wearables [30]. These studies have compared different concepts within their chosen approach and technology concept (e.g., display) to describe what specific communicating form and design will have the most positive effect on interactions with the AV. However, few studies exist that describe and compare multiple external communication concepts across technologies and approaches to come to a more complete overview of user preferences towards external communication systems [6, 26, 31]. Furthermore, the few studies that do exist have a limited sample size [32].

Studies that compare preferences for external communication systems rely on the subjective responses that they trigger among HRUs. The subjective responses of HRUs previously used in studies indicate several evaluation criteria: recognizability, unambiguousness, interaction comfort, intuitive comprehensibility and the feeling of increased safety [6, 26, 31]. These factors closely relate to factors used to describe usability and trust which are considered important factors for Human-Machine Interaction (HMI), user acceptance and the adoption of automated systems and their correct use [33–36]. Lastly, the subjective responses for usability and trust that external communication systems trigger could be dependent of what kind of scenario is encountered. As such, scenario is an additional aspect to be considered while comparing external communication systems. In short, trust and usability are important factors for testing the potential of external communication systems and different scenarios are to be considered.

1.2 Culture

A currently unexplored research direction in external communication is the implication of cultural differences on the design of external communication by AVs. Here, the term culture refers to patterns in how people think, feel and act, which influences how they communicate amongst each other and with technology [37]. Cultural differences are crucial to consider in the development of Human-Computer Interaction (HCI) and HMI because behavior and performance are strongly influenced by culture [38, 39]. Cultural adaptability in HCI and HMI could thus prove valuable. Research in culture and cultural adaptability could aid in generating trust that is required for the successful integration of AVs in our traffic system. To consider culture in the development of external communication, two approaches are proposed in the literature. The first approach is to conceptualize HMI for specific cultures, which is known as localization [37]. With localization different HMIs are designed for different markets to satisfy the different expectations and opportunities. The second approach is to create unity across different vehicle types, brands and cultures by designing HMI following universal principles in order to improve understandability and avoid confusion across cultures [17]. This approach is called internationalization and aims to facilitate universal access. However the question then arises, "What aspects of external communication could be unified, and to what extent?" More specifically, where should localization begin and internationalization end?

1.3 Aim of the Study

This study aims to gain insight in HRUs' preferences for different external communication concepts in three target cultures (i.e. China, Germany and the United States of America). The preferences for the following concepts for external communication systems are researched: Display, LED, emphasized inclination caused by vehicle behavior ("Incline"), Directed Sound Beam ("Sound"), Smart Assistant in Wearables, Smart infrastructure, and Augmented Reality. These concepts for AVs' external communication on HRUs were compared based on the subjective responses describing usability and trust. In addition, the present research aims to explore the perspective and information expectations of HRUs. Through these insights, this study aims to offer to insights in internationalization and localization opportunities for external communication systems. The overall aim of this study can be summarized by the following research question: "Which HRU preferences regarding information needs and concepts for external communication can be identified across and within the cultures of China, Germany and the United States of America?"

2   Method

The explorative study was conducted using a 3×7-mixed factors design with the independent variables of culture and technology concept. Participants represented either the culture China, USA or Germany (between-subjects factor). The within-subject factor of technology concept consisted of seven levels (see Sect. 2.2).

2.1   Participant Selection

A total of 752 participants (48.01% Female) from a consumer panel participated in our study with a mean age of 42.31 years (SD = 13.06 years; min = 18, max = 70).

To ensure that the three target cultures were represented accurately, participants were filtered by applying several selection criteria. Firstly, all participants in the sample were required to be residents of one of the target cultures (i.e. China, Germany and United States). Secondly, for participants holding a nationality of and residency in one of the target cultures, an additional criterion of a minimum residency of 5 years was applied. Thirdly, participants with other nationalities were only included in the sample if they had resided in one of the target cultures all of their lives. Fourth, through the outlier detection of cultural VSM 2013 dimensions scores of each participant, representativeness of the national cultural dimensions scores for each subsample was checked. Lastly, participants could not have a color perception and/or physical impairment as this would hinder the perception of the stimuli or may significantly alter behavior in traffic. Forty-two datasets of participants were excluded based on the selection criteria. The selected 710 participants (47.18% Female) had a mean age of 42.58 years (SD = 13.02 years; min = 18, max = 70).

2.2 Stimuli Selection

The concepts tested in this research were selected based on their described potential in the literature. The first external communication system selected was a display that was implemented in the front of the vehicle and showed dynamic icons [2, 26, 27, 40]. In addition to the display, a concept with a LED bar placed in front of the windshield was selected. This LED bar aims to communicate through indicating the direction and trajectory intent of the AV with animation created by the individual LEDs [17, 23, 25]. Next to these concepts, an adjustable suspension system was included in the tested concepts. This suspensions system (later referred to as Incline) emphasizes the vehicle's natural inclination behavior to increase vehicle cues [7, 8, 12, 41]. Another selected concept relied on acoustics. This concept is referred to as directed sound beam and consists of multiple speakers that direct sound to a desired spatial location, making it possible to deliver spatial, personalized acoustic information [28, 29]. Literature also investigated an external communication concept relying on wearables, which refers to electronic devices offering haptic, acoustic and/or visual feedback. In this case, the wearable used as a communication system is a smartphone that offers personalized information to the HRUs through a pop-up notification [42]. To increase versatility and abide by traffic regulations when being an occupant in the vehicle, wearables have been operationalized as a smart assistant in a smartphone and in-vehicle HMI. Next, light projections on the street facilitated by the technology of multi-lens arrays (MLAs) were added to the line-up of tested concepts [43]. These light projections indicate the intent of AVs through animations and icons projected on the street. It must be noted however, that the projected light has been converted to smart infrastructure as projected light is not yet feasible in full daylight. Similar to light projections, smart infrastructure allows for additional information and adjustable road markings. However, rather than projecting light, the lights are to be embedded in the road. To ensure that the value of the adjusted concepts was not lost, the concepts were adapted without taking away the base concept through which the modality communicated for external communication. Apart from slight alterations to two concepts, one concept was added: Augmented Reality.

This technology holds a high potential due to its intuitive presentation of condensed three-dimensional personalized information [44].

In short, this study explores the potential of the following concepts for external communication systems: Display, LED, emphasized inclination caused by vehicle behavior ("Incline"), Directed Sound Beam ("Sound"), Smart Assistant in Wearables, Smart infrastructure, and Augmented Reality to gain insight in HRUs' preferences for and information and perspective expectations of external communication systems.

To explore the potential of external communication systems, previous study rely on HRUs' subjective responses to factors describing usability and trust. Based on these previous studies, the factors that together test usability are subjective efficiency of use, subjective ease of learning, high memorability and expected performance [6, 26, 31]. The adaptation of the identified factor of trust generated through the improvement of human understanding is described by increased safety and demonstrable willingness to use an information system [45]. As such, the aspects have been tested relating to usability are whether the concept is clear and easy to understand, communicates in an intuitive manner and is easily recognizable. For trust, the aspect tested are whether the external communication system makes HRUs feel safer in traffic and whether HRUs experience higher comfort during and after interaction. These subjective HRUs responses that can be derived and could prove useful indicators for the potential of external communication systems revolve around usability and trust which could depend on the encountered scenario.

2.3   Materials, Procedure and Measures

Participation was voluntary and the questionnaire lasted between 45 to 60 min. Each section of the questionnaire had a short introduction to inform the participant what they could expect. The first section consisted of questions related to demographics that were based on the VSM 2013 model [46] (i.e. age, gender, nationality, current country of residence, educational background and employment). Three questions were added to enable the selection of participants: years of residency in current country of residence, color perception impairments and physical impairments. The second section consisted of questions related to personal attitude towards different topics in their life to investigate culture as described by the VSM 2013 model [46, 47]. Through 24 items, each with a 5-point Likert scale, Hofstede's six cultural dimensions can be calculated, which together describe the characteristics of a national culture.

The different external communication concepts were introduced via a video for each selected concept. Each video showed an AV attempting to communicate with its environment in two driving scenarios: (1) a pedestrian crosswalk and (2) a cooperative driving scenario (see Fig. 1). These were chosen as HRUs rely heavily on communication with drivers in these driving scenarios [6, 8]. All seven videos were presented in a randomized order and adapted to traffic signs and road markings of each culture.

Participants were instructed to turn on the sound before being allowed to continue with the questionnaire. After reading a short explanation about how each video would show an AV attempting to communicate with its environment, the participants were asked to consider for each video what message and how the vehicle wished to communicate. To ensure that participants understood each concept correctly, a short explanation of the

Fig. 1. Visualization of different scenarios and the possible variations within each scenario: *(Top to bottom)* (1) Pedestrian crosswalk scenario with approaching vehicle from left with AR, (2) Cooperative Behavior scenario with approaching vehicle from right with Display

concept was shown after each video. This explanation described how the technology functioned and the intended communication by the presented concept. Participants had to reflect on the perceived and intended communication. During this reflection, the participants had the option to watch the video again. Participants were then asked to rate their preferences for each concept. Specifically, participants were to respond to the multi-measures of trust and usability, and expected performance in (or suitability for) multiple scenarios in order to gain insight in expectations and preferences for external communication. These aspects were all scored on a 7-point Likert scale (1 = "Completely disagree" - 7 = "Completely agree"). Firstly, the expected performance was measured by asking how much participants agreed with whether the different concepts would perform well (or be suitable) in certain driving scenarios (i.e. warning in a safety-critical situation; a non-safety-critical situation, and a scenario unrelated to driving). Secondly, the multi-item measure of trust was tested with two questions: "I would feel safer in traffic" and "I would find this concept comfortable in interaction"). Thirdly, several factors of usability were scored using the following questions: (1) "This concept is clear and easy to understand" (subjective efficiency of use); (2) "This concept communicates in an intuitive manner" (ease of learning); (3) "This concept is easily recognizable" (memorability), and (4) "I think that this concept would work well in different kinds of traffic or weather" (expected performance). This rating process was repeated for all videos. To ensure that the participant was able to recall the concept accurately, the video of the concept was available and could be replayed during the rating process.

The last section of the questionnaire revolved around the information expectations and general evaluation. To measure HRUs' information expectations which information should be communicated by external communication systems, several questions related to intent, awareness and status were asked. All questions were scored on a 5-point scale that was derived from the Kano Model (i.e. (5) "I like it that way", (4) "It must be that way", (3) "I am neutral", (2) "I can live with it that way" or (1) "I dislike it that way") [48]. The questions were formulated as follows: "Please evaluate which information you think should be communicated by the Autonomous Vehicle:"

- *Intent-related questions:* "The AV is going to start moving", "The AV is turning", "The AV is deaccelerating", "What the AVs trajectory is", "The AV is performing a "special maneuver" (e.g. parking)", and "The AV will change its direction"
- *Questions related to awareness:* "The AV has detected you (the human road user)", "The AV is giving right of way", and "The AV is taking right of way"
- *Status-related questions:* "Why the AV is behaving the way it does (traffic jam ahead, parking)", "The speed of the AV", and "The AV is driving autonomously"

To gain insight in expectations related to the communication perspective, participants were asked: "From what perspective should the AV communicate?" The participant was able to choose between (1) "The vehicle will communicate its own status or behavior" (for example, "I brake, I see you") and (2) "The vehicle will communicate its advice for other road users" (for example "You can cross").

3   Results and Discussion

Data analysis was performed with a significance level of $\alpha = .05$ and with the factors of culture as between-subject factor, technology concept as within-subject factor and a hierarchal level for participant. Through ANOVAs and Tukey HSD pairwise comparisons, the effects of culture and technology concepts were examined. A Chi square test was executed to investigate the preferences for perspective.

3.1   Communicating a Warning in Safety-Critical Scenarios

The main effect of culture was significant ($F(2, 287) = 77.55, p < .001$). The Chinese sample ($M = 4.82, SD = 1.94$) scored significantly higher than the American sample ($M = 4.05, SD = 2.19, p < .001$) and the German sample ($M = 3.09, SD = 2.02, p < .001$).

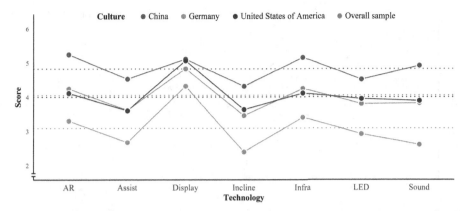

Fig. 2. Means of the subjective suitability of the concepts (lines) and mean across concepts (dotted lines) for overall sample and subsamples for communicating in safety-critical scenarios.

The German and American sample scored significantly different ($p < .001$) as well. The difference in cultural means is visualized with the colored dotted lines in Fig. 2.

Technology shows a main effect on the subjective scores for safety-critical scenarios ($F(2, 945) = 84.98, p < .001$). In safety-critical scenarios, the perceived performance concept of Display was significantly rated the highest than all other technologies ($M = 4.83, SD = 1.80, p < 0.05$). Augmented Reality ($M = 4.24, SD = 2.13$) and Smart Infrastructure ($M = 4.24, SD = 2.12$) were rated second with no significant difference between them ($p = 1$). The technologies of Sound, LED, Assist and Incline had no significant differences ($p > .05$) and were significantly rated lowest.

A significant interaction effect ($F(12, 205) = 9.22, p < .001$) between technology and culture was identified. The most striking result to emerge in relation to the interaction effect is the difference in score for Display. As shown in Fig. 2, America and Germany rated Display significantly higher in comparison to the other technologies. China did not score Display higher and rather rated Display not significantly different as Smart Infrastructure, Augmented Reality and Sound ($p > .05$). These technologies together did have the highest score in China. To summarize, although a consistent universal preference for Display, Augmented Reality and Smart Infrastructure can be identified, China did not indicate an equal preference for Display in comparison to other technologies as the United States and Germany. These differentiating preferences could indicate an opportunity for localization.

3.2 Communicating in Non-safety Critical Scenarios

Culture shows a main effect on subjective scores for non-safety critical scenarios ($F(2, 468) = 128.66, p < .001$). Similar to the previous scores of the safety-critical scenario, China ($M = 5.28, SD = 1.70$) scored higher than the United States ($M = 4.19, SD = 2.17, p < .001$) and Germany ($M = 3.23, SD = 2.03, p < .001$). A difference between German and American scores was present as well ($p < .001$). The mean scores of each culture have been visualized with the colored dotted lines in Fig. 3.

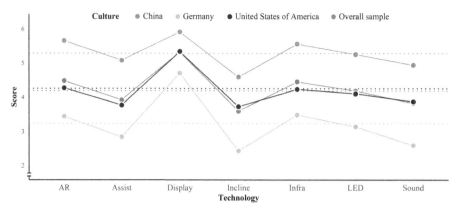

Fig. 3. Means of the subjective suitability of the concepts (lines) and mean across concepts (dotted lines) for overall sample and subsamples for non-safety critical scenarios.

The main effect of technology is significant ($F(2, 1412) = 129.39, p < .001$). The perceived performance concept of Display was rated the highest ($M = 5.33, SD = 1.59$). The second highest score, due to no significant difference ($p = 1$), was given to Augmented Reality ($M = 4.48, SD = 2.14$) and Smart Infrastructure ($M = 4.45, SD = 2.14$). The technology of LED was preferred after these technologies with $M = 4.19$ ($SD = 2.14$), followed by Smart Assistant ($M = 3.92, SD = 2.20$) and Sound ($M = 3.82, SD = 2.19$) with no significant difference identifiable between these technology concepts ($p > .5$). Incline ($M = 3.59, SD = 2.16$) scored significantly lowest ($p > .05$).

The interaction effect of technology and culture was significant ($F(12, 144) = 6.59, p < .001$). Interestingly, the interaction effect did not influence the overall trend in highest expected performance scores for non-safety critical scenarios (see Fig. 3). The preferred technology concepts in each culture remain Display, Augmented Reality, Smart Infrastructure and LED. All in all, results show a consisted trend in preferred technologies across cultures with Display as the most preferred technology, followed by Augmented Reality, Smart Infrastructure and LED for non-safety critical scenarios.

3.3 Communicating in Scenarios Unrelated to Driving

The main effect of culture is significant ($F(2, 288) = 81.92, p < .001$). Once more, China ($M = 4.81, SD = 1.86$) scored higher than the United States ($M = 3.98, SD = 2.21, p < .001$) and Germany ($M = 2.99, SD = 2.01, p < .001$) with a significant difference between these cultures as well ($p < .001$). The colored dotted lines in Fig. 4 show the different mean scores of each culture.

Technology shows a significant main effect on the perceived performance of technology concepts in scenarios unrelated to driving ($F(2, 391) = 37.09, p < .001$). With no significant difference ($p < 0.1$) found, the perceived performance of Display ($M = 4.30, SD = 1.95$), Augmented Reality ($M = 4.22, SD = 2.21$) and Smart Infrastructure ($M = 4.16, SD = 2.19$) received the highest scores. Similarly, Smart Assistant ($M = 4.22, SD = 2.21$), LED ($M = 4.22, SD = 2.21$) and Sound ($M = 4.22, SD = 2.21$) received no significant difference in scores and were rated highest after the previously

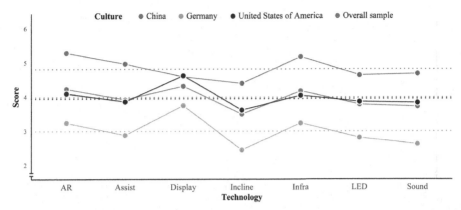

Fig. 4. Means of the subjective suitability of the concepts (lines) and mean across concepts (dotted lines) for overall sample and subsamples for scenarios unrelated to driving.

named technologies. Incline ($M = 3.49$, $SD = 2.18$) scored the lowest for the perceived performance in non-safety critical scenarios. Most technology concept did not score significantly different from each other, which can be seen in Fig. 4.

The interaction effect of technology and culture was significant ($F(12, 191) = 9.07$, $p < .001$). Strikingly, the score for Display by China in comparison to the previous scenarios was considerably lower and was rated here as one of the lowest technologies. In comparison, Display is rated significantly as the highest technology in both America ($M = 4.61$, $SD = 1.90$, $p < .05$) and Germany ($M = 3.74$, $SD = 1.96$, $p < .05$). It must be noted however that all technologies apart from Display and Incline in both Germany and the United States do not score significantly different ($p > 0.05$) from each other. A similar observation can be made in China where Smart Assistant, Display, LED and Sound did not receive a significant difference in score ($p > 0.05$). This observation was made again in China's highest scored technologies: Augmented Reality, Smart Infrastructure and Smart Assistant ($p > 0.05$). All in all, the result of this analysis could indicate less strong preferences for technologies in scenarios unrelated to driving, with the exception of Display and Incline. In other words, in scenarios unrelated to driving localization opportunities may prove valuable if only one technology is used in the external communication system.

3.4 Usability of Technology Concepts

The multi-tem measure of usability (4 items; $\alpha \geq 0.87$) shows a significant main effect of culture ($F(2, 399) = 111.08$, $p < .001$). Similar to the previous scenarios, China ($M = 5.37$, $SD = 1.69$) scored higher than the United States ($M = 4.27$, $SD = 2.19$, $p < .001$) and Germany ($M = 3.50$, $SD = 2.10$, $p < .001$). Here, the cultures of Germany and the United States significantly differ in scores as well ($p < .001$). Figure 5 show the different means for each culture through the colored dotted lines.

A main effect of technology is present ($F(2, 5518) = 512.56$, $p < .001$). Display scored the highest in usability ($M = 5.37$, $SD = 1.63$, followed by Smart Infrastructure ($M = 4.73$, $SD = 2.06$) and Augmented Reality ($M = 4.70$, $SD = 2.10$) with no significant

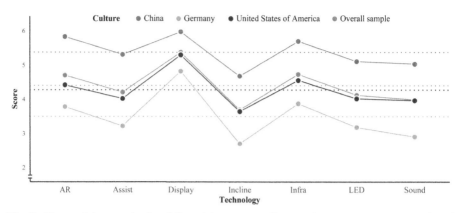

Fig. 5. Means of the perceived usability of the concepts (lines) and mean across concepts (dotted lines) for overall sample and subsamples.

difference between them ($p < 0.5$). The following highest scores of Smart Assistant ($M = 4.20, SD = 2.15$) and LED ($M = 4.12, SD = 2.14$) also had no significant difference in score ($p > 0.1$). Incline ($M = 3.69, SD = 2.20$) scored the lowest.

The interaction effect of technology and culture was significant for subjective usability ($F(12, 400) = 18.60, p < .001$). Although Fig. 5 shows slight differences in overall trends caused by the interaction effect on Display, Smart Infrastructure and Sound, the order based on highest scores remains unchanged. Usability scores indicate that the technology of Display is rated the highest across and within all cultures. With no significant difference between Augmented Reality and Smart Infrastructure found in China ($p < 0.5$), Germany ($p < 0.5$) and the United States ($p = 1$), a trend of preferred technologies seems to be identifiable. This trend within usability closely resembles the scores of non-safety critical scenarios (see Fig. 3). All in all, with similar preference for technologies based on usability scores, universalization could be possible.

3.5   Trust in Technology Concepts

The variables of subjective safety and comfort are items in the multi-item measure of trust (2 items; $\alpha \geq .79$). Trust is influenced by the main effect of culture ($F(2, 39) = 111.03, p < .001$). In line with previous results, China ($M = 5.20, SD = 1.76$) scored higher than the United States ($M = 4.11, SD = 2.25, p < .001$) and Germany ($M = 4.00, SD = 2.10, p < .001$). Here, the cultures of Germany and the United States have a significant difference in scores as well ($p < .001$). See Fig. 6 for the different means for each culture visualized by the colored dotted lines.

A main effect of technology influences trust scores ($F(2, 2516) = 248.31, p < .001$). Across cultures, Display scored the highest ($M = 5.06, SD = 1.82$). Smart Infrastructure ($M = 4.48, SD = 2.014$) and Augmented Reality ($M = 4.36, SD = 2.24$) had no significant difference between them ($p < 0.1$) and were rated highest after Display. The technology of LED ($M = 3.98, SD = 2.19$) was significantly lower than the previously named technologies ($p < .001$). After LED, the technologies of Smart Assistant ($M = 3.79, SD = 2.23$) and Sound ($M = 3.78, SD = 2.23$) were rated with no significant difference

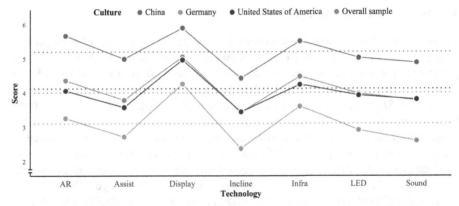

Fig. 6. Means of the trust of the concepts (lines) and mean across concepts (dotted lines) for overall sample and subsamples.

between them ($p < .001$). The significantly lowest score ($p < .001$) was given to the technology of Incline ($M = 3.43$, $SD = 2.21$).

Subjective trust scores show an interaction effect of technology and culture ($F(12, 165) = 8.12$, $p < .001$). Based on the technology results within culture, the technologies of Display, Smart Infrastructure and Augmented Reality were scored highest in usability. The interaction effect did not affect the trend of preference for these technologies. See Fig. 6 for an overview of the subjective usability for each technology and culture.

3.6 Communication Perspective Expectations

There is no significant influence of culture on preferences concerning the perspective of communication by external communication (*Chi square* $= 0.86$, $p > .5$, *df* $=2$) indicating an opportunity for universalization. The preference for the "Egocentric" perspective was preferred significantly more than the "Allocentric" perspective (*Chi square* $= 3.86$, $p < .05$, *df* $= 1$) (see Table 1). In short, the preference for communication perspective is not influenced by culture and the universal preferred perspective is "Egocentric" perspective.

Table 1. Percentage and N concerning preferences for perspective across and within cultures.

	CH (N=251)		GER (N=239)		USA (N = 220)		Overall (N = 710)	
	N	%	N	%	N	%	N	%
Egocentric	145	57.77	151	60.16	135	53.78	431	60.7
Allocentric	106	42.23	88	35.06	85	33.86	279	39.3

3.7 Information Expectations

The information expectations relate to three main areas: communication of intent, awareness and status and were rated on a 5-point scale based on the KANO model. The preferences for information expectations indicate a main effect of culture ($F(2, 50) = 63.10$, $p < .001$). China ($M = 4.32$, $SD = 0.74$) scored higher than the United States ($M = 3.89$, $SD = 1.05$, $p < .001$) and Germany ($M = 3.55$, $SD = 1.12$, $p < .001$). Germany and the United States have a significant difference in scores as well ($p < .001$). The results indicate that German participants tended to score all information aspects in the range of neutral and expected information (3–4). In contrast, China rated all scores between expected and liked (4–5), which could indicate a preference for a larger amount of information. This indication for preferences is supported by HCI guidelines [49].

A main effect of information was found ($F(11, 86) = 19.57$, $p < .001$). The information aspects of Give Way ($M = 4.12$, $SD = 0.92$) and Detection of HRU ($M = 4.09$, $SD = 0.92$), both related to awareness, were indicated as most liked and expected information and did not significantly differ in scores ($p > .5$). The following score of communication the intention of Turning ($M = 3.98$, $SD = 1.01$) did not differ significantly from Detection of HRU ($p > .5$). In addition, the following scores for communicating an intention to Change direction ($M = 3.97$, $SD = 1.00$), to execute a Special maneuver ($M = 3.94$, $SD = 1.03$) and to indicate intention through Trajectory ($M = 3.87$, $SD = 1.05$), and the aspect of Autonomous level ($M = 3.94$, $SD = 1.01$) and Reason for behavior ($M = 3.88$, $SD = 1.01$) related to status (or mode) were not scored significantly different from each other ($p > .05$). Lastly, the lowest rated information aspects of intention to start Braking ($M = 3.86$, $SD = 1.05$) and to start Moving ($M = 3.83$, $SD = 1.03$); of awareness to Take right of way ($M = 3.82$, $SD = 1.15$), and of mode related to current Speed ($M = 3.79$, $SD = 1.11$) did show significantly different scores in comparison to each other and to Reason for behavior and Trajectory ($p > .1$).

An interaction effect of information and culture is present ($F(22, 44) = 4.97$, $p < .001$). The interaction effects becomes apparent on multiple occasions as can be seen in Fig. 7 which show several deviations of trends in universal and cultural preferences. Firstly, where Germany and the United States liked Current Speed significantly less in comparison to other information aspects, China rated the information aspect among the highest. Secondly, the scores of the American sample also show a significantly lower preference for Trajectory. Thirdly, Germany indicated a significantly lower preference for Take Way. Fourthly, China scored Start Moving information significantly lower than other information preferences. In addition, the preference for Detect and Give Way are rated much higher in Germany in comparison to the other information aspects. This difference in rating is not as extreme in China and even less so in the United States of America. These cultural trends differentiate considerably indicating highly differentiating preferences for information. This is the most striking result to emerge from the data related to information expectations and indicate an important finding in the understanding that universalization for external communication may not provide optimized experiences across cultures. In other words, information expectations may require localization as approach in the design of external communication systems to optimize interaction experiences.

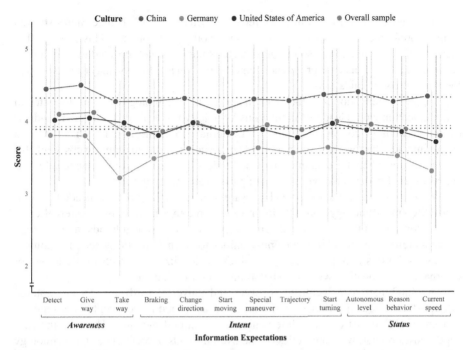

Fig. 7. Mean and SD for information expectations concerning awareness, intent and status to be communicated by external for the overall sample and cultural subsamples.

4   Conclusions

This study aimed to compare concepts for and investigate information expectations of external communication across and within cultures. The aim of the study originated from the want for a large sample study where multiple external communication concepts across technologies and approaches are compared as noted in previous research [6, 26, 31, 32]. Moreover, by executing the study in different cultures, cultural differences in user preferences for these various external communication concepts can be investigated which was yet to be explored for external communication [38, 39]. Based on the results, localization and universalization opportunities for external communication were to be identified.

The results in this paper provide evidence that culture influences information expectations and preferences for external communication concepts.

Although culture significantly impacts technology concept preferences for external communication, universal trends can be identified. Based on the results in usability, trust, safety critical scenario and non-safety critical scenario, the technologies of Display, Smart Infrastructure and Augmented Reality have the highest potential for external communication systems. Although a consistent universal preference can be identified, scenarios unrelated to driving show that culture influences preference so that a difference in preferences can be identified. Where Germany and the United States prefer the technologies previously listed, China indicates a preference for Augmented Reality, Smart

Assistant and Smart Infrastructure. Moreover, Germany and the United States showed a higher preference for Display in comparison to other technologies. This considerably higher preference in comparison to other technologies was not repeated in the Chinese culture. These differences in preferences could show that localization may prove valuable for external communication.

The recommendation for a localization approach in technology concepts for external communication would lead to Autonomous Vehicles with localized hardware to address cultural technology concept preferences for external communication. Localized external communication hardware would not only lead to increased manufacturing costs but also pose problems when, for example, crossing borders. In other words, hardware solutions are considered undesirable when designing external communication.

When considering that localized hardware is undesirable, it may be needed to combine different technology concepts in order to come to a more holistic external communication system that is preferred across cultures. The universal trends in technology concept preferences could indicate opportunities for such a holistic system. In addition, software solutions may provide opportunities for localization which would allow for external communication systems to be optimized for each culture.

The combination of localized software and universal hardware can be captured through the design of external communication as a modular concept. This modular concept could consist of a baseline concept and cultural modules. The baseline concept focuses on hardware and relies on the overall trends in preferences for technology concepts across cultures and other universal trends. The cultural modules are software-oriented and should aim to optimize satisfaction of user expectations and user experience through localization.

Based on the results of this study, the baseline concept should consider Display in combination with Augmented Reality, Smart Infrastructure and/or LED. It must be noted however that Sound may hold value to allow for a multi-modal concept for external communication to stimulate additional senses to vision and support HRUs with vision defects. A universal trend which can be woven into the baseline concept is the universal preference for the "Egocentric" perspective as the perspective in which external communication systems are to communicate. A universal perspective could improve the unambiguousness and universal understanding of external communication when traveling to different cultures and may lower the threshold caused by regulations for the implementation of AVs in different cultures.

The baseline concept should be extended with software-based cultural modules that aim to optimize satisfaction of user expectations and user experience. This opportunity for localization aligns with the cultural preferences for information. More specifically, the influence of culture is most apparent in the differentiating preferences for information. Due to the cultural influence, universalization of communicated information by external communication may not provide optimized experiences in all cultures. As a result, information expectations should be considered an important aspect in the development of the cultural modules to facilitate localization.

5   Limitations and Future Work

The finding of this study provide a positive indication for universalization of the different concepts for external communication. This is desirable to be investigated in more detail in future work, since the different tested concepts may not have been designed optimally for each of the different cultures as the design of the concepts has been based on studies that have been executed in specific or single cultures. This extents to the environment in which the concepts where introduced. The virtual environment shown in the movies does not express the cultures precisely even though markings and signs have been adjusted. Future research should also consider the potential effects of culture on the design of external communication more carefully, for example through the application of HCI-guidelines to adapt each concept for each culture. In addition, future research on external communication can extend the explanations of which universal and culture-specific design guidelines can be identified that positively influence the usability and trust of external communication within each culture.

References

1. Taxonomy and Definitions for Terms Related to Driving Automation Systems for On-Road Motor Vehicles (2018). https://saemobilus.sae.org/content/J3016_201806/
2. Habibovic, A., Andersson, J., Malmsten Lundgren, V., Klingegård, M., Englund, C., Larsson, S.: External vehicle interfaces for communication with other road users? Presented at the (2019)
3. Gao, P., Kaas, H.-W., Mohr, D., Wee, D.: Automotive Revolution–Perspective Towards 2030: How the Convergence of Disruptive Technology-Driven Trends Could Transform the Auto Industry. Adv. Ind. McKinsey Co. (2016)
4. Haghi, A., Ketabi, D., Ghanbari, M., Rajabi, H.: Assessment of human errors in driving accidents; analysis of the causes based on aberrant behaviors. Life Sci. J. 11, 414–420 (2014)
5. Alavi, S.S., Mohammadi, M.R., Souri, H., Kalhori, S.M., Jannatifard, F., Sepahbodi, G.: Personality, driving behavior and mental disorders factors as predictors of road traffic accidents based on logistic regression. Iran. J. Med. Sci. 42, 24 (2017)
6. Ackermann, C., Beggiato, M., Schubert, S., Krems, J.F.: An experimental study to investigate design and assessment criteria: what is important for communication between pedestrians and automated vehicles? Appl. Ergon. 75, 272–282 (2019). https://doi.org/10.1016/j.apergo.2018.11.002
7. Fuest, T., Michalowski, L., Traris, L., Bellem, H., Bengler, K.: Using the driving behavior of an automated vehicle to communicate intentions - a wizard of Oz study. In: 2018 21st International Conference on Intelligent Transportation Systems (ITSC), pp. 3596–3601. IEEE (2018)
8. Risto, M., Emmenegger, C., Vinkhuyzen, E., Cefkin, M., Hollan, J.: Human-vehicle interfaces: the power of vehicle movement gestures in human road user coordination. In: Proceedings of the 9th International Driving Symposium on Human Factors in Driver Assessment, Training, and Vehicle Design: Driving Assessment 2017, pp. 186–192. University of Iowa, Iowa City (2017)
9. Šucha, M., Dostal, D., Risser, R.: Pedestrian-driver communication and decision strategies at marked crossings. Acc. Anal. Prev. 102, 41–50 (2017). https://doi.org/10.1016/j.aap.2017.02.018

10. Mahadevan, K., Somanath, S., Sharlin, E.: Communicating awareness and intent in autonomous vehicle-pedestrian interaction. In: Mandryk, R., Hancock, M., Perry, M., Cox, A. (eds.) CHI 2018: Proceedings of the 2018 CHI Conference on Human Factors in Computing Systems, Montréal, QC, Canada/sponsored by ACM SIGCHI, 21–26 April 2018, pp. 1–12. The Association for Computing Machinery, New York (2018)

11. Matthews, M., Chowdhary, G., Kieson, E.: Intent communication between autonomous vehicles and pedestrians (2017)

12. Dey, D., Martens, M., Eggen, B., Terken, J.: The Impact of vehicle appearance and vehicle behavior on pedestrian interaction with autonomous vehicles. Presented at the October 12 (2017)

13. Tapiro, H., Oron-gilad, T., Hurwitz, D.S.: The critical role of human factors in designing the interaction of autonomous vehicles with pedestrians. In: Australasian College of Road Safety Conference, Gold Coast, Australia (2015)

14. Wagner, P.: traffic control and traffic management in a transportation system with autonomous vehicles. In: Maurer, M., Gerdes, J.C., Lenz, B., Winner, H. (eds.) Autonomous Driving, pp. 301–316. Springer, Heidelberg (2016). https://doi.org/10.1007/978-3-662-48847-8_15

15. Rothenbucher, D., Li, J., Sirkin, D., Mok, B., Ju, W.: Ghost driver: a field study investigating the interaction between pedestrians and driverless vehicles. In: 2016 25th IEEE International Symposium on Robot and Human Interactive Communication (RO-MAN), pp. 795–802. IEEE (2016)

16. Rasouli, A., Kotseruba, I., Tsotsos, J.K.: Agreeing to cross: how drivers and pedestrians communicate. In: 2017 IEEE Intelligent Vehicles Symposium (IV), pp. 264–269. IEEE (2017)

17. Habibovic, A., et al.: Communicating intent of automated vehicles to pedestrians. Front. Psychol. 9 (2018). https://doi.org/10.3389/fpsyg.2018.01336

18. Deb, S., Warner, B., Poudel, S., Bhandari, S.: Identification of external design preferences in autonomous vehicles. In: Proceedings of the 2016 Industrial and Systems Engineering Research Conference, Anaheim, California, pp. 69–44 (2016)

19. Hoff, K.A., Bashir, M.: Trust in automation: integrating empirical evidence on factors that influence trust. Hum. Factors 57, 407–434 (2015)

20. Sheridan, T.B.: Humans and Automation: System Design and Research Issues. Human Factors and Ergonomics Society, Santa Monica (2002)

21. Lee, J.D., See, K.A.: Trust in automation: designing for appropriate reliance. Hum. Factors 46, 50–80 (2004)

22. Parasuraman, R., Riley, V.: Humans and automation: use, misuse, disuse, abuse. Hum. Factors J. Hum. Factors Ergon. Soc. 39, 230–253 (1997). https://doi.org/10.1518/001872097778543886

23. Florentine, E., Ang, M.A., Pendleton, S.D., Andersen, H., Ang, M.H.: Pedestrian notification methods in autonomous vehicles for multi-class mobility-on-demand service. In: Proceedings of the Fourth International Conference on Human Agent Interaction - HAI 2016, pp. 387–392. ACM Press, New York (2016)

24. Krahnstöver, A.Z.: Licht führt!?. Springer Fachmedien Wiesbaden, Wiesbaden (2017). https://doi.org/10.1007/978-3-658-17161-2

25. Löcken, A., Heuten, W., Boll, S.: AutoAmbiCar: using ambient light to inform drivers about intentions of their automated cars. In: Proceedings of the 8th International Conference on Automotive User Interfaces and Interactive Vehicular Applications Adjunct - Automotive'UI 16, pp. 57–62. ACM Press, New York (2016)

26. Fridman, L., Mehler, B., Xia, L., Yang, Y., Facusse, L.Y., Reimer, B.: To walk or not to walk: crowdsourced assessment of external vehicle-to-pedestrian displays, pp. 1–12 (2017)

27. Kitayama, S., Kondou, T., Ohyabu, H., Hirose, M., Narihiro, H., Maeda, R.: Display system for vehicle to pedestrian communication. In: SAE Technical Paper Series, pp. 4–8 (2017)

28. Fortino, A., Eckstein, L., Viehöfer, J., Pampel, J.: Acoustic vehicle alerting systems (AVAS) - regulations, realization and sound design challenges. SAE Int. J. Passeng. Cars Mech. Syst. 9(2016-01-1784) (2016). https://doi.org/10.4271/2016-01-1784
29. Poveda-Martínez, P., Peral-Orts, R., Campillo-Davo, N., Nescolarde-Selva, J., Lloret-Climent, M., Ramis-Soriano, J.: Study of the effectiveness of electric vehicle warning sounds depending on the urban environment. Appl. Acoust. 116, 317–328 (2017). https://doi.org/10.1016/j.apacoust.2016.10.003
30. Meschtscherjakov, A., Tscheligi, M., Fröhlich, P., McCall, R., Riener, A., Palanque, P.: Mobile interaction with and in autonomous vehicles. In: Proceedings of the 19th International Conference on Human-Computer Interaction with Mobile Devices and Services - MobileHCI 2017, pp. 1–6. ACM Press, New York (2017)
31. Clamann, M., Aubert, M., Cummings, M.: Evaluation of vehicle-to-pedestrian communication displays for autonomous vehicles. In: 96th Annual Transportation Research Board Meeting, Washington, DC (2017)
32. Rasouli, A., Tsotsos, J.K.: Autonomous vehicles that interact with pedestrians: a survey of theory and practice. IEEE Trans. Intell. Transp. Syst., 1–19 (2019). https://doi.org/10.1109/TITS.2019.2901817
33. Hjetland, S.M.: Designing for trust in autonomous systems (2015)
34. Körber, M., Baseler, E., Bengler, K.: Introduction matters: manipulating trust in automation and reliance in automated driving. Appl. Ergon. 66, 18–31 (2018)
35. Bevan, N.: Usability is quality of use. In: Advances in Human Factors/Ergonomics, pp. 349–354. Elsevier (1995)
36. Dillon, A.: Cross-cultural differences in automotive HMI design: a comparative study between UK and Indian users' design preferences. J. Usability Stud. 11, 45–65 (2016)
37. Ford, G., Kotzé, P.: Designing usable interfaces with cultural dimensions. Presented at the (2005)
38. Khan, T., Williams, M.: A study of cultural influence in automotive HMI: measuring correlation between culture and HMI usability. SAE Int. J. Passeng. Cars Electron. Electr. Syst. 7(2014-01-0263) (2014). https://doi.org/10.4271/2014-01-0263
39. Heimgärtner, R., Solanki, A., Windl, H.: Cultural user experience in the car—toward a standardized systematic intercultural agile automotive UI/UX design process. In: Meixner, G., Müller, C. (eds.) Automotive User Interfaces. HIS, pp. 143–184. Springer, Cham (2017). https://doi.org/10.1007/978-3-319-49448-7_6
40. Othersen, I., Conti-Kufner, A.S., Dietrich, A., Maruhn, P., Bengler, K.: Designing for automated vehicle and pedestrian communication: Perspectives on eHMIs from older and younger persons. Proc. Hum. Factors Ergon. Soc. Eur., 135–148 (2018)
41. Othersen, I., Cramer, S., Salomon, C.: HMI for external communication – Kann die Fahrzeugbewegung als Kommunikationskanal zwischen einem Fahrzeug und einem Fußgänger dienen? VDI-Berichte, 10(2360), 145–154 (2019)
42. Meschtscherjakov, A., Wanko, L., Batz, F.: LED-A-pillars: displaying distance information on the cars chassis. In: Adjunct Proceedings of the 7th International Conference on Automotive User Interfaces and Interactive Vehicular Applications - AutomotiveUI 2015, pp. 72–77. ACM Press, New York (2015)
43. Future Talk - Evolution of Light
44. Plavšic, M., Duschl, M., Tönnis, M., Bubb, H., Klinker, G.: Ergonomic design and evaluation of augmented reality based cautionary warnings for driving assistance in urban environments. Proc. Intl. Ergon. Assoc. (2009)
45. Fraedrich, E., Lenz, B.: Societal and individual acceptance of autonomous driving. In: Maurer, M., Gerdes, J.C., Lenz, B., Winner, H. (eds.) Autonomous Driving: Technical, Legal and Social Aspects, pp. 621–640. Springer, Heidelberg (2016). https://doi.org/10.1007/978-3-662-48847-8_29

46. Hofstede, G., Minkov, M.: Values survey module 2013, 1–17 (2013). https://doi.org/10.1111/avj.12084
47. Hofstede, G.: Dimensionalizing cultures: the Hofstede model in context. Online Read. Psychol. Cult. **2** (2011). https://doi.org/10.9707/2307-0919.1014
48. Xu, Q., Jiao, R.J., Yang, X., Helander, M., Khalid, H.M., Opperud, A.: An analytical Kano model for customer need analysis. Des. Stud. **30**, 87–110 (2009). https://doi.org/10.1016/j.destud.2008.07.001
49. Heimgärtner, R.: Cultural differences in human computer interaction: results from two online surveys. In: Osswald, A., Stempfhuber, M., Wolff, C. (eds.) Open Innovation. Proceedings of the 10th International Symposium for Information Science, pp. 145–157. UVK, Constance (2007)

Research on Intelligent Design Tools to Stimulate Creative Thinking

Jingwen Xu, Chi-Ju Chao, and Zhiyong Fu[✉]

Department of Information Art and Design, Tsinghua University, Beijing 100084, China
598254196@qq.com, aboutblank329@foxmail.com,
fuzhiyong@tsinghua.edu.cn

Abstract. With the development of artificial intelligence technology, AI has been more and more employed in the process of design, providing unlimited possibilities for future design tools. In order to promote AI technology to better assist design and meet the needs of non-professional designers who want to participate in design, we research on design tools with AI's participation, based on the 'double diamond' design process model, combining reasonable design thinking processes and design tools. Taking the expansion stage of students' thinking as the starting point, we use the WoZ method to conduct experiments and compare the different effects of AI's assistants with different functions in different thinking stages. We explore what role AI should play in the design process and how strong it should interfere with the designers and put forward some ideas for future design tools.

Keywords: Design systems · Artificial intelligence · Design tools · Creative thinking

1 Introduction

1.1 The Popularity of Online Collaboration with Non-designers' Teams

Nowadays, collaboration among multiple people, teams, and even companies has been a common situation. Daily meetings, regular project discussions, progress reports, etc. are common in everyday collaborative activities, and with the development of network technology, online collaboration has become a new form of teamwork, but it is also difficult to make sure that everyone is online in real time every day. The significance of collaboration tools is actually here, helping everyone to communicate seamlessly whenever they have requests, sharing information, and updating progress.

Although different teams have natural differences in terms of number, location, resources, goals, processes, etc., there are similarities in their division of design tasks and daily needs.

1.2 The Design Process is not yet Universal

A correct and reasonable design process can guide the team to design efficiently, but the design process hasn't been widely used at present.

© Springer Nature Switzerland AG 2020
P.-L. P. Rau (Ed.): HCII 2020, LNCS 12192, pp. 661–672, 2020.
https://doi.org/10.1007/978-3-030-49788-0_50

The design process is a set of problem-solving methods formed by the induction and modeling of the designer's thinking. It usually includes a series of explicit steps to provide participants with a thinking frame, which plays an important leading role. Under the guidance of the design process, even non-designers can quickly grasp the design method, so that all staff can participate in the design work in an orderly manner. However, at present, the design process has not been widely used in design work, and non-designers' teams need the guidance of correct and reasonable design process.

1.3 The Trend of Using Artificial Intelligence in Daily Work

We are already in the digital world. Computers are increasingly entering people's work and life. Driven by efficiency, value creation and informatization, artificial intelligence has become a hot topic in technological development. Big data and deep learning technologies also provide important technical support for artificial intelligence. The way of value creation in the future will be efficient information processing, and artificial intelligence will be a powerful force for the development of technology and society. Its historical significance is no less than that of the industrial revolution [1].

2 Related Work

2.1 Design Thinking and Design Process

Design thinking has attracted widespread attention since it was first cited by Peter Rowe, Dean of the Harvard School of Design in 1987 [2]. The use of design thinking can encourage teams to come up with ideas democratically, thereby improving the efficiency of communication between participants, proposing and selecting solutions that are suitable for each participant. The design thinking has also been verified by practice since its inception, which has promoted design positively. The trump majors of top universities in the world such as Harvard University, Stanford University and MIT, the project of Fujitsu Japan called Design Thinking for Future Schools, the "Design Thinking Toolkit" jointly launched by the well-known design company IDEO and Riverdale Country School which is in New York, are all typical examples of the use of design thinking [3].

During the development of design thinking, many models have been proposed [4]. The earliest process expression of design thinking was almost a repetition of the traditional design process. Later, deeper empathy and more specific forms of multidisciplinary collaboration were added. The design process proposed in Herbert Simon's 1969 book *The Sciences of the Artificial* [5] is definition, research, creativity, prototype, selection, implementation and learning, which has become the cornerstone of the design process. Professor Liu Guanzhong of Tsinghua University divided the process of design into six steps, which includes observation, analysis, induction, association, creation and evaluation, progressively building up the model of problem solving [6].

The book *A Designing for Growth: A Design Thinking Tool Kit for Managers* [7] written by Jeanne Liedtka, professor of business administration at the University of Virginia's Darden College, and Tim Ogilvie, founder of innovation consulting firm Peer

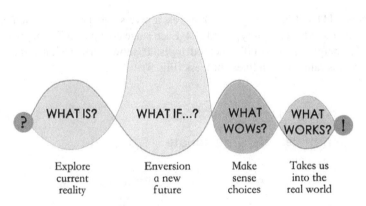

Fig. 1. The process of 4W

Insight, redefines design process to 4 W, that is, design can be seen as the answer to four questions, which are What is? What if? What wows? What works? (Fig. 1).

The British Design Association's double diamond model uses 4D to define the process of design thinking, including Discover, Define, Develop and Deliver. The double diamond chart divides thinking into two phases, namely the divergent phase and the contractive phase. The core is to find the right problems and finding the right solutions. This is a structured design method that is loved and used by many designers (Fig. 2).

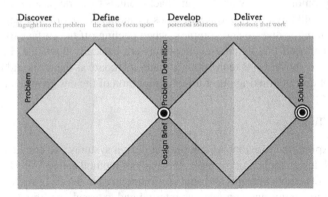

Fig. 2. Double diamond model

The process used by IDEO includes three stages: inspiration, conception, implementation. It also released a set of design method cards [8], covering learning, observation, inquiry and so on. IDEO also developed an environment-based toolkit [9] and the HCD process [10] (human-centered design) was reinterpreted as an acronym for Hear, Create and Deliver.

D.school [11] defines design thinking as a five-stage process, which includes empathize, define, ideate, prototype and test. Each stage contains different goals, implementation principles and specific method tools. It shows that design thinking is an iterative process rather than a linear process (Fig. 3).

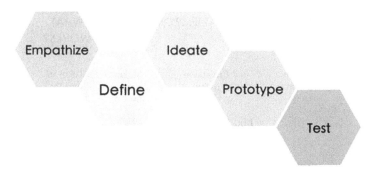

Fig. 3. The design thinking process of d.school.

Design Sprint is a rapid design method proposed by Google. It is mainly applicable to team collaboration tasks that need to come up with design plans in a short time. It includes 6 stages: understanding, definition, divergence, decision, prototype, and verification. This approach allows non-technical designers to understand the methods of designing quickly and promote people with different backgrounds to participate in collaboration.

In a word, although there are many types of models about design and the theory of design thinking is diverse and complex, when peeling off the surface to explore the essence, we can find that its core connotations are basically the same. Based on the model discussed above, we take the double diamond model as the basis and incorporate the concept of design thinking into further exploration of the design system.

2.2 Design Tools

In order to explore the role that AI can play in design systems, we conducted research on traditional design tools. For example, the analysis method called SWOT [12] is used to formulate corporate strategies and analyze competitors in the early strategic planning of product. It integrates and summarizes internal and external conditions from various aspects and analyzes the advantages and disadvantages, opportunities and threats, positioning the product and analyzing the core competitiveness of it among similar products. For another example, Japanese professor Noriaki Kano invented the Kano model, which is a useful tool for classifying and prioritizing user needs, analyzing the impact of user needs on user satisfaction and reflecting the non-linear relationship between product performance and user satisfaction. In the kano model, the quality characteristics of products and services are divided into five types: Basic Quality, Performance Quality, Attractive Quality, Indifferent Quality, Reverse Quality. It classifies user needs and prioritizes multiple function points and can be used during the convergence phase of the design process. Design tools can be divided into figurative and abstract tools, traditional tools and new tools [13]. Material tools are those such as pens, paper, whiteboards, pictures, which are also traditional tools. Abstract tools are those such as words, conceptual framework, theory, brainstorming. AI assistive tools are also new forms of design tools.

2.3 Design Systems and Collaboration Platforms

There exist some systems of design and collaboration. For example, Teambition [14] is a collaborative creation tool for teams. InVision [15] is a prototype design and team collaboration tool. UXPin [16] is a collaboration tool, its team version can add three different member roles into a project, including administrator, creator, and collaborator. Figma [17] is real-time collaborative UI design tool based on browser and cloud, allowing multiple designers to edit a project online at the same time. Zeplin [18] is a standardized design style tool for designers and developers to work together.

However, there are some shortcomings. According to the inspection of these systems, we understand the current situation of this type of systems: firstly, These systems are not good at guiding the design process and the function of collaboration is greater than the function of design assistance. Secondly, AI assistance has not been widely used. Thirdly, these systems are highly specialized and not suitable for non-design professionals.

2.4 AI Technology

AI assistance is one of the focus of our research due to the broad application prospects of AI in recent years. Artificial intelligence technology was first formed in the middle of the last century. In recent years, with the development of cloud computing, intelligent algorithms and other technologies, huge improvements have been made in computing capabilities, data analysis and practical applications [19].

The theory of design process and design methods have already been mature, but the design system only stays as a tool and cannot guide the design well. At the same time, the AI technology which is getting more and more mature has not participated in the design process. We think this is a great opportunity.

3 Concept, Design and Prototype

3.1 Concept

We hope to construct a design system with the participation of AI, so We need more detailed research on the role of AI in designing systems. In this article, we preliminarily propose a design tool that can be combined with the theory of design process and based on the concept of AI.

We gave it four characteristics:

1. Guidance: Design thinking and design process are integrated into the design system, and users are guided by the system to do design tasks based on scientific thinking and process. A scientific and reasonable design process can effectively lead the divergence and convergence of design thinking. We hope that new design tools can use the design process to play the role of guidance for participants.

2. Intelligence: We hope we can combine this system with artificial intelligence and big data. The new design tools we propose will take AI as the core which will intelligently give tips and help in the design process according to the pace of the designers' discussions.
3. Non-interference: Our design tools will not directly interfere with the design process, but will give tips and help during the design process. In this way, it will not unduly interfere with the original design process and the behavior of participants can be observed and the system can be improved more effectively.
4. Modularity: The design process is complicated, but if it is decomposed, the design work can be simplified. Therefore, we need to decompose the design system into modules one by one from top to bottom, and each module implements a subfunction. Modules can be combined or decomposed, and they are independent relatively as well as interacting with each other at the same time. By completing the tasks of each module, the entire design is completed step by step, making the design tasks easy to control and manage.

3.2 Design

The target users of our design are college students who don't major in design. The common characteristics of these students are that they are less familiar with the design process and are often not good at expanding ideas and participating in practice compared with students who have learnt about design.

The final product is a design thinking extension assistant. When students discuss, it will provide students with information that can help them expand their thinking according to the pace of their discussions, eliminating the limit of students' thinking, intelligently controlling the rhythm to better guide the design process. Since the use of images and texts can help people process complex information, our mind extension assistants will provide information in the form of images combined with words [20].

In order to construct this tool, we need to explore the relationship between tool forms and design process. Tools can help students expand their thinking by giving materials, which can be images or texts. The characteristic of images is that it is able to elicit richer and deeper associations, while the texts are characterized by easier understanding and higher efficiency. In the design process, design thinking can be divided into two parts: divergence and convergence. We use the double diamond model [21] to conduct experiments to explore the applicability of different forms of design tool in different phases of the design process.

3.3 Prototype

We made simple prototypes with paper and do experiments. We have two goals in the experiments. First, during the phase of thinking convergence and divergence, which method performs better, and which is more suitable for the entire design process. The second is to discuss the symbiotic relationship [22] between the new design tools with AI's intervention and users, and the relationship between the AI and users. In our prototype, we developed conservative and passive strategies for AI to test the relation.

Our experiments need to collect a lot of pictures and texts. In order to simulate AI's decisions, we collected it manually. At the same time, in order to avoid the interference of subjective factors, the task of collecting materials was distributed to different types of people. Each person collected 10 words/phrases and 10 pictures for the specified requirements formulated for the experiment. We collected a total of 200 words/phrases and 200 pictures (Fig. 4 and Fig. 5).

Fig. 4. Partial pictures

4   Test and Result

4.1   Test

Experimental method: Wizard of OZ. Wizard of Oz (WoZ) is a technique for prototyping and experimenting dynamically with a system's performance that uses a human in the design loop. It was originally developed by HCI researchers in the area of speech and natural language interfaces as a means to understand how to design systems before the underlying speech recognition or response generation systems were mature [23]. We use the WoZ method to simulate the role that AI can play in the design process. We conduct experiments by simulating actual discussion scenarios to collect various feedback during the design process.

Staff assignments: We did 6 groups of experiments, each group had 4 students, and one of the students played the role of AI to participate in the discussion.

Experimental process: The discussion lasted for 1.5 h each time. In the discussion, we introduced the concept of the double diamond model [21]. There were 2 periods of divergence and 2 periods of convergence. The processes include demand exploration, demand definition, brainstorming, and solution derivation (Fig. 6 and Fig. 7).

Fig. 5. Partial words/phrases

Fig. 6. Experimental process

Experimental design: In the experiment, we gave 3 groups of students picture materials and 3 groups of students text materials. "AI" provided participants with information in two forms: passive and active. Passive means that "AI" will give students relevant materials for discussion every 30 s. Active refers to the fact that during the discussion, "AI" will determine whether the frequency of giving information should be increased according to the trend of the discussion.

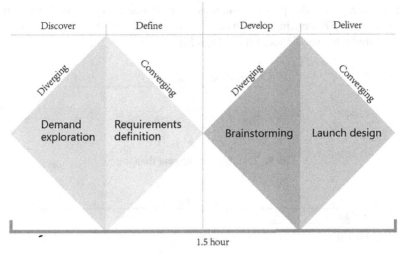

Fig. 7. The phases of the experiment

4.2 Result

Experiments showed that AI' intervention in the design process was helpful. We observed 6 groups of discussions. In the 1.5-h discussion, the stage of thinking divergence took 65 min and the stage of thinking convergence took 25 min.

In general, students were more dependent on the "AI" assistant when their mind diverged and were less dependent on the "AI" assistant when the mind was shrinking. In the discussions of 6 groups, "AI" gave information for a total of 1,332 (180 * 6 + 252) times, of which 960 (127 * 6 + 198) times were given when the mind was diverged, and 372 (53 * 6 + 54) times were given when the mind shrank. When the mind diverged, 28.9% of the materials proposed were discussed, while only 9.7% of the materials were discussed when the mind was contracted. Therefore, we think that this "AI" assistant is more suitable for the divergent thinking stage in the design process (Fig. 8).

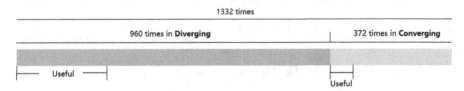

Fig. 8. Divergent thinking and convergent thinking

We use the 45-min divergence of thinking in the experiment as an entry point to explore the performance of texts and pictures in the discussion. On the whole, the performance of the pictures was better. In the three discussion groups that provided texts, "AI" gave materials for a total of 504 times, of which only 24% (121) were used in the discussion, and only 30% of the texts used were converted into specific ideas (11). In the three discussion groups that provided pictures, AI gave materials for a total of 456 times, of which 34.4% (157) were included in the discussion, and 14% (64) of the pictures used were converted into specific ideas (Fig. 9 and Fig. 10).

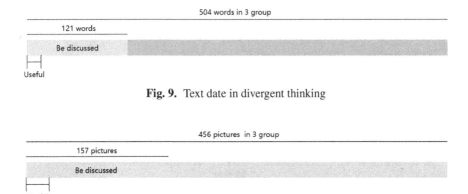

Fig. 9. Text date in divergent thinking

Fig. 10. Picture data in divergent thinking

Let's take one of the groups given texts and one of the groups given pictures as an example to explore the difference between when "AI" is in active and in passive states. When "AI" was passive, pictures were more useful than texts. Both groups received materials for 127 times, of which 44 were pictures and 24 were texts. Through observation, we can easily judge that the picture materials are more likely to cause students' association, and the content of discussion will be deeper and more specific. However, when "AI" was active, the texts' guidance was stronger. "AI" provided 39 words/phrases to one group, of which 18 were used in the discussion, and 25 pictures were provided to the other, of which 10 were used in the discussion (Fig. 11).

Fig. 11. Active vs passive

5   Conclusion and Future Work

Through data analysis and simple interviews with some students after the experiments, we can draw the following conclusions about this study:

1. At the stage of divergent thinking, students rely more on AI thinking assistants. At this stage, it is difficult for students who have not studied design methods to start discussions, and the tips given by the AI assistant can help them start the topic. However, at the stage of thinking convergence, students basically draw conclusions from previously divergent ideas without much help from thinking assistants.
2. Pictures are more helpful for divergent thinking. The texts are too abstract. Although it is easy to understand, students generally think that the texts cannot trigger more associations. Although the picture materials need to be interpreted, they can help students trigger a series of chained associations, which is more helpful for divergent discussions.
3. If the relationship between the AI and the user is different, the behavior of the AI will also be different. When AI is passive, the pictures are more effective and when AI is active, the texts are more effective. Through interviews with several students, we found that this is because the pictures take a certain amount of time to interpret, so if AI is active, it occupies the main role and will interrupt the students' thinking. At this time, the characteristics of texts are more efficient, the continuous output of which can help students compose scenes and enrich the imagination. when the AI is in a passive state, the role of AI is weakened, and students' discussions are dominant, and the pictures can stimulate deeper discussions.
4. AI should not overly interfere with the design process, but it is more suitable for AI to play a role of supporting design. In the experiments about the design process involving AI in this article, we just provided simple materials without other intervention, which has already affected the discussion process of students to some extent. Therefore, we think that when AI participates in the design process, it should not interfere too much with the students ' discussions, so as not to affect the students' design thinking too much.

In general, the combination of AI and design processes will undoubtedly become an important trend in the future. In order to meet this trend, we will continue to explore the possibilities of future design tools. In the future, we need to conduct advance research on the form and intensity of AI's intervention. We will explore how AI is involved in other design processes in more ways in the design process. Since this article mainly discusses the intervention of AI in design thinking, in the future we will study how AI can better participate in the process of thinking convergence. And we will gradually form a more complete AI design system concept through several experiments.

Acknowledgement. This paper is supported by Tsinghua University Teaching Reform Project (2019 autumn DX05_01), Construction of Online Educational Tools and Evaluation System Based on Design Thinking.

References

1. Artificial Intelligence Evolution. Latest Cases and Trends in Artificial Intelligence [DB/OL]. 30 July 2018–27 February 2020 https://www.jianshu.com/p/2df26f9c6411

2. Spitler, L., Talbot, L.: Design thinking as a method of improving communication efficacy, 437–444 (2017). https://doi.org/10.24928/2017/0270

3. Anderson, N., Tims, C., Hashemi, C.H., Xiao, J.: Improving the quality of online learning using design thinking. China Distance Educ. (09), 5–12+95 (2014)

4. Dam, R.F., Teo, Y.S.: Design Thinking: A Quick Overview, February 2020. https://www.int eraction-design.org/literature/article/design-thinking-a-quick-overview?r=iris-liu

5. Simon, H.: The Sciences of the Artificial. MIT Press, Cambridge (1969)

6. Liu, G.: "Science"-innovative design thinking method. In: China Association for Science and Technology. Energy Conservation and Environmental Protection Harmonious Development-Proceedings of the 2007 China Association for Science and Technology (II). Ministry, pp. 449–457 (2007)

7. Liedtka, J., Ogilvie, T.: A Designing for growth: a design thinking tool kit for managers (2011)

8. IDEO, Method Cards, March 2020. https://www.ideo.com/post/method-cards

9. IDEO, Design Kit. The Human-Centered Design Toolkit, March 2020. https://www.ideo.com/post/design-kit

10. Chen, Y.: Analysis of the principles and design process of HCD. Design (22), 136–137 (2016)

11. d.school. The Design Thinking Process, March 2020. http://dschool.stanford.edu/redesigni ngtheater/the-design-thinking-process/

12. Gong, X.: SWOT analysis as a general analysis method for strategic research. J. Xidian Univ. (Soc. Sci. Ed.) (01), 49–52 (2003)

13. Stolterman, E., Pierce, J.: Design tools in practice: Studying the designer-tool relationship in interaction design. In: Proceedings of the Designing Interactive Systems Conference, DIS 2012 (2012). https://doi.org/10.1145/2317956.2317961

14. Teambition. https://www.teambition.com/. Accessed 2020

15. Invision. https://www.invisionapp.com. Accessed 2020

16. uxpin. https://www.uxpin.com/. Accessed 2020

17. figma. https://www.figma.com/. Accessed 2020

18. zeplin. https://zeplin.io/. Accessed 2020

19. Zeng, H.: Application of artificial intelligence technology in education. Electron. Technol. Softw. Eng. (19), 241–242 (2019)

20. Ma, X.: Augmenting text with multiple pictures can facilitate online information processing across language barriers (2014)

21. Council D.: The 'double diamond' design process model. Design Council (2005)

22. Hernandez-Orallo, J., Vold, K.: AI extenders: the ethical and societal implications of humans cognitively extended by AI, pp. 507–513 (2019). https://doi.org/10.1145/3306618.3314238

23. Kelley, J.F.: An empirical methodology for writing user-friendly natural language computer applications. In: Proceedings of the SIGCHI Conference on Human Factors in Computing Systems, pp. 193–196. ACM, New York (1983). https://doi.org/10.1145/800045.801609

Design of Intelligent Public Restrooms in Tourist Cities

Qing Yang[✉], Dan Li, and Ya Tan

School of Design and Art, Beijing Institute of Technology, Zhuhai, People's Republic of China
5305427@qq.com, 381394100@qq.com, 58293171@qq.com

Abstract. As digitization elevated the public's needs, aesthetic and cognitive demands concerning public facilities [1], the design of public restroom in tourist city is facing new challenges. The research, by taking the example of independent public restrooms and those affiliated to other buildings in Kuala Lumpur, analyzes the needs of tourists about city public restroom, and tries to design a query system of intelligent and information-rich public restroom. Focusing on the needs of people from different countries and regions and varying cultural backgrounds, the research proposes the design of a public restroom smart search system which takes the inner and outer space of public restroom into account, and a reasonable evaluation method of public restroom data. The significance of the proposal lies in that it can help provide effective service and showcase the good images of city to tourists on one hand, and on the other hand, provide the managers and designers of public restroom in tourist city a design model for reference, which can help to strengthen the harmonious relationship among people, public restrooms and cities.

Keywords: Public restroom in tourist city · Public restroom design · Intelligent · Interactive design

1 Introduction

The Travel and Tourism Competitiveness Index Report considers infrastructure as one of the vital sectors for modern city design [2]. In recent years, more modern city designs have put public restrooms in eye-catching locations instead of hiding them away from the public. The design of public restrooms in tourist city, which orients people coming from different countries and regions, having diverse cultures and speaking various languages, should explore the needs of users and managers of public restroom, and problems that might emerge which concern public restroom design. By taking a cross-cultural perspective and applying interactive design methodology, the design of public restrooms should cater to the users' physiological, behavioral and emotional needs, and reflect the local environment, culture and prospect [3]. With the development of technology, digitization can make public restroom smarter and information-rich, which facilitates the sustainable development of modern tourist cities and enhances their tourism competitiveness.

© Springer Nature Switzerland AG 2020
P.-L. P. Rau (Ed.): HCII 2020, LNCS 12192, pp. 673–682, 2020.
https://doi.org/10.1007/978-3-030-49788-0_51

2 The Demand of Public Restrooms in Tourist City

Populous area in tourist city has a high demand for public restroom, since tourists from different cultural backgrounds have various needs for public restroom [4]. As the capital of Malaysia, a multi-ethnic and multi-cultural nation, Kuala Lumpur is a popular destination among tourists all over the world for its unique natural scenery and rich culture [5]. For this reason, the research chooses to study the representative independent public restrooms or those affiliated to other buildings in the scenic spots, parks, shopping centers and transportation hubs with a large number of visitors in Kuala Lumpur.

2.1 The Demand of Public Restrooms from the Perspective of the Users

According to universalism, public restroom should cater to the needs of people from different age groups. Thus, the design of public restroom should take the needs of people of various ages into consideration, and try to create a comfortable and user-friendly space [6]. The users of public restrooms in tourist cities include all sorts of people engaging in public activities, such as adults, elders, children, pregnant women, mothers and infants and disabled people. Besides using restrooms, tourists usually also need to wait for their turns, tidy their clothes and other belongings, wait for their companions, smoke, or rest for a while. Public restrooms are built primarily to satisfy people's physiological need and safety need, the most basic needs in Maslow's hierarchy of needs [7]. In recent years, the incorporation of gender-neutral restroom into public restroom in developing countries reflects the humane feature of public restroom revolution [6]. Gender-neutral restroom is designed for the convenience of elders, children and disabled people. For instance, China's CJJ14-2016 Standard for Design of Urban Public Toilets stipulates that the floor area of gender-neutral public restroom should be no less than 6.5 m^2, that facilities such as barrier-free toilet, handrail, wash basins for adults and children, toilet for children, multifunctional children's table, children's safety seat are essential for gender-neutral public restroom [see Fig. 1]. Some developed countries on the other hand are considering meeting higher-level needs of public restroom users. An example is that the smart and user-friendly public toilet in Japan takes users' emotional need into account by incorporating music or the sound of running water to cover the awkward noise made when people use the toilet [8], and by calling restroom "switch room" which implies its function of helping users to change their moods. Thus, the users' experience is improved. Based in Germany, Wall GmbH which built the public restrooms in Berlin kept innovating to enhance users' experience. It has made restroom more than merely a place to meet the public's physiological need. Chanel and Apple have advertised their products on the exterior wall of public restrooms in Berlin. Advertisements can also be seen on the interior wall and facilities, for instance, toilet paper. Wall GmbH has also built a number of high-end pay toilets in Berlin which offer personal care, diapers, music, and the services of cleaning leather shoes, massaging back and so on. Such luxurious toilets, which can be found everywhere in Berlin, have drawn countless tourists [9]. The example of Wall GmbH shows that public restrooms in tourist cities can also act as media to meet the public's higher-level needs.

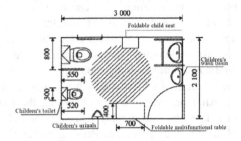

Fig. 1. Gender-neutral public restroom and restroom plan (unit: mm)

2.2 The Problems of Public Restrooms from the Perspective of Tourists in Kuala Lumpur

According to Hofstede (1980), for people who have similar education and life experience, their culture is their similar psychology. But for people from different countries and regions, their psychology shows evident diversity [10]. Hofstede also developed the so-called "Cultural Onion" which indicates that the innermost value shows the strongest diversity among cultures [11]. The diversity of cultures results in people's variant understandings of restroom culture [9, 12]. Malaysia is a country with high religious diversity. The co-existence of miscellaneous cultures demands that the design of public restroom takes the needs of people from different backgrounds into consideration. The design of public restroom in Malaysia prioritizes the habits of Muslim users. So tourists from all-around the world may have troubles using the public restroom designed especially for Muslim people in Kuala Lumpur. The design of the public restroom of Kota Damansara subway station failed to do so, which resulted in inconvenience for users [see Fig. 2].

Fig. 2. Lack of word description for a Muslim

Fig. 3. A water pipe with a shower head

Fig. 4. Peacock totems for the New Year of India prayer room

For instance, due to the lack of word description for a Muslim prayer room, some tourists mistakenly entered the prayer room and then left in embarrassment. The lack of necessary information often causes tourists inconvenience or leaves them an awkward situation. For example, there is a water pipe near the door attached with a shower head in the public restroom, whose use the non-Muslim people usually have no idea about [see Fig. 3]. For another example, when the New Year of India draws near, the exterior of public restrooms of Indian subways would be painted with peacock totems [see Fig. 4] the meaning of which usually no Chinese tourists being interviewed knew.

After interviewing and observing the tourists visiting the five tourist attractions of Kuala Lumpur, namely Petronas Twin Towers, KLCC Park, National Mosque of Malaysia, Batu Caves and Petaling Street, the researcher of this report found the main problems concerning public restrooms of Kuala Lumpur: the tricky location, the long line outside of women's restroom and the unpleasant environment. The male/female ratio of the 45 tourists randomly interviewed is 1:2. For the 28 people who needed to use the washroom, 71% of them didn't know there were public restrooms nearby. For the 21 people who knew that there were restrooms in the nearby malls, 76% of them couldn't find the exact locations. This indicates that the increase of number of public restrooms is far from enough. Public restrooms should also be easily found by the public. Another problem is whether in a mall or a scenic spot, a long line outside of woman's restroom is a common scene while men's restroom is not crowded. One cause of this difference is the physiological differences between men and women. Statistics show that women go to restroom more often and they spend more time in the restroom than men. Another reason lies in the design of toilets. For men's restroom, there are urinals and toilets while for women's restroom, the number of sit toilet far exceeds that of squat toilet as in Fig. 5. Non-Muslims use squat toilets in KL.

Fig. 5. The sit-and-squat ratio in the subway station   **Fig. 6.** The trash bins

Among the Chinese tourists interviewed by the researcher, women prefer to use squat toilet than sit toilet for the sake of their habit and sanitation. So, the limited number of squat toilets partly causes the long lines outside of women's restrooms. Also, for small restrooms, semi-automatic trash bins are used [see Fig. 6]. When the trash bins are broken or unavailable for some reason, this leaves tourists with a disagreeable impression.

A noteworthy phenomenon is that, shown by a survey conducted by the researcher, out of 25 tourists who accessed a public restroom affiliated to a building near the National Mosque of Malaysia, 11 people needed to use the restroom while 14 people waited for their companions. As there was no waiting area, 17 tourists being interviewed expressed their wish for an out-door waiting area near the restroom [see Table 1].

Table 1. The needs for restroom service at different stages

Stages	Tourists (at independent public restrooms)	Tourists (at public restrooms affiliated to other buildings)
Before using washroom	Difficult to find; free of charge	Easy to find by following signs; having long lines, some with charge
While using washroom	Excrete; tidy clothes	Excrete; tidy clothes and other belongs; Touch-up
After using washroom	Leave immediately	Wait for companion, rest for a while (elders, pregnant women)
Feeling	The environment needs to be improved	Users need to stay or rest for a while after using toilet

2.3 The Construction of Public Restrooms from the Perspective of Managers in Kuala Lumpur

As early as in 2001, a restroom sanitation movement rolled out in Malaysia. In 2006, the Kuala Lumpur government built 12 automatic public restrooms, each costing RM 400,000. These restrooms are now already abandoned [see Fig. 7]. However, the unpleasant environment of public restrooms remained till 2016. The public, including the tourists, complained about the public restrooms. The Malaysian government has the environment of public restrooms across the country evaluated every year. Restrooms with good environment score 3 to 5 stars while those with bad environment only score 1 to 2 stars. Of all the Malaysian public restrooms being evaluated in 2016, restrooms with an unsatisfactory environment account for 43.7%, while restrooms score 4 or 5 stars take up merely 4% of the total. The central government of Malaysia takes charge of the construction and maintenance of public restrooms mainly to improve sanitation. Of the public restrooms in Kuala Lumpur studied by the researcher, the sanitation of public restrooms affiliated to other buildings in subways and malls is obviously better than those independent in scenic areas as Figs. 8 shows.

In the respect of public restroom management, the cost of maintenance for either independent restroom or that affiliated to other buildings far exceeds that of construction. For example, the functional life for an independent public restroom is 40 years. If the maintenance cost of it per month is RM 2,200, its maintenance cost for a lifecycle of 40 years is RM 1,057,000, compared to the construction cost of RM 60,000. Public restrooms, like other public facilities, have asset lifecycles [see Fig. 9].

Fig. 7. Automatic public restrooms in 2006   **Fig. 8.** Dependence public restroom

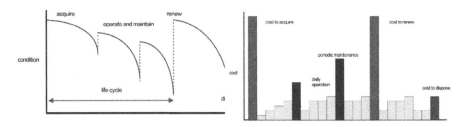

Fig. 9. Extended asset lifecycle

3 Framework for the Design of Intelligent Public Restrooms

In recent years, intelligent search systems have been installed in some public restrooms with a large number of users in a few countries while Kuala Lumpur has not used.

3.1 Intelligent Guidance System for Public Restrooms

The guidance system functions as it obtains real-time information of toilet occupancy and displays this information on screen [13]. The public can access the layout of public restrooms and the occupancy of toilets through the screen on the exterior wall of restroom. The information include: the number of vacancy toilets and occupied toilets, information of the air quality, the density of H_2S and NH_3 etc. [see Fig. 10] [13].

The intelligent public restrooms guidance system consists of sensing, data analyzing, evaluating and displaying sectors [Fig. 11]. The principle of intelligent guidance systems is using detectors to collect the signals of toilet occupation, using collection modules to analyze the signals, transmitting the signals through Ethernet to a UNC mainframe which controls the screens on the doors of independent restrooms and on the exterior wall. The screen on the doors use red or green lights to show toilet availability which is obtained by door lock sensor and that installed on the back wall of restroom [13, 14]. In users' cognition, red means unavailability while green represents availability, just like a traffic light. In this way, users can find an available toilet quickly.

The public restroom intelligent guidance system enables tourists search for information about public restrooms at any time. When people need to use the restroom, they usually can hold on for 7 to 15 min [15]. Since a smart guidance screen can only be

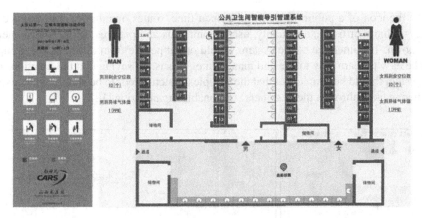

Fig. 10. Display of search system for public restroom

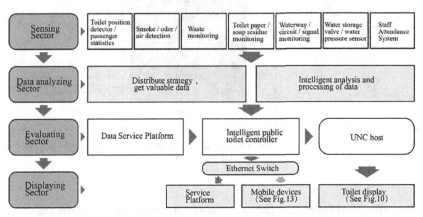

Fig. 11. The framework of intelligent guidance system for public restrooms

seen by people near the public restroom, using a smart phone map APP to find a public restroom is very necessary for people who don't see a restroom. To search "public restroom or Tandas (Malay)" on Google Maps, information about the location of public restrooms, the routes, distance, photos from different angles and user's comments show in Fig. 12. However, the public have no way to access to real-time information of the availability of toilets. Thus, their needs are difficult to meet. To solve this problem, an APP or WeChat public account platform [Fig. 13] which incorporates a public restroom smart search system and which is based on iOS, Android or WP mobile operating system is needed. The public can enter the APP or WeChat public account platform through downloading, scanning code or face scanning. The public restroom map platform and the smart guidance system complement each other by loose coupling. The functions of the platform and guidance system should include map control, real-time positioning, information search, navigation and tracing. To use the full range of functions, a user opens the APP which automatically initiates positioning. Immediately a map showing nearby public restrooms and their distances to the user appears [16, 17]. When the user clicks the icon of a public restroom, the open time, routes, number of visitors, occupancy, volume of tissue, air quality, user's comments and pictures of the environment are shown. The interface of the system should orient people from different age groups and be easy to learn how to use and manipulate. This is not saying that the functions of the system should be simple, but that the complex system needs to be simplified to avoid confusing users through the designers' reasonable organization [18].

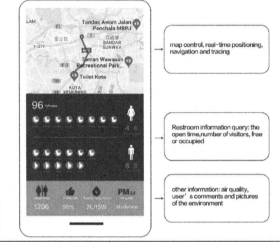

Fig. 12. The interface for entering "public toilets" on Google Map

Fig. 13. The interface design of an intelligent guidance of public restroom APP

3.2   The Design of Inner and Outer Space of Intelligent Public Restrooms

Everyone has experienced queuing, but with varying experience. During the time of waiting for using a toilet, interactive design can turn user's boredom into behavior (from-to structure by Michael Polanyi 1958) [19]. To do this, the exterior wall of public restrooms can be turned into LED screens. As it takes women 3 min on average to use a restroom, soft music could be played, accompanied by moving and static advertising at the unit of 3 min. The advertising material could be public information, for instance guidance on using a toilet, city images and local culture. It could also be commercial, for example, tourist attraction guidance, food guidance and shopping discount information. If users want to access to further information, they can scan the code. At the same time, people in the waiting area can also follow this information. When tourists from different countries and regions scan to enter the system, they first need to choose a language and then choose different information which meets their needs. Within a short period of time, user's agitation converts into an activity through visual, acoustic, tactile, even gustatory cognition. Such a system can help users to save time and be more efficient by enabling them to obtain some useful information during the fragmented waiting time [1].

The mirrors in public restroom including those above the wash basin can be replaced by multi-touch screen, digital screen or projection. Somatosensory interaction with

3-dimentional gestures can also be incorporated into the walls above urinals [20]. These technologies provide restroom users with information about the weather, tourist attractions, food and accommodation, exhibitions and so on. The technologies can also facilitate discussions on social topics, such as current hot issues, entertainment shows and scenic areas in the city, thus creating interactive activities for public restroom users.

3.3 The Method of Evaluating Intelligent Public Restroom Data

For traditional restrooms, daily maintenance involves keeping it clean and tidy, while for smart toilets, in addition to the basic maintenance, three things are needed. First, to make a reasonable evaluation framework and standard for data of smart public restroom. Second, to collect, analyze, use and maintain data, and adjust and revise operation strategy through process evaluation, judge and reflect on the effect of public restroom projects by means of summative evaluation. Third, to increase the asset utilization ratio of public restrooms and increase cost benefit ratio judging from the lifecycle of public facilities so as to continue running and developing public restroom. These three points combined improve the effectiveness and reliability of smart public restrooms.

4 The Significance of the Design of Intelligent Public Restroom in Tourist City

Both local residents and international tourists have high expectation of its public restrooms in tourist city [4]. Tourists want public restrooms to be both comfortable and sanitary, and they expect to have a pleasant experience of using public restrooms. Smart restroom, incorporated with the technologies of the Internet of Things, big data, cloud computing, network transmission and sensor, endows traditional restroom basic capabilities, such as the abilities of real-time sensing, accurately judging and executing, making possible the delicacy management of municipal public restrooms and those for scenic spots and express way service areas in tourist city. The building of a design framework for smart and information-rich public restrooms in tourist cities, on one hand, can provide tourists with satisfactory service, showcase the good image of the city, enrich the scenic spots and bring convenience and comfort to tourists. On the other hand, smart public restroom stands for a design model for the reference of constructing infrastructure in tourist city. In the broad sense, such restrooms show a city's possibility, improve its investment climate, and boost its economy. In the narrow sense, such restrooms perfect city infrastructure, meet the needs of people from different age groups and create comforting space.

5 Conclusion

The aim of designing smart and information-rich public restrooms in tourist cities is to make people's life in the information age more efficient, convenient and science-based [21], is to give people a pleasant experience when they use restrooms, is to help users create a new experience by interactive design. It is vital for tourists to have a nice "encounter" with a public restroom, for the tourist's cognition of an agreeable personal experience can transform into a lasting impression or memory [18]. The elevation from body to visual sense to cognition reflects the relationship among public restrooms, people and cities. When well-designed and constructed, public restrooms can even become a symbol of a city's culture.

References

1. Zhang, H.: Research on the interaction design of urban public facilities. Technol. Commun. **11**, 53–55 (2013)
2. World Economic Forum: Travel and Tourism Competitiveness Report 2017 [EB/OL] (2017). http://reports.weforum.org/travel-and-tourism-competitiveness-report-2017/
3. Yi-yuan, Z.E.N.: On the ideas and strategies of modern urban public toilet design. J. Sichuan Cult. Ind. Vocat. Coll. **2**, 52–55 (2017)
4. Chen, L.: Economic value and cultural identity of landscape design of public toilets in tourist cities. Soc. Sci. **11**, 82–86 (2015)
5. Anuar, A.N.A., Ahmad, H., Jusoh, J.: Policy and tourism development strategy towards tourist friendly destination in Kuala Lumpur. Asian Soc. Sci. **9**(2), 180–190 (2013)
6. Chun-li, X.U., Zhang, X., Ma, J.: Development of urban public toilets design in the information age. Shanxi Architect. **44**(6), 14–16 (2017)
7. Zhang, X., Wang, F.: Research on the interaction design of urban public facilities. Packag. Eng. **31**(10), 30–33 (2010)
8. Li-bo, S.U.: The Perfect Embodiment of Japanese toilet "Humanization" Design. Art Des. **108**, 89–91 (2013)
9. Greed, C., Qu, M., Wang, W.: Inclusive Urban Design: Public Toilets. Machinery Industry Press, Norwalk (2005)
10. Feng, N.: Post-industrial cross-cultural design in digital interactive product design. Art Educ. Res. **11**, 73 (2015)
11. Wang, M.: The differences of cross-cultural design communication and cultural values concepts. Creativity Des. **6**, 61–65 (2012)
12. Jiang, M.: Research on design of cruise atrium space service from cross-cultural perspective. Design **11**, 170–173 (2017)
13. Liu, J.: Intelligent guidance management of public toilets in Taiyuan south station. In: Proceedings of the 13th China Intelligent Transportation Conference. Electronic Industry Press (2018)
14. Shang-qing, R.E.N., Zhao, D.I.N.G.: Overview of intelligent public toilet control system scheme. J. Kaili Univ. **27**(6), 117–119 (2009)
15. Zhou, X.: Study on design of toilet in scenic area. Xi'an University of Architecture and Technology's Master Thesis (2009)
16. Yan, G.A.O.: Development Conception for toilet intelligent query system of mobile terminal equipment. Environ. Sanit. Eng. **23**(3), 61–62 (2015)
17. Qin, J., Shuang, X.: The affective interaction design in mobile social applications based on strong and weak ties. Packag. Eng. **38**(14), 80–84 (2017)
18. Norman, D.A., Zhang, L.: Design Psychology 2-How to Manage Complexity. CITIC Publishing House, Beijing (2011)
19. Tang, L.: Design for body-consciousness. Packag. Eng. **40**(20), 1–8 (2019)
20. Tan, H., Li, W., Tan, Z.: 3D gesture interaction product design of vehicle information system. Packag. Eng. **36**(18), 45–48 (2015)
21. Cukier, K.: Big Data: A Revolution that Will Transform How We Live Work and Think. Zhejiang People Publishing House, Hangzhou (2013)

The Trend Analysis Method of Urban Taxi Order Based on Driving Track Data

Linchao Yang[1], Guozhu Jia[1], Fajie Wei[1], Wenbin Chang[2], and Shenghan Zhou[2(✉)]

[1] School of Economics and Management, Beihang University, Beijing 100191, China
[2] School of Reliablility and Systems Engineering, Beihang University, Beijing 100191, China
zhoush@buaa.edu.cn

Abstract. The paper tries to build the analysis framework to explore the implication trend of the complex taxi order. The online taxi has become one of the important means of urban travel with the popularity of the Internet and smart phones. The analysis of online taxi order may contribute to better understand urban traffic trends and people's living habits. The ride-hailing platform can track every order completely through the client, which provides a basis for the analysis of order trend. With the development of big data analysis methods, it is also possible to analyze the trend of urban ride-hailing orders. The research object of this study is the driving tracking data of online taxi orders in Chengdu in October 2016 provided by Didi Chuxing GAIA Initiative. The month covers the China's National Day holiday. And it is the very typical traffic research scenario. This paper analyzed the change trend of urban online taxi order quantity over time, compared the taxi order quantity trends on workday and weekend, and found that workday and weekend order trends about online taxi have structured differently. In addition, k-means algorithm and DBSCAN algorithm were used to analyze the optimal order-waiting location for online taxi drivers, and the comparison between the two methods was made. It is found that DBSCAN algorithm performs better in analyzing such problems. Didi is the largest ride-hailing platform in China, and Chengdu is one of the mega-cities in southwest China. The analysis based on the data of Didi and Chengdu can provide typical research paradigms for the order analysis of urban taxi to some extent.

Keywords: Online taxi trend · Driving track data · Trend analysis · *K*-means · DBSCAN

1 Introduction

At present, online taxi service has become one of the important role of urban travel. For cities' administrators, the study on the analysis of online taxi order trend has gradually become an important basis and reference for urban traffic planning. For residents, it can help arrange their travel plan properly in advance. For drivers, it is the key to maximizing the benefits of orders. Didi, one of the world's largest ride-hailing platform, has the largest order-tracking data. With the development of data mining technology, it provides an opportunity for the analysis of large-scale urban order-tracking data. Based on the

© Springer Nature Switzerland AG 2020
P.-L. P. Rau (Ed.): HCII 2020, LNCS 12192, pp. 683–698, 2020.
https://doi.org/10.1007/978-3-030-49788-0_52

order data of Didi platform, this paper studies the change trend of urban taxi quantity over time and the recommendation method of optimal order-waiting location for drivers.

The existing research on ride-hailing order is mainly aimed at the long-term or short-term demand forecasting of passenger and the order dispatch methods. In terms of demand forecasting, considering the dependencies in space, time and exogenous factors, Ke, Zheng and Yang et al. [1] proposed the fusion convolutional long short-term memory network (FCL-Net), which get higher prediction accuracy than other learning methods such as time-series prediction model, ANN, XGBoost, LSTM and CNN. Wei, Wang and Wo et al. [2] proposed a hybrid model named ZEST to predict passenger demand for chauffeured car service and the model performed better than 5 other models on the three-month datasets. Chan, Dillon and Singh et al. [3] proposed a novel Neural-Network-Based training model to improve the generalization capabilities of short-term traffic flow forecasting method using a hybrid exponential smoothing and Levenberg–Marquardt algorithm. Cheng, Liu and Wei et al. [4] compared the forecasting performances of the day-to-day travel demand variations in the deep neural network, the stacked long short-term memory (LSTM) network and the feature-level data fusion model, and the LSTM model was proved to be best.

In terms of order dispatch methods, based on the order data of Didi, Jiang, Wang, and Liu et al. [5] constructed a car dispatching network to analyze the evolution of vehicle demand. Xu, Li and Guan et al. [6] presented an large-scale order dispatch algorithm, which improved the on-demand ride-hailing platform's efficiency remarkably and had been deployed in Didi Chuxing. Tang, Wang and Xu et al. [7] proposed a deep value-network learning method based solution for order dispatching and the test result in Didi platform showed that the method significantly improved the driver income and user experience. Wang, Qin and Tang et al. [8] took the ride-hailing order dispatching as a Markov Decision Process and proposed a deep reinforcement learning method with knowledge transfer. To improve the service experience of Mobile Taxi-Hailing (MTH) in flexibility and personalized pricing, Zhao, Xiao and Wu et al. [9] introduced a Reverse-Auction-Based Competitive Order Assignment for MTH Systems.

Through literature research, we can find that there are few studies on ride-hailing order distribution characteristics. Dong, Wang and Hu et al. [10] studied the regularity of residents' pick-up and drop-off activities from the perspective of time and space by connecting residents' travel behavior with their daily activities. Han and Wang [11] analyzed the spatial and temporal distribution characteristics, the hot roads and regions of residents' travel at different times based on taxi GPS data. Qin, Zhou and Wu et al. [12] detected the hotpots of Wuhan and discovered the changing patterns among them by field clustering based on taxi trajectory data. Actually, The trend analysis of ride-hailing order is a typical kind of distribution characteristics analysis.

Based on the driving tracking data of Didi, this paper studies the change trend of urban taxi order quantity over time and the recommendation method of the optimal order-waiting location for drivers. In the trend analysis, the probability distribution was used to analyze the change trends of the weekday order quantity and the weekend order quantity in a day, and the two are compared. In the study of the recommendation method of the optimal order-waiting location for drivers on weekends, k-means algorithm and DBSCAN algorithm were used to cluster the pick-up positions of all orders on weekends, and the center point coordinates were obtained as the recommended location for drivers waiting for orders, and the performances of the two algorithms were compared.

2 Data Sources and Data Description

In order to study the time trend of online taxi order quantity and the recommendation method of the optimal order-waiting location for drivers, we obtained the data from Didi platform. Didi is the largest Chinese online ride-hailing platform. It provides online ride-hailing services for various types of taxis, private car, express trains, etc., which almost covers all large and medium-sized cities in China. Didi platform data is one of the most representative data for studying the trend characteristics of China's online taxi orders. This research obtained the order data of some regions (longitude: 104.0422–104.1296, latitude: 30.56294–30.72775) of Chengdu in October 2016 from Didi platform. The data set is real-time data collected during the order journey. From the beginning of the order, the system automatically records time and location information (longitude and latitude) every 3 s. The data includes order number, driver name, time, longitude coordinates and latitude coordinates. As shown in Table 1, the order number and driver name have been replaced with codes, and the time is standard time.

Table 1. Online booking order records

DIVERS	INDENTS	TIME	LNG	LAT
4b0b3348878c30aa5b7d321e02429dd5	091f6e074483ad1b3f814929246e3f00	1475297342	104.1103	30.67110
4b0b3348878c30aa5b7d321e02429dd5	091f6e074483ad1b3f814929246e3f00	1475297345	104.1101	30.67159
......	......	......	......	......
7d98c920d9c0e8911399d460f08c5f97	92fd683456f28aed9b4c86d91d9994de	1475309262	104.0486	30.69439
7d98c920d9c0e8911399d460f08c5f97	92fd683456f28aed9b4c86d91d9994de	1475309265	104.0486	30.69439
......	......	......	......	......
3a7013bfbbdcb48f7f203ed5d30c8e01	464b015cf95322f3c07df5abb908f61f	1475299354	104.0582	30.65471
3a7013bfbbdcb48f7f203ed5d30c8e01	464b015cf95322f3c07df5abb908f61f	1475299357	104.0583	30.65463
......	......	......	......	......

In order to avoid the impact of the 2016 National Day holiday (October 1 to October 7), this paper mainly uses the data from mid to late October for research. In the third part, we analyze the trend of urban online taxi orders. For the analysis of weekdays, we selected order data on Monday (October 10, October 17), Wednesday (October 12 and October 19) and Friday (October 14th, October 21st), which are more representative because Monday, Wednesday, and Friday represent the various stages of the weekdays. For the analysis of weekend, we chose order data for days of October 15, 22, 29 (Sat.) and October 16, 23, 30 (Sun.). In the fourth part, we analyzed the recommended location of urban online taxi order, using the order data of October 22, 29 (Saturday) and October 23, 30 (Sunday).

Due to the huge amount of data (the daily data contains nearly 50 million records, involving more than 200,000 orders and tens of thousands of drivers), during the research, we randomly extracted part of the order data from the original data for analysis to ensure the rationality of data distribution, which could represent the characteristics of the original data. Because the record of the start point of the order directly reflects the passenger's demands in time and location (the order does not necessarily form an order again at the end point of the order), it is one of the most important data of an order. And a large number of data on the starting point of the order can reflect the change trend of the order in time and location. Due to these reasons, we extracted the data of the order start time from the data of the order tracking record as a sample. The details of the data used in the research are shown in Table 2.

Table 2. Data sample description

Date	Order quantity	Driver quantity
October 10	12838	10232
October 12	12626	10053
October 14	14846	10910
October 15	11298	9013
October 16	13027	9951
October 17	12164	9089
October 19	12051	9734
October 21	11219	9325
October 22	10487	8921
October 23	11105	9301
October 29	10419	8786
October 30	11874	9772

3   Trend Analysis of Urban Online Taxi Orders

Based on the urban online taxi order data of Didi, we use the probability density figure to analyze the trend of online taxi orders at different periods in a day. Firstly, the trend of online taxi orders on typical weekdays and weekends are analyzed day by day. Secondly, the trend of online taxi orders among weekdays and that between Saturday and Sunday are compared respectively. Thirdly, the characteristics of online taxi orders on weekdays and weekends are analyzed and compared.

3.1   Day-by-Day Analysis

In day-by-day analysis, we mainly analyzed the online taxi orders on Monday, Wednesday, Friday, Saturday and Sunday.

As can be seen from Fig. 1, the order quantity on October 10 and October 17 show the same trend over time. Perhaps affected by the travel to work, the number of online taxi orders increases rapidly at 7:00 and reaches its first peak at around 10:00. With the increase of people's travel demands in the daytime, the number reaches the second peak at 15:00, and then reaches the third peak at around 18:00 during off hours. Due to the uncertainty of people's off time and the influence of social interaction, the number of online taxi orders at night shows a slow decline trend.

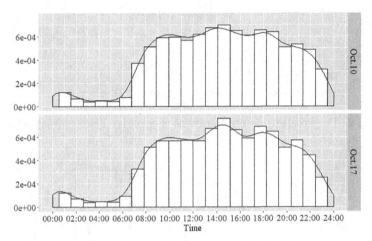

Fig. 1. The probability density distribution of online taxi orders on Monday

Figure 2 shows the probability density distribution of online taxi orders in different time periods on Wednesday. From the figure, we can see that the order distribution also shows three peaks: morning, noon and evening peaks: rapidly increasing in the morning to reach its first peak, reaching its second peak in the afternoon, and turning to decline slowly after 18:00. The order trend in two days is basically the same.

The same as the order trend on Monday and Wednesday, the order trend on Friday has the same features as Fig. 3 shows. Due to the impact of morning rush hours and daytime travel, the number of online taxi orders increases rapidly in the morning, with a peak in the morning and a peak in the noon. The difference is the trend at night. As the next day is Saturday, people's travel demand on Friday night is still very high, which may increase, and there are still a large number of online taxi orders after 22:00.

For Saturday, as Fig. 4 shows, the trend of the number of online taxi is relatively simple and obvious. Since Saturday is a rest day, after 8:00, the number of orders has begun to increase slowly. With the increasing travel demand in daytime, it reaches the peak at 13:00–14: 00, and then basically remains at the level. Besides, there will be a slight increase around 18:00 during the evening peak, and it will slowly decrease at night.

It can be seen from Fig. 5 that the distribution of online taxi orders on three Sundays is basically the same. Since Sunday is also a rest day, people generally travel late. The number of orders start to increases slowly after 8:00, remains stable after reaching the peak in the afternoon, and slowly declines at night.

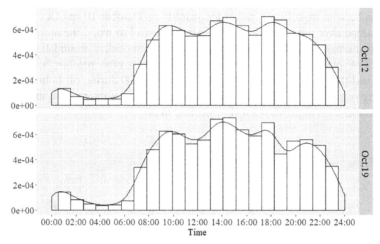

Fig. 2. The probability density distribution of online taxi orders on Wednesday

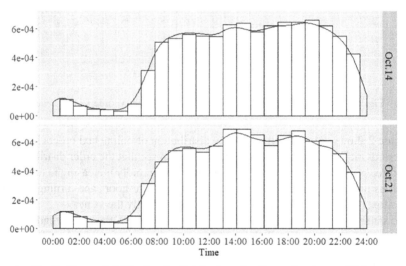

Fig. 3. The probability density distribution of online taxi orders on Friday

Through day-by-day analysis of Monday, Tuesday, Wednesday, Friday, Saturday and Sunday, it can be seen that although on different dates, the trend changing of online taxi orders on a certain day within a week is basically consistent. Therefore, we can combine the data to analyze and compare the changing trend of online taxi orders on different working days and rest days.

3.2 Weekday Comparative Analysis

We combined the data of Monday, Wednesday and Friday of different dates to analyze the trend changing of online taxi orders during the working days and make a comparison.

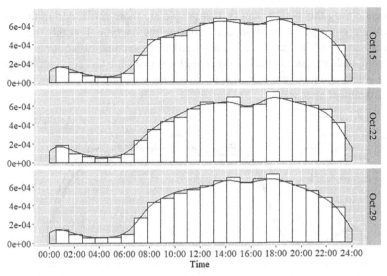

Fig. 4. The probability density distribution of online taxi orders on Saturday

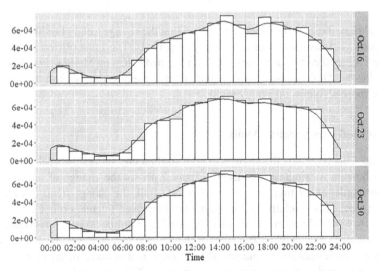

Fig. 5. The probability density distribution of online taxi orders on Sunday

During working days, the trend changes of online taxi orders in each time period in a day are similar, as shown in Fig. 6. First of all, the number of weekday taxi orders shows three kurtosis trends over time, namely "morning peak" for going to the office, "noon peak" for day travel and "evening peak" for getting off work. Second, "morning peak" starts from 6:00, and the number of online taxi orders increases rapidly between 6:00 and 8:00, reaching the peak of "morning peak" between 9:00 and 10:00. Third, the "noon peak" occurs between 13:00 and 16:00 on weekdays. The reason for the increase in online taxi orders during this period may be that people are more active during the daytime, and there are more people who are active during the daytime. Fourthly, the "evening peak" starts around 17:00 and ends around 22:00. The evening rush hour is longer than the morning rush, because the off-hours of each company are less concentrated than the working hours, and there are demands for overtime work and people's social activities and entertainment after work.

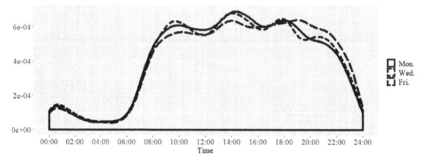

Fig. 6. The probability density distribution of online taxi orders on Weekdays

In addition, it can also be seen from Fig. 6 that the change trend of online taxi orders in different weekdays varies in different time periods. The most obvious is that more orders are occured on Friday during the "evening rush" than on Monday and Wednesday and last longer. This may be caused by the fact that the next day is a rest day, so people schedule more activities on Friday night.

3.3   Weekend Comparative Analysis

Compared with weekday, the change trend of online taxi orders on weekends is more simple, showing a "single peak" trend. Figure 7 shows the change of the number of online taxi orders over time on weekends. First of all, the change trend of online taxi orders on weekends is relatively flat, without the repeated fluctuation in a day. Second, from 6:00 to 12:00, the changing trend of the number of online taxi orders is significantly slower than that of weekdays. The peak of the day appears at 14:00 and 18:00. But the difference is little because people usually choose to rest at home in the morning of the rest day, and schedule activities at noon or evening. Third, it can be clearly seen from Fig. 7 that the peak value of online taxi orders on Saturday occurs at 18:00, at which point the order number distribution is more intensive than that on Sunday. The peak on Sunday was at 14:00, when orders were much higher than on Saturday. It can be seen that people usually choose Saturday evening and Sunday noon for activities, which can also be seen in the peak orders in early Sunday morning.

Based on the analysis of the trend changing of urban online taxi orders, we can roughly draw the following conclusions. Firstly, the trend of urban online taxi orders shows consistent characteristics on a day within a week. Secondly, the change of the order of online taxi in weekdays presents a structure of "three peaks". Thirdly, the number

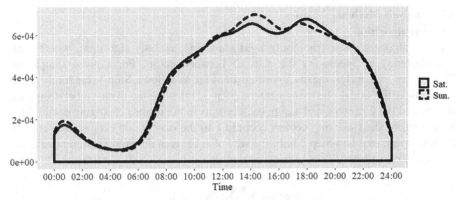

Fig. 7. The probability density distribution of online taxi orders on Weekends

of online taxi orders on weekends varies greatly from that on workday, and peaks at different times on Saturdays and Sundays (18:00 on Saturday and 14:00 on Sunday).

4 The Optimal Order-Waiting Location Analysis on Weekends Based on Clustering

Due to the influence of working place, residential location, commute time and work status, on weekdays, the time and location of people choosing online taxi is relatively fixed. From the perspective of the taxi driver, the location and time for ordering is also relatively fixed. Compared with weekdays, people have more choices about where to travel and when to travel on weekends, which leads to an increase in the uncertainty of taxi drivers getting orders at a certain location. We assume that without the interference of other external factors, the closer the location of the taxi driver is to the taxi passenger, the more favorable it is for him to obtain orders at a lower cost. This assumption is logically valid. At present, the industry usually uses real-time heat maps to provide information on order-concentrated areas to taxi drivers, but this method has one disadvantage: If the driver is far from the heat map recommendation area, when he arrives in the area, the order trend in the area may have changed. We can analyze the location of the historical order reaching point (order starting point) to find the center point in the intensive order area, so that the driver can choose the closest order receiving location to wait for the order, so as to provide historical experience and macro perspective help the drivers. Cluster analysis is a method of classifying different classes into samples, and the samples in each class are very similar. Because we use longitude and latitude information for cluster analysis, the samples in each class are similar in location. We get the center point of these samples, which is the best location for the taxi driver to wait for order. In this part, we use two cluster analysis methods (the k-means algorithm and the DBSCAN algorithm) to analyze the latitude and longitude information of the starting point of the online taxi order on the weekend to obtain the optimal order-waiting location, and the two methods were compared.

4.1 Analysis Methods

k-means algorithm

k-means algorithm is a commonly used clustering analysis algorithm, which realizes clustering by measuring the distance between points. For a given sample set $D = \{x_1, x_2, \ldots, x_m\}$, k-means algorithm steps are as follows. Step 1, randomly select k objects as the initial clustering centers. Step 2, calculate the distance between each objects and the k centers, and assign each objects to the nearest clustering center. Step 3, recalculate the clustering centers according to the existing objects in each classes. Step 4, repeat Step 1 to Step 2 until the sum of squares in the group is minimized, i.e., formula (1) is satisfied.

$$arg \min_s \sum_{i=1}^{k} \sum_{x \in C_i} ||x - \mu_i||^2 \tag{1}$$

Here $C = \{C_1, C_2, \ldots, C_k\}$ represents the sets of classes, $\mu_i = \frac{1}{|C_i|} \sum_{x \in C_i} x$ represents the mean vector of C_i.

DBSCAN algorithm

DBSCAN (Density-based spatial clustering of applications with noise) algorithm is different from k-means algorithm. It is a Density-based clustering method. For a given data set $D = \{x_1, x_2, \ldots, x_m\}$ and parameters (*Eps*, *MinPts*) are defined as follows [13].

Definition 1: The Eps-neighborhood of a point x_j, is defined by $N_{Eps}(x_i) = \{x_j \in D | dist(x_i, x_j) \leq Eps\}$. There are at least a minimum number (*MinPts*) of points in an Eps-neighborhood of that point.

Definition 2: If $x_j \in N_{Eps}(x_i)$ and $|N_{Eps}(x_i)| \geq Minpts$ (core point condition), point x_j is directly density-reachable from point x_i.

Definition 3: If there is a chain of points $p_1, p_2, \ldots, p_n, p_1 = x_i, p_n = x_j$, and point p_{i+1} is directly density-reachable from point p_i, point x_j is density-reachable from point x_i.

Definition 4: Point x_i is density-connected to point x_j, if there is a point x_k such that both, x_i and x_j are density-reachable from x_k.

The DBSCAN algorithm defines a non-empty subset that satisfies the following conditions as a cluster C: 1) $\forall x_i, x_j$, if $x_i \in C$ and x_j is density-reachable from x_i, then $x_j \in C$; 2) $\forall x_i, x_j \in C$, x_j is density-connected to x_i. Therefore, DBSCAN algorithm first selects any core point, and then aggregates samples into corresponding categories according to the sample density.

4.2 The Optimal Order-Waiting Location Comparison on Weekends

We used the starting location data of online taxi orders from three weekends in October (Oct. 15–16, Oct. 22–23, and Oct. 29–30) as samples for analysis. The data included 68,209 orders, distributed at different locations in the sample area, as shown in Fig. 8.

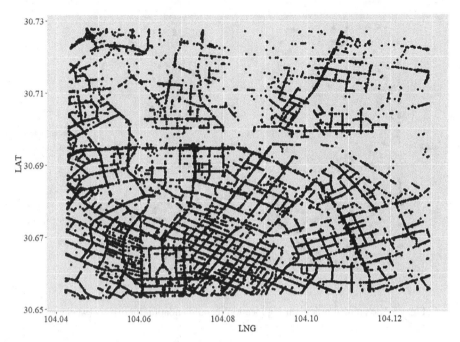

Fig. 8. Start location of weekend online taxi orders

$x_i' = \frac{x_i}{\text{max}x}$ was used to standardize the data. Then k-means algorithm and DBSCAN algorithm were used to cluster the order starting position (the longitude, the latitude) respectively, get taxi drivers' best waiting order positions. The sample area is 152.91 km^2, and tens of thousands of drivers are active in this area every day. Therefore, in the k-means algorithm, take $k = 100$, divide the sample into 100 categories, and get 100 best waiting order positions. In DBSCAN algorithm, the parameters $Eps = 0.000001$ and $MinPts = 50$ were taken. The samples were clustered by cluster analysis, and the result of each class of center point was shown in Fig. 9, where DBSCAN algorithm automatically clustered the samples into 163 classes.

It can be seen from Fig. 9 that the effects of the k-means algorithm and the DBSCAN algorithm are basically the same: the center points of each type obtained basically cover the sample area; and the locations of the cluster center points obtained by the two methods are similar. There are differences only around (104.06, 30.685) and (104.09, 30.690). However, the k-means algorithm needs to set the number of clusters in advance, and the DBSCAN algorithm can automatically cluster based on sample characteristics. In the application of recommending the best waiting position for online booking, it is difficult to estimate the number of recommended points in advance, and the advantages of the DBSCAN algorithm are reflected.

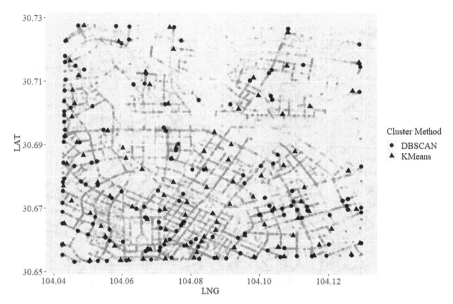

Fig. 9. The optimal order-waiting location recommendation for drivers (weekend)

4.3   Recommended Order-Waiting Location Comparison in Different Time Period

Through the analysis of 3.3, although the number of taxi orders on Saturday and Sunday is basically same, but there are still differences at individual points in time, such as the probability density of the number of orders at 18:00 on Saturday is larger than Sunday, and the probability density of orders at 14:00 on Sunday is greater than Saturday. Therefore, only by analyzing the starting positions of all taxi orders on the weekend, it is not possible to provide a more practical recommendation of the waiting position for the online taxi drivers. We performs a cluster analysis on the starting positions of the order of the ride-hailing order in different time periods on Saturday and Sunday respectively. We can get the cluster center points in different time periods, so as to provide the recommended locations where taxi divers can wait for online orders in different time periods. This recommendation is obviously more practical.

We still use the k-means algorithm and the DBSCAN algorithm for analysis in different time periods: in the k-means algorithm, take $k = 10$; in the DBSCAN algorithm, take parameters $Eps = 0.000001$, $MinPts = 50$. The samples are clustered by cluster analysis. The results of each type of center point are shown in Fig. 10, Fig. 11, Fig. 12 and Fig. 13, and the number of cluster categories of the DBSCANS algorithm is shown in Table 3.

The results indicate that there are great differences in the location of the classified central point, obtained by clustering analysis of the starting position of orders in different time periods. Besides, DBSCAN algorithm can automatically determine the number of clustering categories according to the characteristics of samples, and obtain different

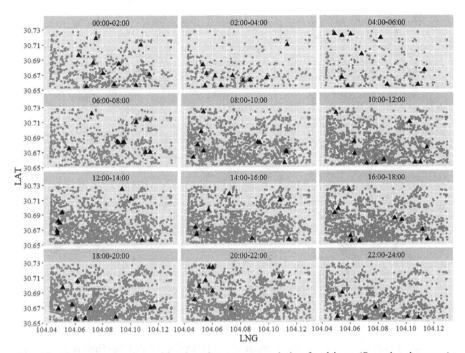

Fig. 10. The optimal order-waiting location recommendation for drivers (Saturday, *k-means*)

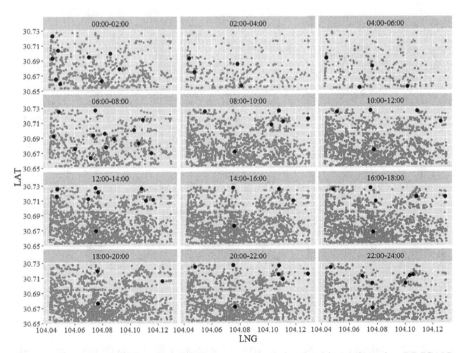

Fig. 11. The optimal order-waiting location recommendation for drivers (Saturday, DBSCAN)

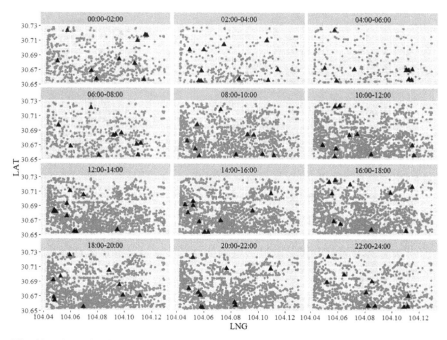

Fig. 12. The optimal order-waiting location recommendation for drivers (Sunday, *k-means*)

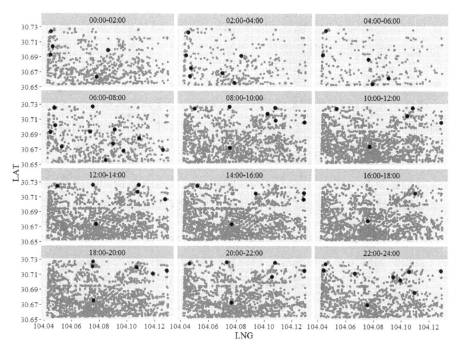

Fig. 13. The optimal order-waiting location recommendation for drivers (Sunday, DBSCAN)

Table 3. The number of recommended cluster for the optimal order-waiting location for online taxi drivers based on DBSCAN

Time periods	Cluster quantity	
	Saturday	Sunday
00:00–02:00	9	5
02:00–04:00	4	7
04:00–06:00	4	5
06:00–08:00	13	12
08:00–10:00	6	7
10:00–12:00	5	5
12:00–14:00	9	6
14:00–16:00	4	5
16:00–18:00	6	2
18:00–20:00	3	6
20:00–22:00	7	6
22:00–24:00	7	9

amounts of waiting-order locations in different time periods, which is more flexible. And because of the basis of density clustering, the classified center can cover all the regions. However, as k-means algorithm needs to specify the clustering category number, its flexibility is inadequate in one more microscopic analysis. Due to the clustering of its nearest neighbors, it is very sensitive to the edge value, so that may be defects in the analysis of nearby waiting order recommendation. From the results of 12:00–14:00 in Fig. 10, k-means algorithm focuses too much on the edge point, where a large number of central points are basically located at the edge of the map, while there is no recommendation of nearby waiting orders in the most concentrated position.

5   Conclusions

This paper mainly discusses two issues: a) the trend of the number of urban online taxi orders over time; b) the recommendation method of optimal order-waiting location on weekend for urban online taxi drivers. Through the analysis of trend changing of the number of online taxi orders, we find that: Firstly, the number changing of online taxi orders during the week reveals a triple-peak structure. Secondly, the number changing of online taxi orders at the weekend approximates a single-peak tendency, while Saturday and Sunday peaked at different moments, which are 18:00 for Saturday and 14:00 for Sunday specifically. In the study of the optimal order-waiting location recommendation method for urban online taxi drivers, we compared k-means algorithm and DBSCAN algorithm and discovered that the DBSCAN algorithm is more suitable for the application of the optimal order-waiting location recommendation.

Acknowledgements. Thanks for the support of data source from Didi Chuxing GAIA Initiative. The study is supported by the National Natural Science Foundation of China (Grant No.71971013 & 71501007) and the Fundamental Research Funds for the Central Universities (YWF-19-BJ-J-330). The study is also sponsored by the Technical Research Foundation (JSZL2016601A004).

References

1. Ke, J.T., Zheng, H.Y., Yang, H., Michael, X.: Short-term forecasting of passenger demand under on-demand ride services: A spatio-temporal deep learning approach. Transp. Res. Part C Emerg. Technol. **85**, 591–608 (2017)
2. Wei, H., Wang, Y.D., Wo, T.Y., et al.: ZEST: a hybrid model on predicting passenger demand for chauffeured car service. In: CIKM 2016, Indianapolis, IN, USA, 24–28 October 2016
3. Chan, K.Y., Dillon, T.S., Singh, J., et al.: Neural-network-based models for short-term traffic flow forecasting using a hybrid exponential smoothing and levenberg-marquardt algorithm. IEEE Trans. Intell. Transp. Syst. **13**(2), 644–654 (2012)
4. Cheng, Q., Liu, Y., Wei, W., et al.: Analysis and forecasting of the day-to-day travel demand variations for large-scale transportation networks: a deep learning approach (2016)
5. Jiang, Z.Y., Wang, Q., Liu, Z.Q., Ma, J.F.: Vehicle demand evolution analysis from the complex network perspective. Phys. A **532**, 121889 (2019)
6. Xu, Z., Li, Z.X., Guan, Q.W., et al.: Large-scale order dispatch in on-demand ride-hailing platforms: a learning and planning approach. In: KDD 2018, London, United Kingdom, 19–23 August 2018
7. Tang, X.C., Wang, Z.D., Xu, Z., et al.: A deep value-network based approach for multi-driver order dispatching. In: KDD 2019, Anchorage, AK, USA, 4–8 August 2019
8. Wang, Z.D., Tony, Q., Tang, X.C., et al.: Deep reinforcement learning with knowledge transfer for online rides order dispatching. In: IEEE International Conference on Data Mining, pp. 617–626 (2018)
9. Zhao, H., Xiao, M., Wu, J., Liu, A., An, B.: Reverse-auction-based competitive order assignment for mobile taxi-hailing systems. In: Li, G., Yang, J., Gama, J., Natwichai, J., Tong, Y. (eds.) DASFAA 2019. LNCS, vol. 11447, pp. 660–677. Springer, Cham (2019). https://doi.org/10.1007/978-3-030-18579-4_39
10. Dong, X.L., Wang, L.Y., Hu, B.B.: Analysis of spatio-temporal distribution characteristics of passenger travel behaviour based on online ride-sharing trajectory data. J. Phys: Conf. Ser. **1187**, 052055 (2019)
11. Han, Y., Wang, W.J.: On the spatial and temporal distribution of resident trip based on taxi GPS data. Geomat. Spat. Inf. Technol. **2**, 87–89 (2018)
12. Qin, K., Zhou, Q., Wu, T., et al.: Hotspots detection from trajectory data based on spatiotemporal data field clustering. ISPRS – Int. Arch. Photogramm. Remote Sensing Spat. Inf. Sci. **XLII-2/W7**, 1319–1325 (2017)
13. Ester, M., Kriegel, H.P., Sander, J., et al.: A density-based algorithm for discovering clusters in large spatial databases with noise. Kdd **96**(34), 226–231 (1996)

Trends in Human-Computer Interaction in the 5G Era: Emerging Life Scenarios with 5G Networks

Jingyu Zhao[1], Andong Zhang[1], Pei-Luen Patrick Rau[1(✉)], Lili Dong[1], and Liang Ge[2]

[1] Tsinghua University, Beijing 100084, China
rp1@mail.tsinghua.edu.cn
[2] Service Laboratory, Huawei Technologies CO., LTD, Shenzhen 518129, China

Abstract. With the development of 5G networks, capacity enhancement, massive connectivity, ultra-high reliability, and low latency are typical use scenarios of information communication technology (ICT). These ICT improvements will also bring large challenges and revolutions to future human-computer interaction (HCI). This study aims to explore the emerging and promising trends and applications of HCI in the 5G era. Eleven experts in HCI and psychology were invited to participate in focus groups and brainstorming studies to collect their perspectives and expectations of future life scenarios and technological applications with 5G networks. Four trends in HCI are identified and proposed in this paper according to their opinions: natural interaction, multimodal display, virtual identity, and cloud data. The four trends are based on the elements of humans, computers, and the interactions between them and the environment. Each identified trend is analyzed in terms of main technological applications, links to 5G development, and future life scenarios. A related trend model was further discussed with respect to user experience and privacy protection.

Keywords: 5G · Human-computer interaction · 5G scenarios

1 Introduction

The fifth generation of mobile communication technology (5G) has made major breakthroughs in data transmission speed, delay, connection, capacity, reliability, and mobility [1]. With the development of 5G networks, capacity enhancement, massive connectivity, ultra-high reliability, and low latency are typical use scenarios in information communication technology (ICT) [2]. These ICT improvements will also bring large challenges and revolutions to future interactions between people and information technology. This change is not limited to hardware, software, interaction paradigms, and new user needs. The emergence of 5G will accelerate the interconnection of all things, promote the construction of smart cities, and realize the development and expansion of digital life, such as intelligent furniture, intelligent transportation, intelligent offices, and other multi-faceted living scenarios [3]. The future will be an era of information explosion. Massive data will be generated during the interactions between people and information technology.

© Springer Nature Switzerland AG 2020
P.-L. P. Rau (Ed.): HCII 2020, LNCS 12192, pp. 699–710, 2020.
https://doi.org/10.1007/978-3-030-49788-0_53

To explore the feasibility of 5G information communication networks in the context of human-computer interaction (HCI), this study focused on the human-center perspective and combined development and daily life applications with 5G technology to imagine future trends in HCI and life scenarios. Eleven experts in HCI and psychology were invited to participate in focus groups and brainstorming studies to collect their perspectives and expectations of future life scenarios and technological applications with 5G networks. The theme of the seminar was the human, computer, and environmental elements involved in the 5G era, including the following: 1) smart devices, emerging smart devices (such as AR glasses, smart speakers, etc.), new interaction paradigms (such as voice interaction, gesture interaction, eye movement, EEG, etc.), and the impact of 5G on hardware and software design; 2) Internet of things (IoT) and big data, data collection, transmission and application, people-centric design of the IoT, 5G impact on the development of the IoT, etc.; 3) artificial intelligence (AI), using AI to understand users through data, the role of AI in the future, the combination of AI and 5G networks, etc.; 4) future life scenarios, data collection, transmission, and applications in different scenarios in the future digital world, as well as data intercommunication between different scenarios.

We conducted open coding of potential trends and life scenarios in the raw discussion record. Axial coding was performed based on the coded topics. Researchers discussed the content of coded transcripts and clustered them into themes. Four dimensions were clustered: natural interactions, multimodal applications, virtual identity, and data in the cloud, and future life scenarios were concluded with each theme. Based on these four themes, we proposed a theoretical HCI trends model for human factor research in the 5G era. In the subsequent sections of this paper, each identified trend is analyzed related to technological application development, and some promising sub-trends are discussed. Human life scenarios in the 5G era are introduced in terms of human needs, emerging smart devices, interaction paradigms, and AI applications. Further discussion is extended to investigate the changes and challenges in user experience, social distance, and technological development in the 5G era.

2 Natural Interaction

2.1 Natural User Interface

A major core of HCI is the input and output of information. The information in the human brain is reflected by the user's physical attributes, including gestures, touch, and voice. These physical attributes of the user are sensed by an input device and converted into information that can be processed by a computer. In current HCI input, in addition to keyboard input, more natural and simple interaction methods have started to attract more attention.

Some researchers have invented the process of tapping with a thumb on the touchpad, in which the user taps the QWERTY keyboard on a virtual screen and receives text feedback on a separate screen, which opens up new possibilities for entering text on ubiquitous computing platforms, such as smart TVs and head mounted displays [4]. Some researchers have invented a pressure-based input method that allows text to be entered with subtle finger movements. Users report that this input is easy to learn and fun to use, demonstrating the feasibility of applying pressure as the primary channel for text input [5]. In addition to the common voice input, Lip-Interact, a lip language input, has also been developed. It uses a front camera to recognize lip movements and uses natural language lip movement coding to recognize mobile phone operation commands. This interaction is not affected by environmental noise and largely alleviates the issues surrounding personal privacy and social norms in the public environment [6].

Although these new types of interaction methods are still being explored in the academic world, attention should also be paid to the exploration of future interaction methods for these more natural and easy interactions in the 5G era.

2.2 Brain-Computer Interface

A brain-computer interface (BCI) builds a direct information transmission path between the brain and the outside world by decoding neural brain activity information in the course of human thinking. BCI provides a new HCI channel that does not rely on peripheral nerves and muscle tissues. BCI hardware is currently divided into invasive and non-invasive types. Intrusive technology requires the acquisition of electroencephalograms (EEGs) by fully implanting electrodes on the surface of the cerebral cortex or inside the brain. However, because it needs to be implanted into the human brain, it is an "invasive" operation, so the research purpose is mainly to help patients with severe disabilities to recover or improve the quality of life.

The Japanese Honda company produced a mind-control robot. Operators can control surrounding robots to perform corresponding actions by imagining their limb movements. According to a study by the University of Rochester in the United States, subjects can control P300 component signals in a virtual reality (VR) scene, such as for switching lights or operating a virtual car. Japanese technology company Neurowear has developed a BCI device called Necomimi Brainwave Cat Ears. This cat ear device can detect human brain waves, and then turn the cat ears to express different emotions. Its similar product, the "Electric Wave Cattail," can control the movement of a cat-like tail device by brainwaves. Emotiv, a neurotechnology company in San Francisco, California, has developed a brain wave compilation device, Emotiv Insight, which can help people with disabilities to control wheelchairs or computers. Some researchers have developed a virtual prosthetic speech system that can decode the brain's speaking intentions, interpret the brain's intended vocal cord movements during speech, and convert them into basically understandable speech using a computer to output speech [7].

In terms of materials for future BCIs, considering biocompatibility, small size, mechanical flexibility, and electronic properties, graphene is one of the most promising candidates for neural interfaces [8]. Although the development of BCI still faces limitations such as ethics, technology, and neurophysiology, this interaction is still expected in the future 5G era.

2.3 Affective Computing

Affective computing refers to identifying a human's emotional signals to create a computer system that can perceive, recognize, and understand human emotions, and can respond intelligently, sensitively, and friendly. That is, to give computers the same observation, understanding, and cognitive ability to produce various emotional characteristics [9].

The research on affective computing mainly includes the collection of affective signals, the analysis, modeling, and recognition of signals, and the research of signal fusion algorithms. The analysis, modeling, and recognition of emotion signals mainly include five methods: facial emotion recognition, speech emotion recognition, limb emotion recognition, language and text recognition, and physiological pattern recognition. At present, traditional research mainly includes text sentiment analysis, speech sentiment analysis, and visual sentiment analysis. These three types of analysis have formed commonly used analysis databases and API interfaces. At the same time, sentiment computing based on massive network data and multi-modal sentiment computing has now emerged. Massive sentiment analysis will be a way to perform sentiment calculations with the support of a large number of networks and various other data in the future, such as judging depression tendency from a person's Weibo data, so as to actively intervene. Multi-modal sentiment computing will make the dimensions in HCI as complete as possible, to judge by multi-result fitting. For example, in the interactions of emotional robots, robots can better recognize human emotions through multi-modal emotion recognition, thereby achieving better interactions.

Affective computing has been applied in classroom teaching, emotional monitoring, medical rehabilitation, and public opinion monitoring. In future digital life scenarios, improved HCI based on emotional computing will also make interactions more vivid and beautiful.

2.4 Life Scenarios

With the increase of capacity and connectivity, natural interactions will become the trend in HCI, especially in AI and the IoT. People will use voice, gestures, touch, EEG, and more natural ways to interact with their smart homes. Their homes will become smart and quiet, which means these devices can perform affective computing and automatically provide more optimal settings. Smart cars can interact with the driver and control the car more safely with a lower loading. Robots of the future will be able to understand user instructions and behave in a more natural way.

3 Multimodal Display

3.1 Virtual Environment Technology

Virtual environment technologies, including VR technology and augmented environment (AR) technology, have attracted increasing academic interest in recent decades. VR technology can build a virtual scene that is totally different from the physical environment, while AR refers to adding virtual information to real physical scenes [10]. Common virtual-environment visual display devices include head-mounted displays, mobile phones, and spatial immersive displays [11]. Holographic projection, a special augmented reality technology, refers to projecting virtual images without the devices mentioned above. Holographic display aims to create a physically pure three-dimensional image, which can be observed from different angles without restriction. Hatsune Miku, a virtual Japanese idol, is the first to adopt holographic technology to show its virtual image in concert [12].

3.2 Multisensory Fusion Technology

With the development of information communication technologies, multimodal information is communicated from person to person, i.e., visual information, auditory information, and tactile information. Multisensory fusion can not only enrich information content but also improve the sense of immersion during usage.

Multisensory fusion technology provides users with immersive experience not only via visual simulation but also auditory and even tactile simulation. The auditory channel is very important when the visual channel is occupied. Take a driving alert system, for instance; it was suggested that auditory alerts contribute to reduced collision rates, shorter reaction times, larger maximum brake pedal force, and larger maximum lane deviation when compared to the baseline condition without a warning [13]. Tactile information can be simulated by an electronic skin system that receives commands wirelessly and then simulates the "touch" with vibration. The user can feel the "touch" by putting a soft and thin device like a patch on the skin [14]. With the development of future technologies, the simulation of VR users' senses might even extend to taste and smell.

For the transmission of multi-channel information, especially when high-definition visual information is included, the speed of transmission is very important. The development of 5G technology brings more possibilities for high-speed, high-quality multimodal information display. The impact model of consumer multisensory perceptual is built form the embodied cognition perspective, multisensory perceptions have indirect positive effect on online purchase intention [15].

3.3 Life Scenarios

Attention to tactile sensation is a hot spot in multi-modal applications. In addition to audiovisual interactions, in the future, when using smart devices, tactile interactions will also become a focus. In terms of entertainment, people can use VR games or VR virtual experiences to obtain a better user experience. AR and MR can be used to assist daily behaviors, such as AR-assisted training and driving. Future autonomous driving may take advantage of the multi-sensory redundancy effect to improve the efficiency and safety of takeovers.

4   Virtual Identity

4.1   Virtual Avatar

Zepeto is a mobile application launched in 2018 and has gained extensive user and media attention in a short time. The core content of Zepeto is social. Different from other traditional social media, Zepeto gives users their own avatars. Users can interact with their avatar, share it through pictures and videos, use the avatar to creative emoticons, take photos with friends' avatars, etc. The popularity of Zepeto reveals people's expectations for their new identities in socializing. Social network users are no longer satisfied with plain text or pictures, and they are expecting a new function to "show off." Various types of virtual technologies, such as face recognition and motion capture, provide the opportunity for each user to create a virtual identity on social networks. Information such as account name, avatar, and personal introduction constitute a simple virtual identity. According to studies on human social attributes, people try to influence their images consciously or unconsciously in their social life, which is called "image management." Image management behaviors occur when people want to show others a good self-image [16, 17]. Therefore, as a social person, people will play their own social roles on social networking services, showing their image characteristics in line with their own expectations and those of others. Virtual, anonymous social media offers users a greater degree of freedom. People can manipulate multiple network personalities to show themselves, making them more humorous, friendly, and cute than they are in reality. One characteristic of network image management is that users can fully control the release of information so that they can present themselves more strategically.

4.2 Virtual Idols

With the development of virtual identity, virtual idols, such as Hatsune Miku, emerge. Idolatry was viewed as a kind of attachment to outstanding figures in fantasies, which are often overpowered or idealized. Fans would be disappointed when the idols show their "true face." For virtual idols, they will never disappoint their fans because they are designed as a "perfect" idol. The most well-known virtual idol is the Japanese virtual singer Hatsune Miku. Miku was born in 2008 and has a sound source library based on the vocal synthesis software VOCALOID released by Yamaha. Using the data of the sound source library, music lovers can make songs and upload them to the Internet. In addition to the speech synthesis technique, virtual 3D image synthesis, such as MikuMikuDance, also enable Miku's fans to create fanworks without restrictions. Fans can create their own virtual idols with the help of virtual technologies and open-source data. Meanwhile, high-quality fanworks also attract more potential fan groups.

On March 9, 2010, Hatsune Miku became the first virtual singer to hold a concert using holographic projection technology, and 2500 concert tickets were snapped up in an instant. More than 30,000 Internet users watched the entire concert through a live webcast. Without a doubt, the virtual entertainment industry represented by virtual idols has large commercial potential. The development of 5G technology will provide more possibilities for this virtual industry.

4.3 Life Scenarios

Everyone could create a virtual avatar in a social network or community communication with image identification and virtual modeling technology and then could use their digital identities to socialize in different scenarios. People could use their avatars for online activities such as entertainment, social networking, and working. Anyone could choose to start their own information rights where others could directly obtain the developed information after identification through AR or other methods. New types of socializing could be conducted through AR and VR.

5 Data in the Cloud

5.1 Big Data in the Cloud

The key characteristics of 5G include capacity enhancement and massive connectivity, which make massive data collection and analysis feasible and reliable. The 4 V definition of big data, volume (amount), variety (type and source), velocity (speed of data transfer), and value (hidden information) is widely recognized [18]. Integrated sensor systems provide users with a large-volume and wide-variety of data with wireless and wearable sensors emerging. Massive data in the cloud eliminates data storage using expansive hardware, and analysis in cloud computing increases the efficiency of the process in software [19].

Big data in the cloud has already been explored in IoT systems, cloud manufacturing, and healthcare service. To deal with multi-dimensional IoT data, researchers proposed an update-and-query efficient index framework based on a key-value store that can support high insert throughput and provide efficient multi-dimensional queries simultaneously [20]. With massive sensor data in cloud manufacturing systems, Hadoop was introduced to conduct sensor data management [21]. Experts believe the adoption of IoT paradigms and data in the cloud in the healthcare field will improve service in the 5G era. They proposed a hybrid model of IoT and cloud computing to manage big data in health service applications consisting of four main components: stakeholders' devices, stakeholders' requests (tasks), cloud broker, and network administrator [22].

5.2 Edge Computing

As cloud computing technology develops, data is increasingly produced at the edge of the network. With the growing quantity of data generated at the edge, the speed of data transportation is becoming a bottleneck for the cloud-based computing paradigm [23]. In this way, edge computing, which refers to the enabling technologies that allow computation to be performed at the network edge so that computing happens near data sources, would be more efficient [24]. If the data is computed through a mobile network edge, the technology is defined as mobile edge computing (MEC) [25]. According to the white paper published by ETSI [26], MEC can be characterized by five points: on-premises, proximity, lower latency, location awareness, and network context information.

Edge computing infrastructures have been recognized to be a solution for high-demand computing power, low latency, and high bandwidth in AR applications [27]. For example, by collecting environmental data, AR combines real and virtual objects handled by MEC. Edge computing also plays a role in content delivery and video acceleration [25]. Users' requests and responses are time-efficient as the edge server is deployed close to the edge devices.

5.3 Life Scenarios

With massive amounts of data produced by each person, all the data could be uploaded to the cloud and synchronized in different mobile digital devices. Edge computing will provide users with rapid feedback on data analysis and fast data transmission. Integrated sensor systems will be combined with smart home devices to collect data on environmental indicators, physiological indicators, movement behavior, etc. Each person generates rich and diverse data. In addition to audio, video, and image data generated by themselves, various sensor data, such as geographical, physiological, and psychological indicators, can be used to construct the data context of the entire life cycle of a person. With the powerful computing capability provided by edge computing, the system analyzes and predicts the habits, rules, and intentions of each person, and provides users with the most needed information services at the right time and place. Through edge computing, the data will be used for warning and monitoring of emergencies and providing timely rescue and emergency guidance. For example, if an elderly person falls or suddenly becomes ill, the system will immediately contact the hospital and their families, and provide practical rescue guidance to those around them. Doctors can search the health cloud and conduct cloud computing with authorized data in the cloud for remote diagnosis and tracking patients' statuses.

6   Discussion

Based on the four trends in HCI in the 5G era, we proposed a trend model. Each basic element of the HCI paradigm is related to potential trends. In terms of the human element, people's virtual identities in social networking and data clouds will complement and co-exist with identities in real life. People will not only exist physically; the data they generate will also become another form of digital existence. As for the computer elements, it is broadly defined and includes many potential devices besides computers. Hardware and software will provide more natural user interfaces for better user experiences and conduct more intelligent computing to recognize emotions. The interaction element will become natural and multimodal. Besides keyboard input and screen display, voice interaction, gesture interaction, haptic interaction, and brain-computer interaction will be popular in the 5G era. All data produced by HCI will be in the cloud, and the environment element will be digitally collected and processed. In addition, 5G high-density coverage allows smart devices to collect more data automatically and provide services proactively (Fig. 1).

In terms of user experience in the 5G era, pragmatic and hedonic aspects are both important. The main benefits of 5G applications are capacity enhancement, massive connectivity, ultra-high reliability, and low latency. These properties result in not only efficient performance but also satisfying entertainment. We could cluster the life scenarios into pragmatic and hedonic aspects with respect to human needs and experience. In natural interactions with a smart home, smart car, or social robot, qualities like object response time and accuracy and the subject's perceived usefulness and easiness play an important role in the user experience. Entertainment elements like perceived enjoyment or cuteness could make an impact on trust and the user experience of intelligent systems or devices. For multimodal displays, one typical scenario is an immersive entertainment

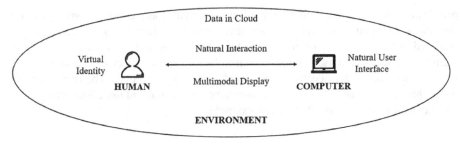

Fig. 1. Trend model

experience. For example, people perceive visual, auditory, and haptic stimuli in a VR environment, and such immersion, provided by 3D modeling and multimodal display, improves the pragmatic user experience. Besides, using multimodal displays for vehicle takeover in smart cars is an efficient improvement. Considering further trends in HCI, pragmatic and hedonic qualities in each dimension need to be identified and discussed to explore related factors and design guidelines.

Looking back at the changes brought by the previous 4G development, we considered the various trends brought by 5G to HCI and continued to imagine the new changes in the future 6G era. From a social point of view, 4G makes real-time face-to-face communication in video calls smoother. In the 6G era, the bandwidth and delay will meet the requirements of real-time projection technology, and the communication between people will be more real. The space limitation will be completely broken. For example, remote operations such as remote surgery with high time accuracy will be more popular and safer. From the perspective of the development of autonomous driving, the latency of 4G networks is as high as tens of milliseconds and unstable, while the latency of 6G will be lower and more stable. At the same time, the bandwidth can even reach the terabyte level. The improvement of network performance can guarantee the usability and safety of human driving. In addition, with the development of smart homes and smart cities, the bottlenecks caused by ICT will be further reduced by 6G technology.

The convergence of digital life and data throughout the life cycle in 5G development trends poses security challenges as well. In the 5G era, data in the cloud will result in abundant monitoring networks and a large amount of personal privacy data. With big data technologies, it is even possible to infer detailed and comprehensive user portraits from non-associated data. The leakage of personal privacy seriously threatens the safety and quality of everyone's life. Although the EU's general data protection regulation is considered as the strictest personal data protection bill in history [28], personal data is still in the possession and control of third parties, and there is a potential risk of leakage. A new type of personal data protection mechanism is needed to allow individuals to control their data by themselves. The acquisition of data requires personal authorization. Similar technologies such as SoLiD and DID have already been applied.

In the 5G era, combined with high-density coverage of base stations, human-centered combined space and time study framework will be developed with high-precision positioning and accurate services. From the perspective of the entire life cycle of people, people are divided into eight-stages according to their ages [29]. As shown in Fig. 2, people at each stage have their own mobility of activity (only at home, two points and one line, three points and three lines, and three points and four lines), and each activity is placed by 5G's precision positioning technology. Based on their position, servers can identify and predict the user's intention, infer the user's next objective, and analyze the potential information service that the user needs. In this way, people can be served according to their location and data. Figure 2 shows the three-point and three-line activities of middle-aged workers from 25–65 as an example.

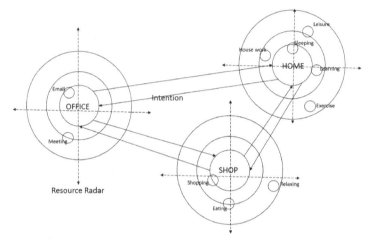

Fig. 2. Mobility of three-point and three-line activities of middle-aged workers from 25–65

At the same time, 5G increases the speed of network uplinks, enabling each information service node to provide external information in real-time and at high speed. Combined with high-performance terminal equipment, a human-centered radar-like information search service will become possible. Combining different locations and different activity intents, we can construct a suitable radar search radius and information preferences, shield and weaken unwanted information, and recommend the information that best meets the user's needs to improve the efficiency of user screening and identification, reduce the search and screening time, and improve user experience.

7   Conclusion

This study invited 11 experts in HCI and psychology to participate in focus groups and brainstorm with the topic of future life scenarios and technological applications with 5G networks to explore the trends in HCI in the 5G era. After open coding and axial coding, four trends of HCI were identified and proposed for technological development and future life scenarios. Natural user interfaces, brain-computer interactions, and affective computing were related to the trend in natural interaction. Virtual environment and multimodal fusion technologies were important parts of the trend in multimodal applications. Humans would have their own virtual avatar, including digital life and virtual images. Big data in the cloud and edge computing became environmental elements of HCI paradigms. The specific life scenarios involved multiple areas such as housekeeping, driving, entertainment, and healthcare and are related to pragmatic and hedonic user experience factors. Personal data protection is still critical in 5G and new technologies are developing. Human-centered combined space and time study frameworks with high-precision positioning and accurate services could be further investigated.

The limitations of this study were that only subjective feedback was collected, and the concluded trends in HCI may be qualitative. However, this study emphasizes experts' perspectives and speculation, and we thought it was acceptable to use these methods. This study reflected potential trend developments and application domains from 11 experts. There is more work required to achieve a deeper perspective.

References

1. Andrews, J.G., et al.: What will 5G be? IEEE J. Sel. Areas Commun. **32**(6), 1065–1082 (2014)
2. Palattella, M.R., et al.: Internet of things in the 5G era: Enablers, architecture, and business models. IEEE J. Sel. Areas Commun. **34**(3), 510–527 (2016)
3. Amy, N.: 5 myths about 5G. IEEE Spectr. (2016)
4. Lu, Y., Yu, C., Yi, X., Shi, Y., Zhao, S.: BlindType: eyes-free text entry on handheld touchpad by leveraging thumb's muscle memory. In: Proceedings of the ACM on Inter-active, Mobile, Wearable and Ubiquitous Technologies, vol. 1, no. 2, pp. 1–24 (2017)
5. Zhong, M., Yu, C., Wang, Q., Xu, X., Shi, Y.: ForceBoard: Subtle text entry leveraging pressure. In: Proceedings of the 2018 CHI Conference on Human Factors in Computing Systems, Montréal, Canada, pp. 1–10. ACM (2018)
6. Sun, K., Yu, C., Shi, W., Liu, L., Shi, Y.: Lip-interact: improving mobile device interaction with silent speech commands. In: Proceedings of the 31st Annual ACM Symposium on User Interface Software and Technology, Berlin, German, pp. 581–593. ACM (2018)
7. Anumanchipalli, G.K., Chartier, J., Chang, E.F.: Speech synthesis from neural decoding of spoken sentences. Nature **568**(7753), 493–498 (2019)
8. Hébert, C., et al.: Flexible graphene solution-gated field-effect transistors: Efficient transducers for micro-electrocorticography. Adv. Func. Mater. **28**(12), 1703976 (2018)
9. Picard, R.W.: Affective Computing. MIT Press, London (2000)
10. Gavish, N.: Evaluating virtual reality and augmented reality training for industrial maintenance and assembly tasks. Interactive Learning Environments **23**(6), 778–798 (2015)
11. Lantz, E.: The future of virtual reality: head mounted displays versus spatially immersive displays (panel). In: Proceedings of the 23rd Annual Conference on Computer Graphics and Interactive Techniques, New York, NY, United States, pp. 485–486. ACM (1996)
12. McLeod, K.: Living in the immaterial world: holograms and spirituality in recent popular music. Popul. Music Soc. **39**(5), 501–515 (2016)
13. Wu, X., Boyle, L.N., Marshall, D., O'Brien, W.: The effectiveness of auditory forward collision warning alerts. Transp. Res. Part F Traffic Psychol. Behav. **59**, 164–178 (2018)
14. Yu, X., et al.: Skin-integrated wireless haptic interfaces for virtual and augmented reality. Nature **575**(7783), 473–479 (2019)
15. Xiao-qing, S., Xi-xiang, S.: The Impact of multisensory perception on consumer's online purchase intention: Embodied cognition perspective. In: Proceedings of 23rd International Conference on Management Science and Engineering Management, Dubai, United Arab Emirates, pp. 200–206. IEEE (2015)
16. Goffman, E.: The Presentation of Self in Everyday Life. Harmondsworth, London (1978)

17. Tedeschi, J.T.: Impression Management Theory and Social Psychological Research. Academic Press, Salt Lake City (2013)
18. Laurila, J.K., et al.: The mobile data challenge: big data for mobile computing research (2012)
19. Hashem, I.A.T., Yaqoob, I., Anuar, N.B., Mokhtar, S., Gani, A., Khan, S.U.: The rise of "big data" on cloud computing: review and open research issues. Inf. Syst. **47**, 98–115 (2015)
20. Ma, Y., et al.: An efficient index for massive IOT data in cloud environment. In: Proceedings of the 21st ACM International Conference on Information and Knowledge Management, Hawaii, USA, pp. 2129–2133. ACM (2012)
21. Bao, Y., Ren, L., Zhang, L., Zhang, X., Luo, Y.: Massive sensor data management framework in cloud manufacturing based on Hadoop. In: Proceedings of 10th International Conference on Industrial Informatics, Beijing, China, pp. 397–401. IEEE (2012)
22. Elhoseny, M., Abdelaziz, A., Salama, A.S., Riad, A.M., Muhammad, K., Sangaiah, A.K.: A hybrid model of Internet of Things and cloud computing to manage big data in health services applications. Future Gener. Comput. Syst. **86**, 1383–1394 (2018)
23. Shi, W., Cao, J., Zhang, Q., Li, Y., Xu, L.: Edge computing: vision and challenges. IEEE Internet Things J. **3**(5), 637–646 (2016)
24. Shi, W., Dustdar, S.: The promise of edge computing. Computer **49**(5), 78–81 (2016)
25. Hu, Y.C., Patel, M., Sabella, D., Sprecher, N., Young, V.: Mobile edge computing—a key technology towards 5G. ETSI White Paper **11**(11), 1–16 (2015)
26. Patel, M., Naughton, B., Chan, C., Sprecher, N., Abeta, S., Neal, A.: Mobile-edge computing introductory technical white paper. White Paper, Mobile-Edge Computing (MEC) Industry Initiative, pp. 1089–7801 (2014)
27. Abbas, N., Zhang, Y., Taherkordi, A., Skeie, T.: Mobile edge computing: a survey. IEEE Internet Things J. **5**(1), 450–465 (2017)
28. Voigt, P., von dem Bussche, A.: The EU general data protection regulation (GDPR). Springer, Cham (2017). https://doi.org/10.1007/978-3-319-57959-7
29. Erikson, E.H.: Childhood and Society, 2nd edn. Norton & Company, New York (1963)

Author Index

Printed in the United States
By Bookmasters